Pluripotent Stem Cell Therapy for Diabetes

Lorenzo Piemonti • Jon Odorico
Timothy J. Kieffer • Valeria Sordi
Eelco de Koning
Editors

Pluripotent Stem Cell Therapy for Diabetes

First Edition

Springer

Editors
Lorenzo Piemonti
Diabetes Research Institute and Clinical Unit of Regenerative Medicine and Transplant
IRCCS San Raffaele Hospital/Vita-Salute San Raffaele University
Milan, Italy

Timothy J. Kieffer
University of British Columbia
Vancouver, BC, Canada

Eelco de Koning
Department of Medicine
Leiden University Medical Center
Leiden, The Netherlands

Jon Odorico
Department of Surgery
University of Wisconsin School of Medicine and Public Health
Madison, WI, USA

Valeria Sordi
IRCCS Ospedale San Raffaele
Diabetes Research Institute
Milan, Italy

ISBN 978-3-031-41942-3 ISBN 978-3-031-41943-0 (eBook)
https://doi.org/10.1007/978-3-031-41943-0

This Springer imprint is published by the registered company Springer Nature Switzerland AG
The registered company address is: Gewerbestrasse 11, 6330 Cham, Switzerland

Paper in this product is recyclable.

To all those who face the daily challenges of type 1 diabetes, this book is dedicated to you. In the pursuit of a world where insulin is no longer a necessity, you have been the unwavering beacon of hope and inspiration.
To the patients who have endured the relentless battle, you have shown remarkable resilience and strength, and your unwavering spirit fuels the drive for a better future.
To the families who have stood by their loved ones, your unwavering support and endless encouragement have been the cornerstones of their journey towards a healthier tomorrow.
To the researchers and scientists who have dedicated their lives, your relentless pursuit of knowledge and breakthroughs has brought us one step closer to an insulin-free world.
To all those who have contributed their time, resources, and expertise, your unwavering belief in the possibility of a brighter future has fuelled the momentum of progress.
As Helen Keller once said, "Alone we can do so little; together we can do so much." It is through our collective efforts, our collaboration, and our shared vision that we can create a world where type 1 diabetes is no longer a barrier. This book stands as a testament to the power of unity, perseverance, and the unwavering belief in a future where treatment options abound.
With heartfelt gratitude and endless admiration, this book is dedicated to the patients, families, and believers, who have fuelled the journey towards an insulin-free world.

Preface

In recent years, the field of diabetes research has witnessed remarkable progress in understanding and exploring novel avenues for the treatment of this chronic disease. Among the various approaches under investigation, beta cell replacement holds great promise as a potential cure for type 1 diabetes. The ability to restore functional beta cells, the insulin-producing powerhouses of the pancreas, could revolutionise the lives of millions affected by this condition. The aim of this book is to offer a comprehensive resource that brings together the vast knowledge and advancements in the field of stem cell therapy for diabetes. From the historical milestones of insulin therapy to the recent breakthroughs in pluripotent stem cell research, this book strives to be a unique compilation that explores the subject matter from various angles. By encompassing a wide range of topics, it serves as a comprehensive guide for researchers, clinicians, and scientists seeking to deepen their understanding of the potential applications of stem cells in regenerative medicine. As we reflect on the centenary of the discovery of insulin, we celebrate the remarkable success story that has saved countless lives over the past century. However, we must also acknowledge the limitations of insulin therapy and renew our commitment to strive for an insulin-free world. The convergence of advancements in stem cell research, tissue engineering, and regenerative medicine presents a timely opportunity to address these limitations and revolutionise diabetes treatment.

Part I sets the foundation by exploring the developmental journey and differentiation of beta cells. It elucidates the potential of human pluripotent stem cells to mimic islet development and discusses the prospects of beta cell replacement as a definitive cure for type 1 diabetes. Additionally, it investigates the derivation of functional human beta cells in vitro and explores key events in islet development that serve as a blueprint for successful in vitro differentiation. Furthermore, it examines methods of enhancing stem cell-derived islets, pushing the boundaries of possibility even further.

Part II delves into the genetic regulatory networks that guide the intricate process of islet development. It unravels the regulatory logic behind biological processes, shedding light on the complex interplay of genes and signalling pathways during pancreas organogenesis. By comprehending these genetic regulatory networks, we can gain a deeper understanding of the intricacies of beta cell development and harness this knowledge for therapeutic applications.

Part III explores the realm of bioengineering approaches for beta cell replacement. It discusses the selection of biocompatible biomaterials for stem cell-derived beta cell transplantation, the design of scaffolds for encapsulating these cells, and the creation of bioengineered vascularised insulin-producing endocrine tissues. Additionally, it examines the innovative concept of 3-D organoids composed of allogeneic mesenchymal stromal and pancreatic islet cells, offering a glimpse into the future of regenerative medicine.

Part IV delves into preclinical models and translational approaches, examining the crucial considerations involved in implant sites for cell-based insulin replacement therapies. It addresses the implementation of genetic safety switches for pluripotent stem cell-derived therapies, mitigating risks and ensuring patient safety. Furthermore, it delves into the concerns surrounding teratoma risk, emphasising the importance of rigorous safety protocols in the development of novel therapeutic approaches.

Finally, Part V provides an insightful glimpse into the clinical horizon of beta cell replacement. It explores the social value of cell-based technologies in type 1 diabetes and sheds light on emerging regulatory perspectives surrounding beta cell replacement products. Drawing on lessons learned from clinical trials of islet transplantation, this part highlights the challenges, successes, and future directions of translating beta cell replacement therapies from the laboratory to the clinic. Moreover, it identifies the minimal stem cell-derived beta cell properties required for successful transplantation and examines ongoing clinical trials involving stem cell-derived insulin-producing cells.

By traversing the diverse terrain of development, bioengineering, preclinical models, and clinical perspectives, this book aims to provide readers with a comprehensive overview of the multifaceted world of beta cell replacement. It is our hope that this collective knowledge will inspire researchers, clinicians, and policymakers alike, fostering collaboration and driving innovation towards a future where type 1 diabetes is no longer a lifelong burden but a conquered challenge.

Milan, Italy — Lorenzo Piemonti
Madison, WI, USA — Jon Odorico
Vancouver, BC, Canada — Timothy J. Kieffer
Milan, Italy — Valeria Sordi
Leiden, The Netherlands — Eelco de Koning

Contents

Part I Development and Differentiation of Beta Cells

Mimicking Islet Development with Human Pluripotent Stem Cells 3
Aubrey L. Faust, Adrian Veres, and Douglas A. Melton

Genetic Regulatory Networks Guiding Islet Development 25
Xin-Xin Yu, Xin Wang, Wei-Lin Qiu, Liu Yang, and Cheng-Ran Xu

Pancreatic Cell Fate Specification: Insights Into Developmental Mechanisms and Their Application for Lineage Reprogramming 49
Sara Gonzalez Ortega, Anna Melati, Victoria Menne, Anna Salowka, Miriam Vazquez Segoviano, and Francesca M. Spagnoli

Factors Influencing In Vivo Specification and Function of Endocrine Cells Derived from Pancreatic Progenitors 67
Nelly Saber and Timothy J. Kieffer

The Promises of Pancreatic Progenitor Proliferation and Differentiation 85
Azuma Kimura and Kenji Osafune

Part II Bioengineering

Selecting Biocompatible Biomaterials for Stem Cell-Derived β-Cell Transplantation 97
Rick de Vries and Aart A. van Apeldoorn

Scaffolds for Encapsulation of Stem Cell-Derived β Cells 123
Rick de Vries and Aart A. van Apeldoorn

Bioengineered Vascularized Insulin Producing Endocrine Tissues 151
Francesco Campo, Alessia Neroni, Cataldo Pignatelli, Juliette Bignard, Ekaterine Berishvili, Lorenzo Piemonti, and Antonio Citro

3 D Organoids of Mesenchymal Stromal and Pancreatic Islet Cells 179
Christof Westenfelder and Anna Gooch

Extracellular Matrix to Support Beta Cell Health and Function. 195
Daniel M. Tremmel, Sara Dutton Sackett, and Jon S. Odorico

Bioactive Materials for Use in Stem Cell Therapies for the Treatment of Type 1 Diabetes 221
Jonathan Hinchliffe and Ipsita Roy

Islet Macroencapsulation: Strategies to Boost Islet Graft Oxygenation. 251
Barbara Ludwig, Carolin Heller, Victoria Sarangova, and Petra B. Welzel

Part III Immunoescape

Immunogenicity of Stem Cell Derived Beta Cells 283
Nicoline H. M. den Hollander and Bart O. Roep

Immune Evasive Stem Cell Islets. 299
Federica Cuozzo, Valeria Sordi, and Lorenzo Piemonti

Islet Immunoengineering 317
Leonor N. Teles, Chris M. Li, Zachary M. Wilkes, Aaron A. Stock, and Alice A. Tomei

Part IV Preclinical Model and Translational Approaches

Considerations Pertaining to Implant Sites for Cell-Based Insulin Replacement Therapies 363
Braulio A. Marfil-Garza, Nerea Cuesta-Gomez, and A. M. James Shapiro

Genetic Safety Switches for Pluripotent Stem Cell-Derived Therapies for Diabetes 403
Dena E. Cohen and Jon S. Odorico

Safety Issues Related to Pluripotent Stem Cell-Based Therapies: Tumour Risk 419
Sanne Hillenius, Joaquin Montilla-Rojo, Thomas F. Eleveld, Daniela C. F. Salvatori, and Leendert H. J. Looijenga

Part V Beta Cell Replacement: Clinical Horizon

An Ethical Perspective on the Social Value of Cell-Based Technologies in Type 1 Diabetes. 461
Dide de Jongh and Eline M. Bunnik

Beta Cell Replacement Cellular Products: Emerging Regulatory Perspectives and Considerations for Program Development 485
Bruce S. Schneider

Lessons Learned from Clinical Trials of Islet Transplantation 499
Thierry Berney, Lionel Badet, Ekaterine Berishvili, Fanny Buron, Philippe Compagnon, Fadi Haidar, Emmanuel Morelon, Andrea Peloso, and Olivier Thaunat

Minimal SC-β-Cell Properties for Transplantation in Diabetic Patients 529
Veronica Cochrane, Yini Xiao, Hasna Maachi, and Matthias Hebrok

Clinical Trials with Stem Cell-Derived Insulin-Producing Cells 547
Ji Lei and James F. Markmann

Modelling of Beta Cell Pathophysiology Using Stem Cell-Derived Islets 573
Tom Barsby, Hossam Montaser, Väinö Lithovius, Hazem Ibrahim, Eliisa Vähäkangas, Sachin Muralidharan, Vikash Chandra, Jonna Saarimäki-Vire, and Timo Otonkoski

Index 599

Contributors

Lionel Badet Department of Urology and Transplantation Surgery, Hospices Civils de Lyon, Lyon, France

Tom Barsby Stem Cells and Metabolism Research Program, Faculty of Medicine, University of Helsinki, Helsinki, Finland

Ekaterine Berishvili Laboratory of Tissue Engineering and Organ Regeneration, Department of Surgery, University of Geneva, Geneva, Switzerland

Cell Isolation and Transplantation Center, Department of Surgery, Geneva University Hospitals and University of Geneva, Geneva, Switzerland

Faculty Diabetes Center, University of Geneva Medical Center, Geneva, Switzerland

Institute of Medical and Public Health Research, Ilia State University, Tbilisi, Georgia

Division of Transplantation, Department of Surgery, University of Geneva Hospitals, Geneva, Switzerland

School of Natural Sciences and Medicine, Ilia State University, Tbilisi, Georgia

Thierry Berney Division of Transplantation, Department of Surgery, University of Geneva Hospitals, Geneva, Switzerland

Department of Transplantation, Nephrology and Clinical Immunology, Hospices Civils de Lyon, Lyon, France

School of Natural Sciences and Medicine, Ilia State University, Tbilisi, Georgia

Juliette Bignard Laboratory of Tissue Engineering and Organ Regeneration, Department of Surgery, University of Geneva, Geneva, Switzerland

Cell Isolation and Transplantation Center, Department of Surgery, Geneva University Hospitals and University of Geneva, Geneva, Switzerland

Faculty Diabetes Center, University of Geneva Medical Center, Geneva, Switzerland

Eline M. Bunnik Erasmus MC, Department of Medical Ethics, Philosophy and History of Medicine, Rotterdam, The Netherlands

Fanny Buron Department of Transplantation, Nephrology and Clinical Immunology, Hospices Civils de Lyon, Lyon, France

Francesco Campo Diabetes Research Institute and Clinical Unit of Regenerative Medicine and Transplant, IRCCS San Raffaele Hospital, Milan, Italy

Vikash Chandra Stem Cells and Metabolism Research Program, Faculty of Medicine, University of Helsinki, Helsinki, Finland

Antonio Citro Diabetes Research Institute and Clinical Unit of Regenerative Medicine and Transplant, IRCCS San Raffaele Hospital, Milan, Italy

Veronica Cochrane Diabetes Center, Department of Medicine, University of California, San Francisco, CA, USA

Dena E. Cohen Regenerative Medical Solutions, Inc, Madison, WI, USA

Philippe Compagnon Division of Transplantation, Department of Surgery, University of Geneva Hospitals, Geneva, Switzerland

Nerea Cuesta-Gomez Alberta Diabetes Institute, Edmonton, AB, Canada

Department of Surgery, University of Alberta, Edmonton, AB, Canada

Federica Cuozzo Diabetes Research Institute, IRCCS San Raffaele Hospital, Milan, Italy

Dide de Jongh Erasmus MC, Department of Medical Ethics, Philosophy and History of Medicine, Department of Nephrology and Transplantation, Rotterdam, The Netherlands

Rick de Vries Maastricht University, MERLN Institute for Technology-Inspired Regenerative Medicine, Maastricht, The Netherlands

Nicoline H. M. den Hollander Department of Internal Medicine, Section Immunomodulation & Regenerative Medicine, Leiden University Medical Center, Leiden, The Netherlands

Thomas F. Eleveld Princess Máxima Center for Pediatric Oncology, Utrecht, The Netherlands

Aubrey L. Faust Department of Stem Cell and Regenerative Biology and Harvard Stem Cell Institute, Harvard University, Boston, MA, USA

Sara Gonzalez Ortega Centre for Gene Therapy and Regenerative Medicine, King's College London, Guy's Hospital, London, UK

Anna Gooch Department of Medicine, University of Utah, Salt Lake City, UT, USA

SymbioCellTech, Salt Lake City, UT, USA

Fadi Haidar Division of Transplantation, Department of Surgery, University of Geneva Hospitals, Geneva, Switzerland

Matthias Hebrok Diabetes Center, Department of Medicine, University of California, San Francisco, CA, USA

Center for Organoid Systems (COS), Technical University Munich (TUM), Garching, Germany

Institute for Diabetes Organoid Technology (IDOT), Helmholtz Center Munich, Helmholtz Diabetes Center (HDC), Neuherberg, Germany

Carolin Heller Department of Medicine III, Paul Langerhans Institute Dresden (PLID) of the Helmholtz Center Munich at the University Hospital Carl Gustav Carus, CRTD/DFG-Center for Regenerative Therapies, Technische Universität Dresden, Dresden, Germany

Leibniz-Institut für Polymerforschung Dresden e.V., Max Bergmann Center of Biomaterials Dresden, Dresden, Germany

Sanne Hillenius Princess Máxima Center for Pediatric Oncology, Utrecht, The Netherlands

Jonathan Hinchliffe Department of Materials Science and Engineering, University of Sheffield, Sheffield, UK

Hazem Ibrahim Stem Cells and Metabolism Research Program, Faculty of Medicine, University of Helsinki, Helsinki, Finland

Timothy J. Kieffer Laboratory of Molecular and Cellular Medicine, Department of Cellular & Physiological Sciences, Life Sciences Institute, University of British Columbia, Vancouver, BC, Canada

School of Biomedical Engineering, University of British Columbia, Vancouver, BC, Canada

Department of Surgery, University of British Columbia, Vancouver, BC, Canada

Azuma Kimura Center for iPS Cell Research and Application (CiRA), Kyoto University, Kyoto, Japan

Ji Lei Division of Transplantation and Center for Transplantation Sciences, Department of Surgery, Massachusetts General Hospital, Harvard Medical School, Boston, MA, USA

Chris M. Li Diabetes Research Institute and Department of Microbiology and Immunology, University of Miami School of Medicine, Miami, FL, USA

Väinö Lithovius Stem Cells and Metabolism Research Program, Faculty of Medicine, University of Helsinki, Helsinki, Finland

Leendert H. J. Looijenga Princess Máxima Center for Pediatric Oncology, Utrecht, The Netherlands

Barbara Ludwig Department of Medicine III, Paul Langerhans Institute Dresden (PLID) of the Helmholtz Center Munich at the University Hospital Carl Gustav Carus, CRTD/DFG-Center for Regenerative Therapies, Technische Universität Dresden, Dresden, Germany

Hasna Maachi Diabetes Center, Department of Medicine, University of California, San Francisco, CA, USA

Braulio A. Marfil-Garza Alberta Diabetes Institute, Edmonton, AB, Canada

Clinical Islet Transplant Program, University of Alberta, Edmonton, AB, Canada

CHRISTUS-LatAm Hub – Excellence and Innovation Center, Monterrey, Mexico

James F. Markmann Division of Transplantation and Center for Transplantation Sciences, Department of Surgery, Massachusetts General Hospital, Harvard Medical School, Boston, MA, USA

Anna Melati Centre for Gene Therapy and Regenerative Medicine, King's College London, Guy's Hospital, London, UK

Douglas A. Melton Department of Stem Cell and Regenerative Biology and Harvard Stem Cell Institute, Harvard University, Boston, MA, USA

Victoria Menne Centre for Gene Therapy and Regenerative Medicine, King's College London, Guy's Hospital, London, UK

Hossam Montaser Stem Cells and Metabolism Research Program, Faculty of Medicine, University of Helsinki, Helsinki, Finland

Joaquin Montilla-Rojo Anatomy and Physiology, Department Clinical Sciences, Faculty of Veterinary Medicine, Utrecht University, Utrecht, The Netherlands

Emmanuel Morelon Department of Transplantation, Nephrology and Clinical Immunology, Hospices Civils de Lyon, Lyon, France

Sachin Muralidharan Stem Cells and Metabolism Research Program, Faculty of Medicine, University of Helsinki, Helsinki, Finland

Alessia Neroni Diabetes Research Institute and Clinical Unit of Regenerative Medicine and Transplant, IRCCS San Raffaele Hospital, Milan, Italy

Jon S. Odorico Transplantation Division, University of Wisconsin-Madison Department of Surgery, Madison, WI, USA

Kenji Osafune Center for iPS Cell Research and Application (CiRA), Kyoto University, Kyoto, Japan

Timo Otonkoski Stem Cells and Metabolism Research Program, Faculty of Medicine, University of Helsinki, Helsinki, Finland

Andrea Peloso Division of Transplantation, Department of Surgery, University of Geneva Hospitals, Geneva, Switzerland

Lorenzo Piemonti Diabetes Research Institute and Clinical Unit of Regenerative Medicine and Transplant, IRCCS San Raffaele Hospital, Milan, Italy

Cataldo Pignatelli Diabetes Research Institute and Clinical Unit of Regenerative Medicine and Transplant, IRCCS San Raffaele Hospital, Milan, Italy

Wei-Lin Qiu School of Basic Medical Sciences, Department of Human Anatomy, Histology, and Embryology, Peking University Health Science Center, Peking-Tsinghua Center for Life Sciences, Peking University, Beijing, China

Bart O. Roep Department of Internal Medicine, Section Immunomodulation & Regenerative Medicine, Leiden University Medical Center, Leiden, The Netherlands

Ipsita Roy Department of Materials Science and Engineering, University of Sheffield, Sheffield, UK

Jonna Saarimäki-Vire Stem Cells and Metabolism Research Program, Faculty of Medicine, University of Helsinki, Helsinki, Finland

Nelly Saber Laboratory of Molecular and Cellular Medicine, Department of Cellular & Physiological Sciences, Life Sciences Institute, University of British Columbia, Vancouver, BC, Canada

Sara Dutton Sackett Transplantation Division, University of Wisconsin-Madison Department of Surgery, Madison, WI, USA

Anna Salowka Centre for Gene Therapy and Regenerative Medicine, King's College London, Guy's Hospital, London, UK

Daniela C. F. Salvatori Anatomy and Physiology, Department Clinical Sciences, Faculty of Veterinary Medicine, Utrecht University, Utrecht, The Netherlands

Victoria Sarangova Department of Medicine III, Hospital Carl Gustav Carus, Technische Universität Dresden, Dresden, Germany

Leibniz-Institut für Polymerforschung Dresden e.V., Max Bergmann Center of Biomaterials Dresden, Dresden, Germany

Bruce S. Schneider Schneider BIO Consultancy LLC, New York, NY, USA

A. M. James Shapiro Clinical Islet Transplant Program, University of Alberta, Edmonton, AB, Canada

Department of Surgery, University of Alberta, Edmonton, AB, Canada

Alberta Diabetes Institute, Edmonton, AB, Canada

Valeria Sordi Diabetes Research Institute, IRCCS San Raffaele Hospital, Milan, Italy

Francesca M. Spagnoli Centre for Gene Therapy and Regenerative Medicine, King's College London, Guy's Hospital, London, UK

Aaron A. Stock Diabetes Research Institute, University of Miami Miller School of Medicine, Miami, FL, USA

Leonor N. Teles Diabetes Research Institute and Department of Biomedical Engineering, University of Miami, Miami, FL, USA

Olivier Thaunat Department of Transplantation, Nephrology and Clinical Immunology, Hospices Civils de Lyon, Lyon, France

Alice A. Tomei Diabetes Research Institute, University of Miami Miller School of Medicine, Miami, FL, USA

Department of Biomedical Engineering, University of Miami Miller School of Medicine, Miami, FL, USA

Department of Surgery, University of Miami Miller School of Medicine, Miami, FL, USA

Daniel M. Tremmel Boston Children's Hospital/Harvard Medical School Department of Cardiac Surgery, Boston, MA, USA

Eliisa Vähäkangas Stem Cells and Metabolism Research Program, Faculty of Medicine, University of Helsinki, Helsinki, Finland

Aart A. van Apeldoorn Maastricht University, MERLN institute for technology-inspired regenerative medicine, Maastricht, The Netherlands

Miriam Vazquez Segoviano Centre for Gene Therapy and Regenerative Medicine, King's College London, Guy's Hospital, London, UK

Adrian Veres Department of Stem Cell and Regenerative Biology and Harvard Stem Cell Institute, Harvard University, Boston, MA, USA

Xin Wang College of Life Sciences, Peking University, Beijing, China

Petra B. Welzel Leibniz-Institut für Polymerforschung Dresden e.V., Max Bergmann Center of Biomaterials Dresden, Dresden, Germany

Christof Westenfelder Department of Medicine, University of Utah, Salt Lake City, UT, USA

SymbioCellTech, Salt Lake City, UT, USA

Zachary M. Wilkes Diabetes Research Institute and Department of Biomedical Engineering, University of Miami, Miami, FL, USA

Yini Xiao Diabetes Center, Department of Medicine, University of California, San Francisco, CA, USA

Cheng-Ran Xu School of Basic Medical Sciences, Department of Human Anatomy, Histology, and Embryology, Peking University Health Science Center, Peking-Tsinghua Center for Life Sciences, Peking University, Beijing, China

Liu Yang School of Basic Medical Sciences, Department of Human Anatomy, Histology, and Embryology, Peking University Health Science Center, Peking-Tsinghua Center for Life Sciences, Peking University, Beijing, China

Xin-Xin Yu School of Basic Medical Sciences, Department of Human Anatomy, Histology, and Embryology, Peking University Health Science Center, Peking-Tsinghua Center for Life Sciences, Peking University, Beijing, China

Part I
Development and Differentiation of Beta Cells

Mimicking Islet Development with Human Pluripotent Stem Cells

Aubrey L. Faust, Adrian Veres, and Douglas A. Melton

1 Characteristics of Diabetes

Glucose homeostasis is maintained by concerted action of endocrine cells across several organs. These cells exert regulatory effects on distant tissues by secreting hormones in response to circulating glucose levels and other metabolic factors. Central among these are insulin-expressing pancreatic beta cells. Beta cells secrete insulin upon sensing elevated circulating glucose levels, and insulin action on liver, muscle, and adipose tissue induces uptake of glucose and synthesis of glycogen or triglycerides. Diseases of insulin insufficiency or insulin resistance – of which there are several mechanistically distinct classes – are referred to as diabetes mellitus.

In type 1 diabetes (T1D), beta cells are lost to a highly specific autoimmune attack that typically begins during childhood. People with T1D are unable to regulate glucose levels without therapeutic intervention once sufficient beta cell mass is lost, typically within a few years of onset. While the initiating cause remains unknown, it is believed to be a failure in or viral infection of beta cells, abetted by a dysfunctional immune system that specifically responds to beta cell antigens. The maladaptive autoimmune response is specific to beta cells, leaving other islet endocrine populations unharmed.

Beta cells lost in T1D do not regenerate over time. In healthy tissue, beta cells are long-lived cells whose mass is maintained by replication rather than differentiation from a stem cell source [17]. Sustained autoimmune attack outpaces any regeneration, continuing until endogenous insulin secretion becomes and remains insufficient to maintain glucose levels within safe bounds. The immune memory persists, and autologous beta cells are still rejected decades after disease onset. This was

A. L. Faust · A. Veres · D. A. Melton (✉)
Department of Stem Cell and Regenerative Biology and Harvard Stem Cell Institute, Harvard University, Boston, MA, USA
e-mail: Melton@vrtx.com

L. Piemonti et al. (eds.), *Pluripotent Stem Cell Therapy for Diabetes*,
https://doi.org/10.1007/978-3-031-41943-0_1

elegantly demonstrated through transplantation of pancreatic tissue from healthy donors into their identical twins with type 1 diabetes [74]. Thus, type 1 diabetes becomes a lifelong disease requiring complex care.

2 Beta Cell Replacement as a Cure for Type 1 Diabetes

Although cases of diabetes have been described for millennia, the first effective therapy was discovered just over a century ago. In the 1920s, Banting and Best successfully isolated insulin from dog pancreases and used it to treat a patient with type 1 diabetes [8]. For the next several decades, insulin therapy primarily involved the use of animal-derived insulin, which had varying degrees of purity [61]. In 1978, the first recombinant human insulin was approved for use, which led to a significant improvement in the safety and efficacy of insulin therapy [55].

In recent years, there have been several enhancements to insulin therapies for type 1 diabetes. These include the development of rapid-acting and long-acting insulin analogs that allow more sophisticated dosing strategies, as well as continuous glucose-monitoring systems paired with insulin pumps. These advances have improved the lives of people with diabetes but do not yet match the blood glucose control achieved by functional beta cells. The resulting imperfect control leads to variation in blood glucose levels beyond normal ranges. Time spent with elevated blood glucose levels causes microvascular damage, with long-term complications including diabetic nephropathy and retinopathy. Finally, the unremitting requirement to monitor and adjust their own physiology using expensive and dangerous medication is a chronic burden on people with T1D.

An alternative treatment approach is to replace not just the hormone product of beta cells, but the cells themselves. Whole pancreas transplantation was first successfully performed in 1966 [33], and the islets within an appropriately secrete insulin and cure diabetes [73]. However, this is an invasive surgery with potential complications driven by the exocrine compartment, leading to the exploration of whether islets could be isolated and engrafted alone. The first demonstration that islets alone can functionally cure diabetes was performed by Paul Lacy in the 1970s. After developing a method for isolating functional islets from exocrine pancreatic tissue, Lacy showed that transplanting islets can restore glucose homeostasis in diabetic animal models [7, 34]. Transplanting cadaveric islets into human patients, however, was less effective until the development of the Edmonton protocol in the early 2000s [65]. This protocol made key advances in islet isolation methodology, the number of transplanted islets, and the choice of immunosuppressive drugs, and drastically increased the success rate of islet transplantation. The challenge of immune rejection that accompanies transplantation of cadaveric or any "non-self" islets–compounded by the autoimmune origin of T1D–is addressed elsewhere.

Presently, transplanted islets can replace insulin injections in most recipients for more than 5 years [66]. The most effective transplant site has been islet injection into the hepatic portal vein, which is performed through a minimally invasive surgery. Furthermore, by embedding islets within liver microvasculature, most of the released insulin acts immediately on the adjacent hepatic tissue and is removed from circulation, as it would be when released from islets within the native pancreas [44]. Effective as this treatment can be, the requirement for immunosuppression and the absence of a standardized and readily available islet supply have limited the use of cadaveric islet transplants.

To generate a limitless supply of human beta cells, two major strategies have been explored. The first is to replace beta cells in vivo by inducing replication of residual beta cells [78] or by transdifferentiating pancreatic, gastric, intestinal, or other cells into insulin-secreting cells [10, 19, 28, 82]. The second strategy is to generate beta cells through the differentiation of human pluripotent stem cells, either by transplanting pancreatic progenitors that differentiate into beta cells in vivo [36], or by transplanting mature beta cells fully differentiated in vitro as though they were islets isolated from an organ donor pancreas [48]. To date, in vitro differentiation of mature beta cells has progressed closer to therapeutic applications than any of the alternatives, including promising early results from a clinical trial by Vertex Pharmaceuticals ("Vertex Provides Updates on Phase 1/2 Clinical Trial of VX-880 for the Treatment of Type 1 Diabetes – Press Release" [77]). While none of these approaches to replenishing beta cells directly address the challenge of autoimmunity, in vitro cells are amenable to engineering as immune evasion approaches advance.

3 Deriving Functional Human Beta Cells In Vitro

Studies of pancreatic development in model organisms described the stages and genes involved in specifying pancreatic endoderm and endocrine differentiation [84]. Those studies formed the basis for subsequent work on the in vitro differentiation of pluripotent stem cells. In 2006, D'Amour and colleagues developed a protocol to progressively specify SOX17+ definitive endoderm, then posterior foregut, then PDX1+ pancreatic endoderm [12]. Numerous publications followed similar protocols, but comparing their results is challenging given a lack of standardized cell type quantification. A notable advance by Nostro et al. employed factors that increased endocrine cell abundance, and this study provided more rigorous quantifications of percentages of cells expressing particular markers, including C-peptide [46]. However, in all these reports, the resulting beta or beta-like cells were not able to secrete insulin in response to glucose challenges in vitro.

Early efforts to induce beta cells from pancreatic progenitors yielded a cell type of ambiguous identity, co-expressing insulin and glucagon. Because this hormone co-expression is not seen in mature islets, these cells were not definitively classified as a specific islet cell type and were instead called "polyhormonal" cells. The hypothesis that these represented a beta cell progenitor was pursued but attempts to mature polyhormonal cells into a mono-hormonal insulin expression state, or to achieve the glucose-responsive insulin secretion expected of a beta cell, proved unsuccessful.

A pivotal advance came in 2014 with the publication of a detailed protocol that described both efficient beta cell differentiation and the first compelling evidence for beta cell function in vitro [48]. Physiological function by sequential glucose-stimulated insulin secretion (GSIS) is significant as that is what leads to glucose homeostasis in vivo. This advance, including all details on inducing factors, timing, and other experimental methods, was shared prior to publication with Rezania et al., who independently verified the findings [56]. The protocol to produce functional stem cell-derived beta cells and islets using human stem cells (essentially described in [48]) was patented and licensed to Semma Therapeutics, now Vertex Pharmaceuticals. The protocol has since been expanded, industrialized, and used in clinical trial.

4 Key Events in Islet Development as a Blueprint for In Vitro Differentiation

Efforts to differentiate beta cells in vitro have relied on an understanding of their normal development in vivo. Endocrine cells comprise approximately 2% of the pancreas by mass and are aggregated in structures called islets, which are interspersed throughout the exocrine tissue. Islets consist of five endocrine cell types. The most abundant are insulin-secreting beta and glucagon-secreting alpha, whose hormone products have opposite metabolic effects and, in tandem, govern glucose homeostasis. The next most abundant are somatostatin-secreting delta cells, which regulate beta and alpha-cell activity through paracrine signaling. Finally, pancreatic-polypeptide-secreting gamma and ghrelin-secreting epsilon are the least abundant. No changes in glycemia or body weight are associated with gamma cell ablation in mice [51], and epsilon cells are absent from the adult pancreases of most species and constitute less than 1% of human islets [81]. Islet structure varies across species. In mice and most mammals, islets have a core-mantle structure with a shell of alpha and other non-beta cells and a beta cell interior [70]. In humans, islet structure is more complex, appearing superficially disorganized but proposed to consist of sheets of alpha and beta cells folded into complex shapes to generate both homotypic and heterotypic contacts between these two cell types [16].

Based on similarities in gene expression patterns, it was hypothesized that pancreatic endocrine cells are derived from neural crest cells. This mistaken view was widely taught and included in many medical texts despite the lack of direct evidence [2]. It is the case that there are many similarities between neurons and islet endocrine cells: both cell types depolarize upon receiving input signals, using the resulting influx of Ca2+ ions to release the contents of secretory granules in specific locations. There are also extensive gene expression commonalities, including many genes that are otherwise largely restricted to neurons. In *Drosophila* and other invertebrates, insulin-expressing cells are in fact neurons [59], spurring speculation that the ancestral cell type may have been an insulin-secreting neuron whose expression programs have subsequently been activated in the gut tube. Regardless of their evolutionary history, the developmental origin of vertebrate islets is decidedly endodermal [53].

The pancreas is generated through branching morphogenesis, with epithelial branches organized into acinar tips and ductal trunks. Endocrine cells develop as scattered cells in the ductal epithelium. After endocrine induction, cells exit the ductal epithelium to form islets. One model to explain islet formation proposes that individual endocrine cells undergo a partial epithelial-to-mesenchymal transition and are delaminated from the ductal epithelium to enter developing islets through preferential adhesion. More recently, an alternative model was proposed whereby large duct segments initiate endocrine induction in concert, initially forming a "peninsula" that subsequently separates into islets as the cells gain mature endocrine identities [67].

At the cellular level, endocrine induction is initiated by the transient expression of Neurog3 [14, 22, 64]. Induction of sufficient levels of Neurog3 to drive endocrine induction requires low levels of Notch signaling [3], with the Notch target Hes1 suppressing transcription of Neurog3 [30, 38]. Downstream of Neurog3, a set of transcription factors (TFs) shared across islet cells maintains endocrine fate; these include Neurod1, Insm1, and Nkx2-2 [63]. After Neurog3 is downregulated, markers of individual cell types including hormones turn on and cell identity becomes apparent. When and how specification of the five different islet cell types occurs, however, is unknown, and this knowledge gap has carried through to in vitro differentiation.

5 Applying Technologies for Cell Characterization and Perturbation

In vitro differentiation poses a practical test for definitions of cell identity: when can we consider a cell type produced in vitro equivalent to its in vivo counterpart? Which genes are definitional to a cell type's identity, and how much can gene expression vary as a consequence of cell state such as age or environment without altering the underlying cell type identity? Perfect concordance in expression of every gene between directed differentiation derivatives and their in vivo counterparts is too high a bar. Differing metabolic or cellular environments must induce differences in gene expression, differences we can safely attribute to state not identity. Similarly, human fetal cells are classified as the same cell types as their adult counterparts, despite an immature state that lacks expression of genes referred to as maturation markers. Consequently, cell function may be a better parameter for comparison of cells formed by in vitro differentiation to their natural counterparts.

In practice, the field progressed despite this puzzle by combining small panels of marker genes with a strong emphasis on functional assays. For beta cells, the bar was set at expression of insulin and the transcription factor NKX6.1, paired with the functional ability to secrete insulin in response to sequential glucose challenges in vitro and separately rescue murine diabetes models after transplantation in vivo. As noted above, these criteria were met in 2014, which produced the first stem cell-derived (SC-) beta cells that performed glucose-stimulated insulin secretion in vitro [48].

Various methods have been deployed to measure the similarity of in vitro and in vivo beta cell gene expression. These methods include biased, hypothesis-based approaches with single-cell resolution (such as immunofluorescence read out by microscopy or flow cytometry) or unbiased, whole-transcriptome approaches (micro-arrays and bulk RNA-seq) limited by an unknown degree of cellular heterogeneity in the input. While these methods gave insight into questions such as whether the resulting beta cells more closely resemble fetal or adult beta cells [27], they left unresolved whether unexpected cell populations might be present, and the extent to which SC-beta transcriptomes might diverge from those in islets.

These limitations changed dramatically with the advent of single-cell RNA sequencing, which enables comprehensive transcriptomic profiling at single-cell resolution. Multiplexed single-cell RNA-seq emerged in 2011, initially with throughputs of fewer than 100 cells [29]. Adapting the molecular biology steps from these techniques to be performed within microscopic droplets created by microfluidics devices enabled the routine application of these techniques to characterize thousands of cells [35, 41]. Applying inDrops, we generated one of the first whole-transcriptome profiles of pancreatic islet cell identity [9]. This provided a reference dataset against which stem cell-derived islet cells can be benchmarked.

During this same period, the application of Cas9 to precisely edit genomes transformed our ability to systematically probe gene function. Forward genetics is an approach to mapping genes that control a phenotype and has been applied at scale in a number of biological contexts for decades. For mammalian genetics, mapping genes that control a phenotype had been accomplished by knocking out one gene at a time or using knockdown libraries of short hairpin RNAs, but variation in on-target efficiency and substantial off-target effects in the latter limited their use as a tool across biology. Genome-wide Cas9 knockout libraries share the concept of shRNA knockdown libraries but benefit from higher on-target and lower off-target activity to provide a new way to systematically study human biology. These approaches depend on next-generation DNA sequencing, which allows simultaneous sequencing of millions of short sequences to reveal which elements of a perturbation library (e.g., specific sgRNAs) drive a phenotype of interest.

6 Stem Cell Differentiation Recapitulates Islet Development

The most commonly used approach for generating a cell type through directed differentiation is to guide cells through successive progenitor states by manipulating signaling pathway activities. While this method succeeded in generating functional beta cells from human pluripotent stem cells, it does not produce beta cells in isolation. Applying droplet-based single-cell RNA-sequencing [76], we determined the identities of the several other cell types that co-develop with SC-beta cells, including SC-alpha and SC-enterochromaffin (EC) populations whose abundance rivals that of SC-beta cells. While the SC-alpha cells may enhance SC-islets due to their complementary role in glucose homeostasis, SC-EC cells are not typically found in islets and are unlikely to contribute to the desired function of SC-islets.

Our reclassification of polyhormonal cells as SC-alpha cells transiently expressing insulin resolved an earlier debate. Previously, population identities in SC-beta differentiations had been approximated by how marker gene immunostaining mapped to the characteristic expression in islet endocrine cell types. Because insulin and glucagon co-expression is not seen in adult islets, the identity and potential of so-called polyhormonal cells was an open question. Demonstrating close concordance between SC-alpha cells and their islet alpha counterparts connects to a broader question of whether in vitro differentiation can derive aberrant cell types that do not map to cells that naturally develop in vivo. In this case, a cell type initially viewed as an in vitro artifact due to insulin and glucagon co-expression was found instead to be a close analog of a canonical cell type and demonstrated to reach this canonical state over time even without additional exogenous factors.

Since our initial single-cell publication, rare populations observed in our and others' data have been classified as analogs of somatostatin-expressing delta and ghrelin-expressing epsilon cells. This demonstrates the derivation of nearly all islet cell types in what is effectively a self-assembly process following directed differentiation to pancreatic endoderm. Pancreatic polypeptide-expressing gamma cells are the sole islet endocrine population that has not been conclusively observed in stem cell differentiation. While this fact remains puzzling, gamma cell function remains a mystery, and their absence in a stem cell-derived islet is unlikely to be detrimental [51].

Non-endocrine cells mature over time in culture into approximations of exocrine acinar and ductal populations. Rather than lingering in a stalled progenitor state, they lose expression of transient developmental markers and gain expression of genes related to their functional roles in secreting and transporting digestive enzymes. Non-endocrine cells are considered undesirable in a graft setting. One reason relates to limited space, especially for SC-islets transplanted in a device. While endocrine cells rarely replicate, exocrine cells retain proliferative potential. Thus, nonendocrine cells occupy space that could be devoted to SC-beta cells instead and are expected to expand further through replication. Second, mature acinar cells secrete proteases that could physically damage the graft. To reduce exocrine cell abundance, we introduced a reaggregation process that depletes these populations (Fig. 1).

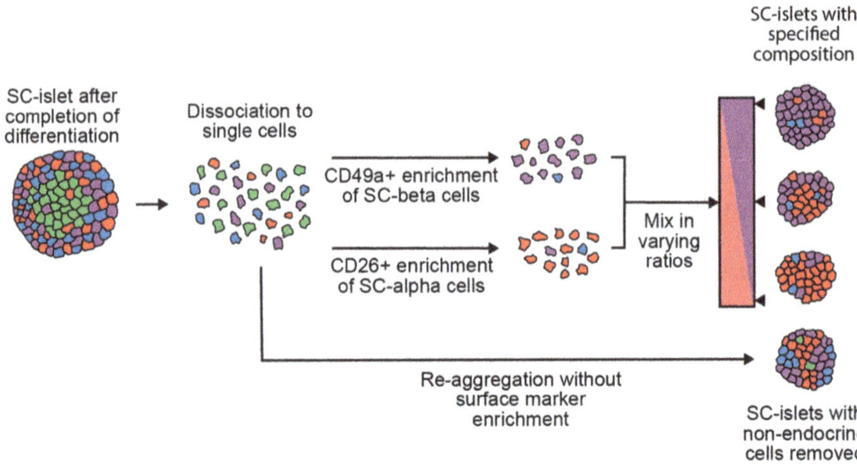

Fig. 1 Controlling SC-islet composition through a combination of surface marker sorting and reaggregation

7 A Piece Out of Place: The Puzzle of Enterochromaffin Cells

Among this near-complete representation of pancreatic exocrine and endocrine cell types, there is the puzzling emergence of enterochromaffin cells. Enterochromaffin cells are the most abundant enteroendocrine cell type and emerge in both the stomach and intestine but have not been identified in the pancreas historically or in recent single-cell sequencing profiles of human and mouse islets. Since our initial description of SC-EC cells in stem cell-derived islets, they have been observed across several protocol variants, cell lines, and research groups [4, 6, 26, 50, 79, 83] and no single-cell RNA-seq study of stem cell-derived beta cell cultures has reported their absence. The mechanisms driving SC-EC differentiation remain elusive. One study identified a compound that gradually reduces SC-EC abundance during extended culture, but this timing suggests selective toxicity rather than developmental control [6].

Two underlying explanations for the presence of enterochromaffin cells in islet-directed differentiation have emerged: (1) incomplete pancreatic specification yields an intestinal progenitor that generates an enteroendocrine cell type, or (2) enterochromaffin identity is a previously undescribed cell state in the developing human pancreas. The first theory is that differentiations contain gastric or intestinal progenitors whose presence is revealed by their non-pancreatic progeny. Because SC-EC cells express the marker CDX2, an intestinal progenitor is more likely than a gastric one. Additionally, the standard method of evaluating whether directed differentiations generate pancreatic endoderm–immunostaining for PDX1–does not exclude the possibility of intestinal identity. Although the loss of PDX1 causes pancreatic agenesis [72], this gene is expressed in adjacent tissues: the duodenum contains Pdx1+/Cdx2+ cells, and the gastric antrum contains Pdx1+/Sox2+ cells [23, 43]. Further confounding progenitor classification, a recent study found that, in a divergence from mouse embryos, the human fetal pancreas contains CDX2+ cells [83]. VIL1, whose promoter is used in mouse models to drive gene expression in the intestinal epithelium [60], is also expressed in the human fetal pancreas [47]. Consequently, marker genes have been insufficient to classify our in vitro progenitors, leaving unresolved the question of whether SC-EC cells derive from intestinal or gastric progenitors. Given the ambiguous picture painted by progenitor markers, an alternative is to consider whether the cell types made in our differentiations are consistent with such a progenitor pool. Other than SC-EC cells researchers have not identified populations that are characteristically gastric or intestinal, but this claim, too, is confounded by marker ambiguity.

A second explanation for the emergence of SC-EC cells is that an equivalent state is present in the human pancreas either during development or under certain physiological conditions. Especially if this represents a divergence between human and mouse, these cells may not have been profiled. Recent evidence supporting this view stems from the discovery of CDX2 expression during human fetal pancreatic development, and enterochromaffin cells may emerge from this progenitor state [83]. From their observation of a possible enterochromaffin population in fetal pancreas and its absence in adult islets, Zhu et al. conclude that this is a transient expression state that resolves toward a beta cell identity, and that such a change would likely occur over time in SC-islets as well. This hypothesis is challenged by data from single-cell sequencing of SC-islets 6 months after transplantation into diabetic mouse models, which found that EC cells persist and even increase their expression of EC identity markers while reducing expression of islet markers [5].

An intermediate explanation is the hypothesis that enterochromaffin cells are a default endocrine cell type in the intestine, stomach, and pancreas, but that conditions during normal pancreatic development efficiently suppress their emergence. Enterochromaffin cells have been observed throughout the stomach and intestine with only subtle gene expression variation across sites [11]. An unusual feature of enterochromaffin cells observed through single-cell sequencing is the overlap of their marker genes with those of the shared expression program of endocrine progenitors. In the stomach, intestine, and pancreas, endocrine induction is driven by transient expression of NEUROG3, and markers that distinguish individual endocrine cell types are upregulated after NEUROG3 turns off. The notable exceptions to this are EC markers, which are co-expressed with NEUROG3 both in the intestine in vivo [20] and in our SC-islet differentiation in vitro.

EC cell emergence may be controlled by a small set of transcription factors that are shared across tissues. In the intestine, EC cells are the only endocrine cell type that does not express Isl1 [20], and intestine-specific deletion of Isl1 increases enterochromaffin abundance while reducing differentiation of all other enteroendocrine cell types [75]. In the adult pancreas, all islet endocrine cells express ISL1. It is possible that efficient induction of ISL1 and similar genes in the pancreas repress EC cell emergence, and that these genes are not induced as highly, rapidly, or efficiently in vitro. Supporting this hypothesis, knockout of ISL1 in our differentiations reduces alpha and beta differentiation relative to EC, and its overexpression achieves the opposite (Veres et al., unpublished). The cause of SC-EC differentiation may be as simple as a developmental event that drives robust ISL1 induction in vivo that is absent in vitro, and identifying and mimicking these cues would enable suppression of the EC fate.

There may be a deep similarity and interconvertibility between beta and EC cells. They are each the most abundant endocrine cell type in their respective tissues, yet they have disjointed tissue distributions. There is also plasticity in both cell types to acquire functional elements of the other. Subsets of beta cells synthesize serotonin during pregnancy [62], which has been described as a reversible cell state but-since serotonin-synthesizing beta cells have not been profiled by scRNA-seq-may yet prove to be transdifferentiation. For EC cells, the evidence is stronger: Foxo1 inhibitors cause them to gain insulin expression and to downregulate genes related to serotonin synthesis and secretion, likely acquiring a beta-like identity [10]. Similarities are also prominent in SC-islet differentiations, with SC-beta and EC cells differentiating simultaneously with gene expression commonalities. Under suitable conditions such as Foxo1 inhibition, SC-EC cells may in the future be induced to transdifferentiate into SC-beta cells. Finally, it is worth noting that the presence of EC cells has not been demonstrated in any assay to adversely affect SC-islet function. These cells can be eliminated or greatly reduced in numbers by alterations to the protocol.

8 Constructing an Islet from Stem Cells

Having demonstrated in vitro derivation of nearly all islet cell types, a question driving the next phase of protocol development is how to achieve the desired proportions of these cells. While only beta cells are destroyed in type 1 diabetes (T1D), paracrine interactions with alpha and other islet cells enhance beta cell function [57]. The ideal composition of an SC-islet is unknown, with questions ranging from the optimal ratio of SC-beta to SC-alpha cells to whether including delta, gamma, or epsilon cells would provide significant functional enhancements. Three general strategies emerge: (1) accept heterogeneous differentiations and instead purify the desired cell types after their emergence, (2) tune the signaling factors to control which cell types differentiate, or (3) tune the genetics of the starting pluripotent stem cells in ways that alter the cell type composition of their differentiated progeny.

In the first consideration, we can compose desired cell type proportions after differentiation by sorting and reassembling cells into an SC-islet. We demonstrated that dissociation and reaggregation favor the adhesion of endocrine cells, removing proliferative non-endocrine cells whose expansion potential is undesirable in grafts [76]. To enrich for only SC-beta cells, we also identified CD49a as a surface marker that selects for SC-beta cells (Fig. 1). More recently, another group identified several surface-targeting antibodies that label SC-beta cells [49], which, if used either alone or in combination with CD49a could increase the yield of beta cells recovered from sorting. Enrichment of SC-alpha cells has also been explored through the development of a differentiation protocol targeted toward generating abundant SC-alpha cells [52] and with the identification of CD26 as a surface marker for further purification [1]. By sorting SC-beta and SC-alpha cells, we can combine them in controlled proportions or even–through sequential reaggregation–approximate the mantle-shell configurations of islets seen in mice and other species (Fig. 1). While differing islet architectures across species offer many templates for the ratio and 3D configuration of beta and alpha cells, optimal parameters have not been explored empirically.

Purifying SC-beta and SC-alpha cells has so far enabled the study of specific immune responses to each cell type [21, 39, 69], but it is very challenging to incorporate into a large-scale beta cell manufacturing process in part because of the loss of desired cells. This applies both to the opportunity cost of generating off-target cells and to incomplete recovery of desired cells during the purification process. Many SC-beta cells are lost after sorting on CD49a, which may stem from recovery during sorting due to a combination of magnetic column saturation and heterogeneous expression levels of CD49a as well as from beta cell death during dissociation, sorting, and reaggregation. Additionally, sorting is a significant manipulation that is very challenging to deploy in a clinical-grade manufacturing pipeline.

Alternatively, to control population emergence during differentiation, the most common approach has been to tune the signaling factors that comprise the differentiation protocol. For stages up to pancreatic progenitor specification or even the induction of endocrine differentiation, many of these factors emulate signals that drive the development of analogous progenitors in vivo. But for the final stage of endocrine development, when we attempt to induce beta cell differentiation without generating other endocrine cell types, the in vivo mechanism also remains elusive. We know of transcription factors that drive bifurcations (e.g., PAX4 for beta and delta vs. ARX for alpha, gamma, and epsilon), and we know that certain populations emerge more frequently at certain time points both in vivo and in vitro, but we are not able to predict this from progenitor gene expression prior to commitment, or to infer upstream signaling activities. Despite this absence of a developmental roadmap, empirically exploring the space of signaling modulators, metabolites, and other media components may generate further refinements to SC-islet composition.

While it is improbable that the potential for signaling modulator control of endocrine ratios has been exhausted, progress using factor-driven approaches has been slow since 2014. While careful studies have shown that inhibition of YAP and canonical WNT each increases SC-beta yield [58, 68], these effects are modest relative to the remaining number of off-target cells. An exception to this characterization is a recent study that claims to make 80% SC-beta cells, a doubling from prior claims [40]. This new protocol extends the length of differentiation and adds ten new small molecules, so the key pathways underlying this effect are unclear. However, it is challenging to compare the reported purity of SC-beta cells to that of other studies since the authors only assess SC-beta identity by expression of NKX6.1

and insulin. Because a subset of SC-EC cells can also express these markers, it will be important for subsequent studies to confirm this advance with either a more extensive marker panel that distinguishes these two cell types, or by scRNA-seq.

An alternative paradigm for controlling differentiation has been explored less: altering the genome of the starting cell populations in ways that shape developmental potential. Through a genome-wide CRISPR knockout screen, we have established the ability of more than 200 genes to alter cell fate during SC-islet differentiation (Veres et al. in press). We highlight the knockout of FBXL14 as a path toward genetically-encoded bias toward SC-beta differentiation and identify several genes with similar effects that could be mutated in parallel to further this advance. While not explored through monoclonal knockout cell lines in our study, genes such as PAX4 that increase SC-alpha cell differentiation could be mutated with the analogous goal to enhance that cell type's differentiation.

While not yet implemented in SC-beta differentiations, another form of genetic control is selection. This is exemplified by a suicide gene that could be inserted downstream of a marker of an undesired cell population, allowing selection against these cells after they emerge. This approach would be more analogous to marker sorting than to signaling factor control, as it shares limitations of potential cell loss and toxicity. While a selection-based strategy would allow population depletion, differentiating fewer of the undesired cells is preferable because substantial cell death within a 3D spheroid could negatively affect adjacent cells. In a limitation shared with marker sorting–a selection approach would not increase the absolute number of desired populations per culture, only their purity.

These strategies offer a roadmap for the optimization of SC-islets. While each has so far been pursued independently, a combination may ultimately provide the ideal control of SC-islet construction.

9 Extending Genetic Control of Islet Cell Fate

Our genome-wide CRISPR knockout screen demonstrates the potential of forward genetics to derive novel gene associations with islet differentiation (Veres et al. in press). Several extensions are feasible with tools already available today (Fig. 2). One such extension is experiments with combinations of perturbations, which could identify epistatic interactions and enable reconstructing the blueprint of relationships among genes. Similarly, experiments that perturb specific moments in time, or specific developmental events, should further refine our understanding of the mechanistic connections between relevant genes. Furthermore, perturbations are not limited to loss-of-function or inhibition. Target genes can also be activated using CRISPRa or overexpressed with pooled open reading frame (ORF) libraries. Pooled application of CRISPRa will benefit from cell line engineering that introduces a single, stably expressed Cas9 transgene to ensure uniform activation of target genes. There could be many genes that are not normally induced during SC-islet development that could perturb this system, and others whose induction or inhibition in

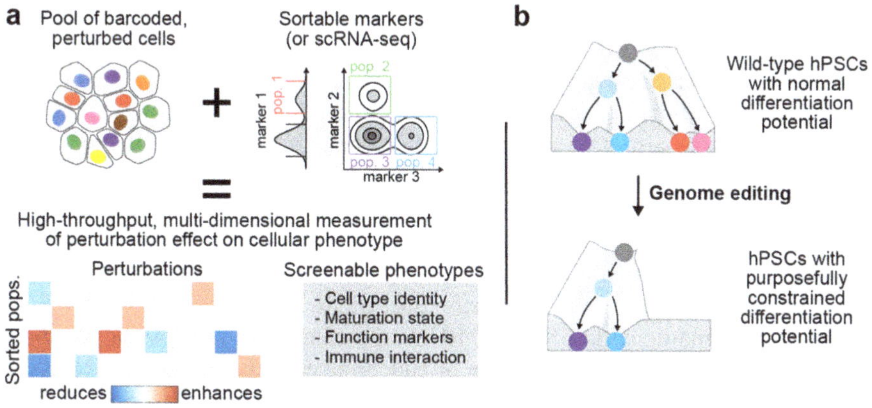

Fig. 2 Controlling SC-islet composition using genetic perturbations. (**a**) Concept of a pooled genetic screen read out through sortable makers as a method of identifying genetic perturbations that control a cellular phenotype, such as differentiation to a given cell type. (**b**) Illustration of a cell line with constrained differentiation potential using the Waddington landscape, with a depiction of a cell line that has lost the ability to differentiate to a particular progenitor and its downstream fates

different timepoints, populations, or at different doses would have considerable and perhaps considerably variable effects. Altogether, this represents a combinatorial explosion of possible choices, and designing new ways to relate them to one another in a mechanistic way will require the design, training, and deployment of sophisticated machine learning tools to guide experimental exploration.

Translating impactful screen associations to engineered cell lines for directed differentiation can take more sophisticated forms than genetically encoding the original perturbation. CRISPR-based techniques such as knockouts, inhibition, and activation allow researchers to pinpoint the effects of specific genetic changes on cell fate decisions. However, more subtle forms of genetic manipulation such as modifying regulatory elements can be used to mimic these effects. For example, disrupting a regulatory element required to induce a gene during a key developmental decision can mimic the effects of a knockout. Similarly, introducing a regulatory element characteristically bound by a lineage-specific transcription factor could induce the expression of a gene at a key decision point and mimic effects of overexpression. An alternative is simply adding the transgene as a 2A fusion to a lineage-specific transcription factor or other marker with a desirable expression pattern. As our ability to design synthetic genetic circuits such as RNA sensors via ADAR [31, 32, 54], engineer novel intracellular signaling pathways [45], and guide cell-cell-interactions [71] continues to develop, such advanced technologies will offer even more sophisticated ways of controlling fate specification and differentiation.

In the limit, a cell line with fully constrained differentiation potential could produce pure cultures of a single islet cell type. If a precise ratio of cells such as SC-beta to SC-alpha cells is preferred, this could be established by mixing two biased cell lines at the start of differentiation rather than relying on imprecisely controlled differentiation to generate this outcome. This exploration of the limits of genetic control on cell fate gives us a way to benchmark our understanding of fate decisions, both revealing and filling gaps in our mechanistic understanding of development.

An important output of genetic studies of SC-islet differentiation will be testing, which conclusions also hold in vivo. The hypothesis in vitro differentiation gains from model organism studies can be repaid in kind. For instance, many reverse genetics studies of conditional transcription factor knockouts in murine islets were performed before the emergence of scRNA-seq. While staining for the five primary islet hormones was common, unbiased cell-type analysis was not available and may have resulted in incomplete or erroneous conclusions. For instance, a study analyzing a mouse conditional knockout of Isl1 in the pancreas noted loss of islet hormones and emergence of a population of endocrine cells that no longer expressed them [18]. These were discussed as an immature islet progenitor, but, based on the strong increase in enterochromaffin cells from our ISL1 knockouts in vitro, it is likely that this murine model generated the same. Revisiting historical conclusions with unbiased cell profiling may reveal unexpected cell types and challenge models of cellular plasticity. Further, in vitro differentiation offers a window into the mechanisms of human islet development, which were previously unmeasurable and may diverge from those of model organisms in important ways.

10 Enhancing Stem Cell-Derived Islets

The next challenges in developing stem cell-derived islets as a cell therapy relates to their survival, safety, and efficacy following transplantation. These challenges range from the interaction of SC-islet cells with other cells (the immune system and vasculature), their environment (hypoxia, glucose, and other metabolites), and even their own genome (tumorigenesis).

Engineering immune evasion for stem cell-derived islets could allow for allogeneic transplantation without the need for immunosuppression or encapsulation. Because of the autoimmune nature of T1D, even autologous stem cell-derived beta cells would be rejected by the recipient. This was demonstrated in cases where patients with T1D received hemi-pancreas transplants from their healthy, identical twin siblings, and rejected these grafts in less than a year [74]. Deriving universal, hypoimmunogenic stem cell lines would also enable standardization in manufacturing stem cell therapies, avoiding the challenges, costs, and inconsistencies of deriving and differentiating patient-specific iPSCs. Current approaches are focused on layering strategies to block both the innate and adaptive immune systems [15, 24], and to induce immune tolerance [21].

Another important consideration is engineering safeguards. Cells generated in vitro have the potential to acquire oncogenic mutations [37, 80], and consequent growth advantages can cause mutant populations to expand. While oncogenic risk can be reduced through cell culture conditions optimized to minimize mutational load and mutant cell advantage and by genotyping and karyotyping, the field is also working to engineer safeguards.

One concept in this space is suicide genes, which can kill graft cells in response to an orally deliverable and bioavailable compound with limited toxicity to host cells. To remove undesirable cells without eliminating the entire graft, suicide genes can be engineered in tandem with pluripotency markers, or – especially in cases like SC-islets where the target cells proliferate rarely – with cell cycle genes [42]. Separately, kill switches linked to essential genes can be used to eliminate all graft cells as a response to tumorigenesis or another adverse event. The relevance of such safety mechanisms increases in the context of hypoimmunogenic lines that would not simply be killed by withdrawing immunosuppressive drugs. Yet the introduction of suicide genes may itself introduce new issues such as the induction of an immune reaction. Overall, it may be best to focus on the fewest number of genetic modifications that enable physiological control and use other methods to provide safeguards against these additional concerns.

Engrafted SC-islets face challenges in adapting to their new environment. One way to enhance SC-islet survival may be the inclusion of supporting cells not normally found in islets. Parathyroid tissue engrafts and vascularizes more readily than islets, and recent work suggesting that parathyroid tissue can confer these pro-engraftment effects on co-transplanted stem cell-derived islets is being tested in a clinical trial. This, however, further complicates the development of a new medicine because it involves incorporating at least one additional cell type.

Engrafted SC-islets also face a critical challenge in assuming their functional role as regulators of circulating glucose levels. A central hypothesis is that increasing SC-beta cell maturity in vitro improves their ability to control glucose levels in vivo. While SC-beta cells have been shown to function in vitro since 2014 [48], there are still differences between SC-beta cells and human islets in the ratio of secretion after incubation in high versus low glucose media, the magnitude of insulin secretion, and the curve of insulin secretion levels over time in dynamic GSIS assays. Transcriptomic studies, too, have consistently concluded that beta cells derive in vitro beta cells more closely resemble their fetal than adult counterparts [6, 27], but the degree of expression of maturation-associated genes is variable. Over several weeks of in vitro culture in minimal media without signaling modulators, the magnitude of GSIS and expression of maturation markers increases [76]. Several mechanisms for the gap in function between SC-beta cells and islets have been identified, including a bottleneck in glycolysis, response to amino acids, and the absence of circadian rhythms [1, 13, 25]. Increasing the maturity of SC-beta cells in vitro may enhance their capacity to assume control of glucose levels rapidly after transplantation. On the other hand, in practice, SC-islets do mature effectively in vivo posttransplantation and adopt a fully functional state. Thus, the issue of obtaining more advanced SC-beta cells in vitro may be primarily useful for preclinical studies on drug interactions.

Insulin was a central molecule in the past century of biology and medicine. From its discovery and application to successfully treat a then-fatal disease, to its star-studded history as the first sequenced peptide and the first recombinant therapeutic product, insulin has driven and showcased key technological advances. Stem cell-derived beta cells, a biological layer of abstraction above insulin, are following in the footsteps of the hormone itself. The production of functional beta cells in vitro has advanced close to its objective, and we have now seen proofs-of-concept in prototyping the use of these very stem cell-derived cells to achieve insulin-free control of diabetes in at least one person for nearly a year at the last update ("Vertex Provides Updates on Phase 1/2 Clinical Trial of VX-880 for the Treatment of Type 1 Diabetes – Press Release" [77]). The next prominent challenges are manufacturing scale and immunological avoidance: both will require significant progress on the basic biology of the inherent phenomena as well as on our ability to realize these discoveries in effective, enduring, and broadly available therapy.

References

1. Alvarez-Dominguez, Juan R., Julie Donaghey, Niloofar Rasouli, Jennifer H. R. Kenty, Aharon Helman, Jocelyn Charlton, Juerg R. Straubhaar, Alexander Meissner, and Douglas A. Melton. 2020. "Circadian Entrainment Triggers Maturation of Human In Vitro Islets." *Cell Stem Cell* 26 (1): 108–122.e10.
2. Andrew, A., B. Kramer, and B. B. Rawdon. 1998. "The Origin of Gut and Pancreatic Neuroendocrine (APUD) Cells--the Last Word?" *The Journal of Pathology* 186 (2): 117–18.
3. Apelqvist, A., H. Li, L. Sommer, P. Beatus, D. J. Anderson, T. Honjo, M. Hrabe de Angelis, U. Lendahl, and H. Edlund. 1999. "Notch Signalling Controls Pancreatic Cell Differentiation." *Nature* 400 (6747): 877–81.
4. Augsornworawat, Punn, Nathaniel J. Hogrebe, Matthew Ishahak, Erica Marquez, Marlie M. Maestas, Mason D. Schmidt, Daniel A. Veronese-Paniagua, et al. 2022. "Defining the Chromatin and Transcriptional Landscape of Stem Cell-Derived Islets." *BioRxiv*. https://doi.org/10.1101/2022.02.26.482126.
5. Augsornworawat, Punn, Kristina G. Maxwell, Leonardo Velazco-Cruz, and Jeffrey R. Millman. 2020. "Single-Cell Transcriptome Profiling Reveals β Cell Maturation in Stem Cell-Derived Islets after Transplantation." *Cell Reports* 32 (8): 108067.
6. Balboa, Diego, Tom Barsby, Väinö Lithovius, Jonna Saarimäki-Vire, Muhmmad Omar-Hmeadi, Oleg Dyachok, Hossam Montaser, et al. 2022. "Functional, Metabolic and Transcriptional Maturation of Human Pancreatic Islets Derived from Stem Cells." *Nature Biotechnology* 40 (7): 1042–55.
7. Ballinger, W. F., and P. E. Lacy. 1972. "Transplantation of Intact Pancreatic Islets in Rats." *Surgery* 72 (2): 175–86.
8. Banting, F. G., C. H. Best, J. B. Collip, W. R. Campbell, and A. A. Fletcher. 1922. "Pancreatic Extracts in the Treatment of Diabetes Mellitus." *Canadian Medical Association Journal* 12 (3): 141–46.
9. Baron, Maayan, Adrian Veres, Samuel L. Wolock, Aubrey L. Faust, Renaud Gaujoux, Amedeo Vetere, Jennifer Hyoje Ryu, et al. 2016. "A Single-Cell Transcriptomic Map of the Human and Mouse Pancreas Reveals Inter- and Intra-Cell Population Structure." *Cell Systems* 3 (4): 346–360.e4.
10. Bouchi, Ryotaro, Kylie S. Foo, Haiqing Hua, Kyoichiro Tsuchiya, Yoshiaki Ohmura, P. Rodrigo Sandoval, Lloyd E. Ratner, Dieter Egli, Rudolph L. Leibel, and Domenico Accili. 2014. "FOXO1 Inhibition Yields Functional Insulin-Producing Cells in Human Gut Organoid Cultures." *Nature Communications* 5 (June): 4242.
11. Busslinger, Georg A., Bas L. A. Weusten, Auke Bogte, Harry Begthel, Lodewijk A. A. Brosens, and Hans Clevers. 2021. "Human Gastrointestinal Epithelia of the Esophagus, Stomach, and Duodenum Resolved at Single-Cell Resolution." *Cell Reports* 34 (10): 108819.

12. D'Amour, Kevin A., Anne G. Bang, Susan Eliazer, Olivia G. Kelly, Alan D. Agulnick, Nora G. Smart, Mark A. Moorman, Evert Kroon, Melissa K. Carpenter, and Emmanuel E. Baetge. 2006. "Production of Pancreatic Hormone--Expressing Endocrine Cells from Human Embryonic Stem Cells." *Nature Biotechnology* 24 (11): 1392.
13. Davis, Jeffrey C., Tiago C. Alves, Aharon Helman, Jonathan C. Chen, Jennifer H. Kenty, Rebecca L. Cardone, David R. Liu, Richard G. Kibbey, and Douglas A. Melton. 2020. "Glucose Response by Stem Cell-Derived β Cells in Vitro Is Inhibited by a Bottleneck in Glycolysis." *Cell Reports* 31 (6): 107623.
14. Desgraz, Renaud, and Pedro L. Herrera. 2009. "Pancreatic Neurogenin 3-Expressing Cells Are Unipotent Islet Precursors." *Development* 136 (21): 3567–74.
15. Deuse, Tobias, Xiaomeng Hu, Alessia Gravina, Dong Wang, Grigol Tediashvili, Chandrav De, William O. Thayer, et al. 2019. "Hypoimmunogenic Derivatives of Induced Pluripotent Stem Cells Evade Immune Rejection in Fully Immunocompetent Allogeneic Recipients." *Nature Biotechnology*, 1.
16. Dolenšek, Jurij, Marjan Slak Rupnik, and Andraž Stožer. 2015. "Structural Similarities and Differences between the Human and the Mouse Pancreas." *Islets* 7 (1): e1024405.
17. Dor, Yuval, Juliana Brown, Olga I. Martinez, and Douglas A. Melton. 2004. "Adult Pancreatic Beta-Cells Are Formed by Self-Duplication Rather than Stem-Cell Differentiation." *Nature* 429 (6987): 41–46.
18. Du, Aiping, Chad S. Hunter, Johanna Murray, Daniel Noble, Chen-Leng Cai, Sylvia M. Evans, Roland Stein, and Catherine Lee May. 2009. "Islet-1 Is Required for the Maturation, Proliferation, and Survival of the Endocrine Pancreas." *Diabetes* 58 (9): 2059–69.
19. Furuyama, Kenichiro, Simona Chera, Léon van Gurp, Daniel Oropeza, Luiza Ghila, Nicolas Damond, Heidrun Vethe, et al. 2019. "Diabetes Relief in Mice by Glucose-Sensing Insulin-Secreting Human α-Cells." *Nature* 567 (7746): 43–48.
20. Gehart, Helmuth, Johan H. van Es, Karien Hamer, Joep Beumer, Kai Kretzschmar, Johanna F. Dekkers, Anne Rios, and Hans Clevers. 2019. "Identification of Enteroendocrine Regulators by Real-Time Single-Cell Differentiation Mapping." *Cell*.
21. Gerace, Dario, Quan Zhou, Jennifer Hyoje-Ryu Kenty, Adrian Veres, Elad Sintov, Xi Wang, Kyle R. Boulanger, Hongfei Li, and Douglas A. Melton. 2023. "Engineering Human Stem Cell-Derived Islets to Evade Immune Rejection and Promote Localized Immune Tolerance." *Cell Reports. Medicine* 4 (1): 100879.
22. Gradwohl, Gérard, Andrée Dierich, Marianne LeMeur, and François Guillemot. 2000. "Neurogenin3 Is Required for the Development of the Four Endocrine Cell Lineages of the Pancreas." *Proceedings of the National Academy of Sciences* 97 (4): 1607–11.
23. Guz, Y., M. R. Montminy, R. Stein, J. Leonard, L. W. Gamer, C. V. Wright, and G. Teitelman. 1995. "Expression of Murine STF-1, a Putative Insulin Gene Transcription Factor, in Beta Cells of Pancreas, Duodenal Epithelium and Pancreatic Exocrine and Endocrine Progenitors during Ontogeny." *Development* 121 (1): 11–18.
24. Han, Xiao, Mengning Wang, Songwei Duan, Paul J. Franco, Jennifer Hyoje-Ryu Kenty, Preston Hedrick, Yulei Xia, et al. 2019. "Generation of Hypoimmunogenic Human Pluripotent Stem Cells." *Proceedings of the National Academy of Sciences of the United States of America* 116 (21): 10441–46.
25. Helman, Aharon, Andrew L. Cangelosi, Jeffrey C. Davis, Quan Pham, Arielle Rothman, Aubrey L. Faust, Juerg R. Straubhaar, David M. Sabatini, and Douglas A. Melton. 2020. "A Nutrient-Sensing Transition at Birth Triggers Glucose-Responsive Insulin Secretion." *Cell Metabolism* 31 (5): 1004–1016.e5.
26. Hiyoshi, Hideyuki, Kensuke Sakuma, Noriko Tsubooka-Yamazoe, Shinya Asano, Taisuke Mochida, Junji Yamaura, Shuhei Konagaya, et al. 2022. "Characterization and Reduction of Non-Endocrine Cells Accompanying Islet-like Endocrine Cells Differentiated from Human IPSC." *Scientific Reports* 12 (1): 4740.
27. Hrvatin, Sinisa, Charles W. O'Donnell, Francis Deng, Jeffrey R. Millman, Felicia Walton Pagliuca, Philip DiIorio, Alireza Rezania, David K. Gifford, and Douglas A. Melton. 2014. "Differentiated Human Stem Cells Resemble Fetal, Not Adult, β Cells." *Proceedings of the National Academy of Sciences* 111 (8): 3038–43.

28. Huang, Xiaofeng, Wei Gu, Jiaoyue Zhang, Ying Lan, Jonathan L. Colarusso, Sanlan Li, Christoph Pertl, et al. 2022. “Restoring Glucose Homeostasis with Stomach-Derived Human Insulin-Secreting Organoids.” *BioRxiv*. https://doi.org/10.1101/2022.12.15.520488.
29. Islam, Saiful, Una Kjällquist, Annalena Moliner, Pawel Zajac, Jian-Bing Fan, Peter Lönnerberg, and Sten Linnarsson. 2011. “Characterization of the Single-Cell Transcriptional Landscape by Highly Multiplex RNA-Seq.” *Genome Research* 21 (7): 1160–67.
30. Jensen, J., E. E. Pedersen, P. Galante, J. Hald, R. S. Heller, M. Ishibashi, R. Kageyama, F. Guillemot, P. Serup, and O. D. Madsen. 2000. “Control of Endodermal Endocrine Development by Hes-1.” *Nature Genetics* 24 (1): 36–44.
31. Jiang, Kaiyi, Jeremy Koob, Xi Dawn Chen, Rohan N. Krajeski, Yifan Zhang, Verena Volf, Wenyuan Zhou, et al. 2022. “Programmable Eukaryotic Protein Synthesis with RNA Sensors by Harnessing ADAR.” *Nature Biotechnology*, October. https://doi.org/10.1038/s41587-022-01534-5.
32. Kaseniit, K. Eerik, Noa Katz, Natalie S. Kolber, Connor C. Call, Diego L. Wengier, Will B. Cody, Elizabeth S. Sattely, and Xiaojing J. Gao. 2022. “Modular, Programmable RNA Sensing Using ADAR Editing in Living Cells.” *Nature Biotechnology*, October. https://doi.org/10.1038/s41587-022-01493-x.
33. Kelly, W. D., R. C. Lillehei, F. K. Merkel, Y. Idezuki, and F. C. Goetz. 1967. “Allotransplantation of the Pancreas and Duodenum along with the Kidney in Diabetic Nephropathy.” *Surgery* 61 (6): 827–37.
34. Kemp, C. B., M. J. Knight, D. W. Scharp, P. E. Lacy, and W. F. Ballinger. 1973. “Transplantation of Isolated Pancreatic Islets into the Portal Vein of Diabetic Rats.” *Nature* 244 (5416): 447.
35. Klein, Allon M., Linas Mazutis, Ilke Akartuna, Naren Tallapragada, Adrian Veres, Victor Li, Leonid Peshkin, David A. Weitz, and Marc W. Kirschner. 2015. “Droplet Barcoding for Single-Cell Transcriptomics Applied to Embryonic Stem Cells.” *Cell* 161 (5): 1187–1201.
36. Kroon, Evert, Laura A. Martinson, Kuniko Kadoya, Anne G. Bang, Olivia G. Kelly, Susan Eliazer, Holly Young, et al. 2008. “Pancreatic Endoderm Derived from Human Embryonic Stem Cells Generates Glucose-Responsive Insulin-Secreting Cells in Vivo.” *Nature Biotechnology* 26 (4): 443.
37. Kuijk, Ewart, Myrthe Jager, Bastiaan van der Roest, Mauro D. Locati, Arne Van Hoeck, Jerome Korzelius, Roel Janssen, et al. 2020. “The Mutational Impact of Culturing Human Pluripotent and Adult Stem Cells.” *Nature Communications* 11 (1): 2493.
38. Lee, J. C., S. B. Smith, H. Watada, J. Lin, D. Scheel, J. Wang, R. G. Mirmira, and M. S. German. 2001. “Regulation of the Pancreatic Pro-Endocrine Gene Neurogenin3.” *Diabetes* 50 (5): 928–36.
39. Leite, Nayara C., Elad Sintov, Torsten B. Meissner, Michael A. Brehm, Dale L. Greiner, David M. Harlan, and Douglas A. Melton. 2020. “Modeling Type 1 Diabetes In Vitro Using Human Pluripotent Stem Cells.” *Cell Reports* 32 (2): 107894.
40. Liu, Haisong, Ronghui Li, Hsin-Kai Liao, Zheying Min, Chao Wang, Yang Yu, Lei Shi, et al. 2021. “Chemical Combinations Potentiate Human Pluripotent Stem Cell-Derived 3D Pancreatic Progenitor Clusters toward Functional β Cells.” *Nature Communications* 12 (1): 3330.
41. Macosko, Evan Z., Anindita Basu, Rahul Satija, James Nemesh, Karthik Shekhar, Melissa Goldman, Itay Tirosh, et al. 2015. “Highly Parallel Genome-Wide Expression Profiling of Individual Cells Using Nanoliter Droplets.” *Cell* 161 (5): 1202–14.
42. Martin, Renata M., Jonas L. Fowler, M. Kyle Cromer, Benjamin J. Lesch, Ezequiel Ponce, Nobuko Uchida, Toshinobu Nishimura, Matthew H. Porteus, and Kyle M. Loh. 2020. “Improving the Safety of Human Pluripotent Stem Cell Therapies Using Genome-Edited Orthogonal Safeguards.” *Nature Communications* 11 (1): 2713.
43. McCracken, Kyle W., Eitaro Aihara, Baptiste Martin, Calyn M. Crawford, Taylor Broda, Julie Treguier, Xinghao Zhang, John M. Shannon, Marshall H. Montrose, and James M. Wells. 2017. “Wnt/β-Catenin Promotes Gastric Fundus Specification in Mice and Humans.” *Nature* 541 (7636): 182–87.

44. Meier, Juris J., Irene Hong-McAtee, Ryan Galasso, Johannes D. Veldhuis, Antoinette Moran, Bernhard J. Hering, and Peter C. Butler. 2006. "Intrahepatic Transplanted Islets in Humans Secrete Insulin in a Coordinate Pulsatile Manner Directly into the Liver." *Diabetes* 55 (8): 2324–32.
45. Morsut, Leonardo, Kole T. Roybal, Xin Xiong, Russell M. Gordley, Scott M. Coyle, Matthew Thomson, and Wendell A. Lim. 2016. "Engineering Customized Cell Sensing and Response Behaviors Using Synthetic Notch Receptors." *Cell* 164 (4): 780–91.
46. Nostro, M. Cristina, Farida Sarangi, Shinichiro Ogawa, Audrey Holtzinger, Barbara Corneo, Xueling Li, Suzanne J. Micallef, et al. 2011. "Stage-Specific Signaling through TGFβ Family Members and WNT Regulates Patterning and Pancreatic Specification of Human Pluripotent Stem Cells." *Development* 138 (5): 861–71.
47. Olaniru, Oladapo Edward, Ulrich Kadolsky, Shichina Kannambath, Heli Vaikkinen, Kathy Fung, Pawan Dhami, and Shanta J. Persaud. 2023. "Single-Cell Transcriptomic and Spatial Landscapes of the Developing Human Pancreas." *Cell Metabolism* 35 (1): 184–199.e5.
48. Pagliuca, Felicia W., Jeffrey R. Millman, Mads Gürtler, Michael Segel, Alana Van Dervort, Jennifer Hyoje Ryu, Quinn P. Peterson, Dale Greiner, and Douglas A. Melton. 2014. "Generation of Functional Human Pancreatic β Cells in Vitro." *Cell* 159 (2): 428–39.
49. Parent, Audrey V., Sudipta Ashe, Gopika G. Nair, Mei-Lan Li, Jessica Chavez, Jennifer S. Liu, Yongping Zhong, Philip R. Streeter, and Matthias Hebrok. 2022. "Development of a Scalable Method to Isolate Subsets of Stem Cell-Derived Pancreatic Islet Cells." *Stem Cell Reports* 17 (4): 979–92.
50. Pellegrini, Silvia, Raniero Chimienti, Giulia Maria Scotti, Francesca Giannese, Dejan Lazarevic, Fabio Manenti, Gaia Poggi, et al. 2021. "Transcriptional Dynamics of Induced Pluripotent Stem Cell Differentiation into β Cells Reveals Full Endodermal Commitment and Homology with Human Islets." *Cytotherapy* 23 (4): 311–19.
51. Perez-Frances, Marta, Léon van Gurp, Maria Valentina Abate, Valentina Cigliola, Kenichiro Furuyama, Eva Bru-Tari, Daniel Oropeza, et al. 2021. "Pancreatic Ppy-Expressing γ-Cells Display Mixed Phenotypic Traits and the Adaptive Plasticity to Engage Insulin Production." *Nature Communications* 12 (1): 4458.
52. Peterson, Quinn P., Adrian Veres, Lihua Chen, Michael Q. Slama, Jennifer H. R. Kenty, Shaimaa Hassoun, Matthew R. Brown, et al. 2020. "A Method for the Generation of Human Stem Cell-Derived Alpha Cells." *Nature Communications* 11 (1): 2241.
53. Pictet, R. L., L. B. Rall, P. Phelps, and W. J. Rutter. 1976. "The Neural Crest and the Origin of the Insulin-Producing and Other Gastrointestinal Hormone-Producing Cells." *Science* 191 (4223): 191–92.
54. Qian, Yongjun, Jiayun Li, Shengli Zhao, Elizabeth A. Matthews, Michael Adoff, Weixin Zhong, Xu An, et al. 2022. "Programmable RNA Sensing for Cell Monitoring and Manipulation." *Nature* 610 (7933): 713–21.
55. Quianzon, Celeste C., and Issam Cheikh. 2012. "History of Insulin." *Journal of Community Hospital Internal Medicine Perspectives* 2 (2). https://doi.org/10.3402/jchimp.v2i2.18701.
56. Rezania, Alireza, Jennifer E. Bruin, Payal Arora, Allison Rubin, Irina Batushansky, Ali Asadi, Shannon O'Dwyer, et al. 2014. "Reversal of Diabetes with Insulin-Producing Cells Derived in Vitro from Human Pluripotent Stem Cells." *Nature Biotechnology* 32 (11): 1121–33.
57. Rodriguez-Diaz, Rayner, R. Damaris Molano, Jonathan R. Weitz, Midhat H. Abdulreda, Dora M. Berman, Barbara Leibiger, Ingo B. Leibiger, et al. 2018. "Paracrine Interactions within the Pancreatic Islet Determine the Glycemic Set Point." *Cell Metabolism* 27 (3): 549–558.e4.
58. Rosado-Olivieri, Edwin A., Kendall Anderson, Jennifer H. Kenty, and Douglas A. Melton. 2019. "YAP Inhibition Enhances the Differentiation of Functional Stem Cell-Derived Insulin-Producing β Cells." *Nature Communications* 10 (1): 1464.
59. Rulifson, Eric J., Seung K. Kim, and Roel Nusse. 2002. "Ablation of Insulin-Producing Neurons in Flies: Growth and Diabetic Phenotypes." *Science* 296 (5570): 1118–20.
60. Rutlin, M., D. Rastelli, W. T. Kuo, J. A. Estep, A. Louis, M. M. Riccomagno, J. R. Turner, and M. Rao. 2020. "The Villin1 Gene Promoter Drives Cre Recombinase Expression in

Extraintestinal Tissues." *Cellular and Molecular Gastroenterology and Hepatology* 10 (4): 864–867.e5.
61. Schernthaner, G. 1993. "Immunogenicity and Allergenic Potential of Animal and Human Insulins." *Diabetes Care* 16 Suppl 3 (December): 155–65.
62. Schraenen, A., K. Lemaire, G. de Faudeur, N. Hendrickx, M. Granvik, L. Van Lommel, J. Mallet, et al. 2010. "Placental Lactogens Induce Serotonin Biosynthesis in a Subset of Mouse Beta Cells during Pregnancy." *Diabetologia* 53 (12): 2589–99.
63. Schreiber, Valérie, Reuben Mercier, Sara Jiménez, Tao Ye, Emmanuel García-Sánchez, Annabelle Klein, Aline Meunier, et al. 2021. "Extensive NEUROG3 Occupancy in the Human Pancreatic Endocrine Gene Regulatory Network." *Molecular Metabolism* 53 (November): 101313.
64. Schwitzgebel, V. M., D. W. Scheel, J. R. Conners, J. Kalamaras, J. E. Lee, D. J. Anderson, L. Sussel, J. D. Johnson, and M. S. German. 2000. "Expression of Neurogenin3 Reveals an Islet Cell Precursor Population in the Pancreas." *Development* 127 (16): 3533–42.
65. Shapiro, A. M. James, Jonathan R. T. Lakey, Edmond A. Ryan, Gregory S. Korbutt, Ellen Toth, Garth L. Warnock, Norman M. Kneteman, and Ray V. Rajotte. 2000. "Islet Transplantation in Seven Patients with Type 1 Diabetes Mellitus Using a Glucocorticoid-Free Immunosuppressive Regimen." *The New England Journal of Medicine* 343 (4): 230–38.
66. Shapiro, A. M. James, Marta Pokrywczynska, and Camillo Ricordi. 2017. "Clinical Pancreatic Islet Transplantation." *Nature Reviews. Endocrinology* 13 (5): 268.
67. Sharon, Nadav, Raghav Chawla, Jonas Mueller, Jordan Vanderhooft, Luke James Whitehorn, Benjamin Rosenthal, Mads Gürtler, et al. 2019. "A Peninsular Structure Coordinates Asynchronous Differentiation with Morphogenesis to Generate Pancreatic Islets." *Cell.*
68. Sharon, Nadav, Jordan Vanderhooft, Juerg Straubhaar, Jonas Mueller, Raghav Chawla, Quan Zhou, Elise N. Engquist, Cole Trapnell, David K. Gifford, and Douglas A. Melton. 2019. "Wnt Signaling Separates the Progenitor and Endocrine Compartments during Pancreas Development." *Cell Reports* 27 (8): 2281–2291.e5.
69. Sintov, Elad, Igor Nikolskiy, Victor Barrera, Jennifer Hyoje-Ryu Kenty, Alexander S. Atkin, Dario Gerace, Shannan J. Ho Sui, Kyle Boulanger, and Douglas A. Melton. 2022. "Whole-Genome CRISPR Screening Identifies Genetic Manipulations to Reduce Immune Rejection of Stem Cell-Derived Islets." *Stem Cell Reports* 17 (9): 1976–90.
70. Steiner, Donald J., Abraham Kim, Kevin Miller, and Manami Hara. 2010. "Pancreatic Islet Plasticity: Interspecies Comparison of Islet Architecture and Composition." *Islets* 2 (3): 135–45.
71. Stevens, Adam J., Andrew R. Harris, Josiah Gerdts, Ki H. Kim, Coralie Trentesaux, Jonathan T. Ramirez, Wesley L. McKeithan, et al. 2023. "Programming Multicellular Assembly with Synthetic Cell Adhesion Molecules." *Nature* 614 (7946): 144–52.
72. Stoffers, D. A., N. T. Zinkin, V. Stanojevic, W. L. Clarke, and J. F. Habener. 1997. "Pancreatic Agenesis Attributable to a Single Nucleotide Deletion in the Human IPF1 Gene Coding Sequence." *Nature Genetics* 15 (1): 106–10.
73. Sutherland, D. E., R. W. Gruessner, D. L. Dunn, A. J. Matas, A. Humar, R. Kandaswamy, S. M. Mauer, et al. 2001. "Lessons Learned from More than 1,000 Pancreas Transplants at a Single Institution." *Annals of Surgery* 233 (4): 463–501.
74. Sutherland, D. E., R. Sibley, X. Z. Xu, A. Michael, A. M. Srikanta, F. Taub, J. Najarian, and F. C. Goetz. 1984. "Twin-to-Twin Pancreas Transplantation: Reversal and Reenactment of the Pathogenesis of Type I Diabetes." *Transactions of the Association of American Physicians* 97: 80–87.
75. Terry, Natalie A., Erik R. Walp, Randall A. Lee, Klaus H. Kaestner, and Catherine Lee May. 2014. "Impaired Enteroendocrine Development in Intestinal-Specific Islet1 Mouse Mutants Causes Impaired Glucose Homeostasis." *American Journal of Physiology. Gastrointestinal and Liver Physiology* 307 (10): G979–91.
76. Veres, Adrian, Aubrey L. Faust, Henry L. Bushnell, Elise N. Engquist, Jennifer Hyoje-Ryu Kenty, George Harb, Yeh-Chuin Poh, et al. 2019. "Charting Cellular Identity during Human in Vitro β-Cell Differentiation." *Nature* 569 (7756): 368–73.

77. "Vertex Provides Updates on Phase 1/2 Clinical Trial of VX-880 for the Treatment of Type 1 Diabetes – Press Release." 2022. Vertex Pharmaceutical Newsroom. 2022. https://news.vrtx.com/news-releases/news-release-details/vertex-provides-updates-phase-12-clinical-trial-vx-880-treatment?_ga=2.159281658.1976707806.1676298239-1596731475.1676298239.
78. Wang, Peng, Juan-Carlos Alvarez-Perez, Dan P. Felsenfeld, Hongtao Liu, Sharmila Sivendran, Aaron Bender, Anil Kumar, et al. 2015. "A High-Throughput Chemical Screen Reveals That Harmine-Mediated Inhibition of DYRK1A Increases Human Pancreatic Beta Cell Replication." *Nature Medicine* 21 (4): 383.
79. Weng, Chen, Jiajia Xi, Haiyan Li, Jian Cui, Anniya Gu, Sisi Lai, Konstantin Leskov, Luxin Ke, Fulai Jin, and Yan Li. 2020. "Single-Cell Lineage Analysis Reveals Extensive Multimodal Transcriptional Control during Directed Beta-Cell Differentiation." *Nature Metabolism* 2 (12): 1443–58.
80. Werbowetski-Ogilvie, Tamra E., Marc Bossé, Morag Stewart, Angelique Schnerch, Veronica Ramos-Mejia, Anne Rouleau, Tracy Wynder, et al. 2009. "Characterization of Human Embryonic Stem Cells with Features of Neoplastic Progression." *Nature Biotechnology* 27 (1): 91–97.
81. Wierup, Nils, Frank Sundler, and R. Scott Heller. 2014. "The Islet Ghrelin Cell." *Journal of Molecular Endocrinology* 52 (1): R35–49.
82. Zhou, Qiao, Juliana Brown, Andrew Kanarek, Jayaraj Rajagopal, and Douglas A. Melton. 2008. "In Vivo Reprogramming of Adult Pancreatic Exocrine Cells to Beta-Cells." *Nature* 455 (7213): 627–32.
83. Zhu, Han, Gaowei Wang, Kim-Vy Nguyen-Ngoc, Dongsu Kim, Michael Miller, Georgina Goss, Jenna Kovsky, et al. 2022. "Improving Stem Cell-Derived Pancreatic Islets Using Single-Cell Multiome-Inferred Regulomes." *BioRxiv*. https://doi.org/10.1101/2022.09.25.509355.
84. Zorn, Aaron M., and James M. Wells. 2009. "Vertebrate Endoderm Development and Organ Formation." *Annual Review of Cell and Developmental Biology* 25 (1): 221–51.

Genetic Regulatory Networks Guiding Islet Development

Xin-Xin Yu, Xin Wang, Wei-Lin Qiu, Liu Yang, and Cheng-Ran Xu

1 Introduction

The pancreas is a compound organ containing endocrine and exocrine compartments that regulate glucose homeostasis and nutrition metabolism, respectively. The endocrine function is executed by islets, which are mini-organs composed primarily of four hormone-secreting endocrine cell types in adults (insulin-secreting β cells, glucagon-secreting α cells, somatostatin-secreting δ cells, and pancreatic polypeptide-secreting PP cells), as well as endothelial cells, neurons, mesenchymal cells and immune cells. The exocrine function is carried out by acinar cells that secrete digestive enzymes and ductal cells that transport pancreatic juices into the duodenum.

Diabetes mellitus is a chronic metabolic disease characterized by endocrine β-cell dysfunction or insulin resistance that causes substantial socioeconomic and medical burdens worldwide. Patients with diabetes suffer from hyperglycemia and consequent serious complications, such as kidney failure, stroke, heart disease and vision loss [6, 90, 112, 121]. The current treatments to improve diabetes symptoms include exogenous insulin supplementation or the transplantation of cadaveric pancreas or islets, although they suffer from a lack of precise physiological regulation

Xin-Xin Yu, Xin Wang and Wei-Lin Qiu contributed equally.

X.-X. Yu · W.-L. Qiu · L. Yang · C.-R. Xu (✉)
School of Basic Medical Sciences, Department of Medical Genetics, Peking University Health Science Center, Peking-Tsinghua Center for Life Sciences, Peking University, Beijing, China
e-mail: cxu@pku.edu.cn

X. Wang
School of Life Sciences, Peking University, Beijing, China

L. Piemonti et al. (eds.), *Pluripotent Stem Cell Therapy for Diabetes*,
https://doi.org/10.1007/978-3-031-41943-0_2

of blood glucose and donor shortages, respectively. The regeneration of significant quantities of β cells or islet organs from human embryonic stem cells (hESCs) or human induced pluripotent stem cells (hiPSCs) is a promising cell therapy modality for diabetes and has attracted considerable interest.

To generate functional islet cells from hESCs or hiPSCs, induction methods are continuously optimized by adjusting signaling pathways to recapitulate in vivo genetic programs of pancreatic lineage differentiation [19, 68, 98, 116, 120, 133, 135, 167]. During cell differentiation, environmental signals derived from the niche (input) trigger intracellular pathways that alter the epigenetic state and genetic regulatory network (GRN), leading to lineage specification or the establishment of particular cell states (output). Thus, the GRN functions as an intermediary between exogenous signal inputs and phenotypic outputs. In this paper, we summarize the GRNs involved in pancreatic lineage development in vivo to advance pancreatic islet regeneration from stem cells in vitro.

2 GRN Facilitates Comprehension of the Regulatory Logic of Biological Processes

GRN is defined as a collection of hierarchical regulatory networks among regulators and their potential targets in particular biological processes, such as dynamic cell fate switching and the maintenance of cellular homeostasis [101]. A GRN is composed of two typical network properties, nodes and edges. Nodes refer to genes and gene products, such as transcription factors (TFs) and miRNAs, and in some cases also to the cis-regulatory elements of genes. Edges represent active or repressed regulatory relationships among genes that are mediated by TF–DNA or protein–protein interactions [151]. A subset of highly connected neighboring nodes is called a module, and genes in the same module are considered to have similar functions or expression patterns [21]. Uncovering the GRNs underlying diverse biological processes enhances the understanding of developmental processes and mechanisms of disease onset. With the technological developments over the past decades, analytical methods for GRNs are evolving toward higher throughput and higher accuracy. In the following, we briefly describe the stages of analytical methods development for inferring GRNs, as well as their advantages and disadvantages (Fig. 1, Table 1).

Stage I: GRNs were constructed based on experimental systems for gene perturbation assays and cis-regulatory analysis [94]. In brief, direct or indirect gene–gene interactions are implied by altering the expression of certain regulatory genes by gene deletion or overexpression and then identifying subsequent changes in expression patterns using techniques such as microarray analysis or quantitative PCR. In some instances, the physical interaction between a given TF and its putative target DNA sequence(s) is further validated by cis-regulatory analysis, such as the electrophoresis mobility shift assay (EMSA) or luciferase assay [40, 63, 156]. This method for inferring regulatory relationships among signals, TFs and genes is relatively accurate, albeit a low-throughput and time consuming approach [93, 94].

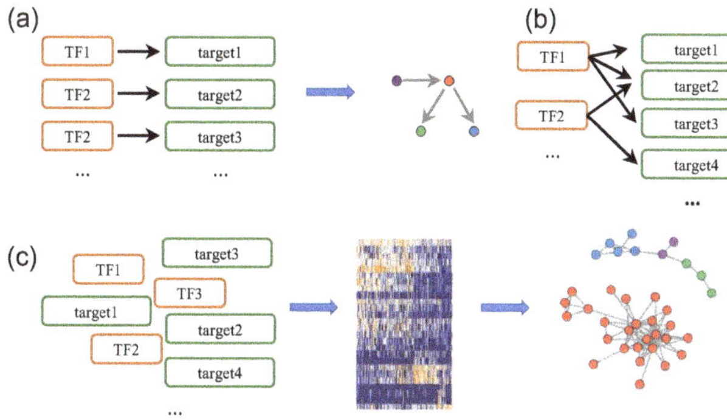

Fig. 1 Strategies for building GRNs. (**a**) Diagram showing the strategy for building a GRN based on data from gene perturbation assays and cis-regulatory analysis. (**b**) Diagram showing the strategy for building a GRN based on ChIP-seq data from multiple TFs. (**c**) Diagram showing the strategy for building a GRN based on RNA-seq, and single-cell multiomics data

Table 1 Approaches for inferring GRNs

	Approaches	Advantages	Disadvantages
Gene perturbation	Gene knockout Gene overexpression	Accurate	Low throughput; time-consuming
Transcriptome analysis	Bulk-cell RNA-seq	High-throughput	High false-positive rate; lack of regulatory hierarchy
	ScRNA-seq	High-throughput; more accurate than bulk-cell RNA-seq; predicts regulatory hierarchy	High false-positive rate; fewer gene counts than bulk-cell RNA-seq
	ScRNA-seq & CRISPR screen	High-throughput; more reliable and efficient than RNA-seq	Off-target effects and quantified error rates of CRISPR screen
Epigenome analysis	ChIP-seq	Efficiently infers TF targets	Requires specific antibodies; difficult to correlate cis-regulatory elements with target genes
	ATAC-seq	High-throughput; no antibodies required	Dependency on Motif database; difficult to correlate cis-regulatory elements with target genes

Stage II: With the advent of high-throughput sequencing techniques, such as RNA-sequencing (RNA-seq) and chromatin immunoprecipitation sequencing (ChIP-seq), and the development of computational tools for the analysis of the generated data, GRNs can be built from a genome-scale perspective.

RNA-seq permits the identification of relationships among thousands of genes from a variety of samples or developmental stages. Gene coexpression networks (GCNs) are typically employed to construct artificial GRNs based on gene expression profiles and are typically mined by the WGCNA [88] and ARACEN [103] algorithms. In a GCN, each gene is considered a node, and the correlation coefficients between genes are used as the weights of edges, which can be calculated using Pearson correlation, Spearman correlation, mutual information or proportionality measures [119, 132]. In GRNs, the directed edges could reflect the regulatory hierarchy from upstream to downstream; however, the edges in GCNs are undirected and thus only reflect relationships but ignore their causality. To overcome this drawback, some algorithms introduce directional inference to generate directed artificial GRNs. The inference can be based on statistical features inside the dataset, as in the GENIE3 [72] and DREAM4 [58] algorithms, or on external information about TF binding sites or from time series, as in SCENIC [9] and MapReduce [1] algorithms.

ChIP-seq may efficiently identify genome-wide TF-binding or chromatin-modified loci and infer the downstream targets of a particular TF. GRNs can subsequently be utilized to delineate the directed network of TFs and their respective targets [12, 124, 169]. Recently, the novel CUT&RUN and CUT&Tag ChIP-seq technologies have been developed as powerful tools to capture TF-bound DNA more efficiently than traditional ChIP-seq technologies [83, 84, 109]. Nevertheless, the establishment of GRNs that rely on TF binding site information still requires a substantial amount of ChIP-seq experimental data, the quality of which is largely dependent on the quality of the antibodies used in the ChIP experiments.

Stage III: The rapid development of single-cell technologies and Assay for Transposase-Accessible Chromatin with high-throughput sequencing (ATAC-seq) has brought about the era of single-cell multiomics. Single-cell multiomics approaches provide a higher-resolution and more comprehensive understanding of organismal developmental processes. Although single-cell RNA-seq (scRNA-seq) detects fewer genes than bulk-cell RNA-seq, it is able to build GRNs more accurately because it can identify cell types and cell heterogeneity that are obscured in bulk-cell RNA-seq. As a result, scRNA-seq facilitates the elimination of false coexpression relationships suggested by cell heterogeneity and the detection of weaker GCN modules (e.g., the cell cycle). In addition, scRNA-seq can provide directional inference for GRNs that reference time-series information derived from pseudotime and RNA velocity analyses during development [87]. Algorithms for building artificial GRNs from scRNA-seq datasets were developed, such as SCODE [108], GRISLI [17] and Scribe [131]. The combination of scRNA-seq and CRISPR screening of gene perturbation, such as perturb-seq [43], CROP-seq [39] and TAP-seq [141], balances the reliability and efficiency of GRN construction by conducting perturbation experiments on many genes simultaneously. Nevertheless, the systematic false positives in these high-throughput experiments remain a challenge to the generation of more reliable GRNs.

ATAC-seq has been combined with TF motif analysis to infer putative regulatory relationships between TFs and target genes [48, 70, 99]. The accessible chromatin regions identified by ATAC-seq can be used to infer binding sites for TFs with known motif information, which is more efficient than identifying TF binding sites with ChIP-seq experiments but requires further validation of the accuracy of the inferred binding sites. Notably, one common issue with ATAC-seq and ChIP-seq analyses is the difficulty in correlating cis-regulatory elements with their actual target genes. To address this issue, the Cicero algorithm was developed to predict cis-regulatory interactions by identifying regions of coaccessibility within the scATAC-seq dataset [128].

Overall, GRN facilitates the grasp of regulatory logic under various contexts, including developmental processes and disease states, and provides new insights for the identification of essential genes. GRN can also predict potential interactions from large-scale datasets by inferring gene–gene relationships, and this information can enhance the efficiency of experimental design.

3 Developmental Pathway of Pancreatic Lineages

Based on genetic investigations and lineage tracing studies in animal models conducted over the past few decades, we have obtained a framework for a pancreatic lineage hierarchy, but there are still many unanswered concerns [24, 89, 150]. Fortunately, recent breakthroughs in single-cell technologies have enhanced our understanding of the molecular characteristics, intermediate cell populations and developmental trajectories of pancreatic lineages during organogenesis [32, 171, 183].

The pancreas is derived from dorsal and ventral definitive endoderm (DE) domains that form during gastrulation and regionalizes into the foregut, foregut lip, midgut and hindgut domains [96]. Previous studies have suggested that the pancreas originates from the DE foregut region [187]; however, our recent study redefined the molecular characteristics and borders of the DE regions using scRNA-seq and indicated that the pancreatic progenitors originate from the DE midgut region at embryonic day (E) 8.5 in mouse [96] (Fig. 2). Subsequently, the dorsal and ventral pancreatic progenitors follow distinct developmental pathways to develop into the same cell type, multipotent progenitor (MP) cells [96, 173]. In rodents, MP cells are heterogeneous and are classified as MP-early and MP-late cells at the developmental time points of E9.5 and E10.5, respectively [182]. The MP-early cells process into the first wave of α cells (α-1st cells) and MP-late cells, which further develop into tip cells (Fig. 2). Tip cells remain multipotent [186] and generate acinar and

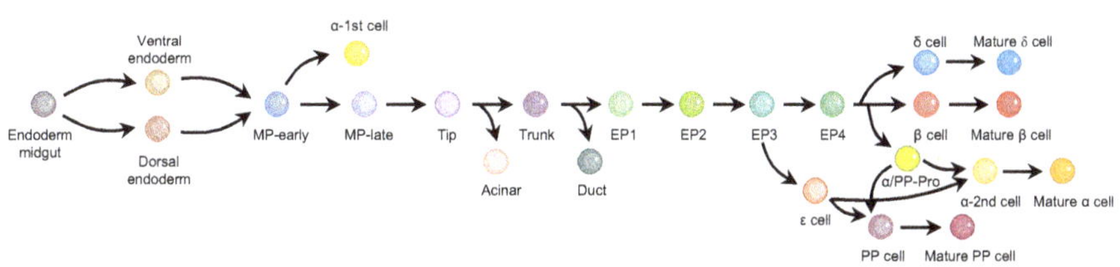

Fig. 2 Pancreatic lineage development model. MP, multipotent progenitor; EP, endocrine progenitor; α-1st cell, the first-wave α cell; α-2nd cell, the second-wave α cell; α/PP-Pro, α/PP-progenitor

trunk cells. Tip and trunk cells constitute the epithelial microlumen structures and are morphologically located in the distal and central regions of the microlumen, respectively. Bipotent trunk cells develop into duct and endocrine progenitor (EP) cells. EP cells are heterogeneous [25, 31, 45, 138, 143, 165, 182] and can be divided into four consecutive stages (EP1–4) [182], which have distinct potentials for certain endocrine lineage specifications (Fig. 2). In mice, EP3 cells generate ε cells, which are progenitors of α cells and PP cells [16, 181] and are rarely observed in adults [11, 35]; EP4 cells at E13.5–E15.5 differentiate mainly into β cells and α/PP progenitor cells, while at E16.5–E18.5, they develop mainly into β cells and δ cells [80, 181]. Thereafter, the endocrine lineages mature into functional cells.

The pancreatic developmental trajectory is generally conserved between mice and humans [181]. Nonetheless, two-wave transitions of endocrine cells are detected during mouse pancreas development but not in humans [76, 168]. Most studies on endocrinogenesis have focused on the second wave of the endocrine transition, and the first wave of transition remains to be further investigated. Moreover, whether ε cells differentiate into α cells and PP cells in humans needs to be further explored. Overall, a comprehensive understanding of the pancreatic lineage developmental route is necessary for elucidating and comprehending the regulatory mechanisms governing the sequential determination of cell fate. Signals governing pancreatic organogenesis have been excellently reviewed elsewhere [24, 65, 89]. Here, we focus on the GRNs involved in different developmental stages of the pancreas.

4 GRNs During Pancreas Organogenesis

4.1 GRNs Controlling MP Generation

At mouse E8.5–9.0 and human embryonic Carnegie stage 10 (CS10), the anterior intestinal portal forms, which is a hallmark of dorsal and ventral endoderm differentiation [41, 55]. ScRNA-seq shows that the endoderm is regionalized before differentiation to specific organs [62, 117], comprising more than ten subpopulations [96]. Subsequently, the pancreatic buds form and protrude at E9.5 in mice and at CS13 in humans [55, 77]. Later, at E11.5 in mice and 6–7 weeks post conception in humans, the dorsal pancreas (DP) and ventral pancreas (VP) fuse into a single organ [89, 122].

Pdx1, Sox9 and Ptf1a are the three indispensable core TFs that regulate the development of both DP and VP [81, 82, 89, 100, 144, 149] (Fig. 3a). *Pdx1* is widely expressed in the midgut and posterior foregut and is responsible for the development of not only the pancreas but also the gastrointestinal tract [118, 157, 161]. Onecut1 directly activates the promoter of *Pdx1*, and knockout of *Onecut1* resulted in downregulation of *Pdx1* expression, leading to disturbed pancreas development [75]. Double knockout of *Foxa1* and *Foxa2* also downregulated *Pdx1* expression in the endoderm and led to pancreas hypoplasia, and ChIP-seq showed that Foxa1 and Foxa2 bound to the enhancers of *Pdx1* [53]. In addition, Pdx1 binds

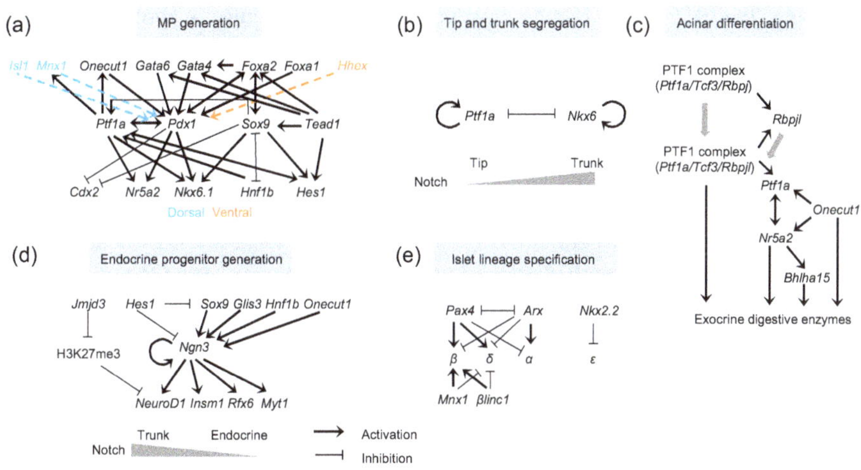

Fig. 3 GRNs controlling different developmental stages. (**a**) MP generation; (**b**) Tip and trunk segregation; (**c**) Acinar differentiation; (**d**) Endocrine progenitor generation; (**e**) Islet lineage specification

to its own enhancer but does not affect its initial expression in the endoderm [118]. *Pdx1* regulates the expression of *Nr5a2*, which plays a key role in the expansion of MP cells, by targeting its promoter [13, 61]. *Sox9* is expressed in pancreatic buds, and its loss leads to pancreatic hypoplasia [127, 146]. *Pdx1* and *Sox9* do not affect each other's expression, but deletion of either resulted in absent or reduced expression of *Ptf1a* and *Nkx6.1*, as well as upregulation of *Cdx2* in pancreatic bud, suggesting that Sox9 cooperates with Pdx1 [149]. It was shown that there is feedback regulation between *Sox9* and *Foxa2* or *Hnf1b*. Knockdown of *Sox9* resulted in the downregulation of *Foxa2* but upregulation of *Hnf1b*, and *Sox9* was downregulated or upregulated by *Foxa2* or *Hnf1b* knockdown, respectively [100]. Moreover, *Sox9* upregulates the Notch pathway and *Hes1* expression [146], which further affects endocrine development [78]. *Ptf1a* is expressed in pancreatic bud and acinar cells, later than *Pdx1* [30]. Ptf1a maintains a high level of *Pdx1* expression by binding to the *Pdx1* enhancer [172] and targets *Mnx1*, *Onecut1*, *Nkx6.1* and *Nr5a2* in MP cells [162]. Ptf1a activates *Dll1* and *Hes1*, which initiate MP cell proliferation [8], and Hes1 maintains *Ptf1a*, forming a positive feedback loop.

Other TFs function in the GRNs also play key roles at the MP stage, such as *Gata4/6*, *Tead1* and *Hnf1b* (Fig. 3a). *Gata4/6* are important for early pancreatic development, and their double knockout by Pdx1-Cre converted the DP and VP to gastrointestinal fate [175, 176]. In humans, heterozygous haploinsufficiency of *GATA6* causes pancreatic agenesis [10], and a missense or a point mutation in *GATA4* can cause diabetes [147]. By summarizing the published TF regulatory relationships and analyzing the basic modules of the GRN, a coherent feed-forward loop comprising *Foxa2*, *Gata4* and *Pdx1* was found, and this network may contribute to the specification of pancreatic fate [14]. Tead1 is a TF involved in YAP-Hippo signaling. In CS16–CS18 human embryos, when the pancreas is about to form an epithelial structure, ChIP-seq showed TEAD binds to the enhancers of several pancreatic development-related TFs, such as *Gata4/6*, *Foxa2*, *Sox9* and *Hes1* [33]. *Hnf1b* is indispensable for pancreas organogenesis, and its absence leads to a complete cessation of VP formation and the inability of the nascent dorsal bud to undergo continuous growth [66]. *Hnf1b* knockout abolishes the expression of *Ptf1a*, but the expression of *Mnx1* and *Pdx1* in the dorsal bud is maintained [66].

In addition to the commonalities mentioned above, there are significant differences in the regulatory mechanisms between DP and VP (Fig. 3a). One of the most important TFs during DP development is *Mnx1*, whose disruption leads to the complete absence of DP but not VP, suggesting that *Mnx1* activates *Pdx1* in the dorsal endoderm by an unknown mechanism [64]. *Mnx1* and *Pdx1* independently repressed Shh signaling to maintain the DP fate in chicken endoderm [57]. Recently, scRNA-seq and gene coexpression network analysis showed that *Mnx1* forms a module with other TFs, such as *Bhlhe41*, *Jun*, *Lmx1b*, *Nfatc2*, *Sim1*, *Sox21*, *Spdef* and *Trp63*, that may regulate DP development [96]. Another key TF in DP development is *Isl1*, whose deletion leads to failure of DP bud formation but not of VP development [7]. During VP development, *Hhex* plays an important role by maintaining the proliferation rate of the anterior lip and positioning the VP away from liver-inducing signals [29]. Notably, a scRNA-seq study showed that the Pdx1$^+$ cells in the DP and VP of E10.5 mice are heterogeneous and are divided into two populations, characterized by low and high Pdx1 expression [95]. The dorsal and ventral Pdx1-high cells are MP cells and are identical at the transcriptome level; however, the Pdx1-low populations in the DP represent premature pancreatic endocrine cells, whereas those in the VP represent tripotential progenitor cells, which give rise to the pancreas, liver and biliary tract [95, 173]. Several TFs were shown to function in ventral cell fate competition. Deletion of *Ptf1a* led to biliary expansion [30]; *Sox17* and *Hes1* knockout caused ectopic ventral pancreas differentiation at the biliary position [51, 155]. Knockout of the histone acetyltransferase *P300* and histone methyltransferase *Ezh2* led to expansion of the VP bud under hepatic compensation [174].

4.2 GRNs for Tip Cell Specification

Following MP cell formation, the pancreas is morphologically divided into tip and trunk regions. The tip cells are pluripotent and can differentiate into bipotent trunk and acinar cells [182, 186]. *Ptf1a* and *Nkx6* (*Nkx6.1* and *Nkx6.2*) are coexpressed in MP cells and then restricted to the tip and trunk, respectively, to maintain cell identity (Fig. 3b). Although *Ptf1a* and *Nkx6.1* positively autoregulate their own gene expression [73, 107], they establish a double-negative feedback network to repress each other's expression [139]. In addition, it has been suggested that the Notch signaling pathway facilitates trunk cell fate commitment by activating *Nkx6.1*, the cis-regulatory element of which is bound by *Rbpj*, a Notch signaling component gene [2].

Upon tip to acinar differentiation, *Ptf1a*, *Tcf3* and *Rbpj* form the trimeric pancreas transcription factor 1 (PTF1) complex and activate the expression of the *Rbpj* paralog *Rbpjl*, which replaces Rbpj to form the *Ptf1a/Tcf3/Rbpjl* complex in acinar cells (Fig. 3c). The PTF1 complex autoregulates the expression of *Ptf1a* and *Rbpjl* [26, 105–107] and activates exocrine digestive enzymes, including *Amy1*, *Ctrb1* and *Cpa1* [26, 50, 107]. In addition, *Nr5a2* is induced by *Ptf1a* in MP cells [162] and establishes a mutually reinforcing regulatory interaction with *Ptf1a* [61]; together, they coregulate acinar differentiation by cobinding the cis-regulatory elements of target genes [61, 69]. *Bhlha15* is activated by *Nr5a2* [61] and regulates acinar cell polarity and exocrine function [126, 134]. Furthermore, ChIP-seq and RNA-seq analyses showed that *Onecut1* directly regulates the expression of *Pft1a*, *Nr5a2* and exocrine digestive enzymes and that the loss of *Onecut1* results in acinar and ductal cell dysplasia [86].

4.3 GRNs Governing the EP Generation

Genetic tracing analysis demonstrated that all endocrine lineages are derived from *Ngn3*$^{+}$ EP cells [56, 60, 67, 80]. *Ngn3* is the key regulator in the initiation of endocrinogenesis [168], and its deletion results in the absence of all pancreatic endocrine lineages [56]. The expression of *Ngn3* is regulated by Notch signaling [148]. Low Notch activity promotes *Sox9* expression, which activates *Ngn3* expression [100, 145, 148], while high Notch activity induces *Hes1*, which represses *Sox9* and *Ngn3* [92] and prevents endocrine differentiation (Fig. 3d). *Glis3* [85], *Hnf1b* [102] and *Onecut1* [74, 102] also activate *Ngn3* expression (Fig. 3d).

In recent years, scRNA-seq has facilitated insights into the developmental pathways, cell heterogeneity and regulatory mechanisms involved in endocrine lineage differentiation [32, 36, 49, 114, 163, 171, 183]. The inferred order of gene expression along developmental trajectories reflects the potential regulatory relationships between genes. Several studies have revealed heterogeneity in mouse and human EP cells, with specific gene expression patterns for each EP state [25, 31, 45, 138, 143, 165, 181, 182], and have shown that a series of TFs are expressed in a cascade during the process of sequential EP differentiation, which may be regulated by sequential GCNs [25, 31, 45, 138, 143, 165, 181, 182]. However, the functions of most genes in GCNs remain to be further investigated.

Multiomics analyses are increasingly being applied in studies of developmental regulation. Integrated ChIP-seq and RNA-seq analyses revealed that the removal of H3K27me3 during EP generation activated key TFs, such as *NeuroD1*, which further established primed and active enhancers for downstream genes. Loss of the H3K27me3 demethylase *Jmjd3* repressed the efficiency of endocrine cell formation [180] (Fig. 3d). Combined scRNA-seq and ATAC-seq analyses revealed that *Ngn3* is a putative pioneer TF during pancreatic endocrine specification [45] and that E14.5 and E16.5 EP cells are in different states, with E16.5 EP cells more likely to generate β cells [138].

4.4 GRNs Governing Endocrine Lineage Specification

Ngn3 is transiently expressed in endocrine progenitors for approximately 12–24 hours [20, 27, 111]. Individual *Ngn3*$^+$ cells are unipotent for a specific endocrine cell type [42], suggesting that the generation of each endocrine lineage requires unique and rapid changes in GRNs.

Direct targets of Ngn3 include *Ngn3* itself [46], *NeuroD1* [71, 113], *Insm1* [110], *Rfx6* [152, 154] and *Myt1* [170], all of which promote the generation of all types of pancreas endocrine lineages [54, 110, 113, 152, 154, 170] (Fig. 3d). *Pax6* [137, 158] and *Isl1* [44] also regulate endocrine differentiation.

It was found that some TFs act as molecular switches to determine the cell fates of specific endocrine lineages. Double-negative feedback regulation of Pax4 and Arx plays an important role in β-cell, δ-cell and α-cell differentiation (Fig. 3e). The loss of *Pax4* results in a decrease in the number of β cells/δ cells and an increase in the number of α cells [153]. In contrast, *Arx* deletion leads to a decrease in the number of α cells and an increase in the number of β cells/δ cells [38]. The loss of both *Pax4* and *Arx* results in the virtually exclusive generation of δ cells [37]. *Nkx2.2* functions independently of *Pax4*, and deletion of *Nkx2.2* in mice led to the replacement of β/α/PP cells by ε cells [129, 159] (Fig. 3e). The TF *Mnx1* and the long noncoding RNA *βlinc1* also control the balance between β cells and δ cells; the loss of *Mnx1* or *βlinc1* leads to a decreased number of β cells and an increased number of δ cells [123] (Fig. 3e). In addition, gain- and loss-of-function experiments showed that *Pdx1* [52, 178] and *Nkx6.1* [140] play important roles in β-cell differentiation and identity maintenance.

Notably, a comparison of gene coexpression networks between humans and mice revealed that there are species-specific regulators of endocrine lineage specification, such as *Pax6* (specifically downregulated in human ε cells), *MafB* (specifically downregulated in human ε cells), *Irx2* (specifically upregulated in mouse ε cells) and *Mef2c* (specifically upregulated in mouse δ cells) [181]. Therefore, species differences should be considered when studying human pancreatic development in animal models.

4.5 GRNs Regulating Endocrine Maturation

Newly generated endocrine cells are functionally immature and must undergo further development. Because β-cell dysfunction is closely related to the pathophysiology of diabetes, most studies have focused on the mechanisms of β-cell maturation rather than on other endocrine cell types [104, 130]. However, there is mutual functional regulation between β cells and other endocrine cells, and the full function of islets depends on the interactions between various endocrine cell types. Therefore, other endocrine cell maturation mechanisms also deserve to be investigated comprehensively.

β-cell maturation occurs postnatally as a coordinated process of cell cycle exit, transcriptional network changes and enhanced functionality [22]. Many extrinsic triggers (signals derived from nutritional and hormonal changes) and intrinsic regulators (transcriptional regulation, epigenetic regulation, and circadian modulation) serve to promote the functional maturity of β cells, resulting in the generation of metabolic networks and machinery (e.g., energy-sensing machinery, mitochondrial metabolism and metabolite trafficking networks) to respond to glucose stimulation and secrete insulin [22]. As several excellent reviews have summarized the roles of environmental signals [22, 136], including mTOR, AMPK, TGF-β and WNT pathways, in the acquisition and maintenance of the β-cell maturation state, the focus of this chapter is on genetic network regulation during β cell maturation.

Several TFs essential for endocrine lineage specification, such as *Foxa2*, *Pdx1*, *NeuroD1* and *Rfx6* [52, 91, 113, 152, 154, 178], retain their functions during β-cell maturation. Foxa2 and Pdx1 co-occupy cis-regulatory regions of mature β cell-specific genes, including *MafA* and *Glut2*, and promote β-cell maturation [23]. ChIP-seq analysis shows that PDX1-targeting genes are largely preserved from fetal to adult stages, suggesting that PDX1 coordinates with various cofactors to play different roles at different developmental stages [177]. *NeuroD1* is required for the acquisition and maintenance of β-cell function [59] and mediates the effect of cyclic AMP by targeting the histone acetyltransferases CBP/P300 to the enhancer elements of β cell-specific genes [164, 185]. Insm1 cobinds with NeuroD1 and Foxa2 at the cis-regulatory regions of genes associated with mature β cells, and its deletion in mice results in β cells with a disordered phenotype characterized by reduced expression of *Pdx1*, *Glut2*, *Ucn3* and *MafA*, comparable to the phenotype of immature β cells [79]. *Rfx6* also participates in the maturation of β cells by regulating Ca^{2+} channels and repressing the expression of β cell disallowed genes, which should be repressed in mature β cells [34, 125].

Some TFs are expressed in differentiated β cells and regulate the maturation process. *MafA* is not expressed during endocrine specification, and the upregulation of its expression is a hallmark of mature β cells. Moreover, *MafA* has been shown to regulate postnatal maturation of rodent and human β cells, and *MafA* deletion impairs insulin secretion and alters gene expression in β cells, downregulating several TFs (*Pdx1*, *NeuroD1* and *Nkx6.1*), functional genes (*Glut2*, *Pcsk1* and *ZnT8*) and DNA methyltransferases (*Dnmt1* and *Dnmt3a*) and upregulating the β cell disallowed gene *Slc16a1* [3, 4, 115]. The defined coregulators of MafA include Pdx1, Foxa2, NeuroD1, Mll3 and Mll4, which are histone 3 lysine 4 (H3K4) methyltransferases [142, 160]. Thyroid hormone (T3) has been demonstrated to promote β-cell maturation in rodents and humans in part by acting through MafA [3, 5]. T3 has been applied to boost *MAFA* expression in order to generate hESC-induced β-like cells [133]. An orphan nuclear receptor and mitochondrial gene regulator, estrogen-related receptor γ (ERRγ), has been shown to regulate postnatal β-cell maturation and drive the formation of a transcriptional network for mitochondrial oxidative metabolism in mice, and its overexpression can drive the maturation of β-like cells derived from iPSCs [179].

Two recent scRNA-seq studies profiled the dynamic transcriptional changes at multiple developmental stages during mouse β-cell maturation and identified molecular features that distinguish immature from mature β cells and proliferative from quiescent β cells [130, 184]. Intriguingly, mature states of proliferative and quiescent β cells are synchronous [130]. Early postnatal β-cell proliferation is influenced by metabolic alterations, such as changes in amino acid metabolism and mitochondrial reactive oxygen species levels, and by a network of nutrient-responsive TFs involving the *Jun/Fos* family [184]. A scRNA-seq analysis of human islets spanning multiple decades of life (juvenile, young adult, and middle-aged) has been performed; although the maturation process of β cells was not elaborated in this study, the dataset could serve as a resource for analyzing the maturation of human endocrine lineages [47]. Notably, α-cell maturation at the single-cell transcriptome level has been dissected and was found to be predominantly associated with the downregulation of gene expression, a regulatory strategy distinct from the upregulation of numerous genes during β-cell maturation [130].

Multiomics studies provide invaluable insights into the GRNs that function during maturation. Using a combination of RNA-seq and ChIP-seq of histone modifications (H3K4me3, H3K27ac, and H3K27me3), one study uncovered age-dependent genomic alterations associated with human α-cell and β-cell maturation [15]. Several TFs are specifically increased with age in β cells, including *SIX2* and *SIX3*, which are expressed solely in adult β cells in humans but not in mice [15]. *SIX2* and *SIX3* have been shown to coordinately regulate β-cell maturation by acting on distinct gene sets, with *SIX2* being required for the maintenance of functional genes and *SIX3* repressing the genes expressed in fetal or neonatal β cells and other endocrine cells to achieve advanced maturation [15, 28]. *SIX2* also drives the functional maturation of stem cell-derived β-like cells [166]. Integrated analysis of the transcriptome and the histone landscape revealed changes in novel signaling pathways (such as adipokine signaling, Th1 pathway, and HOTAIR pathway) and histone modifications in the binding motifs of key TFs (such as *Gata1*, *Foxo1*, *E2f1*, *E2f3*, and *MafB*) during maturation of intact rat islets [97]. However, this study was performed on whole islets and did not distinguish specific regulators of each endocrine lineage.

5 Conclusions and Perspectives

Benefiting from numerous in vivo animal studies of pancreatic lineage development, in vitro approaches to generate functional β-like cells from hESCs or hiPSCs have been developed, and significant breakthroughs have been achieved in recent years [19, 68, 98, 116, 120, 133, 135, 167]. This in vitro differentiation system is widely utilized as a platform for investigating human pancreatic development and diseases. Although the induced β-like cells express some signature genes of primary mature β cells (e.g., *PDX1*, *NKX6.1*, *MAFA*, and *INS*) and are able to secrete insulin in response to glucose stimulation and relieve diabetes in animal models, they are still functionally immature, with transcriptional and metabolic features similar to those of fetal β cells, and they require in vivo transplantation for further maturation [18, 19]. Furthermore, off-target and nonendocrine lineages, as well as polyhormonal cells and enterochromaffin cells, are generated as byproducts, which substantially affects the induction efficiency and may adversely affect the function of β-like cells. The primary reasons for these issues include but are not limited to (1) species-specific differences in the development of pancreatic lineages between humans and rodents, (2) inadequate knowledge of human pancreas development due to the difficulty in obtaining human pancreas tissues from various life stages, and (3) bias in the differentiation potential of hESC and hiPSC lines due to variations in cell origin and manufacturing procedures.

To define and characterize in vitro induced pancreatic lineages as well as to enhance induction efficiency, it is essential to decode the continuous developmental pathways and genetic programs of human pancreatic lineages. The combination of single-cell omics technologies and bioinformatic analysis methods provides an unprecedented opportunity to address this process using scarce human samples. By constructing in vivo GRNs of pancreatic lineages, we can evaluate and optimize the cellular properties at each step of induction, thereby avoiding the cumulative bias introduced by the use of inferior progenitor cells during long-term culture, facilitating the derivation of β cells with genetic characteristics more similar to their primary counterparts and potentially reducing or eliminating the development of nontarget cells. In addition, the panel of marker genes currently employed to indicate the cell identity of each induced cell state should be replaced with core genes from the GCNs.

Acknowledgments We thank the members of the Xu laboratory for their advice and comments. We apologize that we were unable to cite many studies owing to space limitations.

Competing Interests The authors declare no competing or financial interests.

References

1. Abduallah, Y., Turki, T., Byron, K., Du, Z., Cervantes-Cervantes, M., and Wang, J.T. (2017). MapReduce Algorithms for Inferring Gene Regulatory Networks from Time-Series Microarray Data Using an Information-Theoretic Approach. Biomed Res Int *2017*, 6261802.
2. Afelik, S., Qu, X., Hasrouni, E., Bukys, M.A., Deering, T., Nieuwoudt, S., Rogers, W., Macdonald, R.J., and Jensen, J. (2012). Notch-mediated patterning and cell fate allocation of pancreatic progenitor cells. Development *139*, 1744–1753.
3. Aguayo-Mazzucato, C., Dilenno, A., Hollister-Lock, J., Cahill, C., Sharma, A., Weir, G., Colton, C., and Bonner-Weir, S. (2015). MAFA and T-3 Drive Maturation of Both Fetal Human Islets and Insulin-Producing Cells Differentiated From hESC. Journal of Clinical Endocrinology & Metabolism *100*, 3651–3659.
4. Aguayo-Mazzucato, C., Koh, A., El Khattabi, I., Li, W.C., Toschi, E., Jermendy, A., Juhl, K., Mao, K., Weir, G.C., Sharma, A., *et al.* (2011). Mafa expression enhances glucose-responsive insulin secretion in neonatal rat beta cells. Diabetologia *54*, 583–593.
5. Aguayo-Mazzucato, C., Zavacki, A.M., Marinelarena, A., Hollister-Lock, J., El Khattabi, I., Marsili, A., Weir, G.C., Sharma, A., Larsen, P.R., and Bonner-Weir, S. (2013). Thyroid

Hormone Promotes Postnatal Rat Pancreatic beta-Cell Development and Glucose-Responsive Insulin Secretion Through MAFA. Diabetes *62*, 1569–1580.
6. Aguilar, D. (2016). Heart Failure, Diabetes Mellitus, and Chronic Kidney Disease: A Clinical Conundrum. Circ Heart Fail *9*.
7. Ahlgren, U., Pfaff, S.L., Jessell, T.M., Edlund, T., and Edlund, H. (1997). Independent requirement for ISL1 in formation of pancreatic mesenchyme and islet cells. Nature *385*, 257–260.
8. Ahnfelt-Ronne, J., Jorgensen, M.C., Klinck, R., Jensen, J.N., Fuchtbauer, E.M., Deering, T., MacDonald, R.J., Wright, C.V., Madsen, O.D., and Serup, P. (2012). Ptf1a-mediated control of Dll1 reveals an alternative to the lateral inhibition mechanism. Development *139*, 33–45.
9. Aibar, S., Gonzalez-Blas, C.B., Moerman, T., Huynh-Thu, V.A., Imrichova, H., Hulselmans, G., Rambow, F., Marine, J.C., Geurts, P., Aerts, J., *et al.* (2017). SCENIC: single-cell regulatory network inference and clustering. Nat Methods *14*, 1083–1086.
10. Allen, H.L., Flanagan, S.E., Shaw-Smith, C., De Franco, E., Akerman, I., Caswell, R., International Pancreatic Agenesis, C., Ferrer, J., Hattersley, A.T., and Ellard, S. (2011). GATA6 haploinsufficiency causes pancreatic agenesis in humans. Nat Genet *44*, 20–22.
11. Andralojc, K.M., Mercalli, A., Nowak, K.W., Albarello, L., Calcagno, R., Luzi, L., Bonifacio, E., Doglioni, C., and Piemonti, L. (2009). Ghrelin-producing epsilon cells in the developing and adult human pancreas. Diabetologia *52*, 486–493.
12. Angelini, C., and Costa, V. (2014). Understanding gene regulatory mechanisms by integrating ChIP-seq and RNA-seq data: statistical solutions to biological problems. Front Cell Dev Biol *2*, 51.
13. Annicotte, J.S., Fayard, E., Swift, G.H., Selander, L., Edlund, H., Tanaka, T., Kodama, T., Schoonjans, K., and Auwerx, J. (2003). Pancreatic-duodenal homeobox 1 regulates expression of liver receptor homolog 1 during pancreas development. Mol Cell Biol *23*, 6713–6724.
14. Arda, H.E., Benitez, C.M., and Kim, S.K. (2013). Gene regulatory networks governing pancreas development. Dev Cell *25*, 5–13.
15. Arda, H.E., Li, L.Y., Tsai, J., Torre, E.A., Rosli, Y., Peiris, H., Spitale, R.C., Dai, C.H., Gu, X.Y., Qu, K., *et al.* (2016). Age-Dependent Pancreatic Gene Regulation Reveals Mechanisms Governing Human beta Cell Function. Cell Metabolism *23*, 909–920.
16. Arnes, L., Hill, J.T., Gross, S., Magnuson, M.A., and Sussel, L. (2012). Ghrelin expression in the mouse pancreas defines a unique multipotent progenitor population. PLoS One *7*, e52026.
17. Aubin-Frankowski, P.C., and Vert, J.P. (2020). Gene regulation inference from single-cell RNA-seq data with linear differential equations and velocity inference. Bioinformatics *36*, 4774–4780.
18. Augsornworawat, P., Maxwell, K.G., Velazco-Cruz, L., and Millman, J.R. (2020). Single-Cell Transcriptome Profiling Reveals beta Cell Maturation in Stem Cell-Derived Islets after Transplantation. Cell Rep *32*, 108067.
19. Balboa, D., Barsby, T., Lithovius, V., Saarimäki-Vire, J., Omar-Hmeadi, M., Dyachok, O., Montaser, H., Lund, P.-E., Yang, M., Ibrahim, H., *et al.* (2022). Functional, metabolic and transcriptional maturation of human pancreatic islets derived from stem cells. Nature Biotechnology.
20. Bankaitis, E.D., Bechard, M.E., and Wright, C.V.E. (2015). Feedback control of growth, differentiation, and morphogenesis of pancreatic endocrine progenitors in an epithelial plexus niche. Genes Dev *29*, 2203–2216.
21. Barabasi, A.L., and Oltvai, Z.N. (2004). Network biology: understanding the cell's functional organization. Nat Rev Genet *5*, 101–113.
22. Barsby, T., and Otonkoski, T. (2022). Maturation of beta cells: lessons from in vivo and in vitro models. Diabetologia *65*, 917–930.
23. Bastidas-Ponce, A., Roscioni, S.S., Burtscher, I., Bader, E., Sterr, M., Bakhti, M., and Lickert, H. (2017). Foxa2 and Pdx1 cooperatively regulate postnatal maturation of pancreatic beta-cells. Molecular Metabolism *6*, 524–534.
24. Bastidas-Ponce, A., Scheibner, K., Lickert, H., and Bakhti, M. (2017). Cellular and molecular mechanisms coordinating pancreas development. Development *144*, 2873–2888.

25. Bastidas-Ponce, A., Tritschler, S., Dony, L., Scheibner, K., Tarquis-Medina, M., Salinno, C., Schirge, S., Burtscher, I., Bottcher, A., Theis, F.J., *et al.* (2019). Comprehensive single cell mRNA profiling reveals a detailed roadmap for pancreatic endocrinogenesis. Development *146*.
26. Beres, T.M., Masui, T., Swift, G.H., Shi, L., Henke, R.M., and MacDonald, R.J. (2006). PTF1 is an organ-specific and notch-independent basic helix-loop-helix complex containing the mammalian suppressor of hairless (RBP-J) or its paralogue, RBP-L. Mol Cell Biol *26*, 117–130.
27. Beucher, A., Martin, M., Spenle, C., Poulet, M., Collin, C., and Gradwohl, G. (2012). Competence of failed endocrine progenitors to give rise to acinar but not ductal cells is restricted to early pancreas development. Dev Biol *361*, 277–285.
28. Bevacqua, R.J., Lam, J.Y., Peiris, H., Whitener, R.L., Kim, S., Gu, X.Y., Friedlander, M.S.H., and Kim, S.K. (2021). SIX2 and SIX3 coordinately regulate functional maturity and fate of human pancreatic beta cells. Genes Dev *35*.
29. Bort, R., Martinez-Barbera, J.P., Beddington, R.S., and Zaret, K.S. (2004). Hex homeobox gene-dependent tissue positioning is required for organogenesis of the ventral pancreas. Development *131*, 797–806.
30. Burlison, J.S., Long, Q., Fujitani, Y., Wright, C.V., and Magnuson, M.A. (2008). Pdx-1 and Ptf1a concurrently determine fate specification of pancreatic multipotent progenitor cells. Dev Biol *316*, 74–86.
31. Byrnes, L.E., Wong, D.M., Subramaniam, M., Meyer, N.P., Gilchrist, C.L., Knox, S.M., Tward, A.D., Ye, C.J., and Sneddon, J.B. (2018). Lineage dynamics of murine pancreatic development at single-cell resolution. Nat Commun *9*, 3922.
32. Carrano, A.C., Mulas, F., Zeng, C., and Sander, M. (2017). Interrogating islets in health and disease with single-cell technologies. Mol Metab *6*, 991–1001.
33. Cebola, I., Rodriguez-Segui, S.A., Cho, C.H., Bessa, J., Rovira, M., Luengo, M., Chhatriwala, M., Berry, A., Ponsa-Cobas, J., Maestro, M.A., *et al.* (2015). TEAD and YAP regulate the enhancer network of human embryonic pancreatic progenitors. Nat Cell Biol *17*, 615–626.
34. Chandra, V., Albagli-Curiel, O., Hastoy, B., Piccand, J., Randriamampita, C., Vaillant, E., Cave, H., Busiah, K., Froguel, P., Vaxillaire, M., *et al.* (2014). RFX6 Regulates Insulin Secretion by Modulating Ca2+ Homeostasis in Human beta Cells. Cell Reports *9*, 2206–2218.
35. Chanoine, J.P., and Wong, A.C. (2004). Ghrelin gene expression is markedly higher in fetal pancreas compared with fetal stomach: effect of maternal fasting. Endocrinology *145*, 3813–3820.
36. Cheung, L.Y.M., and Rizzoti, K. (2020). Cell population characterization and discovery using single-cell technologies in endocrine systems. J Mol Endocrinol *65*, R35–R51.
37. Collombat, P., Hecksher-Sorensen, J., Broccoli, V., Krull, J., Ponte, I., Mundiger, T., Smith, J., Gruss, P., Serup, P., and Mansouri, A. (2005). The simultaneous loss of Arx and Pax4 genes promotes a somatostatin-producing cell fate specification at the expense of the alpha- and beta-cell lineages in the mouse endocrine pancreas. Development *132*, 2969–2980.
38. Collombat, P., Mansouri, A., Hecksher-Sorensen, J., Serup, P., Krull, J., Gradwohl, G., and Gruss, P. (2003). Opposing actions of Arx and Pax4 in endocrine pancreas development. Genes Dev *17*, 2591–2603.
39. Datlinger, P., Rendeiro, A.F., Schmidl, C., Krausgruber, T., Traxler, P., Klughammer, J., Schuster, L.C., Kuchler, A., Alpar, D., and Bock, C. (2017). Pooled CRISPR screening with single-cell transcriptome readout. Nat Methods *14*, 297–301.
40. Davidson, E.H., Rast, J.P., Oliveri, P., Ransick, A., Calestani, C., Yuh, C.H., Minokawa, T., Amore, G., Hinman, V., Arenas-Mena, C., *et al.* (2002). A provisional regulatory gene network for specification of endomesoderm in the sea urchin embryo. Dev Biol *246*, 162–190.
41. de Bakker, B.S., de Jong, K.H., Hagoort, J., de Bree, K., Besselink, C.T., de Kanter, F.E., Veldhuis, T., Bais, B., Schildmeijer, R., Ruijter, J.M., *et al.* (2016). An interactive three-dimensional digital atlas and quantitative database of human development. Science *354*.
42. Desgraz, R., and Herrera, P.L. (2009). Pancreatic neurogenin 3-expressing cells are unipotent islet precursors. Development *136*, 3567–3574.

43. Dixit, A., Parnas, O., Li, B., Chen, J., Fulco, C.P., Jerby-Arnon, L., Marjanovic, N.D., Dionne, D., Burks, T., Raychowdhury, R., *et al.* (2016). Perturb-Seq: Dissecting Molecular Circuits with Scalable Single-Cell RNA Profiling of Pooled Genetic Screens. Cell *167*, 1853–1866 e1817.
44. Du, A.P., Hunter, C.S., Murray, J., Noble, D., Cai, C.L., Evans, S.M., Stein, R., and May, C.L. (2009). Islet-1 is Required for the Maturation, Proliferation, and Survival of the Endocrine Pancreas. Diabetes *58*, 2059–2069.
45. Duvall, E., Benitez, C.M., Tellez, K., Enge, M., Pauerstein, P.T., Li, L., Baek, S., Quake, S.R., Smith, J.P., Sheffield, N.C., *et al.* (2022). Single-cell transcriptome and accessible chromatin dynamics during endocrine pancreas development. Proc Natl Acad Sci U S A *119*, e2201267119.
46. Ejarque, M., Cervantes, S., Pujadas, G., Tutusaus, A., Sanchez, L., and Gasa, R. (2013). Neurogenin3 cooperates with Foxa2 to autoactivate its own expression. J Biol Chem *288*, 11705–11717.
47. Enge, M., Arda, H.E., Mignardi, M., Beausang, J., Bottino, R., Kim, S.K., and Quake, S.R. (2017). Single-Cell Analysis of Human Pancreas Reveals Transcriptional Signatures of Aging and Somatic Mutation Patterns. Cell *171*, 321–330 e314.
48. Fiers, M., Minnoye, L., Aibar, S., Bravo Gonzalez-Blas, C., Kalender Atak, Z., and Aerts, S. (2018). Mapping gene regulatory networks from single-cell omics data. Brief Funct Genomics *17*, 246–254.
49. Fu, H., Sun, H., Kong, H., Lou, B., Chen, H., Zhou, Y., Huang, C., Qin, L., Shan, Y., and Dai, S. (2021). Discoveries in Pancreatic Physiology and Disease Biology Using Single-Cell RNA Sequencing. Front Cell Dev Biol *9*, 732776.
50. Fujikura, J., Hosoda, K., Kawaguchi, Y., Noguchi, M., Iwakura, H., Odori, S., Mori, E., Tomita, T., Hirata, M., Ebihara, K., *et al.* (2007). Rbp-j regulates expansion of pancreatic epithelial cells and their differentiation into exocrine cells during mouse development. Dev Dyn *236*, 2779–2791.
51. Fukuda, A., Kawaguchi, Y., Furuyama, K., Kodama, S., Horiguchi, M., Kuhara, T., Koizumi, M., Boyer, D.F., Fujimoto, K., Doi, R., *et al.* (2006). Ectopic pancreas formation in Hes1-knockout mice reveals plasticity of endodermal progenitors of the gut, bile duct, and pancreas. J Clin Invest *116*, 1484–1493.
52. Gannon, M., Ables, E.T., Crawford, L., Lowe, D., Offield, M.F., Magnuson, M.A., and Wright, C.V.E. (2008). pdx-1 function is specifically required in embryonic cells to generate appropriate numbers of endocrine cell types and maintain glucose homeostasis. Dev Biol *314*, 406–417.
53. Gao, N., LeLay, J., Vatamaniuk, M.Z., Rieck, S., Friedman, J.R., and Kaestner, K.H. (2008). Dynamic regulation of Pdx1 enhancers by Foxa1 and Foxa2 is essential for pancreas development. Genes Dev *22*, 3435–3448.
54. Gierl, M.S., Karoulias, N., Wende, H., Strehle, M., and Birchmeier, C. (2006). The zinc-finger factor Insm1 (IA-1) is essential for the development of pancreatic beta cells and intestinal endocrine cells. Genes Dev *20*, 2465–2478.
55. Gittes, G.K. (2009). Developmental biology of the pancreas: a comprehensive review. Dev Biol *326*, 4–35.
56. Gradwohl, G., Dierich, A., LeMeur, M., and Guillemot, F. (2000). neurogenin3 is required for the development of the four endocrine cell lineages of the pancreas. Proc Natl Acad Sci U S A *97*, 1607–1611.
57. Grapin-Botton, A., Majithia, A.R., and Melton, D.A. (2001). Key events of pancreas formation are triggered in gut endoderm by ectopic expression of pancreatic regulatory genes. Genes Dev *15*, 444–454.
58. Greenfield, A., Madar, A., Ostrer, H., and Bonneau, R. (2010). DREAM4: Combining genetic and dynamic information to identify biological networks and dynamical models. PLoS One *5*, e13397.
59. Gu, C.Y., Stein, G.H., Pan, N., Goebbels, S., Hornberg, H., Nave, K.A., Herrera, P., White, P., Kaestner, K.H., Sussel, L., *et al.* (2010). Pancreatic beta Cells Require NeuroD to Achieve and Maintain Functional Maturity. Cell Metabolism *11*, 298–310.

60. Gu, G., Dubauskaite, J., and Melton, D.A. (2002). Direct evidence for the pancreatic lineage: NGN3+ cells are islet progenitors and are distinct from duct progenitors. Development *129*, 2447–2457.
61. Hale, M.A., Swift, G.H., Hoang, C.Q., Deering, T.G., Masui, T., Lee, Y.K., Xue, J., and MacDonald, R.J. (2014). The nuclear hormone receptor family member NR5A2 controls aspects of multipotent progenitor cell formation and acinar differentiation during pancreatic organogenesis. Development *141*, 3123–3133.
62. Han, L., Chaturvedi, P., Kishimoto, K., Koike, H., Nasr, T., Iwasawa, K., Giesbrecht, K., Witcher, P.C., Eicher, A., Haines, L., *et al.* (2020). Single cell transcriptomics identifies a signaling network coordinating endoderm and mesoderm diversification during foregut organogenesis. Nature Communications *11*, 4158.
63. Hardy, K., Mansfield, L., Mackay, A., Benvenuti, S., Ismail, S., Arora, P., O'Hare, M.J., and Jat, P.S. (2005). Transcriptional networks and cellular senescence in human mammary fibroblasts. Mol Biol Cell *16*, 943–953.
64. Harrison, K.A., Thaler, J., Pfaff, S.L., Gu, H., and Kehrl, J.H. (1999). Pancreas dorsal lobe agenesis and abnormal islets of Langerhans in Hlxb9-deficient mice. Nat Genet *23*, 71–75.
65. Hashemitabar, M., and Heidari, E. (2019). Redefining the signaling pathways from pluripotency to pancreas development: In vitro beta-cell differentiation. J Cell Physiol *234*, 7811–7827.
66. Haumaitre, C., Barbacci, E., Jenny, M., Ott, M.O., Gradwohl, G., and Cereghini, S. (2005). Lack of TCF2/vHNF1 in mice leads to pancreas agenesis. Proc Natl Acad Sci U S A *102*, 1490–1495.
67. Heller, R.S., Jenny, M., Collombat, P., Mansouri, A., Tomasetto, C., Madsen, O.D., Mellitzer, G., Gradwohl, G., and Serup, P. (2005). Genetic determinants of pancreatic epsilon-cell development. Dev Biol *286*, 217–224.
68. Hogrebe, N.J., Augsornworawat, P., Maxwell, K.G., Velazco-Cruz, L., and Millman, J.R. (2020). Targeting the cytoskeleton to direct pancreatic differentiation of human pluripotent stem cells. Nat Biotechnol.
69. Holmstrom, S.R., Deering, T., Swift, G.H., Poelwijk, F.J., Mangelsdorf, D.J., Kliewer, S.A., and MacDonald, R.J. (2011). LRH-1 and PTF1-L coregulate an exocrine pancreas-specific transcriptional network for digestive function. Genes Dev *25*, 1674–1679.
70. Hu, X., Hu, Y., Wu, F., Leung, R.W.T., and Qin, J. (2020). Integration of single-cell multi-omics for gene regulatory network inference. Comput Struct Biotechnol J *18*, 1925–1938.
71. Huang, H.P., Liu, M., El-Hodiri, H.M., Chu, K., Jamrich, M., and Tsai, M.J. (2000). Regulation of the pancreatic islet-specific gene BETA2 (neuroD) by neurogenin 3. Mol Cell Biol *20*, 3292–3307.
72. Huynh-Thu, V.A., Irrthum, A., Wehenkel, L., and Geurts, P. (2010). Inferring regulatory networks from expression data using tree-based methods. PLoS One *5*.
73. Iype, T., Taylor, D.G., Ziesmann, S.M., Garmey, J.C., Watada, H., and Mirmira, R.G. (2004). The transcriptional repressor Nkx6.1 also functions as a deoxyribonucleic acid context-dependent transcriptional activator during pancreatic beta-cell differentiation: Evidence for feedback activation of the nkx6.1 gene by Nkx6.1. Mol Endocrinol *18*, 1363–1375.
74. Jacquemin, P., Durviaux, S.M., Jensen, J., Godfraind, C., Gradwohl, G., Guillemot, F., Madsen, O.D., Carmeliet, P., Dewerchin, M., Collen, D., *et al.* (2000). Transcription factor hepatocyte nuclear factor 6 regulates pancreatic endocrine cell differentiation and controls expression of the proendocrine gene ngn3. Mol Cell Biol *20*, 4445–4454.
75. Jacquemin, P., Lemaigre, F.P., and Rousseau, G.G. (2003). The Onecut transcription factor HNF-6 (OC-1) is required for timely specification of the pancreas and acts upstream of Pdx-1 in the specification cascade. Dev Biol *258*, 105–116.
76. Jennings, R.E., Berry, A.A., Kirkwood-Wilson, R., Roberts, N.A., Hearn, T., Salisbury, R.J., Blaylock, J., Piper Hanley, K., and Hanley, N.A. (2013). Development of the human pancreas from foregut to endocrine commitment. Diabetes *62*, 3514–3522.
77. Jennings, R.E., Berry, A.A., Strutt, J.P., Gerrard, D.T., and Hanley, N.A. (2015). Human pancreas development. Development *142*, 3126–3137.

78. Jensen, J., Pedersen, E.E., Galante, P., Hald, J., Heller, R.S., Ishibashi, M., Kageyama, R., Guillemot, F., Serup, P., and Madsen, O.D. (2000). Control of endodermal endocrine development by Hes-1. Nat Genet *24*, 36–44.
79. Jia, S., Ivanov, A., Blasevic, D., Mueller, T., Purfuerst, B., Sun, W., Chen, W., Poy, M.N., Rajewsky, N., and Birchmeier, C. (2015). Insm1 cooperates with Neurod1 and Foxa2 to maintain mature pancreatic beta-cell function. Embo Journal *34*, 1417–1433.
80. Johansson, K.A., Dursun, U., Jordan, N., Gu, G., Beermann, F., Gradwohl, G., and Grapin-Botton, A. (2007). Temporal control of neurogenin3 activity in pancreas progenitors reveals competence windows for the generation of different endocrine cell types. Dev Cell *12*, 457–465.
81. Jonsson, J., Carlsson, L., Edlund, T., and Edlund, H. (1994). Insulin-promoter-factor 1 is required for pancreas development in mice. Nature *371*, 606–609.
82. Kawaguchi, Y., Cooper, B., Gannon, M., Ray, M., MacDonald, R.J., and Wright, C.V. (2002). The role of the transcriptional regulator Ptf1a in converting intestinal to pancreatic progenitors. Nat Genet *32*, 128–134.
83. Kaya-Okur, H.S., Janssens, D.H., Henikoff, J.G., Ahmad, K., and Henikoff, S. (2020). Efficient low-cost chromatin profiling with CUT&Tag. Nature Protocols.
84. Kaya-Okur, H.S., Wu, S.J., Codomo, C.A., Pledger, E.S., Bryson, T.D., Henikoff, J.G., Ahmad, K., and Henikoff, S. (2019). CUT&Tag for efficient epigenomic profiling of small samples and single cells. Nat Commun *10*, 1930.
85. Kim, Y.S., Kang, H.S., Takeda, Y., Hom, L., Song, H.Y., Jensen, J., and Jetten, A.M. (2012). Glis3 regulates neurogenin 3 expression in pancreatic beta-cells and interacts with its activator, Hnf6. Mol Cells *34*, 193–200.
86. Kropp, P.A., Zhu, X., and Gannon, M. (2019). Regulation of the Pancreatic Exocrine Differentiation Program and Morphogenesis by Onecut 1/Hnf6. Cell Mol Gastroenterol Hepatol *7*, 841–856.
87. La Manno, G., Soldatov, R., Zeisel, A., Braun, E., Hochgerner, H., Petukhov, V., Lidschreiber, K., Kastriti, M.E., Lönnerberg, P., Furlan, A., *et al.* (2018). RNA velocity of single cells. Nature.
88. Langfelder, P., and Horvath, S. (2008). WGCNA: an R package for weighted correlation network analysis. BMC Bioinformatics *9*, 559.
89. Larsen, H.L., and Grapin-Botton, A. (2017). The molecular and morphogenetic basis of pancreas organogenesis. Semin Cell Dev Biol *66*, 51–68.
90. Lau, L.H., Lew, J., Borschmann, K., Thijs, V., and Ekinci, E.I. (2019). Prevalence of diabetes and its effects on stroke outcomes: A meta-analysis and literature review. J Diabetes Investig *10*, 780–792.
91. Lee, C.S., Sund, N.J., Behr, R., Herrera, P.L., and Kaestner, K.H. (2005). Foxa2 is required for the differentiation of pancreatic alpha-cells. Dev Biol *278*, 484–495.
92. Lee, J.C., Smith, S.B., Watada, H., Lin, J., Scheel, D., Wang, J., Mirmira, R.G., and German, M.S. (2001). Regulation of the pancreatic pro-endocrine gene neurogenin3. Diabetes *50*, 928–936.
93. Levine, M., and Davidson, E.H. (2005). Gene regulatory networks for development. Proc Natl Acad Sci U S A *102*, 4936–4942.
94. Li, E., and Davidson, E.H. (2009). Building developmental gene regulatory networks. Birth Defects Res C Embryo Today *87*, 123–130.
95. Li, L.C., Qiu, W.L., Zhang, Y.W., Xu, Z.R., Xiao, Y.N., Hou, C., Lamaoqiezhong, Yu, P., Cheng, X., and Xu, C.R. (2018). Single-cell transcriptomic analyses reveal distinct dorsal/ ventral pancreatic programs. EMBO Rep *19*.
96. Li, L.C., Wang, X., Xu, Z.R., Wang, Y.C., Feng, Y., Yang, L., Qiu, W.L., Yang, L., Yu, X.X., Gu, J., *et al.* (2021). Single-cell patterning and axis characterization in the murine and human definitive endoderm. Cell Res *31*, 326–344.
97. Lien, Y.C., Lu, X.M., Won, K.J., Wang, P.Z., Osei-Bonsu, W., and Simmons, R.A. (2021). The Transcriptome and Epigenome Reveal Novel Changes in Transcription Regulation During Pancreatic Rat Islet Maturation. Endocrinology *162*.

98. Liu, H., Li, R., Liao, H.-K., Min, Z., Wang, C., Yu, Y., Shi, L., Dan, J., Hayek, A., Martinez Martinez, L., *et al.* (2021). Chemical combinations potentiate human pluripotent stem cell-derived 3D pancreatic progenitor clusters toward functional β cells. Nature Communications *12*, 3330.
99. Lowe, E.K., Cuomo, C., Voronov, D., and Arnone, M.I. (2019). Using ATAC-seq and RNA-seq to increase resolution in GRN connectivity. Methods Cell Biol *151*, 115–126.
100. Lynn, F.C., Smith, S.B., Wilson, M.E., Yang, K.Y., Nekrep, N., and German, M.S. (2007). Sox9 coordinates a transcriptional network in pancreatic progenitor cells. Proc Natl Acad Sci U S A *104*, 10500–10505.
101. Macneil, L.T., and Walhout, A.J. (2011). Gene regulatory networks and the role of robustness and stochasticity in the control of gene expression. Genome Res *21*, 645–657.
102. Maestro, M.A., Boj, S.F., Luco, R.F., Pierreux, C.E., Cabedo, J., Servitja, J.M., German, M.S., Rousseau, G.G., Lemaigre, F.P., and Ferrer, J. (2003). Hnf6 and Tcf2 (MODY5) are linked in a gene network operating in a precursor cell domain of the embryonic pancreas. Hum Mol Genet *12*, 3307–3314.
103. Margolin, A.A., Nemenman, I., Basso, K., Wiggins, C., Stolovitzky, G., Dalla Favera, R., and Califano, A. (2006). ARACNE: an algorithm for the reconstruction of gene regulatory networks in a mammalian cellular context. BMC Bioinformatics *7 Suppl 1*, S7.
104. Masson, M.H., Poisson, C., Guerardel, A., Mamin, A., Philippe, J., and Gosmain, Y. (2014). Foxa1 and Foxa2 Regulate alpha-Cell Differentiation, Glucagon Biosynthesis, and Secretion. Endocrinology *155*, 3781–3792.
105. Masui, T., Long, Q., Beres, T.M., Magnuson, M.A., and MacDonald, R.J. (2007). Early pancreatic development requires the vertebrate Suppressor of Hairless (RBPJ) in the PTF1 bHLH complex. Genes Dev *21*, 2629–2643.
106. Masui, T., Swift, G.H., Deering, T., Shen, C.C., Coats, W.S., Long, Q.M., Elsasser, H.P., Magnuson, M.A., and MacDonald, R.J. (2010). Replacement of Rbpj With Rbpjl in the PTF1 Complex Controls the Final Maturation of Pancreatic Acinar Cells. Gastroenterology *139*, 270–280.
107. Masui, T., Swift, G.H., Hale, M.A., Meredith, D.M., Johnson, J.E., and MacDonald, R.J. (2008). Transcriptional autoregulation controls pancreatic Ptf1a expression during development and adulthood. Mol Cell Biol *28*, 5458–5468.
108. Matsumoto, H., Kiryu, H., Furusawa, C., Ko, M.S.H., Ko, S.B.H., Gouda, N., Hayashi, T., and Nikaido, I. (2017). SCODE: an efficient regulatory network inference algorithm from single-cell RNA-Seq during differentiation. Bioinformatics *33*, 2314–2321.
109. Meers, M.P., Bryson, T.D., Henikoff, J.G., and Henikoff, S. (2019). Improved CUT&RUN chromatin profiling tools. Elife *8*.
110. Mellitzer, G., Bonne, S., Luco, R.F., Van De Casteele, M., Lenne-Samuel, N., Collombat, P., Mansouri, A., Lee, J., Lan, M., Pipeleers, D., *et al.* (2006). IA1 is NGN3-dependent and essential for differentiation of the endocrine pancreas. The EMBO journal *25*, 1344–1352.
111. Miyatsuka, T., Li, Z.M., and German, M.S. (2009). Chronology of Islet Differentiation Revealed By Temporal Cell Labeling. Diabetes *58*, 1863–1868.
112. Moss, S.E., Klein, R., and Klein, B.E. (1994). Ten-year incidence of visual loss in a diabetic population. Ophthalmology *101*, 1061–1070.
113. Naya, F.J., Huang, H.P., Qiu, Y.H., Mutoh, H., DeMayo, F.J., Leiter, A.B., and Tsai, M.J. (1997). Diabetes, defective pancreatic morphogenesis, and abnormal enteroendocrine differentiation in BETA2/NeuroD-deficient mice. Genes Dev *11*, 2323–2334.
114. Ngara, M., and Wierup, N. (2022). Lessons from single-cell RNA sequencing of human islets. Diabetologia *65*, 1241–1250.
115. Nishimura, W., Takahashi, S., and Yasuda, K. (2015). MafA is critical for maintenance of the mature beta cell phenotype in mice. Diabetologia *58*, 566–574.
116. Nostro, M.C., Sarangi, F., Yang, C., Holland, A., Elefanty, A.G., Stanley, E.G., Greiner, D.L., and Keller, G. (2015). Efficient generation of NKX6-1+ pancreatic progenitors from multiple human pluripotent stem cell lines. Stem Cell Reports *4*, 591–604.

117. Nowotschin, S., Setty, M., Kuo, Y.Y., Liu, V., Garg, V., Sharma, R., Simon, C.S., Saiz, N., Gardner, R., Boutet, S.C., *et al.* (2019). The emergent landscape of the mouse gut endoderm at single-cell resolution. Nature *569*, 361–367.
118. Offield, M.F., Jetton, T.L., Labosky, P.A., Ray, M., Stein, R.W., Magnuson, M.A., Hogan, B.L., and Wright, C.V. (1996). PDX-1 is required for pancreatic outgrowth and differentiation of the rostral duodenum. Development *122*, 983–995.
119. Omranian, N., and Nikoloski, Z. (2017). Computational Approaches to Study Gene Regulatory Networks. Methods Mol Biol *1629*, 283–295.
120. Pagliuca, F.W., Millman, J.R., Gurtler, M., Segel, M., Van Dervort, A., Ryu, J.H., Peterson, Q.P., Greiner, D., and Melton, D.A. (2014). Generation of functional human pancreatic beta cells in vitro. Cell *159*, 428–439.
121. Palsson, R., and Patel, U.D. (2014). Cardiovascular complications of diabetic kidney disease. Adv Chronic Kidney Dis *21*, 273–280.
122. Pan, F.C., and Brissova, M. (2014). Pancreas development in humans. Curr Opin Endocrinol Diabetes Obes *21*, 77–82.
123. Pan, F.C., Brissova, M., Powers, A.C., Pfaff, S., and Wright, C.V.E. (2015). Inactivating the permanent neonatal diabetes gene Mnx1 switches insulin-producing beta-cells to a delta-like fate and reveals a facultative proliferative capacity in aged beta-cells. Development *142*, 3637–3648.
124. Pavesi, G. (2017). ChIP-Seq Data Analysis to Define Transcriptional Regulatory Networks. Adv Biochem Eng Biotechnol *160*, 1–14.
125. Piccand, J., Strasser, P., Hodson, D.J., Meunier, A., Ye, T., Keime, C., Birling, M.C., Rutter, G.A., and Gradwohl, G. (2014). Rfx6 Maintains the Functional Identity of Adult Pancreatic beta Cells. Cell Reports *9*, 2219–2232.
126. Pin, C.L., Rukstalis, J.M., Johnson, C., and Konieczny, S.F. (2001). The bHLH transcription factor Mist1 is required to maintain exocrine pancreas cell organization and acinar cell identity. J Cell Biol *155*, 519–530.
127. Piper, K., Ball, S.G., Keeling, J.W., Mansoor, S., Wilson, D.I., and Hanley, N.A. (2002). Novel SOX9 expression during human pancreas development correlates to abnormalities in Campomelic dysplasia. Mech Dev *116*, 223–226.
128. Pliner, H.A., Packer, J.S., McFaline-Figueroa, J.L., Cusanovich, D.A., Daza, R.M., Aghamirzaie, D., Srivatsan, S., Qiu, X., Jackson, D., Minkina, A., *et al.* (2018). Cicero Predicts cis-Regulatory DNA Interactions from Single-Cell Chromatin Accessibility Data. Mol Cell *71*, 858–871 e858.
129. Prado, C.L., Pugh-Bernard, A.E., Elghazi, L., Sosa-Pineda, B., and Sussel, L. (2004). Ghrelin cells replace insulin-producing beta cells in two mouse models of pancreas development. Proceedings of the National Academy of Sciences of the United States of America *101*, 2924–2929.
130. Qiu, W.L., Zhang, Y.W., Feng, Y., Li, L.C., Yang, L., and Xu, C.R. (2017). Deciphering Pancreatic Islet beta Cell and alpha Cell Maturation Pathways and Characteristic Features at the Single-Cell Level. Cell Metab *25*, 1194–1205 e1194.
131. Qiu, X., Rahimzamani, A., Wang, L., Ren, B., Mao, Q., Durham, T., McFaline-Figueroa, J.L., Saunders, L., Trapnell, C., and Kannan, S. (2020). Inferring Causal Gene Regulatory Networks from Coupled Single-Cell Expression Dynamics Using Scribe. Cell Syst *10*, 265–274 e211.
132. Quinn, T.P., Richardson, M.F., Lovell, D., and Crowley, T.M. (2017). propr: An R-package for Identifying Proportionally Abundant Features Using Compositional Data Analysis. Sci Rep *7*, 16252.
133. Rezania, A., Bruin, J.E., Arora, P., Rubin, A., Batushansky, I., Asadi, A., O'Dwyer, S., Quiskamp, N., Mojibian, M., Albrecht, T., *et al.* (2014). Reversal of diabetes with insulin-producing cells derived in vitro from human pluripotent stem cells. Nat Biotechnol *32*, 1121–1133.
134. Rukstalis, J.M., Kowalik, A., Zhu, L.Q., Lidington, D., Pin, C.L., and Konieczny, S.F. (2003). Exocrine specific expression of Connexin32 is dependent on the basic helix-loop-helix transcription factor Mist1. J Cell Sci *116*, 3315–3325.

135. Russ, H.A., Parent, A.V., Ringler, J.J., Hennings, T.G., Nair, G.G., Shveygert, M., Guo, T., Puri, S., Haataja, L., Cirulli, V., *et al.* (2015). Controlled induction of human pancreatic progenitors produces functional beta-like cells in vitro. The EMBO journal *34*, 1759–1772.
136. Salinno, C., Cota, P., Bastidas-Ponce, A., Tarquis-Medina, M., Lickert, H., and Bakhti, M. (2019). beta-Cell Maturation and Identity in Health and Disease. International Journal of Molecular Sciences *20*.
137. Sander, M., Neubuser, A., Kalamaras, J., Ee, H.C., Martin, G.R., and German, M.S. (1997). Genetic analysis reveals that PAX6 is required for normal transcription of pancreatic hormone genes and islet development. Genes Dev *11*, 1662–1673.
138. Scavuzzo, M.A., Hill, M.C., Chmielowiec, J., Yang, D., Teaw, J., Sheng, K., Kong, Y., Bettini, M., Zong, C., Martin, J.F., *et al.* (2018). Endocrine lineage biases arise in temporally distinct endocrine progenitors during pancreatic morphogenesis. Nat Commun *9*, 3356.
139. Schaffer, A.E., Freude, K.K., Nelson, S.B., and Sander, M. (2010). Nkx6 Transcription Factors and Ptf1a Function as Antagonistic Lineage Determinants in Multipotent Pancreatic Progenitors. Dev Cell *18*, 1022–1029.
140. Schaffer, A.E., Taylor, B.L., Benthuysen, J.R., Liu, J.X., Thorel, F., Yuan, W.P., Jiao, Y., Kaestner, K.H., Herrera, P.L., Magnuson, M.A., *et al.* (2013). Nkx6.1 Controls a Gene Regulatory Network Required for Establishing and Maintaining Pancreatic Beta Cell Identity. PLoS Genet *9*, 15.
141. Schraivogel, D., Gschwind, A.R., Milbank, J.H., Leonce, D.R., Jakob, P., Mathur, L., Korbel, J.O., Merten, C.A., Velten, L., and Steinmetz, L.M. (2020). Targeted Perturb-seq enables genome-scale genetic screens in single cells. Nat Methods *17*, 629–635.
142. Scoville, D.W., Cyphert, H.A., Liao, L., Xu, J., Reynolds, A., Guo, S., and Stein, R. (2015). MLL3 and MLL4 Methyltransferases Bind to the MAFA and MAFB Transcription Factors to Regulate Islet beta-Cell Function. Diabetes *64*, 3772–3783.
143. Sean, d.l.O., Liu, Z., Sun, H., Yu, S.K., Wong, D.M., Chu, E., Rao, S.A., Eng, N., Peixoto, G., Bouza, J., *et al.* (2022). Single-Cell Multi-Omic Roadmap of Human Fetal Pancreatic Development. bioRxiv, 2022.2002.2017.480942.
144. Servitja, J.M., and Ferrer, J. (2004). Transcriptional networks controlling pancreatic development and beta cell function. Diabetologia *47*, 597–613.
145. Seymour, P.A., Freude, K.K., Dubois, C.L., Shih, H.P., Patel, N.A., and Sander, M. (2008). A dosage-dependent requirement for Sox9 in pancreatic endocrine cell formation. Dev Biol *323*, 19–30.
146. Seymour, P.A., Freude, K.K., Tran, M.N., Mayes, E.E., Jensen, J., Kist, R., Scherer, G., and Sander, M. (2007). SOX9 is required for maintenance of the pancreatic progenitor cell pool. Proc Natl Acad Sci U S A *104*, 1865–1870.
147. Shaw-Smith, C., De Franco, E., Lango Allen, H., Batlle, M., Flanagan, S.E., Borowiec, M., Taplin, C.E., van Alfen-van der Velden, J., Cruz-Rojo, J., Perez de Nanclares, G., *et al.* (2014). GATA4 mutations are a cause of neonatal and childhood-onset diabetes. Diabetes *63*, 2888–2894.
148. Shih, H.P., Kopp, J.L., Sandhu, M., Dubois, C.L., Seymour, P.A., Grapin-Botton, A., and Sander, M. (2012). A Notch-dependent molecular circuitry initiates pancreatic endocrine and ductal cell differentiation. Development *139*, 2488–2499.
149. Shih, H.P., Seymour, P.A., Patel, N.A., Xie, R., Wang, A., Liu, P.P., Yeo, G.W., Magnuson, M.A., and Sander, M. (2015). A Gene Regulatory Network Cooperatively Controlled by Pdx1 and Sox9 Governs Lineage Allocation of Foregut Progenitor Cells. Cell Rep *13*, 326–336.
150. Shih, H.P., Wang, A., and Sander, M. (2013). Pancreas organogenesis: from lineage determination to morphogenesis. Annu Rev Cell Dev Biol *29*, 81–105.
151. Singh, A.J., Ramsey, S.A., Filtz, T.M., and Kioussi, C. (2018). Differential gene regulatory networks in development and disease. Cell Mol Life Sci *75*, 1013–1025.
152. Smith, S.B., Qu, H.Q., Taleb, N., Kishimoto, N.Y., Scheel, D.W., Lu, Y., Patch, A.M., Grabs, R., Wang, J.H., Lynn, F.C., *et al.* (2010). Rfx6 directs islet formation and insulin production in mice and humans. Nature *463*, 775–780.

153. Sosa-Pineda, B., Chowdhury, K., Torres, M., Oliver, G., and Gruss, P. (1997). The Pax4 gene is essential for differentiation of insulin-producing beta cells in the mammalian pancreas. Nature *386*, 399–402.
154. Soyer, J., Flasse, L., Raffelsberger, W., Beucher, A., Orvain, C., Peers, B., Ravassard, P., Vermot, J., Voz, M.L., Mellitzer, G., *et al.* (2010). Rfx6 is an Ngn3-dependent winged helix transcription factor required for pancreatic islet cell development. Development *137*, 203–212.
155. Spence, J.R., Lange, A.W., Lin, S.C., Kaestner, K.H., Lowy, A.M., Kim, I., Whitsett, J.A., and Wells, J.M. (2009). Sox17 regulates organ lineage segregation of ventral foregut progenitor cells. Dev Cell *17*, 62–74.
156. Stathopoulos, A., Van Drenth, M., Erives, A., Markstein, M., and Levine, M. (2002). Whole-genome analysis of dorsal-ventral patterning in the Drosophila embryo. Cell *111*, 687–701.
157. Stoffers, D.A., Zinkin, N.T., Stanojevic, V., Clarke, W.L., and Habener, J.F. (1997). Pancreatic agenesis attributable to a single nucleotide deletion in the human IPF1 gene coding sequence. Nat Genet *15*, 106–110.
158. StOnge, L., SosaPineda, B., Chowdhury, K., Mansouri, A., and Gruss, P. (1997). Pax6 Is required for differentiation of glucagon-producing alpha-cells in mouse pancreas. Nature *387*, 406–409.
159. Sussel, L., Kalamaras, J., Hartigan-O'Connor, D.J., Meneses, J.J., Pedersen, R.A., Rubenstein, J.L.R., and German, M.S. (1998). Mice lacking the homeodomain transcription factor Nkx2.2 have diabetes due to arrested differentiation of pancreatic beta cells. Development *125*, 2213–2221.
160. Tennant, B.R., Robertson, A.G., Kramer, M., Li, L., Zhang, X., Beach, M., Thiessen, N., Chiu, R., Mungall, K., Whiting, C.J., *et al.* (2013). Identification and analysis of murine pancreatic islet enhancers. Diabetologia *56*, 542–552.
161. Thomas, I.H., Saini, N.K., Adhikari, A., Lee, J.M., Kasa-Vubu, J.Z., Vazquez, D.M., Menon, R.K., Chen, M., and Fajans, S.S. (2009). Neonatal diabetes mellitus with pancreatic agenesis in an infant with homozygous IPF-1 Pro63fsX60 mutation. Pediatr Diabetes *10*, 492–496.
162. Thompson, N., Gesina, E., Scheinert, P., Bucher, P., and Grapin-Botton, A. (2012). RNA profiling and chromatin immunoprecipitation-sequencing reveal that PTF1a stabilizes pancreas progenitor identity via the control of MNX1/HLXB9 and a network of other transcription factors. Mol Cell Biol *32*, 1189–1199.
163. Tritschler, S., Theis, F.J., Lickert, H., and Bottcher, A. (2017). Systematic single-cell analysis provides new insights into heterogeneity and plasticity of the pancreas. Mol Metab *6*, 974–990.
164. Van de Velde, S., Wiater, E., Tran, M., Hwang, Y.S., Cole, P.A., and Montminy, M. (2019). CREB Promotes Beta Cell Gene Expression by Targeting Its Coactivators to Tissue-Specific Enhancers. Mol Cell Biol *39*.
165. van Gurp, L., Muraro, M.J., Dielen, T., Seneby, L., Dharmadhikari, G., Gradwohl, G., van Oudenaarden, A., and de Koning, E.J.P. (2019). A transcriptomic roadmap to alpha- and beta-cell differentiation in the embryonic pancreas. Development *146*.
166. Velazco-Cruz, L., Goedegebuure, M.M., Maxwell, K.G., Augsornworawat, P., Hogrebe, N.J., and Millman, J.R. (2020). SIX2 Regulates Human beta Cell Differentiation from Stem Cells and Functional Maturation In Vitro. Cell Reports *31*.
167. Velazco-Cruz, L., Song, J., Maxwell, K.G., Goedegebuure, M.M., Augsornworawat, P., Hogrebe, N.J., and Millman, J.R. (2019). Acquisition of Dynamic Function in Human Stem Cell-Derived beta Cells. Stem Cell Reports *12*, 351–365.
168. Villasenor, A., Chong, D.C., and Cleaver, O. (2008). Biphasic Ngn3 expression in the developing pancreas. Dev Dyn *237*, 3270–3279.
169. Wade, J.T. (2015). Mapping Transcription Regulatory Networks with ChIP-seq and RNA-seq. Adv Exp Med Biol *883*, 119–134.
170. Wang, S., Zhang, J., Zhao, A., Hipkens, S., Magnuson, M.A., and Gu, G. (2007). Loss of Myt1 function partially compromises endocrine islet cell differentiation and pancreatic physiological function in the mouse. Mech Dev *124*, 898–910.

171. Wang, Y.J., and Kaestner, K.H. (2019). Single-Cell RNA-Seq of the Pancreatic Islets--a Promise Not yet Fulfilled? Cell Metab *29*, 539–544.
172. Wiebe, P.O., Kormish, J.D., Roper, V.T., Fujitani, Y., Alston, N.I., Zaret, K.S., Wright, C.V., Stein, R.W., and Gannon, M. (2007). Ptf1a binds to and activates area III, a highly conserved region of the Pdx1 promoter that mediates early pancreas-wide Pdx1 expression. Mol Cell Biol *27*, 4093–4104.
173. Willnow, D., Benary, U., Margineanu, A., Vignola, M.L., Konrath, F., Pongrac, I.M., Karimaddini, Z., Vigilante, A., Wolf, J., and Spagnoli, F.M. (2021). Quantitative lineage analysis identifies a hepato-pancreato-biliary progenitor niche. Nature.
174. Xu, C.R., Cole, P.A., Meyers, D.J., Kormish, J., Dent, S., and Zaret, K.S. (2011). Chromatin "prepattern" and histone modifiers in a fate choice for liver and pancreas. Science *332*, 963–966.
175. Xuan, S., Borok, M.J., Decker, K.J., Battle, M.A., Duncan, S.A., Hale, M.A., Macdonald, R.J., and Sussel, L. (2012). Pancreas-specific deletion of mouse Gata4 and Gata6 causes pancreatic agenesis. J Clin Invest *122*, 3516–3528.
176. Xuan, S., and Sussel, L. (2016). GATA4 and GATA6 regulate pancreatic endoderm identity through inhibition of hedgehog signaling. Development *143*, 780–786.
177. Yang, X., Raum, J.C., Kim, J., Yu, R., Yang, J., Rice, G., Li, C., Won, K.J., Stanescu, D.E., and Stoffers, D.A. (2022). A PDX1 cistrome and single-cell transcriptome resource of the developing pancreas. Development *149*.
178. Yang, Y.P., Thorel, F., Boyer, D.F., Herrera, P.L., and Wright, C.V.E. (2011). Context-specific alpha-to-beta-cell reprogramming by forced Pdx1 expression. Genes Dev *25*, 1680–1685.
179. Yoshihara, E., Wei, Z., Lin, C.S., Fang, S., Ahmadian, M., Kida, Y., Tseng, T., Dai, Y., Yu, R.T., Liddle, C., *et al.* (2016). ERR gamma Is Required for the Metabolic Maturation of Therapeutically Functional Glucose-Responsive beta Cells. Cell Metabolism *23*, 622–634.
180. Yu, X.X., Qiu, W.L., Yang, L., Li, L.C., Zhang, Y.W., and Xu, C.R. (2018). Dynamics of chromatin marks and the role of JMJD3 during pancreatic endocrine cell fate commitment. Development *145*.
181. Yu, X.X., Qiu, W.L., Yang, L., Wang, Y.C., He, M.Y., Wang, D., Zhang, Y., Li, L.C., Zhang, J., Wang, Y., *et al.* (2021). Sequential progenitor states mark the generation of pancreatic endocrine lineages in mice and humans. Cell Res *31*, 886–903.
182. Yu, X.X., Qiu, W.L., Yang, L., Zhang, Y., He, M.Y., Li, L.C., and Xu, C.R. (2019). Defining multistep cell fate decision pathways during pancreatic development at single-cell resolution. EMBO J *38*.
183. Yu, X.X., and Xu, C.R. (2020). Understanding generation and regeneration of pancreatic beta cells from a single-cell perspective. Development *147*.
184. Zeng, C., Mulas, F., Sui, Y., Guan, T., Miller, N., Tan, Y., Liu, F., Jin, W., Carrano, A.C., Huising, M.O., *et al.* (2017). Pseudotemporal Ordering of Single Cells Reveals Metabolic Control of Postnatal beta Cell Proliferation. Cell Metab *25*, 1160–1175 e1111.
185. Zhang, L., Sheng, C., Zhou, F., Zhu, K., Wang, S., Liu, Q., Yuan, M., Xu, Z., Liu, Y., Lu, J., *et al.* (2021). CBP/p300 HAT maintains the gene network critical for beta cell identity and functional maturity. Cell Death Dis *12*, 476.
186. Zhou, Q., Law, A.C., Rajagopal, J., Anderson, W.J., Gray, P.A., and Melton, D.A. (2007). A multipotent progenitor domain guides pancreatic organogenesis. Dev Cell *13*, 103–114.
187. Zorn, A.M., and Wells, J.M. (2009). Vertebrate endoderm development and organ formation. Annu Rev Cell Dev Biol *25*, 221–251.

Pancreatic Cell Fate Specification: Insights Into Developmental Mechanisms and Their Application for Lineage Reprogramming

Sara Gonzalez Ortega, Anna Melati, Victoria Menne, Anna Salowka, Miriam Vazquez Segoviano, and Francesca M. Spagnoli

1 Introduction

The pancreas is both an exocrine and endocrine gland derived from the endoderm germ layer [65, 83]. The exocrine part of the pancreas is composed of acini and ducts, which form a network of tubes to transport the digestive enzymes, synthesized by acinar cells, to the duodenum. The endocrine compartment is organized in the so-called islets of Langerhans, composed of five main cell types that are responsible for the production and release of hormones that regulate blood glucose levels [65, 74]. The most abundant endocrine cell types are the β-cells and α-cells, which produce insulin and glucagon, respectively [65, 74]. Insufficient β-cell mass or impaired β-cell function lead to dysregulation of glucose metabolism and development of diabetes, which currently affects more than 400 million people worldwide [61, 74]. The two major types of diabetes are type 1 diabetes (T1D) and type 2 diabetes (T2D) [47]. In individuals with T1D, insulin-producing β-cells are destroyed as a result of an autoimmune-mediated attack, requiring constant exogenous insulin administration [47, 61]. Despite the availability of insulin as a treatment to temporarily lower blood glucose levels, this remedy cannot avoid either the acute dangers of hypoglycemia or the long-term complications of hyperglycemia. Eventually, the challenge of curing T1D centers on the replacement of a functional β-cell mass. Islet transplantation has provided proof-of-principle that cell replacement therapy can effectively cure patients, leading to insulin independence for

Authors "Sara Gonzalez Ortega, Anna Melati, Victoria Menne, Anna Salowka, and Miriam Vazquez Segoviano" are Joint first authors.

S. G. Ortega · A. Melati · V. Menne · A. Salowka · M. V. Segoviano · F. M. Spagnoli (✉)
Centre for Gene Therapy and Regenerative Medicine,
King's College London, Guy's Hospital, London, UK
e-mail: francesca.spagnoli@kcl.ac.uk

L. Piemonti et al. (eds.), *Pluripotent Stem Cell Therapy for Diabetes*,
https://doi.org/10.1007/978-3-031-41943-0_3

5 years when combined with proper immunomodulatory regimens [11, 28, 81]. However, the clinical application of this approach is extremely limited by the scarcity of cadaveric donor tissue, and alternative sources of insulin-producing β-cells are therefore required [28, 61]. In recent years, significant progress has been made in developing culture conditions for the directed differentiation of human pluripotent stem cells (PSCs) into pancreatic progenitors and β-like cells that secrete insulin and are responsive to glucose to a certain extent [10, 25, 64, 72, 75]. Notably, the transplantation of human ESC-derived insulin-producing surrogates for cadaveric islets has recently entered clinical trials [28].

The differentiation protocols for generating pancreatic cells from human PSCs (either ESC or iPSCs) closely mimic in vivo pancreatic embryonic development through sequential changes of in vitro culture conditions and step-wise addition of signaling factors and cytokines [10, 25, 64, 72, 75]. Alternatively, adult differentiated somatic cells can be used as starting cell source, instead of PSCs, and induced to acquire a pancreatic β-cell fate upon ectopic expression of key developmental factors [44, 48], in a so-called direct lineage reprogramming approach. Both directed differentiation of human PSCs and lineage reprogramming relies on the understanding of the genetic regulators and mechanisms governing pancreas and islet development. Below, we discuss recent developments in the characterization of transcription factor (TF) networks governing cell fate decisions during pancreas organogenesis. We also consider the possible implications of these findings for direct lineage reprogramming and how they might advance this strategy into an effective treatment of diabetes.

2 Recent Insights Into Pancreas Development and Endocrine Fate Specification

All different pancreatic cell types arise from a common Pdx1$^+$ progenitor population (PP) that becomes specified as a result of the interplay between extrinsic signals from the surrounding microenvironment and intrinsic genetic determinants [27, 48, 83]. Numerous excellent reviews have been written on pancreatic development [14, 65, 68, 74, 83] and should be consulted for a more comprehensive view. Here, we summarize the main events of this multi-step process in Fig. 1 and review the most recent progress in elucidating regulatory mechanisms underlying pancreatic fate specification and differentiation, with a special emphasis on cell fate allocation within the endocrine lineage.

2.1 Pancreas Lineage Allocation and Specification

Pancreas organogenesis starts upon the induction of *Pdx1/PDX1* in the foregut endoderm at embryonic day (E) 8.5 in mouse and at approximately 29 days post conception (dpc) in human [65, 68]. Specified pancreatic progenitors expand in the

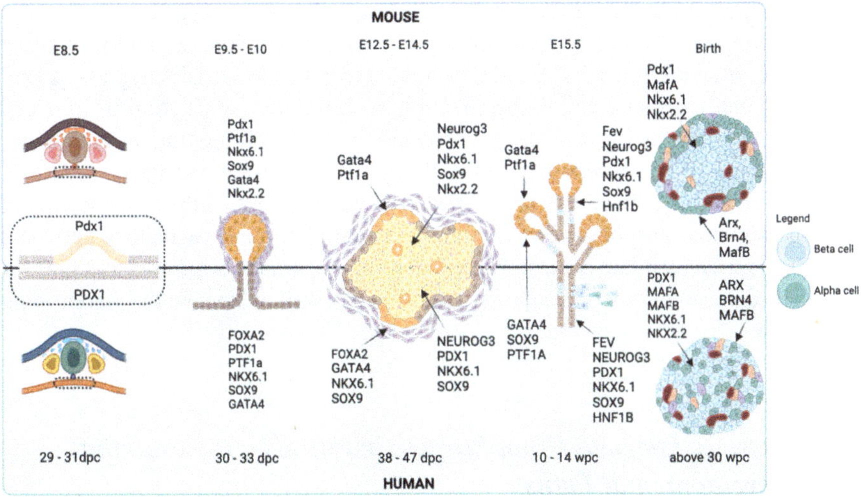

Fig. 1 Multistep cell fate decisions during pancreatic development. Schematic representation of the developing pancreas at indicated developmental stages in mouse (top) and human (bottom) embryos. Sub-set of TFs typifying distinct stages and lineages are displayed. (Created with BioRender.com)

surrounding mesenchyme and start expressing a plethora of TFs that maintain their multipotent state (*e.g.,* Sox9, Nkx6.1, Prox1, and Hnf1b) between E10.5 and 12.5 in the mouse embryo. Subsequently, the pancreatic epithelium undergoes important remodeling events, which result in the formation of a tubular network called the plexus, with a proximo-distal 'tip and trunk' domain organization [12, 14, 48, 53, 65, 69, 83, 85]. Elongating branch tips are formed at the periphery of the pancreatic epithelium, whereas the center contains the luminal plexus and corresponds to trunk domains. This proximo-distal architecture coincides with cell fate restriction of pancreatic progenitors and subsequent differentiation, whereby cells at the 'tip' adopt an acinar differentiation program, while cells within the 'trunk' contribute to the ductal and endocrine cell lineages [48, 65, 83]. A key event during endocrine development is the activation of the TF Neurogenin 3 (Neurog3) in a sub-set of pancreatic progenitor cells within the trunk region [65, 68] (Fig. 1). *Neurog3* expression marks endocrine progenitor (EP) cells and is concomitant with a series of morphogenetic events (e.g. cytoskeleton remodeling, apical narrowing and basal ward cell movement) [9, 12, 59], which ultimately lead to egression of the EPs from the epithelium [40] and clustering next to the forming nascent islets of Langerhans [53, 65]. At the same time, the accumulation of *Neurog3* levels in EP promotes the exit from the cell cycle and their differentiation into all hormone-producing islet cells [65, 68]. Over the last decades, many of the TFs promoting islet differentiation downstream of Neurog3 have been reported (*e.g.,* Neurod1, Mafa) [53, 65, 74]; however, the mechanisms regulating lineage segregation in the endocrine pancreas are still elusive. Likewise, the coordination between cellular and molecular events during pancreatic endocrine fate specification remains to be defined. The vesicle trafficking protein Syt13 has been recently identified as a connecting piece in these events, acting as a morphogenetic driver downstream of Neurog3 and promoting β-cell fate allocation in EPs [9]. Deletion of *Syt13* in mouse EPs impairs proper endocrine cell egression and β-specification, resulting in a shift from β- to α-cell fate [9]. These results suggest that the acquisition of a β-cell program might require precise timing of EP occupancy within the trunk epithelium and faster or slower egression results in the acquisition of an alternative α-cell fate. Further dissecting of the link between morphogenesis signaling and intrinsic gene-regulatory networks (GRN) that drive cell fate acquisition will continue to improve current strategies for islet cell-replacement therapy starting from PSC as well as lineage reprogramming (see below).

2.2 Lessons Learned From Single-Cell Analyses: Resolving Developmental Paths

In recent years, fast-evolving single-cell sequencing technologies, including single-cell RNA-seq (scRNA-seq), single-nucleus RNA-seq (snRNA-seq), scATAC-seq, have allowed investigation of developmental processes in the pancreas at unprecedented detail, providing insights into cellular heterogeneity in both mouse and human pancreas as well as transcriptional programs regulating endocrine lineage specification [13, 15, 19, 29, 33, 39, 63, 79, 88, 90, 97]. Thanks to scRNA-seq studies, the developmental paths within the mouse endocrine lineage have started to be resolved at a high level of granularity. Four subtypes of Neurog3$^+$ EP cells were annotated based on the differences in their transcriptome and epigenome [19, 79, 97]. These EP states possess distinct development potentials displaying a bias toward α-cells before E14.5, whereas late EPs preferentially differentiate into β-cells [79, 97]. Such bias is consistent with previous observations in transgenic mouse models that showed a differential stage-dependent output of islet cell types from Neurog3$^+$ EPs [49]. Yet, the direct genetic program downstream of Neurog3/NEUROG3 and the mechanisms that are responsible for producing distinct islet cell types only start to be unveiled.

Sc-transcriptomics complemented by spatial analyses provided us with the first cellular atlases of the human fetal and neonatal pancreas [29, 39, 63, 88, 97]. These datasets have been particularly helpful for reconstructing the human endocrine lineage through pseudotemporal ordering of endocrine cells from all developmental timepoints as in a 'virtual genetic lineage tracing' [29, 97]. Four distinct putative human EP populations were inferred displaying an earlier fate restriction, as a precommitted state between α and β lineages, compared to the mouse [29, 97]. This trajectory is consistent with another study that utilized mitochondrial genome variants within adult α- and β-cells as endogenous lineage-tracing markers in humans [56]. To further elucidate cellular lineage hierarchies in the human pancreas and experimentally validate predictions generated in silico, future experiments will benefit from methods for lineage barcoding to human pancreas modeled ex vivo with human PSC cultures [16] or in vivo using single-cell sequencing of mtDNA mutations as a marker of cell lineages [73].

2.3 *Lessons Learned From Single-Cell Analyses: Building GRNs*

Understanding how gene expression programs are controlled requires identifying regulatory relationships between TFs and their target genes, that can be organised and modeled as a hierarchical network, so-called a GRN [2, 67]. The GRNs define cell-type-specific transcriptional states, which in turn underlie cellular morphology and function [67]. During development, GRNs can adapt in response to external signals and other influences, resulting in a series of different developing cell states [2]. Single-cell measurements of transcriptome and epigenome states have been a major driving force behind recent advances in GRN mapping [35]. ScRNA-seq data alone can be used for inferring GRNs, but their combination with sc-epigenome profiling, such as scATAC-seq, allows to map GRNs with higher accuracy [35]. Ahead of the 'single-cell era', Arda et al. compiled a set of GRNs underlying main pancreatic cell states during development through extensive data mining, which integrated mutants analysis, gene expression and enhancer analyses, and biochemical studies of gene regulation [4]. The recent release of single-cell profiles from mouse and human embryonic pancreas has accelerated the uncovering of TF networks and GRN configurations governing cell states across pancreatic development. Duvall et al. [33] combined scRNA-seq and ATAC-seq analysis to map gene regulatory networks that define pancreatic endocrine lineage. By focusing the analysis on Neurog3$^+$ cells isolated from E15.5 and E17.5 mouse pancreas, they succeeded in defining networks of TFs either in common or highly specific to single lineages (β, α, or δ). Moreover, TFs that follow an ON-OFF pattern as cells differentiate were identified; for example, Xbp1, Id2, and Creb3 were found to be abundant in duct cells, while their levels decreased in Neurog3$^+$ progenitors and then increased again upon endocrine fate differentiation [33]. Further insights into the NEUROG3 direct genetic program and GRNs controlling islet cell development came from studies conducted in hESC-derived pancreatic cells as a model for human pancreas development [3, 80]. Genome-wide NEUROG3-bound regions were identified, and the analysis of NEUROG3 occupancy supports its role in activating the endocrine program in PP cells, uncovering potential co-regulation of target genes by FOXA TFs [3, 80]. Interestingly, Schreiber et al. also suggested the role of NEUROG3 in priming genes essential for later human β-cell maturation and function [80].

Downstream of the Neurog3, scRNA-seq identified the TF Fev as a marker of an intermediate EP population and upstream of differentiated, hormone-expressing endocrine cells [15, 19, 79, 97]. In a more recent study, snATAC-seq was used to analyze chromatin accessibility at single-cell resolution in Fev+ mouse endocrine cells [30]. Integration of the chromatin data with scRNA-seq datasets enabled the construction of cell-type-specific GRNs, which have started to shed light on how Fev may govern cell fate decisions. GRN configuration identified both known and novel regulators of the ductal-to-EP cell state transition as well as of pro-α and pro-β cell fates in the mouse [30]. Notably, FEV-expressing cells were also found in human fetal endocrine pancreas and human PSC-derived EPs [29].

Because of the limited access to human fetal tissues, in vitro stem cell-derived pancreatic cell types are being used more and more for modeling human pancreas development [7, 39, 52, 68, 102]. Data integration has shown a good correlation between the transcriptional profiles of EP cells derived from PSC differentiation in vitro and those endogenous counterparts from human fetal pancreas [29, 102]. Thus, in vitro PSC-derived models not only enable the map GRNs and underlying cellular identities in the human pancreas but also provide an accessible platform for directly testing GRN components and nodes through CRISPR-Cas9 genome editing approaches. For instance, CRISPR-perturbation functional analysis in hESC showed that FEV is not solely a marker of β-cell progenitors but also plays a role in regulating endocrine cell specification in the developing human pancreas [29]. Another successful implementation of such a combinatorial approach in an hESC platform led to the identification of a safeguard mechanism in humans that sustains pancreatic GRNs while restricting the activation of alternative programs, such as liver and duodenum [95]. In this context, HHEX was reported to cooperate with FOXA1/2 and GATA4, two TFs with pioneer factor activity, to promote pancreas specification [95]. More recently, sc-multiome-inferred GRNs have been comprehensively characterized for each step of SC-derived islet differentiation protocols, unveiling uncharacterized candidate cell fate regulators, such as EBF1 in β-cells and CDX2 in enterochromaffin-like cells [7, 102]. As a next step, these approaches should be expanded to connect the core GRN with the essential cues of the extra-cellular environment to predict if, and under what conditions, a cell state transition can be encouraged, as already done in other contexts [70]. Indeed, single-cell studies present an exciting opportunity to further decipher intercellular signaling behind cell–cell communication, which is crucial for β-cell differentiation [6, 29, 39, 63], and then integrate the information within GRN circuits.

In sum, we now have a greater understanding of the TFs governing cell states across pancreatic development, representing the building blocks of GRNs. The characterization of GRNs has started to pinpoint both known and novel candidate regulators of pancreatic endocrine cell identity, providing a resource for further investigation of their role in establishing or preserving cellular identity.

3 Direct Reprogramming for Pancreatic Cells

Direct lineage reprogramming is the conversion of one specialized cell type to another without the need for a pluripotent intermediate [44, 82]. This approach offers the possibility to overcome the limited regenerative capacities of the pancreas, providing a source of pancreatic β-like cells for diabetes cell therapies, but also a model for studying cell identity [44, 76, 101]. Reprogramming of one specialized cell type into another can be achieved by ectopically expressing defined TFs, which are relevant for the acquisition of the desired cell fate [44, 76, 82]. Thus, a strategy underlying successful lineage reprogramming has its fundaments in developmental biology and is based on the knowledge of the transcriptional networks underlying the establishment and maintenance of cellular identity. Re-wiring core TFs within a GRN can impose a new molecular program and, thereby, a new cellular identity onto terminally differentiated cells [54].

3.1 Direct Lineage Reprogramming: Inside the Pancreas

To date, most of the efforts to generate pancreatic β-cells by lineage reprogramming have focused on the conversion between cell types within the pancreas [24, 48, 84]. A pioneering study by the Melton group identified the combination of three key pancreatic TFs [Pdx1, Neurog3, MafA (referred to as PNM)], with well-known roles at subsequent stages along the pancreatic β-cell lineage, as sufficient for converting acinar cells into β-like cells in vivo in the mouse pancreas [100]. PNM-reprogrammed cells displayed a close resemblance in morphology and gene expression to endogenous β-cells, being able to ameliorate hyperglycemia in diabetic mouse models [100]. Since then, several groups have continued to use either the PNM alone or in combination with other TFs to reprogram pancreatic as well as non-pancreatic cells into β-like cells (Fig. 2). Besides the acini [100], additional cellular sources explored so far include pancreatic endocrine cells (e.g., α-cells, δ-cells) [23, 24, 87, 94], pancreatic duct cells [8, 55, 78], other endoderm-derivatives, such as the liver [20, 76], intestine [22, 86], and antral stomach cells [5], as well as others of non-endodermal origin, like fibroblasts [37].

Inducing cell interconversion within pancreatic islets (e.g., α-cells to β-cells) is a particularly attractive strategy for various reasons, including the common origin of islet cells from EPs, their anatomical proximity, and similar functionality [24, 77]. Especially, the α-cells are among the most promising targets due to their similarity in gene expression and chromatin modifications to the β-cells [1, 18], but also because few α-cells are required in an islet for maintaining glucose homeostasis and reducing glucagon levels may be beneficial for the treatment of diabetes [23, 43, 87]. Moreover, ATAC-seq showed that the α-cell genome is remarkably accessible, suggesting that α-cells are epigenetically poised to become β-cells but are prevented from acquiring a β-cell transcriptional program by repressive regulators [1, 60]. In line with these observations, α-cells have been reported to spontaneously convert into insulin-producing cells in vivo following extreme experimental (>99%) β-cell ablation in the mouse pancreas [87].

Recent work by Xiao et al. [94] showed that α-cells can be reprogrammed into functional β-like cells in vivo upon ectopic expression of *Pdx1* and *MafA* (PM) via

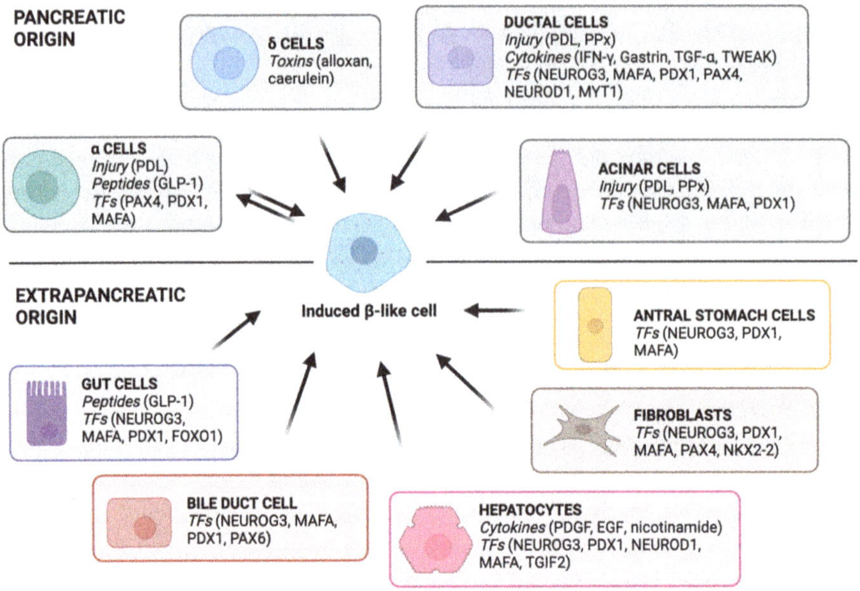

Fig. 2 Overview of recently implemented cell types and strategies in lineage reprogramming methods for pancreatic β-cell generation. (Created with BioRender.com)

an adeno-associated viral vector (AAV) delivered into the pancreatic ducts of diabetic mice. The AAV-PM delivery resulted in increased glucose responsiveness and normalized the glucose levels for the following 4 months [94]. Notably, the transgenic mouse model used in this study allowed the authors to perform lineage tracing, providing compelling evidence that the newly formed β-cells arose from α-cells. Finally, they also showed proof-of-concept evidence for α-to-β conversion in vitro in human islets following the treatment with a high-dose of streptozotocin, which was reported to ablate a large proportion of human β-cells [94]. Subsequent studies have further highlighted the advantages of using AAVs as an efficient and long-term gene delivery strategy to induce the conversion of α-to-β-cells [41, 42]. For example, the combination of an AAV vector serotype 8 (AAV8) with the Glucagon promoter allowed to target the expression of *Pdx1* and *MafA* cassette specifically to α-cells in diabetic mice [42].

Overall, these recent findings represent a significant step forward in lineage reprogramming as a treatment for diabetic patients, and concerns over the translation into the clinic are slowly being tackled. For instance, the novel delivery process via an endoscope of the AAV-PM cassette into pancreatic ducts, after having shown promise in animal models [41, 94], has been now licensed by the gene therapy developer Genprex and will move into phase 1 clinical trial. In a similar vein, Furuyama et al. [38] directly tested α-to-β-cells plasticity in human islet cells. Upon forced expression of *PDX1* and *MAFA* via adenoviral transduction, the authors reported the conversion of human α-cells into insulin-positive cells with about 35%

efficiency [38]. Furthermore, the reprogrammed human α-cells were transplanted into a diabetic immunodeficient mouse model and further studied by transcriptomic and proteomic analyses. The reprogrammed cells showed a hybrid state maintaining the expression of α-cell markers, while producing and processing insulin, which was enough to improve glucose tolerance in diabetic mouse [38].

Recent studies have also shown that α-to-β-cell reprogramming can be enhanced by modulating cellular plasticity with extrinsic factors or boosting β cell-proliferative capacity [45]. For instance, the forced expression of *Mycl* in adult mice results in an expansion of functional insulin-secreting cells due to both α-to-β conversion and replication of the mature β-cells [45]. Importantly, *MYCL* was also shown to stimulate cell proliferation in vitro in human adult cadaveric islets [45].

3.2 Direct Lineage Reprogramming: Outside the Pancreas

The use of non-pancreatic cells with endodermal origin as an alternative cellular source for generating insulin-producing cells offers some important advantages compared to pancreatic cells, including easier accessibility and larger availability (Fig. 2). For instance, the intestinal epithelium is a highly regenerative tissue with cells undergoing continuous renewal, which is different from the very low turnover rate of the pancreas [51]. In addition, the intestine contains an enteroendocrine cell population, which shares a set of TFs, such as Neurog3, with the endocrine pancreas [22]. Interestingly, recent single-cell sequencing and GRN analyses unveiled endogenous *INSULIN* gene expression in a cell population of the human fetal intestine, further highlighting its therapeutic potential as a source for generating β-cells [34]. Conversely, enteroendocrine cells that synthesize serotonin have also been found in the fetal human pancreas, in addition to the gut, resembling a transient fetal pre-β-cell compartment [102]. Thus, the induction of insulin-expressing enteroendocrine cells is a promising avenue for cell therapy in diabetes.

To date, Pdx1, MafA, Neurog3, and Foxo1 have been reported as the major TFs capable of driving gut-to-pancreas lineage reprogramming. Genetic ablation of *Foxo1* in Neurog3$^+$ enteroendocrine cells in mice gave rise to insulin-producing and glucose-responsive cells that express markers of mature β-cells in the gut [86]. Follow-up studies from the same group showed a conserved role for FOXO1 in humans. For instance, FOXO1 inhibition in human intestinal organoids resulted in the generation of insulin-positive cells as well as the expression of terminally differentiated β-cell markers [17]. More recently, they reported that cells with Paneth/goblet features and positive for INSULIN, which is present in the human fetal intestine, can be reprogrammed into intestinal β-like cells upon genetic or pharmacologic FOXO1 ablation [32].

Ectopic expression of PMN was shown to induce the formation of β like-cells, which secreted insulin and coalesced into "neoislets" below the crypt base in the gut of adult mice [22]. Similar results were obtained by forcing the expression of the same TFs in human intestinal organoids [22]. Likewise, the same set of TFs was employed to induce lineage reprogramming in the stomach, which also harbors hormone-secreting enteroendocrine cells [5]. Also, this reprogramming strategy was successfully translated into a human model [46]. Taking advantage of human stomach-derived organoids, the authors showed that sequential activation of NEUROG3, PDX1, and MAFA induces insulin-positive cells displaying β-cells hallmarks and capable of restoring glucose homeostasis of diabetic mice upon transplantation [46].

Lastly, when considering the potential for clinical application, the accessibility and highly regenerative capacity of the liver make hepatic cells another attractive starting source for generating β-cells [76]. Numerous reviews have been written on the adult plasticity between these two cell types and should be consulted for a more comprehensive view [76, 98]. Very recent studies using quantitative approaches and lineage tracing analysis [93] have unveiled significant plasticity in the ventral foregut during development, leading to a revision of the lineage relationship between liver and pancreas progenitors. A multipotent subpopulation of hepato-pancreato-biliary (HPB) progenitors has been identified in the mouse with the ability to give rise not only to pancreas and gall bladder cells but also to liver cells [93]. These findings have opened up the possibility of exploiting the HPB plasticity for future reprogramming strategies.

Altogether, these studies clearly support the potential of direct lineage reprogramming for generating β-like cells in vitro as well as in vivo, starting from cells inside and outside the pancreas. Reprogramming of more easily accessible cell types outside of the pancreas, such as cells in the digestive tract or the liver, could be more attractive from a clinical and surgical perspective.

4 Concluding Remarks and Future Directions

Even though rewriting of cell identity is possible, and newly induced cells can mitigate disease phenotypes in animal models [54, 82], important challenges remain to be overcome for accelerating a transition to clinical application, such as low conversion rates and heterogeneity of the resulting populations, which often comprise cells partially differentiated or incompletely specified. The availability of single-cell transcriptomics and new approaches, such as Tagging/Barcoding coupled with next-generation sequencing, has greatly improved our ability to study direct reprogramming in vitro [50, 62] as well as in vivo [57] and will help to address the current limitations. Indeed, single-cell analyses have started defining high-resolution maps of the journey that cells undergo during conversion, from which one can define 'productive' and 'non-productive' conversion paths and identify barriers to changing cell identity. For instance, scRNA-seq of in vivo PMN-induced acinar-to-β-cell conversion identified p53 as a barrier factor in the initial stage of reprogramming and the DNA methyltransferase Dnmt3a as a subsequent barrier [57]. Interestingly, p53 seems to act as a common barrier for mammalian cell regeneration, also in neurons and iPSCs, representing a common target for enhancing adult mammalian cell regeneration [57, 82]. In addition, sc-transcriptomics provides comprehensive molecular profiles of the reprogrammed cells, assessing the level of cellular authenticity and maturity that can be reached by the induced cells and to what extent they resemble the in vivo counterparts. Expanding the analysis to the epigenome, using scATAC-seq, we will also answer the question to which extent the epigenetic memory of the source cell is retained in reprogrammed cells. This will help in understanding if and how such a residual memory might interfere with the new function of a reprogrammed cell.

Single-cell and bulk-transcriptome data have also fuelled the development of computational strategies for identifying putative reprogramming factors [31, 50, 71, 89]; complementary and, perhaps in certain contexts, more powerful than a developmental biology-informed approach. Notably, recent algorithms that have been developed to predict combinations of reprogramming factors consider not only the initial and final cell states but also intermediate cell states or trajectories during cell reprogramming [89]. Besides computational approaches, CRISPR-Cas genome screening can also be used for defining successful reprogramming strategies. Liu et al. [58] developed an elegant approach to identify factors promoting neuronal fate in an unbiased fashion using a CRISPR-activation-based screening of TFs and other regulators in mouse ESCs. A core cluster of six pro-neural genes *Ngn1, Zeb1, Tcf15, Foxo1, Ezh2*, and *Brn2* was identified from this screening; some of these factors also yielded successful reprogramming of fibroblasts into neurons [58]. A similar unbiased approach to systematically identify regulators of pancreatic fate specification sufficient for direct reprogramming has yet to be pursued.

An alternative strategy to direct lineage reprogramming is to induce the desired cell fate by overexpressing lineage-specific TFs in PSCs, in a so-called forward programming approach [36]. This method combines the advantages of directed differentiation and lineage reprogramming, enabling scalable and rapid generation of human cell types in a more efficient manner. Successful examples have been reported in the context of the hematopoietic, muscle, and neuronal lineages [36, 66]. For instance, the combined expression of the TFs Ascl1, Brn2, and Myt1l is not only sufficient to convert mouse fibroblasts into neurons [91] but also efficiently drives neuronal specification from hPSCs [82]. More recently, the same group showed that forward programming of hPSCs with the TF NGN2 induces rapid neuronal differentiation with nearly 100% yield and purity in less than 2 weeks [99]. Similar TF-programming approaches from hPSCs have not been attempted in the context of the pancreatic lineage; this might be a strategy to pursue for enhancing the generation of mature, functional β-cells.

In addition to the right choice of TFs and cellular sources, successful translation of direct reprogramming from basic research into the clinic needs the development of adequate gene delivery systems. AAV vectors are leading delivery platforms currently used in pre-clinical and clinical studies for safe and efficient in vivo gene delivery [92]. Local AAV delivery of reprogramming TFs to pancreatic ducts has succeeded in inducing conversion to β-like cells [41, 94]. However, AAV serotypes that selectively target pancreatic cells upon systemic delivery still need to be developed. Another, even more promising, TFs-delivery platform for in vitro and in vivo direct reprogramming is the use of synthetic mRNAs coupled to lipid nanoparticles (LPN) [21]. The success of mRNA vaccines during the COVID-19 pandemic has greatly accelerated the development of mRNA-based medicine. In vitro synthesized mRNA presents multiple advantages for direct reprogramming applications, including immediate but transitory mechanism of action, non-integrative properties, and safe and relatively simple manufacturing [21]. A proof-of-concept study demonstrated that direct conversion of human pancreatic duct-derived cells into insulin-producing cells can be achieved using a single synthetic modified-mRNA encoding for the MAFA [26]. Recent improvement in LNPs, which further shield the mRNA from enzymatic degradation and facilitates cellular entry, will accelerate future clinical applications of TF-based reprogramming approaches [96].

In conclusion, we are in an exciting new age of direct lineage reprogramming. Ongoing concerted efforts in the fields of development, stem cell and computational biology will enable rapid advances in our ability to genetically access and rewrite cellular identity. Further delineation of GRNs defining pancreatic cell lineages will be crucial for designing synthetic engineering approaches to generate β-cells or islet tissues and move the field closer to clinical application.

Acknowledgments Our apologies go to all authors whose important work could not be mentioned due to space limitations. We gratefully acknowledge the financial support of the Wellcome Trust PhD program "Advanced therapies for regenerative medicine" (grant number 218461/Z/19/Z). FMS is supported by the Wellcome Trust grant 'Probing cellular heterogeneity in the pancreatic microenvironment' (grant number 221807/Z/20/Z). We thank Dr. Fay Minty for her support of the program.

Conflict of Interest Statement The authors declare no competing interests.

References

1. Ackermann, A. M., Wang, Z., Schug, J., Naji, A. & Kaestner, K. H. 2016. Integration of ATAC-seq and RNA-seq identifies human alpha cell and beta cell signature genes. *Molecular Metabolism,* 5, 233–244.
2. Almeida, N., Chung, M. W. H., Drudi, E. M., Engquist, E. N., Hamrud, E., Isaacson, A., Tsang, V. S. K., Watt, F. M. & Spagnoli, F. M. 2021. Employing core regulatory circuits to define cell identity. *Embo j,* 40, e106785.
3. Alvarez-Dominguez, J. R., Donaghey, J., Rasouli, N., Kenty, J. H. R., Helman, A., Charlton, J., Straubhaar, J. R., Meissner, A. & Melton, D. A. 2020. Circadian Entrainment Triggers Maturation of Human In Vitro Islets. *Cell Stem Cell,* 26, 108–122.e10.
4. Arda, H. E., Benitez, C. M. & Kim, S. K. 2013. Gene regulatory networks governing pancreas development. *Dev Cell,* 25, 5–13.
5. Ariyachet, C., Tovaglieri, A., Xiang, G., Lu, J., Shah, M. S., Richmond, C. A., Verbeke, C., Melton, D. A., Stanger, B. Z., Mooney, D., Shivdasani, R. A., Mahony, S., Xia, Q., Breault, D. T. & Zhou, Q. 2016. Reprogrammed Stomach Tissue as a Renewable Source of Functional β Cells for Blood Glucose Regulation. *Cell Stem Cell,* 18, 410-421.
6. Armingol, E., Officer, A., Harismendy, O. & Lewis, N. E. 2021. Deciphering cell–cell interactions and communication from gene expression. *Nature Reviews Genetics,* 22, 71–88.

7. Augsornworawat, P., Hogrebe, N. J., Ishahak, M., Schmidt, M. D., Marquez, E., Maestas, M. M., Veronese-Paniagua, D. A., Gale, S. E., Miller, J. R., Velazco-Cruz, L. & Millman, J. R. 2023. Single-nucleus multi-omics of human stem cell-derived islets identifies deficiencies in lineage specification. *Nat Cell Biol.* 25, 904–916.
8. Azzarelli, R., Rulands, S., Nestorowa, S., Davies, J., Campinoti, S., Gillotin, S., Bonfanti, P., Göttgens, B., Huch, M., Simons, B. D. & Philpott, A. 2018. Neurogenin3 phosphorylation controls reprogramming efficiency of pancreatic ductal organoids into endocrine cells. *Sci Rep.,* 18, 15374.
9. Bakhti, M., Bastidas-Ponce, A., Tritschler, S., Czarnecki, O., Tarquis-Medina, M., Nedvedova, E., Jaki, J., Willmann, S. J., Scheibner, K., Cota, P., Salinno, C., Boldt, K., Horn, N., Ueffing, M., Burtscher, I., Theis, F. J., Coskun, Ü. & Lickert, H. 2022. Synaptotagmin-13 orchestrates pancreatic endocrine cell egression and islet morphogenesis. *Nature Communications,* 13, 4540.
10. Balboa, D., Barsby, T., Lithovius, V., Saarimäki-Vire, J., Omar-Hmeadi, M., Dyachok, O., Montaser, H., Lund, P. E., Yang, M., Ibrahim, H., Näätänen, A., Chandra, V., Vihinen, H., Jokitalo, E., Kvist, J., Ustinov, J., Nieminen, A. I., Kuuluvainen, E., Hietakangas, V., Katajisto, P., Lau, J., Carlsson, P. O., Barg, S., Tengholm, A. & Otonkoski, T. 2022. Functional, metabolic and transcriptional maturation of human pancreatic islets derived from stem cells. *Nat Biotechnol,* 40, 1042–1055.
11. Balzano, G., Maffi, P., Nano, R., Mercalli, A., Melzi, R., Aleotti, F., De Cobelli, F., Magistretti, P., Scavini, M., Secchi, A., Falconi, M. & Piemonti, L. 2019. Diabetes-free survival after extended distal pancreatectomy and islet auto transplantation for benign or borderline/malignant lesions of the pancreas. *Am J Transplant,* 19, 920–928.
12. Bankaitis, E. D., Bechard, M. E., Gu, G., Magnuson, M. A. & Wright, C. V. E. 2018. ROCK-nmMyoII, Notch and Neurog3 gene-dosage link epithelial morphogenesis with cell fate in the pancreatic endocrine-progenitor niche. *Development,* 145:dev162115.
13. Baron, M., Veres, A., Wolock, Samuel L., Faust, Aubrey L., Gaujoux, R., Vetere, A., Ryu, Jennifer H., Wagner, Bridget K., Shen-Orr, Shai S., Klein, Allon M., Melton, Douglas A. & Yanai, I. 2016. A Single-Cell Transcriptomic Map of the Human and Mouse Pancreas Reveals Inter- and Intra-cell Population Structure. *Cell Systems,* 3, 346–360.e4.
14. Bastidas-Ponce, A., Scheibner, K., Lickert, H. & Bakhti, M. 2017. Cellular and molecular mechanisms coordinating pancreas development. *Development,* 144, 2873–2888.
15. Bastidas-Ponce, A., Tritschler, S., Dony, L., Scheibner, K., Tarquis-Medina, M., Salinno, C., Schirge, S., Burtscher, I., Böttcher, A., Theis, F. J., Lickert, H. & Bakhti, M. 2019. Comprehensive single cell mRNA profiling reveals a detailed roadmap for pancreatic endocrinogenesis. *Development*, 146(12):dev173849.
16. Biddy, B. A., Kong, W., Kamimoto, K., Guo, C., Waye, S. E., Sun, T. & Morris, S. A. 2018. Single-cell mapping of lineage and identity in direct reprogramming. *Nature,* 564, 219–224.
17. Bouchi, R., Foo, K. S., Hua, H., Tsuchiya, K., Ohmura, Y., Sandoval, P. R., Ratner, L. E., Egli, D., Leibel, R. L. & Accili, D. 2014. FOXO1 inhibition yields functional insulin-producing cells in human gut organoid cultures. *Nature Communications,* 5, 4242.
18. Bramswig, N. C., Everett, L. J., Schug, J., Dorrell, C., Liu, C., Luo, Y., Streeter, P. R., Naji, A., Grompe, M. & Kaestner, K. H. 2013. Epigenomic plasticity enables human pancreatic alpha to beta cell reprogramming. *J Clin Invest,* 123, 1275–84.
19. Byrnes, L., Wong, D., Subramaniam, M., Meyer, N., Gilchrist, C., Knox, S., Tward, A., Ye, C. & Sneddon, J. 2018. Lineage dynamics of murine pancreatic development at single-cell resolution. *Nat Commun.,* 9, 3922.
20. Cerdá-Esteban, N., Naumann, H., Ruzittu, S., Mah, N., Pongrac, I. M., Cozzitorto, C., Hommel, A., Andrade-Navarro, M. A., Bonifacio, E. & Spagnoli, F. M. 2017. Stepwise reprogramming of liver cells to a pancreas progenitor state by the transcriptional regulator Tgif2. *Nature Communications, 8, 14127.*
21. Chabanovska, O., Galow, A. M., David, R. & Lemcke, H. 2021. mRNA – A game changer in regenerative medicine, cell-based therapy and reprogramming strategies. *Adv Drug Deliv Rev,* 179, 114002.

22. Chen, Y. J., Finkbeiner, S. R., Weinblatt, D., Emmett, M. J., Tameire, F., Yousefi, M., Yang, C., Maehr, R., Zhou, Q., Shemer, R., Dor, Y., Li, C., Spence, J. R. & Stanger, B. Z. 2014. De novo formation of insulin-producing "neo-β cell islets" from intestinal crypts. *Cell Reports, 6, 1046-1058.*
23. Chera, S., Baronnier, D., Ghila, L., Cigliola, V., Jensen, J. N., Gu, G., Furuyama, K., Thorel, F., Gribble, F. M., Reimann, F. & Herrera, P. L. 2014. Diabetes recovery by age-dependent conversion of pancreatic δ-cells into insulin producers. *Nature, 514, 503-507.*
24. Cigliola, V., Thorel, F., Chera, S. & Herrera, P. L. 2016. Stress-induced adaptive islet cell identity changes. *Diabetes Obes Metab,* 18 Suppl 1, 87–96.
25. Cogger, K. F., Sinha, A., Sarangi, F., McGaugh, E. C., Saunders, D., Dorrell, C., Mejia-Guerrero, S., Aghazadeh, Y., Rourke, J. L., Screaton, R. A., Grompe, M., Streeter, P. R., Powers, A. C., Brissova, M., Kislinger, T. & Nostro, M. C. 2017. Glycoprotein 2 is a specific cell surface marker of human pancreatic progenitors. *Nat Commun,* 8, 331.
26. Corritore, E., Lee, Y.-S., Pasquale, V., Liberati, D., Hsu, M.-J., Lombard, C. A., Van Der Smissen, P., Vetere, A., Bonner-Weir, S., Piemonti, L., Sokal, E. & Lysy, P. A. 2016. V-Maf Musculoaponeurotic Fibrosarcoma Oncogene Homolog A Synthetic Modified mRNA Drives Reprogramming of Human Pancreatic Duct-Derived Cells Into Insulin-Secreting Cells. *Stem Cells Translational Medicine,* **5,** 1525–1537.
27. Cozzitorto, C. & Spagnoli, F. M. 2019. Pancreas organogenesis: The interplay between surrounding microenvironment(s) and epithelium-intrinsic factors. *Curr Top Dev Biol.,* 132, 221-256.
28. de Klerk, E. & Hebrok, M. 2021. Stem Cell-Based Clinical Trials for Diabetes Mellitus. *Front Endocrinol (Lausanne),* 12**,** 631463.
29. De La O, S., Liu, Z., Sun, H., Yu, S. K., Wong, D. M., Chu, E., Rao, S. A., Eng, N., Peixoto, G., Bouza, J., Shen, Y., Knox, S. M., Tward, A. D., Gloyn, A. L. & Sneddon, J. B. 2022. Single-Cell Multi-Omic Roadmap of Human Fetal Pancreatic Development. *bioRxiv.* 2022.02.17.480942.
30. De La O, S., Yao, X., Chang, S., Liu, Z. & Sneddon, J. B. 2023. Single-cell chromatin accessibility of developing murine pancreas identifies cell state-specific gene regulatory programs. *Mol Metab,* 73**,** 101735.
31. Dixit, A., Parnas, O., Li, B., Chen, J., Fulco, C. P., Jerby-Arnon, L., Marjanovic, N. D., Dionne, D., Burks, T., Raychowdhury, R., Adamson, B., Norman, T. M., Lander, E. S., Weissman, J. S., Friedman, N. & Regev, A. 2016. Perturb-Seq: Dissecting Molecular Circuits with Scalable Single-Cell RNA Profiling of Pooled Genetic Screens. *Cell,* 167**,** 1853–1866.e17.
32. Du, W., Wang, J., Kuo, T., Wang, L., Mckimpson, W. M., Son, J., Watanabe, H., Kitamoto, T., Lee, Y., Creusot, R. J., Ratner, L. E., Mccune, K., Chen, Y. W., Grubbs, B. H., Thornton, M. E., Fan, J., Sultana, N., Diaz, B. S., Balasubramanian, I., Gao, N., Belvedere, S. & Accili, D. 2022. Pharmacological conversion of gut epithelial cells into insulin-producing cells lowers glycemia in diabetic animals. *J Clin Invest.* 132(24):e162720.
33. Duvall, E., Benitez, C. M., Tellez, K., Enge, M., Pauerstein, P. T., Li, L., Baek, S., Quake, S. R., Smith, J. P., Sheffield, N. C., Kim, S. K. & Arda, H. E. 2022. Single-cell transcriptome and accessible chromatin dynamics during endocrine pancreas development. *Proceedings of the National Academy of Sciences of the United States of America.* 119(26): e2201267119.
34. Egozi, A., Llivichuzhca-Loja, D., Mccourt, B. T., Bahar Halpern, K., Farack, L., An, X., Wang, F., Chen, K., Konnikova, L. & Itzkovitz, S. 2021. Insulin is expressed by enteroendocrine cells during human fetal development. *Nature Medicine, 27, 2104-2107.*
35. Fiers, M., Minnoye, L., Aibar, S., Bravo González-Blas, C., Kalender Atak, Z. & Aerts, S. 2018. Mapping gene regulatory networks from single-cell omics data. *Brief Funct Genomics,* 17**,** 246–254.
36. Flitsch, L. J., Laupman, K. E. & Brüstle, O. 2020. Transcription Factor-Based Fate Specification and Forward Programming for Neural Regeneration. *Frontiers in cellular neuroscience.* Front Cell Neurosci. 14, 121.

37. Fontcuberta-Pisunyer, M., García-Alamán, A., Prades, È., Téllez, N., Alves-Figueiredo, H., Ramos-Rodríguez, M., Enrich, C., Fernandez-Ruiz, R., Cervantes, S., Clua, L., Ramón-Azcón, J., Broca, C., Wojtusciszyn, A., Montserrat, N., Pasquali, L., Novials, A., Servitja, J.-M., Vidal, J., Gomis, R. & Gasa, R. 2023. Direct reprogramming of human fibroblasts into insulin-producing cells using transcription factors. *Communications Biology, 6,1-15.*
38. Furuyama, K., Chera, S., Van Gurp, L., Oropeza, D., Ghila, L., Damond, N., Vethe, H., Paulo, J. A., Joosten, A. M., Berney, T., Bosco, D., Dorrell, C., Grompe, M., Ræder, H., Roep, B. O., Thorel, F. & Herrera, P. L. 2019. Diabetes relief in mice by glucose-sensing insulin-secreting human α-cells. *Nature, 567, 43-48.*
39. Gonçalves, C. A., Larsen, M., Jung, S., Stratmann, J., Nakamura, A., Leuschner, M., Hersemann, L., Keshara, R., Perlman, S., Lundvall, L., Thuesen, L. L., Hare, K. J., Amit, I., Jørgensen, A., Kim, Y. H., Del Sol, A. & Grapin-Botton, A. 2021. A 3D system to model human pancreas development and its reference single-cell transcriptome atlas identify signaling pathways required for progenitor expansion. *Nature Communications 2021 12, 1-17.*
40. Gouzi, M., Kim, Y. H., Katsumoto, K., Johansson, K. & Grapin-Botton, A. 2011. Neurogenin3 initiates stepwise delamination of differentiating endocrine cells during pancreas development. *Dev Dyn,* 240, 589–604.
41. Guo, P., Wiersch, J., Xiao, X. & Gittes, G. 2019. Simplified Purification of AAV and Delivery to the Pancreas by Intraductal Administration. *Methods in Molecular Biology,* 1950, 373-387.
42. Guo, P., Zhang, T., Lu, A., Shiota, C., Huard, M., Whitney, K. E. & Huard, J. 2023. Specific reprogramming of alpha cells to insulin-producing cells by short glucagon promoter-driven Pdx1 and MafA. *Molecular Therapy, 28, 355-365.*
43. Habener, J. F. & Stanojevic, V. 2013. Alpha cells come of age. *Trends in endocrinology and metabolism, 24, 153-163.*
44. Heinrich, C., Spagnoli, F. M. & Berninger, B. 2015. In vivo reprogramming for tissue repair. *Nature Cell Biology, 17, 204-211.*
45. Hirano, M., So, Y., Tsunekawa, S., Kabata, M., Ohta, S., Sagara, H., Sankoda, N., Taguchi, J., Yamada, Y., Ukai, T., Kato, M., Nakamura, J., Ozawa, M., Yamamoto, T. & Yamada, Y. 2022. MYCL-mediated reprogramming expands pancreatic insulin-producing cells. *Nature Metabolism, 4, 254-268.*
46. Huang, X., Gu, W., Zhang, J., Lan, Y., Colarusso, J. L., Li, S., Pertl, C., Lu, J., Kim, H., Zhu, J., Breault, D. T., Sévigny, J. & Zhou, Q. 2023. Stomach-derived human insulin-secreting organoids restore glucose homeostasis. *Nature Cell Biology, 25, 778–786.*
47. IDF. The IDF Diabetes Atlas 2021 Report. *IDF Diabetes Atlas,* 10th.
48. Isaacson, A. & Spagnoli, F. M. 2021. Pancreatic cell fate specification: insights into developmental mechanisms and their application for lineage reprogramming. *Current Opinion in Genetics & Development, 70, 32-39.*
49. Johansson, K. A., Dursun, U., Jordan, N., Gu, G., Beermann, F., Gradwohl, G. & Grapin-Botton, A. 2007. Temporal control of neurogenin3 activity in pancreas progenitors reveals competence windows for the generation of different endocrine cell types. *Dev Cell,* 12, 457–65.
50. Kamimoto, K., Adil, M. T., Jindal, K., Hoffmann, C. M., Kong, W., Yang, X. & Morris, S. A. 2023. Gene regulatory network reconfiguration in direct lineage reprogramming. *Stem Cell Reports,* 18, 97–112.
51. Kopp, J. L., Grompe, M. & Sander, M. 2016. Stem cells versus plasticity in liver and pancreas regeneration. *Nat Cell Biol,* 18, 238–45.
52. Krentz, N. A. J., Lee, M. Y. Y., Xu, E. E., Sproul, S. L. J., Maslova, A., Sasaki, S. & Lynn, F. C. 2018. Single-Cell Transcriptome Profiling of Mouse and hESC-Derived Pancreatic Progenitors. *Stem Cell Reports,* 11, 1551–1564.
53. Larsen, H. L. & Grapin-Botton, A. 2017. The molecular and morphogenetic basis of pancreas organogenesis. *Seminars in Cell & Developmental Biology, 66, 51-68.*
54. Leaman, S., Marichal, N. & Berninger, B. 2022. Reprogramming cellular identity in vivo. *Development,* 149(4):dev200433.

55. Lee, J., Sugiyama, T., Liu, Y., Wang, J., Gu, X., Lei, J., Markmann, J. F., Miyazaki, S., Miyazaki, J., Szot, G. L., Bottino, R. & Kim, S. K. 2013. Expansion and conversion of human pancreatic ductal cells into insulin-secreting endocrine cells. *Elife,* 2, e00940.
56. Li, L., Yufeng, Z., Weizhou, Q., Yao, L., Yingkun, Z., Fanghe, L., Cenxi, L., Guangxing, L., Yanling, S., Jia, S., Chaoyong, Y. & Jin, L. 2021. Single-cell transcriptome lineage tracing of human pancreatic development identifies distinct developmental trajectories of alpha and beta cells. *bioRxiv*, https://doi.org/10.1101/2021.01.14.426320
57. Liu, G., Li, Y., Li, M., Li, S., He, Q., Liu, S., Su, Q., Chen, X., Xu, M., Zhang, Z.-N., Shao, Z. & Li, W. 2023. Charting a high-resolution roadmap for regeneration of pancreatic beta cells by in vivo transdifferentiation from adult acinar cells. *Science Advances,* 9, eadg2183.
58. Liu, Y., Yu, C., Daley, T. P., Wang, F., Cao, W. S., Bhate, S., Lin, X., Still, C., Liu, H., Zhao, D., Wang, H., Xie, X. S., Ding, S., Wong, W. H., Wernig, M. & Qi, L. S. 2018. CRISPR Activation Screens Systematically Identify Factors that Drive Neuronal Fate and Reprogramming. Cell Stem Cell, 3, 758-771.e8.
59. Löf-Öhlin, Z. M., Nyeng, P., Bechard, M. E., Hess, K., Bankaitis, E., Greiner, T. U., Ameri, J., Wright, C. V. & Semb, H. 2017. EGFR signalling controls cellular fate and pancreatic organogenesis by regulating apicobasal polarity. *Nat Cell Biol,* 19, 1313–1325.
60. Mawla, A. M., van der Meulen, T. & Huising, M. O. 2023. Chromatin accessibility differences between alpha, beta, and delta cells identifies common and cell type-specific enhancers. *BMC Genomics,* 24, 202.
61. Migliorini, A., Nostro, M. C. & Sneddon, J. B. 2021. Human pluripotent stem cell-derived insulin-producing cells: A regenerative medicine perspective. *Cell Metabolism, 33, 721-731.*
62. Morris, S. A. 2016. Direct lineage reprogramming via pioneer factors; a detour through developmental gene regulatory networks. *Development,* 143, 2696–705.
63. Olaniru, O. E., Kadolsky, U., Kannambath, S., Vaikkinen, H., Fung, K., Dhami, P. & Persaud, S. J. 2023. Single-cell transcriptomic and spatial landscapes of the developing human pancreas. *Cell Metabolism, 35, 184-199.e5.*
64. Pagliuca FW, Millman JR, Gürtler M, Segel M, Van Dervort A, Ryu JH, et al 2014. Generation of functional human pancreatic β cells in vitro. *Cell,* 159, 428–39.
65. Pan, F. C. & Wright, C. 2011. Pancreas organogenesis: from bud to plexus to gland. *Developmental dynamics*, 240, 530-565.
66. Pawlowski, M., Ortmann, D., Bertero, A., Tavares, J. M., Pedersen, R. A., Vallier, L. & Kotter, M. R. N. 2017. Inducible and Deterministic Forward Programming of Human Pluripotent Stem Cells into Neurons, Skeletal Myocytes, and Oligodendrocytes. *Stem Cell Reports.* 8, 803-812.
67. Peter, I. S. & Davidson, E. H. 2016. Implications of Developmental Gene Regulatory Networks Inside and Outside Developmental Biology. *Curr Top Dev Biol,* 117, 237–51.
68. Petersen, M. B. K., Gonçalves, C. A. C., Kim, Y. H. & Grapin-Botton, A. 2018. Recapitulating and Deciphering Human Pancreas Development From Human Pluripotent Stem Cells in a Dish. *Curr Top Dev Biol,* 129, 143–190.
69. Pierreux, C. E., Cordi, S., Hick, A. C., Achouri, Y., Ruiz De Almodovar, C., Prévot, P. P., Courtoy, P. J., Carmeliet, P. & Lemaigre, F. P. 2010. Epithelial: Endothelial cross-talk regulates exocrine differentiation in developing pancreas. *Developmental Biology, 347, 216-227.*
70. Prochazka, L., Benenson, Y. & Zandstra, P. W. 2017. Synthetic gene circuits and cellular decision-making in human pluripotent stem cells. *Current Opinion in Systems Biology,* 5, 93–103.
71. Rackham, O. J. L., Firas, J., Fang, H., Oates, M. E., Holmes, M. L., Knaupp, A. S., Suzuki, H., Nefzger, C. M., Daub, C. O., Shin, J. W., Petretto, E., Forrest, A. R. R., Hayashizaki, Y., Polo, J. M., Gough, J. & The, F. C. 2016. A predictive computational framework for direct reprogramming between human cell types. *Nature Genetics,* 48, 331–335.
72. Rezania A, Bruin JE, Arora P, Rubin A, Batushansky I, Asadi A, et al. 2014. Reversal of diabetes with insulin-producing cells derived in vitro from human pluripotent stem cells. *Nat Biotechnol.,* 32, 1121–33.

73. Rodrigo, A. G., Raúl, A. R. H., Jiazhen, R., Maureen, D., Tatiana, A. K., Amanda, B. M., Gary, W. F., Nancy, R. Z. & Sydney, M. S. 2023. Clonal cell states link Barrett's esophagus and esophageal adenocarcinoma. *bioRxiv*, 2023.01.26.525564.
74. Romer, A. I. & Sussel, L. 2015. Pancreatic islet cell development and regeneration. *Current Opinion in Endocrinology, Diabetes and Obesity,* 22, 255–64.
75. Russ Ha, P. A., Ringler JJ, Hennings TG, Nair GG, Shveygert M, et al. 2015. Controlled induction of human pancreatic progenitors produces functional beta-like cells in vitro. *EMBO J,* 34, 1759–72.
76. Ruzittu, S., Willnow, D. & Spagnoli, F. M. 2020. Direct Lineage Reprogramming: Harnessing Cell Plasticity between Liver and Pancreas. *Cold Spring Harb Perspect Biol,* 12(7):a035626.
77. Saleh, M., Gittes, G. K. & Prasadan, K. 2021. Alpha-to-beta cell trans-differentiation for treatment of diabetes. *Biochemical Society Transactions, 49, 2539–2548.*
78. Sancho, R., Gruber, R., Gu, G. & Behrens, A. 2014. Loss of Fbw7 Reprograms Adult Pancreatic Ductal Cells into α, δ, and β Cells. *Cell Stem Cell,* 15, 139–53.
79. Scavuzzo, M. A., Hill, M. C., Chmielowiec, J., Yang, D., Teaw, J., Sheng, K., Kong, Y., Bettini, M., Zong, C., Martin, J. F. & Borowiak, M. 2018. Endocrine lineage biases arise in temporally distinct endocrine progenitors during pancreatic morphogenesis. *Nat Commun,* 9, 3356.
80. Schreiber, V., Mercier, R., Jiménez, S., Ye, T., García-Sánchez, E., Klein, A., Meunier, A., Ghimire, S., Birck, C., Jost, B., de Lichtenberg, K. H., Honoré, C., Serup, P. & Gradwohl, G. 2021. Extensive NEUROG3 occupancy in the human pancreatic endocrine gene regulatory network. *Molecular Metabolism,* 53, 101313.
81. Shapiro AMJ, Pokrywczynska M & Ricordi C. 2017. Clinical pancreatic islet transplantation. *Nat Rev Endocrinol.,* 13, 268–277.
82. Shelby, H., Shelby, T. & Wernig, M. 2022. Somatic Lineage Reprogramming. *Cold Spring Harb Perspect Biol.,* 14, a040808.
83. Shih, H. P., Wang, A. & Sander, M. 2013. Pancreas Organogenesis: From Lineage Determination to Morphogenesis. *Annual Review of Cell and Developmental Biology,* 29, 81–105.
84. Stanger, B. Z. & Hebrok, M. 2013. Control of Cell Identity in Pancreas Development and Regeneration. *Gastroenterology,* 144, 1170–1179.
85. Sznurkowska, M. K., Hannezo, E., Azzarelli, R., Rulands, S., Nestorowa, S., Hindley, C. J., Nichols, J., Göttgens, B., Huch, M., Philpott, A. & Simons, B. D. 2018. Defining Lineage Potential and Fate Behavior of Precursors during Pancreas Development. *Dev Cell,* 46, 360–375.
86. Talchai, C., Xuan, S., Kitamura, T., Depinho, R. A. & Accili, D. 2012. Generation of functional insulin-producing cells in the gut by Foxo1 ablation. *Nature Genetics, 44, 406-412.*
87. Thorel, F., Népote, V., Avril, I., Kohno, K., Desgraz, R., Chera, S. & Herrera, P. L. 2010. Conversion of adult pancreatic alpha-cells to beta-cells after extreme beta-cell loss. *Nature, 464, 1149-1154.*
88. Tosti, L., Hang, Y., Debnath, O., Tiesmeyer, S., Trefzer, T., Steiger, K., Ten, F. W., Lukassen, S., Ballke, S., Kühl, A. A., Spieckermann, S., Bottino, R., Ishaque, N., Weichert, W., Kim, S. K., Eils, R. & Conrad, C. 2021. Single-Nucleus and In Situ RNA–Sequencing Reveal Cell Topographies in the Human Pancreas. *Gastroenterology, 160, 1330-1344.e11.*
89. Tran, A., Yang, P., Yang, J. Y. H. & Ormerod, J. 2022. Computational approaches for direct cell reprogramming: from the bulk omics era to the single cell era. *Briefings in Functional Genomics,* 21, 270–279.
90. van Gurp, L., Muraro, M. J., Dielen, T., Seneby, L., Dharmadhikari, G., Gradwohl, G., van Oudenaarden, A. & de Koning, E. J. P. 2019. A transcriptomic roadmap to α- and β-cell differentiation in the embryonic pancreas. *Development,* 146(12):dev173716.
91. Vierbuchen, T., Ostermeier, A., Pang, Z. P., Kokubu, Y., Südhof, T. C. & Wernig, M. 2010. Direct conversion of fibroblasts to functional neurons by defined factors. *Nature, 463, 1035-1041.*

92. Wang, D., Tai, P. W. L. & Gao, G. 2019. Adeno-associated virus vector as a platform for gene therapy delivery. *Nat Rev Drug Discov,* 18, 358–378.
93. Willnow, D., Benary, U., Margineanu, A., Vignola, M. L., Konrath, F., Pongrac, I. M., Karimaddini, Z., Vigilante, A., Wolf, J. & Spagnoli, F. M. 2021. Quantitative lineage analysis identifies a hepato-pancreato-biliary progenitor niche. *Nature, 597, 87-91.*
94. Xiao, X., Guo, P., Shiota, C., Zhang, T., Coudriet, G. M., Fischbach, S., Prasadan, K., Fusco, J., Ramachandran, S., Witkowski, P., Piganelli, J. D. & Gittes, G. K. 2018. Endogenous Reprogramming of Alpha Cells into Beta Cells Induced by Viral Gene Therapy Reverses Autoimmune Diabetes. *Cell Stem Cell, 22, 78.*
95. Yang, D., Cho, H., Tayyebi, Z., Shukla, A., Luo, R., Dixon, G., Ursu, V., Stransky, S., Tremmel, D. M., Sackett, S. D., Koche, R., Kaplan, S. J., Li, Q. V., Park, J., Zhu, Z., Rosen, B. P., Pulecio, J., Shi, Z. D., Bram, Y., Schwartz, R. E., Odorico, J. S., Sidoli, S., Wright, C. V., Leslie, C. S. & Huangfu, D. 2022a. CRISPR screening uncovers a central requirement for HHEX in pancreatic lineage commitment and plasticity restriction. *Nature Cell Biology 2022, 24, 1064-1076.*
96. Yang, L., Gong, L., Wang, P., Zhao, X., Zhao, F., Zhang, Z., Li, Y. & Huang, W. 2022b. Recent Advances in Lipid Nanoparticles for Delivery of mRNA. *Pharmaceutics,* 14.
97. Yu, X. X., Qiu, W. L., Yang, L., Wang, Y. C., He, M. Y., Wang, D., Zhang, Y., Li, L. C., Zhang, J., Wang, Y. & Xu, C. R. 2021. Sequential progenitor states mark the generation of pancreatic endocrine lineages in mice and humans. *Cell Research, 31, 886.*
98. Zaret, K. S. & Grompe, M. 2008. Generation and regeneration of cells of the liver and pancreas. *Science,* 322, 1490–4.
99. Zhang, Y., Pak, C. H., Han, Y., Ahlenius, H., Zhang, Z., Chanda, S., Marro, S., Patzke, C., Acuna, C., Covy, J., Xu, W., Yang, N., Danko, T., Chen, L., Wernig, M. & Südhof, T. C. 2013. Rapid single-step induction of functional neurons from human pluripotent stem cells. *Neuron, 78, 785-798.*
100. Zhou, Q., Brown, J., Kanarek, A., Rajagopal, J. & Melton, D. A. 2008. In vivo reprogramming of adult pancreatic exocrine cells to beta-cells. *Nature, 455, 627-632.*
101. Zhou, Q. & Melton, Douglas A. 2018. Pancreas regeneration. *Nature,* 557, 351–358.
102. Zhu, H., Wang, G., Nguyen-Ngoc, K. V., Kim, D., Miller, M., Goss, G., Kovsky, J., Harrington, A. R., Saunders, D. C., Hopkirk, A. L., Melton, R., Powers, A. C., Preissl, S., Spagnoli, F. M., Gaulton, K. J. & Sander, M. 2023. Understanding cell fate acquisition in stem-cell-derived pancreatic islets using single-cell multiome-inferred regulomes. *Dev Cell,* 58, 727–743.e11.

Factors Influencing In Vivo Specification and Function of Endocrine Cells Derived from Pancreatic Progenitors

Nelly Saber and Timothy J. Kieffer

1 Introduction

An estimated 537 million people were living with diabetes in 2021 [35]. This number is projected to increase to 643 million by 2030 and 783 million by 2045 [35]. Diabetes is characterized by high blood glucose levels or hyperglycemia as a result of insulin deficiency or dysfunction. Around 90% of all diabetes cases are type 2, whereas 10% are type 1. In type 1 diabetes (T1D), the autoimmune destruction of insulin-producing β cells in the pancreas leads to hyperglycemia. People living with T1D require exogenous insulin to control their blood glucose levels. However, due to the challenges of dosing insulin appropriately, they often experience periods of hypoglycemia and hyperglycemia. Islet transplantation can successfully maintain blood glucose levels within the appropriate range without exogenous insulin [18, 49, 82, 92, 93]. Unfortunately, there are not enough donors to provide islets for the millions of people living with diabetes. Human pluripotent stem cells (hPSCs) can serve as an unlimited supply of cells that can be differentiated into pancreatic cells to treat diabetes. Differentiation protocols have been guided primarily by studies of pancreas development in model systems [20, 21]. Cell products at the pancreatic endoderm cell (PEC) stage can differentiate into mature endocrine cells following

N. Saber
Laboratory of Molecular and Cellular Medicine, Department of Cellular & Physiological Sciences, Life Sciences Institute, University of British Columbia, Vancouver, BC, Canada

T. J. Kieffer (✉)
Laboratory of Molecular and Cellular Medicine, Department of Cellular & Physiological Sciences, Life Sciences Institute, University of British Columbia, Vancouver, BC, Canada

School of Biomedical Engineering, University of British Columbia, Vancouver, BC, Canada

Department of Surgery, University of British Columbia, Vancouver, BC, Canada
e-mail: tim.kieffer@ubc.ca

L. Piemonti et al. (eds.), *Pluripotent Stem Cell Therapy for Diabetes*,
https://doi.org/10.1007/978-3-031-41943-0_4

implant in rodents [40] and humans [73, 94]. However, the specific factors that influence the development of these cells in vivo are still largely unknown. Given that human embryonic stem cell (hESC)-derived PECs are currently being used in clinical trials, it is important to understand how different factors in the recipients may influence the differentiation and maturation of these PECs into endocrine cells. Here, we discuss some of the factors that can affect the differentiation and maturation of hESC-derived PECs in vivo. For example, sex and site of implantation can impact the development of glucose-responsive, insulin-producing cells in rodents. Mice that received hESC-derived PECs under the kidney capsule displayed glucose-responsive, insulin secretion faster in females than males [83, 84]. Furthermore, when hESC-derived PECs were implanted subcutaneously in macroencapsulation devices in mice, they became glucose-responsive, insulin-producing cells faster than PECs implanted under the kidney capsule or in the gonadal fat pad [84]. Macroencapsulation device grafts also displayed more endocrine rather than ductal cells and did not develop large cysts in contrast to cells implanted in the kidney or fat [84]. Therefore, it is important to consider factors in the recipient such as sex and implant site that can affect the endocrine specification and function of hESC-derived PECs when used to treat people with diabetes.

2 Pancreas Development in Mice and Humans

Given the current use of hPSC-derived PECs in clinical trials, it is important to understand what factors in the recipient may influence the endocrine specification and function of these cells in vivo. We found that the sex of the recipient, thyroid hormone levels, and the site of implantation can affect the acquisition of glucose-stimulated insulin secretion in PECs in mice. Furthermore, macroencapsulation devices appear to promote the differentiation of PECs toward endocrine as opposed to ductal cells and minimize the formation of cysts. Future clinical studies should watch for sex-specific differences in PEC performance and additional studies are required to investigate why hESC-derived PECs implanted in female mice become glucose responsive faster than in male mice. However, as rodent models are often used to study the development of hPSCs into β cells in vivo, differences between humans and rodents in pancreas development and function should be considered when interpreting and translating these data to clinical applications.

Much of what is known about pancreas development is from studying nonhuman models including chicks, fish, Xenopus (frogs), rats, and mice. This is mostly due to the relative ease of availability, as well as the inability to control for age, genetic differences, the method of procurement, or pathophysiology of human donors prior to procurement of the pancreas [99]. In mammalian development, upon fertilization of the egg, the newly formed zygote undergoes multiple cell divisions or cleavage resulting in the blastocyst. The blastocyst contains the inner cell mass, which forms the embryo, and trophectoderm cells, which becomes the extraembryonic tissues that provide protection and nutrients to the embryo. The inner cell mass undergoes gastrulation, involving the movement and reorganization of cells into the three germ layers: ectoderm, mesoderm, and endoderm. The ectoderm is the outermost germ layer that gives rise to the skin and nervous system whereas the mesoderm, the middle layer, forms muscle and the circulatory and skeletal systems. The definitive endoderm is the innermost germ layer that becomes the lining of the respiratory and digestive systems and their associated organs. In contrast, the primitive endoderm from the extraembryonic tissues becomes the visceral and parietal endoderm. Nodal, part of the transforming growth factor (TGF) β family, is important for the development of endoderm and mesoderm in vertebrates with high levels promoting endoderm formation and low levels inducing mesoderm formation [20, 104].

The mammalian pancreas arises from the definitive endoderm, which begins as a flat sheet of cells that is transformed into the primitive gut tube and undergoes anterior-posterior patterning, resulting in the foregut, midgut, and hindgut regions [104]. The midgut and hindgut regions give rise to the small and large intestines. The anterior foregut becomes the esophagus, stomach, thyroid, and lungs, whereas the posterior foregut develops into the duodenum, liver, and pancreas. Evagination of the posterior foregut endoderm into the surrounding mesenchyme on embryonic day (E) 9.5 in mice [103] and 26 days post-conception (dpc) in humans [72] results in the ventral and dorsal pancreatic buds. Most of the pancreas including the upper part of the head, neck, body, and tail is derived from the dorsal pancreatic bud whereas the inferior part of the head and the uncinate process stem from the ventral bud. The whole pancreas is formed when the ventral pancreatic bud migrates posteriorly and fuses with the dorsal bud upon gut rotation around 56 dpc to form the whole pancreas [72]. Reduced expression of cardiac fibroblast growth factor (FGF) leads to the formation of the ventral pancreas in mice [24, 91]. In contrast, expression of retinoic acid in mice [61] and inhibition of sonic hedgehog (SHH) signaling through notochord factors such as activin and FGF2 in mice and chicks influence dorsal pancreas formation [31]. Both ventral and dorsal pancreatic buds contain multipotent progenitor cells that give rise to exocrine, ductal, and endocrine cells [103]. These multipotent cells express key pancreatic progenitor transcription factors pancreatic duodenal homeobox 1 (*Pdx1*) and basic helix-loop-helix protein pancreas-specific transcription factor 1A *(Ptf1a)* as well as avian myelocytomatosis viral oncogene homolog *(cMyc)* and carboxypeptidase A *(Cpa)* [103]. Branching of the pancreatic epithelium around E11.5 in mice results from the proliferation of multipotent progenitor cells away from the center of the buds [103], which is dependent on mesenchymal FGF10 signaling [5]. Similar to mice, *FGF7* and *FGF10* are expressed in mesenchymal cells in the human fetal pancreas at 6–9 gestational weeks and promote the proliferation of pancreatic epithelial cells [101]. The tips of the branches eventually develop into exocrine cells, while the trunks are composed of ductal and endocrine cells [103]. The trunks become solely made up of ductal cells once the endocrine cells leave the epithelium [103].

Developmental processes for islet formation are largely conserved, but there are notable differences. Pancreas development occurs over just 10 days in mice compared to several months in humans [56]. The endocrine cell composition and organization of islets also differs considerably between species [95]. β cells make up the majority of rodent islet cells whereas human islets consist of fewer β cells and more α and δ cells [63]. One study found that human islets contained approximately 54% β cells, 34% α cells, and 10% δ cells, whereas mouse islets were composed of approximately 75% β cells, 19% α cells, and 6% δ cells [12]. Moreover, rodent islets have a characteristic β cell-rich core with α and δ cells along the periphery, whereas adult human islets appear to be a random mix of cells [99]. However, small human islets display the characteristic layout of rodent islets, and larger islets may be composed of small groups of cell clusters that resemble rodent islets [11]. The orchestration of endocrine cell formation also differs between species. While there is a single phase of expression of the master regulator neurogenin 3 *(NGN3)*, a transcription factor that is essential for endocrine cell development [81], peaking at 10–14 weeks post-conception (wpc) in humans [36, 85], there is a biphasic wave of *Ngn3* expression in mice from E8.5 to E11.0 and E12.0 [98]. At E9.5, α and PP cells are observed in the mouse pancreas, followed by β and δ cells at E11.5 and E13.5, respectively [34]. In humans, β cells appear first at 7.5 wpc followed by α and δ cells a week later and PP cells at 10 wpc [72]. Cells immunoreactive for both insulin and glucagon are detected in the human fetal pancreas from 9 to 21 weeks, with a peak between 11 and 13 weeks, but are rare in adult pancreas [78]. These polyhormonal cells are typically missing the β cell transcription factors *PDX1*, NK6 homeobox 1 *(NKX6.1)*, and v-maf musculoaponeurotic fibrosarcoma oncogene homolog A *(MAFA)* yet express the α cell transcription factor aristaless-related homeobox *(ARX)* [78]. Therefore, some mature α cells may develop from a temporary population of cells that co-express insulin and glucagon [66, 75, 78].

3 The Role of Transcription Factors in Endocrine Cell Specification

Transcription factors play a pivotal role in pancreas development and function. The controlled expression of transcription factors during development influences ductal, exocrine, and endocrine cell fate as well as further endocrine cell specification into β, α, δ, ε, and PP cells. As discussed above, endocrine cells are derived from common progenitor cells that express the transcription factors *PDX1, PTF1A, NKX6.1*, and *NGN3*. Mutations in key pancreatic transcription factors such as these result in pancreas malformation, dysfunction, and disease.

The importance of PDX1 in pancreas development was established with observations that mutations in the *Pdx1* gene resulted the absence of a pancreas in mice [37, 67] and humans [90, 97]. PDX1 is first detected in mice at E8.5 [70]. In humans, PDX1 is detectable in the dorsal foregut endoderm at 29 dpc following the inhibition of SHH signaling [31], which occurs at 25–27 dpc [36]. Sex-determining region Y-box 17 *(SOX17)* is also detectable in the pancreas where SHH is repressed, whereas forkhead box protein A2 *(FOXA2)* is expressed in all endodermal epithelial cells at 27 dpc prior to *PDX1* expression in humans [36]. FOXA2 appears to regulate *Pdx1* expression in mice [26]. At 30–33 dpc in humans, *PDX1* and sex-determining region Y-box 9 *(SOX9)* are both expressed in the dorsal and ventral pancreatic bud cells along with GATA binding protein 4 (GATA4) [36]. Similar to humans, *Sox9* is expressed in *Pdx1*-positive cells in the dorsal and ventral pancreas around E10.5 in mice and regulates the expression of *Ngn3* [46]. SOX9 immunoreactivity is eventually lost in cells that become highly immunoreactive for NGN3 or insulin but remains present in ductal cells in the human pancreas [36]. NKX6.1 is also observed at 30–33 dpc in humans, whereas SOX17 is no longer present in the pancreas at this time [36]. *Nkx6.1* is expressed in most epithelial cells in the mouse pancreas as early as E10.5, which becomes restricted to insulin-expressing and ductal cells by E15.5 [88]. Mice with a homozygous mutation for *Nkx6.1* displayed a significant reduction in insulin-expressing cells but no difference in glucagon-, somatostatin-, or PP-expressing cells compared to wildtype mice [88], demonstrating the importance of NKX6.1 in the specification of endocrine cells toward a β cell fate. Increased expression of *NGN3* from 47 dpc to 9–10 wpc in humans correlates with an increase in insulin-expressing cells [36]. The number of cells immunoreactive for NGN3 peak at 10–14 wpc, declining at 18 wpc until it is no longer detected at 35–41 wpc in the human fetal pancreas [85]. Additionally, expression of *FOXA2, NKX6.1*, and NK2 homeobox 2 *(NKX2.2)* is observed in groupings of β cells in humans at 10 wpc [36]. Also, at this time, NKX2.2 is confined to β cells whereas, at 14 wpc, NKX6.1 is co-expressed with insulin in β cells and SOX9 in ductal cells in the human pancreas [36]. β, α, δ, and PP cells are all present in human fetal islets from 12 to 13 wpc and onwards [72]. Mice with a homozygous mutation for *Ngn3* completely lack β, α, δ, and PP cells, develop diabetes, and die 1–3 days following birth [28]. *Ngn3* is also important for the induction of several endocrine cell-related transcription factors including LIM homeobox protein islet 1 *(Isl1)*, paired box 4 *(Pax4)* and 6 *(Pax6)*, neuronal differentiation 1 *(Neurod1)* [28], and *Arx* [19] in mice. β and α cell development also requires expression of v-maf musculoaponeurotic fibrosarcoma oncogene homolog B *(MAFB)* and *MAFA* [59]. Mice with a homozygous *MAFB* mutation displayed lower numbers of insulin- and glucagon-expressing cells compared to wildtype mice throughout development [3]. Moreover, insulin-expressing cells lacked expression of *Pdx1, Nkx6.1,* and glucose transporter 2 *(Glut2)* [3]. In contrast, *MAFA* deficient mice displayed a reduced ratio of β to α cells, decreased expression of insulin 1, insulin 2, and *Glut2* as well as impaired glucose-stimulated insulin secretion compared to wildtype mice [102]. Normally, *Mafb* is co-expressed with *Nkx6.1* and insulin until *Pdx1* expression is increased and *Mafa* becomes present in mouse β cells as well [65]. Both *Mafb* and *Mafa* [51] as well as *Pdx1* [68], *Pax6* [87], and *Neurod1* [64] appear to be involved in the transcription of insulin in mouse β cells. *Mafb* is co-expressed with glucagon in the absence of *Nkx6.1* in mouse α cells [65]. Eventually, adult mouse β cells lose *Mafb* expression while *Mafa* remains whereas *Mafb* continues to be expressed in α cells [65]. However, unlike in mice, *MAFB* continues to be expressed in both adult human β and α cells [22, 78].

Within the mouse pancreas, peptide hormone urocortin 3 (UCN3) is found exclusively in β cells and is associated with β cell maturation [9] but it is present in both β and α cells in humans [57]. Previous studies demonstrated that UCN3 administration stimulates insulin and glucagon secretion in rodents, ultimately leading to increased plasma glucose levels [42] and high fat diet-fed or aged *Ucn3*-null mice display improved glucose tolerance compared to matched wildtype controls [43]. It was later discovered that UCN3 is co-secreted with insulin following a glucose stimulus and helps facilitate the release of somatostatin from δ cells to decrease the production of insulin and glucagon, perhaps serving to maintain the glycemic set point [58]. Collectively, these examples highlight that there are significant differences in islet formation and function between animal models and humans.

4 Regulation and Acquisition of Glucose-Stimulated Insulin Secretion

Although the endocrine pancreas makes up a small portion of the organ, it plays a large role in maintaining blood glucose homeostasis. Blood glucose levels are normally maintained between 4 and 6 mM in humans by pancreatic production of insulin and glucagon [79, 80]. While humans have a glycemic set point of ~5 mM, mice have a higher glycemic set point of ~8 mM which would be considered hyperglycemic in humans [80]. Islets control the glycemic set point as transplanting the islets from different species (e.g., humans, monkeys, pigs) into healthy or streptozotocin (STZ)-induced diabetic mice results in the recipient achieving the glycemic set point of the islet donor [23, 80]. Diabetic mice that received human islets under the kidney capsule maintained the human glycemic set point even after mouse islets were transplanted in the eye [80], most likely due to human islets having a lower glucose threshold for insulin secretion than mouse islets [33]. Overall, these data demonstrate that the pancreatic islets control the glycemic set point, for which there are species differences.

While glucose is a potent stimulator of insulin release in adult β cells, human and rodent fetal β cells respond poorly to glucose and produce more insulin from an amino acid stimulus such as arginine and leucine [4, 10, 32, 60]. Insulin-expressing cells are found at E11.5 in mice [34] and 7.5 wpc [72] in humans, but glucose-stimulated insulin secretion is not observed until after birth. Due to the lack of availability of human fetal cells and the longer maturation period of human β cells compared to rodents, many studies investigating the acquisition of glucose-stimulated insulin secretion in β cells used rodent models. Glucose-stimulated insulin release from neonatal rat islets was less than half that obtained from 3 month old islets [7]. It was also found that insulin release is stimulated at lower levels of glucose in neonatal mouse islets compared to older islets [9]. Therefore, mature β cells have a higher glucose threshold for insulin release than immature β cells [9]. This switch from immature to mature β cells occurs during weaning in mice when they transition from a fat-rich milk to a carbohydrate-rich chow diet [96]. Weaning leads to decreased insulin production in response to low glucose and increased production during high glucose, which results in enhanced glucose-stimulated insulin secretion [96]. Changes in the sensitivity of mechanistic target of rapamycin complex 1 (mTORC1) to nutrients regulate the change from amino acid-stimulated to glucose-stimulated production of insulin [32]. mTORC1 has been shown to play a role in β cell proliferation and insulin secretion [6]. In adult human and mouse β cells, amino acids alone cannot fully activate mTORC1 and require the addition of glucose to achieve peak activity levels whereas in fetal β cells, only amino acids are required to fully activate mTORC1 [32]. Therefore, the acquisition of glucose-stimulated insulin secretion in human and rodent β cells following birth appears to correspond with a change in the nutritional environment.

5 Maturation of Human Embryonic Stem Cell-Derived β Cells In Vivo

As previously noted, the limited access of human fetal cells and the longer period of β cell development and maturation compared to rodent cells make it difficult to study the changes that occur in β cells throughout human gestation and following birth. hPSCs provide a model to study β cell development and maturation as well as a source of transplantable cells that could be used to treat diabetes. hPSCs can be differentiated through the stages of pancreas development including definitive endoderm, primitive gut tube, posterior foregut, and pancreatic endoderm in cell culture format. Once the pancreatic endoderm is achieved, the cells are competent to finish maturation into functional islet cells in vivo following implant, whether in rodents [14, 40, 76] or humans [73, 94]. Moreover, hESC-derived PECs can prevent or ameliorate hyperglycemia in STZ-induced [14, 40, 74] or diet-induced diabetic mice [16]. hPSCs can be differentiated further in vitro to stages including pancreatic endocrine precursors, immature β cells, and maturing β cells that are also capable of treating hyperglycemia in diabetic mouse models including STZ-induced [77] and Akita mice, which have a mutation in the insulin gene [69]. Some of the advantages of implanting hESC-derived cells at the pancreatic endoderm stage rather than the maturing β cell stage are that they have a lower oxygen consumption rate and thus are more resistant to a hypoxic environment [71], it takes less time (~12 days vs ~27–42) to differentiate hESCs into PECs than β cells in vitro [89], and they are less expensive to produce. hESC-PECs are similar to 6- to 9-week human fetal pancreas tissue which is composed mostly of pancreatic epithelial cells and some endocrine hormone-expressing cells including insulin and glucagon [78]. As demonstrated by Kroon et al. and others [14, 40, 76], hESC-derived PECs require a several-month-long maturation period in vivo to differentiate into glucose-responsive insulin-producing cells. The signals that promote their differentiation in vivo are largely unknown and identifying them may be the key to better understanding what drives human β cell development and maturation. Clinical trials (ClinicalTrials.gov Identifiers: NCT02239354 and NCT03163511) are currently using hESC-PECs contained within macroencapsulation devices that are implanted subcutaneously to treat T1D [73, 94]. Therefore, it is important to understand the factors that these cells may be exposed to when implanted that could affect their differentiation into mature functional islet cells.

Besides changes in the nutritional environment, other factors have been found to be important for the development of glucose responsivity. For example, thyroid hormone enhances the expression of *Mafa* and glucose-stimulated insulin secretion in immature rat islets in vitro [1, 2]. Building upon this discovery, we found that the addition of thyroid hormone triiodothyronine (T3) promotes the formation of functional β cells in vitro [77]. Interestingly, when we compared the maturation of hESC-derived PECs in rats and mice, we found that these cells developed glucose-stimulated insulin secretion faster in the rats, and rats had higher levels of T3 than mice [15]. Additionally, hESC-derived PECS implanted in chronic hypothyroid mice had impaired glucose-stimulated insulin secretion and elevated arginine-stimulated plasma glucagon levels compared to the same cells implanted in control euthyroid mice [17]. hESC-derived grafts from chronic hypothyroid mice displayed fewer cells immunoreactive for insulin, heterogenous MAFA expression, and increased immunoreactivity for glucagon and ghrelin compared to grafts from euthyroid mice [17]. In fact, we recently found that high levels of circulating thyroid hormones in human and mouse recipients of PECs were correlated with lower circulating glucagon levels in humans and faster maturation of PECs into insulin-producing cells in mice suggesting that thyroid hormones may promote the differentiation of these cells towards a β cell rather than an α cell fate [74]. Since up to one-third of people living with T1D also have thyroid dysfunction [38], these studies are clinically relevant and highlight the impact of the recipient on the differentiation of hESC-derived PECS in vivo as chronic thyroid hormone deficiency may promote the differentiation of these cells toward an α and ε fate rather than β cells [17].

We assessed the effect of hyperglycemia on the differentiation of hESC-derived PECs by implanting these cells in STZ-induced diabetic and healthy control mice [14]. We found that the hyperglycemic environment enhanced the differentiation of hESC-derived PECs into glucose-responsive insulin-producing cells as mice that received either low- or high-dose STZ injection displayed higher fed levels of human C-peptide compared to control mice that did not receive STZ [14]. Similarly, another group found that STZ-induced diabetic mice implanted with human induced pluripotent stem cell-derived pancreatic endocrine cells displayed increased plasma human C-peptide levels sooner compared to nondiabetic mice implanted withe these cells [61]. Glucose may also play a role in the development of β cells in vitro as mouse E13.5 pancreases cultured with glucose concentrations of 0, 0.75, 1.25, 2.5, 5, or 10 mM for 7 days displayed increased insulin expression in a dose-dependent manner [29]. Furthermore, in the absence of glucose, *Neurod1* expression was decreased in correlation with insulin compared to mouse E13.5 pancreases cultured with 10 mM glucose [29]. Since the majority of people living with diabetes have type 2, which is the result of insulin dysfunction rather than deficiency like type 1, we implanted hESC-derived PECs in high fat diet-fed mice as a model of type 2 diabetes (T2D) [16]. We observed that high fat diets did not affect the maturation of these cells in vivo and hESC-derived PECs improved glucose tolerance after 24 weeks post-implantation [16]. Furthermore, high fat diet-fed mice that received both hESC-derived PECs and diabetes drugs displayed reduced body weight in which treatment with sitagliptin or metformin also ameliorated hyperglycemia after 12 weeks [16]. Overall, it has been shown that several factors in the recipient can impact the differentiation and maturation of hESC-derived PECs in vivo but one caveat with all these studies is that only male rodents were used.

Female rodents are often avoided in research due to the concern that the estrous cycle may induce variability [54], but given that there are notable sex differences in diabetes incidence and pancreatic islet function [25, 52, 53], it is important to understand how the sex of the recipient may affect the development of hPSC-derived glucose-responsive insulin-secreting cells in vivo. The estimated prevalence of diabetes is slightly lower in women (10.2%) than men (10.8%) aged 20–79 years, and in 2021, there were approximately 17.7 million more men living with diabetes than women [35]. While most autoimmune diseases are characterized by a female predominance, T1D is characterized by a male predominance in Caucasians [53]. For example, the incidence of T1D was higher in Swedish males (15.9 per 100,000/year) compared to females (8.6 per 100,000/year) from ages 15 to 34 [8]. One study found that girls with T1D had higher serum C-peptide levels, especially early in puberty, than boys with T1D suggesting they had more preserved β cell function [86]. Another study observed that adolescents with T1D had lower serum levels of estrogen activity than non-diabetic adolescents [50]. Interestingly, estrogen treatment of female nonobese diabetic mice, which spontaneously develop autoimmune insulin dependent (i.e., type 1) diabetes, has been shown to delay the development of diabetes [27]. There is also a sex difference in the prevalence of T2D [53]. There are more men with T2D before the age of 60 while there are more women with T2D after the age of 60 [100]. Females tend to have greater insulin sensitivity, likely because of higher circulating levels of estrogen. Premenopausal women have enhanced insulin sensitivity compared to age-matched men when normalized to lean mass [55]. However, insulin sensitivity decreases following menopause or ovariectomy [55]. Ovariectomized rats that received treatment with a placebo rather than estrogen pellets displayed significantly lower insulin sensitivity [39]. Taken together, these studies suggest that female gonadal hormones (e.g., estrogen) may play a protective role against both T1D and T2D.

Female islets may also function better than male islets as glucose-stimulated insulin secretion was higher in isolated islets from female human pancreas donors compared to males [30]. Female human islets also appear to have more beneficial effects than male islets in islet transplantation. For example, transplant recipients that received islets from at least one female human pancreas donor displayed longer graft survival than recipients that received islets from all male donors [41]. Additionally, female islets from human pancreas donors with T2D displayed enhanced glucose-stimulated insulin secretion compared to males [13]. The beneficial effects of female human islets on graft survival could be due to the presence of a higher percentage of β cells compared to male islets [48] or due to the increased resilience of female β cells to islet stressors such as endoplasmic reticulum stress compared to males [13]. Future studies need to be conducted to determine whether female hPSC-derived islets function better than males as often studies and differentiation protocols currently utilize only male stem cell lines.

Female transplant recipients also displayed longer graft survival than male recipients of islets from at least one female human pancreas donor [41]. The beneficial effects of female recipients on graft survival could be due to the presence of estrogen. For example, STZ-induced hyperglycemic mice implanted with human islets displayed improved blood glucose and islet engraftment when treated with 17 β-estradiol compared to mice treated with a vehicle [44]. We implanted male hESC-derived PECs in male and female mice and found that these cells developed glucose-stimulated insulin secretion faster in females than males in two different cohorts of mice [83]. Therefore, it appears that the sex of the recipient may impact the differentiation and maturation of hESC-derived PECs in vivo, and female sex hormones may promote the development of glucose-stimulated insulin secretion. We are currently investigating whether estrogen promotes the differentiation and maturation of hESC-derived PECs in vitro.

Another factor that could play a role in the differentiation and maturation of hESC-derived PECs in vivo is the site of implantation. While there are differences in sex hormones between males and females that could explain why hESC-derived PECs may have developed glucose-stimulated insulin secretion faster in females than males, we also observed increased adipose tissue surrounding the kidney capsule, the site of implantation, in female immunodeficient SCID-beige mice compared to males [83]. Additionally, it has been observed that hESC-derived PECs implanted in the gonadal fat pad of mice developed glucose-stimulated human C-peptide secretion faster than when the cells were implanted in the subcutaneous site or under the kidney capsule [40]. Therefore, it is possible that adipose tissue may promote the development of hESC-derived glucose-responsive insulin-secreting cells, possibly through signaling factors such as adipokines [45]. We compared the differentiation and maturation of hESC-derived PECs in the gonadal fat pad, kidney capsule, and subcutaneously within macroencapsulation devices in both male and female mice [84]. Regardless of sex, hESC-derived PECs developed glucose-stimulated insulin secretion the fastest when implanted subcutaneously within macroencapsulation devices compared to under the kidney capsule or in the gonadal fat pad [84]. Furthermore, unlike cells implanted into fat or kidney, those within macroencapsulation devices were mostly immunoreactive for endocrine rather than ductal markers and did not display cysts [84]. The restrictive nature of the macroencapsulation devices may prevent the development of off-target (i.e., non-endocrine) cells as it was discovered that cell confinement promotes the differentiation of endocrine cells from hESC-derived PECs whereas cell spreading leads to ductal cell formation [47]. We also found in a third cohort of mice that hESC-derived PECs implanted under the kidney capsule developed glucose-stimulated insulin secretion faster in females than males, although we have yet to identify the mechanism [84]. Interestingly, cells implanted under the kidney capsule in females displayed higher plasma levels of arginine-stimulated glucagon and GLP-1 compared to males and other implantation sites [84]. Perhaps the presence of more glucagon in the kidney capsule grafts at an earlier timepoint post-implantation in females may have enhanced glucose-stimulated insulin secretion. This study demonstrates the benefits of using macroencapsulation devices when implanting hESC-derived PECs since off-target cell differentiation was minimized and glucose-stimulated insulin secretion was established sooner relative to the non-encapsulated implants. Additionally, the ability to easily remove these cells if any issues arise is another advantage of macroencapsulation devices that is particularly relevant in clinical trials using PSCs that have the potential to form off-target cells. It remains to be determined how the site of implantation may impact the differentiation and maturation of more differentiated cells. We have previously shown that hESC-derived maturing β cells develop glucose-stimulated human C-peptide secretion faster when implanted under the kidney capsule of female mice compared to males [83] but have not directly compared these cells in different implantation sites. However, it is evident that both male and female subjects should be included in studies investigating the differentiation and maturation of hPSC-derived PECs in vivo.

6 Conclusion

We have demonstrated that the sex of the recipient, thyroid hormone levels, and the site of implantation are factors that can affect the development of glucose responsivity in PECs. Additionally, the implant site, specifically the use of macroencapsulation devices, can influence the differentiation of PECs toward an endocrine rather than a ductal cell fate. Macroencapsulation devices also prevent the formation of large cysts and can be removed if any issues arise with the hPSC-derived cells. While we only used male hPSC-derived cells in our studies, it would be interesting to see whether factors in the recipient affect female hPSC-derived cells in a similar manner and whether the use of female cells will display beneficial effects in the recipients. Future studies need to be conducted to determine the mechanism behind why hESC-derived PECs develop glucose responsivity faster in females than males when implanted under the kidney capsule in mice. However, as most studies investigating the development of hPSCs into β cells in vivo take place in rodent models, we must continue to be cognizant of the differences between humans and rodents when analyzing these data and inferring findings to clinical situations. Additional trials will be needed to determine what factors in the recipient may influence the endocrine specification and function of hPSC-derived cells in humans.

References

1. Aguayo-Mazzucato, C. *et al.* (2011) 'Mafa expression enhances glucose-responsive insulin secretion in neonatal rat beta cells', *Diabetologia*. Diabetologia, 54(3), pp. 583–593. https://doi.org/10.1007/S00125-010-2026-Z.
2. Aguayo-Mazzucato, C. *et al.* (2013) 'Thyroid Hormone Promotes Postnatal Rat Pancreatic β-Cell Development and Glucose-Responsive Insulin Secretion Through MAFA', *Diabetes*. American Diabetes Association, 62(5), pp. 1569–1580. https://doi.org/10.2337/DB12-0849.

3. Artner, I. *et al.* (2007) 'MafB is required for islet beta cell maturation', *Proceedings of the National Academy of Sciences of The United States of America*. Proc Natl Acad Sci U S A, 104(10), pp. 3853–3858. https://doi.org/10.1073/PNAS.0700013104.
4. Asplund, K. (1973) 'Dynamics of insulin release from the foetal and neonatal rat pancreas', *European journal of clinical investigation*. Eur J Clin Invest, 3(4), pp. 338–344. https://doi.org/10.1111/J.1365-2362.1973.TB00360.X.
5. Bhushan, A. *et al.* (2001) 'Fgf10 is essential for maintaining the proliferative capacity of epithelial progenitor cells during early pancreatic organogenesis', *Development (Cambridge, England)*. Development, 128(24), pp. 5109–5117. https://doi.org/10.1242/DEV.128.24.5109.
6. Blandino-Rosano, M. *et al.* (2017) 'Loss of mTORC1 signalling impairs β-cell homeostasis and insulin processing', *Nature communications*. Nat Commun, 8. https://doi.org/10.1038/NCOMMS16014.
7. Bliss, C. R. and Sharp, G. W. G. (1992) 'Glucose-induced insulin release in islets of young rats: time-dependent potentiation and effects of 2-bromostearate', *The American journal of physiology*. Am J Physiol, 263(5 Pt 1). https://doi.org/10.1152/AJPENDO.1992.263.5.E890.
8. Blohmé, G. *et al.* (1992) 'Male predominance of type 1 (insulin-dependent) diabetes mellitus in young adults: results from a 5-year prospective nationwide study of the 15-34-year age group in Sweden', *Diabetologia*. Diabetologia, 35(1), pp. 56–62. https://doi.org/10.1007/BF00400852.
9. Blum, B. *et al.* (2012) 'Functional beta-cell maturation is marked by an increased glucose threshold and by expression of urocortin 3', *Nature Biotechnology 2012 30:3*. Nature Publishing Group, 30(3), pp. 261–264. https://doi.org/10.1038/nbt.2141.
10. Bonner-Weir, S., Aguayo-Mazzucato, C. and Weir, G. C. (2016) 'Dynamic development of the pancreas from birth to adulthood', *Upsala Journal of Medical Sciences*, 121(2), pp. 155–158. https://doi.org/10.3109/03009734.2016.1154906.
11. Bonner-Weir, S., Sullivan, B. A. and Weir, G. C. (2015) 'Human Islet Morphology Revisited: Human and Rodent Islets Are Not So Different After All', *The journal of histochemistry and cytochemistry : official journal of the Histochemistry Society*. J Histochem Cytochem, 63(8), pp. 604–612. https://doi.org/10.1369/0022155415570969.
12. Brissova, M. *et al.* (2005) 'Assessment of human pancreatic islet architecture and composition by laser scanning confocal microscopy', *Journal of Histochemistry and Cytochemistry*. SAGE PublicationsSage CA: Los Angeles, CA, 53(9), pp. 1087–1097. https://doi.org/10.1369/JHC.5C6684.2005/ASSET/IMAGES/LARGE/10.1369_JHC.5C6684.2005-FIG2.JPEG.
13. Brownrigg, G. P. *et al.* (2023) 'Sex differences in islet stress responses support female β cell resilience', *Molecular metabolism*. Mol Metab, 69, p. 101678. https://doi.org/10.1016/J.MOLMET.2023.101678.
14. Bruin, J. E. *et al.* (2013) 'Maturation and function of human embryonic stem cell-derived pancreatic progenitors in macroencapsulation devices following transplant into mice', *Diabetologia*, 56(9), pp. 1987–1998. https://doi.org/10.1007/s00125-013-2955-4.
15. Bruin, J. E., Asadi, A., *et al.* (2015a) 'Accelerated Maturation of Human Stem Cell-Derived Pancreatic Progenitor Cells into Insulin-Secreting Cells in Immunodeficient Rats Relative to Mice', *Stem Cell Reports*. The Authors, 5(6), pp. 1081–1096. https://doi.org/10.1016/j.stemcr.2015.10.013.
16. Bruin, J. E., Saber, N., *et al.* (2015b) 'Treating diet-induced diabetes and obesity with human embryonic stem cell-derived pancreatic progenitor cells and antidiabetic drugs', *Stem cell reports*. Stem Cell Reports, 4(4), pp. 605–620. https://doi.org/10.1016/J.STEMCR.2015.02.011.
17. Bruin, J. E. *et al.* (2016) 'Hypothyroidism impairs human stem cell-derived pancreatic progenitor cell maturation in mice', *Diabetes*. American Diabetes Association Inc., 65(5), pp. 1297–1309. https://doi.org/10.2337/db15-1439.
18. Chetboun, M. *et al.* (2023) 'Association between primary graft function and 5-year outcomes of islet allogeneic transplantation in type 1 diabetes: a retrospective, multicentre, observational cohort study in 1210 patients from the Collaborative Islet Transplant Registry', *The lancet. Diabetes & endocrinology*. Lancet Diabetes Endocrinol. https://doi.org/10.1016/S2213-8587(23)00082-7.

19. Collombat, P. *et al.* (2003) 'Opposing actions of Arx and Pax4 in endocrine pancreas development', *Genes & development*. Genes Dev, 17(20), pp. 2591–2603. https://doi.org/10.1101/GAD.269003.
20. D'Amour, K. A. *et al.* (2005) 'Efficient differentiation of human embryonic stem cells to definitive endoderm', *Nature Biotechnology 2005 23:12*. Nature Publishing Group, 23(12), pp. 1534–1541. https://doi.org/10.1038/nbt1163.
21. D'Amour, K. A. *et al.* (2006) 'Production of pancreatic hormone-expressing endocrine cells from human embryonic stem cells', *Nature biotechnology*. Nat Biotechnol, 24(11), pp. 1392–1401. https://doi.org/10.1038/NBT1259.
22. Dai, C. *et al.* (2012) 'Islet-enriched gene expression and glucose-induced insulin secretion in human and mouse islets', *Diabetologia*. Diabetologia, 55(3), pp. 707–718. https://doi.org/10.1007/S00125-011-2369-0.
23. Davalli, A. M. *et al.* (1995) 'Function, mass, and replication of porcine and rat islets transplanted into diabetic nude mice', *Diabetes*. Diabetes, 44(1), pp. 104–111. https://doi.org/10.2337/DIAB.44.1.104.
24. Deutsch, G. *et al.* (2001) 'A bipotential precursor population for pancreas and liver within the embryonic endoderm', *Development (Cambridge, England)*. Development, 128(6), pp. 871–881. https://doi.org/10.1242/DEV.128.6.871.
25. Gannon, M. *et al.* (2018) 'Sex differences underlying pancreatic islet biology and its dysfunction', *Molecular Metabolism*. Elsevier, 15, p. 82. https://doi.org/10.1016/J.MOLMET.2018.05.017.
26. Gao, N. *et al.* (2008) 'Dynamic regulation of Pdx1 enhancers by Foxa1 and Foxa2 is essential for pancreas development', *Genes & development*. Genes Dev, 22(24), pp. 3435–3448. https://doi.org/10.1101/GAD.1752608.
27. Gourdy, P. *et al.* (2016) 'Estrogen Therapy Delays Autoimmune Diabetes and Promotes the Protective Efficiency of Natural Killer T-Cell Activation in Female Nonobese Diabetic Mice', *Endocrinology*. Endocrinology, 157(1), pp. 258–267. https://doi.org/10.1210/EN.2015-1313.
28. Gradwohl, G. *et al.* (2000) 'neurogenin3 is required for the development of the four endocrine cell lineages of the pancreas', *Proceedings of the National Academy of Sciences of The United States of America*. Proc Natl Acad Sci U S A, 97(4), pp. 1607–1611. https://doi.org/10.1073/PNAS.97.4.1607.
29. Guillemain, G. *et al.* (2007) 'Glucose is necessary for embryonic pancreatic endocrine cell differentiation', *Journal of Biological Chemistry*. Elsevier, 282(20), pp. 15228–15237. https://doi.org/10.1074/jbc.M610986200.
30. Hall, E. *et al.* (2014) 'Sex differences in the genome-wide DNA methylation pattern and impact on gene expression, microRNA levels and insulin secretion in human pancreatic islets', *Genome biology*. BioMed Central, 15(12), p. 522. https://doi.org/10.1186/S13059-014-0522-Z/FIGURES/10.
31. Hebrok, M., Kim, S. K. and Melton, D. A. (1998) 'Notochord repression of endodermal Sonic hedgehog permits pancreas development', *Genes & development*. Genes Dev, 12(11), pp. 1705–1713. https://doi.org/10.1101/GAD.12.11.1705.
32. Helman, A. *et al.* (2020) 'A Nutrient-Sensing Transition at Birth Triggers Glucose-Responsive Insulin Secretion', *Cell metabolism*. NLM (Medline), 31(5), pp. 1004–1016. https://doi.org/10.1016/J.CMET.2020.04.004.
33. Henquin, J. C., Dufrane, D. and Nenquin, M. (2006) 'Nutrient control of insulin secretion in isolated normal human islets', *Diabetes*. Diabetes, 55(12), pp. 3470–3477. https://doi.org/10.2337/DB06-0868.
34. Herrera, P. L. *et al.* (1991) 'Embryogenesis of the murine endocrine pancreas; early expression of pancreatic polypeptide gene', *Development*. The Company of Biologists, 113(4), pp. 1257–1265. https://doi.org/10.1242/DEV.113.4.1257.
35. IDF Diabetes Atlas | Tenth Edition (2021). Available at: https://diabetesatlas.org/ (Accessed: 30 June 2022).
36. Jennings, R. E. *et al.* (2013) 'Development of the human pancreas from foregut to endocrine commitment', *Diabetes*, 62(10), pp. 3514–3522. https://doi.org/10.2337/db12-1479.

37. Jonsson, J. *et al.* (1994) 'Insulin-promoter-factor 1 is required for pancreas development in mice', *Nature*. Nature, 371(6498), pp. 606–609. https://doi.org/10.1038/371606A0.
38. Kadiyala, R., Peter, R. and Okosieme, O. E. (2010) 'Thyroid dysfunction in patients with diabetes: clinical implications and screening strategies', *International journal of clinical practice*. Int J Clin Pract, 64(8), pp. 1130–1139. https://doi.org/10.1111/J.1742-1241.2010.02376.X.
39. Kawakami, M. *et al.* (2018) 'Estrogen replacement enhances insulin-induced AS160 activation and improves insulin sensitivity in ovariectomized rats', *American journal of physiology. Endocrinology and metabolism*. Am J Physiol Endocrinol Metab, 315(6), pp. E1296–E1304. https://doi.org/10.1152/AJPENDO.00131.2018.
40. Kroon, E. *et al.* (2008) 'Pancreatic endoderm derived from human embryonic stem cells generates glucose-responsive insulin-secreting cells in vivo', *Nature Biotechnology*, 26(4), pp. 443–452. https://doi.org/10.1038/nbt1393.
41. Lemos, J. R. N. *et al.* (2022) 'Prolonged Islet Allograft Function is Associated With Female Sex in Patients After Islet Transplantation', *The Journal of clinical endocrinology and metabolism*. J Clin Endocrinol Metab, 107(3), pp. E973–E979. https://doi.org/10.1210/CLINEM/DGAB787.
42. Li, C. *et al.* (2003) 'Urocortin III Is Expressed in Pancreatic β-Cells and Stimulates Insulin and Glucagon Secretion', *Endocrinology*. Oxford Academic, 144(7), pp. 3216–3224. https://doi.org/10.1210/EN.2002-0087.
43. Li, C. *et al.* (2007) 'Urocortin 3 regulates glucose-stimulated insulin secretion and energy homeostasis', *Proceedings of the National Academy of Sciences of The United States of America*. National Academy of Sciences, 104(10), pp. 4206–4211. https://doi.org/10.1073/PNAS.0611641104/ASSET/B436EB73-E023-4FEA-A278-022CAAE99824/ASSETS/GRAPHIC/ZPQ0060752580005.JPEG.
44. Liu, S. *et al.* (2013) 'Oestrogens improve human pancreatic islet transplantation in a mouse model of insulin deficient diabetes', *Diabetologia*, 56(2), pp. 370–381. https://doi.org/10.1007/s00125-012-2764-1.
45. Lo, J. C. *et al.* (2014) 'Adipsin is an adipokine that improves β cell function in diabetes', *Cell*. Cell Press, 158(1), pp. 41–53. https://doi.org/10.1016/j.cell.2014.06.005.
46. Lynn, F. C. *et al.* (2007) 'Sox9 coordinates a transcriptional network in pancreatic progenitor cells', *Proceedings of the National Academy of Sciences of The United States of America*. Proc Natl Acad Sci U S A, 104(25), pp. 10500–10505. https://doi.org/10.1073/PNAS.0704054104.
47. Mamidi, A. *et al.* (2018) 'Mechanosignalling via integrins directs fate decisions of pancreatic progenitors', *Nature*. Nature Publishing Group, 564(7734), pp. 114–118. https://doi.org/10.1038/s41586-018-0762-2.
48. Marchese, E. *et al.* (2015) 'Enumerating β-Cells in Whole Human Islets: Sex Differences and Associations With Clinical Outcomes After Islet Transplantation', *Diabetes Care*. American Diabetes Association, 38(11), p. e176. https://doi.org/10.2337/DC15-0723.
49. Marfil-Garza, B. A. *et al.* (2022) 'Pancreatic islet transplantation in type 1 diabetes: 20-year experience from a single-centre cohort in Canada', *The lancet. Diabetes & endocrinology*. Lancet Diabetes Endocrinol, 10(7), pp. 519–532. https://doi.org/10.1016/S2213-8587(22)00114-0.
50. Martínez, D. *et al.* (2016) 'Oestrogen activity of the serum in adolescents with Type 1 diabetes', *Diabetic medicine : a journal of the British Diabetic Association*. Diabet Med, 33(10), pp. 1366–1373. https://doi.org/10.1111/DME.13078.
51. Matsuoka, T. *et al.* (2003) 'Members of the large Maf transcription family regulate insulin gene transcription in islet beta cells', *Molecular and cellular biology*. Mol Cell Biol, 23(17), pp. 6049–6062. https://doi.org/10.1128/MCB.23.17.6049-6062.2003.
52. Mauvais-Jarvis, F. (2015) 'Sex differences in metabolic homeostasis, diabetes, and obesity', *Biology of Sex Differences*. Biology of Sex Differences, 6(1), pp. 1–9. https://doi.org/10.1186/s13293-015-0033-y.

53. Mauvais-Jarvis, F. (2018) 'Gender differences in glucose homeostasis and diabetes', *Physiology & behavior*. NIH Public Access, 187, p. 20. https://doi.org/10.1016/J.PHYSBEH.2017.08.016.
54. Mauvais-Jarvis, F., Arnold, A. P. and Reue, K. (2017) 'A Guide for the Design of Preclinical Studies on Sex Differences in Metabolism', *Cell metabolism*. Cell Metab, 25(6), pp. 1216–1230. https://doi.org/10.1016/J.CMET.2017.04.033.
55. Mauvais-Jarvis, F., Clegg, D. J. and Hevener, A. L. (2013) 'The role of estrogens in control of energy balance and glucose homeostasis', *Endocrine reviews*. Endocr Rev, 34(3), pp. 309–338. https://doi.org/10.1210/ER.2012-1055.
56. McKnight, K. D., Wang, P. and Kim, S. K. (2010) 'Deconstructing Pancreas Development to Reconstruct Human Islets from Pluripotent Stem Cells', *Cell stem cell*. NIH Public Access, 6(4), p. 300. https://doi.org/10.1016/J.STEM.2010.03.003.
57. van der Meulen, T. *et al.* (2012) 'Urocortin 3 Marks Mature Human Primary and Embryonic Stem Cell-Derived Pancreatic Alpha and Beta Cells', *PLOS ONE*. Public Library of Science, 7(12), p. e52181. https://doi.org/10.1371/JOURNAL.PONE.0052181.
58. Van Der Meulen, T. *et al.* (2015) 'Urocortin3 mediates somatostatin-dependent negative feedback control of insulin secretion', *Nature medicine*. NIH Public Access, 21(7), p. 769. https://doi.org/10.1038/NM.3872.
59. van der Meulen, T. and Huising, M. O. (2015) 'Role of transcription factors in the transdifferentiation of pancreatic islet cells', *Journal of Molecular Endocrinology*, 54(2), pp. R103–R117. https://doi.org/10.1530/JME-14-0290.
60. Milner, R. D. and Ashworth, M. A. (1971) 'Insulin release from human fetal pancreas in vitro.', *Archives of Disease in Childhood*. BMJ Publishing Group Ltd, 46(249), pp. 737–737. https://doi.org/10.1136/ADC.46.249.737-D.
61. Mochida, T. *et al.* (2020) 'Insulin-Deficient Diabetic Condition Upregulates the Insulin-Secreting Capacity of Human Induced Pluripotent Stem Cell-Derived Pancreatic Endocrine Progenitor Cells After Implantation in Mice', Diabetes, 69(4), pp. 634–646. https://doi.org/10.2337/db19-0728
62. Molotkov, A., Molotkova, N. and Duester, G. (2005) 'Retinoic acid generated by Raldh2 in mesoderm is required for mouse dorsal endodermal pancreas development', *Developmental dynamics : an official publication of the American Association of Anatomists*. Dev Dyn, 232(4), pp. 950–957. https://doi.org/10.1002/DVDY.20256.
63. Nair, G. and Hebrok, M. (2015) 'Islet formation in mice and men: Lessons for the generation of functional insulin-producing β cells from human pluripotent stem cells'. https://doi.org/10.1016/j.gde.2015.03.004.
64. Naya, F. J., Stellrecht, C. M. M. and Tsai, M. J. (1995) 'Tissue-specific regulation of the insulin gene by a novel basic helix-loop-helix transcription factor', *Genes & development*. Genes Dev, 9(8), pp. 1009–1019. https://doi.org/10.1101/GAD.9.8.1009.
65. Nishimura, W. *et al.* (2006) 'A switch from MafB to MafA expression accompanies differentiation to pancreatic beta-cells', *Developmental biology*. Dev Biol, 293(2), pp. 526–539. https://doi.org/10.1016/J.YDBIO.2006.02.028.
66. Nostro, M. C. *et al.* (2015) 'Efficient generation of NKX6-1+ pancreatic progenitors from multiple human pluripotent stem cell lines', *Stem Cell Reports*. Cell Press, 4(4), pp. 591–604. https://doi.org/10.1016/j.stemcr.2015.02.017.
67. Offield, M. F. *et al.* (1996) 'PDX-1 is required for pancreatic outgrowth and differentiation of the rostral duodenum', *Development (Cambridge, England)*. Development, 122(3), pp. 983–995. https://doi.org/10.1242/DEV.122.3.983.
68. Ohlsson, H., Karlsson, K. and Edlund, T. (1993) 'IPF1, a homeodomain-containing transactivator of the insulin gene', *The EMBO journal*. EMBO J, 12(11), pp. 4251–4259. https://doi.org/10.1002/J.1460-2075.1993.TB06109.X.
69. Pagliuca, F. W. *et al.* (2014) 'Generation of functional human pancreatic β cells in vitro', *Cell*. Cell Press, 159(2), pp. 428–439. https://doi.org/10.1016/j.cell.2014.09.040.
70. Pan, F. C. and Wright, C. (2011) 'Pancreas organogenesis: From bud to plexus to gland', *Developmental Dynamics*, 240(3), pp. 530–565. https://doi.org/10.1002/dvdy.22584.

71. Pepper, A. R. *et al.* (2017) 'Transplantation of Human Pancreatic Endoderm Cells Reverses Diabetes Post Transplantation in a Prevascularized Subcutaneous Site', *Stem Cell Reports*. ElsevierCompany., 8(6), pp. 1689–1700. https://doi.org/10.1016/j.stemcr.2017.05.004.
72. Piper, K. *et al.* (2004) 'Beta cell differentiation during early human pancreas development', *Journal of Endocrinology*. BioScientifica, 181(1), pp. 11–23. https://doi.org/10.1677/JOE.0.1810011.
73. Ramzy, A. *et al.* (2021) 'Implanted pluripotent stem-cell-derived pancreatic endoderm cells secrete glucose-responsive C-peptide in patients with type 1 diabetes', *Cell stem cell*. Cell Stem Cell, 28(12), pp. 2047–2061.e5. https://doi.org/10.1016/J.STEM.2021.10.003.
74. Ramzy, A. *et al.* (2023) 'Thyroid hormone levels correlate with the maturation of implanted pancreatic endoderm cells in patients with type 1 diabetes', The Journal of Clinical Endocrinology & Metabolism. Advance online publication. https://doi.org/10.1210/clinem/dgad499
75. Rezania, A. *et al.* (2011) 'Production of functional glucagon-secreting α-cells from human embryonic stem cells', *Diabetes*. Diabetes, 60(1), pp. 239–247. https://doi.org/10.2337/db10-0573.
76. Rezania, A. *et al.* (2012) 'Maturation of human embryonic stem cell-derived pancreatic progenitors into functional islets capable of treating pre-existing diabetes in mice', *Diabetes*. Diabetes, 61(8), pp. 2016–2029. https://doi.org/10.2337/db11-1711.
77. Rezania, A. *et al.* (2014) 'Reversal of diabetes with insulin-producing cells derived in vitro from human pluripotent stem cells', *Nature Biotechnology*. Nature Publishing Group, 32(11), pp. 1121–1133. https://doi.org/10.1038/nbt.3033.
78. Riedel, M. J. *et al.* (2012) 'Immunohistochemical characterisation of cells co-producing insulin and glucagon in the developing human pancreas', *Diabetologia*, 55(2), pp. 372–381. https://doi.org/10.1007/s00125-011-2344-9.
79. Röder, P. V. *et al.* (2016) 'Pancreatic regulation of glucose homeostasis', *Experimental & Molecular Medicine*. Korean Society for Biochemistry and Molecular Biology, 48(3), p. e219. https://doi.org/10.1038/EMM.2016.6.
80. Rodriguez-Diaz, R. *et al.* (2018) 'Paracrine Interactions within the Pancreatic Islet Determine the Glycemic Set Point', *Cell Metabolism*. Elsevier Inc., 27(3), pp. 549-558.e4. https://doi.org/10.1016/j.cmet.2018.01.015.
81. Rukstalis, J. M. and Habener, J. F. (2009) 'Neurogenin3: a master regulator of pancreatic islet differentiation and regeneration', *Islets*. Islets, 1(3), pp. 177–184. https://doi.org/10.4161/ISL.1.3.9877.
82. Ryan, E. A. *et al.* (2001) 'Clinical outcomes and insulin secretion after islet transplantation with the edmonton protocol', *Diabetes*, 50(4), pp. 710–719. https://doi.org/10.2337/diabetes.50.4.710.
83. Saber, N. *et al.* (2018) 'Sex Differences in Maturation of Human Embryonic Stem Cell-Derived β Cells in Mice', *Endocrinology*, 159(4), pp. 1827–1841. https://doi.org/10.1210/en.2018-00048.
84. Saber, N. *et al.* (2023) 'The impact of different implantation sites and sex on the differentiation of human pancreatic endoderm cells into insulin-secreting cells in vivo', *Diabetes*. American Diabetes Association. https://doi.org/10.2337/DB22-0692.
85. Salisbury, R. J. *et al.* (2014) 'The window period of NEUROGENIN3 during human gestation', *Islets*. Landes Bioscience, 6(3). https://doi.org/10.4161/19382014.2014.954436/SUPPL_FILE/KISL_A_954436_SM1789.DOCX.
86. Samuelsson, U. *et al.* (2013) 'Residual beta cell function at diagnosis of type 1 diabetes in children and adolescents varies with gender and season', *Diabetes/metabolism research and reviews*. Diabetes Metab Res Rev, 29(1), pp. 85–89. https://doi.org/10.1002/DMRR.2365.
87. Sander, M. *et al.* (1997) 'Genetic analysis reveals that PAX6 is required for normal transcription of pancreatic hormone genes and islet development', *Genes & development*. Genes Dev, 11(13), pp. 1662–1673. https://doi.org/10.1101/GAD.11.13.1662.
88. Sander, N. *et al.* (2000) 'Homeobox gene Nkx6.1 lies downstream of Nkx2.2 in the major pathway of β-cell formation in the pancreas', *Development*. The Company of Biologists, 127(24), pp. 5533–5540. https://doi.org/10.1242/DEV.127.24.5533.

89. Schulz, T. C. (2015) 'Concise Review: Manufacturing of Pancreatic Endoderm Cells for Clinical Trials in Type 1 Diabetes', *Stem Cells Translational Medicine*. Oxford Academic, 4(8), pp. 927–931. https://doi.org/10.5966/SCTM.2015-0058.
90. Schwitzgebel, V. M. *et al.* (2003) 'Agenesis of human pancreas due to decreased half-life of insulin promoter factor 1', *The Journal of clinical endocrinology and metabolism*. J Clin Endocrinol Metab, 88(9), pp. 4398–4406. https://doi.org/10.1210/JC.2003-030046.
91. Serls, A. E. *et al.* (2005) 'Different thresholds of fibroblast growth factors pattern the ventral foregut into liver and lung', *Development (Cambridge, England)*. Development, 132(1), pp. 35–47. https://doi.org/10.1242/DEV.01570.
92. Shapiro, A. M. J. *et al.* (2000) 'Islet transplantation in seven patients with type 1 diabetes mellitus using a glucocorticoid-free immunosuppressive regimen', *The New England journal of medicine*. N Engl J Med, 343(4), pp. 230–238. https://doi.org/10.1056/NEJM200007273430401.
93. Shapiro, A. M. J. *et al.* (2006) 'International trial of the Edmonton protocol for islet transplantation', *The New England journal of medicine*. N Engl J Med, 355(13), pp. 1318–1330. https://doi.org/10.1056/NEJMOA061267.
94. Shapiro, A. M. J. *et al.* (2021) 'Insulin expression and C-peptide in type 1 diabetes subjects implanted with stem cell-derived pancreatic endoderm cells in an encapsulation device', *Cell reports. Medicine*. Cell Rep Med, 2(12). https://doi.org/10.1016/J.XCRM.2021.100466.
95. Steiner, D. J. *et al.* (2010) 'Pancreatic islet plasticity: interspecies comparison of islet architecture and composition', *Islets*. Islets, 2(3). https://doi.org/10.4161/ISL.2.3.11815.
96. Stolovich-Rain, M. *et al.* (2015) 'Weaning triggers a maturation step of pancreatic β cells', *Developmental cell*. Dev Cell, 32(5), pp. 535–545. https://doi.org/10.1016/J.DEVCEL.2015.01.002.
97. Thomas, I. H. *et al.* (2009) 'Neonatal diabetes mellitus with pancreatic agenesis in an infant with homozygous IPF-1 Pro63fsX60 mutation', *Pediatric diabetes*. NIH Public Access, 10(7), p. 492. https://doi.org/10.1111/J.1399-5448.2009.00526.X.
98. Villasenor, A., Chong, D. C. and Cleaver, O. (2008) 'Biphasic Ngn3 expression in the developing pancreas', *Developmental dynamics : an official publication of the American Association of Anatomists*. Dev Dyn, 237(11), pp. 3270–3279. https://doi.org/10.1002/DVDY.21740.
99. Walker, J. T. *et al.* (2021) 'The Human Islet: Mini-Organ With Mega-Impact', *Endocrine reviews*. Endocr Rev, 42(5), pp. 605–657. https://doi.org/10.1210/ENDREV/BNAB010.
100. Wild, S. *et al.* (2004) 'Global prevalence of diabetes: estimates for the year 2000 and projections for 2030', *Diabetes care*. Diabetes Care, 27(5), pp. 1047–1053. https://doi.org/10.2337/DIACARE.27.5.1047.
101. Ye, F., Duvillié, B. and Scharfmann, R. (2005) 'Fibroblast growth factors 7 and 10 are expressed in the human embryonic pancreatic mesenchyme and promote the proliferation of embryonic pancreatic epithelial cells', *Diabetologia*. Springer, 48(2), pp. 277–281. https://doi.org/10.1007/S00125-004-1638-6/FIGURES/4.
102. Zhang, C. *et al.* (2005) 'MafA Is a Key Regulator of Glucose-Stimulated Insulin Secretion', *Molecular and Cellular Biology*. Taylor & Francis, 25(12), p. 4969. https://doi.org/10.1128/MCB.25.12.4969-4976.2005.
103. Zhou, Q. *et al.* (2007) 'A multipotent progenitor domain guides pancreatic organogenesis', *Developmental cell*. Dev Cell, 13(1), pp. 103–114. https://doi.org/10.1016/J.DEVCEL.2007.06.001.
104. Zorn, A. M. and Wells, J. M. (2009) 'Vertebrate endoderm development and organ formation', *Annual review of cell and developmental biology*. Annu Rev Cell Dev Biol, 25, pp. 221–251. https://doi.org/10.1146/ANNUREV.CELLBIO.042308.113344.

The Promises of Pancreatic Progenitor Proliferation and Differentiation

Azuma Kimura and Kenji Osafune

Abbreviations

EGF	Epidermal growth factor
ESC	Embryonic stem cell
FGF	Fibroblast growth factor
GFR	growth factor reduced
iPSC	induced pluripotent stem cell
KGF	Keratinocyte growth factor
NKX6.1	NK6 homeobox 1
PDX1	Pancreatic and duodenal homeobox 1
PP	Pancreatic progenitor
PSC	Pluripotent stem cell
RA	Retinoic acid
SOX9	SRY-box transcription factor 9

1 Introduction

One of the major challenges that remain for establishing cell therapy for clinical applications is the manufacturing of billions and billions of cells required for transplantation. Despite successes in generating pancreatic cells from human pluripotent

Azuma Kimura and Kenji Osafune contributed equally with all other contributors.

A. Kimura · K. Osafune (✉)
Center for iPS Cell Research and Application (CiRA), Kyoto University, Kyoto, Japan
e-mail: osafu@cira.kyoto-u.ac.jp

L. Piemonti et al. (eds.), *Pluripotent Stem Cell Therapy for Diabetes*,
https://doi.org/10.1007/978-3-031-41943-0_5

stem cells (hPSCs) such as human embryonic stem cells (hESCs) and human induced pluripotent stem cells (hiPSCs), practical usage of stem cell-derived pancreatic β cells will only be feasible when cell preparation procedures are capable of meeting such incredible demands. Pioneers in stem cell biology have previously succeeded in developing culture methods that maintain hPSCs at the pluripotent state with indefinite self-renewal capacity [33, 34]. As a result, growth factors, medium conditions, and culture systems have been studied and optimized extensively for massive propagation of undifferentiated hPSCs. Several groups have now established scalable systems to generate hPSC-derived pancreatic cells as two- or three-dimensional cultures by first preparing vast numbers of hPSCs and subsequently inducing their differentiation [22, 30]. Although this way of generating hPSC-derived pancreatic cells seems fairly plausible, maintaining hPSCs at such large scales can be burdensome for several reasons. First, it requires large quantities of costly medium to preserve hPSCs in the undifferentiated state. Second, it involves several steps of precise manipulation to direct hPSCs to differentiate into pancreatic lineage cells. Third, it may ultimately result in low cell yield and quality because the end products are likely to contain unwanted cell types due to inefficiencies at each differentiation step. Therefore, a system that can be fine-tuned and circumvent these limitations may prove to be more enticing for clinical applications of hPSC-derived pancreatic cells.

Having the ability to make large quantities of hPSC-derived pancreatic progenitors (PPs) opens the door to stable, cost-effective, time-saving, and high-quality supplies of transplantable pancreatic cells for diabetes treatment. Several groups have since attempted to expand intermediate cells—cells in the early stages of differentiation, namely definitive endoderm and primitive gut tube cells—in a chemically defined medium or in the presence of organ-matched feeder cells [4, 8, 31]. In this chapter, we will take a PP-centric view of pancreatic lineage cell preparation systems and explore early pancreatic cell development and the roles of various signaling pathways regulating PP proliferation and differentiation. Looking forward, we will also consider the feasibility of these methods for therapeutic applications, drug screening, and modeling of human pancreas development.

2 Development of Pancreatic Progenitors

Human pancreas development begins when a sheet of endoderm epithelium expresses the transcription factor pancreatic and duodenal homeobox 1 (PDX1), first detected in the presumptive pancreatic endoderm at Carnegie stage (CS) 12 (29–31 days post-conception) [9], and evaginates to form the dorsal and ventral pancreatic buds from the posterior foregut [24]. PDX1-expressing PPs are multipotent and can give rise to exocrine cells, ductal cells, and endocrine cells including insulin-secreting β cells, glucagon-secreting α cells, and somatostatin-secreting δ cells [5, 7]. Because PPs undergo extensive self-renewal following the emergence of pancreatic buds, it is thus this developmental stage that stem cell biologists are attempting to recapitulate in vitro for the expansion of hPSC-derived PPs. By CS13 (30–33 days post-conception), both dorsal and ventral pancreatic buds have formed apparent structures, and microlumen formation begins [9]. PPs therefore undergo either a symmetrical or asymmetrical cell division, resulting in complex cell-fate decisions and branching morphogenesis during this time [13, 36]. *Pdx1* knockout mice or a patient carrying a frameshift mutation in the *PDX1* gene have been shown to lack pancreatic tissues, indicating that PDX1 activity is absolutely required for PP pool expansion [11, 32]. As such, the expression of PDX1 in PPs is crucial for maintaining their proliferative and differentiation potentials.

Another transcription factor, NK6 homeobox 1 (NKX6.1), is activated following the expression of PDX1. NKX6.1 is considered as the principal endocrine and trunk progenitor marker and has been suggested to be the key transcription factor required for the proper activation of endocrine commitment in PPs [21, 26, 27]. *Nkx6.1* deletion prevents the generation of functional β cells in mice [28, 29]; therefore, NKX6.1 is believed to play essential roles, especially during β cell development [2]. However, PPs expressing both PDX1 and NKX6.1 are also considered multipotent. hPSC differentiation protocols established to date for β cell generation have focused on activating NKX6.1 following PDX1 expression; hence, these two seemingly different stages of cells are designated as PP1 (PDX1-positive) and PP2 (PDX1 and NKX6.1 double-positive) stages in pancreatic differentiation protocols [22, 25]. This chapter therefore considers cells expressing either PDX1 alone or both PDX1 and NKX6.1 as the PPs that stem cell biologists are attempting to expand in vitro.

3 hPSC-Derived Pancreatic Progenitor Proliferation

Trott and colleagues were the first to describe a long-term culture of self-renewing hPSC-derived multipotent PPs [35]. They established a culture platform that enables PP expansion derived from multiple genetically diverse hPSCs. PPs expressing PDX1 and SRY-box transcription factor 9 (SOX9) were expanded for more than 25 passages on mouse embryonic fibroblast feeder cells, 3T3-J2, in the presence of signaling growth factors and molecules including retinoic acid (RA), epidermal growth factor (EGF), fibroblast growth factor 10 (FGF10), SB431542 (a TGF-β pathway inhibitor), and DAPT (a Notch pathway inhibitor) (Fig. 1). Passaged PPs maintained the expression of key pancreatic markers including PDX1, SOX9, ONECUT1, and GATA6 without inducing premature endocrine commitment. For example, PDX1 expression was maintained in up to 90% of the expanded PPs without any changes in chromosomal integrity. A transcriptomic comparison with their in vivo counterpart indicated that the expanded PPs, which stained negative for NKX6.1, closely resemble embryonic PP cells between CS12 and CS13. In fact, the described culture platform did not maintain or upregulate NKX6.1 expression during expansion. Nevertheless, the expanded PPs possessed differentiation potential into PDX1 and NKX6.1 double-positive multipotent PPs after exposure to EGF, RA, and FGF7, which indicated that the expansion of PPs represents an alternative

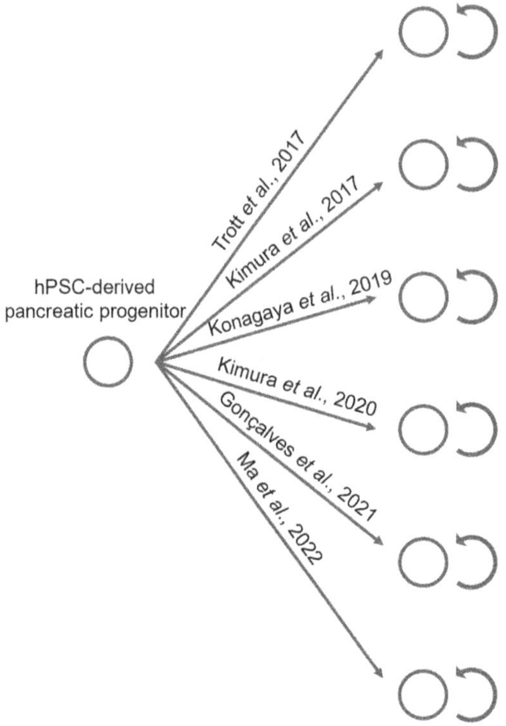

Factors	Feeder cells
RA EGF FGF10 SB431542 DAPT	3T3-J2
KGF LDN193189 EGF AT7867	-
EGF, RA, KGF RSPO1, SANT1 LDN193189 CHIR99021 SB431542	-
KGF EGF Nicotinamide	3T3-Wnt7a/b
FGF2 GFR-Matrigel	-
EGF FGF2 RepSox I-BET151	CF1

Fig. 1 Factors and feeder cells used for the expansion of hPSC-derived pancreatic progenitors

system that possibly circumvents the need to repeatedly induce differentiation of hPSCs into PPs.

We recently attempted to identify proliferation inducers for hPSC-derived PPs and reported a small molecule-based expansion of PPs [15]. In this work, we performed a chemical screen and identified AT7867, which induced the proliferation of PDX1-expressing PPs without causing DNA damage. The culture system used also included three other factors: keratinocyte growth factor (KGF), LDN193189 (a BMP pathway inhibitor), and EGF, but even AT7867 alone exhibited a strong proliferative effect on PPs. Interestingly, AT7867 promoted the proliferation of only PDX1-expressing PPs but no other cell types such as the definitive endoderm. Furthermore, the expanded PPs retained the potential to differentiate into PDX and NKX6.1 double-positive cells as well as insulin-producing β cells. Small molecules are typically available off-the-shelf and stable over time, thus making such systems cost-effective and fine-tunable, while co-culture systems may be influenced by variabilities in the supporting cells between passages, ultimately affecting the proliferation and differentiation of the expanded PPs. Additionally, co-culture systems face potential contamination problems by the feeder cells, which hamper them as a reliable source of hPSC-derived pancreatic cells for clinical applications.

More recently, Konagaya and Iwata reported a xenogenic-free, three-dimensional culture system for the expansion of PDX1- and SOX9-expressing PPs in the presence of EGF, RA, KGF, R-spondin1, SANT-1 (a Hedgehog pathway inhibitor), LDN193189, CHIR99021 (a WNT pathway inhibitor), and SB431542 [16]. The cell aggregates in that study expanded in size in agarose gel plates and were not only passaged but survived through freeze-thaw cycles. NKX6.1 expression was induced, albeit heterogeneously, and the doubling time of PPs in this system was much longer (~180 hours, compared to ~65 hours, as reported by Trott and colleagues). This may have been due to the difference in the culture formats, specifically monolayer versus three-dimensional cultures. Importantly, their culture system included chemically defined conditions, which help to prevent potential contamination with undesirable materials such as xenogenic cells, proteins, and extracellular matrix.

In recent years, additional methods for PP expansion using small molecules, feeder cells, and/or three-dimensional culture formats have been reported. We have further attempted to identify the mechanisms underlying small molecule-induced PP proliferation using transcriptome analysis and knockdown experiments. Although AT7867 is an ATP-competitive inhibitor against AKT and p70S6K, the treatment did not reduce phosphorylation levels of AKT in the PPs. Intrigued by this result, we analyzed gene expression of PPs treated with AT7867 and found that it had upregulated *WNT7B* expression [14]. Knocking down *WNT7B* expression abolished the compound's proliferative effects, thus indicating that AT7867 facilitates proliferation by regulating *WNT7B* expression in PPs. Co-culturing with mouse embryonic fibroblast cell line, NIH3T3, expressing mouse Wnt7a/b or human embryonic kidney cell line, HEK293, expressing human WNT7A/B promoted the proliferation of hPSC-derived PPs, thus confirming that they were stimulated by WNT7A/B signaling for expansion. Phosphoproteome analysis showed that AT7867 inhibited Ying Yang 1 (YY1) phosphorylation, which is known to play a multifaceted role in various biological processes. In addition, we observed that *YY1* knockdown reduced *WNT7B* expression, suggesting that it plays a role in regulating PP proliferation upstream of WNT7B.

Previously, Gonçalves and colleagues dissected the human fetal pancreas between 7 and 10 weeks post-conception through single-cell transcriptome analysis [6]. They used this atlas to benchmark hPSC-derived PPs and showed that a subpopulation of the PPs resembled human fetal progenitors transcriptomically. The PPs, embedded in growth factor reduced (GFR)-Matrigel (mouse-derived extracellular matrix) with a medium containing FGF2 and a Rho kinase inhibitor, were passaged more than 20 times with a doubling time of roughly 65 hours, resulting in a 10^{20}-fold increase in PP number. Although NKX6.1 expression was low and heterogenous, the PPs were successfully differentiated into β cells upon induction into the endocrine lineage.

Ma and colleagues recently identified a small molecule, I-BET151, which promotes hPSC-derived PP expansion while retaining the expression of both PDX1 and NKX6.1 [18]. Their system also utilized feeder cells as supporting cells, and the PPs expanded more than 10^{20}-fold within 35 passages in the presence of EGF, FGF2, and RepSox (a TGF-β pathway inhibitor) in the culture medium. Pancreatic β cells differentiated from the expanded PPs responded to blood glucose levels and ameliorated diabetes when transplanted under the kidney capsule of diabetic mice. Altogether, these studies have successfully demonstrated that PPs can be induced to proliferate and are amenable to further differentiation into β cells in vitro, affirming that PPs can serve as a renewable and expandable cell source for diabetes treatment.

4 Signaling Pathways Governing Pancreatic Progenitor Proliferation and Differentiation

Genetic studies using rodent models have advanced our understanding of pancreatic development enormously. Naturally, current protocols that directly differentiate hPSCs into pancreatic lineage cells were largely established based upon those findings. For example, FGF10 and RA that are expressed by the surrounding pancreatic mesenchyme are important for inducing PDX1 expression and the expansion of the pancreatic buds [3, 17, 19, 37]. Additional signaling pathways have been shown to play critical roles during human pancreas development including FGF, RA, Hedgehog, EGF, WNT, Notch, and Hippo, yet novel insights are still being uncovered using human fetuses and hPSC-derived pancreatic tissues [12, 23].

For instance, transcriptome analysis of the human fetal pancreas revealed that FGF2 and FGF9 are the most abundant FGFs available at CS12, whereas FGF7 and FGF10 were reported as the primary FGF ligands during pancreas development in mice [6]. These findings highlight a species difference between humans and mice and strongly suggest the use of appropriate ligands for differentiating and expanding PPs. In addition, the same study has identified other EGF ligands that may be involved in human pancreas development, namely transforming growth factor α (TGF-α), EGF containing fibulin extracellular matrix protein 1 (EFEMP1), serine peptidase inhibitor Kazal type 1 (SPINK1), and amphiregulin (AREG). Although EGF receptor (EGFR) is expressed by the human fetal pancreatic epithelium, we and others have confirmed that EGF addition into the culture medium does not or only mildly promote PP proliferation [15, 18]. Conversely, whether and to what extent TGF-α, EFEMP1, SPINK1, and AREG are involved in activating EGFR remains an open question.

WNT signaling is another important pathway for pancreas development. However, studies on the source and identity of WNT ligands during human pancreas development have been limited. Using hPSC-derived PPs as a model, we identified WNT7B as a PP mitogen, and our findings indicated that WNT7B operates via the non-canonical WNT pathway; while Wnt7b in contrast activates the canonical Wnt/β-catenin signaling pathway during mouse pancreatic bud formation [1, 14]. In line with this, several reports have shown that activating the canonical WNT pathway in PPs reduces PDX1 expression [6, 14]. These data together suggest a species difference in signaling between WNT ligands in the context of pancreas development.

Lastly, our understanding of the mechanisms of endocrine lineage cell development remains incomplete. Inhibition of Notch signaling is often used to direct PPs to differentiate into endocrine progenitor cells [10, 20]. However, no protocols to date have been able to differentiate PPs into β cells with high efficiency and purity. One possibility is that Notch signaling requires cell-to-cell contact-based signaling and thus may be difficult to fully recapitulate in vitro. Given that PP expansion systems are amenable to high-throughput screening assays with chemical compounds, RNA interference, and CRISPR-based endogenous gene modulations, PPs derived from hPSCs will likely prove to be a powerful tool for dissecting mechanisms of human pancreas development, not only limited to PP maintenance and expansion, but also three-dimensional branching morphogenesis and endocrine specification. It is with much hope that we are now closer than ever to the day that homogenous β cell induction methods are established.

5 Conclusion

There is no doubt that the expansion of PPs in vitro has propelled hPSC-derived pancreatic cell therapy for diabetes to the next stage. However, PPs differentiate asynchronously in vitro under current protocols, resulting in heterogenous cultures—particularly toward endocrine lineages. The ability to synchronize differentiation is thus a sine qua non for generating hPSC-derived β cells with high efficiency and purity. As such, there remain many hurdles associated with the quantity and quality of transplantable cells before these technologies can be applied in clinical settings.

Acknowledgments We thank Dr. Kelvin Hui for his comments on the manuscript. This work was supported by the Japan Society for the Promotion of Science (JSPS) through its Grant-in-Aid for Scientific Research (B) (JSPS KAKENHI Grant Number 21H02979) to K.O. and by the Japan Agency for Medical Research and Development (AMED) through its research grant "Core Center for iPS Cell Research (Grant Number 22bm0104001h0010), R&D Program of Regenerative/Cellular Medicine and Gene Therapies from the Basic to Nonclinical Phase (Grant Number 22bm1123002h0001), Research Center Network for Realization of Regenerative Medicine" to K.O. and by the iPS Cell Research Fund.

References

1. Afelik, S., Pool, B., Schmerr, M., Penton, C. & Jensen, J. 2015. Wnt7b is required for epithelial progenitor growth and operates during epithelial-to-mesenchymal signaling in pancreatic development. *Dev Biol,* 399, 204–17.
2. Aigha, I. I. & Abdelalim, E. M. 2020. NKX6.1 transcription factor: a crucial regulator of pancreatic β cell development, identity, and proliferation. *Stem Cell Res Ther,* 11, 459.
3. Bhushan, A., Itoh, N., Kato, S., Thiery, J. P., Czernichow, P., Bellusci, S. & Scharfmann, R. 2001. Fgf10 is essential for maintaining the proliferative capacity of epithelial progenitor cells during early pancreatic organogenesis. *Development,* 128, 5109–17.

4. Cheng, X., Ying, L., Lu, L., Galvão, A. M., Mills, J. A., Lin, H. C., Kotton, D. N., Shen, S. S., Nostro, M. C., Choi, J. K., Weiss, M. J., French, D. L. & Gadue, P. 2012. Self-renewing endodermal progenitor lines generated from human pluripotent stem cells. *Cell Stem Cell,* 10, 371–84.
5. Gannon, M., Herrera, P. L. & Wright, C. V. 2000. Mosaic Cre-mediated recombination in pancreas using the pdx-1 enhancer/promoter. *Genesis,* 26, 143–4.
6. Gonçalves, C. A., Larsen, M., Jung, S., Stratmann, J., Nakamura, A., Leuschner, M., Hersemann, L., Keshara, R., Perlman, S., Lundvall, L., Thuesen, L. L., Hare, K. J., Amit, I., Jørgensen, A., Kim, Y. H., Del Sol, A. & Grapin-Botton, A. 2021. A 3D system to model human pancreas development and its reference single-cell transcriptome atlas identify signaling pathways required for progenitor expansion. *Nat Commun,* 12, 3144.
7. Gu, G., Dubauskaite, J. & Melton, D. A. 2002. Direct evidence for the pancreatic lineage: NGN3+ cells are islet progenitors and are distinct from duct progenitors. *Development,* 129, 2447–57.
8. Hannan, N. R., Fordham, R. P., Syed, Y. A., Moignard, V., Berry, A., Bautista, R., Hanley, N. A., Jensen, K. B. & Vallier, L. 2013. Generation of multipotent foregut stem cells from human pluripotent stem cells. *Stem Cell Reports,* 1, 293–306.
9. Jennings, R. E., Berry, A. A., Kirkwood-Wilson, R., Roberts, N. A., Hearn, T., Salisbury, R. J., Blaylock, J., Piper Hanley, K. & Hanley, N. A. 2013. Development of the human pancreas from foregut to endocrine commitment. *Diabetes,* 62, 3514–22.
10. Jennings, R. E., Berry, A. A., Strutt, J. P., Gerrard, D. T. & Hanley, N. A. 2015. Human pancreas development. *Development,* 142, 3126–37.
11. Jonsson, J., Carlsson, L., Edlund, T. & Edlund, H. 1994. Insulin-promoter-factor 1 is required for pancreas development in mice. *Nature,* 371, 606–9.
12. Jørgensen, M. C., Ahnfelt-Rønne, J., Hald, J., Madsen, O. D., Serup, P. & Hecksher-Sørensen, J. 2007. An illustrated review of early pancreas development in the mouse. *Endocr Rev,* 28, 685–705.
13. Kim, Y. H., Larsen, H. L., Rué, P., Lemaire, L. A., Ferrer, J. & Grapin-Botton, A. 2015. Cell cycle-dependent differentiation dynamics balances growth and endocrine differentiation in the pancreas. *PLoS Biol,* 13, e1002111.
14. Kimura, A., Toyoda, T., Iwasaki, M., Hirama, R. & Osafune, K. 2020. Combined Omics Approaches Reveal the Roles of Non-canonical WNT7B Signaling and YY1 in the Proliferation of Human Pancreatic Progenitor Cells. *Cell Chem Biol,* 27, 1561–1572.e7.
15. Kimura, A., Toyoda, T., Nishi, Y., Nasu, M., Ohta, A. & Osafune, K. 2017. Small molecule AT7867 proliferates PDX1-expressing pancreatic progenitor cells derived from human pluripotent stem cells. *Stem Cell Res,* 24, 61–68.
16. Konagaya, S. & Iwata, H. 2019. Chemically defined conditions for long-term maintenance of pancreatic progenitors derived from human induced pluripotent stem cells. *Sci Rep,* 9, 640.
17. Kumar, M., Jordan, N., Melton, D. & Grapin-Botton, A. 2003. Signals from lateral plate mesoderm instruct endoderm toward a pancreatic fate. *Dev Biol,* 259, 109–22.
18. Ma, X., Lu, Y., Zhou, Z., Li, Q., Chen, X., Wang, W., Jin, Y., Hu, Z., Chen, G., Deng, Q., Shang, W., Wang, H., Fu, H., He, X., Feng, X. H. & Zhu, S. 2022. Human expandable pancreatic progenitor-derived β cells ameliorate diabetes. *Sci Adv,* 8, eabk1826.
19. Martín, M., Gallego-Llamas, J., Ribes, V., Kedinger, M., Niederreither, K., Chambon, P., Dollé, P. & Gradwohl, G. 2005. Dorsal pancreas agenesis in retinoic acid-deficient Raldh2 mutant mice. *Dev Biol,* 284, 399–411.
20. Memon, B. & Abdelalim, E. M. 2020. Stem Cell Therapy for Diabetes: Beta Cells versus Pancreatic Progenitors. *Cells,* 9.
21. Nostro, M. C., Sarangi, F., Yang, C., Holland, A., Elefanty, A. G., Stanley, E. G., Greiner, D. L. & Keller, G. 2015. Efficient generation of NKX6-1+ pancreatic progenitors from multiple human pluripotent stem cell lines. *Stem Cell Reports,* 4, 591–604.
22. Pagliuca, F. W., Millman, J. R., Gürtler, M., Segel, M., Van Dervort, A., Ryu, J. H., Peterson, Q. P., Greiner, D. & Melton, D. A. 2014. Generation of functional human pancreatic β cells in vitro. *Cell,* 159, 428–39.

23. Pan, F. C. & Wright, C. 2011. Pancreas organogenesis: from bud to plexus to gland. *Dev Dyn,* 240, 530–65.
24. Piper, K., Brickwood, S., Turnpenny, L. W., Cameron, I. T., Ball, S. G., Wilson, D. I. & Hanley, N. A. 2004. Beta cell differentiation during early human pancreas development. *J Endocrinol,* 181, 11–23.
25. Rezania, A., Bruin, J. E., Arora, P., Rubin, A., Batushansky, I., Asadi, A., O'dwyer, S., Quiskamp, N., Mojibian, M., Albrecht, T., Yang, Y. H., Johnson, J. D. & Kieffer, T. J. 2014. Reversal of diabetes with insulin-producing cells derived in vitro from human pluripotent stem cells. *Nat Biotechnol,* 32, 1121–33.
26. Rezania, A., Bruin, J. E., Xu, J., Narayan, K., Fox, J. K., O'Neil, J. J. & Kieffer, T. J. 2013. Enrichment of human embryonic stem cell-derived NKX6.1-expressing pancreatic progenitor cells accelerates the maturation of insulin-secreting cells in vivo. *Stem Cells,* 31, 2432–42.
27. Russ, H. A., Parent, A. V., Ringler, J. J., Hennings, T. G., Nair, G. G., Shveygert, M., Guo, T., Puri, S., Haataja, L., Cirulli, V., Blelloch, R., Szot, G. L., Arvan, P. & Hebrok, M. 2015. Controlled induction of human pancreatic progenitors produces functional beta-like cells in vitro. *EMBO J,* 34, 1759–72.
28. SANDER, M., SUSSEL, L., CONNERS, J., Scheel, D., Kalamaras, J., Dela Cruz, F., Schwitzgebel, V., Hayes-Jordan, A. & German, M. 2000. Homeobox gene Nkx6.1 lies downstream of Nkx2.2 in the major pathway of beta-cell formation in the pancreas. *Development,* 127, 5533–40.
29. Schaffer, A. E., Taylor, B. L., Benthuysen, J. R., Liu, J., Thorel, F., Yuan, W., Jiao, Y., Kaestner, K. H., HERRERA, P. L., Magnuson, M. A., May, C. L. & Sander, M. 2013. Nkx6.1 controls a gene regulatory network required for establishing and maintaining pancreatic Beta cell identity. *PLoS Genet,* 9, e1003274.
30. Schulz, T. C., Young, H. Y., Agulnick, A. D., Babin, M. J., Baetge, E. E., Bang, A. G., Bhoumik, A., Cepa, I., CESARIO, R. M., Haakmeester, C., Kadoya, K., Kelly, J. R., Kerr, J., Martinson, L. A., Mclean, A. B., Moorman, M. A., Payne, J. K., Richardson, M., Ross, K. G., Sherrer, E. S., Song, X., Wilson, A. Z., Brandon, E. P., Green, C. E., Kroon, E. J., Kelly, O. G., D'Amour, K. A. & Robins, A. J. 2012. A scalable system for production of functional pancreatic progenitors from human embryonic stem cells. *PLoS One,* 7, e37004.
31. Sneddon, J. B., Borowiak, M. & Melton, D. A. 2012. Self-renewal of embryonic-stem-cell-derived progenitors by organ-matched mesenchyme. *Nature,* 491, 765–8.
32. Stoffers, D. A., Zinkin, N. T., Stanojevic, V., Clarke, W. L. & Habener, J. F. 1997. Pancreatic agenesis attributable to a single nucleotide deletion in the human IPF1 gene coding sequence. *Nat Genet,* 15, 106–10.
33. Takahashi, K., Tanabe, K., Ohnuki, M., Narita, M., Ichisaka, T., Tomoda, K. & Yamanaka, S. 2007. Induction of pluripotent stem cells from adult human fibroblasts by defined factors. *Cell,* 131, 861–72.
34. Thomson, J. A., Itskovitz-Eldor, J., Shapiro, S. S., Waknitz, M. A., Swiergiel, J. J., Marshall, V. S. & Jones, J. M. 1998. Embryonic stem cell lines derived from human blastocysts. *Science,* 282, 1145–7.
35. Trott, J., Tan, E. K., Ong, S., Titmarsh, D. M., Denil, S. L. I. J., Giam, M., Wong, C. K., Wang, J., Shboul, M., Eio, M., Cooper-White, J., Cool, S. M., Rancati, G., Stanton, L. W., Reversade, B. & Dunn, N. R. 2017. Long-Term Culture of Self-renewing Pancreatic Progenitors Derived from Human Pluripotent Stem Cells. *Stem Cell Reports,* 8, 1675–1688.
36. Villasenor, A., Chong, D. C., Henkemeyer, M. & Cleaver, O. 2010. Epithelial dynamics of pancreatic branching morphogenesis. *Development,* 137, 4295–305.
37. Ye, F., Duvillié, B. & Scharfmann, R. 2005. Fibroblast growth factors 7 and 10 are expressed in the human embryonic pancreatic mesenchyme and promote the proliferation of embryonic pancreatic epithelial cells. *Diabetologia,* 48, 277–81.

Part II
Bioengineering

Selecting Biocompatible Biomaterials for Stem Cell-Derived β-Cell Transplantation

Rick de Vries and Aart A. van Apeldoorn

1 Clinical Islet Transplantation

Currently, clinical islet transplantation (CIT) is the most promising minimal invasive therapy to treat the most severe cases of type 1 diabetes, in which exogenous insulin administration can no longer be effectively used to control blood glucose levels. During CIT, the pancreas of a deceased donor is harvested, pancreatic islets are isolated and subsequently infused into the liver of the patient through the portal vein [1, 2]. A milestone in the advancement of islet transplantation was the publication of the Edmonton Protocol in the year 2000, which, among others, dictates the usage of a specific immunosuppressive regimen and an islet dose of 11,000 IEQ/kg, and was quickly adopted throughout the field [2]. However, despite great progress in isolation and transplantation protocols in the last two decades, less than 40% of patients show insulin independence 5 years after islet transplantation, which further reduces to 30% after 10 years of transplantation [3]. Furthermore, CIT is associated with a loss of 60% of transplanted islets within the first days post-implantation, which explains the need for an average of two to three donors to cure one type 1 diabetes patient [4, 5]. This marked decrease of islet mass is, among others, caused by mechanical stress, the presence of an instant blood-mediated immune response (IBMIR) within the liver and a lack of oxygen flow to the islets due to impaired vascularization, and long-term allo- and autoimmunity (Fig. 1) [5–8]. Transplanted islets are known to revascularize in roughly 14 days, but even after 3 months, intrahepatic implanted islets experience a low oxygen tension <10 mmHg [9, 10]. A lack

R. de Vries · A. A. van Apeldoorn (✉)
Maastricht University, MERLN Institute for Technology-Inspired Regenerative Medicine, Maastricht, the Netherlands
e-mail: a.vanapeldoorn@maastrichtuniversity.nl

L. Piemonti et al. (eds.), *Pluripotent Stem Cell Therapy for Diabetes*,
https://doi.org/10.1007/978-3-031-41943-0_6

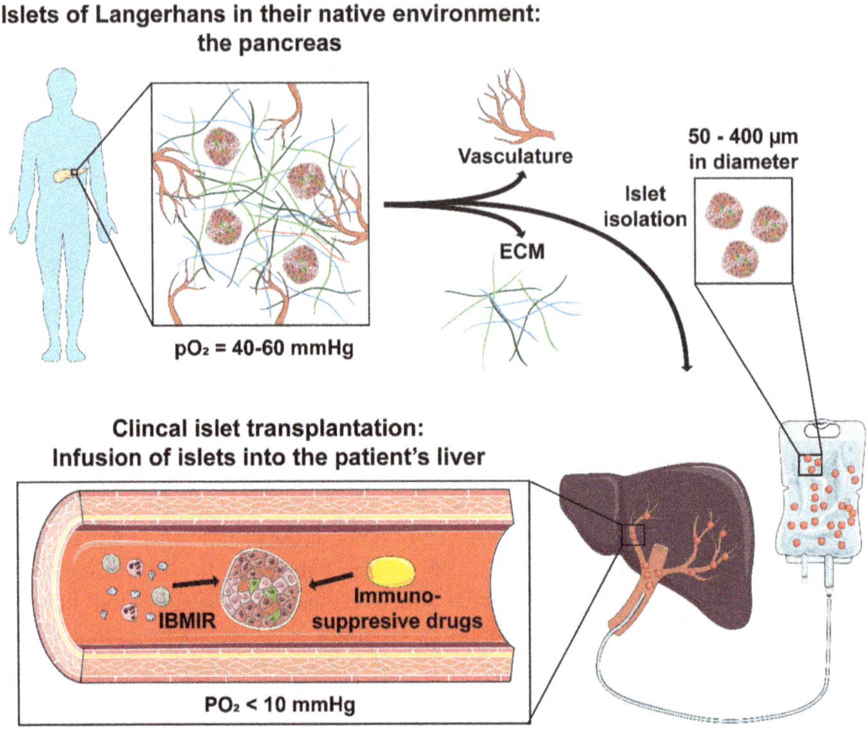

Fig. 1 Stress factors for islets during clinical islet transplantation. Islets naturally reside within the pancreas, where they are surrounded by a specific extracellular matrix (ECM), a dense capillary network, and an oxygen tension of 40–60 mmHg. During islet isolation, the vasculature and ECM are removed, making the islets depend on diffusion to obtain oxygen and nutrients. Islets are infused into the liver through the portal vein in the current CIT procedure, where the islets then embolize in the local vasculature. However, the local milieu surrounding these islets is highly different from the pancreas, with a different ECM composition, relatively high lipid and glucose concentrations, a relatively low oxygen tension, high concentrations of immunosuppressive drugs, and an instant-blood mediated inflammatory reaction (IBMIR)

of oxygen will result in the formation of necrotic cores and ultimately cell death, attributing to the progressive loss of islets after transplantation [11]. Moreover, it is known that glucose responsiveness of β-cells highly depends on local oxygen levels [12, 13]. Under hypoxic conditions, insulin secretion is expected to be reduced by several orders of magnitude compared to islets within their native pancreatic environment [14]. Islets therefore require an implantation site that supports swift restoration of oxygen tension. Improving CIT could potentially lead to better maintenance of glycemia, diminished or absent need for insulin therapy, and less risk of long-term complications.

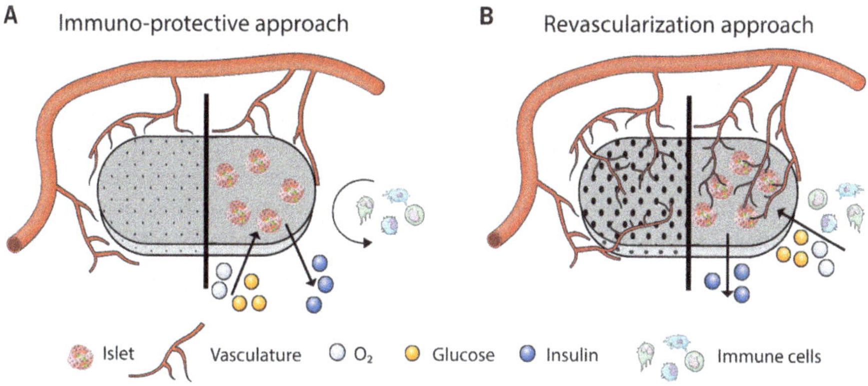

Fig. 2 Schematic overview of the principles from macro-encapsulation devices following either the immunoprotective or revascularization approach. (**a**) Immunoprotective approach in which islets are encapsulated in a barrier that allows free diffusion of glucose, oxygen, and insulin, while blocking immune cells. Over time, vasculature will grow over the device surface, but islets will still depend on diffusion through the barrier to obtain nutrients. (**b**) Revascularization approach in which islets are encapsulated within a device consisting of a macroporous layer that allows free diffusion of glucose, oxygen and insulin directly after transplantation, as well as blood vessel ingrowth over time

2 Strategies for Extrahepatic Transplantation

Bioengineering approaches can be used to offer novel implantation sites to islets that potentially improve the outcomes of CIT. Two different bioengineering approaches include either encapsulation of islets in a "closed" immunoprotective device, or encapsulation of islets in an "open" device which supports revascularization after transplantation. The advantages and limitations of these different strategies are discussed in the following sections and summarized in Fig. 2.

2.1 *The Immunoprotective Barrier Strategy*

CIT relies on the systemic administration of immunosuppressive drugs to prevent loss of the transplanted islets due to a recurrence of the auto- and alloimmune reaction against the donor tissue and β-cells specifically. However, this chronic immunosuppressive therapy comes with significant side effects such as opportunistic infections, a significantly higher risk of specific malignancies, and loss of islet mass [15]. An appealing alternative for immunosuppressive drugs would therefore be an immunoprotective barrier strategy, where a passive barrier separates encapsulated islets from the host immune system. This barrier has a certain molecular size cut-off, allowing diffusion of insulin, glucose, nutrients, and oxygen through the barrier, while, at the same time, immune cells are blocked. Immunoprotective strategies can follow three different encapsulation approaches. In short, nano-encapsulation involves the application of a nanometer-range thick polymer coating on the outside of islets that shields the patient's immune system [16, 17]. A second encapsulation approach involves the microencapsulation of islets by grouping individual or small groups of islets in microcapsules, typically hydrogels [18, 19]. The third approach is macro-encapsulation, where up to 100,000 islets or cell aggregates are co-transplanted into one or more devices [18, 19]. The prevention of direct blood access for transplanted islets is the most limiting factor in the immunoprotective barrier strategy. Transplanted islets therefore are solely dependent on diffusion to obtain oxygen and nutrients and sensing of fluctuating glucose concentrations. Protection against the host immune system is therefore traded for low oxygen levels, which limits the survival of islets.

2.2 *The Revascularization Strategy*

Islets are removed from their native environment by enzymatic digestion and mechanical disruption during islet isolation, removing most of the native ECM and vasculature. Upon transplantation, islets solely depend on diffusion to get sufficient nutrients and oxygen during the first 14 days after transplantation [9, 10]. The combined effect of disrupting islet ECM, exposure to inflammatory cytokines, endotoxins, and hypoxia, due to a lack of vascularization, results in significant degradation of the islet graft [4, 5, 11, 20]. Revascularization strategies therefore aim to promote the reconnection of islets to the host vasculature as quickly as possible, to restore oxygen tension and thereby maintain islet viability and functionality after transplantation. Most importantly, the maximum distance of any cell from its nearest capillary rarely exceeds 200 μm and is usually less than 100 μm [21, 22]. It is therefore vital that delivery devices are as thin and as small as possible to reduce both the diffusion and vascular ingrowth distance into the device. Effective use of a revascularization strategy can help boost angiogenesis, lead to a reduction in the amount of islets (and therefore donor organs) needed for transplantation, potentially offer better glycemic maintenance, leading to a reduction or absence of insulin therapy, and potentially lower the risk of long-term complications. Although the use of immunosuppressive drugs may still be essential for macro-encapsulation devices utilizing a revascularization strategy, the islets can also be transplanted in a more favorable environment than the liver.

3 Stem Cell-Derived β-Cells

Besides the negative effects of a suboptimal transplantation site, donor shortage and the need for lifelong expensive immunosuppressive therapy are also major limitations that restrict CIT to the most severe cases of type 1 diabetes [18]. However, alternative cell sources have gained more and more attention in the last couple of years. Specifically, stem cell-derived β-cells (SCs, either derived from human embryonic stem cells or induced pluripotent stem cells) have gained interest as they allow for the generation of insulin-producing cells via a strict cell culture regimen mimicking β-cell development, and several clinical trials have started to test their clinical potential [23]. The great benefit of culturing SCs is that cell lines and cell aggregate sizes can be standardized, and work can be planned in advance, removing dependency on donor organs, potentially leading to the widespread application of β-cell replacement therapy. However, SCs cannot (yet) be fully differentiated into mature β-cells in vitro, posing a certain risk that still undifferentiated stem cells may form teratomas after transplantation [24]. Important to mention is that no teratoma formation was observed in recent clinical studies, but there are still regulatory concerns [25, 26]. Moreover, given that these SC-derived β-cells are not fully mature, their insulin secretion capacity does not compare to primary islets [27]. This also may lead to the fact that islet delivery devices require a larger cell loading capacity of SC-derived β-cells compared to regular pancreatic islets. In its turn, the immature nature of SCs may also be a benefit, as some have described that SCs may be less sensitive to hypoxia, even though it is currently not clear whether or not this effect is lost upon the maturation of the cells [28, 29].

Very recently, others have published on immune-evasive human islet-like organoids derived from induced pluripotent stem cells [30, 31]. The possibility of transplanting insulin-producing cells without the need for immunosuppressive drugs would be a game-changer in the islet transplantation field, as immunosuppressive drugs are toxic to islets and can have considerable side effects [15, 32]. Especially interesting are future combinations of immune-evasive β-cells and a revascularization approach, which would offer the best of both encapsulation strategies, potentially offering both high viability of islets without the need for immunosuppressive drugs. However, commercial-scale protocols for the manufacturing of SCs are currently still under development and remain challenging [33, 34]. Nevertheless, the great potential of these alternative cell sources, especially when combined with bioengineering approaches, will likely obtain a central role in the future of the field of β-cell replacement strategies.

4 Biomaterials for Encapsulation of SCs

In essence, the same principles for biomaterial selection for islets can be applied to SCs. A biomaterial is "a material intended to interface with biological systems to evaluate, treat, augment or replace any tissue, organ or function of the body" [35]. An extensive evaluation of different biomaterials used for islet encapsulation strategies has been discussed elsewhere [36]. Instead, this chapter will focus on characteristics of biomaterials that influence their biocompatibility and how biomaterials can be adjusted to improve their biocompatibility in light of β-cell replacement therapy. Biocompatibility is defined by The Williams Dictionary of Biomaterials as

"The ability of a material to perform with an appropriate host response in a specific application" [37]. Given that there is a wide variety of different islet encapsulation strategies, it is impossible to give a general indication of how an optimal biocompatible islet encapsulation tissue response should look like. Any implanted biomedical device will face a foreign body response, which starts directly after implantation when plasma proteins cover the medical device [38]. A subsequent acute inflammation starts when monocytes migrate toward the implant and adhere to the surface through interactions with the adsorbed proteins. The monocytes differentiate toward macrophages and release cytokines that regulate the tissue response [39]. Following acute inflammation, chronic inflammation is characterized by the presence of monocytes, macrophages, and foreign body giant cells (fused macrophages) and usually takes about two weeks [38]. Subsequently, granulation tissue, the precursor to fibrous capsule formation, can be identified by the presence of macrophages, the infiltration of fibroblasts, and neovascularization [40]. The final stage of the foreign body response usually starts three to four weeks after device implantation and can be recognized by collagen deposition around the implant, forming a fibrous capsule [41–43]. This fibrous capsule is characterized by an avascular "dense scar plate" with excess collagen fibers aligned parallel to the implant surface [44]. Biomaterial adherent macrophages can secrete proteins that modulate fibrosis and the formation of the fibrous capsule. The fibrous capsule is a common problem for medical devices, and is especially cumbersome for β-cell delivery devices as it leads to long diffusion distances for nutrients and oxygen, finally leading to deprivation and subsequent loss of the islet graft [45]. It is therefore essential to prevent evocation of a strong immune response to realize a positive transplantation outcome.

Traditionally, islets are frequently encapsulated in either hydrogels or seeded into hard polymer-based devices. Hydrogels are a class of biomaterials capable of absorbing and retaining large quantities of water [46]. Hydrogels are polymeric materials that hold a three-dimensional structure due to crosslinks between polymeric fibers. The strength of the network can often be tuned, ranging from soft, almost liquid-like appearances to hard, almost solid materials. Hydrogels can be used to encapsulate cells, by which hydrogels mimic the function of extracellular matrix [47]. An ideal hydrogel would therefore mimic the native environment of the encapsulated cells. Most hydrogels are permeable to oxygen, nutrients, and other water-soluble compounds, making them an interesting option for the fabrication of complex tissue constructs [48]. However, hydrogels also have limitations. Natural hydrogels generally have weak mechanical properties, synthetic hydrogels lack major biological components to allow for cell interactions, while hydrogels are commonly used for immunoprotective islet delivery methods [49]. Varieties of different hydrogels have been tried for islet encapsulation and implantation, with alginate being the leading biomaterial for hydrogel encapsulation [50–52]. There are general issues with hydrogel encapsulation, including biocompatibility and foreign body responses, tuning of biodegradability and stability, and maintaining proper islet viability and insulin release, at the same time, indicating that the optimal hydrogel formulation is yet to be determined [36].

In contrast to hydrogel-based cell delivery devices, hard polymer-based devices are usually made from biomaterials that induce a minimal foreign body response and have a proven track record in other clinical applications. Common biomaterials are for instance poly(glycolic acid) (PLA), poly(lactic acid) (PLA), and their copolymer poly(lactic acid-*co*-glycolic acid) (PLG) [53–59], polydimethylsiloxane (PDMS) [60, 61] polycaprolactone (PCL) [62], expanded polytetrafluoroethylene (ePTFE) [63, 64], polytetrafluoroethylene (PTFE) [65] and block copolymers of poly(ethylene oxide terephthalate) (PEOT) and poly(butylene terephthalate) (PBT) [66]. The main advantage of solid versus hydrogel-based islet delivery devices is their mechanical strength and retrievability after implantation. Both the immunoprotective and revascularization strategies can be implemented with hard polymer-based devices, depending on the device design.

5 Criteria That Influence Biocompatibility of Biomaterials

Multiple criteria are important when selecting biomaterials for SCs encapsulation devices. The importance of these material characteristics may differ between encapsulation strategies, meaning that a biomaterial that displays an appropriate tissue response in an immunoprotective device may not necessarily be the best biomaterial for the device following the revascularization strategy. Among others, the biocompatibility of materials for encapsulation devices will depend on chemical composition, porosity, functionalization, topography and mechanical properties, degradation properties, and compatibility to sterilization techniques, and will be discussed in the following sections and summarized in Figs. 3 and 4. While the different topics discussed in the following sections are applicable to islets and SC-derived β-cells, the basic principle applies to any cell type. However, intrinsic differences between cell types may alter the order of importance of these factors, as for instance SCs are known to be more hypoxia tolerant compared to primary islets [28, 29]. Also important to keep in mind is that the acceptance of the human body toward a biomaterial is not dictated by just one of these biomaterial characteristics but is the result of a complex interplay between all of these parameters at the same time.

5.1 Biomaterial Composition

The surface chemistry of a biomaterial dictates the degree and rate of protein adsorption, which is considered a vital component of the initial host tissue response capable of further activation of the immune system (Fig. 3) [39, 67]. The wettability (hydrophilicity/hydrophobicity) of biomaterials has long been considered an important parameter controlling the attraction of monocytes, macrophages, and lymphocytes [68]. The formation of a layer of water molecules attracted by a hydrophilic surface is believed to reduce protein adsorption, limiting cellular adhesion on these

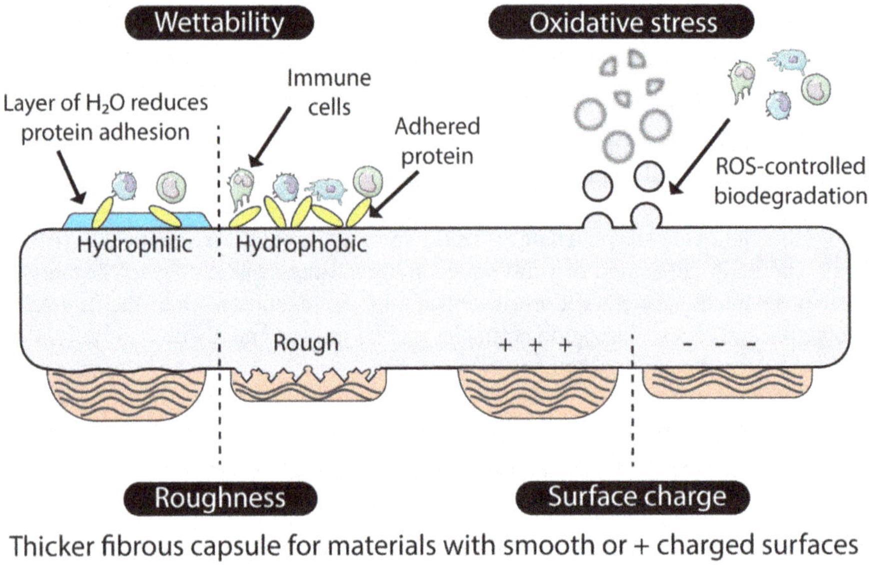

Fig. 3 Schematic overview of the different surface chemistry characteristics that influence the host tissue response of biomaterials. Wettability mainly influences the adherence of proteins, which further facilitates the adherence of immune cells such as macrophages to the biomaterials. The presence of a layer of water molecules is believed to reduce protein adhesion and therefore dampen the immune response. The adhered immune cells try to degrade any foreign body through an oxidative attack of the biomaterial by the release of reactive oxygen species (ROS). While the goal of this attack is to degrade the biomaterial, the surrounding tissue becomes inflamed, exacerbating the immune response to the biomaterial. Moreover, both smooth surfaces and positively charged biomaterials have been reported to induce a thicker fibrous capsule surrounding biomaterials, although some contradicting results have also been reported. These contradictions are likely caused by the fact that each of these four factors influence the host tissue response and are challenging to observe independent from one another

surfaces [69]. However, the hydrophilicity of biomaterials alone is not a good predictor of biological response to biomaterials. Despite the general trend that more proteins are adsorbed on hydrophobic materials, it is essential to consider specific structural features or side groups that influence protein adsorption [70, 71]. For instance, the fibrotic capsule and cell infiltration were compared for microparticles exposing $-CH_3$, CF_x, -OH, $-NH_2$, and –COOH groups [72]. Microparticles coated with –OH and $-NH_2$ groups elucidated a stronger immune response, while $-CF_x$ and –COOH groups elucidated the least inflammatory response. This is also linked to the surface charge of a biomaterial. Materials with positively charged surfaces (cationic surfaces) have been shown to elicit a pro-inflammatory response while negatively charged surfaces (anionic surfaces) have been shown to reduce macrophage adhesion and the formation of foreign body giant cells [73, 74]. However, contradicting results have also been published where cationic surfaces lead to an anti-inflammatory response [75]. The incorporation of zwitterionic groups has been

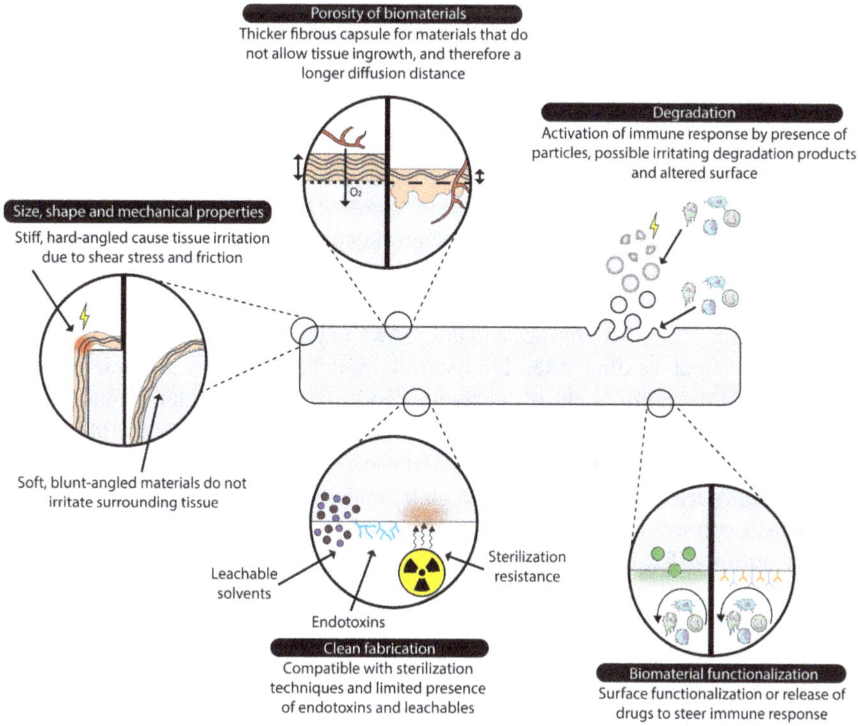

Fig. 4 The influence of biomaterial porosity, degradation, size, shape and mechanical properties, clean fabrication techniques, and functionalization on the local host tissue response. Biomaterials with pores that allow tissue ingrowth generally show a thinner fibrous capsule and short oxygen diffusion distance, while biomaterials with pores that block tissue ingrowth show a relatively thick fibrous capsule and long oxygen diffusion distance. A combination of size, shape, and mechanical properties of cell delivery devices can cause tissue irritation and subsequent activation of an inflammatory response. Clean fabrication techniques are vital, as the presence of leachable solvents and endotoxins can activate the immune system. Sterilization techniques can also alter the physicochemical properties of materials, so the biocompatibility of biomaterials should always be determined after sterilization. The immune response toward biomaterials can be altered by functionalization of the material, either by dampening the immune response through surface functionalization or by local release of immunosuppressive drugs. The degradation products can activate the immune system, especially when these particles are tissue irritating. Moreover, the altered biomaterial surface also triggers activation of the immune system

shown to reduce the inflammatory response, formation of a thinner fibrotic capsule, and shift of macrophages toward the anti-inflammatory M2 phenotype [76].

Next, surface roughness also influences the host tissue response, as it has been shown that proteins adsorb differently on surfaces with a different roughness [77]. The magnitude of the surface roughness is important, as proteins with dimensions that are in the same order as the surface roughness are projected to be conformationally altered upon adsorption [40]. The roughness of the material is then believed to locally alter surface wettability or surface charge, influencing further protein adsorption [40]. In general, it has been shown that smoother surfaces of the same material induce a thicker fibrous capsule compared to rough surfaces [78, 79]. However, important to realize is the fact that rough hydrophobic surfaces are prone to the formation of gas nucleoli within pores, which may initiate an immune response [80].

Finally, an often-overlooked biomaterial characteristic is the indirect induction of oxidative stress of biomaterials on encapsulated cells [81–83]. Oxidative stress occurs when the production of oxidants, mainly reactive oxygen species (ROS), nitrogen species (RNS), and lipid peroxidation (LPO) products, surpasses the antioxidant capacity of cells or tissues and has shown to play a critical role in inflammation, fibrosis, and healing after biomaterial implantation [84]. Especially ROS contributes to the recruitment of leukocytes and macrophages, and stimulation of oxidants often leads to progressive oxidant production, exacerbating the inflammation response [85, 86]. Implanted biomaterials can stimulate oxidant formation as immune cells perform oxidative attacks on biomaterials and their degradation products, which can even lead to chronic inflammation and loss of the biomaterial's biocompatibility [87, 88]. Others have also demonstrated that coating biomaterials with antioxidants diminish the fibrous capsule thickness and recruitment of foreign body giant cells [89]. Besides activation of the immune system, oxidative stress is also an important factor during β-cell transplantation, as β-cells are particularly sensitive to oxidative stress due to their intrinsically low antioxidant capacity [90]. Contributing to the cellular demise are remnants of degraded β-cells, which trigger the immune system, leading to a more severe immune reaction, on top of the already stressed β-cells, increasing the risk of device failure [91]. Screening biomaterials for their anti-oxidative or low oxidative stress-inducing properties is therefore an appealing strategy for the selection of promising biomaterial candidates.

Determining the degree of influence of these surface composition parameters on the host immune response is difficult, as the aforementioned biomaterial properties are not only linked but also act in unison on the surrounding tissue. For instance, surface wettability is influenced by both surface roughness and charge [92]. Another study indicated that the impact of surface roughness depends on the surface wettability. For hydrophilic materials, blood platelet adherence was lower on rough surfaces compared to smooth surfaces, but in contrast, for hydrophobic materials, blood platelet adherence was higher on rough surfaces compared to smooth surfaces [93]. This might also explain why contradicting results are obtained by different research groups, where roughness is sometimes considered to induce either pro-inflammatory or anti-inflammatory responses, as reviewed elsewhere [94]. Surface composition therefore clearly has an influence on the host tissue response, but it is currently not possible to determine what the ideal composition should be. Moreover, it should be noted that the influence of composition of biomaterials between studies is difficult to compare as important material parameters (such as architecture and porosity as discussed in the following sections) often also differ between experiments, complicating the comparison. In an ideal setting, the biocompatibility of biomaterials should be compared by comparing samples that only differ in their chemical composition.

5.2 Porosity of Biomaterials

Already back in 1960, it was reported that the porosity of a biomaterial had a greater impact on the host tissue response than the material composition [95]. In the case of vascular grafts, its biocompatibility, measured by low thrombogenicity and avoidance of calcification over time, is directly related to the tissue ingrowth [44, 95]. In addition, the host tissue reaction of five different PET hernia meshes differing in pore size and pore shape were compared in a porcine model [96]. It turned out that the pore size (ranging between 1–5 mm) and pore shape (either square, diamond, or hexagon) significantly influenced tissue ingrowth and integration in surrounding tissue, while mesh density (g/m^2) had no influence.

In the case of immunoprotective barriers for encapsulation of SCs, the barrier molecular cut-off size dictates the effectiveness of the barrier. Cellular immunity against T-lymphocytes can be realized with a cut-off in the range of several micrometers, while humoral immunity can be achieved by pore radii in the range of 10 nm [15]. Moreover, the host tissue response toward biomaterials leads to the secretion of islet-destructive cytokines, such as tumor necrosis factor (TNF)-α, interleukin (IL)-1β, and interferon (INF)-γ, which have hydrodynamic radii similar to insulin, indicating the difficulty in designing proper barriers for immunoprotective devices [97–102]. Immunoprotective devices prevent tissue ingrowth, which leads to fibrous encapsulation of the device with a relatively thick fibrous layer compared to devices that allow tissue ingrowth [44]. In addition, this dense fibrous layer is avascular, leading to an added barrier for oxygen diffusion (Fig. 4). In contrast, revascularization strategies utilize a macro-porous design that allows for host tissue ingrowth and neovascularization within the device, facilitating direct contact with the encapsulated beta cells [44].

Revascularization strategies often require macro-porous thermoplastic polymers. In the late 1990s, several animal studies were performed with a device consisting of two membranes that were sealed along their edges, except for a loading port [103]. Cell penetration-preventing membranes (0.02 μm pore size) were laminated to membranes with pore sizes ranging between 0.8 and 8 μm to determine the influence of pore size on vascularization ingrowth. Interestingly, outer membranes with a pore size of 5 μm showed 80–100 fold more vasculature structures than membranes with a pore size of 0.02 μm for at least a year in a subcutaneous rat model. Another study using polyurethane device indicated that an increased number of microvessels and a thinner fibrous capsule were detected around porous devices compared to solid devices [43]. More recently, other studies have indicated that pore sizes of 30–40 μm are recommended to obtain maximum stimulation of neovascularization [104, 105]. Others have also indicated that this pore range is associated with relative high numbers of infiltrating macrophages, but that these macrophages hold relative high portion of anti-inflammatory M2 macrophages [106]. Taken together, the pore dimensions and shape can play a key role in both the fibrous layer formation and vascular ingrowth into hard polymer cell delivery device. The absence of tissue ingrowth leads to the formation of a relative thick, avascular fibrous layer that impedes the survival and functionality of encapsulated SCs.

5.3 Topography and Mechanical Properties

Shape is an important biomaterial characteristic that can direct a local tissue response. One study described the foreign body response against medical-grade polymers extruded into rods with circular-, triangular-, and pentagonal-shaped cross-sections [107]. From these geometries, the circular rods evoked the mildest foreign body response, followed by pentagonal rods and then triangular rods. Others have also indicated that a spherical shape is more favorable than a disc shape [108]. In general, biomaterials with sharp edges are therefore considered less biocompatible, but there is no consensus on an ideal geometry [108].

Besides shape, the device size highly influences the local tissue response to biomaterials. Only a limited number of studies have evaluated the influence of biomaterial size on modulating the foreign body response [108–110]. For instance, the size of PLGA microspheres seems to influence the foreign body response, with a higher cellular and macrophage influx and thicker fibrous capsules surrounding small microspheres. No foreign body giant cells were detected for small microspheres (6 μm diameter), and some were detected around large microspheres (30 μm diameter) [109]. In addition, spherical materials that are 1.5 mm or larger in diameter have been shown to be more biocompatible than smaller-sized counterparts for hydrogels, ceramics, metals, and plastics [108]. This effect was independent of the total implanted surface area and visible in all the material types mentioned above. Moreover, silica nanoparticles with diameters between 30–1000 nm induced lysosomal destabilization, cell death, and pro-inflammatory factors compared to nanoparticles with diameters between 3000 and 10,000 nm [110]. It is vital that the encapsulation devices are as thin as possible to reduce the diffusion distance (and potentially vasculature ingrowth distance), as a lack of oxygen would result in the formation of necrotic cores and, ultimately, cell death [11]. The remnants of these dead cells will trigger the immune system, leading to a more severe immune reaction and an increased risk of implant failure [91]. Moreover, the utilization of thinner implants has also been shown to elicit a thinner fibrous capsule [43, 74]. Others have therefore described that subcutaneous implants should preferably have a small width, an ergonomic shape with well-rounded edges, and should be made of a smooth material to prevent wound irritation [111].

Furthermore, it is known that pressure, friction, and shear stress can be imposed on tissue depending on the design and shape of a medical device [42, 111, 112]. A relatively high friction and shear stress may force movement on a biomaterial and lead to a chronic inflammatory response toward the biomaterial [111, 112]. The outcome is suboptimal wound healing resulting in scar tissue formation [113, 114]. Moreover, tissues located at highly motile areas of the body, such as subcutaneous tissue in the extremities, are subjected to higher shear stresses compared to areas with less movement, such as subcutaneous tissue located in the abdomen, indicating that the implantation region can influence the local tissue response to an implantable medical device [42]. Next to biomaterial shape, mechanical properties are a crucial element that can affect the engraftment of a device. Bioengineered cell delivery devices should provide sufficient mechanical stability to allow easy surgical handling and device retrieval when required but be pliable and flexible enough to follow the anatomical contours of our body. However, a mismatch between biomechanical properties and tissue, hard vs soft or stiff vs elastic, can increase inflammation due to continuous mechanical irritation of the tissue [40, 41, 115].

5.4 *Influence of Biomaterial Degradation*

Absorbable or highly degradable biomaterials should temporarily provide structural support while the surrounding tissue is healing after surgery. In this case, the retrievability of such a device is compromised, but the absence of long-term presence of the biomaterial should avoid chronic inflammation, late infection, or possibly pain [44]. Absorbable biomaterials initiate the same foreign body response as non-absorbable biomaterials, with an even more robust early inflammatory response due to the presence of (partially degraded) particles released from the bulk biomaterial [44]. Once a non-degradable biomaterial is removed, its surrounding fibrous capsule disappears [116]. A similar effect is seen once an absorbable biomaterial is dissolved [117]. Degradation can be realized through hydrolytic, oxidation, or enzymatic mechanisms [118]. The immune system can influence degradation as biomaterial adherent macrophages and foreign body giant cells attempt to degrade biomaterials by releasing ROS and degrading enzymes [40, 119, 120]. Both the degradation products and changes in surface characteristics as a result of degradation of the biomaterial can activate the immune system [38]. The properties of degradation products themselves can influence the local immune response. For instance, PLGA, a common degradable biomaterial for tissue engineering [121], has acidic degradation products that are believed to trigger an immune response [76, 122]. Finally, the degradation rate of biomaterials seems to be influenced by their surrounding tissue, as the local composition of immune cells differs for different implantation sites [119]. This implies that the same biomaterial can elucidate different degradation profiles depending on the local tissue composition. One should therefore be conscious of the intended implant location, the degradation profile, and degradation products once selecting a degradable biomaterial.

5.5 *Functionalization of Biomaterials*

Biomaterial properties can be altered, or functionalized, to further improve their tolerability within the human body or to induce a desired tissue reaction. This can be realized by either altering the surface chemistry of biomaterials or by incorporating specific drugs. The idea behind the latter this is that only tissue in the direct surrounding of the biomaterial needs to be influenced rather than systemically inducing tolerance or dampening the entire immune system with for instance immunosuppressive drugs like tacrolimus or rapamycin [123]. Systemic administration of these factors is usually associated with off-target effects that result in unwanted side effects [124]. Local delivery of immunomodulatory factors is therefore hypothesized to improve long-term clinical outcome, while not subjecting a patient to systemic therapy.

Altering the biomaterial surface is a viable strategy to improve its biocompatibility, as the surface chemistry of a biomaterial dictates the degree of protein adsorption, as explained above. Immune-modulating surface chemistries can be engineered through numerous different techniques including non-covalent deposition, adsorption of biomolecules, or covalent cross-linking of functional groups such as silanes, thiols, or specific biomolecules [125, 126].

Immune cells involved in the allogenic immune reaction can be regulated by using hydrogel microcapsules functionalized with Fas ligand (FasL) coupled to streptavidin (SA-FasL). FasL can extend allogeneic islet survival and allograft tolerance by inducing apoptosis in T effector cells supported by T_{reg} cells [127]. When combined with acute rapamycin treatment, the FasL microcapsules facilitated graft survival over 100 days and reversed diabetes in mice [127]. Most interesting, naked human islets that were co-transplanted with SA-FasL presenting hydrogel microcapsules have shown robust glycemic control over 6 months in immunocompetent non-human primates [128]. Another strategy to regulate the immune response is to use biotinylated poly (ethylene glycol) (PEG) microcapsules conjugated with programmed death-ligand 1 (PD-L1) coupled via streptavidin which can increase T_{reg} numbers locally, resulting in improved islet survival in BALB/c mice [129].

Alternatively, heparin has anti-inflammatory properties by blocking the cell adhesion molecule P-selectin on activated platelets [130]. Heparin is highly soluble in water, so it can be deposited onto a suitable substrate by dipping the sample in an aqueous heparin solution. However, P-selectin is highly expressed in the early and middle stages of inflammation but not in late-stage inflammation, leading to loss of immunomodulatory effects over time [130]. In addition, non-covalent deposition of biomolecules on a substrate depends on the local Van der Waals forces as well as electrostatic and hydrophobic interactions [131]. The challenge is that poor affinity of heparin to a biomaterial results in detachment of heparin from the biomaterial surface, and diffusion into the surrounding tissue, losing its positive effect [131]. It may therefore be required to modify a biomaterial to enable non-covalently long-term binding of heparin and other biomolecules to obtain a sustained effect [132].

A specific class of surface modifications includes coatings with extracellular matrix (ECM) molecules. The predicted benefits of ECM are twofold: (1) ECM plays a role in islet cell mass preservation as disconnecting β-cells from islet ECM may induce apoptosis, and (2) the role of ECM in maintaining β-cell function [11, 133]. Given that remnants of dead cells trigger the immune system and exacerbate an inflammatory response [91], one strategy to improve the biocompatibility of a biomaterial is to promote islet survival by creating a supportive microenvironment. Cell delivery devices supplemented with ECM molecules have been shown to enhance the function and survival of pancreatic islets via mechanisms that are largely unknown [134–139]. Interestingly, only the outer perimeter of islets needs to be in contact with ECM in order to improve islet viability and functionality [140]. Numerous studies have shown the positive effect of co-culturing islets with (combinations of) ECM compounds such as laminin, vitronectin, collagen I, collagen IV, fibronectin, and fibrinogen in cell culture, but there is yet no consensus on what the optimal ECM composition should be, as was excellently reviewed elsewhere [133].

Instead of surface modification of biomaterials, incorporation of a compound in the bulk biomaterial is another viable option that can realize local release of immunomodulatory drugs over time. Transforming growth factor-β1 (TGF-β1) is an anti-inflammatory cytokine that, however, can have adverse effects on graft fibrosis when administered systemically [141]. TGF-β1 has a versatile role within the immune system, making it an interesting target for drug delivery [123]. Microporous scaffolds with bulk incorporation of TGF-β1 have been shown to mitigate the immune response toward encapsulated islets [142]. Dexamethasone is another attractive compound due to its ability to promote the anti-inflammatory M2 phenotype of macrophages [143]. The dexamethasone concentration has to be moderated, as excessive dosages of the compound can impair survival and functionality of islets [143]. Dexamethasone has been incorporated into several cell encapsulation strategies, including PDMS devices, PLG micro-micelles, and nanofiber gels [143–145]. Another example is tacrolimus, an immunosuppressive drug often systemically used for islet transplantation to impede T-cell activation, which has been incorporated into the bulk of degradable PLG microparticles that were incorporated into Matrigel [146].

The duration of most local delivery strategies is limited to 30–60 days, which only leads to a delay in the onset of a fibrotic response [147]. An interesting alternative was developed by Farah et al., where a controlled-drug delivery system was designed based on the crystal formation of drugs targeted toward colony-stimulating factor-1 receptor (CSF1R) expressed in monocytes and macrophages [147]. The crystals were demonstrated to suppress the foreign body response for up to 1.3 years in rodents and 6 months in non-human primates, extensively allowing islets to engraft and revascularize [147]. Altogether, biomaterial functionalization approaches have shown efficacy in small animal models, but there are still limitations to address. This includes depletion of the incorporated compound over time and/or the possibility of only influencing a part of the immune response [123]. These limitations will have to be addressed before continuing to clinical testing. In addition, biomaterial functionalization can only do so much, and the basic principle of the device should already support islet survival and functionality.

5.6 Influence of Sterilization Techniques and Clean Fabrication

To ensure the safety of medical devices, one should prove that devices are sterile and free of endotoxins and residual solvents [148]. Medical devices should be free from micro-organisms to prevent infection of tissues and unwanted host tissue responses after device implantation. Numerous sterilization techniques are currently available, of which steam sterilization (autoclavation), gamma irradiation, electron beam sterilization, and ethylene oxide treatment are the most commonly applied [149]. However, sterilization techniques can also significantly alter biomaterial properties. For instance, commonly used clinically approved thermoplastic polymers are known to be modified by gamma irradiation [150]. Depending, among others, on the polymer structure, the dose, and duration of gamma irradiation, polymer chains can be ionized by gamma irradiation, leading either to chain crosslinking or scission of covalent bonds, altering the physicochemical properties of polymers [151]. It is therefore vital to determine the foreign body response of biomaterials after they have been treated with a specific sterilization technique to facilitate the translation into future clinical trials. Besides sterility, endotoxins should be limited as much as possible during medical device fabrication. Parts of the outer membrane of gram-negative bacteria hold endotoxins, which are released in substantial amounts during the destruction of bacterial cell walls by either antibiotics or the host immune system [152]. Endotoxins can strongly interact with most biomaterials and represent an important issue, as small amounts of endotoxin can elicit a strong inflammatory reaction that may overrule any biological effect of the biomaterial itself [152]. Another important aspect of device safety is residual solvents. Manufacturing of medical devices often involves the use of solvents, but some residual solvents may be left behind in the end product. Solvents are often toxic, have no therapeutic importance, and could thus negatively affect the quality of a medical device, and should therefore be limited as much as possible [153]. Altogether, the robustness of a biomaterial against sterilization techniques and preventing the inclusion of endotoxins and residual solvents in fabrication procedures are important aspects that influence the selection of suitable biomaterials and could determine the biosafety of a cell delivery device.

5.7 *Graft Monitoring*

Traditionally, the function of transplanted islets can be determined through several methods. Hepatically delivered islets can be monitored through intraportal needle biopsies, which have a limited practical value because of low islet sampling rates [154]. The mean fluctuations in serum glucose concentrations measure glucose metabolic instability and have been shown to decrease greatly upon islet transplantation [2]. Evidently, insulin therapy requirement is linked with glucose homeostasis and graft function. However, it fails to represent an early graft rejection [155]. Instead, C-peptide levels are conventionally determined as both C-peptide and insulin are produced in a 1:1 ratio from the proinsulin molecule in pancreatic β-cells. This allows the determination of graft function even when a patient is treated with exogenous insulin. Another common marker for graft function is hemoglobin A1c (HbA1c), a glycated form of hemoglobin whose frequency is directly proportional to blood sugar levels. Most importantly, it is a reliable long-term marker of blood sugar levels as it lasts longer than maximum blood sugar levels and has therefore become the benchmark for assessing long-term glycemic control [156]. Because of the cumulative nature of the HbA1c marker, it does not provide valuable information on early graft rejection [155]. The functionality of the islet graft can be evaluated by challenging the graft with a glucose bolus through either an oral glucose tolerance test (OGTT) or intravenous glucose tolerance test (IVGTT), but these interventions are not routinely performed in the clinic [155]. Instead, mixed meal tolerance tests are routinely used to determine insulin secretion, but they are intensive and invasive to perform [157].

The current lack of suitable non-invasive methods to monitor graft viability and function in the liver is a major limitation of clinical islet transplantation [158]. In recent years, research has focused on innovative imaging methods to track islet morphology, location, distribution, and function to predict and track islet graft survival in vivo at the transplant location. Clinically applicable imaging techniques that can track transplanted islets include high-frequency ultrasonography (US), positron emission tomography (PET), magnetic resonance imaging (MRI), X-ray, computed tomography (CT), and single-photon emission tomography (SPECT), as has been excellently reviewed elsewhere [158–161]. The detection sensitivity, toxicity, and long-term stability of imaging labels are important factors to consider for a successful translation of these techniques into the clinics. Currently, imaging modalities can detect the location and distribution of islet grafts but are limited to the very early post-transplantation period and can only visualize clusters of islets due to limits in resolution [162]. Islet imaging and tracking could play an important role in detecting (early) post-transplantation complications and is therefore an important parameter to consider when designing a bioengineering approach and selecting potential implantation sites. Moreover, it is also important for the selection of suitable biomaterials, as they should preferably not hinder the imaging techniques used. Some biomaterials have intrinsic radioactive properties, while in other cases, radiopaque markers can be mixed into biomaterials without influencing their biocompatibility [163, 164]. Evaluation of biomaterials that hold (intrinsic) radioactive properties in future screenings may lead to biomaterials that are both biocompatible and easy to monitor after transplantation.

6 Conclusions

Type 1 diabetes is a life-threatening disease distinguished by an autoimmune response toward insulin-producing β-cells within pancreatic islets. A β-cell replacement strategy following a bioengineering approach through encapsulation of islets, or stem cell-derived β-cells, in cell delivery devices can be used to support cells or offer novel implantation sites. However, the selection of biomaterials for cell delivery devices that evoke an acceptable host immune response is challenging due to the high complexity of the immune system. There is a multitude of biomaterial characteristics that influence the immune system. The importance of some of these features may differ between encapsulation strategies. For instance, a biomaterial that displays an appropriate tissue response in an immunoprotective device may not necessarily be the most ideal biomaterial for devices that follow a revascularization strategy, and vice versa. Important biomaterial characteristics include their chemical composition, porosity, topography, mechanical properties, degradation characteristics, and functionalization. Furthermore, the acceptance of the human body toward a biomaterial is not dictated by just one of these parameters but is the result of a complex interplay between all these parameters at the same time. Besides these material characteristics, it is highly important that fabrication procedures of these biomaterials are as clean as possible to guarantee sterility and minimize the presence of endotoxins and residual solvents. Finally, it is recommended that the field of cell delivery devices should also focus on biomaterials that allow non-invasive monitoring of encapsulated cells or be tailored in such a way that non-invasive monitoring is possible. Collectively, this chapter highlights the importance and large variety of biomaterial characteristics that influence the biocompatibility and suitability of biomaterials used for β-cell replacement therapy.

References

1. Ricordi, C., *Islet transplantation: a brave new world.* Diabetes, 2003. **52**(7): p. 1595–603.
2. Shapiro , A.M.J., et al., *Islet Transplantation in Seven Patients with Type 1 Diabetes Mellitus Using a Glucocorticoid-Free Immunosuppressive Regimen.* New England Journal of Medicine, 2000. **343**(4): p. 230–238.
3. Vantyghem, M.C., et al., *Ten-Year Outcome of Islet Alone or Islet After Kidney Transplantation in Type 1 Diabetes: A Prospective Parallel-Arm Cohort Study.* Diabetes Care, 2019. **42**(11): p. 2042–2049.
4. Eich, T., et al., *Positron Emission Tomography: A Real-Time Tool to Quantify Early Islet Engraftment in a Preclinical Large Animal Model.* Transplantation, 2007. **84**(7): p. 893–898.
5. Biarnes, M., et al., *Beta-cell death and mass in syngeneically transplanted islets exposed to short- and long-term hyperglycemia.* Diabetes, 2002. **51**(1): p. 66–72.
6. Davalli, A.M., et al., *Vulnerability of islets in the immediate posttransplantation period. Dynamic changes in structure and function.* Diabetes, 1996. **45**(9): p. 1161–7.
7. Carlsson, P.-O., *Influence of microenvironment on engraftment of transplanted β-cells.* Upsala Journal of Medical Sciences, 2011. **116**(1): p. 1–7.
8. Delaune, V., et al., *Intraportal islet transplantation: the impact of the liver microenvironment.* Transpl Int, 2017. **30**(3): p. 227–238.
9. Morini, S., et al., *Revascularization and remodelling of pancreatic islets grafted under the kidney capsule.* Journal of Anatomy, 2007. **210**(5): p. 565–577.
10. Olsson, R., et al., *Increased numbers of low-oxygenated pancreatic islets after intraportal islet transplantation.* Diabetes, 2011. **60**(9): p. 2350–2353.
11. Komatsu, H., et al., *Oxygen environment and islet size are the primary limiting factors of isolated pancreatic islet survival.* PLOS ONE, 2017. **12**(8): p. e0183780.
12. Buchwald, P., *A local glucose-and oxygen concentration-based insulin secretion model for pancreatic islets.* Theor Biol Med Model, 2011. **8**: p. 20.
13. Dionne, K.E., C.K. Colton, and M.L. Yarmush, *Effect of hypoxia on insulin secretion by isolated rat and canine islets of Langerhans.* Diabetes, 1993. **42**(1): p. 12–21.
14. Dionne, K.E., C.K. Colton, and M. Lyarmush, *Effect of Hypoxia on Insulin Secretion by Isolated Rat and Canine Islets of Langerhans.* Diabetes, 1993. **42**(1): p. 12–21.

15. Barkai, U., A. Rotem, and P. de Vos, *Survival of encapsulated islets: More than a membrane story.* World J Transplant, 2016. **6**(1): p. 69–90.
16. Ernst, A.U., et al., *Nanotechnology in cell replacement therapies for type 1 diabetes.* Adv Drug Deliv Rev, 2019. **139**: p. 116–138.
17. Opara, A., et al., *Islet cell encapsulation – Application in diabetes treatment.* Experimental Biology and Medicine, 2021. **246**(24): p. 2570–2578.
18. Desai, T. and L.D. Shea, *Advances in islet encapsulation technologies.* Nature Reviews Drug Discovery, 2017. **16**(5): p. 338–350.
19. Farina, M., et al., *Cell encapsulation: Overcoming barriers in cell transplantation in diabetes and beyond.* Advanced Drug Delivery Reviews, 2019. **139**: p. 92–115.
20. Gaber, A.O. and D. Fraga, *Advances in long-term islet culture.* Cell Biochemistry and Biophysics, 2004. **40**(3): p. 49–54.
21. Krogh, A., *The supply of oxygen to the tissues and the regulation of the capillary circulation.* The Journal of physiology, 1919. **52**(6): p. 457–474.
22. Place, T.L., F.E. Domann, and A.J. Case, *Limitations of oxygen delivery to cells in culture: An underappreciated problem in basic and translational research.* Free Radical Biology and Medicine, 2017. **113**: p. 311–322.
23. de Klerk, E. and M. Hebrok, *Stem Cell-Based Clinical Trials for Diabetes Mellitus.* Frontiers in endocrinology, 2021. **12**: p. 631463–631463.
24. Lee, A.S., et al., *Tumorigenicity as a Clinical Hurdle for Pluripotent Stem Cell Therapies.* Nature medicine, 2013. **19**(8): p. 998–1004.
25. Hentze, H., et al., *Teratoma formation by human embryonic stem cells: evaluation of essential parameters for future safety studies.* Stem Cell Res, 2009. **2**(3): p. 198–210.
26. Ramzy, A., et al., *Implanted pluripotent stem-cell-derived pancreatic endoderm cells secrete glucose-responsive C-peptide in patients with type 1 diabetes.* Cell Stem Cell, 2021. **28**(12): p. 2047–2061.e5.
27. Kieffer, Timothy J., *Closing in on Mass Production of Mature Human Beta Cells.* Cell Stem Cell, 2016. **18**(6): p. 699–702.
28. Hyder, A., C. Laue, and J. Schrezenmeir, *Variable Responses of Islet Cells of Different Ages and Species to Hypoxia 11Supported by the grant of "Stiftung Inseltransplantation" to Ch. Laue and by the Egyptian government grant to A. Hyder.* Transplantation Proceedings, 1998. **30**(2): p. 578–580.
29. Emamaullee, J.A., et al., *Neonatal Porcine Islets Exhibit Natural Resistance to Hypoxia-Induced Apoptosis.* Transplantation, 2006. **82**(7): p. 945–952.
30. Tahbaz, M. and E. Yoshihara, *Immune Protection of Stem Cell-Derived Islet Cell Therapy for Treating Diabetes.* Frontiers in Endocrinology, 2021. **12**.
31. Yoshihara, E., et al., *Immune-evasive human islet-like organoids ameliorate diabetes.* Nature, 2020. **586**(7830): p. 606–611.
32. Johnson, J.D., et al., *Different Effects of FK506, Rapamycin, and Mycophenolate Mofetil on Glucose-Stimulated Insulin Release and Apoptosis in Human Islets.* Cell Transplantation, 2009. **18**(8): p. 833–845.
33. Lee, B., et al., *Challenges and Solutions for Commercial Scale Manufacturing of Allogeneic Pluripotent Stem Cell Products.* Bioengineering (Basel, Switzerland), 2020. **7**(2): p. 31.
34. Ellis, C., J.G. Lyon, and G.S. Korbutt, *Optimization and Scale-up Isolation and Culture of Neonatal Porcine Islets: Potential for Clinical Application.* Cell Transplantation, 2016. **25**(3): p. 539–547.
35. O'Brien, F.J., *Biomaterials & scaffolds for tissue engineering.* Materials Today, 2011. **14**(3): p. 88–95.
36. de Vries, R., et al., *Chapter 33 - Bioengineering, biomaterials, and β-cell replacement therapy*, in *Transplantation, Bioengineering, and Regeneration of the Endocrine Pancreas*, G. Orlando, et al., Editors. 2020, Academic Press. p. 461–486.
37. Williams, D.F., *The Williams Dictionary of Biomaterials.* 1999: Liverpool University Press.
38. Anderson, J.M., A. Rodriguez, and D.T. Chang, *FOREIGN BODY REACTION TO BIOMATERIALS.* Seminars in immunology, 2008. **20**(2): p. 86–100.

39. Anderson, J.M., *Biological Responses to Materials.* Annual Review of Materials Research, 2001. **31**(1): p. 81–110.
40. Klopfleisch, R. and F. Jung, *The pathology of the foreign body reaction against biomaterials.* Journal of Biomedical Materials Research Part A, 2017. **105**(3): p. 927–940.
41. Sheikh, Z., et al., *Macrophages, Foreign Body Giant Cells and Their Response to Implantable Biomaterials.* Materials (Basel, Switzerland), 2015. **8**(9): p. 5671–5701.
42. Ibrahim, M., et al., *Characterization of the Foreign Body Response to Common Surgical Biomaterials in a Murine Model.* Eur J Plast Surg, 2017. **40**(5): p. 383–392.
43. Ward, W.K., et al., *The effect of microgeometry, implant thickness and polyurethane chemistry on the foreign body response to subcutaneous implants.* Biomaterials, 2002. **23**(21): p. 4185–4192.
44. Jordan, S.W., et al., *Implant Porosity and the Foreign Body Response.* Plastic and Reconstructive Surgery, 2018. **141**(1): p. 103e–112e.
45. Cayabyab, F., L.R. Nih, and E. Yoshihara, *Advances in Pancreatic Islet Transplantation Sites for the Treatment of Diabetes.* Frontiers in Endocrinology, 2021. **12**.
46. Ozbolat, I.T., *Scaffold-Based or Scaffold-Free Bioprinting: Competing or Complementing Approaches?* Journal of Nanotechnology in Engineering and Medicine, 2015. **6**(2): p. 024701-024701-6.
47. Ahmed, E.M., et al., *An innovative method for preparation of nanometal hydroxide superabsorbent hydrogel.* Carbohydr Polym, 2013. **91**(2): p. 693–8.
48. Thomas, B.H., et al., *Hydrophilic-hydrophobic hydrogels for cartilage replacement.* J Mech Behav Biomed Mater, 2009. **2**(6): p. 588–95.
49. Zhu, J. and R.E. Marchant, *Design properties of hydrogel tissue-engineering scaffolds.* Expert Rev Med Devices, 2011. **8**(5): p. 607–26.
50. Cantarelli, E., A. Citro, and L. Piemonti, *Pancreatic Islet Transplantation Technologies: State of the Art of Micro- and Macro-Encapsulation.* CURRENT TRANSPLANTATION REPORTS, 2017. **4**(3): p. 169–183.
51. Vaithilingam, V., S. Bal, and B.E. Tuch, *Encapsulated Islet Transplantation: Where Do We Stand?* The review of diabetic studies : RDS, 2017. **14**(1): p. 51–78.
52. de Vos, P., et al., *Alginate-based microcapsules for immunoisolation of pancreatic islets.* Biomaterials, 2006. **27**(32): p. 5603–5617.
53. Dufour, J.M., et al., *Development of an ectopic site for islet transplantation, using biodegradable scaffolds.* Tissue Eng, 2005. **11**(9-10): p. 1323–31.
54. Salvay, D.M., et al., *Extracellular matrix protein-coated scaffolds promote the reversal of diabetes after extrahepatic islet transplantation.* Transplantation, 2008. **85**(10): p. 1456–64.
55. Yap, W.T., et al., *Collagen IV-modified scaffolds improve islet survival and function and reduce time to euglycemia.* Tissue Eng Part A, 2013. **19**(21-22): p. 2361–72.
56. Blomeier, H., et al., *Polymer Scaffolds as Synthetic Microenvironments for Extrahepatic Islet Transplantation.* Transplantation, 2006. **82**(4): p. 452–459.
57. Hlavaty, K.A., et al., *Enhancing human islet transplantation by localized release of trophic factors from PLG scaffolds.* Am J Transplant, 2014. **14**(7): p. 1523–32.
58. Kheradmand, T., et al., *Permanent Protection of PLG Scaffold Transplanted Allogeneic Islet Grafts in Diabetic Mice Treated with ECDI-fixed Donor Splenocyte Infusions.* Biomaterials, 2011. **32**(20): p. 4517–4524.
59. Gibly, R., et al., *Porous Scaffolds Support Extrahepatic Human Islet Transplantation, Engraftment, and Function in Mice.* Vol. 22. 2012.
60. Brady, A.-C., et al., *Proangiogenic Hydrogels Within Macroporous Scaffolds Enhance Islet Engraftment in an Extrahepatic Site.* Tissue Engineering. Part A, 2013. **19**(23-24): p. 2544–2552.
61. Pedraza, E., et al., *Macroporous Three Dimensional PDMS Scaffolds for Extrahepatic Islet Transplantation.* Cell transplantation, 2013. **22**(7): p. 1123–1135.
62. Nyitray, C.E., et al., *Polycaprolactone Thin-Film Micro- and Nanoporous Cell-Encapsulation Devices.* ACS Nano, 2015. **9**(6): p. 5675–5682.

63. Sörenby, A.K., et al., *Macroencapsulation protects against sensitization after allogeneic islet transplantation in rats.* Transplantation, 2006. **82**(3): p. 393–7.
64. Sörenby, A.K., et al., *Preimplantation of an immunoprotective device can lower the curative dose of islets to that of free islet transplantation: studies in a rodent model.* Transplantation, 2008. **86**(2): p. 364–6.
65. Magisson, J., et al., *Safety and function of a new pre-vascularized bioartificial pancreas in an allogeneic rat model.* Journal of Tissue Engineering, 2020. **11**: p. 2041731420924818.
66. Buitinga, M., et al., *Micro-fabricated scaffolds lead to efficient remission of diabetes in mice.* Biomaterials, 2017. **135**: p. 10–22.
67. Major, M.R., et al., *The Foreign Body Response: At the Interface of Surgery and Bioengineering.* Plastic and Reconstructive Surgery, 2015. **135**(5): p. 1489–1498.
68. MacEwan, M.R., et al., *Monocyte/lymphocyte interactions and the foreign body response: In vitro effects of biomaterial surface chemistry.* Journal of Biomedical Materials Research Part A, 2005. **74A**(3): p. 285–293.
69. Harding, J.L. and M.M. Reynolds, *Combating medical device fouling.* Trends in Biotechnology, 2014. **32**(3): p. 140–146.
70. Alexander, M.R. and P. Williams, *Water contact angle is not a good predictor of biological responses to materials.* Biointerphases, 2017. **12**(2): p. 02c201.
71. Vogler, E.A., *Protein adsorption in three dimensions.* Biomaterials, 2012. **33**(5): p. 1201–1237.
72. Kamath, S., et al., *Surface chemistry influences implant-mediated host tissue responses.* J Biomed Mater Res A, 2008. **86**(3): p. 617–26.
73. Khandwekar, A. and C.K. Rho, *Modulation of cellular responses on engineered polyurethane implants.* J Biomed Mater Res A, 2012. **100**(9): p. 2211–22.
74. Goswami, D., et al., *Design Considerations for Macroencapsulation Devices for Stem Cell Derived Islets for the Treatment of Type 1 Diabetes.* Advanced Science, 2021. **8**(16): p. 2100820.
75. Kakizawa, Y., et al., *Precise manipulation of biophysical particle parameters enables control of proinflammatory cytokine production in presence of TLR 3 and 4 ligands.* Acta Biomater, 2017. **57**: p. 136–145.
76. Mariani, E., et al., *Biomaterials: Foreign Bodies or Tuners for the Immune Response?* International journal of molecular sciences, 2019. **20**(3): p. 636.
77. den Braber, E.T., et al., *Orientation of ECM protein deposition, fibroblast cytoskeleton, and attachment complex components on silicone microgrooved surfaces.* J Biomed Mater Res, 1998. **40**(2): p. 291–300.
78. Rosengren, A., et al., *Tissue reactions to polyethylene implants with different surface topography.* J Mater Sci Mater Med, 1999. **10**(2): p. 75–82.
79. Anderson, J.M. and A.K. McNally, *Biocompatibility of implants: lymphocyte/macrophage interactions.* Seminars in Immunopathology, 2011. **33**(3): p. 221–233.
80. Ward, C.A., et al., *Reduction in platelet adhesion to biomaterials by removal of gas nuclei.* Trans Am Soc Artif Intern Organs, 1974. **20a**: p. 77–84.
81. Sthijns, M., et al., *Oxidative stress in pancreatic alpha and beta cells as a selection criterion for biocompatible biomaterials.* Biomaterials, 2021. **267**: p. 120449.
82. Sthijns, M., et al., *The response of three-dimensional pancreatic alpha and beta cell co-cultures to oxidative stress.* PLoS One, 2022. **17**(3): p. e0257578.
83. Eguchi, N., et al., *Anti-Oxidative Therapy in Islet Cell Transplantation.* Antioxidants (Basel), 2022. **11**(6).
84. Mouthuy, P.-A., et al., *Biocompatibility of implantable materials: An oxidative stress viewpoint.* Biomaterials, 2016. **109**: p. 55–68.
85. Wattamwar, P.P. and T.D. Dziubla, *Modulation of the Wound Healing Response Through Oxidation Active Materials*, in *Engineering Biomaterials for Regenerative Medicine: Novel Technologies for Clinical Applications*, S.K. Bhatia, Editor. 2012, Springer New York: New York, NY. p. 161–192.
86. Lugrin, J., et al., *The role of oxidative stress during inflammatory processes.* Biol Chem, 2014. **395**(2): p. 203–30.

87. Ozmen, I., M. Naziroglu, and R. Okutan, *Comparative study of antioxidant enzymes in tissues surrounding implant in rabbits.* Cell Biochem Funct, 2006. **24**(3): p. 275–81.
88. Lin, T.H., et al., *Chronic inflammation in biomaterial-induced periprosthetic osteolysis: NF-κB as a therapeutic target.* Acta Biomater, 2014. **10**(1): p. 1–10.
89. Udipi, K., et al., *Modification of inflammatory response to implanted biomedical materials in vivo by surface bound superoxide dismutase mimics.* J Biomed Mater Res, 2000. **51**(4): p. 549–60.
90. Gerber, P.A. and G.A. Rutter, *The Role of Oxidative Stress and Hypoxia in Pancreatic Beta-Cell Dysfunction in Diabetes Mellitus.* Antioxid Redox Signal, 2017. **26**(10): p. 501–518.
91. Groot, M.d., et al., *Response of Encapsulated Rat Pancreatic Islets to Hypoxia.* Cell Transplantation, 2003. **12**(8): p. 867–875.
92. Puah, L.S., et al., *Influence of Surface Charge on Wetting Kinetics.* Langmuir, 2010. **26**(22): p. 17218–17224.
93. Zingg, W., et al., *Platelet adhesion to smooth and rough hydrophobia and hydrophilic surfaces under conditions of static exposure and laminar flow.* Biomaterials, 1981. **2**(3): p. 156–158.
94. Abaricia, J.O., et al., *Control of innate immune response by biomaterial surface topography, energy, and stiffness.* Acta Biomaterialia, 2021. **133**: p. 58–73.
95. Wesolowski, S.A., et al., *Porosity: primary determinant of ultimate fate of synthetic vascular grafts.* Surgery, 1961. **50**: p. 91–6.
96. Lake, S.P., et al., *Pore size and pore shape--but not mesh density--alter the mechanical strength of tissue ingrowth and host tissue response to synthetic mesh materials in a porcine model of ventral hernia repair.* J Mech Behav Biomed Mater, 2015. **42**: p. 186–97.
97. Montolio, M., et al., *Interleukin-1beta and inducible form of nitric oxide synthase expression in early syngeneic islet transplantation.* J Endocrinol, 2007. **192**(1): p. 169–77.
98. Cetkovic-Cvrlje, M. and D.L. Eizirik, *TNF-α and IFN-γ potentiate the deleterious effects of IL-1β on mouse pancreatic islets mainly via generation of nitric oxide.* Cytokine, 1994. **6**(4): p. 399–406.
99. Jensen, S.S., et al., *Insulin diffusion and self-association characterized by real-time UV imaging and Taylor dispersion analysis.* Journal of Pharmaceutical and Biomedical Analysis, 2014. **92**: p. 203–210.
100. Lu C, D.R., Li L., *The effect of interleukin-1 (IL-1) concentration on single cell NF-kB activation in a gradient-generating microfluidic device.* 2011.
101. Uthaiah, R.C., et al., *IIGP1, an Interferon-γ-inducible 47-kDa GTPase of the Mouse, Showing Cooperative Enzymatic Activity and GTP-dependent Multimerization.* Journal of Biological Chemistry, 2003. **278**(31): p. 29336–29343.
102. Oliva, A., J. Fariña, and M.a. Llabrés, *Development of two high-performance liquid chromatographic methods for the analysis and characterization of insulin and its degradation products in pharmaceutical preparations.* Journal of Chromatography B: Biomedical Sciences and Applications, 2000. **749**(1): p. 25–34.
103. Brauker, J.H., et al., *Neovascularization of synthetic membranes directed by membrane microarchitecture.* Journal of Biomedical Materials Research, 1995. **29**(12): p. 1517–1524.
104. Sharkawy, A.A., et al., *Engineering the tissue which encapsulates subcutaneous implants. II. Plasma-tissue exchange properties.* J Biomed Mater Res, 1998. **40**(4): p. 586–97.
105. Madden, L.R., et al., *Proangiogenic scaffolds as functional templates for cardiac tissue engineering.* Proc Natl Acad Sci U S A, 2010. **107**(34): p. 15211–6.
106. Bryers, J.D., C.M. Giachelli, and B.D. Ratner, *Engineering biomaterials to integrate and heal: the biocompatibility paradigm shifts.* Biotechnol Bioeng, 2012. **109**(8): p. 1898–911.
107. Matlaga, B.F., L.P. Yasenchak, and T.N. Salthouse, *Tissue response to implanted polymers: The significance of sample shape.* Journal of Biomedical Materials Research, 1976. **10**(3): p. 391–397.
108. Veiseh, O., et al., *Size- and shape-dependent foreign body immune response to materials implanted in rodents and non-human primates.* Nature materials, 2015. **14**(6): p. 643–651.
109. Zandstra, J., et al., *Microsphere size influences the foreign body reaction.* Eur Cell Mater, 2014. **28**: p. 335–47.

110. Kusaka, T., et al., *Effect of silica particle size on macrophage inflammatory responses.* PLoS One, 2014. **9**(3): p. e92634.
111. Audouard, E., et al., *Optimized Protocol for Subcutaneous Implantation of Encapsulated Cells Device and Evaluation of Biocompatibility.* Frontiers in bioengineering and biotechnology, 2021. **9**: p. 620967–620967.
112. Gefen, A., et al., *Device-related pressure ulcers: SECURE prevention.* Journal of Wound Care, 2020. **29**(Sup2a): p. S1–S52.
113. Rosen RD, M.B., *Wound Dehiscence.* 2022, Tresure Island (FL): StatPearls Publishing.
114. Chun, J.J., et al., *Causes of Surgical Wound Dehiscence: A Multicenter Study.* J Wound Manag Res, 2018. **14**(2): p. 74–79.
115. Hilborn, J. and L.M. Bjursten, *A new and evolving paradigm for biocompatibility.* J Tissue Eng Regen Med, 2007. **1**(2): p. 110–9.
116. Thomson, H.G., *The fate of the pseudosheath pocket around silicone implants.* Plast Reconstr Surg, 1973. **51**(6): p. 667–71.
117. Delbeke, L.O., et al., *Histologic reaction to four synthetic microsutures in the rabbit.* Fertil Steril, 1983. **40**(2): p. 248–52.
118. Tamaríz, E. and A. Rios-Ramirez. *Biodegradation of Medical Purpose Polymeric Materials and Their Impact on Biocompatibility.* 2013.
119. Reid, B., et al., *PEG hydrogel degradation and the role of the surrounding tissue environment.* J Tissue Eng Regen Med, 2015. **9**(3): p. 315–8.
120. Antmen, E., N.E. Vrana, and V. Hasirci, *The role of biomaterials and scaffolds in immune responses in regenerative medicine: macrophage phenotype modulation by biomaterial properties and scaffold architectures.* Biomaterials Science, 2021. **9**(24): p. 8090–8110.
121. Stratton, S., et al., *Bioactive polymeric scaffolds for tissue engineering.* Bioact Mater, 2016. **1**(2): p. 93–108.
122. Nair, A. and L. Tang, *Influence of scaffold design on host immune and stem cell responses.* Semin Immunol, 2017. **29**: p. 62–71.
123. Wang, X., et al., *Local Immunomodulatory Strategies to Prevent Allo-Rejection in Transplantation of Insulin-Producing Cells.* Advanced Science, 2021. **8**(17): p. 2003708.
124. Dang, T.T., et al., *Spatiotemporal effects of a controlled-release anti-inflammatory drug on the cellular dynamics of host response.* Biomaterials, 2011. **32**(19): p. 4464–70.
125. Kharbikar, B.N., G.S. Chendke, and T.A. Desai, *Modulating the foreign body response of implants for diabetes treatment.* Adv Drug Deliv Rev, 2021. **174**: p. 87–113.
126. Bose, S., et al., *A retrievable implant for the long-term encapsulation and survival of therapeutic xenogeneic cells.* Nature Biomedical Engineering, 2020.
127. Headen, D.M., et al., *Local immunomodulation Fas ligand-engineered biomaterials achieves allogeneic islet graft acceptance.* Nat Mater, 2018. **17**(8): p. 732–739.
128. Lei, J., et al., *FasL microgels induce immune acceptance of islet allografts in nonhuman primates.* Science Advances, 2022. **8**(19): p. eabm9881.
129. Coronel, M.M., et al., *Immunotherapy via PD-L1-presenting biomaterials leads to long-term islet graft survival.* Sci Adv, 2020. **6**(35): p. eaba5573.
130. Tu, Z., et al., *Design of therapeutic biomaterials to control inflammation.* Nature Reviews Materials, 2022.
131. Nijhuis, A.W.G., S.C.G. Leeuwenburgh, and J.A. Jansen, *Wet-Chemical Deposition of Functional Coatings for Bone Implantology.* Macromolecular Bioscience, 2010. **10**(11): p. 1316–1329.
132. Kolar, M., et al., *Covalent Binding of Heparin to Functionalized PET Materials for Improved Haemocompatibility.* Materials (Basel), 2015. **8**(4): p. 1526–1544.
133. Child, A., E.J. Larkin, and M.J. Fontaine, *Chapter 22 - Co-encapsulation of ECM proteins to enhance pancreatic islet cell function*, in *Transplantation, Bioengineering, and Regeneration of the Endocrine Pancreas*, G. Orlando, et al., Editors. 2020, Academic Press. p. 307–313.
134. Llacua, A., et al., *Extracellular matrix components supporting human islet function in alginate-based immunoprotective microcapsules for treatment of diabetes.* J Biomed Mater Res A, 2016. **104**(7): p. 1788–96.

135. Llacua, L.A., M.M. Faas, and P. de Vos, *Extracellular matrix molecules and their potential contribution to the function of transplanted pancreatic islets.* Diabetologia, 2018. **61**(6): p. 1261–1272.
136. Llacua, L.A., B.J. de Haan, and P. de Vos, *Laminin and collagen IV inclusion in immunoisolating microcapsules reduces cytokine-mediated cell death in human pancreatic islets.* J Tissue Eng Regen Med, 2018. **12**(2): p. 460–467.
137. Stendahl, J.C., D.B. Kaufman, and S.I. Stupp, *Extracellular Matrix in Pancreatic Islets: Relevance to Scaffold Design and Transplantation.* Cell transplantation, 2009. **18**(1): p. 1–12.
138. Hadavi, E., et al., *Fibronectin and Collagen IV Microcontact Printing Improves Insulin Secretion by INS1E Cells.* Tissue Engineering Part C: Methods, 2018. **24**(11): p. 628–636.
139. Hadavi, E., et al., *Microwell scaffolds using collagen-IV and laminin-111 lead to improved insulin secretion of human islets.* Tissue Engineering Part C: Methods, 2019. **25**(2): p. 71–81.
140. Tremmel, D.M., et al., *A human pancreatic ECM hydrogel optimized for 3-D modeling of the islet microenvironment.* Scientific Reports, 2022. **12**(1): p. 7188.
141. Border, W.A. and N.A. Noble, *Transforming growth factor beta in tissue fibrosis.* N Engl J Med, 1994. **331**(19): p. 1286–92.
142. Liu, J.M.H., et al., *Transforming growth factor-beta 1 delivery from microporous scaffolds decreases inflammation post-implant and enhances function of transplanted islets.* Biomaterials, 2016. **80**: p. 11–19.
143. Jiang, K., et al., *Local release of dexamethasone from macroporous scaffolds accelerates islet transplant engraftment by promotion of anti-inflammatory M2 macrophages.* Biomaterials, 2017. **114**: p. 71–81.
144. Kuppan, P., et al., *Co-localized immune protection using dexamethasone-eluting micelles in a murine islet allograft model.* American Journal of Transplantation, 2020. **20**(3): p. 714–725.
145. Webber, M.J., et al., *Controlled release of dexamethasone from peptide nanofiber gels to modulate inflammatory response.* Biomaterials, 2012. **33**(28): p. 6823–32.
146. Pathak, S., et al., *Single synchronous delivery of FK506-loaded polymeric microspheres with pancreatic islets for the successful treatment of streptozocin-induced diabetes in mice.* Drug Deliv, 2017. **24**(1): p. 1350–1359.
147. Farah, S., et al., *Long-term implant fibrosis prevention in rodents and non-human primates using crystallized drug formulations.* Nat Mater, 2019. **18**(8): p. 892–904.
148. *Medical Device Regulations*, in *Regulation (EU) 2017/45* E.P.a.t.c.o.m. devices, Editor. 2017. p. 228.
149. Qiu, Q.Q., W.Q. Sun, and J. Connor, *4.410 - Sterilization of Biomaterials of Synthetic and Biological Origin*, in *Comprehensive Biomaterials*, P. Ducheyne, Editor. 2011, Elsevier: Oxford. p. 127–144.
150. Harrell, C.R., et al., *Risks of Using Sterilization by Gamma Radiation: The Other Side of the Coin.* International journal of medical sciences, 2018. **15**(3): p. 274–279.
151. Naikwadi, A.T., et al., *Gamma Radiation Processed Polymeric Materials for High Performance Applications: A Review.* Frontiers in Chemistry, 2022. **10**.
152. *Endotoxins—the Invisible Companion in Biomaterials Research.* Tissue Engineering Part B: Reviews, 2013. **19**(5): p. 391–402.
153. Sojitra, C., et al., *Development and validation of residual solvent determination by headspace gas chromatography in Imatinib Mesylate API.* SN Applied Sciences, 2019. **1**(3): p. 233.
154. Toso, C., et al., *Histologic Graft Assessment After Clinical Islet Transplantation.* Transplantation, 2009. **88**(11): p. 1286–1293.
155. Pileggi, A., et al., *Factors influencing Islet of Langerhans graft function and monitoring.* Clin Chim Acta, 2001. **310**(1): p. 3–16.
156. Kohnert, K.-D., et al., *Utility of different glycemic control metrics for optimizing management of diabetes.* World journal of diabetes, 2015. **6**(1): p. 17–29.
157. Besser, R.E., et al., *Lessons from the mixed-meal tolerance test: use of 90-minute and fasting C-peptide in pediatric diabetes.* Diabetes Care, 2013. **36**(2): p. 195–201.
158. Paredes-Juarez, G.A., P. de Vos, and J.W.M. Bulte, *Recent progress in the use and tracking of transplanted islets as a personalized treatment for type 1 diabetes.* Expert Rev Precis Med Drug Dev, 2017. **2**(1): p. 57–67.

159. Arifin, D.R. and J.W.M. Bulte, *Imaging of pancreatic islet cells.* Diabetes/Metabolism Research and Reviews, 2011. **27**(8): p. 761–766.
160. Low, G., et al., *Role of Imaging in Clinical Islet Transplantation.* RadioGraphics, 2010. **30**(2): p. 353–366.
161. Kim, D. and H.-S. Jun, *In Vivo Imaging of Transplanted Pancreatic Islets.* Frontiers in Endocrinology, 2018. **8**(382).
162. Arifin, D.R. and J.W.M. Bulte, *In Vivo Imaging of Pancreatic Islet Grafts in Diabetes Treatment.* Frontiers in Endocrinology, 2021. **12**.
163. Sneha, K.R. and G.S. Sailaja, *Intrinsically radiopaque biomaterial assortments: a short review on the physical principles, X-ray imageability, and state-of-the-art developments.* Journal of Materials Chemistry B, 2021. **9**(41): p. 8569–8593.
164. Bortoluzzi, E.A., et al., *Tissue response to white mineral aggregate-based cement containing barium sulfate as alternative radiopacifier: A randomized controlled animal study.* Microsc Res Tech, 2021. **84**(4): p. 705–711.

Scaffolds for Encapsulation of Stem Cell-Derived β Cells

Rick de Vries and Aart A. van Apeldoorn

1 Strategies for Extrahepatic Transplantation

While clinical islet transplantation currently involves the intrahepatic implantation site, bioengineering approaches offer novel sites that could potentially improve the outcome of clinical islet transplantation and could further support islets by tailoring their microenvironment. The two main bioengineering approaches include either encapsulation of islets in an immunoprotective graft, or in a device that supports revascularization after transplantation. The manner of encapsulation also allows one to further subdivide encapsulation techniques into nano-, micro-, and macroencapsulation devices. The advantages and limitations of the different approaches and techniques are discussed in the following sections, and summarized in Table 1. While these encapsulation strategies are generally applicable to primary islets, the same (dis)advantages are relevant for stem cell-derived β-cells.

2 Nanoencapsulation

Nanoencapsulation is a technique that involves the application of a thin (nanometer range) polymer coating on the outside of islets that shields the islets from the patient's immune system [1–4]. This allows for small transplantation volumes and short diffusion distances for oxygen and nutrients. The polymer capsule surrounding the islets can also be functionalized for further support of islet viability and functionality. However, nanoencapsulation is notorious for a lack of

R. de Vries · A. A. van Apeldoorn (✉)
Maastricht University, MERLN Institute for Technology-Inspired Regenerative Medicine, Maastricht, The Netherlands
e-mail: a.vanapeldoorn@maastrichtuniversity.nl

L. Piemonti et al. (eds.), *Pluripotent Stem Cell Therapy for Diabetes*,
https://doi.org/10.1007/978-3-031-41943-0_7

Table 1 (Dis)advantages of immune-protective barrier and revascularization strategies

Transplantation strategy	Advantages	Disadvantages
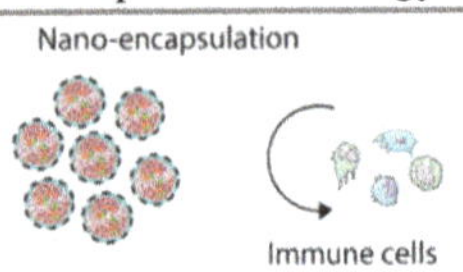	Small transplantation volume	Lack of long-term stability
	Suitable for variety of implantation sites	Incomplete coating elicits immune response
	Short diffusion distances	Difficult to retrieve- Islets depend on diffusion of O_2
	No need for immunosuppressives	
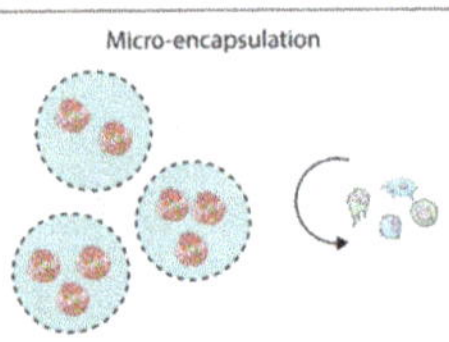	Faster diffusion of oxygen, insulin and nutrients compared to macro-encapsulation devices	Suitable for limited number of implantation sites
	No need for immunosuppressives	Unreliable long-term stability
	Breach of a microcapsule does not affect the whole tissue graft	Clumping of microcapsules leads to hypoxic conditions
		Islets depend on diffusion of O_2
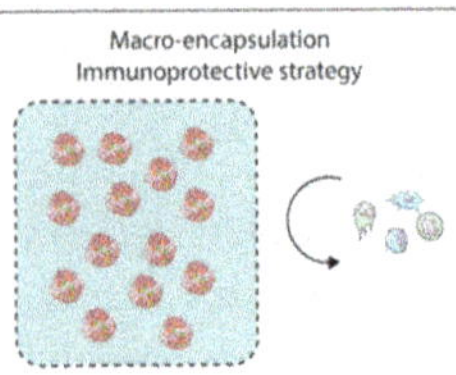	High stability	Least number of suitable implantation sites
	Physical protection for islets	High risk for hypoxia-related islet mass loss
	Easy to retrieve	Any breach of device leads to entire graft loss
	No need for immunosuppressives	Islets depend on diffusion of O_2
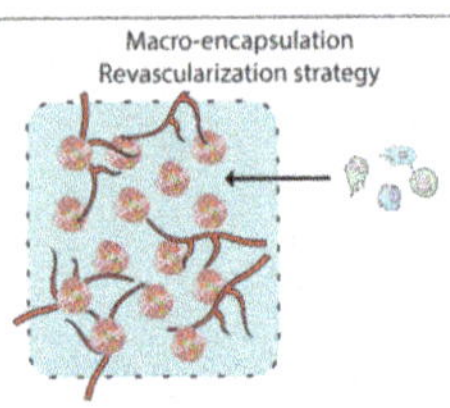	High stability	Least number of suitable implantation sites
	Physical protection for islets	Initially a high risk for hypoxia-related islet mass loss
	Easy to retrieve	Need for immunosuppressives
	High supply of O_2 after re-vascularization (7–14 days post implantation)	

long-term stability of the coating, and it is sometimes difficult to achieve a perfect coating of islets without gaps, triggering an immune response due to direct exposure of the islets to the host tissue [5]. Moreover, nanoencapsulated islets are difficult to retrieve, like naked islets, which all contribute to the fact that nanoencapsulation is currently not routinely performed in the clinic. Nevertheless, a clinical trial (NCT00260234) was initiated where two recipients were subcutaneously transplanted with nanoencapsulated allogeneic islets without long-term immunosuppression [6]. The full islet dose target of the clinical trial was 33,000 IEQ/kg body weight, but both recipients only received half the targeted does. It should therefore come as no surprise that patients did not achieve insulin independence after 4 months of implantation [6]. The recipients experienced a decrease in the number of hyperglycemic and hypoglycemic events, but C-peptide levels were lower than anticipated, and the clinical trial was terminated [5, 6].

3 Microencapsulation

The second encapsulation technique is microencapsulation of islets, which involves encapsulation of individual or small groups of islets in microcapsules. The technique was first used to treat diabetic rats in 1980 [7]. Important advances have been made since, with improvements in the field of biomaterial device design, reduction in the capsule diameter and mitigation of the immune response induced by the microcapsules [6]. Microcapsules are often made from hydrogels and typically range from 100 to 1000 μm in diameter [4]. A benefit of the microencapsulation strategy compared to the immunoprotective macroencapsulation approaches discussed further below is that only a single (group of) islet(s) is affected and attacked by the immune system when a microcapsule breaks. Therefore, only islets in that particular capsule are affected, while the other microencapsulated islets will not be affected. In contrast, the entire tissue graft is at risk once an immunoprotective macroencapsulation device is damaged. In theory, the relatively larger surface-to-volume ratio of microcapsules make them less susceptible to oxygen and nutrient diffusion issues [8]. However, the maximum diffusion distance for effective oxygen and nutrient diffusion is between 100 and 200 μm [9, 10], indicating that microcapsules with diameters above 400 μm will negatively influence survival and functionality of encapsulated islets. In addition, implantation of any biomaterial will elicit the formation of a fibrous capsule surrounding the device, increasing the diffusion distance even further [11]. Other disadvantages of microcapsules include unreliable long-term stability of hydrogels, challenging conditions for live imaging and tracking, difficulty in retrieval of all microcapsules post-transplantation, and hypoxia due to slow diffusion of oxygen and nutrients through the hydrogel, especially when microcapsules have diameters larger than 400 μm [4, 12, 13]. Furthermore, incomplete encapsulation, or only partial coverage of islets with hydrogel, is another challenge for microencapsulation systems. Incomplete capsules can induce severe inflammatory responses resulting in fibrosis of undamaged microcapsules and lead to necrosis of transplanted islets [14]. A major limitation to translate microencapsulated islets into the clinic is the difficulty in upscaling the strategy with large-scale microcapsule production systems. Microencapsulating clinically relevant numbers of islets cannot be performed in a reasonable amount of time, which may induce hypoxic stress and eventually loss of islet function and viability [15]

A wide range of different biomaterials have been used to fabricate microcapsules, including collagen, agarose, chitosan and gelatin, but most microcapsules used are alginate-based [5, 13]. Fabrication of alginate microcapsules is relatively easy to achieve under mild physiological conditions, making alginate an interesting biomaterial candidate for microcapsules [5]. However, alginate does not have a small enough hydrogel network to allow significant permselectivity against immune cells and factors that can destroy encapsulated cells [16]. Therefore, a range of different polycation coatings have been evaluated for alginate-based microcapsules over the past decades to improve their mechanical stability and immunoisolating efficacy [16]. These synthetic coatings, however, may affect the biocompatibility of alginate microcapsules [6, 17–19]. Surface roughness, surface charge, protein adsorption and hydrogel elasticity play a key role in the foreign body response toward microcapsules [16, 20]. Despite the fact that the optimal combination of these parameters has yet to be established, microencapsulated islets and β-cells have shown to be a highly successful combination to reverse diabetes in small animal models [21–25]. However, it is equally important to also test the same strategy in preclinical large animals, since it has been shown that the foreign body response toward microcapsules highly depends on the animal model used [26]. In contrast to the highly successful rodent studies, outcomes of large animal studies such as dogs and non-human primates, are sometimes highly promising but often show only very limited and inconsistent successes [27–29]. This indicates the importance of testing cell delivery strategies in human-sized models.

3.1 Extrahepatic Transplantation Sites for Microencapsulation Approaches

A relatively large number of donor islets is required to achieve insulin independence in patients which interferes with the transplantation of microencapsulated islets into the liver as it would pose a high risk of thrombosis [5]. Microcapsules have been transplanted at various sites, but by far the most frequent implantation sites used are the intraperitoneal space and peritoneal cavity [5, 13, 16, 17, 30] (Fig. 1). Oxygen levels in the peritoneal cavity are generally lower compared to the pancreas leading to decrease in islet survival and function [13, 18]. In order to compensate for this limitation, a larger number of encapsulated islets is required to reach normoglycemia compared to CIT [31]. Animal studies in rats, dogs and pigs (all quadrupeds) have shown promising results after intraperitoneal injection of microcapsules [30]. However, outcomes from studies with non-human primates and first-phase clinical trials in humans (bipeds) were less consistent. In quadrupeds, the microcapsules remain scattered throughout the peritoneal cavity, while in bipeds, they tend to cluster together on the pelvic floor due to gravity, leading to increased levels of hypoxia [6, 32]. The lack of control over confinement of microcapsules also provides a challenge in case they need to be retrieved in case of adverse effects or replacement. The omentum has been used in small animal models [33, 34] and some large animal

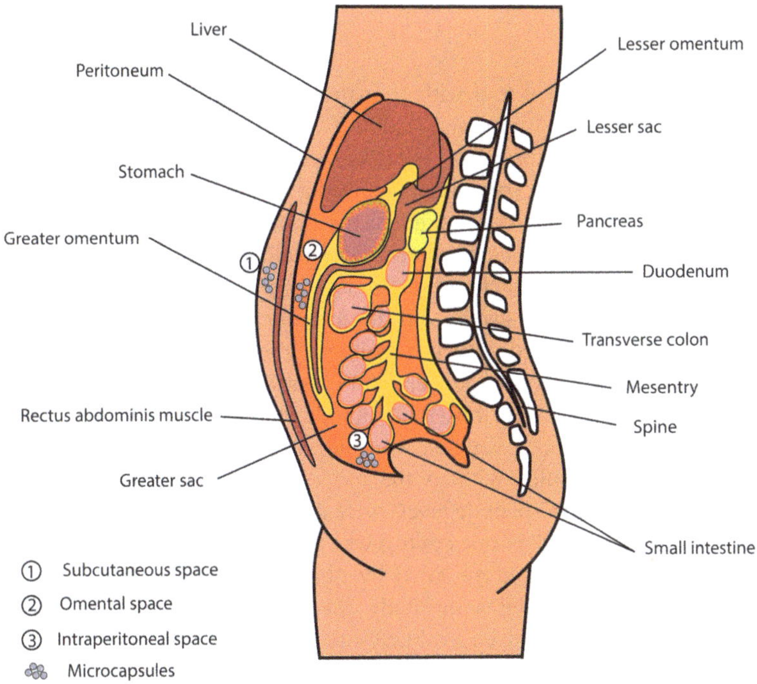

Fig. 1 Extrahepatic implantation sites for microencapsulating islet delivery devices. The size of microencapsulating islet delivery devices limits the possible implantation sites, with the intraperitoneal space being the most common implantation site. The subcutaneous space is not necessarily located on top of the abdomen, and the greater omentum is often folded into a pouch, preventing movement of the microcapsules throughout the greater sac. Both omental and intraperitoneal spaces require opening the peritoneum, which is associated with increased surgical complications compared to the subcutaneous space

models with some success [29, 30]. Parts of the omentum can surgically be reshaped into a smaller pouch that provides a dedicated space to house microencapsulated islets, which is highly vascularized and enables delivery of insulin to the portal circulation similar to the pancreas and liver and confines the capsules to a specific location improving retrievability [30, 35].

As an alternative, the subcutaneous space offers a large surface area, is easy accessible, and can be accessed via a minimal invasive intervention, but generally has poor oxygen tension, while revascularization is slow and the foreign body response is high [36, 37]. Improving the oxygen supply will be required to facilitate microcapsule-based islet transplantation at the subcutaneous site. Nevertheless, the real clinical potential of this implantation site is yet to be determined, as most studies have only been performed in small animals.

3.2 Clinical Trials

The first successful clinical trial with alginate-based microencapsulated islets was performed in 1994 [38]. A total of 678,000 IEQ were transplanted into the peritoneal space, leading to a dose 9957 islets/kg, while the projected target dose was 20,000 islets/kg. Therefore, the patient received a supplemental dose of 5000 islets/kg. Patients underwent low-dose maintenance immunosuppression treatment with 50 mg daily administration of cyclosporine and azathioprine, and patients were free from exogeneous insulin injections for 9 months. Following this promising trial, human donor islets were encapsulated in alginate microcapsules coated with a polycation layer and transplanted without immunosuppression in 2006 [39]. Although, patients still had to use exogeneous insulin to control their blood glucose, no signs of immune sensitization were observed. Another clinical trial was conducted in 2009 in which four diabetic patients were intraperitoneally implanted with islet-encapsulated microcapsules without the use of immunosuppression [40]. Even though initially elevated C-peptide levels were detected, this effect was lost 4 weeks post-transplantation and glycemic control was not improved. Microcapsules recovered after 16 months were surrounded by fibrous tissue and contained necrotic islets. Most notably, an interesting study was performed by Jacobs-Tulleneers-Thevissen et al., which involved evaluation of similar microcapsules in mice and humans [23]. Encapsulated human islets were capable to cure diabetic mice even with low β-cell purity islet preparations, while only a poor rise in C-peptide could be measured after transplantation in humans. The discrepancy between small animal models and clinical outcomes stresses the importance of using large preclinical animal models for evaluation of novel transplantation strategies.

Besides human islets, another endeavor called the "Diabecell" approach by Living Cell Technologies Company was used to transplant 5000–10,000 IEQ/kg body weight of microencapsulated porcine islets in patients in 2007 [5]. The alginate microcapsules were coated with a polycation layer to improve immunoprotective properties, and were capable to reduce exogenous insulin requirements for six out of eight patients for up to 8 months. An individual case report of the Diabecell technology describes the intraperitoneal transplantation of encapsulated neonatal porcine cells [41]. Exogenous insulin requirements were reduced by 30% for the first 12 weeks but returned to pre-transplantation levels within a year. Interestingly, capsules retrieved after 9.5 years still displayed some viable islet cells. Similarly, a phase II clinical study using Diabecell technology in 16 patients demonstrated a reduction in insulin dosage. Although none of the clinical trials with porcine tissue have resulted in patients being free from exogeneous insulin, these studies showed that the technique is safe [42].

In summary, microencapsulation of islets has been extensively studied over the past decades. Some promising studies have indicated the potential of microencapsulation for islet transplantation. Despite the promising improvements in biomaterials and encapsulation strategies, clinical translation has been slow and unsuccessful in terms of consistently abolishing the need for exogenous insulin administration for type 1 diabetic patients. Important limitations for clinical translation of microencapsulation include difficulties in retrievability of microcapsules, exacerbated hypoxia due to aggregation of microcapsules within the intraperitoneal space, and difficulty in upscaling toward large-scale microcapsule production systems. Future studies should focus on the testing of microencapsulation strategies in large preclinical animal models.

4 Macroencapsulation

The third encapsulation technique is macroencapsulation of islets, whereby up to 100–100,000 islets are transplanted into one or several macroencapsulating devices. The devices allow for easy retrieval of the graft in case of graft failure or complications. The earliest macroencapsulation approaches were developed in the 1950s [43]. The basic configuration of macroencapsulation devices is an immunoprotective barrier surrounding a large number of islets encapsulated in the delivery device. Even though polymeric membranes can offer physical protection to islets and block host immune cells, the relatively high islet density within macroencapsulation devices usually leads to hypoxia-related cell death. Unfortunately, most biomaterials exacerbate this problem since they often induce a fibrotic immune response in which the implant is gradually covered by a layer of connective fibrous tissue, increasing the diffusion distance for oxygen and nutrients toward the encapsulated islets [44]. Nevertheless, the retrievability of macroencapsulation devices makes them especially interesting for stem-cell derived β-cells, which are currently still believed to have a tumorigenic potential [45]. Several different macroencapsulation islet delivery devices have been developed over the years which utilize either the immunoprotective and revascularization strategy or one of each. Macroencapsulation devices can be categorized depending on their underlying working mechanisms, such as improved local oxygen delivery, pre-vascularization, immunoprotective membrane-based devices, intravascular islet delivery devices, and more exotic fiber-reinforced hydrogel approaches and 3D printing approaches. Each of these different configurations will be discussed in the following sections and are summarized in Table 2.

4.1 Extrahepatic Transplantation Sites for Macroencapsulation Approaches

In the field, several extrahepatic implantation sites have been suggested including the kidney capsule, the bone marrow, spleen, the greater omentum, the lung, the gastric submucosal space, skeletal muscle, subcutaneous space, pre-peritoneal space, peritoneal cavity, epidydimal fat pad and the vascular system, each with their

Table 2 Advantages and limitations of several macroencapsulation approaches

Strategy	Advantages	Limitations	Most relevant devices	Key references
Local oxygen delivery	High oxygenation of islets	Daily replenishment required	βAir	[60, 62, 64]
				NCT02064309
	Immunoprotective	Questionable long term stability	OxySite	[68, 70]
Pre-vascularization/'deviceless'	High vascularization surrounding islets	No large animal data	Cell Pouch	[80]
		Questionable effect for clinically relevant dosages		NCT01652911
		Not immunoprotective		NCT03513939
		Requires two surgeries		
Membrane-based devices	Use of clinically approved biomaterials	Immunoprotective devices show limited diffusion of insulin	Theracyte (PEC-Encap & PEC-Direct)	[90–92]
				NCT02239354
				NCT03163511
	Could be used for immunoprotective or revascularization strategy	Limited large animal data	Encellin	[95, 96]
			MailPan	[77]
			Microwell device	[97, 98]
Intravascular devices	High oxygenation of islets	High risk of complications	iBAP	[101]
	Immunoprotective			
Fiber-reinforced hydrogels	Immunoprotective	Tissue adhesion	Islet Sheet	[102]
	Mechanically stable	Clinically challenging dimensions of upscaled devices	NEED	[106]
			TRAFFIC	[105]
			NICE	[107]
3D (bio)printing	Immunoprotective	Islets do not show functionality	Direct bioprinting	[110, 112]
	Easy upscaling	Lack of *in vivo* data	NICHE 1	[116]
	High control over distribution of islets and support cells/factors		NICHE 2	[117]

own advantages and limitations [46–55]. However, the considerable size of macroencapsulation strategies only leaves a few options, with the greater omentum, peritoneal cavity, pre-peritoneal space, and subcutaneous space being often used as implantation sites (Fig. 2). The subcutaneous space is often considered due to easy

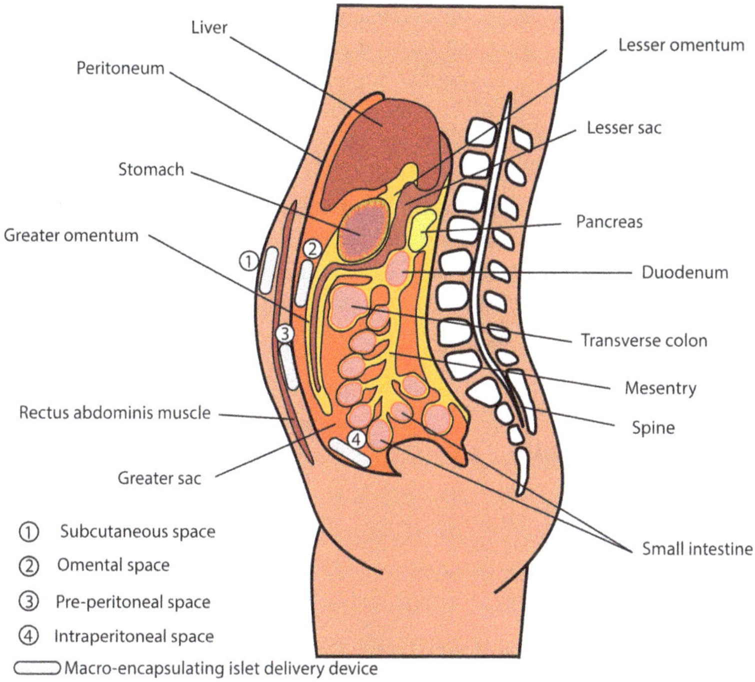

Fig. 2 Extrahepatic implantation sites for macroencapsulating islet delivery devices. The size of macroencapsulating islet delivery devices limits the possible implantation sites, with the subcutaneous space, omental space, pre-peritoneal space and intraperitoneal space being the most common implantation sites. The subcutaneous space is not necessarily located on top of the abdomen, and the greater omentum is often folded into a pouch, preventing movement of the device throughout the greater sac. Both omental and intraperitoneal spaces require opening the peritoneum, which is associated with increased surgical complications compared to the subcutaneous and pre-peritoneal space

and minimal invasive access, the opportunity to easily monitor implants over time, and the possibility to recover implants if needed [51]. Nonetheless, hypoxia, poor revascularization of transplanted islets, and a relatively strong inflammatory response are common problems associated with subcutaneous implantation [17]. The peritoneal cavity (also known as the intraperitoneal space) offers an interesting implantation site due to the available space to house a large number of islets or devices. Nevertheless, it is also associated with limited revascularization of transplanted islets, delayed glucose responsiveness and chronic hypoxic stress [56]. More and more groups have shown interest in the greater omentum, which is a large apron-like layer of white adipose tissue that hangs down the stomach and covers the small intestines [57]. It is the first anatomic structure observed when the abdominal cavity is opened anteriorly. During islet transplantation, the greater omentum is often reshaped into a pouch, preventing (encapsulated) islets from migrating into the greater sac. Given that the abdominal wall and peritoneum must be opened, intraperitoneal and omental islet transplantation are associated with an elevated risk of surgical complications including internal hernia and adhesion ileus [57, 58]. The pre-peritoneal site is the space just in front of the peritoneum and is therefore not associated with a high risk of surgical complications. Even though this site lays deeper in the body than the subcutaneous site, its relatively superficial location allows for easy transplantation, recovery and monitoring of islet delivery devices without violating the peritoneal space [59]. Nevertheless, to date, only few devices have been implanted at the pre-peritoneal site, leading to a relative gap in knowledge and experience with this site for islet delivery. Important to note is that certain implantation sites could work very well in combination with a certain encapsulation strategy, while devices using another encapsulation strategy implanted at the same location may fail, complicating the choice for a universal optimal implantation site.

5 Macroencapsulation Devices

5.1 Local Oxygen Delivery

5.1.1 βAir

Beta-O_2 developed the βAir device to supply external oxygen to encapsulated islets. The disc-shaped device consists of three major components (1) a 500–600 μm thick alginate hydrogel slab containing islets, (2) a gas chamber with inlet and outlet ports, and (3) a 25 μm thick polytetrafluoroethylene (PTFE) membrane separating the alginate slab from the host tissue (Fig. 3a) [60]. Islets within the alginate slab receive oxygen by diffusion through a silicon-rubber membrane. The device was implanted subcutaneously in rats with access ports placed on the dorsal side of the animal between the scapula, connected by short polyurethane tubes with access ports used for refilling oxygen [61]. Unfortunately, this βAir device requires daily replenishing of the gas mixture, thereby still requiring continuous monitoring by the patient. The βAir device was placed pre-peritoneal in non-human primates with porcine islets and showed adequate glucose-regulated insulin secretion without the need for immunosuppressive medication [62]. Importantly, a recent study showed that usage of hyper-physiological levels of pO_2 allows high islet loading densities within the βAir device, effectively reducing the device dimensions [63]. A case report for this device in a single patient reported that islets retained function for the 10-month study duration, with a modest reduction in exogenous insulin [64]. A phase 1 clinical safety study was conducted where four patients were each implanted with one or two devices (holding 155,000–180,000 IEQs) for 3–6 months [65]. Implantation of the device was safe for the patients and supported islet survival, but unfortunately, it showed low circulating C-peptide levels with no impact on metabolic control [66]. Additional studies will need to show the clinical benefit of using the βAir devices for clinical islet transplantation in humans.

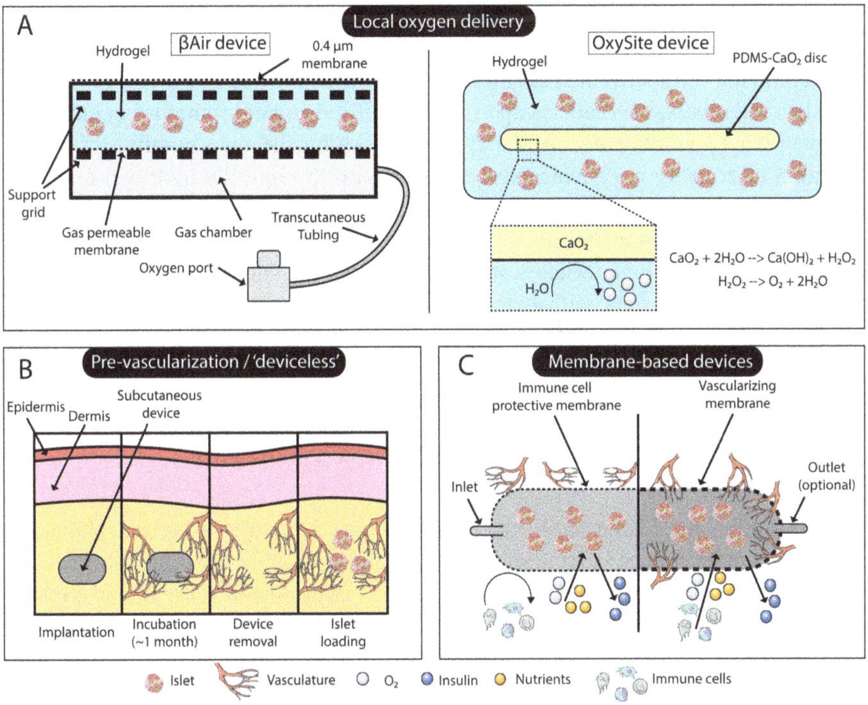

Fig. 3 General approach for macroencapsulating devices following the local oxygen delivery approach, the pre-vascularization approach and membrane-based devices. (**a**) The local oxygen delivery is split between external oxygen supply (βAir device) and internal oxygen supply (OxySite device), while islets are encapsulated into a hydrogel. (**b**) The pre-vascularization approach requires the implantation of a device, and the subsequent formation of a foreign body response resulting in vascularization of the space. The device is removed after approximately a month, and islets are loaded in the pre-vascularized space. (**c**) Membrane-based device can either be immunoprotective, where immune cells and vasculature cannot penetrate a membrane with relatively small openings. On the other side, membrane-based devices can be used for a revascularization approach with relatively large pore openings which allow vasculature (and immune cells) to enter the device

5.2 *Oxygen Generating Biomaterials*

5.2.1 OxySite

Others have tried to increase the islet packing density of macroencapsulating islet delivery devices by enhancing local oxygen supply, mainly by entrapping oxygen-generating biomaterials within a device (Fig. 3a). The incorporation of microbeads consisting of a H_2O_2/PVP core and PLGA shell in PDMS disks was able to improve the function of β-cells during in vitro culture [67]. In a similar but different way hydrolytically reactive oxygen-generating biomaterials based on a mixture of CaO_2 and PDMS forming what the authors coined as the OxySite device are being developed to locally deliver oxygen to β-cells [68, 69]. These OxySite devices were shown to enhance the proliferation of insulinoma β-cells for more than 3 weeks under hypoxic culture conditions [70]. Human islets transplanted with the OxySite device into the intraperitoneal cavity in diabetic mice showed improved viability of the islets and reversal of diabetes up to 30 days [68]. Another group further enhanced the working of silicone-encapsulated CaO_2 disks by the incorporation of hemoglobulin, effectively improving oxygen diffusivity through the hydrogel and neutralizing reactive oxygen species, which are harmful side products produced by the oxygen-generating biomaterials [71]. Nevertheless, the long-term durability of oxygen-generating biomaterials is still being optimized. This approach may serve as a bridge to supply islets with extra oxygen within the early-post-transplantation period when the grafts have not yet been vascularized.

5.3 Pre-vascularization or 'Device-Less' Strategies

Various studies have been published over the years describing islet transplantation approaches within a pre-vascularized site (mainly the subcutaneous site) or pre-vascularized device, and in general have shown enhanced cell survival and engraftment (Fig. 3b) [72–79]. Pre-vascularization approaches initially often implant an easily retrievable polymer device, which triggers a foreign body response and, finally, neovascularization surrounding the construct. After approximately a month, the device is removed, and islets are transplanted in the created cavity. A leading example of this pre-vascularization approach is the Sernova Cell Pouch which is implanted at a subcutaneous site [80]. The Cell Pouch is not immunoisolating. The device is inserted under the skin for 30 days to enable vascular integration with the device. Subsequently, polymer rods are removed, leaving channels that can be filled with islets, which can also be microencapsulated prior to transplantation. Mouse syngeneic islets implanted within Cell Pouch devices into the subcutaneous space have been shown to induce normoglycaemia for up to 100 days [80]. A 3-year phase I/II clinical study using this device started in 2012 but was terminated after recruiting three patients [81, 82]. There were no detectable C-peptide levels, no decrease in insulin requirements nor insulin independence, indicative that the device required modifications. A new phase I/II clinical study was initiated in 2018, which still requires the recruitment of patients [83].

Another approach utilizes a trocar to subcutaneously implant a catheter to initiate a foreign body response and formation of a hollow space lined with neovasculature over the course of 4 weeks. Transplantation of islets into such a space enabled the reversal of diabetes in rodent models for over a year [74]. This method was recently improved by utilizing a methacrylic acid-coated silicone tube to promote vascularization of the subcutaneous pocket in rodent models [76]. Instead of naked islets, the space was filled with a collagen hydrogel containing a mixture of islets and human umbilical vein endothelial cells (HUVECs), which lead to reversal of diabetes for up to 70 days in rats. Important to realize is that the addition of non-clinically approved cells such as HUVECs will make the translation toward the clinic more difficult and may not be approved for clinical use in the near future.

Another device following the pre-vascularization strategy was developed by Pileggi et al., which described a cylindrical stainless-steel mesh (pores of 450 μm) with two polytetrafluoroethylene (PTFE) stoppers at the extremities [72]. Devices were subcutaneously implanted in rats and allowed to engraft and pre-vascularize over a 40-day period. A small cutaneous incision was used to remove one of the PTFE stoppers, rodent islets were seeded in the device, and another PTFE stopper was used to seal the device. The devices showed to be biocompatibility and allowed long-term graft function by diabetes reversal up to 160 days. Despite promising rodent outcomes, debate has started within the islet transplantation field that pre-vascularization of the subcutaneous site could be helpful to reduce the islet dose and still get a metabolic effect, but it is not sufficient to support high cell seeding densities required for reasonably effective clinical-sized devices in humans [17]. Large animal studies resembling the human situation with upscaled devices are necessary to evaluate these pre-vascularization strategies.

5.4 *Membrane-Based Devices*

5.4.1 Theracyte Device

In general, membrane-based devices are created by sealing two membranes to each other in such a way that they form a pouch, leaving an opening for cell seeding (Fig. 3c). The most renowned membrane-based device is the immunoisolating Theracyte™ device, originally developed by Baxter Healthcare Corporation and intended to be subcutaneously implanted [84–86]. The inner membrane has a pore diameter of 0.4 μm to protect the islets against invading immune cells, while the outer membrane has 5 μm pores to support tissue infiltration and promote angiogenesis surrounding the device. The macroporous outer layer was found to be critically important in preventing a fibrotic response and attracting blood vessels toward the immunoisolating membrane [87]. Newer generations of the Theracyte™ device, known as Encaptra® devices, are only made from immunoisolating membranes to protect the transplanted cells from direct interaction with immune cells, while allowing oxygen and nutrients to pass. Encaptra® devices are typically loaded with SCs, called pancreatic endoderm cells (PECs), that have been shown to be able to normalize blood glucose levels in diabetic mice after approximately 120 days [88]. The transplantation of SCs with the help of a Theracyte device has demonstrated its potential in supporting their survival and subsequent maturation toward mature β-cells [89]. The combination of PEC-01 cells, the first generation of PECs, and the Encaptra® device was called PEC-Encap or VC-01™, and was used for a phase I/II clinical trial in 2014 to test the safety, tolerability, and efficacy of the combination product (NCT02239354). Some of these devices still contained insulin-positive cells after a 2-year implantation period, but unfortunately cell survival was suboptimal due to a foreign body response toward the Encaptra® devices, and as a result, proper insulin secretion was not observed [90]. In light of discussion in the field and these suboptimal outcomes, the company switched strategies and launched a follow-up clinical trial in 2017 to investigate safety, tolerability and C-peptide production of PEC-01s in a non-immunoprotective device called VC-02, or PEC-Direct (clinicaltrials.gov: NCT03163511). The VC-02 device is a 3 cm × 9 cm × 1 mm macroencapsulating device that includes large openings that facilitate the ingrowth of blood vessels into the device. Although this approach requires the use of immunosuppressive drugs, β-cells were observed in 63% of VC-02 devices, while 35% of devices showed C-peptide secretion within 26 weeks post-transplantation [91, 92]. Patients showed 20% reduced insulin requirement, 13% more time in normal blood glucose range, <7.0% stable HbA1c levels, and improved hypoglycemic awareness [91]. Most reported adverse effects were due to the implantation of the device (28%), or side effects of the immunosuppressive drugs (34%) [92]. No safety concerns were reported; specifically, no teratoma formation was seen, nor was any metastatic growth outside of the VC-02 devices detected [92]. Each patient received 250–500 million PEC-01s, which translates to ~5300 IEQ/kg body weight, but the secreted C-peptide levels were only 1/10th to 1/30th of that required for insulin independence [93, 94]. Either transplantation of multiple devices, upscaling of the devices, or increasing cell survival within the devices are therefore needed to reach full clinical potential.

5.4.2 Encellin Device

The Encellin cell encapsulation device (CED) consists of a nanoporous, thin polycaprolactone (PCL) film created with a nano-templating technique. Micro- and nanoporous (30–100 nm pore size) films are spun cast on silicon wafers. PCL devices are assembled by heat-sealing two PCL thin films at the outer borders, except for a small opening that allows cell seeding. After seeding of cells, the opening is also heat-sealed for a period of 10 s. The heat sealing leads to some loss of encapsulated cells due to generated heat. Nevertheless, the devices showed good viability of the encapsulated cells for up to 90 days in an allogeneic mouse model, with limited foreign body response and rapid revascularization [95]. More recently, a 6-month mouse study showed long-term survival and glucose responsiveness of the encapsulated SCs transplanted at the intraperitoneal cavity using the Encellin devices [96]. However, regardless of these promising outcomes, reversal of diabetes with Encellin devices still needs to be proven.

5.4.3 MailPan Device

Defymed designed the MailPan (macroencapsulation of pancreatic islets) device, which is another endeavor of a polymer membrane encapsulation aiming to prevent immune cells from entering the device while allowing diffusion of oxygen, nutrients, glucose and insulin in and out of the device [77]. The device consists of a planar pouch formed with two polyester membranes that are heat-sealed at their periphery. Its key feature is an inlet/outlet system made from two silicone catheters permanently connected to two titanium subcutaneous injection ports. Therefore, the device can be implanted and engrafted before islets are seeded in the device, and the cells could potentially be replaced without retrieving the entire device, avoiding repeated surgery. The MailPan device was inserted in the peritoneal cavity of rats and sutured to the peritoneum and abdominal muscles, while the injection ports were placed under the skin. Interestingly, the device showed oxygenation levels around 1% O_2 1 week post-transplantation, which increased to 6% O_2 over a 6-week period. Transplantation of a rat β-cell line with the MailPan device within diabetic rats led to a normalization of blood glucose levels after 2 months of transplantation. Pre-clinical studies with large animals will have to show the clinical feasibility of the approach. Nevertheless, Defymed started a collaboration with Semma Therapeutics to use the MailPan technology in combination with stem cell-derived insulin-secreting cells. Pre-clinical large animal studies of this combined product are currently ongoing.

5.4.4 Microwell-Array Device

Our group has previously published a macroencapsulation strategy involving an open microwell-array islet delivery device, which was successfully used to reverse chemically induced diabetes in mice [97, 98]. The device consists of two thin, porous polymer films which together form a thin envelope to retain islets or β-cell aggregates. One film is imprinted with a dense array of microwells, a feature unique to this islet delivery implant, which can capture individual islets and retain them in a specific spot. The other film acts as a lid, finally entrapping the islets seeded within the microwells. Overall, the device provides physical protection to the islets while the pores in the sheets enable revascularization, allow for fast diffusion of oxygen and nutrients in the early post-transplantation period, and promote the swift ingrowth of vasculature to prevent hypoxia-related cell death. The microwell structure ensures that islets can be captured in individual microwells, ensuring a uniform distribution of islets throughout the device while preventing islet aggregation, thereby promoting the proper supply of oxygen and nutrients to the entrapped cells. This first-generation microwell-array islet delivery device, however, had a few disadvantages. The device was constructed from a specific block copolymer composition based on polyethylene-glycol-terephthalate and polybutylene-terephthalate, which has not (yet) been approved for clinical use. Moreover, concerns have been raised that this material may induce necrosis in transplanted cells [99]. An additional disadvantage is that this first prototype microwell-array islet delivery device was closed by suturing a lid on top, which is a cumbersome, unstandardized procedure. Moreover, the device consisted of two thin film membranes, which lack mechanical stability and therefore potentially allow folding of the devices, a problem which might exacerbate upon increasing the dimensions [100]. First tests in diabetic mice showed that the open device concept was very effective in reaching normoglycemia within a week in mice [97]. A new generation device was recently developed from a different clinically approved biomaterial. In light of future clinical use, a combination of computational modeling and in vitro studies on oxygen distribution, and in vivo experiments in small and large animals were used to further optimize the design and led to the manufacturing of devices that can hold clinically relevant number of islets (unpublished results).

5.5 *Intravascular Bioartificial Pancreas Devices (iBAPs)*

Intravascular devices have been studied to avoid hypoxia and associated necrosis of islets and prevent the delay of insulin response often experienced by other macroencapsulation strategies. In the first stage, a hollow fiber device is located outside the bloodstream, were the blood flow causes the first passage of ultrafiltrate from the bloodstream to the external islet graft. However, this approach failed due to excessive blood clotting in the lumen of the hollow fibers [101]. Others have therefore focused on an intravascular perfusing bioartifical pancreas device (iBAP) (Fig. 4a). Vascular perfusion devices are advantageous in creating high oxygen tension of the arterial blood exposed to islets; however, iBAP devices require surgery for the placing of shunts, are prone to vascular thrombosis and are exposed to a potential risk in diversion of large volume of blood from a distal extremity, limiting the development and application of BAP devices [102]. One specific iBAP device contained an acrylic housing with islets separated from the common iliac artery and vein through a semipermeable membrane. However, clotting and thrombosis occurred at either the site of anastomosis or the junction of the graft, leading immediate failure of the device [103]. In general, the polymer membranes used in iBAP devices produced inadequate levels of ultrafiltrate, resulting in insufficient convective mass transport, which on its turn also severely restricted convective oxygen transport and accompanying loss of islet viability and function [104]. However, new technological advances have allowed for the creation of silicon nanopore membranes (SNM) and subsequent SNM-based iBAP devices that enable relatively high islet viability (>60%) and an insulin response within 15 minutes post glucose stimulation at clinically relevant densities up to 11,400 IEQ/cm^2 in vitro [105]. A porcine study showed that transplantation of islets with an SNM-based iBAP device with pores of 10 nm led to high islet viability (>85%) at a clinically relevant islet density of 5700 IEQ/cm^2 with detectable C-peptide levels and hemocompatibility under convection [105]. This study stresses that iBAP devices could be scaled-up and tested in large animal models.

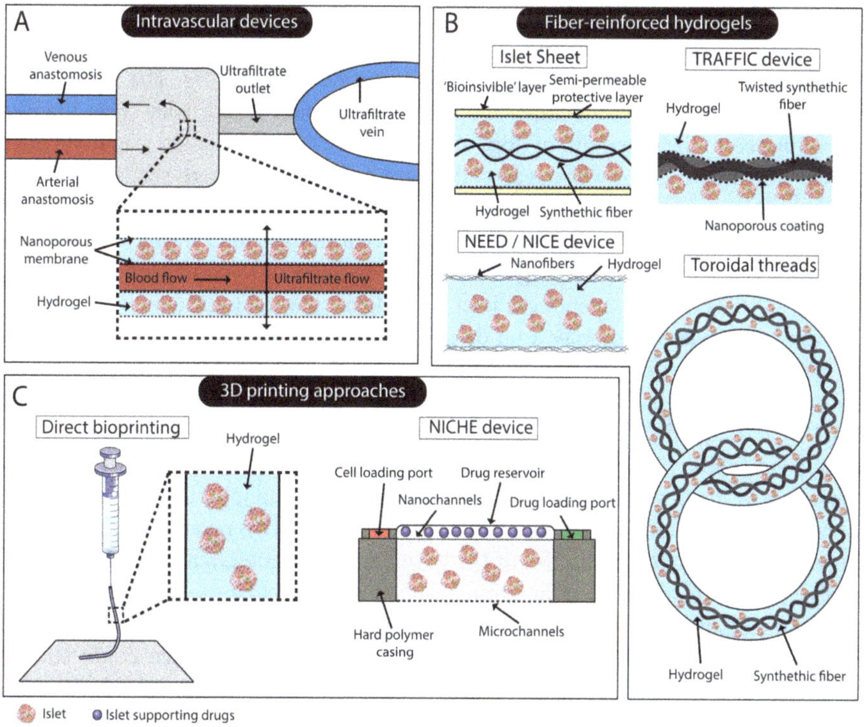

Fig. 4 General approach for macroencapsulating devices following the intravascular approach, the fiber-reinforced hydrogel approach or a 3D printing approach. (**a**) Intravascular devices are anastomosed to a vein and artery, while ultrafiltration through a nanoporous membrane provides hydrogel-encapsulated islets with oxygen and nutrients. Insulin secreted by islets is collected through an ultrafiltrate outlet anastomosed to another vein. (**b**) The Islet Sheet device is a pancake-shaped device where islets are encapsulated within a hydrogel that is reinforced with a synthetic fiber. The semipermeable protective layer and 'bioinvisible' layer are optional. The TRAFFIC device holds a fiber-like shape and consists of a twisted, synthetic fiber encapsulated in a hydrogel casing. The NEED or NICE devices are also fiber-like shaped devices, which have an outer layer of nanofibers and an inner core of hydrogel-encapsulated islets. Toroidal threads consist of 3D-printed, interconnected rings that are subsequently encapsulated within a hydrogel, finally forming interconnected toroidal structures. (**c**) One 3D printing approach entails direct bioprinting, where islets are directly printed within a hydrogel. Another approach is 3D printing of a hard polymer construct and subsequent loading of islets (and optionally islet-supporting drugs) in the construct, such as the NICHE device

5.6 *Fiber-Reinforced Hydrogels*

Several macroencapsulation devices aim to improve the retrieval of hydrogel-encapsulated islets by the inclusion of reinforcement within the hydrogel. So-called Islet Sheets are composed of highly purified alginate with a reinforcing mesh that encapsulates islets and optionally a 'bioinvisible' film that should mitigate the foreign body response toward the device (Fig. 4b) [106]. Human islets encapsulated within Islet Sheet devices were shown to be viable and functional after 8 weeks of in vitro culture and after subcutaneous implantation in rats [107]. The same group also utilized the islet sheet principle in a canine model where islet sheet devices were sutured to the omentum of dogs and showed reversal of diabetes up to 84 days [106]. A similar system was developed where porcine islets were seeded on a collagen matrix to create a cell monolayer before encapsulating the islets within an alginate slab. These islet sheet devices were then transplanted into the abdominal subcutaneous tissue of cynomolgus monkeys and showed to correct for diabetes up to 6 months [108]. However, these islet sheet devices were associated with a strong immune response, which may limit their clinical application.

Another endeavor is the impregnation of a coated polymer tread with an alginate hydrogel, forming a thread-reinforced alginate fiber called the TRAFFIC device [109]. The TRAFFIC device is easily upscalable, retrievable with a minimally invasive laparoscopic procedure and allows for a relatively short oxygen diffusion distance due to the threaded core. The device provided immunoprotection to either rat or human islets after transplantation in mice, and showed to reverse diabetes for up to several months. Retrievability of the device was evaluated in a scaled-up version of the TRAFFIC after transplantation at the peritoneal cavity in dogs. Despite the promising results, minor tissue adhesion occasionally occurred, resulting in partial detachment of the hydrogel from the thread. The complete device could still be retrieved after excising the adhered tissue, and utilization of a twist-folded threat instead of a straight thread did seem to improve the hydrogel-thread adhesion, but it does indicate the Achilles heel of the TRAFFIC device. The same group also developed another fiber-reinforced hydrogel approach in which highly porous, electrospun nanofibers were impregnated with a hydrogel precursor solution and subsequently crosslinked to form nanofiber-enabled hydrogel devices (NEEDs) [110]. The devices retain their hydrogel properties (such as biocompatibility) and nanofiber properties (such as mechanical robustness). Islet-loaded NEEDs were transplanted into the peritoneal cavity and showed reversal of diabetes for at least 8 weeks in mice with only minimal fibrosis. A next generation of this was called a nanofiber-integrated cell encapsulation (NICE) device [111]. The 100 μm thick nanofibrous porous skins were 500 μm – 3 mm in diameter, consisted of 270 nm thick fibers and contained interconnected pores in the order of 1 μm. The devices restored normoglycemia in diabetic mice up to 200 days, and a scaled up version of the NICE device was transplanted and fully retrieved from the peritoneal cavity in dogs with a laparoscopic procedure. Even though empty devices showed a fibrous layer thickness below 5 μm after 4 weeks of transplantation at the intraperitoneal site in mice, dogs showed minor adhesions between the NICE device and either the peritoneal wall or the liver after 2 weeks of transplantation. The technique itself allows for easy upscaling of the fibers, however the loading density of the devices is relatively low. The dogs were transplanted with a device of 17 cm in length, holding 2500 SCs aggregates. Based on these data, a NICE device holding 300,000 SC aggregates would be more than 20 m long. Such dimensions would be clinically challenging to implant and may lead to serious complications such as wide-spread adhesions or constriction of organs. Future diabetic, large animal studies will need to show the clinical suitability of the TRAFFIC, NEED and NICE device.

A similar upscalable approach was used by Ernst et al., where interconnected, toroidal-shaped threads of flexible resin were 3D printed, impregnated with a hydrogel precursor solution and crosslinked [112]. The high flexibility and recovery to its original design are benefits of the device. To show therapeutic potential, islets were transplanted into the peritoneal cavity of diabetic mice. The devices maintained glucose-responsive, and most animals showed reversal of diabetes over 84 days. Despite these promising results, the flexible resin is not a clinically approved material that will therefore considerably delay the approval of this device toward the clinic. In addition, organ entanglement was occasionally observed which should be highly avoided to prevent severe side complications. Future large animal studies are needed to determine the clinical translatability of the strategy.

5.7 3D (Bio)printing Approaches

Besides the classical polymeric approach, 3D (bio)printing has also become a potential technique for macroencapsulation of pancreatic islets in recent years, as excellently reviewed by others [113]. Interestingly, hydrogels can also be used for direct 3D bioprinting, which combines the advantages of micro- and macroencapsulation (Fig. 4c). A 3D printed construct would ideally facilitate a maximized surface area/volume ratio for optimal diffusion of oxygen and nutrients, while also utilizing a superordinate macrostructure for easier retrieval and stability. The technique also allows for easy upscaling of the constructs to clinically relevant sizes and printing designs can easily be altered to desired shapes. 3D printing is a relatively new islet encapsulation technique with only a few experienced groups. Despite promising viability results from 3D bioprinted pancreatic islets, an impaired diffusion of insulin through the hydrogels likely delays or prevents the release of insulin from the hydrogels [114–118]. This highlights that a careful selection of hydrogel formulations for 3D bioprinting is of utmost importance. Moreover, the selected hydrogels should be of general manufacturing grade (GMP) level to ensure easy translation toward the clinic, but the availability of different GMP grade hydrogel is currently scarce. Future developments into clinical-grade hydrogels may offer suitable hydrogels for 3D bioprinting of islets.

Next to direct bioprinting, a polymer islet encapsulation can be 3D printed and subsequently loaded with islets. Daoud et al. showed long-term viability and insulin release in in vitro cultured human islets in 3D printed constructs [119]. Sabek et al. developed the NICHE device, aiming to obtain an extensive vascularized environment prior to islet transplantation. The 3D printed discoidal NICHE device consists of 50 μm thick polylactic acid (PLA) fibers, creating inner layers with microreservoirs of 300 × 300 μm to prevent aggregation of encapsulated islets, and covered with an outer layer of 100 × 100 μm squares to prevent islet loss and allow for the ingrowth of vasculature [120, 121]. Devices were then transplanted subcutaneously in nude mice and left to vascularize over a 4-week period. Subsequently, printed devices were intracutaneously injected with 2000 IEQ of human pancreatic islets through a loading port (1 mm diameter). Despite this relatively high islet dose, devices required refilling after a 12-week period with an additional 2000 IEQ to obtain insulin release levels similar to kidney capsule control samples. The next generation of the device (NICHE-2) holds the same design with the exception that a drug reservoir is included for sustained local release of immunosuppressive drugs through a nanofluidic silicon membrane [16, 122]. The NICHE-2 device also allows for transcutaneous drug replenishment, or the addition of islet supporting factors such as growth factors, nutrients, or oxygen. Although completed studies only involve rodent studies, porcine and NHP models are under investigation [16]. More recently, the two printing approaches were combined by printing of hydrogel inside or on top of a 3D-printed polymeric backbone [123, 124], which showed promising in vitro data. In addition, support cells such as HUVECs or MSCs or support structures such as ECM-(mimicking) molecules or growth factors could easily be included in these approaches. However, one should be cautious about utilizing these additional compounds as they may prevent an easy translation to the clinic when the support cells and/or molecules are not clinically approved. Despite the promising 3D (bio)printing results, in vivo data is often still lacking for 3D printing approaches, which makes it difficult to determine the actual clinical potential of these techniques.

6 What Do Patients Want Themselves?

Recently, a short online survey held among type 1 diabetes patients was reported, of which the majority (>80%) of the population had a duration of diabetes exceeding 5 years and were treated with constant glucose monitoring and insulin pumps [125]. Interestingly, the subcutaneous implantation site was their top choice (53%), followed by intra-abdominal implantation (30%), and the least attractive site was a device outside of their body. Considering the device size and shape, a microencapsulation technology was preferred by 48% of patients, followed by macroencapsulation strategies (32% of patients), which seems to overlap with the fact that 58% of patients would like the device to be as small as possible, 25% of patients do not care about visibility, while 16% preferred the device not to be visible at all. Finally, 83% of patients are open to the use of induced pluripotent stem cell-derived β-cells. Unfortunately, allogeneic stem cell-derived β-cells (or stem cell bank derived β-cells), which have the benefit of off-the-shelve availability, were not taken along in this study.

7 Concluding Remarks

Islet encapsulation techniques enable the use of extrahepatic transplantation sites. However, each encapsulation approach (either nanoencapsulation, microencapsulation, or macroencapsulation) have their benefits and disadvantages. Nanoencapsulation allows for small transplantation volumes and short diffusion distances for oxygen and nutrients. However, nanoencapsulation is notorious for occasional incomplete coating of islets and has a lack of long-term stability, which leads to a loss of the islet graft in the long-term. Alternatively, microencapsulation strategies can be used to isolate small groups of islets in microcapsules. The relative large, required volume limits the number of suitable implantation sites to the subcutaneous space, omentum, and intraperitoneal space. The relatively large surface to volume ratio of microcapsules should provide improved oxygen diffusion toward islets, but sedimentation of microcapsules in the often-used intraperitoneal space have indicated severe hypoxic conditions. In addition, microcapsules are difficult to retrieve, which may limit their suitability for stem cell-derived β-cells.

As for the current macroencapsulation strategies previously published, local oxygen delivery is an interesting strategy that should improve local islet viability. Current devices either require daily oxygen replenishment and therefore still require constant monitoring (βAir device) or long-term local oxygen delivery through oxygen-generating biomaterials which has yet to be evaluated in large animal models (OxySite device). Despite the promising rodent models, debate has started that the pre-vascularization strategy is only helpful to reduce the islet dose for diabetes in small animal models, but it is not sufficient to support islets within human sized devices. Viacyte has shown that there is a clear clinical potential for membrane-based strategies with their VC-02 device. Future improvements in insulin production of their PEC-01 SCs will determine the success of their islet transplantation strategy. Other membrane-based devices such as the Encellin and MailPan devices have yet to be evaluated in (diabetic) large animal studies. Vascular perfusion devices (or intravascular devices) are advantageous by offering a high oxygen tension to islets. However, these devices require extensive surgery and have a high risk of potential complications, limiting their development and application. Strategies involved with fiber-reinforced hydrogels such as the Islet Sheet device, NEED device, TRAFFIC device and NICE device can easily be upscaled, but device dimensions holding clinically relevant islet numbers become unpractical and may lead to organ constriction. Finally, 3D (bio)printing may offer interesting approaches in the future. However production of 3D printed devices in a good manufacturing practice (GMP)-grade environment is still very challenging. In addition, other hurdles to overcome include standardization of the production process (especially challenging due to the vast number of different 3D printers out there) and the lack of proper compatible materials for printing β-cells. Most important, the current lack of sufficient in vivo data prevents a realistic prediction of clinical translatability of this approach.

There is a clear need for evaluation of encapsulation devices within large animal models before the translation toward the clinic can be made. Rodent models often do not translate to large animals and human needs [6]. When an encapsulation device fails in rodents, it will almost never succeed in larger animals. On the other hand, if the encapsulation device is successful in rodent models, there is still a great risk that the device could fail in larger animals. Moreover, β-cell delivery strategies need to adhere to the strict safety regulations for medical devices or even advanced medicinal therapy products in case stem cell-derived β-cells are used. This means that the use of non-clinically approved biomaterials, support cells and/or additional compounds to realize proper islet function in animal models likely delays or even prevent strategies to reach the clinic. Besides these regulatory issues, strategies can also fail to reach the clinic due to the use of technology, which is too complex or just simply too expensive compared to current therapy. Nevertheless, several promising and perhaps also more reasonable encapsulation strategies have been engineered and are currently under evaluation in human clinical trials, showing that β-cell replacement therapy supported by a dedicated medical device strategy may soon become a reality.

References

1. Zhi, Z.l., et al., *Nano-scale encapsulation enhances allograft survival and function of islets transplanted in a mouse model of diabetes.* Diabetologia, 2012. **55**(4): p. 1081–1090.
2. Krol, S., et al., *Multilayer nanoencapsulation. New approach for immune protection of human pancreatic islets.* Nano Lett, 2006. **6**(9): p. 1933–9.
3. Ernst, A.U., et al., *Nanotechnology in cell replacement therapies for type 1 diabetes.* Adv Drug Deliv Rev, 2019. **139**: p. 116–138.
4. Opara, A., et al., *Islet cell encapsulation – Application in diabetes treatment.* Experimental Biology and Medicine, 2021. **246**(24): p. 2570–2578.
5. Espona-Noguera, A., et al., *Review of Advanced Hydrogel-Based Cell Encapsulation Systems for Insulin Delivery in Type 1 Diabetes Mellitus.* Pharmaceutics, 2019. **11**(11).
6. Scharp, D.W. and P. Marchetti, *Encapsulated islets for diabetes therapy: History, current progress, and critical issues requiring solution.* Advanced Drug Delivery Reviews, 2014. **67–68**: p. 35–73.
7. Lim, F. and A. Sun, *Microencapsulated islets as bioartificial endocrine pancreas.* Science, 1980. **210**(4472): p. 908–910.
8. Cao, R., et al., *Mathematical predictions of oxygen availability in micro- and macro-encapsulated human and porcine pancreatic islets.* J Biomed Mater Res B Appl Biomater, 2020. **108**(2): p. 343–352.
9. Krogh, A., *The supply of oxygen to the tissues and the regulation of the capillary circulation.* The Journal of physiology, 1919. **52**(6): p. 457–474.
10. Place, T.L., F.E. Domann, and A.J. Case, *Limitations of oxygen delivery to cells in culture: An underappreciated problem in basic and translational research.* Free radical biology & medicine, 2017. **113**: p. 311–322.
11. Anderson, J.M., *Biological Responses to Materials.* Annual Review of Materials Research, 2001. **31**(1): p. 81–110.
12. Smith, K.E., R.C. Johnson, and K.K. Papas, *Update on cellular encapsulation.* Xenotransplantation, 2018. **25**(5): p. e12399.

13. Desai, T. and L.D. Shea, *Advances in islet encapsulation technologies.* Nature Reviews Drug Discovery, 2017. **16**(5): p. 338–350.
14. Bhujbal, S.V., et al., *A novel multilayer immunoisolating encapsulation system overcoming protrusion of cells.* Scientific Reports, 2014. **4**(1): p. 6856.
15. Jr, W.F.K. and E.C. Opara, *Immunoisolation techniques for islet cell transplantation.* Expert Opinion on Biological Therapy, 2002. **2**(5): p. 503–511.
16. Farina, M., et al., *Cell encapsulation: Overcoming barriers in cell transplantation in diabetes and beyond.* Advanced Drug Delivery Reviews, 2019. **139**: p. 92–115.
17. Papas, K.K., et al., *Oxygenation strategies for encapsulated islet and beta cell transplants.* Adv Drug Deliv Rev, 2019. **139**: p. 139–156.
18. Colton, C.K., *Oxygen supply to encapsulated therapeutic cells.* Adv Drug Deliv Rev, 2014. **67–68**: p. 93–110.
19. Ma, M., et al., *Core–Shell Hydrogel Microcapsules for Improved Islets Encapsulation.* Advanced Healthcare Materials, 2013. **2**(5): p. 667–672.
20. Strand, B.L., et al., *Poly-L-Lysine Induces Fibrosis on Alginate Microcapsules via the Induction of Cytokines.* Cell Transplant, 2001. **10**(3): p. 263–275.
21. Espona-Noguera, A., et al., *Type 1 Diabetes Mellitus reversal via implantation of magnetically purified microencapsulated pseudoislets.* Int J Pharm, 2019. **560**: p. 65–77.
22. Duvivier-Kali, V.F., et al., *Complete protection of islets against allorejection and autoimmunity by a simple barium-alginate membrane.* Diabetes, 2001. **50**(8): p. 1698–705.
23. Jacobs-Tulleneers-Thevissen, D., et al., *Sustained function of alginate-encapsulated human islet cell implants in the peritoneal cavity of mice leading to a pilot study in a type 1 diabetic patient.* Diabetologia, 2013. **56**(7): p. 1605–14.
24. Souza, Y.E., et al., *Islet transplantation in rodents. Do encapsulated islets really work?* Arq Gastroenterol, 2011. **48**(2): p. 146–52.
25. Vegas, A.J., et al., *Long-term glycemic control using polymer-encapsulated human stem cell–derived beta cells in immune-competent mice.* Nature Medicine, 2016. **22**: p. 306.
26. Vaithilingam, V., et al., *Effect of prolonged gelling time on the intrinsic properties of barium alginate microcapsules and its biocompatibility.* J Microencapsul, 2011. **28**(6): p. 499–507.
27. Sun, Y., et al., *Normalization of diabetes in spontaneously diabetic cynomologus monkeys by xenografts of microencapsulated porcine islets without immunosuppression.* J Clin Invest, 1996. **98**(6): p. 1417–22.
28. Elliott, R.B., et al., *Intraperitoneal alginate-encapsulated neonatal porcine islets in a placebo-controlled study with 16 diabetic cynomolgus primates.* Transplant Proc, 2005. **37**(8): p. 3505–8.
29. Bochenek, M.A., et al., *Alginate encapsulation as long-term immune protection of allogeneic pancreatic islet cells transplanted into the omental bursa of macaques.* Nature Biomedical Engineering, 2018. **2**(11): p. 810–821.
30. Vaithilingam, V., S. Bal, and B.E. Tuch, *Encapsulated Islet Transplantation: Where Do We Stand?* Rev Diabet Stud, 2017. **14**(1): p. 51–78.
31. Qi, M., et al., *Encapsulation of human islets in novel inhomogeneous alginate-ca2+/ba2+ microbeads: in vitro and in vivo function.* Artif Cells Blood Substit Immobil Biotechnol, 2008. **36**(5): p. 403–20.
32. De Vos, P., et al., *Why do microencapsulated islet grafts fail in the absence of fibrotic overgrowth?* Diabetes, 1999. **48**(7): p. 1381–1388.
33. Kobayashi, T., et al., *Survival of microencapsulated islets at 400 days posttransplantation in the omental pouch of NOD mice.* Cell Transplant, 2006. **15**(4): p. 359–65.
34. Pareta, R., et al., *Long-term function of islets encapsulated in a redesigned alginate microcapsule construct in omentum pouches of immune-competent diabetic rats.* Pancreas, 2014. **43**(4): p. 605–13.
35. Damyar, K., et al., *An overview of current advancements in pancreatic islet transplantation into the omentum.* Islets, 2021. **13**(5–6): p. 115–120.
36. Vaithilingam, V., et al., *Coencapsulation of Target Effector Cells With Mesenchymal Stem Cells Reduces Pericapsular Fibrosis and Improves Graft Survival in a Xenotransplanted Animal Model.* Cell Transplant, 2016. **25**(7): p. 1299–317.

37. Mochizuki, Y., et al., *Co-Microencapsulation of Islets and MSC CellSaics, Mosaic-Like Aggregates of MSCs and Recombinant Peptide Pieces, and Therapeutic Effects of Their Subcutaneous Transplantation on Diabetes.* Biomedicines, 2020. **8**(9).
38. Soon-Shiong, P., et al., *Insulin independence in a type 1 diabetic patient after encapsulated islet transplantation.* The Lancet, 1994. **343**(8903): p. 950–951.
39. Calafiore, R., et al., *Microencapsulated pancreatic islet allografts into nonimmunosuppressed patients with type 1 diabetes: first two cases.* Diabetes Care, 2006. **29**(1): p. 137–8.
40. Tuch, B.E., et al., *Safety and viability of microencapsulated human islets transplanted into diabetic humans.* Diabetes Care, 2009. **32**(10): p. 1887–9.
41. Elliott, R.B., et al., *Live encapsulated porcine islets from a type 1 diabetic patient 9.5 yr after xenotransplantation.* Xenotransplantation, 2007. **14**(2): p. 157–61.
42. Matsumoto, S., et al., *Clinical Benefit of Islet Xenotransplantation for the Treatment of Type 1 Diabetes.* EBioMedicine, 2016. **12**: p. 255–262.
43. Algire, G.H., J.M. Weaver, and R.T. Prehn, *Growth of cells in vivo in diffusion chambers. I. Survival of homografts in immunized mice.* J Natl Cancer Inst, 1954. **15**(3): p. 493–507.
44. Schweicher, J., C. Nyitray, and T.A. Desai, *Membranes to achieve immunoprotection of transplanted islets.* Frontiers in bioscience (Landmark edition), 2014. **19**: p. 49–76.
45. Lee, A.S., et al., *Tumorigenicity as a Clinical Hurdle for Pluripotent Stem Cell Therapies.* Nature medicine, 2013. **19**(8): p. 998–1004.
46. Liu, Z., et al., *Pig-to-Primate Islet Xenotransplantation: Past, Present, and Future.* Cell Transplant, 2017. **26**(6): p. 925–947.
47. Smink, A.M., M.M. Faas, and P. de Vos, *Toward engineering a novel transplantation site for human pancreatic islets.* Diabetes, 2013. **62**(5): p. 1357–64.
48. Stokes, R.A., et al., *Transplantation sites for human and murine islets.* Diabetologia, 2017. **60**(10): p. 1961–1971.
49. Cantarelli, E., et al., *Transplant Site Influences the Immune Response After Islet Transplantation: Bone Marrow Versus Liver.* Transplantation, 2017. **101**(5): p. 1046–1055.
50. Chen, X., et al., *The epididymal fat pad as a transplant site for minimal islet mass.* Transplantation, 2007. **84**(1): p. 122–5.
51. Uematsu, S.S., et al., *The Optimization of the Prevascularization Procedures for Improving Subcutaneous Islet Engraftment.* Transplantation, 2018. **102**(3): p. 387–395.
52. Carlsson, P.-O., et al., *Markedly Decreased Oxygen Tension in Transplanted Rat Pancreatic Islets Irrespective of the Implantation Site.* Diabetes, 2001. **50**(3): p. 489–495.
53. Itoh, T., et al., *The Spleen Is an Ideal Site for Inducing Transplanted Islet Graft Expansion in Mice.* PLOS ONE, 2017. **12**(1): p. e0170899.
54. Sakata, N., G. Yoshimatsu, and S. Kodama, *The Spleen as an Optimal Site for Islet Transplantation and a Source of Mesenchymal Stem Cells.* Int J Mol Sci, 2018. **19**(5).
55. Lau, H., et al., *Exploring Insulin Production Following Alveolar Islet Transplantation (AIT).* International journal of molecular sciences, 2021. **22**(19): p. 10185.
56. Zhu, H., et al., *Selection of Implantation Sites for Transplantation of Encapsulated Pancreatic Islets.* Tissue Eng Part B Rev, 2018.
57. Sakata, N., G. Yoshimatsu, and S. Kodama, *White Adipose Tissue as a Site for Islet Transplantation.* Transplantology, 2020. **1**(2): p. 55–70.
58. Sammour, T., et al., *Peritoneal Damage: The Inflammatory Response and Clinical Implications of the Neuro-Immuno-Humoral Axis.* World Journal of Surgery, 2010. **34**(4): p. 704–720.
59. McDermott, B., et al., *Developing a morphomics framework to optimize implant site-specific design parameters for islet macroencapsulation devices.* J R Soc Interface, 2021. **18**(185): p. 20210673.
60. Barkai, U., et al., *Enhanced oxygen supply improves islet viability in a new bioartificial pancreas.* Cell Transplant, 2013. **22**(8): p. 1463–76.
61. Neufeld, T., et al., *The Efficacy of an Immunoisolating Membrane System for Islet Xenotransplantation in Minipigs.* PLOS ONE, 2013. **8**(8): p. e70150.
62. Ludwig, B., et al., *Favorable outcome of experimental islet xenotransplantation without immunosuppression in a nonhuman primate model of diabetes.* Proceedings of the National Academy of Sciences of the United States of America, 2017. **114**(44): p. 11745–11750.

63. Evron, Y., et al., *Long-term viability and function of transplanted islets macroencapsulated at high density are achieved by enhanced oxygen supply.* Scientific Reports, 2018. **8**(1): p. 6508.
64. Ludwig, B., et al., *Transplantation of human islets without immunosuppression.* Proceedings of the National Academy of Sciences, 2013. **110**(47): p. 19054–19058.
65. *An Open Label, Pilot Investigation, to Assess the Safety and Efficacy of Transplantation of Macro-encapsulated Human Islets Within the Bioartificial Pancreas Beta-Air in Patients With Type 1 Diabetes Mellitus.* https://ClinicalTrials.gov/show/NCT02064309.
66. Carlsson, P.-O., et al., *Transplantation of macroencapsulated human islets within the bioartificial pancreas βAir to patients with type 1 diabetes mellitus.* American journal of transplantation: official journal of the American Society of Transplantation and the American Society of Transplant Surgeons, 2018. **18**(7): p. 1735–1744.
67. McReynolds, J., et al., *Modeling spatial distribution of oxygen in 3d culture of islet beta-cells.* Biotechnology Progress, 2017. **33**(1): p. 221–228.
68. Coronel, M.M., et al., *Oxygen generating biomaterial improves the function and efficacy of beta cells within a macroencapsulation device.* Biomaterials, 2019. **210**: p. 1–11.
69. Coronel, M.M., R. Geusz, and C.L. Stabler, *Mitigating hypoxic stress on pancreatic islets via in situ oxygen generating biomaterial.* Biomaterials, 2017. **129**: p. 139–151.
70. Pedraza, E., et al., *Preventing hypoxia-induced cell death in beta cells and islets via hydrolytically activated, oxygen-generating biomaterials.* Proceedings of the National Academy of Sciences, 2012. **109**(11): p. 4245–4250.
71. Mouré, A., et al., *Extracellular hemoglobin combined with an O2-generating material overcomes O2 limitation in the bioartificial pancreas.* Biotechnology and Bioengineering, 2019. **116**(5): p. 1176–1189.
72. Pileggi, A., et al., *Reversal of diabetes by pancreatic islet transplantation into a subcutaneous, neovascularized device.* Transplantation, 2006. **81**(9): p. 1318–24.
73. Kaufman-Francis, K., et al., *Engineered vascular beds provide key signals to pancreatic hormone-producing cells.* PLoS One, 2012. **7**(7): p. e40741.
74. Pepper, A.R., et al., *A prevascularized subcutaneous device-less site for islet and cellular transplantation.* Nat Biotechnol, 2015. **33**(5): p. 518–23.
75. Luan, N.M. and H. Iwata, *Long-Term Allogeneic Islet Graft Survival in Prevascularized Subcutaneous Sites Without Immunosuppressive Treatment.* American Journal of Transplantation, 2014. **14**(7): p. 1533–1542.
76. Vlahos, A.E., et al., *A scalable device-less biomaterial approach for subcutaneous islet transplantation.* Biomaterials, 2021. **269**: p. 120499.
77. Magisson, J., et al., *Safety and function of a new pre-vascularized bioartificial pancreas in an allogeneic rat model.* Journal of Tissue Engineering, 2020. **11**: p. 2041731420924818.
78. Sakurai, T., et al., *The Efficient Prevascularization Induced by Fibroblast Growth Factor 2 With a Collagen-Coated Device Improves the Cell Survival of a Bioartificial Pancreas.* Pancreas, 2004. **28**(3): p. e70–e79.
79. *Seeding of Pancreatic Islets into Prevascularized Tissue Engineering Chambers.* Tissue Engineering Part A, 2009. **15**(12): p. 3823–3833.
80. Pepper, A.R., et al., *Diabetes Is Reversed in a Murine Model by Marginal Mass Syngeneic Islet Transplantation Using a Subcutaneous Cell Pouch Device.* Transplantation, 2015. **99**(11): p. 2294–2300.
81. *A Phase I/II Study of the Safety and Efficacy of Sernova's Cell PouchTM for Therapeutic Islet Transplantation.* https://ClinicalTrials.gov/show/NCT01652911.
82. Gala-Lopez, B., et al., *Subcutaneous clinical islet transplantation in a prevascularized subcutaneous pouch - Preliminary experience.* CELLR4, 2016. **4**(5).
83. *A Safety, Tolerability and Efficacy Study of Sernova's Cell Pouch™ for Clinical Islet Transplantation.* https://ClinicalTrials.gov/show/NCT03513939.
84. Sörenby, A.K., et al., *Preimplantation of an immunoprotective device can lower the curative dose of islets to that of free islet transplantation: studies in a rodent model.* Transplantation, 2008. **86**(2): p. 364–6.

85. Sörenby, A.K., et al., *Macroencapsulation protects against sensitization after allogeneic islet transplantation in rats.* Transplantation, 2006. **82**(3): p. 393–7.
86. Kumagai-Braesch, M., et al., *The TheraCyte™ Device Protects against Islet Allograft Rejection in Immunized Hosts.* Cell Transplantation, 2013. **22**(7): p. 1137–1146.
87. Tibell, A., et al., *Survival of Macroencapsulated Allogeneic Parathyroid Tissue One Year after Transplantation in Nonimmunosuppressed Humans.* Cell Transplantation, 2001. **10**(7): p. 591–599.
88. Rezania, A., et al., *Maturation of human embryonic stem cell-derived pancreatic progenitors into functional islets capable of treating pre-existing diabetes in mice.* Diabetes, 2012. **61**(8): p. 2016–2029.
89. Kirk, K., et al., *Human embryonic stem cell derived islet progenitors mature inside an encapsulation device without evidence of increased biomass or cell escape.* Stem Cell Research, 2014. **12**(3): p. 807–814.
90. HENRY, R.R., et al., *Initial Clinical Evaluation of VC-01TM Combination Product—A Stem Cell–Derived Islet Replacement for Type 1 Diabetes (T1D).* Diabetes, 2018. **67**(Supplement_1).
91. Ramzy, A., et al., *Implanted pluripotent stem-cell-derived pancreatic endoderm cells secrete glucose-responsive C-peptide in patients with type 1 diabetes.* Cell Stem Cell, 2021. **28**(12): p. 2047–2061.e5.
92. Shapiro, A.M.J., et al., *Insulin expression and C-peptide in type 1 diabetes subjects implanted with stem cell-derived pancreatic endoderm cells in an encapsulation device.* Cell Reports Medicine, 2021. **2**(12): p. 100466.
93. Ramachandran, K., H.H. Huang, and L. Stehno-Bittel, *A Simple Method to Replace Islet Equivalents for Volume Quantification of Human Islets.* Cell Transplant, 2015. **24**(7): p. 1183–94.
94. Jeyam, A., et al., *Clinical Impact of Residual C-Peptide Secretion in Type 1 Diabetes on Glycemia and Microvascular Complications.* Diabetes Care, 2020. **44**(2): p. 390–398.
95. Nyitray, C.E., et al., *Polycaprolactone Thin-Film Micro- and Nanoporous Cell-Encapsulation Devices.* ACS Nano, 2015. **9**(6): p. 5675–82.
96. Chang, R., et al., *Nanoporous Immunoprotective Device for Stem-Cell-Derived β-Cell Replacement Therapy.* ACS Nano, 2017. **11**(8): p. 7747–7757.
97. Buitinga, M., et al., *Micro-fabricated scaffolds lead to efficient remission of diabetes in mice.* Biomaterials, 2017. **135**: p. 10–22.
98. Buitinga, M., et al., *Microwell scaffolds for the extrahepatic transplantation of islets of Langerhans.* PLoS One, 2013. **8**(5): p. e64772.
99. Smink, A.M., et al., *Selection of polymers for application in scaffolds applicable for human pancreatic islet transplantation.* Biomed Mater, 2016. **11**(3): p. 035006.
100. Hadavi, E., et al., *In vitro degradation profiles and in vivo biomaterial-tissue interactions of microwell array delivery devices.* J Biomed Mater Res B Appl Biomater, 2021. **109**(1): p. 117–127.
101. Chick, W.L., et al., *Artificial pancreas using living beta cells:. effects on glucose homeostasis in diabetic rats.* Science, 1977. **197**(4305): p. 780–2.
102. Song, S. and S. Roy, *Progress and Challenges in Macroencapsulation Approaches for Type 1 Diabetes (T1D) Treatment: Cells, Biomaterials, and Devices.* Biotechnology and bioengineering, 2016. **113**(7): p. 1381–1402.
103. Maki, T., et al., *Treatment of Diabetes by Xenogeneic Islets Without Immunosuppression: Use of a Vascularized Bioartificial Pancreas.* Diabetes, 1996. **45**(3): p. 342–347.
104. Burczak, K., et al., *Protein permeation through poly(vinyl alcohol) hydrogel membranes.* Biomaterials, 1994. **15**(3): p. 231–8.
105. Song, S., et al., *An intravascular bioartificial pancreas device (iBAP) with silicon nanopore membranes (SNM) for islet encapsulation under convective mass transport.* Lab on a chip, 2017. **17**(10): p. 1778–1792.
106. Storrs, R., et al., *Preclinical development of the Islet Sheet.* Ann N Y Acad Sci, 2001. **944**: p. 252–66.

107. Lamb, M., et al., *Function and Viability of Human Islets Encapsulated in Alginate Sheets: In Vitro and in Vivo Culture.* Transplantation Proceedings, 2011. **43**(9): p. 3265–3266.
108. Dufrane, D., R.-M. Goebbels, and P. Gianello, *Alginate Macroencapsulation of Pig Islets Allows Correction of Streptozotocin-Induced Diabetes in Primates up to 6 Months Without Immunosuppression.* Transplantation, 2010. **90**(10): p. 1054–1062.
109. An, D., et al., *Designing a retrievable and scalable cell encapsulation device for potential treatment of type 1 diabetes.* Proceedings of the National Academy of Sciences, 2018. **115**(2): p. E263.
110. An, D., et al., *Developing robust, hydrogel-based, nanofiber-enabled encapsulation devices (NEEDs) for cell therapies.* Biomaterials, 2015. **37**: p. 40–48.
111. Wang, X., et al., *A nanofibrous encapsulation device for safe delivery of insulin-producing cells to treat type 1 diabetes.* Sci Transl Med, 2021. **13**(596).
112. Ernst, A.U., L.H. Wang, and M. Ma, *Interconnected Toroidal Hydrogels for Islet Encapsulation.* Adv Healthc Mater, 2019. **8**(12): p. e1900423.
113. Gurlin, R.E., J.A. Giraldo, and E. Latres, *3D Bioprinting and Translation of Beta Cell Replacement Therapies for Type 1 Diabetes.* Tissue Eng Part B Rev, 2021. **27**(3): p. 238–252.
114. Duin, S., et al., *3D Bioprinting of Functional Islets of Langerhans in an Alginate/Methylcellulose Hydrogel Blend.* Advanced Healthcare Materials, 2019. **8**(7): p. 1801631.
115. Liu, X., et al., *Development of a Coaxial 3D Printing Platform for Biofabrication of Implantable Islet-Containing Constructs.* Advanced Healthcare Materials, 2019. **8**(7): p. 1801181.
116. Marchioli, G., et al., *Fabrication of three-dimensional bioplotted hydrogel scaffolds for islets of Langerhans transplantation.* Biofabrication, 2015. **7**(2): p. 025009.
117. Kim, J., et al., *3D cell printing of islet-laden pancreatic tissue-derived extracellular matrix bioink constructs for enhancing pancreatic functions.* Journal of Materials Chemistry B, 2019. **7**(10): p. 1773–1781.
118. Song, J. and J.R. Millman, *Economic 3D-printing approach for transplantation of human stem cell-derived β-like cells.* Biofabrication, 2016. **9**(1): p. 015002–015002.
119. Daoud, J.T., et al., *Long-term in vitro human pancreatic islet culture using three-dimensional microfabricated scaffolds.* Biomaterials, 2011. **32**(6): p. 1536–42.
120. Farina, M., et al., *3D Printed Vascularized Device for Subcutaneous Transplantation of Human Islets.* Biotechnol J, 2017. **12**(9).
121. Sabek, O.M., et al., *Three-dimensional printed polymeric system to encapsulate human mesenchymal stem cells differentiated into islet-like insulin-producing aggregates for diabetes treatment.* Journal of Tissue Engineering, 2016. **7**: p. 2041731416638198.
122. Sih, J., et al., *Characterization of nanochannel delivery membrane systems for the sustained release of resveratrol and atorvastatin: new perspectives on promoting heart health.* Analytical and Bioanalytical Chemistry, 2013. **405**(5): p. 1547–1557.
123. Hwang, D.G., et al., *A 3D bioprinted hybrid encapsulation system for delivery of human pluripotent stem cell-derived pancreatic islet-like aggregates.* Biofabrication, 2021. **14**(1): p. 014101.
124. Idaszek, J., et al., *Alginate-based tissue-specific bioinks for multi-material 3D-bioprinting of pancreatic islets and blood vessels: A step towards vascularized pancreas grafts.* Bioprinting, 2021. **24**: p. e00163.
125. Mohammadi, M.R., et al., *Preferences of Type 1 Diabetic Patients on Devices for Islet Transplantation.* Cell Transplant, 2020. **29**: p. 963689720952343.

Bioengineered Vascularized Insulin Producing Endocrine Tissues

Francesco Campo, Alessia Neroni, Cataldo Pignatelli, Juliette Bignard, Ekaterine Berishvili, Lorenzo Piemonti, and Antonio Citro

The authors Ekaterine Berishvili, Lorenzo Piemonti, and Antonio Citro have equally contributed to this work as last author

1 Shifting the Paradigm from In Vivo to Ex Vivo Islet Engraftment

Intrahepatic islet transplantation is the current clinical beta cell replacement strategy for type 1 diabetes (T1D) patients. However, the reduced islet engraftment upon transplantation is one of the major limitations that hinders the success rate of the procedure [1, 2]. In native context, the pancreatic islets are embedded in their extracellular matrix (ECM), surrounded by a dense vascular network. This configuration is well known as "vascularized endocrine niche" and represents a specific

The authors Francesco Campo, Alessia Neroni and Cataldo Pignatelli have equally contributed to this work as first author

F. Campo · A. Neroni · L. Piemonti (✉)
Diabetes Research Institute and Clinical Unit of Regenerative Medicine and Transplant, IRCCS San Raffaele Hospital, Milan, Italy

University Vita-Salute San Raffaele, Milan, Italy
e-mail: piemonti.lorenzo@hsr.it

C. Pignatelli · A. Citro (✉)
Diabetes Research Institute and Clinical Unit of Regenerative Medicine and Transplant, IRCCS San Raffaele Hospital, Milan, Italy
e-mail: citro.antonio@hsr.it

J. Bignard
Laboratory of Tissue Engineering and Organ Regeneration, Department of Surgery, University of Geneva, Geneva, Switzerland

Cell Isolation and Transplantation Center, Department of Surgery, Geneva University Hospitals and University of Geneva, Geneva, Switzerland

Faculty Diabetes Center, University of Geneva Medical Center, Geneva, Switzerland

L. Piemonti et al. (eds.), *Pluripotent Stem Cell Therapy for Diabetes*,
https://doi.org/10.1007/978-3-031-41943-0_8

micro-environment that is mandatory for the pancreatic endocrine tissue to daily absolve its function [3]. In physiological condition, this fine-tuned activity is regulated and controlled, from the metabolic point of view, by a defined microarchitecture based on to cell-to-cell and cell-to-ECM interactions [2]. In contrast, upon intrahepatic transplantation, the islet microenvironment is characterized by a severe and extensive inflammatory processes that are caused by the two major steps of the procedures: (i) the isolation and (ii) the transplantation [4].

During the isolation process, islets lost their native microenvironment due to a combination of enzymatic and mechanical steps, resulting in an inevitably inflamed preparation [5–9]. Numerous studies have reported that immediately following isolation, islets release a significant quantity of inflammatory chemo/cytokines, which could negatively impact the early transplantation phase [4, 10–12]. In this context, several approaches have been developed to tackle the inflammation before or after implantation without substantial improvement in the final clinical outcome [4, 13–16].

In vivo, islets required time to remodel the hepatic parenchyma and "engineer" de novo their own specific microenvironment [17]. During this timeframe, islets are subjected to several insults related to the surgery and the immunological reaction - i.e., ischemic reperfusion injury, instant blood-mediated inflammatory reaction (IBMIR), innate/adaptive immune response - that substantially reduce the effectiveness of the procedure [4, 8, 18]. For several years, groups aimed to ameliorate the islet engraftment, function and survival by modulating the events upon implantation [19–24]. The clinical translation of this approach shifted the overall implantable mass to 10.000 IEq/kg to compensate for the loss of the islets in the early transplantation phase. Consequently, to satisfy the ratio patient/islet demand, the isolation procedure moved from 1 to 2 donor pancreas for each transplantation without the definition of an effective immunosuppressive regimen able to achieve stable long-term insulin independence [23–27].

As of now, bioengineering approaches are intended to overcome the classical limitation of islet transplantation. There is a progressive interest in engineering the islet niche micro-environment ex vivo prior to in vivo implantation (Fig. 1). These approaches are growing on the basis of the different limitations that the islet faced during isolation and engraftment time in vivo: (i) the avascular phase; (ii) the ECM interaction and (iii) the mechanical and inflammatory stress [2].

E. Berishvili (✉)
Laboratory of Tissue Engineering and Organ Regeneration, Department of Surgery, University of Geneva, Geneva, Switzerland

Cell Isolation and Transplantation Center, Department of Surgery, Geneva University Hospitals and University of Geneva, Geneva, Switzerland

Faculty Diabetes Center, University of Geneva Medical Center, Geneva, Switzerland

School of Natural Sciences and Medicine, Ilia State University, Tbilisi, Georgia
e-mail: ekaterine.Berishvili@unige.ch

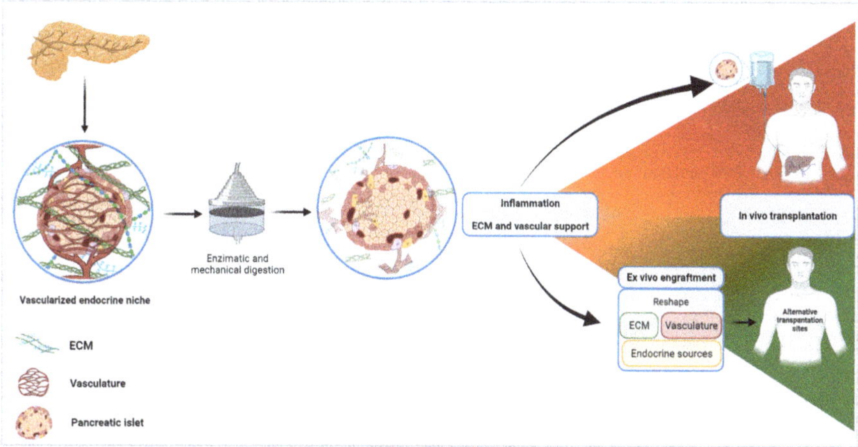

Fig. 1 Shifting the paradigm from in vivo to ex vivo engraftment. In the physiological state, pancreatic islets are embedded in the vascularized endocrine niche (i.e., extracellular matrix (ECM) and vasculature essential for their survival and function. During the isolation process, the microenvironment is disrupted, resulting in inflammation and the loss of ECM and vascular support. When infused in vivo (hepatic site), the islets face further challenges and require time to adapt. Conversely, by engineering the microenvironment ex vivo through the reshape of ECM, vasculature or using alternative endocrine sources, it is possible to significantly enhance the engraftment and functionality of the graft once it is implanted in vivo in alternative sites

More specifically, bioengineering approaches are developing new strategies that used alternative cell sources and native or synthetic materials to reshape the niche architecture in order to provide to engrafting cells an environment where they can be safely harbored maximizing the cell-to-scaffold ratio [28]. Of note, there is an increasing awareness that engineering an endocrine pancreas demands meticulously designed, with particular attention on recreating a customized endocrine microenvironment that considers not only the endocrine and vasculature compartments but also the three-dimensional microstructure conferred by the ECM organization. Several attempts demonstrated that the fine design of these three components ex vivo may drive the generation of a safe, controlled and organized endocrine environment that may improve the endocrine engraftment and function in vivo [29–31]. To this end, new approaches aim to generate a bioengineered scaffold resembling, as much as possible, the microstructures of the native 3D endocrine niche architecture, including the ECM composition and disposition, as well as the vasculature to provide the endocrine components with a suitable native-like microenvironment (Fig. 2).

The generation of an endocrine vascularized pancreas platform prior to implantation could enable the ex vivo assembly of the three essential building blocks – ECM-like microarchitecture, vasculature, and endocrine cells – promoting the connection between the endocrine and vascular compartments. From a clinical perspective, the ability to ex vivo assemble the endocrine and vascular compartments with a native-like scaffold has the potential to: (1) prevent inflammatory events that

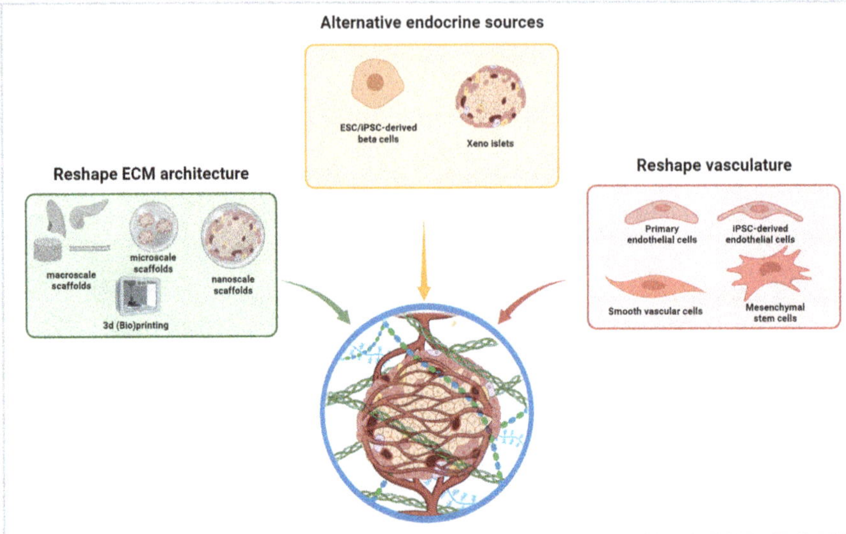

Fig. 2 Reshape the vascularized endocrine niche. Schematic overview on how to ex vivo engineer an insulin endocrine tissue: ECM architecture (green box) by using macro-micro-nano-scale scaffolds or 3D-bioprinting; vasculature (red box) regenerating ex vivo vascular network with several cell types; exploiting alternative endocrine sources (yellow box)

may affect the endocrine components in the peri-transplant phase, and maximize the number of endocrine cells capable of engraftment; (2) enable rapid connection to the host vascular network, thanks to an established vascularization of the scaffold; and (3) promote rapid graft function upon implantation. Thus, it is crucial to comprehend the methods to bioengineer vascularized scaffolds for recreating the pancreatic endocrine niche, including how the building blocks can be assembled. The following paragraphs explore approaches for reshaping the ECM architecture, the vasculature and endocrine compartment to bioengineer a vascularized endocrine insulin producing tissue.

2 Reshaping the Architecture

2.1 The ECM Relevance in Bioengineering the Vascularized Endocrine Pancreas

ECM is a key driver in controlling islet fate, particularly in terms of survival and function. The predominant ECM proteins, in native islet niche, are collagen type I and IV [32], laminin isoforms 111, 511, 521, 411 whose expression changes during endocrine tissue maturation [33] and fibronectin [34]. These proteins contribute to islet development, maturation, maintenance of the capsule, survival by promoting the expression of the anti-apoptotic BCL-2 protein [35] and insulin secretion [36]. Therefore, bioengineering tools aimed to create an adequate, supportive ECM structure with a physiological composition and proper architecture that mimics the native endocrine niche to design functional insulin-producing tissue ex vivo [8]. From a bioengineering perspective, ECM may represent a tunable shield that not only provide mechanical support to cells but also protects the graft from the immune system by acting as a physical barrier between endocrine tissues and recipient immune cells [37]. Herein, we reviewed the most up-to-date strategies for bioengineering insulin-producing endocrine tissue by reshaping the ECM proteins content using biomimetic 3D scaffolds casted with different techniques as whole organ decellularization and recellularization, 3D (bio)-printing, micro and nano-encapsulation.

3D scaffolds can be fabricated exploiting biomaterials of different origin as synthetic, naturally derived, or biological biomaterials. Among synthetic biomaterials relevant for pancreas bioengineering are Polyethilenglycol (PEGs), Polyester as poly(lactic acid) (PLA, PDLA, PLLA, PDLLA), poly(glycolic acid) (PGA), poly(lactic acid-co-glycolic acid) (PLGA), and poly(caprolactone) (PCL), silicon-based biomaterials as Polydimethylsiloxane (PDMS), or Polytetrafluoroethylene (PFTE) [38]. Synthetic biomaterials offer enhanced reproducibility and control on composition, tunable properties [39]. PEG is a very versatile biomaterial in terms of physicochemical properties, but it lacks cell binding sites [39]. However, the molecule itself has several functional reaction groups which can be easily exploited in specific chemical reaction to further modify and functionalized the biomaterial linking growth factors, such as laminin or collagen-derived peptides, through click-chemistry [40] to improve beta cell survival, viability and function [35]. Synthetic scaffolds can be further modified to improve local immuno-tolerance of the graft in an allo/xenogeneic setting via immobilization of immunomodulatory ligands as FASL, PDL1 or immunosuppressive, anti-inflammatory compounds as CTLA4-Ig, TGF-Beta as extensively reviewed by Wang et al. [41]. Naturally derived biomaterials offer some control on stiffness, degradation rates; however, they are limited by structural and composition variability. They can be subdivided in polysaccharide based as alginate, chitosan or hyaluronic acid, or protein based as collagen, fibronectin or fibrin [38].

Additionally, 3D scaffolds can be classified according to their scale and the number of endocrine cells embedded in macro-encapsulating devices housing many islets, micro-devices containing a small number of islets, or nano-encapsulating devices housing a single islet or cell [37]. Each different category has been designed to solve a particular limit: the lack of vascularization, the lack of ECM support, an impaired oxygen and nutrient diffusion with the final aim to improve the endocrine function and the immune isolation of the graft.

2.2 Macroscale Scaffolds

Macroscale scaffolds are designed with a reshaped architecture that supports large-scale engraftment and contains all cellular graft components. They can be designed with native ECM or synthetic/naturally derived polymer, with the goal to physically concealing the endocrine graft against to the immunological response and improving cell interaction and engraftment. Viacyte was one of the pioneers of macroscale scaffolds tested in preclinical and clinical settings. They evaluated a synthetic macro-device called VC-01 in their first trial (NCT02239354), which encapsulated embryonic stem cell (SC)-derived islet-like clusters in a double layer of PFTE membrane. The first layer was permeable, promoting tissue integration and nutrient and oxygen exchange, while the inner one was cell impermeable but still capable to favor the exchange of nutrients and oxygen. However, the trial was suspended because of a low cell survival and engraftment rate due to the foreign body reaction (FBR) reaction, which clogged the membrane and limited oxygen/nutrient exchange [42].

Additional attempts at the macroscale device tried a combinatorial approach by incorporating a synthetic polymer-based device with a protein-based hydrogel ECM. Recently, Paez-Mayorga et al. reported the generation of a refillable dual-reservoir device called NICHE. This device is composed of a central cell reservoir, closed by a 100 μm-pore grid, loaded with a hydrogel containing mesenchymal stem cells to facilitate angiogenesis, tissue penetration, and engraftment of endocrine cells, and an external U-shaped drug reservoir capable of releasing immunosuppressive compounds in a controlled manner. The authors demonstrated that after double implantation of allogenic pancreatic islets, the in vivo pre-vascularized NICHE supported their complete vascularization and engraftment, reversed diabetes within 40 days in an immune-competent rat model and maintained euglycemic conditions for over 150 days [43]. This approach was also tested in monkeys for safety purposes, indicating a future potential application in a clinical setting. Another relevant macroscale scaffold was published by Liang et al. It was designed to improve the engraftment eliminating hypoxia in the avascular phase using oxygen-generating microbeads named OxySite, which were incorporated in the PDMS. The resulting scaffolds generated local oxygenation for up 20 days. The integration of this oxygen-generating material leads to an improved graft function in a syngeneic model compared to control implants [44]. An et al. developed a hydrogel nanofiber-enabled encapsulation macro-device (NEEDs) in which a highly porous electrospun nanofibrous membrane of pre-made tubular or planar device is impregnated with liquid hydrogel precursor exploiting water-based properties. Following successful crosslinking, the researchers created a novel device that combines the biocompatibility of a hydrogel with the robustness of nanofibers, resulting in a structure with properties of both materials. As a proof-of-concept study, rat pancreatic islets were encapsulated in the device and transplanted into diabetic mice. Within 2 days of transplantation, the diabetic condition was reversed, and this was maintained for the 8-week follow-up period. Upon retrieval of the device, viable and functional islets were found, demonstrating the efficacy of the device in promoting islet survival and function [45]. Also, An et al. developed the TRAFFIC (Thread-Reinforced Alginate Fiber for Islet Encapsulation) device, which incorporates a tough polymer thread to provide mechanical resistance, as well as a uniform, strongly adhered alginate hydrogel layer for the encapsulation of beta cells. The TRAFFIC system is significant for its scalability to a clinically relevant size, as a longer thread can accommodate a larger islet mass, and for its high retrievability through minimally invasive laparoscopic procedures [46].

Recently, significant progress has been made in the field of macroscale scaffolds using 3D whole decellularized organ scaffolds. These decellularized organs are obtained by constantly detergents perfusion in the native vascular tree, which allows for the preservation of both ultrastructure and vascular network architecture [47]. Furthermore, the main ECM components with their native geometrical organ distribution are also preserved. With the well-preserved native architecture and ECM composition, the organ can be re-endothelialized, thereby regenerating a vascular network. This innovative approach holds tremendous potential for creating functional bioengineered organs for transplantation.

This approach has been applied since 2010 to various organs such as heart [48], kidney [49] lung [50] and pancreas [51] with the final aim to engineer an artificial organ recellularizing those scaffolds with several cell sources. In the field of endocrine pancreas bioengineering, whole organ decellularization was first applied to rodent pancreata, which were recellularized with murine islets, demonstrating the procedure's feasibility and showing improved insulin secretion after glucose stimuli [52]. Later, clinically relevant-sized pancreata from pig [53] and humans [51] were decellularized and recellularized with human islets and primary endothelial cells. However, considering the native structural organization of the pancreas – with only 5% endocrine tissue, receiving 10–15% of the entire organ blood flow, scattered in the dominant exocrine tissue 95% – this data underlines the organ's architectural complexity, pointing out the presence of exocrine and endocrine compartments with various features. However, from a bioengineering standpoint, the regeneration of the endocrine microenvironment using a scaffold derived from the pancreas poses a significant challenge. The endocrine side represents only a small portion of the organ structure, and pancreatic implantation is a major invasive surgery.

To this end, studies have shown that reprogramming decellularized organ function using different cell types in an ex vivo repurposing fashion may improve cell engraftment and function in a controlled microenvironment [29, 47–50]. Indeed, several groups tried to re-purpose a whole organ scaffold from per se highly vascularized tissues, such as the spleen [54], pericardium [55], kidney [56], liver [57] and lung [30] to the endocrine pancreas. In particular, the decellularized lung has been re-purposed as it is characterized by two anatomically independent compartments for air and blood volume, also showing an overlap of 22 ECM proteins with pancreas. Alveoli are readily accessible via the large airways and embedded in the pulmonary vasculature, mirroring the dense vascular network of the endocrine pancreatic tissue. In 2019, Citro et al. used a decellularized rat lung left lobe scaffold to bioengineer a vascularized islet organ (VIO) [30], using HUVEC and rodents islets, seeded respectively through pulmonary vessel system and trachea. The VIO, transplanted subcutaneously in immunocompromised diabetic mice, achieve function in just 5 days. They demonstrated the great impact of native organ ECM in promoting faster engraftment and vascularization ex vivo and an accelerated endocrine function in vivo. Following the same idea, a new version of the platform was proposed with the integration of immature neonatal porcine islets (NPIs) and healthy subject-derived endothelial cells, named blood outgrowth endothelial cells (BOEC). The platform fosters in vitro NPIs maturation and in vivo primary function after transplantation for over 18 weeks [30].

On the same path, it is possible to create macroscale devices using native organ ECM, which can be re-designed according to biological need. Whole decellularized organ can be enzymatically digested to generate thermal-triggered hydrogels (ECM-derived scaffolds). Enzymatic digestion has been applied to several decellularized tissues as extensively reviewed by Saldin et al. [58], obtaining decellularized ECM (dECM) hydrogels. Recently, Jiang et al. generated a dECM hydrogel from porcine lung, bladder and pancreas [59]. Encapsulation of both rodents and human islets in those hydrogels resulted in an improved function compared to batch-matched native islets cultured in standard culture condition [59]. More recently Sackett et al. [60] successfully generated and characterized a dECM hydrogel from human pancreas suitable for cell culture and transplantation. These studies demonstrate that human islets embedded in dECM hydrogels have increased insulin secretion in response to glucose stimuli, better mitochondrial function, and lower apoptosis levels compared to standard culture conditions [61]. However, dECM hydrogels tend to lose some native components, have poor mechanical properties, and degrade rapidly in vivo [62, 63]. Therefore, they can be reshaped and functionalized with bio-inert materials of both synthetic or natural origin such as alginate to reduce degradability and increase stiffness [64]. Nonetheless, dECM hydrogels showed excellent potential in terms of production and future clinical applicability, but they have a strong intrinsic limit: the loss of native organ ultrastructure and architecture.

To overcome this limitation, a 3D bioprinting approach may be used to reshape the organ structure architecture with more precise stiffness, geometrical spatial cell-to-ECM contact, and function. 3D bioprinting involves the deposition of a biomaterial, alone or with cells, layer-by-layer in a desired position to create a 3D tissue-like structure [65]. Bioink is the term used for any polymeric biomaterial with cells and potentially in a single set up you can simultaneously use several bioinks with different cells loaded. Based on the layer deposition methods, bioprinters can be classified into four major groups: extrusion-based, inkjet, stereolithography based, and light based [66]. Among these groups, extrusion-based are the most used in beta cell replacement because they showed good flexibility in terms of spatial resolution, printable bioinks, and viability of the cells after the extrusion since they have slow printing speed minimizing the risk of damaging cells [67].

Several groups tried to recreate and rebuild the ECM structure using 3D bioprinters coupled with naturally derived materials such as alginate, gelatin or synthetic ones such as poly-lactic acid (PLA) printing rodents, human islets and iPSCs-derived beta cells [68]. Of note, Hwang et al. developed a hybrid system composed of a macroporous polymer capsule with a stagger-type membrane and a nanoporous 3D printed dECM hydrogel containing islet-like cell aggregates. These aggregates remained viable and maintained their insulin-producing function [69]. In 2019 Kim et al. reported one of the first attempt in bioprinting a dECM hydrogel for beta cell replacement. They developed a dECM pancreas in which hiPSC-derived beta cells were bioprinted, showing an upregulation of insulin secretion and an enhanced maturation [70]. Bioprinting technology is becoming increasingly relevant in the field of beta cell replacement, although its application remains limited. A significant advantage of this technology is its unique capability to integrate different cell types and biomaterials simultaneously. However, some major challenges still need to be addressed, such as the need to overcome the vasculature issue for the perfusion of larger constructs, and to improve printing speed and yield for higher cell survival and viability.

2.3 *Microscale Scaffolds*

Microscale scaffolds have the potential to reshape the architecture of the endocrine niche by maximizing the cell-to-scaffold ratio [28]. This enables better oxygenation and nutrient diffusion, as well as protecting isolated islets from host immune reactions by providing a physical barrier [28]. Various biomaterials can be used to fabricate microscale scaffolds, including synthetic, naturally derived, and biological materials as ECM-derived one. Different fabrication techniques, such as batch emulsion systems, microfluidic emulsion systems, and electrohydrodynamic (EHD) spraying, are used based on the starting biomaterial [71–73]. In batch emulsion systems, a pre-polymer is solved in acqueous solution and mixed with an oil phase, then the exposition to an external crosslinker induces microparticles formation. In microfluidic emulsion techniques, aqueous cell suspension is pumped with a defined flow rate in a microchannel where it encounters an oil phase, allowing its hydrophobic-forces and shear stress-driven encapsulation, assuring a higher control on particle size [72]. In EHD spraying, a viscous water-based solution containing the cells is extruded upon the application of high voltage at the nozzle, allowing the formation of droplets, which encapsulate cells [73]. There are some limitations to each technique, such as poor control over scaffold size with batch emulsion systems [74], while in microfluidic emulsion ones and in EHD there is a better control over scaffold size with a higher reproducibility, scalability and high throughput [75].

Recently, a new microencapsulation technique named core-shell spheriphication has been introduced and applied to both synthetic and naturally derived biomaterials. Harrington et al. developed a nonemulsion-based method to generate hydrogel microsphere compatible with slow-hardening biomaterials [76].

Among synthetic biomaterials, the most relevant for microencapsulation are PEGs [76, 77], and PLGA [38, 78]. They were largely used thanks to their high reproducibility and tunable mechanical/physical/chemical properties obtaining good results in promoting beta cell survival, function and engraftment after transplantation [38]. Harrington et al. microencapsulated canine pancreatic islets in PEG-based scaffolds further modified with diacrylate (PEGDA). They tested not only the function of encapsulated islets, which were able to revert diabetes up to 16 weeks, but also the ability of PEGDA micro-scaffolds to protect the graft after transplantation in NOD mice presenting adaptive immune cells [76]. In another study, pancreatic islets were seeded on PLGA-based micro-scaffolds and then transplanted in immunocompromised mouse models. After imaging, pancreatic islets remained localized in scaffolds, survived and were able to reverse diabetes compared to control groups [79]. To improve their function, PLGA scaffolds were also coated with ECM-derived proteins such as collagen type IV, laminin and fibronectin [80] or other synthetic molecules to improve biocompatibility, hydrophilicity and cell adhesion [81].

Among naturally derived biomaterials alginate is considered the best option due to its natural origin, cost-effectiveness and tunable physicochemical properties [82], and it was largely applied to several cell sources from rodent to human pancreatic islets and also with SC-derived beta cells. Hu et al. demonstrated firstly that an alginate microscale scaffolds modified with low-methoxy-pectin can significantly reduce innate immune cell activation both in vitro and in vivo in xenogeneic setting, hampering peri-capsular fibrosis formation [83]. Bochenek et al. advanced to a clinically relevant animal model using non-human primates to test various microscale scaffolds modified with alginate. They identified a compound that can significantly reduce FBR and fibrosis, thereby protecting the graft. The graft remained viable and metabolically active for up to 4 months without requiring immunosuppression [84]. Lastly, J. Vegas et al. encapsulated SC-derived beta cells in alginate derivatives, transplanting them into the peritoneal space of immunocompetent diabetic mouse models. Recipients treated with alginate-encapsulated SC-derived beta cells maintain normoglycemia up to 6 months in the absence of immunosuppression [85]. All those studies demonstrated the feasibility of the microscale scaffolds with successful transplantation in rodents and non-human primates with optimal metabolic activity and immune protection of the graft.

However, microscale systems present some limitation: insufficient mass transfer due to scaffold geometry, disproportional graft volume, and un-physiological intracapsule microenvironment. The mass transfer of oxygen and nutrients remains a key challenge in encapsulation field since the availability of oxygen primary and then of nutrients is essential for islet survival and function [86]. Indeed, various oxygenating materials [87] and devices [88, 89] were proposed with optimal outcomes in preclinical setting leading also to a clinical trial (NCT02064309) which partially confirmed preclinical results. The mass transfer issue is strictly linked to the scaffold geometry. While a spherical geometry is typically used for microscale scaffolds due to the casting techniques mentioned previously, recent evidence suggests that other shapes, such as rods or toroid, may offer more benefits in terms of larger surface area or smaller diffusion distances [90, 91]. However, a larger exchange surface or volume of the graft can make transplantation more challenging, as an appropriate transplantation site must be found to accommodate larger-sized graft [92].

Additionally, the intra-capsule microenvironment might be insufficient to guarantee all the cell-ECM interactions necessary for the correct function of the encapsulated endocrine cells. Therefore, a key challenge is to identify factors that can promote islet survival once added to the intra-capsule microenvironment. Quin et al. reported that the combinatorial addiction of both solubilized ECM components as collagen type IV and RGD peptide and necroptosis inhibitor factor in intra-capsular micro-environment of alginate-based scaffold was able to reduce inflammatory stress, cytokine overproduction and prevent suppression of FAK/AKT pathway contributing to higher survival and enhanced longevity of the pancreatic islets after microencapsulation [93].

2.4 Nanoscale Scaffolds

Nanoscale devices are becoming a valuable alternative in beta cell replacement bio-engineering, offering two main types of scaffolds: (1) conformational coating, which consists of a biocompatible hydrogel covering the cell and forming a cross-linked structure on the cell surface [71]; (2) layer-by-layer (LBL) encapsulation which deposits alternate layers of polyanions and polycations using the negative surface of the pancreatic islets [94]. Both systems offer the advantage of a small graft volume, making it suitable for transplantation in various sites. Additionally, their significant benefits are attributed to the durability and perm-selectivity of the coating [95]. The durable coating increases cell viability by protecting individual encapsulated islets from mechanical stress, while perm-selectivity allows for the selective passage of specific biomolecules or nutrients that promote islet survival, safeguarding the graft from immune cell destruction [95].

In 2014 Tomei et al., firstly described a conformational coating approach, which allows islet encapsulation in a thin layer of hydrogel. Coated islets exhibited no delay in GSIS compared to uncoated batch-matched islets in vitro and a similar function in vivo when transplanted in diabetic syngeneic mouse models. Later, in 2018, Manzoli et al. proved the feasibility of the graft immune-isolation: mismatched murine islets previously encapsulated with PEG, were transplanted in an allogenic mouse model, which achieve normoglycemia for over 100 days [96]. Additionally, Stabler et al. proposed a conformational coating approach where pancreatic islets were coated with PEG polymer to mitigate host-graft reaction and transplanted in non-human primates in a reduced immunosuppressive regimen. Recipients of coated islets showed an improved early C-peptide and glycemic control as well as a reduced insulin requirement [97].

Another study by Syed et al. used LBL deposition of chitosan coating on human pancreatic islets. They tested islet function both ex vivo and in vivo. Ex vivo, embedded human islets preserve their insulin secretion performance in response to glucose stimuli. In vivo, encapsulated islets transplanted under the kidney capsule of diabetic C57Bl/JI mice allowed long-term normoglycemia with minimal infiltration of immune cells into the graft [98]. De Toni et al. compared PEG conformational coated and alginate microencapsulated batch-matched islets after retrieval upon transplantation in mice, measuring several parameters such as the size of the islets and capsule, as well as the capsule permeability to generate an in-silico simulation through Comsol multiphysics software [99]. They demonstrated that insulin response to glucose stimuli is dependent on capsule size, the presence of a perm-selective material can promote a delay in insulin secretion and a minimal capsule size is needed for physiological function. Interestingly, as the in silico model predicted, in microscale capsule, glucose remain for longer time causing an overexposure of islet to glucose which release insulin for more prolonged time with a non-physiological insulin secretion in vitro and possibly leading to a more frequent hypoglycemia event in vivo [99].

To date, nano-encapsulation seems a promising approach for beta cell replacement. However further studies are needed to assess the contribution of this nano-scaffold to the survival, engraftment, and function of islets.

Currently, each category of scaffolds has its own set of advantages and disadvantages. As we move from the macro to the nanoscale architecture, the surface area available for oxygen and nutrient exchange is maximized in tandem. However, this also means, from a clinical point of view, that the final size of the graft is increased, requiring thousands of micro or nano-devices. In addition, the choice of SC-derived sources for engineering insulin-producing tissue raises concerns about tracking and retrieving the graft, which should be easily removable and traceable. Therefore, a delicate balance must be struck between scale, proper architecture, suitable starting biomaterials for ECM remodeling, and efficient bioengineering of insulin-producing tissue while safeguarding it from the host immune response.

3 Reshaping the Vasculature

3.1 The Relevance of the Pre-vascularization in Bioengineering the Vascularized Endocrine Pancreas

The vasculature within endocrine tissue is crucial in maintaining proper physiological function of the islets [100]. In the field of bioengineering, a functional vascular (re)generation presents a significant challenge [100]. Therefore, a promising strategy that is independent of scaffold size and composition is the pre-vascularization of the graft. The integration of functional vascular networks ex vivo into bioengineered systems is crucial to recapitulate the endocrine pancreatic niche, and to provide the endocrine cells with adequate oxygenation and physiological nutrients [100]. In this context, bioengineering approaches are taking advantage of the possibility to generate modular scaffold ex vivo that can integrate reshaped vascular structure in a safe and controlled environment prior to the transplantation. This need is strongly supported by experimental data that showed that the delivery of angiogenic factors during transplantation is insufficient to rapidly reconstitute the specialized three-dimensional islet niche vasculature, thereby hampering engraftment success rates [101].

Since the 1990, both in vitro and in vivo investigations focused on recreating a vascular network using: (1) human endothelial cells (ECs) to line the lumen of the bioengineered vascular structures; (2) human perivascular cells to support EC function and provide perivascular stability to the newly generating vascular networks; and (3) scaffolds to provides ECM support and physical cues for cells to correctly interact and reorganize in a functional vascular network [102].

3.2 *Sources of Cells for the Generation of Vasculature Network*

3.2.1 Human Primary Endothelial Cells

For a long time, research studies predominantly utilized fully committed adult human ECs isolated from vascular tissue. In the early 1970s, the first isolated and in vitro cultured ECs were obtained from human umbilical veins (HUVECs) [103] which proved to be a pivotal EC model for both vascular biology and vascular network bioengineering. The first attempt at ex vivo vascular bioengineering was released by Black et al. in 1998, that used HUVECs to recreate a network resembling a human capillary in a tissue-engineered skin equivalent [104]. In this direction, a step forward was released by Schechner et al., who defined the conditions for assembling HUVECs into a capillary network within a three-dimensional hydrogel, demonstrating that a bioengineered human vascular network could connect with the host circulatory system after surgical implantation in immune-deficient mice [105].

Scaffold bioengineering for beta cell replacement has deeply investigated the potential role of the vascularization as a supporting strategy to promote endocrine tissue engraftment. In this path, HUVEC cells were the first, and one of the most, used ECs to demonstrate proof-of-concept of both ex vivo or in vivo scaffold based islet treatment. In this regard, Song et al. demonstrated that HUVEC can self-assemble within a fibrin-based ECM scaffold to form microvascular meshes that generate a controlled tubular microvascular structure. When this concept-technology was used in the presence of rat pancreatic islets, it significantly improved vascularization of subcutaneously transplanted endocrine cells and corrected diabetes in recipient mice for up to 3 months [106]. In the same path, native organ ECM scaffold-based technology may be used in beta cell replacement. Citro et al. in 2019 first and in 2023 later, showed that native decellularized organ ECM can be used as an approach for vascular bioengineering in T1D field. Also in this case, a self-assembly approach, with HUVEC [30] first and BOEC [29] later, in the presence of (im) mature islets showed that the pre-vascularization strategy is able, not only to improve the islet engraftment and function ex vivo but also improve the endocrine function in vivo for up to 18 weeks after implantation [29, 30]. As mentioned previously, in islet niche bioengineering, the vasculature can be reshaped by either modifying the main vascular structure or by modifying the endothelial cell repertoire of the islet prior to implantation. In this context, Wassmer et al. demonstrated the potential of dis- and re-aggregating pancreatic islets in combination with supporting cells such as HUVECs and human amniotic epithelial cells. Through the crosstalk generated between HUVECs and pancreatic islets, the graft's revascularization was facilitated after transplantation, promoting the viability and engraftment of the islets themselves [107].

In bioengineering approaches, the heterogeneity of ECs continues to be the subject of intense investigation based on the crucial evidence that the endothelium regulates multiple regenerative processes in an organ/tissue-specific manner. After successfully isolating ECs from the umbilical cord, other human tissues were used to obtain vessel/organ-specific endothelial cells as ECs obtained from skin [108] and adipose tissue [109] small diameter veins and the microvasculature. Alternatively, to HUVEC, the most studied primary human ECs were the dermal microvascular endothelial cells (HDMECs). As shown by Nör et al., those ECs have great potential, once embedded in a hydrogel, to organize themselves into functional micro vessels and to create anastomoses with the host vascular system 10 days after transplantation [110]. As previously reported, primary isolated ECs represent a valuable option to reshape the endocrine vasculature with a mature functional commitment.

3.2.2 Human Endothelial Primary Progenitor Cells

In bioengineering perspective to reshape the vasculature in a clinically relevant manner, progenitor and stem cell sources have been taken into account and considered as alternative sources to autologous, tissue-specific, human ECs. One alternative emerged in 1990 with the discovery of progenitor cells that circulate in human peripheral blood and differentiate in culture into ECs [111]. Research groups used to refer to them as endothelial progenitor cells (EPC) or BOEC. To date, they are known with a wide consensus as endothelial colony-forming cells (ECFC). The identification of ECFC represents a promising opportunity for clinical applications as this population of ECs can be obtained in a non-invasive manner, reducing the immunogenicity of a potential bioengineered device [112].

ECFCs were extensively characterized over the past 18 years with the identification of a robust endothelial phenotype [113]. ECFCs express all regular EC markers (e.g., VE-Cadherin, CD31, vWF), they uptake low-density lipoproteins (e.g., LDL-Ac) and bind to specific lectins with high affinity (e.g., UEA -1). Also, from the functional point of view, they can form vascular networks both in vitro and in vivo [114]. Lastly, ECFCs maintain an endothelial identity for a prolonged time in culture, suggesting their phenotype stability.

The first use of human ECFCs in bioengineering vascular networks was reported by Wu et al. In this study, cord blood-derived ECFCs together with human smooth muscle cells (SMCs) was seeded ex vivo in three-dimensional polyglycolic acid-poly-lactic acid (PGA-PLLA) scaffolds forming human micro vessels which were uniformly distributed throughout the construct. Thus, this proof-of-concept study indicated that human ECFCs are suitable for creating microvascular networks within tissue-engineered constructs [115].

Furthermore, the ability of human ECFCs to self-assemble into perfusable vascular networks in vivo was evaluated in 2007 by Melero-Martin et al. The group incorporated human ECFCs and SMCs, as supporting cells, derived from human saphenous veins, into matrigel and injected subcutaneously into immune-deficient nude mice. After 1 week, an extensive network of luminal structures was unambiguously coated by human ECFCs and contained murine erythrocytes, indicating the formation of functional anastomoses with the host vasculature [116].

Concordantly to the previous findings, Yoder et al. reported the ability of human ECFCs to create a perfused blood vessel network after implantation in NOD/SCID mice using a collagen/fibronectin hydrogel construct. This study was critical to point out the ECFCs ability to form vascular networks in vivo without any exogenous supporting cells as perivascular cells [113]. However, subsequent studies showed that, in order to achieve higher microvascular density, it is mandatory to add, to the reshaped vasculature structure, mural cells to obtain a functional complete vascular architecture [117–119].

In 2011, Kang et al. demonstrated that a bioengineered vascular network with human ECFCs in mice could be explanted and reconnected in secondary mice, re-establishing perfusion. This feature could extend the potential applications of this cell-based technology to transplant large tissue-engineered constructs [120].

Regarding the bioengineering of insulin-producing endocrine tissues, Bellofatto et al. generated functional, pre-vascularized endocrine constructs adopting amniotic membrane-derived hydrogel, islets, and BOECs to be transplanted in diabetic hosts. The authors found that amnion-derived hydrogel seeded with endocrine pancreatic tissue and ECs could be used for functional, pre-vascularized endocrine construct bioengineering [121].

Concerning the bioengineering of endocrine tissue, Citro et al. upgrade the VIO platform mentioned before [30] to create a system with a personalized vascular network using ECFCs derived from the peripheral blood of healthy donors. In this work, the researchers have identified alternative xeno sources of endocrine cells and BOECs as an endothelial primary cell source, to engineer a native organ vascular network.

Although isolation efficiency may be affected by donor-related colony-forming ability, ECFCs may offer several important advantages over other sources of ECs such as: (1) can be easily isolated from healthy and diabetic subjects [122], (2) have a high proliferative capacity and can promote neovascularization in vivo in immune-deficient mice [123], (3) have already been proposed for the treatment of hemophilia and as a tool in gene therapy approaches [124], (4) can deliver therapeutic agents directly into the bloodstream due to their direct connection to the vascular system. Thus, using ECSFSs will increase the biocompatibility of the generated scaffold as the vascular compartment can be composed of autologous endothelial cells [29].

ECFCs have emerged as a highly promising cell type for vascular bioengineering; however, their limited distribution may be attributed to their low frequency in adults. In fact, ECFCs constitute a small subpopulation of cells circulating in human peripheral blood, estimated to be around 0.05–0.2 cells/ml, which is 15 times lower than the frequency observed in umbilical cord blood [125]. Moreover, the limitation in adult ECFC isolation can be attributed to their low frequency and lack of a unique set of distinctive cellular markers [126]. In addition, there were concerns about donor-to-donor variability, and several studies have demonstrated the absence of ECFC in a substantial fraction of healthy and non-healthy donors [127]. Of note, ECFCs derived from umbilical cord blood do not present the same challenges as adult ECFCs. In culture, umbilical ECFCs are significantly more frequent with an extended lifespan than adult ECFCs. Although umbilical cord blood ECFCs may have a better isolation efficiency and improved plasticity compared to adult one, the use of those cells is limited in bioengineering perspective since most of the patients do not have access to their own umbilical blood, with the ultimate goal of enabling the development of autologous cell therapies [102]. This limitation could be overcome by genetically modifying [128] the ECFCs, from human or xeno origin [129], making them invisible to the immune system. This would allow the use of these cells even on HLA-incompatible patient.

3.2.3 Pluripotent Stem Cells-Derived Endothelial Cells

To reshape the vascular side of the islet niche microenvironment, ECs derived from human induced pluripotent stem cells (iPSC) are rising a great interest in bioengineering perspective as alternative sources for autologous vascular cells.

Thanks to the Yamanaka protocol nowadays is possible to reprogram human fibroblast into iPSCs by promoting the expression of a defined set of transcription factors opening to the possibility to produce patient-specific ECs for regenerative medicine [130]. In this direction, in 2011, Rufaihah et al. demonstrated that human iPSC-derived ECs may successfully work in vivo when transplanted into ischemic hind limbs of immune-deficient mice. Cells were not only successfully incorporated into the host vasculature, but also significantly ameliorated the local blood flow owing to the anastomosis created [131].

In order to reshape the endocrine niche for diabetes approach, Samuel et al. showed that blood vessels could also be generated in vivo with human ECs obtained from iPSCs derived from T1D patients, suggesting the feasibility and potential application of these cells for future clinical translation [132].

In this direction, authors showed the potential of the EC derived from iPSCs in bioengineering and beta cell replacement. Minglin Ma et al. engineered microvascular networks from human iPSC-derived ECs (iPSC-ECs) to achieve customized microvasculature, demonstrating the feasibility of this approach, as a proof of concept, for the treatment of T1D; indeed, they subcutaneously transplanted a microvascular iPSC based EC meshes device combined with rat islets enabling correction of diabetes in SCID-Beige mice for up to 3 months [106].

Despite the results, several obstacles remain before iPSC-ECs become a clinical reality. For example, uncertainty about their potential tumorigenicity, long-term consequences of possible genetic and epigenetic alterations, and issues with their immunogenicity [133]. To minimize risks, adequate controls over the differentiation process should be implemented as routine screening for undesired mutations or practical purification steps at the end to eliminate possible undifferentiated cells [134]. Furthermore, the risk of malignancy should be minimized by preserving genomic stability during cell expansion. It is possible to screen these tumor types with tumorigenicity assays with the final tissue product before using iPSC [135]. Of note, iPSC-ECs might become a helpful tool in the study of tissue specificity. Increasing evidence indicates that the endothelium regulates multiple processes in a tissue-specific manner [136]. For example, tissue-specific EC-derived factors stimulate self-renewal and in situ expansion of stem cells residing in the lung, liver, bone, and neural tissues, contributing to the regeneration of these tissues upon injury [137]. Therefore, to recapitulate the full complexity of human tissues, bioengineers may need to use tissue-specific primary ECs. Nolan et al. showed that after engraftment in the liver and kidney, immature ECs underwent an in vivo education process, acquiring structural and phenotypic attributes of tissue-resident native ECs (i.e., liver ECs and glomerular ECs, respectively) [138]. Consequently, iPSC-ECs may undergo a tissue-specific education/maturation process, which bioengineers could exploit to recapitulate EC heterogeneity. Thus, iPSCs-ECs have excellent potential to bioengineer a personalized vascular network.

4 Reshaping the Endocrine Side: Alternative Sources and Their Application

Although insulin administration is the primary treatment for T1D, it may not always be sufficient to maintain an adequate glycemic control, leading to the development of severe complications. In such situations, replacing the damaged beta cells can provide a more effective solution, enabling real-time insulin regulation. Pancreas or pancreatic islets transplantation are currently considered the best therapeutic options when T1D develops into severe secondary complications. However, both pancreas and islet transplantation, face challenges, such as a shortage of organ donors and the need for lifelong immunosuppression. This prompted researchers to explore various strategies, such as co-transplantation of islets with accessory cells and establishing an alternative beta cell source. Pluripotent stem cells (PSC), adult SC, and animals are all being utilized in this pursuit.

Various cell types have been shown to positively impact islet function in co-culture studies. These findings have led to the development of combined structures that merge native islet cells or other insulin-producing cells with supportive accessory cells. This innovative approach enhances islet cell viability, functionality, and offers immune protection, thereby improving overall treatment outcomes.

Co-transplantation of islets alongside various cell types, including fibroblasts, ECs, and amniotic epithelial cells, has demonstrated enhanced functionality, survival, and revascularization of the transplanted islets [139–141].

Recent studies have focused on generating islet spheroids by combining insulin-producing cells with accessory cells. This approach has shown promising results in enhancing functionality, matrix restoration, vascularization, and signaling.

Lebreton et al. have uncovered the significant potential of human amniotic epithelial cells (hAECs) by incorporating them into islet organoids. Acting as a shield, those cells markedly substantially improving engraftment and graft function in a diabetic mouse model [142]. Due to their cytoprotective properties, employing hAECs in islet transplants has the potential to prevent inflammation-induced islet destruction [142]. Wassmer et al. demonstrated further demonstrated the benefits of pre-vascularization by integrating ECs into the organoids. Assembled pre-vascularized islet organoids (PIOs) exhibited enhanced in vitro function and superior engraftment due to improved vascularization compared to native islets [143].

A significant problem with islet cells is their restricted availability. The most promising sources for acquiring insulin-producing tissue in vast quantities are SC-derived beta cells and xenogeneic islets of porcine-origin. The isolation of bone marrow or adipose tissue derived mesenchymal stem cells (MSCs) [144], as well as PSC, such as embryonic stem cells (ESCs) [137] and iPSCs [145] have opened up new therapeutic approaches for the treatment of various diseases. These cells hold great promise, particularly for conditions such as type 1 diabetes, where the loss of a single cell type does not require transplanting back to its original location to restore its function. ESCs and iPSCs are particularly attractive due their ability to self-renew and differentiate into various cell types, unlike to adult SCs, which have limited proliferation capacity while maintaining an undifferentiated state, and tend to differentiate into specific adult tissue cells [146]. Therefore, one of the most advanced strategies in research is focused on differentiating PSC into beta cells.

In 2006, the first in vitro protocol was developed to differentiate ESCs into insulin-producing cells. However, these cells showed limited responsiveness to glucose stimulation [147]. A potential solution involved transplanting ESCs derived pancreatic endoderm cells, into immunodeficient mice, which led to spontaneous in vivo differentiation and maturation. After 3 months, human insulin levels increased, and mice were protected against hyperglycemia [148].

Since then, multiple groups have explored effective approaches to differentiate ESCs into the insulin-producing beta cells, tackling challenges such as attaining homogeneity and functional maturity. In 2014, Rezania et al. developed a successful in vitro differentiation protocol. The insulin secretion of the differentiated cells was nearly identical to that of human islets. As a result, diabetes was successfully reversed in mice within 2 months after transplantation [149]. Despite these encouraging results, the use of ESCs necessitates the destruction of a human embryo, giving rise to ethical dilemma. Furthermore, the immune rejection remains a significant concern in the context of transplantation [150]. To address these issues [145], in parallel to ESCs research, the differentiation of iPSCs derived from healthy or diabetic individuals into insulin-secreting cells has been explored. The Melton group's three-dimensional cell culture strategy facilitated the transplantation of mature and functional SC-derived beta cells in NRG-Akita mice, restoring glycemic control within two weeks [151]. This protocol, using iPSCs from diabetic patient, suggests potential for personalized diabetes treatment [152]. Belmonte's group improved the differentiation protocol by developing a chemical formula to boost the efficiency of PSC-derived NKX6.1+/INS+ beta-cells, emphasizing the importance of 3D-progenitor cells during differentiation [153]. More recently, a novel differentiation protocol published on STAR protocol journal by Otokonski et al. has successfully demonstrated the generation of functional SC-islets [142]. Integrated analysis revealed enhanced in vitro beta cell functionality. Despite persistent metabolic and transcriptomic differences between PSC-derived and primary beta cells, PSC-derived beta cells offer considerable for diabetes research and treatment [154]. Nonetheless, challenges such as PSC tumorigenicity, constrain clinical applications. Transplantation of PSC-derived beta cells can lead to a tumor formation, due to residual undifferentiated cells. Pellegrini et al. suggest using clinical-grade anti-CD30 treatment to mitigate iPSCs-derived beta cell tumorigenicity [155].

These remarkable studies have led to clinical trials, initiated by two companies, ViaCyte and Vertex. ViaCyte has conducted trials on T1D patients, using human ESCs-derived pancreatic progenitors encapsulated in a PTFE device, as previously described [146]. More recently, Vertex has begun a trial where fully differentiated hESC-derived islet clusters are transplanted intraportally. In this study, patients with unstable T1D received VX-880, a SC-derived, fully matured pancreatic islet cell treatment without encapsulation, necessitating immunosuppression (NCT04786262). The first three patients showed improved glycemic control and no severe side effects. Currently, Vertex is launching an encapsulation device to use with VX-880.

Porcine islet xenotransplantation presents a compelling alternative to human islet transplantation. Owing to the absence of human major histocompatibility complex (MHC) antigens, pig islets may preclude allo-sensitization and MHC-restricted autoimmunity while exhibiting physiological and metabolic functions akin to those of human islets [143]. Researchers have extensively investigated various sources of porcine islets, including fetal and neonatal porcine cell clusters, as well as adult islets [29, 156–158]. In recent preclinical studies, pig islets transplanted into non-human primates restored glucose homeostasis and maintained normoglycemia [159]. A multitude of clinical trials have been completed or are currently in progress to assess the safety and efficacy of porcine islet transplantation in humans [160]. Genetically engineered pigs expressing human complement regulatory proteins can avert hyperacute rejection. Additionally, various immunosuppressive strategies, such as tolerance induction through chimerism or mixed chimerism and combination therapy with immunosuppressive drugs, have been devised to prevent acute and chronic rejection. The availability of multi-transgenic animals has recently culminated in the remarkable reports of pig-to-human heart and kidney transplants [161, 162]. Consequently, the prospects for xenogeneic islets to serve as the source of insulin-producing tissue have never been more promising.

In conclusion, the treatment of T1D has seen remarkable progress in recent years with the development of novel therapies and alternative sources of insulin-producing tissue. Stem cell technology and porcine islet xenotransplantation have shown great promise in clinical trials, but challenges such as immune rejection, tumorigenicity, and the risk of PERV transmission must be addressed to ensure their safety and efficacy. Continued research and development in these areas will be essential in achieving long-term success and improving the lives of those affected by type 1 diabetes.

Acknowledgments FC and AN have contributed to the writing of the work in partial fulfillment of the requirements for obtaining the PhD degree at Vita-Salute San Raffaele University, Milano, Italy. CP has contributed to the writing and reviewing the manuscript. JB and EB have contributed to the writing of the manuscript. LP and AC have written and reviewed the manuscript and supervised the work. This work was supported by grants from the European Commission (Horizon 2020 Framework Program; VANGUARD grant 874700); Juvenile Diabetes Research Foundation (JDRF; grant 3-SRA-2022-1155-S-B); Fondazione Italiana Diabete (FID); "SOStegno 70 Insieme ai ragazzi diabetici Associazione Onlus" (project "Beta is better"), and fundraising campaign "Un brutto t1po".

References

1. Chetboun, M. *et al.* Association between primary graft function and 5-year outcomes of islet allogeneic transplantation in type 1 diabetes: a retrospective, multicentre, observational cohort study in 1210 patients from the Collaborative Islet Transplant Registry. *Lancet Diabetes Endocrinol.* (2023). doi:https://doi.org/10.1016/S2213-8587(23)00082-7
2. Pignatelli, C., Campo, F., Neroni, A., Piemonti, L. & Citro, A. Bioengineering the Vascularized Endocrine Pancreas: A Fine-Tuned Interplay Between Vascularization, Extracellular-Matrix-Based Scaffold Architecture, and Insulin-Producing Cells. *Transpl. Int.* **35**, (2022).
3. Almaça, J., Caicedo, A. & Landsman, L. Beta cell dysfunction in diabetes: the islet microenvironment as an unusual suspect. *Diabetologia* **63**, 2076–2085 (2020).
4. Citro, A., Cantarelli, E. & Piemonti, L. Anti-inflammatory strategies to enhance islet engraftment and survival. *Curr. Diab. Rep.* **13**, 733–744 (2013).
5. Wang, M., Crager, M. & Pugazhenthi, S. Modulation of apoptosis pathways by oxidative stress and autophagy in cells. *Exp. Diabetes Res.* **2012**, (2012).
6. Jansson, L. & Carlsson, P. O. Graft vascular function after transplantation of pancreatic islets. *Diabetologia* **45**, 749–763 (2002).
7. Keshtkar, S. *et al.* Protective effect of nobiletin on isolated human islets survival and function against hypoxia and oxidative stress-induced apoptosis. *Sci. Rep.* **9**, 1–13 (2019).
8. Citro, A. & Ott, H. C. Can We Re-Engineer the Endocrine Pancreas? *Current Diabetes Reports* **18**, (2018).
9. Piemonti, L., Guidotti, L. G. & Battaglia, M. Modulation of early inflammatory reactions to promote engraftment and function of transplanted pancreatic islets in autoimmune diabetes. *Adv. Exp. Med. Biol.* **654**, 725–747 (2010).
10. Melzi, R. *et al.* Role of CCL2/MCP-1 in islet transplantation. *Cell Transplant.* **19**, 1031–1046 (2010).
11. Piemonti, L. *et al.* Human pancreatic islets produce and secrete MCP-1/CCL2: Relevance in human islet transplantation. *Diabetes* **51**, 55–65 (2002).
12. Nano, R. *et al.* Human pancreatic islet preparations release HMGB1: (Ir)relevance for graft engraftment. *Cell Transplant.* **22**, 2175–2186 (2013).
13. Dugnani, E. & Citro, A. Filling the gap to improve islet engraftment and survival using anti-inflammatory approaches. in *Transplantation, Bioengineering, and Regeneration of the*

Endocrine Pancreas: Volume 1 741–750 (Academic Press, 2019). doi:https://doi.org/10.1016/B978-0-12-814833-4.00059-9
14. Citro, A., Cantarelli, E., Pellegrini, S., Dugnani, E. & Piemonti, L. Anti-Inflammatory Strategies in Intrahepatic Islet Transplantation: A Comparative Study in Preclinical Models. *Transplantation* **102**, 240–248 (2018).
15. Citro, A. *et al.* CCL2/MCP-1 and CXCL12/SDF-1 blockade by L-aptamers improve pancreatic islet engraftment and survival in mouse. *Am. J. Transplant.* **19**, 3131–3138 (2019).
16. Maffi, P. *et al.* Targeting CXCR1/2 does not improve insulin secretion after pancreatic islet transplantation: A phase 3, double-blind, randomized, placebo-controlled trial in type 1 diabetes. *Diabetes Care* **43**, 710–718 (2020).
17. Brissova, M. *et al.* Islet Microenvironment, Modulated by Vascular Endothelial Growth Factor-A Signaling, Promotes β Cell Regeneration, *Cell Metab.* **19**, 498–511 (2014).
18. Delaune, V., Berney, T., Lacotte, S. & Toso, C. Intraportal islet transplantation: the impact of the liver microenvironment. *Transpl. Int.* **30**, 227–238 (2017).
19. Bluestone, J. A. & Tang, Q. Solving the Puzzle of Immune Tolerance for β-Cell Replacement Therapy for Type 1 Diabetes. *Cell Stem Cell* **27**, 505–507 (2020).
20. Berney, T., Andres, A., Toso, C., Majno, P. & Squifflet, J. P. mTOR Inhibition & Clinical Transplantation: Pancreas & Islet. *Transplantation* **102**, S30–S31 (2017).
21. Samojlik, M. M. & Stabler, C. L. Designing biomaterials for the modulation of allogeneic and autoimmune responses to cellular implants in Type 1 Diabetes. *Acta Biomaterialia* **133**, 87–101 (2021).
22. Pepper, A. R., Bruni, A. & Shapiro, A. M. J. Clinical islet transplantation: Is the future finally now? *Current Opinion in Organ Transplantation* **23**, 428–439 (2018).
23. Shapiro, A. M. J. *et al.* Islet Transplantation in Seven Patients with Type 1 Diabetes Mellitus Using a Glucocorticoid-Free Immunosuppressive Regimen. *N. Engl. J. Med.* **343**, 230–238 (2000).
24. Shapiro, A. M. J. *et al.* International trial of the Edmonton protocol for islet transplantation. *N. Engl. J. Med.* **355**, 1318–1330 (2006).
25. Brandhorst, D., Brandhorst, H., Acreman, S., Abraham, A. & Johnson, P. R. V. High concentrations of etanercept reduce human islet function and integrity. *J. Inflamm. Res.* **14**, 599–610 (2021).
26. Marfil-Garza, B. A., Shapiro, A. M. J. & Kin, T. Clinical islet transplantation: Current progress and new frontiers. *J. Hepatobiliary. Pancreat. Sci.* **28**, 243–254 (2021).
27. Shapiro, A. M. J., Pokrywczynska, M. & Ricordi, C. Clinical pancreatic islet transplantation. *Nat. Rev. Endocrinol.* **13**, 268–277 (2017).
28. de Vos, P., Faas, M. M., Strand, B. & Calafiore, R. Alginate-based microcapsules for immunoisolation of pancreatic islets. *Biomaterials* **27**, 5603–5617 (2006).
29. Citro, A. *et al.* Directed self-assembly of a xenogeneic vascularized endocrine pancreas for type 1 diabetes. *Nat. Commun.* **14**, 878 (2023).
30. Citro, A. *et al.* Biofabrication of a vascularized islet organ for type 1 diabetes. *Biomaterials* **199**, 40–51 (2019).
31. Wassmer, C.-H. *et al.* Bio-Engineering of Pre-Vascularized Islet Organoids for the Treatment of Type 1 Diabetes. *Transpl. Int.* **35**, 7 (2022).
32. Stendahl, J. C., Kaufman, D. B. & Stupp, S. I. Extracellular matrix in pancreatic islets: relevance to scaffold design and transplantation. *Cell Transplant.* **18**, 1–12 (2009).
33. Cross, S. E. *et al.* Key Matrix Proteins Within the Pancreatic Islet Basement Membrane Are Differentially Digested During Human Islet Isolation. *Am. J. Transplant.* **17**, 451–461 (2017).
34. Llacua, L. A., Faas, M. M. & de Vos, P. Extracellular matrix molecules and their potential contribution to the function of transplanted pancreatic islets. *Diabetologia* **61**, 1261–1272 (2018).
35. Weber, L. M., Hayda, K. N., Haskins, K. & Anseth, K. S. The effects of cell-matrix interactions on encapsulated β-cell function within hydrogels functionalized with matrix-derived adhesive peptides. *Biomaterials* **28**, 3004–3011 (2007).
36. Aamodt, K. I. & Powers, A. C. Signals in the pancreatic islet microenvironment influence β-cell proliferation. *Diabetes, Obesity and Metabolism* **19**, 124–136 (2017).

37. Desai, T. & Shea, L. D. Advances in islet encapsulation technologies. *Nature Reviews Drug Discovery* **16**, 338–350 (2017).
38. Hunckler, M. D. & García, A. J. Engineered Biomaterials for Enhanced Function of Insulin-Secreting β-Cell Organoids. *Adv. Funct. Mater.* **30**, 1–15 (2020).
39. Lutolf, M. P. & Hubbell, J. A. Synthetic biomaterials as instructive extracellular microenvironments for morphogenesis in tissue engineering. *Nat. Biotechnol.* **23**, 47–55 (2005).
40. Devaraj, N. K. & Finn, M. G. Introduction: Click Chemistry. *Chem. Rev.* **121**, 6697–6698 (2021).
41. Wang, X. *et al.* Local Immunomodulatory Strategies to Prevent Allo-Rejection in Transplantation of Insulin-Producing Cells. *Adv. Sci.* **8**, 1–19 (2021).
42. Henry, R. R. *et al.* Initial Clinical Evaluation of VC-01TM Combination Product—A Stem Cell–Derived Islet Replacement for Type 1 Diabetes (T1D). *Diabetes* **67**, 138-OR (2018).
43. Paez-Mayorga, J. *et al.* Implantable niche with local immunosuppression for islet allotransplantation achieves type 1 diabetes reversal in rats. *Nat. Commun.* **13**, 7951 (2022).
44. Liang, J. P. *et al.* Engineering a macroporous oxygen-generating scaffold for enhancing islet cell transplantation within an extrahepatic site. *Acta Biomater.* **130**, 268–280 (2021).
45. An, D. *et al.* Developing robust, hydrogel-based, nanofiber-enabled encapsulation devices (NEEDs) for cell therapies. *Biomaterials* **37**, 40–48 (2015).
46. An, D. *et al.* Designing a retrievable and scalable cell encapsulation device for potential treatment of type 1 diabetes. *Proc. Natl. Acad. Sci. U. S. A.* **115**, E263–E272 (2017).
47. Guyette, J. P. *et al.* Perfusion decellularization of whole organs. **9**, 1451–1468 (2014).
48. Ott, H. C. *et al.* Perfusion-decellularized matrix: Using nature's platform to engineer a bioartificial heart. *Nat. Med.* **14**, 213–221 (2008).
49. Song, J. J. *et al.* Regeneration and experimental orthotopic transplantation of a bioengineered kidney. *Nat. Med.* **19**, 646–651 (2013).
50. Ott, H. C. *et al.* Regeneration and orthotopic transplantation of a bioartificial lung. *Nat. Med.* **16**, 927–933 (2010).
51. Peloso, A. *et al.* The human pancreas as a source of protolerogenic extracellular matrix scaffold for a new-generation bioartificial endocrine pancreas. *Ann. Surg.* **264**, 169–179 (2016).
52. Napierala, H. *et al.* Engineering an endocrine Neo-Pancreas by repopulation of a decellularized rat pancreas with islets of Langerhans. *Sci. Rep.* **7**, 1–12 (2017).
53. Sayed-Hadi Mirmalek-Sani, Orlando, G. *et al.* Porcine pancreas extracellular matrix as a platform for endocrine pancreas bioengineering. **34**, 5488–5495 (2014).
54. Vishwakarma, S. K. *et al.* Molecular dynamics of pancreatic transcription factors in bioengineered humanized insulin producing neoorgan. *Gene* **675**, 165–175 (2018).
55. Wang, X., Wang, K., Zhang, W., Qiang, M. & Luo, Y. A bilaminated decellularized scaffold for islet transplantation: Structure, properties and functions in diabetic mice. *Biomaterials* **138**, 80–90 (2017).
56. Willenberg, B. J. *et al.* Repurposed biological scaffolds: Kidney to pancreas. *Organogenesis* **11**, 47–57 (2015).
57. Goh, S. K., Bertera, S., Richardson, T. & Banerjee, I. Repopulation of decellularized organ scaffolds with human pluripotent stem cell-derived pancreatic progenitor cells. *Biomed. Mater.* **18**, (2023).
58. Saldin, L. T., Cramer, M. C., Velankar, S. S., White, L. J. & Badylak, S. F. Extracellular matrix hydrogels from decellularized tissues: Structure and function. *Acta Biomater.* **49**, 1–15 (2017).
59. Jiang, K. *et al.* 3-D physiomimetic extracellular matrix hydrogels provide a supportive microenvironment for rodent and human islet culture. *Biomaterials* **198**, 37–48 (2019).
60. Sackett, S. D. *et al.* Extracellular matrix scaffold and hydrogel derived from decellularized and delipidized human pancreas. *Sci. Rep.* **8**, 1–16 (2018).
61. Tremmel, D. M. *et al.* A human pancreatic ECM hydrogel optimized for 3-D modeling of the islet microenvironment. *Sci. Rep.* **12**, 7188 (2022).
62. Salg, G. A. *et al.* The emerging field of pancreatic tissue engineering: A systematic review and evidence map of scaffold materials and scaffolding techniques for insulin-secreting cells. *Journal of Tissue Engineering* **10**, (2019).

63. Hussein, K. H. *et al.* New insights into the pros and cons of cross-linking decellularized bio-artificial organs. *International Journal of Artificial Organs* **40**, 136–141 (2017).
64. Li, X., Sun, Q., Li, Q., Kawazoe, N. & Chen, G. Functional Hydrogels With Tunable Structures and Properties for Tissue Engineering Applications . *Frontiers in Chemistry* **6**, (2018).
65. Bishop, E. S. *et al.* 3-D bioprinting technologies in tissue engineering and regenerative medicine: Current and future trends. *Genes Dis.* **4**, 185–195 (2017).
66. Kang, H. W., Kengla, C., Lee, S. J., Yoo, J. J. & Atala, A. 3-D organ printing technologies for tissue engineering applications. *Rapid Prototyp. Biomater. Princ. Appl.* 236–253 (2014). doi:https://doi.org/10.1533/9780857097217.236
67. Espona-Noguera, A. *et al.* Review of advanced hydrogel-based cell encapsulation systems for insulin delivery in type 1 diabetes mellitus. *Pharmaceutics* **11**, (2019).
68. Gurlin, R. E., Giraldo, J. A. & Latres, E. 3D Bioprinting and Translation of Beta Cell Replacement Therapies for Type 1 Diabetes. *Tissue Eng. - Part B Rev.* **27**, 238–252 (2021).
69. Hwang, D. G. *et al.* A 3D bioprinted hybrid encapsulation system for delivery of human pluripotent stem cell-derived pancreatic islet-like aggregates. *Biofabrication* **14**, (2021).
70. Kim, J. *et al.* 3D cell printing of islet-laden pancreatic tissue-derived extracellular matrix bioink constructs for enhancing pancreatic functions. *J. Mater. Chem. B* **7**, 1773–1781 (2019).
71. Ghasemi, A., Akbari, E. & Imani, R. An Overview of Engineered Hydrogel-Based Biomaterials for Improved β-Cell Survival and Insulin Secretion. *Front. Bioeng. Biotechnol.* **9**, 686 (2021).
72. Li, W. *et al.* Microfluidic fabrication of microparticles for biomedical applications. *Chem. Soc. Rev.* **47**, 5646–5683 (2018).
73. Daly, A. C., Riley, L., Segura, T. & Burdick, J. A. Hydrogel microparticles for biomedical applications. *Nat. Rev. Mater.* **5**, 20–43 (2020).
74. Ding, S., Serra, C. A., Vandamme, T. F., Yu, W. & Anton, N. Double emulsions prepared by two–step emulsification: History, state-of-the-art and perspective. *J. Control. Release* **295**, 31–49 (2019).
75. Wang, J. *et al.* Droplet Microfluidics for the Production of Microparticles and Nanoparticles. *Micromachines* **8**, (2017).
76. Harrington, S., Ott, L., Karanu, F., Ramachandran, K. & Stehno-Bittel, L. A versatile microencapsulation platform for hyaluronic acid and polyethylene glycol. *Tissue Eng. - Part A* **27**, 153–164 (2021).
77. Harrington, S., Karanu, F., Ramachandran, K., Williams, S. J. & Stehno-Bittel, L. PEGDA microencapsulated allogeneic islets reverse canine diabetes without immunosuppression. *PLoS One* **17**, 1–20 (2022).
78. Skoumal, M. *et al.* Localized immune tolerance from FasL-functionalized PLG scaffolds. 271–281 (2020). doi:https://doi.org/10.1016/j.biomaterials.2018.11.015.Localized
79. Herman Blomeier, Xiaomin Zhang, Christopher Rives, Marcela Brissova, Elizabeth Hughes, Marshall Baker, Alvin C. Powers4 Dixon B. Kaufman, Lonnie D. Shea, and W. L. L. J. Polymer Scaffolds as Synthetic Microenvironments for Extrahepatic Islet Transplantation. *Bone* **23**, 1–7 (2008).
80. Salvay, D. M. *et al.* Extracellular Matrix Protein-Coated Scaffolds Promote the Reversal of Diabetes After Extrahepatic Islet Transplantation. *Transplantation* **85**, 1456–1464 (2008).
81. Zhang, M. *et al.* Study on the Effect of PDA-PLGA Scaffold Loaded With Islet Cells for Skeletal Muscle Transplantation in the Treatment of Diabetes. *Front. Bioeng. Biotechnol.* **10**, 1–12 (2022).
82. Rosiak, P., Latanska, I., Paul, P., Sujka, W. & Kolesinska, B. Modification of Alginates to Modulate Their Physic-Chemical Properties and Obtain Biomaterials with Different Functional Properties. *Molecules* **26**, (2021).
83. Hu, S. *et al.* Toll-like receptor 2-modulating pectin-polymers in alginate-based microcapsules attenuate immune responses and support islet-xenograft survival. *Biomaterials* **266**, (2021).
84. Bochenek, M. A. *et al.* Alginate encapsulation as long-term immune protection of allogeneic pancreatic islet cells transplanted into the omental bursa of macaques. *Nat. Biomed. Eng.* **2**, 810–821 (2018).

85. Vegas, A. J. *et al.* Long-term glycemic control using polymer-encapsulated human stem cell-derived beta cells in immune-competent mice. *Nat. Med.* **22**, 306–311 (2016).
86. Komatsu, H., Kandeel, F. & Mullen, Y. Impact of Oxygen on Pancreatic Islet Survival. *Pancreas* **47**, 533–543 (2018).
87. Pedraza, E., Coronel, M. M., Fraker, C. A., Ricordi, C. & Stabler, C. L. Preventing hypoxia-induced cell death in beta cells and islets via hydrolytically activated, oxygen-generating biomaterials. *Proc. Natl. Acad. Sci. U. S. A.* **109**, 4245–4250 (2012).
88. Barkai, U. *et al.* Enhanced oxygen supply improves islet viability in a new bioartificial pancreas. *Cell Transplant.* **22**, 1463–1476 (2013).
89. An, D. *et al.* An Atmosphere-Breathing Refillable Biphasic Device for Cell Replacement Therapy. *Adv. Mater.* **31**, 1–8 (2019).
90. Chen, Y., Nguyen, D. T., Kokil, G. R., Wong, Y. X. & Dang, T. T. Microencapsulated islet-like microtissues with toroid geometry for enhanced cellular viability. *Acta Biomater.* **97**, 260–271 (2019).
91. An, D. *et al.* Mass production of shaped particles through vortex ring freezing. *Nat. Commun.* **7**, 1–10 (2016).
92. Fotino, N., Fotino, C. & Pileggi, A. Re-engineering islet cell transplantation. *Pharmacological Research* **98**, 76–85 (2015).
93. Qin, T. *et al.* Inclusion of extracellular matrix molecules and necrostatin-1 in the intracapsular environment of alginate-based microcapsules synergistically protects pancreatic β cells against cytokine-induced inflammatory stress. *Acta Biomater.* **146**, 434–449 (2022).
94. Krol, S., Baronti, W. & Marchetti, P. Nanoencapsulated human pancreatic islets for β-cell replacement in Type 1 diabetes. *Nanomedicine (London, England)* **15**, 1735–1738 (2020).
95. Youn, W. *et al.* Single-Cell Nanoencapsulation: From Passive to Active Shells. *Adv. Mater.* **32**, 1907001 (2020).
96. Manzoli, V. *et al.* Immunoisolation of murine islet allografts in vascularized sites through conformal coating with polyethylene glycol. *Am. J. Transplant. Off. J. Am. Soc. Transplant. Am. Soc. Transpl. Surg.* **18**, 590–603 (2018).
97. Stabler, C. L. *et al.* Transplantation of PEGylated islets enhances therapeutic efficacy in a diabetic nonhuman primate model. *Am. J. Transplant.* **20**, 689–700 (2020).
98. Syed, F. *et al.* Conformal coating by multilayer nano-encapsulation for the protection of human pancreatic islets: In-vitro and in-vivo studies. *Nanomedicine Nanotechnology, Biol. Med.* **14**, 2191–2203 (2018).
99. De Toni, T. *et al.* Parallel Evaluation of Polyethylene Glycol Conformal Coating and Alginate Microencapsulation as Immunoisolation Strategies for Pancreatic Islet Transplantation. *Front. Bioeng. Biotechnol.* **10**, 1–16 (2022).
100. Atala, A., Kasper, F. K. & Mikos, A. G. Engineering complex tissues. *Sci. Transl. Med.* **4**, 160rv12 (2012).
101. Said, S. S., Pickering, J. G. & Mequanint, K. Advances in growth factor delivery for therapeutic angiogenesis. *J. Vasc. Res.* **50**, 35–51 (2013).
102. Wang, K., Lin, R. Z. & Melero-Martin, J. M. Bioengineering human vascular networks: trends and directions in endothelial and perivascular cell sources. *Cell. Mol. Life Sci.* **76**, 421–439 (2019).
103. Gimbrone, M. A. J., Cotran, R. S. & Folkman, J. Human vascular endothelial cells in culture. Growth and DNA synthesis. *J. Cell Biol.* **60**, 673–684 (1974).
104. Black, A. F., Berthod, F., L'heureux, N., Germain, L. & Auger, F. A. In vitro reconstruction of a human capillary-like network in a tissue-engineered skin equivalent. *FASEB J. Off. Publ. Fed. Am. Soc. Exp. Biol.* **12**, 1331–1340 (1998).
105. Schechner, J. S. *et al.* In vivo formation of complex microvessels lined by human endothelial cells in an immunodeficient mouse. *Proc. Natl. Acad. Sci. U. S. A.* **97**, 9191–9196 (2000).
106. Song, W. *et al.* Engineering transferrable microvascular meshes for subcutaneous islet transplantation. *Nat. Commun.* **10**, 4602 (2019).
107. Wassmer, C.-H. *et al.* Bio-Engineering of Pre-Vascularized Islet Organoids for the Treatment of Type 1 Diabetes. *Transpl. Int.* **0**, 7 (2022).

108. Davison, P. M., Bensch, K. & Karasek, M. A. Isolation and growth of endothelial cells from the microvessels of the newborn human foreskin in cell culture. *J. Invest. Dermatol.* **75**, 316–321 (1980).
109. Kern, P. A., Knedler, A. & Eckel, R. H. Isolation and culture of microvascular endothelium from human adipose tissue. *J. Clin. Invest.* **71**, 1822–1829 (1983).
110. Nör, J. E. *et al.* Engineering and characterization of functional human microvessels in immunodeficient mice. *Lab. Invest.* **81**, 453–463 (2001).
111. Lin, Y., Weisdorf, D. J., Solovey, A. & Hebbel, R. P. Origins of circulating endothelial cells and endothelial outgrowth from blood. *J. Clin. Invest.* **105**, 71–77 (2000).
112. Medina, R. J. *et al.* Endothelial Progenitors: A Consensus Statement on Nomenclature. *Stem Cells Transl. Med.* **6**, 1316–1320 (2017).
113. Yoder, M. C. *et al.* Redefining endothelial progenitor cells via clonal analysis and hematopoietic stem/progenitor cell principals. *Blood* **109**, 1801–1809 (2007).
114. Au, P. *et al.* Differential in vivo potential of endothelial progenitor cells from human umbilical cord blood and adult peripheral blood to form functional long-lasting vessels. *Blood* **111**, 1302–1305 (2008).
115. Wu, X. *et al.* Tissue-engineered microvessels on three-dimensional biodegradable scaffolds using human endothelial progenitor cells. *Am. J. Physiol. Heart Circ. Physiol.* **287**, H480–7 (2004).
116. Melero-Martin, J. M. *et al.* In vivo vasculogenic potential of human blood-derived endothelial progenitor cells. *Blood* **109**, 4761–4768 (2007).
117. Sieminski, A. L., Hebbel, R. P. & Gooch, K. J. Improved microvascular network in vitro by human blood outgrowth endothelial cells relative to vessel-derived endothelial cells. *Tissue Eng.* **11**, 1332–1345 (2005).
118. Ren, X. *et al.* Engineering pulmonary vasculature in decellularized rat and human lungs. *Nat. Biotechnol.* **33**, (2015).
119. Jain, R. K. Molecular regulation of vessel maturation. *Nat. Med.* **9**, 685–693 (2003).
120. Kang, K.-T., Allen, P. & Bischoff, J. Bioengineered human vascular networks transplanted into secondary mice reconnect with the host vasculature and re-establish perfusion. *Blood* **118**, 6718–6721 (2011).
121. Bellofatto, K., Lebreton, F., Hanna, R., Fonseca, L. M., Bignard, J., Galvan, V., Peloso, A., Berney, Thierry., Compagnon, P., VANGUARD Consortium; Berishvili, E. 228.1: Hydrogel-based, prevascularized, retrievable endocrine construct to treat Type 1 Diabetes. *Transplantation* **107(10S2):p 57 (2023).** https://doi.org/10.1097/01.tp.0000994048.59940.dd
122. Coppens, V. *et al.* Human blood outgrowth endothelial cells improve islet survival and function when co-transplanted in a mouse model of diabetes. *Diabetologia* **56**, 382–390 (2013).
123. Hoshi, R. A. *et al.* The blood and vascular cell compatibility of heparin-modified ePTFE vascular grafts. *Biomaterials* **34**, 30–41 (2013).
124. Olgasi, C. *et al.* Efficient and safe correction of hemophilia A by lentiviral vector-transduced BOECs in an implantable device. *Mol. Ther. Methods Clin. Dev.* **23**, 551–566 (2021).
125. Ingram, D. A. *et al.* Identification of a novel hierarchy of endothelial progenitor cells using human peripheral and umbilical cord blood. *Blood* **104**, 2752–2760 (2004).
126. Mund, J. A., Estes, M. L., Yoder, M. C., Ingram, D. A. J. & Case, J. Flow cytometric identification and functional characterization of immature and mature circulating endothelial cells. *Arterioscler. Thromb. Vasc. Biol.* **32**, 1045–1053 (2012).
127. Rignault-Clerc, S. *et al.* Functional late outgrowth endothelial progenitors isolated from peripheral blood of burned patients. *Burns* **39**, 694–704 (2013).
128. Dudek, A. Z. *et al.* Systemic inhibition of tumour angiogenesis by endothelial cell-based gene therapy. *Br. J. Cancer* **97**, 513–522 (2007).
129. Shradhanjali, A. *et al.* Characterization of Blood Outgrowth Endothelial Cells (BOEC) from Porcine Peripheral Blood. *J. Vis. Exp.* **2022**, 1–14 (2022).
130. Park, I.-H. *et al.* Reprogramming of human somatic cells to pluripotency with defined factors. *Nature* **451**, 141–146 (2008).

131. Rufaihah, A. J. *et al.* Endothelial cells derived from human iPSCS increase capillary density and improve perfusion in a mouse model of peripheral arterial disease. *Arterioscler. Thromb. Vasc. Biol.* **31**, e72–9 (2011).
132. Samuel, R. *et al.* Generation of functionally competent and durable engineered blood vessels from human induced pluripotent stem cells. *Proc. Natl. Acad. Sci. U. S. A.* **110**, 12774–12779 (2013).
133. Wu, S. M. & Hochedlinger, K. Harnessing the potential of induced pluripotent stem cells for regenerative medicine. *Nat. Cell Biol.* **13**, 497–505 (2011).
134. Wimmer, R. A., Leopoldi, A., Aichinger, M., Kerjaschki, D. & Penninger, J. M. Generation of blood vessel organoids from human pluripotent stem cells. *Nat. Protoc.* **14**, 3082–3100 (2019).
135. Stevens, K. R. & Murry, C. E. Human Pluripotent Stem Cell-Derived Engineered Tissues: Clinical Considerations. *Cell Stem Cell* **22**, 294–297 (2018).
136. Cleaver, O. & Melton, D. A. Endothelial signaling during development. *Nat. Med.* **9**, 661–668 (2003).
137. Thomson, J. A. *et al.* Embryonic stem cell lines derived from human blastocysts. *Science* **282**, 1145–1147 (1998).
138. Nolan, D. J. *et al.* Molecular signatures of tissue-specific microvascular endothelial cell heterogeneity in organ maintenance and regeneration. *Dev. Cell* **26**, 204–219 (2013).
139. Lebreton, F. *et al.* Shielding islets with human amniotic epithelial cells enhances islet engraftment and revascularization in a murine diabetes model. *Am. J. Transplant.* **20**, 1551–1561 (2020).
140. Takahashi, Y., Sekine, K., Kin, T., Takebe, T. & Taniguchi, H. Self-Condensation Culture Enables Vascularization of Tissue Fragments for Efficient Therapeutic Transplantation. *Cell Rep.* **23**, 1620–1629 (2018).
141. Matsushima, H. *et al.* Human Fibroblast Sheet Promotes Human Pancreatic Islet Survival and Function In Vitro. *Cell Transpl.* **25**, 1525–1537 (2016).
142. Barsby, T. *et al.* Differentiating functional human islet-like aggregates from pluripotent stem cells. *STAR Protoc* **3**, 101711 (2022).
143. Liu, Z. *et al.* Pig-to-Primate Islet Xenotransplantation: Past, Present, and Future. *Cell Transpl.* **26**, 925–947 (2017).
144. Pittenger, M. F. *et al.* Mesenchymal stem cell perspective: cell biology to clinical progress. *NPJ Regen Med* **4**, 22 (2019).
145. Takahashi, K. & Yamanaka, S. Induction of pluripotent stem cells from mouse embryonic and adult fibroblast cultures by defined factors. *Cell* **126**, 663–676 (2006).
146. Yamanaka, S. Pluripotent Stem Cell-Based Cell Therapy-Promise and Challenges. *Cell Stem Cell* **27**, 523–531 (2020).
147. D'Amour, K. A. *et al.* Production of pancreatic hormone-expressing endocrine cells from human embryonic stem cells. *Nat Biotechnol* **24**, 1392–1401 (2006).
148. Kroon, E. *et al.* Pancreatic endoderm derived from human embryonic stem cells generates glucose-responsive insulin-secreting cells in vivo. *Nat. Biotechnol.* **26**, 443–452 (2008).
149. Rezania, A. *et al.* Reversal of diabetes with insulin-producing cells derived in vitro from human pluripotent stem cells. *Nat. Biotechnol.* **32**, 1121–1133 (2014).
150. Volarevic, V. *et al.* Ethical and Safety Issues of Stem Cell-Based Therapy. *Int J Med Sci* **15**, 36–45 (2018).
151. Pagliuca, F. W. *et al.* Generation of functional human pancreatic β cells in vitro. *Cell* **159**, 428–439 (2014).
152. Millman, J. R. *et al.* Generation of stem cell-derived β-cells from patients with type 1 diabetes. *Nat. Commun.* **7**, (2016).
153. Liu, H. *et al.* Chemical combinations potentiate human pluripotent stem cell-derived 3D pancreatic progenitor clusters toward functional beta cells. *Nat Commun* **12**, 3330 (2021).
154. Balboa, D. *et al.* Functional, metabolic and transcriptional maturation of human pancreatic islets derived from stem cells. *Nat Biotechnol* **40**, 1042–1055 (2022).
155. Pellegrini, S. *et al.* Treating iPSC-Derived beta Cells with an Anti-CD30 Antibody-Drug Conjugate Eliminates the Risk of Teratoma Development upon Transplantation. *Int J Mol Sci* **23**, (2022).

156. Vizzardelli, C. *et al.* Neonatal porcine pancreatic cell clusters as a potential source for transplantation in humans: characterization of proliferation, apoptosis, xenoantigen expression and gene delivery with recombinant AAV. *Xenotransplantation* **9**, 14–24 (2002).
157. Ellis, C., Lyon, J. G. & Korbutt, G. S. Optimization and Scale-up Isolation and Culture of Neonatal Porcine Islets: Potential for Clinical Application. *Cell Transpl.* **25**, 539–547 (2016).
158. Nagaraju, S., Bottino, R., Wijkstrom, M., Trucco, M. & Cooper, D. K. Islet xenotransplantation: what is the optimal age of the islet-source pig? *Xenotransplantation* **22**, 7–19 (2015).
159. Kim, J. M. *et al.* Long-term porcine islet graft survival in diabetic non-human primates treated with clinically available immunosuppressants. *Xenotransplantation* **28**, e12659 (2021).
160. Matsumoto, S. *et al.* Long-term follow-up for the microbiological safety of clinical microencapsulated neonatal porcine islet transplantation. *Xenotransplantation* **27**, e12631 (2020).
161. Calvin Kagan Muhammad Haq, Muhammad Mohiuddin, Susie N Hong-Zohlman, Manjula Ananthram, Charles C Hong, Vincent Y See, Stephen Shorofsky, Bartley Griffith and Timm Dickfeld, R. S. A. Abstract 12072: EKG Appearance and Evolution of Baseline EKG-Characteristics in the Worldwide First Genetically Modified Porcine-to-Human Xenotransplant ("Pig Heart-in-Human Body"). *Circulation* (2022).
162. Porrett, P. M. *et al.* First clinical-grade porcine kidney xenotransplant using a human decedent model. *Am J Transpl.* **22**, 1037–1053 (2022).

3D Organoids of Mesenchymal Stromal and Pancreatic Islet Cells

Christof Westenfelder and Anna Gooch

1 Background

1.1 Type I Diabetes Mellitus

Type 1 diabetes mellitus (T1DM) is an autoimmune disorder characterized by a lack of insulin production due to autoimmune destruction of insulin-producing beta cells, in turn leading to the lifelong need for exogenous, injected insulin and frequent blood glucose monitoring. An estimated eight to nine million individuals are currently living with the disease worldwide, and ~500,000 new cases in children and adults are diagnosed each year [1]. Poor glycemic control due to nonphysiologic and inadequate insulin dosing can lead to significant morbidities such as cardiac disease, blindness, strokes, and other micro- and macro-vascular complications, as well as premature death. Insidiously, insulin injections themselves also contribute to complications such as hypoglycemic episodes and importantly, insulin toxicity. Since nearly 50% of endogenously produced insulin is inactivated by the liver, nonphysiologic, subcutaneous delivery of significantly higher insulin doses by the diabetic patient can lead to peripheral insulin toxicity, insulin resistance, and potentially an increase in cancer risk [2–5].

C. Westenfelder (✉)
University of Utah, Department of Medicine, Salt Lake City, UT, USA

SymbioCellTech, Salt Lake City, UT, USA
e-mail: christof.westenfelder@hsc.utah.edu

A. Gooch
SymbioCellTech, Salt Lake City, UT, USA

L. Piemonti et al. (eds.), *Pluripotent Stem Cell Therapy for Diabetes*,
https://doi.org/10.1007/978-3-031-41943-0_9

2 Current Challenges and Progress in Cellular Therapeutics

The need for multiple daily insulin injections and blood glucose checks, the often young age of the newly afflicted, and the possible serious complications of T1DM have prompted intensive research with a focus on cellular therapeutics that would provide an endogenous and curative source of insulin (see Table 1).

2.1 Islet and Pancreas Transplants

Currently, autologous and allogeneic islet and pancreas transplants are two such available options. However, allogeneic islet transplants are hampered by a number of obstacles, such as expense, scarcity of suitable organs, and the chronic, lifelong need for potentially harmful antirejection drugs [6, 7]. Thus, islet transplants remain rare, with only ~4400 having been performed worldwide between 2000 and 2020 [6]. While they are considered standard of care in Europe, and elsewhere in the world for diabetic patients with hypoglycemic unawareness, in the USA, they are still considered experimental and subject to regulation as a biologic rather than an organ or tissue transplant. As a consequence, islet transplants are not eligible for insurance reimbursement in the United States, and thus can be prohibitively expensive and are only rarely performed [7].

2.2 Improvement of Islet Transplant Outcomes

In recent years, steps have been taken to increase the efficiency and viability of transplanted islets so that fewer are required, steps that would increase the available donor pool. Importantly, there are well-documented phenotypical donor and islet quality differences that have been demonstrated to correlate with the outcome of islet and pancreas transplants. In order to account for such differences and to maximize the possibility of successful outcomes with as few donors as possible, as well as to conserve resources, guidelines and screening systems, such as the North American Islet Donor Score (NAIDS) for both donors and islets, are now utilized [8–13]. The NAIDS system scores donors based on a number of characteristics including body weight, body surface area, body mass index, use of vasopressors, certain blood chemistries, age, and HbA1c levels [10]. Other studies have demonstrated that human islet viability, purity, size, morphology, and integrity correlate well with in vivo function, with viability >96%, islet diameter of >200 μm for more than 10% of the islets, a solid, spherical shape, and a well-rounded border being highly predictive of a successful clinical transplant outcome, also pretested in STZ-diabetic NOD/SCID mice [13, 14]. However, even with the application of such stringent islet donor quality criteria, only a sufficient islet yield for islet transplants

Table 1 Advantages and disadvantages of biotherapies that provide an endogenous and curative source of insulin

Biotherapy	Advantages	Disadvantages
Islet transplants	Often curative; Reduces hypoglycemic events Improves hypoglycemic unawareness Physiologic delivery	Requires immune-suppression Often >1 donor, repeatedly Donor scarcity Significant loss of islets from isolation and engraftment
Pancreas transplant	1 donor Curative Physiologic delivery	Requires immune-suppression Donor scarcity High morbidity with procedure
iPSC- and ESC-derived islet replacement therapies	Potentially curative Potentially unlimited supply Physiologic delivery	Require immune-suppression Potential for teratoma formation Not retrievable (portal system administration)
Open encapsulation devices for islets or iPSC- and ESC-derived therapies	Can be curative; Reduces hypoglycemic events Improves hypoglycemic unawareness Retrievable Revascularization	Requires immune-suppression, Invasive Limited implantation sites Foreign body reaction Nonphysiologic delivery Potential for cell migration out of capsule (limited retrievability)
Closed encapsulation devices for islets or iPSC- and ESC-derived therapies	Potentially curative Retrievable No immune-suppression	Foreign body reaction Islet amyloid accumulation Escape of autoantigens Invasive Limited implantation sites Nonphysiologic delivery
Islet transplants modified by MSCs	Often curative; Reduces hypoglycemic events Physiologic delivery Fewer islet donors needed Better islet survival	Requires immune-suppression Donor scarcity
3D organoids of MSCs and culture-expanded islet cells with omental engraftment	Potentially curative Auto- and alloimmune isolation (no immune-suppression needed) Better engraftment and survival Minimally invasive Donor pool expansion Physiologic delivery	Requires cadaveric islet donors

is reported by some centers in the range of 30–50% [10]. Nevertheless, the use of the NAIDS and other scoring criteria has the capacity to facilitate the a priori avoidance of actually harvesting islets from suboptimal donors, which is cost and time-saving [10, 11, 15–17]. However, since still significant numbers of islet preparations fail to meet the quality criteria that such scoring systems apply, it is inevitable that such "suboptimal" islet preparations may be discarded and thus may potentially represent a waste of a highly valuable resource if found to be effective for novel cell-based therapies of T1DM that rely on culture-expanded islet cells rather than whole, fresh islets. However, even with these screening methods, islets from multiple donors are often needed for successful islet transplantation [8, 11–13, 18, 19].

Another approach taken in an ongoing clinical trial (NCT03746769) is the administration of gastrin to islet transplant recipients, which has shown excellent results to date. Qualified participants will receive treatment with a single islet transplant and two rounds of gastrin treatment (twice daily injections for 30 days) with transplant and again 6 months later. Furthermore, multiple studies have demonstrated that the co-administration of mesenchymal stem cells (MSCs) does support the function of islet allografts. As summarized in Fig. 1, MSCs have well-characterized immune-modulatory, anti-inflammatory, and complex trophic activities [20–33]. In studies where MSCs were cocultured or co-administered with islets or where islets were coated with them prior to administration, outcomes were improved, demonstrating that the inclusion of MSCs in islet transplantation can help reduce the number of needed islet donors [34–44].

2.3 Scarcity of Suitable Islet Donors

Several approaches have been taken to try to overcome the issue of scarcity of suitable organ donors. One very promising approach is to avoid the use of donated islets entirely, replacing them with human, induced pluripotent (iPSCs) or embryonic

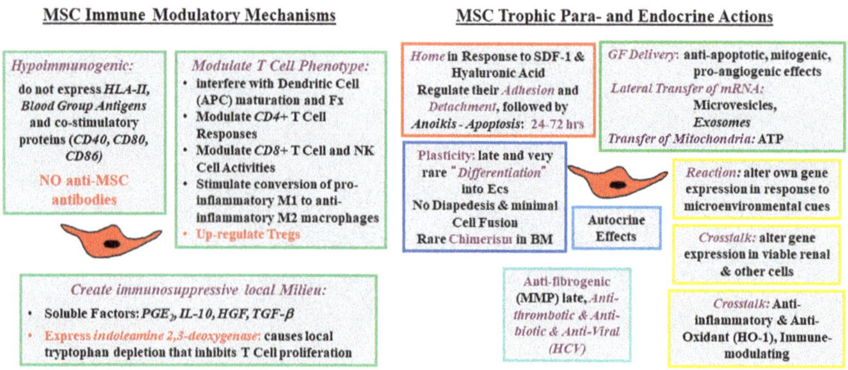

Fig. 1 MSC immune-modulatory and trophic actions

stem cells (ESCs) that are differentiated in vitro to produce insulin and other islet hormones [45–47]. Thus far, these have proven to still require multiple in vitro cell differentiation steps that also address potential teratoma formation by residual undifferentiated iPSCs and ECSs. Yet ideally, if further developed, this approach would provide a clonal, insulin-producing cell line that would function in vivo physiologically [47]. The most successful iPSC-based therapeutic (VX-880) to date has been administered to a handful of individuals with both very promising and some mixed results (*Vertex Pharmaceuticals*, press release). VX-880 is delivered into the portal system of the liver, and while providing physiologic delivery of insulin, such a transplant site offers no opportunity to retrieve the cells should the need arise. These issues together with the prevention of teratoma and teratocarcinoma formation, are addressed by these novel SC-based biotherapies [48, 49]. iPSC-based therapies like allogeneic islet transplants still require the lifelong use of potentially harmful antirejection drugs, although efforts are being made to mitigate these drawbacks through gene editing and Treg-based strategies [47, 50].

The successful culture expansion of islets has long been known to facilitate the production of high numbers of islet cells that have therapeutic potential [51, 52]. Since the expansion of hormone-producing islet cells results in their progressive differentiation, mediated by cell cycle entry and reversible EMT, the obtained cells must be redifferentiated to reestablish physiological hormone production in vivo [51–55]. In vitro redifferentiation of these cells is effective, yet this approach has only been tested in a few preclinical studies [51].

2.4 Immune Isolation of Islet Transplants

Encapsulation systems that use, e.g., alginate, have long been tested as a possible solution to provide allo-and autoimmune isolation of allogeneic islet and iPSC-based transplants. These can be susceptible to foreign body reactions and secondary failure of the graft [45, 56–59]. There are data from two promising preclinical encapsulation technologies that provide glycemic control and that lack the foreign body reaction, while one of these is also retrievable [60, 61]. However, two recent reports from an ongoing Phase I/II clinical trial conducted at seven centers (United States, Canada), sponsored by *ViaCyte/Vertex,* describe the results of open, subcutaneously placed encapsulation devices (VC-02) containing human ESC-derived pancreatic hormone-producing cells. The utilized devices are retrievable but open to facilitate vascularization and therefore require antirejection drugs. Neither trial site reported achievement of insulin independence and reestablishment of euglycemia, but insulin expression and meal-sensitive C-peptide release were demonstrated. Adverse Events were related to involved surgical procedures and the use of antirejection drugs [62, 63].

Most recently, it was shown that editing of primary human islet cells to the hypoimmune HLA class I- and II-negative and CD47 overexpressing phenotype, followed by their reaggregation into pseudo-islets generated an effective, anitirejection drug-independent therapy for T1DM in humanized autoimmune diabetic mice [64].

2.5 *Use of Novel, Prevascularized Organoids*

The significant losses of transplanted islets that are caused because of hypoxic stress that is caused during their isolation has been addressed by the engineering of sophisticated prevascularized organoids [65, 66]. These preclinical proof-of-concept studies were highly effective in supporting effective engraftment and essential vascularization.

2.6 *Induction of Immune Tolerance with Tregs*

Several ongoing and completed clinical trials have demonstrated that the administration, at the time of islet transplant, of polyclonal Tregs can improve the subsequent function of allogeneic islets [47, 67, 68]. Of note is the fact that the administration of MSCs will increase Treg numbers in the microenvironment of transplanted islets, which is considered a contributing factor to the beneficial actions of co-transplanted MSCs [69].

2.7 *Prevention of Teratoma/Teratocarcinoma Formation*

Potential oncogenic transformation by residual undifferentiated iPSCs and ECSs in novel stem cell-based biotherapies of T1DM is an essential safety issue [70]. Use of chemical inhibitors [71], Insertion of suicide genes [72], and most recently, the in vitro use of an anti-CD30 antibody linked to the antitubulin agent monomethyl auristatin E was shown to eliminate the oncogenic potential of undifferentiated iPSC, a technique that may have substantial translational value [73].

2.8 *Islet Hormone Delivery, Extrahepatic Implantation Sites, and Retrievability*

Functioning, allogeneic islet transplants provide physiological hormone delivery into the portal system of the liver, paralleling that of the pancreas. However, a significant portion of islets is lost during islet isolation due to ischemic stress, apoptosis, and, postintraportal implantation, due to the instant blood-mediated inflammatory reaction [IBMIR], low oxygen pressure, and chronic exposure to high concentrations of intraportally delivered antirejection drugs [74]. Surgical retrieval of intraportal grafts is not possible, which can become a safety concern.

Preclinical and some clinical data from subcutaneous, under kidney capsule, intramuscular, intraocular, epididimal fat, and other implantation sites of both islet or insulin-producing cells show generally adequate glycemic control [75, 76]. However, like s.c. insulin injections, there is the possibility for the development of peripheral insulin toxicity because the physiological inactivation, by up to 50%, of intra-portally delivered insulin does not occur [2, 77]. The long-term significance of this issue requires further investigation. Most of these implantation sites allow the retrieval of the transplant if needed. The omentum has been demonstrated in several pre- and clinical studies to be a suitable islet transplantation site that provides adequate oxygenation, vascularization and its venous system delivers islet hormones into the portal system as the pancreas does. If warranted, the omental transplants can be retrieved [75].

2.9 *Metabolic Control Achieved with Mono Hormonal (Insulin) Versus Polyhormonal (Insulin, Glucagon, SST, PPY, and Ghrelin) Cell Transplants*

In the normal islet, complex paracrine interactions between the hormone-producing cells have been identified, together resulting in physiological glycemic control [78]. Whether the novel iPSC-derived, by differentiation, primarily insulin-producing cells result, long-term, in suboptimal metabolic control and the secondary development of end-organ complications remains to be demonstrated in the ongoing clinical trials.

3 Summary

Novel and promising approaches in the field of stem cell therapy for T1DM are being developed and tested globally. *The main issues that need to be addressed are*:

(1) Donor scarcity, if applicable; (2) Durable auto- and alloimmune isolation of allogeneic biotherapies that optimally avoid the need of anti-rejection drugs and provide durable insulin independence and optimal metabolic control; (3) Protection of transplanted cells from inflammation- and hypoxia-induced apoptosis; (4) Adequate Angiogenesis; (5) Optimal implantation site; (6) Retrievability; (7) Physiological Insulin and other islet hormone delivery into the hepatic portal system; (8) Absent tumorigenicity.

As shown below, these critical issues are addressed by the formation and therapeutic use of 3D organoids of mesenchymal stromal and pancreatic islet cells.

4 Technology: 3D Organoids of Mesenchymal Stromal and Pancreatic Islet Cells

4.1 Rationale

As discussed above, and due to their paracrine trophic and immune-modulatory properties summarized in Fig. 1, mesenchymal stromal cells can provide protection to whole and to dissociated islets, and thus are excellent cellular tools for overcoming the issues of allo- and autoimmune isolation, protection from inflammation and hypoxia, and promotion of adequate angiogenesis stated in the above conclusion [20–33].

As further summarized above, cells within islets can be culture-expanded but will undergo potentially significant dedifferentiation as a result, and to some extent, exhibit reduced endocrine cell phenotype. However, as also summarized above, dedifferentiation or transdifferentiation are processes seen in type 2 diabetes, a reversible human disease. Thus, de/transdifferentiation and redifferentiation of islet cells is a natural phenomenon [79].

4.2 Hypothesis

With the above-detailed concerns and limitations in mind, i.e., lack of organ donors, auto- and alloimmune rejection of grafts, retrievability, efficiency, physiologic hormone delivery, and cumbersome technology, we chose not to pursue an improvement of current islet transplantation technology, but rather a new cellular therapeutic strategy. As illustrated in Fig. 2, we hypothesized that inclusion of high numbers of healthy MSCs, when aggregated with culture-expanded islet cells (ICs) could provide adequate allo- and autoimmune protection to those cells, and that the pro-angiogenic, pro-survival, anti-inflammatory, and other pleotropic, cell supportive functions of MSCs could provide a protective microenvironment where the islet cell component could resume its physiologic, endocrine functions, providing a diabetic subject with needed insulin and other islet hormones.

This bioengineered therapeutic is called "Neo-Islets" (NIs) and is distinct in two ways from previous work done where to-be-transplanted islets were either coated or co-administered with MSCs [34, 35, 37–43]. First, the pleotropic actions of MSCs are greatly amplified by increasing the intrinsic numbers of MSCs in NIs to approximately 50% from the 2% typically found in native islets [80]. Second, the islet cells used in NIs are culture-expanded and partially dedifferentiated, as detailed in the sections below.

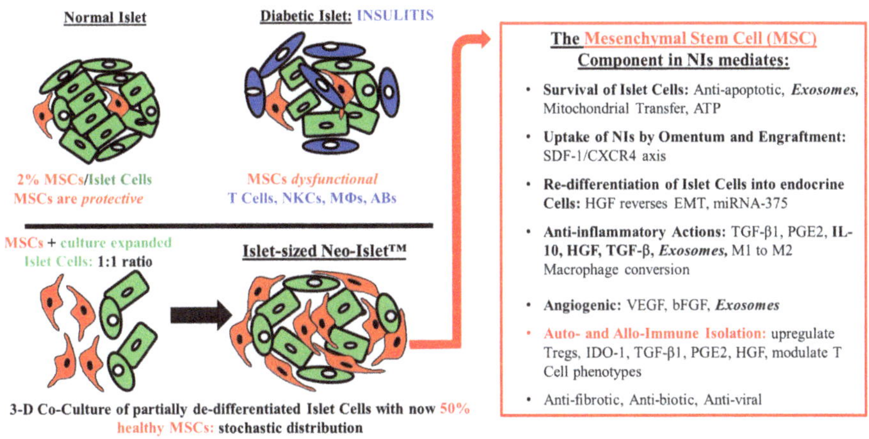

Fig. 2 Rationale for the use of MSCs in high numbers in NIs

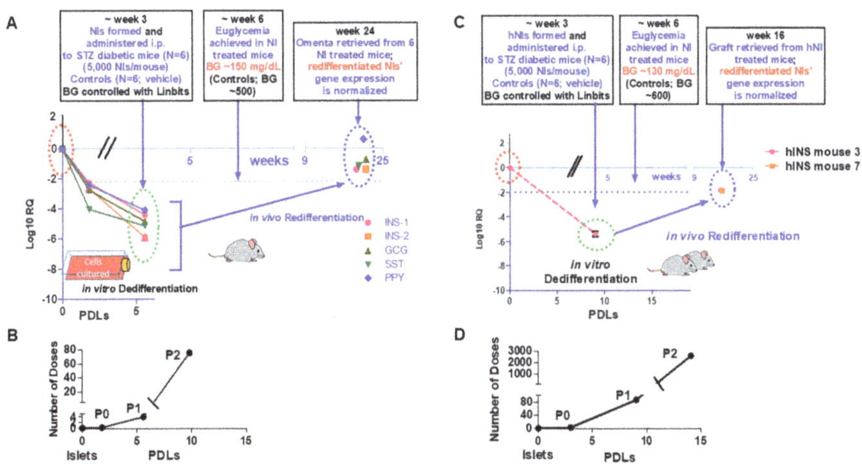

Fig. 3 In vitro dedifferentiation of cultured murine (**a**) or human (**c**) Islet Cells (ICs)and their in vivo redifferentiation as the endocrine component of i.p.-administered NIs (**a**, **c**). Culture expansion of ICs allows multiple doses to be produced from a single islet donor (**b**, **d**)

To date, NIs have been created using cells from mice, dogs, and humans. NIs from any of these species can restore euglycemia in mice [69, 81]. Culture expansion of ICs such that endocrine gene expression is down-regulated but not lost in vitro does allow for redifferentiation in vivo while simultaneously allowing multiple therapeutic doses to be created from one donor (See Fig. 3).

The MSC component allows for the allogeneic use of NIs without the need for antirejection drugs, as has been demonstrated in both NOD mice and spontaneously diabetic pet dogs [69, 81, 82]. In the latter study, which is being conducted under an

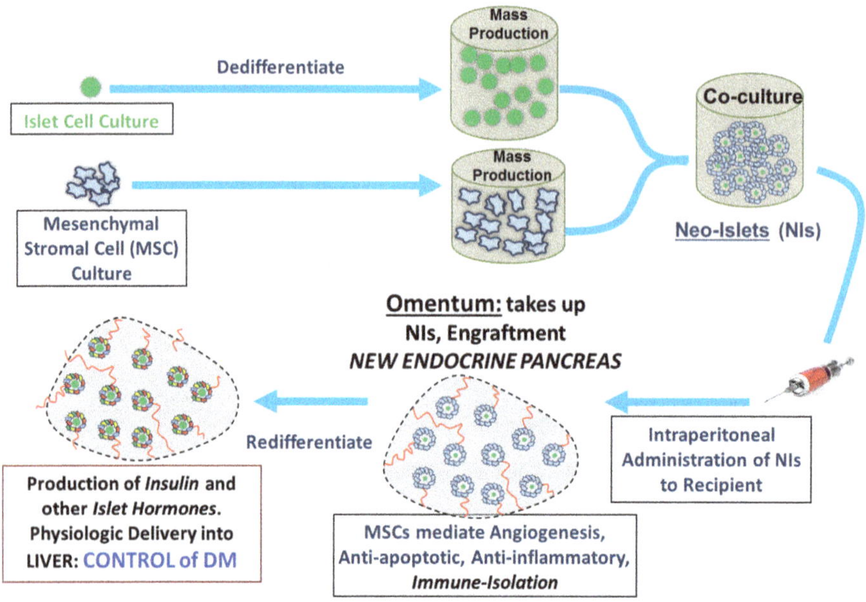

Fig. 4 Process of Neo-Islet formation, scale-up, delivery, and function. Neo-Islets are safe, simple, efficient, and less expensive

FDA-guided INAD and has been ongoing for more than 3 years, allogeneically administered NIs have resulted in a durably reduced need for insulin by up to 50% and simultaneously improved glycemic control as demonstrated by significantly reduced blood glucose, fructosamine, and HbA1c levels. No adverse events to date have been attributed to NI administration. This promising, safe therapy is easily scaled (see Fig. 4), and an IND-enabling study is currently being conducted in order to facilitate translation of this technology to the human clinic in the near future.

Acknowledgments and Competing Interests CW is a consultant, and AG is a full-time employee of SymbioCellTech, LLC. Both CW and AG hold shares in and are members of the Board of Directors of SymbioCellTech, LLC. SymbioCellTech, LLC funded this work in its entirety. We acknowledge the excellent laboratory work conducted by the members of our lab, Ping Zhang, Zhuma Hu, and Sabiha Chowdhury, as well as our collaborating veterinarians, Julie Fischer and Natasha LoySon (both at Veterinary Specialty Hospital in San Diego, CA), Rance Sellon (Washington State University, Pullman, WA), and Melanie Thompson (Mountain West Veterinary Specialty Hospital (Layton, UT). Neo-Islets are a patented technology in the US, Canada, Japan, and Europe.

References

1. Gregory GA, Robinson TIG, Linklater SE, Wang F, Colagiuri S, de Beaufort C, et al. Global incidence, prevalence, and mortality of type 1 diabetes in 2021 with projection to 2040: a modelling study. Lancet Diabetes Endocrinol. 2022;10: 741–760. https://doi.org/10.1016/S2213-8587(22)00218-2

2. Gregory JM, Kraft G, Scott MF, Neal DW, Farmer B, Smith MS, et al. Insulin Delivery Into the Peripheral Circulation: A Key Contributor to Hypoglycemia in Type 1 Diabetes. Diabetes. 2015;64: 3439–3451. https://doi.org/10.2337/db15-0071
3. Edgerton DS, Moore MC, Gregory JM, Kraft G, Cherrington AD. Importance of the route of insulin delivery to its control of glucose metabolism. Am J Physiol Metab. 2021;320: E891–E897. https://doi.org/10.1152/ajpendo.00628.2020
4. Gregory JM, Cherrington AD, Moore DJ. The Peripheral Peril: Injected Insulin Induces Insulin Insensitivity in Type 1 Diabetes. Diabetes. 2020;69: 837–847. https://doi.org/10.2337/dbi19-0026
5. Zhong W, Mao Y. Daily Insulin Dose and Cancer Risk Among Patients With Type 1 Diabetes. JAMA Oncol. 2022;8: 1356. https://doi.org/10.1001/jamaoncol.2022.2960
6. Berney T, Andres A, Bellin MD, de Koning EJP, Johnson PR V, Kay TWH, et al. A Worldwide Survey of Activities and Practices in Clinical Islet of Langerhans Transplantation. Transpl Int. 2022;35: 10507. https://doi.org/10.3389/ti.2022.10507
7. Witkowski P, Philipson LH, Kaufman DB, Ratner LE, Abouljoud MS, Bellin MD, et al. The demise of islet allotransplantation in the United States: A call for an urgent regulatory update. Am J Transplant. 2021;21: 1365–1375. https://doi.org/10.1111/ajt.16397
8. Henquin J-C. Influence of organ donor attributes and preparation characteristics on the dynamics of insulin secretion in isolated human islets. Physiol Rep. 2018;6: e13646. https://doi.org/10.14814/phy2.13646
9. Wang L-J, Kin T, O'Gorman D, Shapiro AMJ, Naziruddin B, Takita M, et al. A Multicenter Study: North American Islet Donor Score in Donor Pancreas Selection for Human Islet Isolation for Transplantation. Cell Transplant. 2016;25: 1515–1523. https://doi.org/10.3727/096368916X691141
10. Gołębiewska JE, Bachul PJ, Wang L, Matosz S, Basto L, Kijek MR, et al. Validation of a New North American Islet Donor Score for Donor Pancreas Selection and Successful Islet Isolation in a Medium-Volume Islet Transplant Center. Cell Transplant. 2019;28: 185–194. https://doi.org/10.1177/0963689718816989
11. Bertuzzi F, Ricordi C. Prediction of Clinical Outcome in Islet Allotransplantation. Diabetes Care. 2007;30: 410–417. https://doi.org/10.2337/dc06-1233
12. Kandeel F, El-Shahawy M, Singh G, Dafoe DC, Isenberg JS, Riggs AD. Towards a Rational Balanced Pancreatic and Islet Allocation Schema. Cell Transplant. 2021;30: 9636897211057130. https://doi.org/10.1177/09636897211057130
13. Salgado M, Gonzalez N, Medrano L, Rawson J, Omori K, Qi M, et al. Semi-Automated Assessment of Human Islet Viability Predicts Transplantation Outcomes in a Diabetic Mouse Model. Cell Transplant. 2020;29: 096368972091944. https://doi.org/10.1177/0963689720919444
14. Komatsu H, Qi M, Gonzalez N, Salgado M, Medrano L, Rawson J, et al. A Multiparametric Assessment of Human Islets Predicts Transplant Outcomes in Diabetic Mice. Cell Transplant. 2021;30: 096368972110522. https://doi.org/10.1177/09636897211052291
15. Hilling DE, Bouwman E, Terpstra OT, Marang-Van De Mheen PJ. Effects of Donor-, Pancreas-, and Isolation-Related Variables on Human Islet Isolation Outcome: A Systematic Review. Cell Transplant. 2014;23: 921–928. https://doi.org/10.3727/096368913X666412
16. Silverman LI, Flanagan F, Rodriguez-Granrose D, Simpson K, Saxon LH, Foley KT. Identifying and Managing Sources of Variability in Cell Therapy Manufacturing and Clinical Trials. Regen Eng Transl Med. 2019;5: 354–361. https://doi.org/10.1007/s40883-019-00129-y
17. Yeh C-C, Wang L-J, Mcgarrigle JJ, Wang Y, Liao C-C, Omami M, et al. Effect of Manufacturing Procedures on Human Islet Isolation from Donor Pancreata Standardized by the North American Islet Donor Score. Cell Transplant. 2017;26: 33–44. https://doi.org/10.3727/096368916X692834
18. Barton FB, Rickels MR, Alejandro R, Hering BJ, Wease S, Naziruddin B, et al. Improvement in outcomes of clinical islet transplantation: 1999–2010. Diabetes Care. 2012;35: 1436–1445. https://doi.org/10.2337/dc12-0063

19. Polonsky KS. The Past 200 Years in Diabetes. N Engl J Med. 2012;367: 1332–1340. https://doi.org/10.1056/NEJMra1110560
20. Inverardi L, Lanzoni G, Dominguez-Bendala J, Ricordi C. MSCs for Diabetes. Mesenchymal Stromal Cells. New York, NY: Springer New York; 2013. pp. 571–597. https://doi.org/10.1007/978-1-4614-5711-4_33
21. DelaRosa O, Sánchez-Correa B, Morgado S, Ramírez C, del Río B, Menta R, et al. Human Adipose-Derived Stem Cells Impair Natural Killer Cell Function and Exhibit Low Susceptibility to Natural Killer-Mediated Lysis. Stem Cells Dev. 2012;21: 1333–1343. https://doi.org/10.1089/scd.2011.0139
22. Spees JL, Lee RH, Gregory CA. Mechanisms of mesenchymal stem/stromal cell function. Stem Cell Res Ther. 2016;7: 125. https://doi.org/10.1186/s13287-016-0363-7
23. Tögel F, Hu Z, Weiss K, Isaac J, Lange C, Westenfelder C. Administered mesenchymal stem cells protect against ischemic acute renal failure through differentiation-independent mechanisms. Am J Physiol Renal Physiol. 2005;289: F31–42. https://doi.org/10.1152/ajprenal.00007.2005
24. Togel F, Weiss K, Yang Y, Hu Z, Zhang P, Westenfelder C. Vasculotropic, paracrine actions of infused mesenchymal stem cells are important to the recovery from acute kidney injury. Am J Physiol Ren Physiol. 2007;292: F1626–F1635. https://doi.org/10.1152/ajprenal.00339.2006
25. Gooch AM, Westenfelder C. MSCs for Renal Repair. In: Hematti P, Keating A, editors. Mesenchymal Stromal Cells. New York, NY: Springer New York; 2013. pp. 599–615. https://doi.org/10.1007/978-1-4614-5711-4_34
26. Burr SP, Dazzi F, Garden O a. Mesenchymal stromal cells and regulatory T cells: the Yin and Yang of peripheral tolerance? Immunol Cell Biol. 2013;91: 12–18. https://doi.org/10.1038/icb.2012.60
27. English K. Mechanisms of mesenchymal stromal cell immunomodulation. Immunol Cell Biol. 2013;91: 19–26. https://doi.org/10.1038/icb.2012.56
28. Kim Y-H, Wee Y-M, Choi M-Y, Lim D-G, Kim S-C, Han D-J. Interleukin (IL)-10 induced by CD11b(+) cells and IL-10-activated regulatory T cells play a role in immune modulation of mesenchymal stem cells in rat islet allografts. Mol Med. 2011;17: 697–708. https://doi.org/10.2119/molmed.2010.00098
29. Spaggiari GM, Moretta L. Cellular and molecular interactions of mesenchymal stem cells in innate immunity. Immunol Cell Biol. 2013;91: 27–31. https://doi.org/10.1038/icb.2012.62
30. LeBlanc K, Davies LC. Mesenchymal stromal cells and the innate immune response. Immunol Lett. 2015;168: 140–146. https://doi.org/10.1016/j.imlet.2015.05.004
31. Caplan AI, Sorrell JM. The MSC curtain that stops the immune system. Immunol Lett. 2015;168: 136–139. https://doi.org/10.1016/j.imlet.2015.06.005
32. Caplan AI. Adult Mesenchymal Stem Cells: When, Where, and How. Stem Cells Int. 2015;2015: 1–6. https://doi.org/10.1155/2015/628767
33. Davis TA, Anam K, Lazdun Y, Gimble JM, Elster EA. Adipose-Derived Stromal Cells Promote Allograft Tolerance Induction. Stem Cells Transl Med. 2014;3: 1444–1450. https://doi.org/10.5966/sctm.2014-0131
34. Duprez IR, Johansson U, Nilsson B, Korsgren O, Magnusson PU. Preparatory studies of composite mesenchymal stem cell islets for application in intraportal islet transplantation. Ups J Med Sci. 2011;116: 8–17. https://doi.org/10.3109/03009734.2010.524320
35. Borg DJ, Weigelt M, Wilhelm C, Gerlach M, Bickle M, Speier S, et al. Mesenchymal stromal cells improve transplanted islet survival and islet function in a syngeneic mouse model. Diabetologia. 2014;57: 522–31. https://doi.org/10.1007/s00125-013-3109-4
36. Hubber EL, Rackham CL, Jones PM. Protecting islet functional viability using mesenchymal stromal cells. Stem Cells Transl Med. 2021;10: 674–680. https://doi.org/10.1002/sctm.20-0466
37. Cavallari G, Olivi E, Bianchi F, Neri F, Foroni L, Valente S, et al. Mesenchymal stem cells and islet cotransplantation in diabetic rats: improved islet graft revascularization and function by human adipose tissue-derived stem cells preconditioned with natural molecules. Cell Transplant. 2012;21: 2771–2781. https://doi.org/10.3727/096368912X637046

38. Rackham CL, Chagastelles PC, Nardi NB, Hauge-Evans a C, Jones PM, King a JF. Co-transplantation of mesenchymal stem cells maintains islet organisation and morphology in mice. Diabetologia. 2011;54: 1127–1135. https://doi.org/10.1007/s00125-011-2053-4
39. Rackham CL, Dhadda PK, Le Lay AM, King AJF, Jones PM. Preculturing Islets With Adipose-Derived Mesenchymal Stromal Cells Is an Effective Strategy for Improving Transplantation Efficiency at the Clinically Preferred Intraportal Site. Cell Med. 2014;7: 37–47. https://doi.org/10.3727/215517914X680047
40. Carlsson P-O, Schwarcz E, Korsgren O, Le Blanc K. Preserved β-Cell Function in Type 1 Diabetes by Mesenchymal Stromal Cells. Diabetes. 2015;64: 587–592. https://doi.org/10.2337/db14-0656
41. Rickels MR, Stock PG, de Koning EJP, Piemonti L, Pratschke J, Alejandro R, et al. Defining outcomes for β-cell replacement therapy in the treatment of diabetes: a consensus report on the Igls criteria from the IPITA/EPITA opinion leaders workshop. Transpl Int. 2018;31: 343–352. https://doi.org/10.1111/tri.13138
42. Rackham CL, Hubber EL, Czajka A, Malik AN, King AJF, Jones PM. Optimizing beta cell function through mesenchymal stromal cell-mediated mitochondria transfer. Stem Cells. 2020;38: 574–584. https://doi.org/10.1002/stem.3134
43. Sordi V, Piemonti L. Mesenchymal Stem Cells as Feeder Cells for Pancreatic Islet Transplants. Rev Diabet Stud. 2010;7: 132–143. https://doi.org/10.1900/RDS.2010.7.132
44. Koehler N, Buhler L, Egger B, Gonelle-Gispert C. Multipotent Mesenchymal Stromal Cells Interact and Support Islet of Langerhans Viability and Function. Front Endocrinol (Lausanne). 2022;13: 822191. https://doi.org/10.3389/fendo.2022.822191
45. Schulz TC, Young HY, Agulnick AD, Babin MJ, Baetge EE, Bang AG, et al. A scalable system for production of functional pancreatic progenitors from human embryonic stem cells. PLoS One. 2012;7: e37004. https://doi.org/10.1371/journal.pone.0037004
46. Pagliuca FW, Millman JR, Gürtler M, Segel M, Van Dervort A, Ryu JH, et al. Generation of Functional Human Pancreatic β Cells In Vitro. Cell. 2014;159: 428–439. https://doi.org/10.1016/j.cell.2014.09.040
47. Loretelli C, Assi E, Seelam AJ, Ben Nasr M, Fiorina P. Cell therapy for type 1 diabetes. Expert Opin Biol Ther. 2020;20: 887–897. https://doi.org/10.1080/14712598.2020.1748596
48. Lezmi E, Benvenisty N. The Tumorigenic Potential of Human Pluripotent Stem Cells. Stem Cells Transl Med. 2022;11: 791–796. https://doi.org/10.1093/stcltm/szac039
49. Benabdallah B, Désaulniers-Langevin C, Colas C, Li Y, Rousseau G, Guimond J V., et al. Natural Killer Cells Prevent the Formation of Teratomas Derived From Human Induced Pluripotent Stem Cells. Front Immunol. 2019;10. https://doi.org/10.3389/fimmu.2019.02580
50. Sintov E, Nikolskiy I, Barrera V, Hyoje-Ryu Kenty J, Atkin AS, Gerace D, et al. Whole-genome CRISPR screening identifies genetic manipulations to reduce immune rejection of stem cell-derived islets. Stem cell reports. 2022;17: 1976–1990. https://doi.org/10.1016/j.stemcr.2022.08.002
51. Efrat S. Mechanisms of adult human β-cell in vitro dedifferentiation and redifferentiation. Diabetes Obes Metab. 2016;18 Suppl 1: 97–101. https://doi.org/10.1111/dom.12724
52. Moreno-Amador JL, Téllez N, Marin S, Aloy-Reverté C, Semino C, Nacher M, et al. Epithelial to mesenchymal transition in human endocrine islet cells. PLoS One. 2018;13: e0191104. https://doi.org/10.1371/journal.pone.0191104
53. Wei J-J, Tang L, Chen L-L, Xie Z-H, Ren Y, Qi H-G, et al. Mesenchymal Stem Cells Attenuates TGF-β1-Induced EMT by Increasing HGF Expression in HK-2 Cells. Iran J Public Health. 2021;50: 908–918. https://doi.org/10.18502/ijph.v50i5.6108
54. Iismaa SE, Kaidonis X, Nicks AM, Bogush N, Kikuchi K, Naqvi N, et al. Comparative regenerative mechanisms across different mammalian tissues npj Regen Med. 2018;3: 6. https://doi.org/10.1038/s41536-018-0044-5
55. Dong L-H, Jiang Y-Y, Liu Y-J, Cui S, Xia C-C, Qu C, et al. The anti-fibrotic effects of mesenchymal stem cells on irradiated lungs via stimulating endogenous secretion of HGF and PGE2. Sci Rep. 2015;5: 8713. https://doi.org/10.1038/srep08713

56. Tuch BE, Keogh GW, Williams LJ, Wu W, Foster JL, Vaithilingam V, et al. Safety and Viability of Microencapsulated Human Islets Transplanted Into Diabetic Humans. Diabetes Care. 2009;32: 1887–1889. https://doi.org/10.2337/dc09-0744
57. Weir GC. Islet encapsulation: advances and obstacles. Diabetologia. 2013;56: 1458–1461. https://doi.org/10.1007/s00125-013-2921-1
58. Ludwig B, Reichel A, Steffen A, Zimerman B, Schally A V, Block NL, et al. Transplantation of human islets without immunosuppression. Proc Natl Acad Sci. 2013;110: 19054–19058. https://doi.org/10.1073/pnas.1317561110
59. Vegas AJ, Veiseh O, Gürtler M, Millman JR, Pagliuca FW, Bader AR, et al. Long-term glycemic control using polymer-encapsulated human stem cell–derived beta cells in immune-competent mice. Nat Med. 2016;22: 306–311. https://doi.org/10.1038/nm.4030
60. Stock AA, Gonzalez GC, Pete SI, De Toni T, Berman DM, Rabassa A, et al. Performance of islets of Langerhans conformally coated via an emulsion cross-linking method in diabetic rodents and nonhuman primates. Sci Adv. 2022;8: eabm3145. https://doi.org/10.1126/sciadv.abm3145
61. Wang X, Maxwell KG, Wang K, Bowers DT, Flanders JA, Liu W, et al. A nanofibrous encapsulation device for safe delivery of insulin-producing cells to treat type 1 diabetes. Sci Transl Med. 2021;13. https://doi.org/10.1126/scitranslmed.abb4601
62. Shapiro AMJ, Thompson D, Donner TW, Bellin MD, Hsueh W, Pettus J, et al. Insulin expression and C-peptide in type 1 diabetes subjects implanted with stem cell-derived pancreatic endoderm cells in an encapsulation device. Cell reports Med. 2021;2: 100466. https://doi.org/10.1016/j.xcrm.2021.100466
63. Ramzy A, Thompson DM, Ward-Hartstonge KA, Ivison S, Cook L, Garcia R V, et al. Implanted pluripotent stem-cell-derived pancreatic endoderm cells secrete glucose-responsive C-peptide in patients with type 1 diabetes. Cell Stem Cell. 2021;28: 2047–2061.e5. https://doi.org/10.1016/j.stem.2021.10.003
64. Hu X, Gattis C, Olroyd AG, Friera AM, White K, Young C, et al. Human hypoimmune primary pancreatic islets avoid rejection and autoimmunity and alleviate diabetes in allogeneic humanized mice. Sci Transl Med. 2023;15: eadg5794. https://doi.org/10.1126/scitranslmed.adg5794
65. Wassmer C-H, Lebreton F, Bellofatto K, Perez L, Cottet-Dumoulin D, Andres A, et al. Bio-Engineering of Pre-Vascularized Islet Organoids for the Treatment of Type 1 Diabetes. Transpl Int. 2022;35. https://doi.org/10.3389/ti.2021.10214
66. Citro A, Moser PT, Dugnani E, Rajab TK, Ren X, Evangelista-Leite D, et al. Biofabrication of a vascularized islet organ for type 1 diabetes. Biomaterials. 2019;199: 40–51. https://doi.org/10.1016/j.biomaterials.2019.01.035
67. Marfil-Garza BA, Hefler J, Bermudez De Leon M, Pawlick R, Dadheech N, Shapiro AMJ. Progress in Translational Regulatory T Cell Therapies for Type 1 Diabetes and Islet Transplantation. Endocr Rev. 2021;42: 198–218. https://doi.org/10.1210/endrev/bnaa028
68. Raffin C, Vo LT, Bluestone JA. Treg cell-based therapies: challenges and perspectives. Nat Rev Immunol. 2020;20: 158–172. https://doi.org/10.1038/s41577-019-0232-6
69. Westenfelder C, Gooch A, Hu Z, Ahlstrom J, Zhang P. Durable Control of Autoimmune Diabetes in Mice Achieved by Intraperitoneal Transplantation of "Neo-Islets," Three-Dimensional Aggregates of Allogeneic Islet and "Mesenchymal Stem Cells". Stem Cells Transl Med. 2017;6: 1631–1643. https://doi.org/10.1002/sctm.17-0005
70. Ben-David U, Benvenisty N. The tumorigenicity of human embryonic and induced pluripotent stem cells. Nat Rev Cancer. 2011;11: 268–77. https://doi.org/10.1038/nrc3034
71. Ben-David U, Gan Q-F, Golan-Lev T, Arora P, Yanuka O, Oren YS, et al. Selective Elimination of Human Pluripotent Stem Cells by an Oleate Synthesis Inhibitor Discovered in a High-Throughput Screen. Cell Stem Cell. 2013; 1–13. https://doi.org/10.1016/j.stem.2012.11.015
72. Schuldiner M, Itskovitz-Eldor J, Benvenisty N. Selective ablation of human embryonic stem cells expressing a "suicide" gene. Stem Cells. 2003;21: 257–65. https://doi.org/10.1634/stemcells.21-3-257

73. Pellegrini S, Zamarian V, Landi E, Cospito A, Lombardo MT, Manenti F, et al. Treating iPSC-Derived β Cells with an Anti-CD30 Antibody–Drug Conjugate Eliminates the Risk of Teratoma Development upon Transplantation. Int J Mol Sci. 2022;23: 9699. https://doi.org/10.3390/ijms23179699
74. Marfil-Garza BA, Imes S, Verhoeff K, Hefler J, Lam A, Dajani K, et al. Pancreatic islet transplantation in type 1 diabetes: 20-year experience from a single-centre cohort in Canada. lancet Diabetes Endocrinol. 2022;10: 519–532. https://doi.org/10.1016/S2213-8587(22)00114-0
75. Deng H, Zhang A, Pang DRR, Xi Y, Yang Z, Matheson R, et al. Bioengineered omental transplant site promotes pancreatic islet allografts survival in non-human primates. Cell reports Med. 2023;4: 100959. https://doi.org/10.1016/j.xcrm.2023.100959
76. Addison P, Fatakhova K, Rodriguez Rilo HL. Considerations for an Alternative Site of Islet Cell Transplantation. J Diabetes Sci Technol. 2020;14: 338–344. https://doi.org/10.1177/1932296819868495
77. Pørksen N, Grøfte T, Greisen J, Mengel A, Juhl C, Veldhuis JD, et al. Human insulin release processes measured by intraportal sampling. Am J Physiol Endocrinol Metab. 2002;282: E695–702. https://doi.org/10.1152/ajpendo.00516.2000
78. Heaton ES, Jin S. Importance of multiple endocrine cell types in islet organoids for type 1 diabetes treatment. Transl Res. 2022;250: 68–83. https://doi.org/10.1016/j.trsl.2022.06.014
79. Wang W, Zhang C. Targeting β-cell dedifferentiation and transdifferentiation: opportunities and challenges. Endocr Connect. 2021;10: R213–R228. https://doi.org/10.1530/EC-21-0260
80. Hayden MR, Yang Y, Habibi J, Bagree S V, Sowers JR. Pericytopathy: Oxidative Stress and Impaired Cellular Longevity in the Pancreas and Skeletal Muscle in Metabolic Syndrome and Type 2 Diabetes. Oxid Med Cell Longev. 2010;3: 290–303. https://doi.org/10.4161/oxim.3.5.13653
81. Westenfelder C, Hu Z, Zhang P, Gooch A. Intraperitoneal administration of human "Neo-Islets", 3-D organoids of mesenchymal stromal and pancreatic islet cells, normalizes blood glucose levels in streptozotocin-diabetic NOD/SCID mice: Significance for clinical trials. Fiorina P, editor. PLoS One. 2021;16: e0259043. https://doi.org/10.1371/journal.pone.0259043
82. Gooch A, Zhang P, Hu Z, Loy Son N, Avila N, Fischer J, et al. Interim report on the effective intraperitoneal therapy of insulin-dependent diabetes mellitus in pet dogs using "Neo-Islets," aggregates of adipose stem and pancreatic islet cells (INAD 012-776). PLoS One. 2019;14: e0218688. https://doi.org/10.1371/journal.pone.0218688

Extracellular Matrix to Support Beta Cell Health and Function

Daniel M. Tremmel, Sara Dutton Sackett, and Jon S. Odorico

1 The Extracellular Matrix

The extracellular matrix (ECM) is a network of proteins and polysaccharides that stimulates cells through structural and biochemical interactions. ECM is an essential part of the tissue environment and influences cell behavior directly by binding cell surface receptors, such as integrins, through regulation of growth factor diffusion and presentation, and through mechanical stimulation [1]. Through these mechanisms, ECM plays a significant role in cell health and identity. The process of isolating islets from the pancreas involves digestion with collagenase and neutral protease, which destroys much of the native islet ECM, affecting islet health [2–4]. Since beta cell replacement is used as a treatment for diabetic patients, understanding the role of ECM on islet and beta cell survival and function could help optimize the treatment. In 2-D culture, non-tissue specific ECM scaffolds like Matrigel or purified ECM proteins are sometimes used to support cell adherence and growth, but in 3-D organoid culture and in engineered tissues, ECM may be optimally tailored to the cell or tissue type [5]. This is especially important given the significant differences in ECM composition among different organs and tissues [10].

There are many types of proteins categorized as 'ECM', and the different families and types of ECM molecules and their potential roles in pancreatic islets have been extensively summarized in previous reviews [6–9]. These structural proteins that constitute the scaffolding around cells, or the "core matrisome" can broadly be

D. M. Tremmel (✉)
Department of Cardiac Surgery, Boston Children's Hospital/Harvard Medical School, Boston, MA, USA
e-mail: daniel.tremmel@childrens.harvard.edu

S. D. Sackett · J. S. Odorico
Department of Surgery, Transplantation Division, University of Wisconsin-Madison, Madison, WI, USA

L. Piemonti et al. (eds.), *Pluripotent Stem Cell Therapy for Diabetes*,
https://doi.org/10.1007/978-3-031-41943-0_10

categorized as collagens, ECM glycoproteins, and ECM proteoglycans [10]. Collagen is the most abundant protein in the body, and the predominant structural ECM component in most tissues [11]. There are 28 different types of collagens, which can either be homo- or heterotrimers of three collagen α chains assembled into a triple helix structure. Collagen proteins can be classified as fibril-forming collagens (Collagen types I, II, III, V and XI) and non-fibril forming collagens, which form networks (Collagen IV), short chains (Collagens VIII, X), beaded filaments (Collagen VI), or support the formation of more complex structures along collagen fibrils (fibril-associated collagens with interruptions in their triple helices, or FACITs) (Collagen types IX, XII, XIV, XIX, XXI) [12]. The ECM glycoprotein family includes ECM proteins that have covalently attached glycan (oligosaccharide) chains, added during post-translational modification (PTM). Glycan chains are short, branched, and comprise a relatively small percentage of the mass of the glycoprotein. The ECM glycoprotein group includes laminins, nidogens, fibronectin, elastin, and many other ECM proteins [10]. The laminin family consists of distinct protein chains, categorized as α (five types), β (four types), and γ (three types) that form heterotrimers (one α, one β, and one γ chain) named based on their composition ($\alpha 1$, $\beta 1$, $\gamma 1$ is called Laminin-111). Laminins form a structural mesh-like network with Collagen IV and nidogens to construct the basement membrane [13]. Some types of collagens are also glycoproteins. ECM Proteoglycans, on the other hand, are ECM proteins with long, linear amino sugar, or glycosaminoglycan (GAG) chains. GAGs are large negatively charged molecules that comprise a high percentage of the mass of the proteoglycan and contribute to the structure of the ECM. There are 20 ECM proteoglycans, including perlecan, biclycan, decorin, and osteoglycin, among others. Hyaluronic acid (HA) is a linear non-sulfated GAG polymer with no core protein, instead it is formed from disaccharide units at the plasma membrane and forms an essential structural component of the ECM [14]. Because it has no protein component, HA is not represented in proteomic studies of the ECM.

In many studies with human islets, ECM has been found to affect beta cell proliferation, differentiation, survival, and insulin secretion dynamics [6]. The culture of islets with individual purified matrix components has been shown to positively affect islet-specific gene expression, further underscoring their importance in islet health [15–17]. Following islet transplantation, host cells are recruited to the graft; endothelial cells, pericytes and Schwann cells establish neurovascular connections with the graft [18], and fibroblast-like stromal cells are recruited to restore the ECM of the transplanted islets [19]. Stem cell-derived islets (SC-islets) are transplanted in a similar manner as primary islets, and undergo maturation post-transplantation, following exposure to these same *in vivo* factors, including the tissue ECM environment [20–22]. Furthermore, integrins, a group of cell surface receptors that can be activated by specific ECM protein motifs, are signaled by ECM to polarize both primary and SC-islets to secrete insulin toward the vasculature and improve glucose-responsive function [23–25]. Because SC-islets express very little endogenous ECM [26], incorporating ECM into SC-islet culture or transplantation may improve function and differentiation outcomes.

2 Pancreas and Islet ECM

For decades, ECM has been shown to have an important role in the health and function of pancreatic islets. Aspects of the ECM composition of the pancreas and islets of many species had been characterized through the use of immunostaining and other targeted approaches (reviewed in [6–9]). However, in vitro studies of islets cultured with ECM have used purified ECM components that were only putatively comparable to the native islet ECM. As *in vitro* islet culture moves from 2-D adherent methods toward more complex customized 3-D systems that better resemble the native cell environment [27], a more complete understanding of the pancreas matrisome is necessary to appropriately engineer these models. Historical analyses from immunohistochemistry studies, using data originating from a variety of different species, have determined that islet ECM contains collagens I, III, IV, V and VI, as well as laminins and fibronectin [6, 28, 29]. Because most native islet ECM is destroyed during isolation, it has been difficult to interrogate isolated islets using proteomic methods to further characterize islet-specific ECM. To begin characterizing ECM relevant to human islet biology at a proteomic level, recent studies have directly, comprehensively, and quantitatively measured expression of ECM proteins in the human pancreas.

In 2021, two published studies quantified the composition of the human pancreas matrisome. Asthana et al. utilized a decellularization and ECM solubilization protocol alongside untreated human pancreas tissue to identify and quantify the proteins that make up the pancreas matrisome [30]. Combining mass spectrometry and multiplex ELISA-based assays, they identify a total of 314 matrisome proteins among all of their pancreas donors and samples. Most importantly, this study demonstrated that decellularization biases the ECM proteins extracted from tissues, generally towards having high collagen preservation and lower retention of other ECM components [30]. The protein composition is further affected by the digestion and solubilization of the ECM to create a gel. In their study, 67 ECM proteins were quantified in the native (non-decellularized) human pancreas, including 37 core ECM proteins (16 ECM glycoproteins, 14 collagen chains, and 7 proteoglycans) and 30 ECM-associated proteins (15 ECM regulators, 12 ECM-affiliated proteins and 3 secreted factors). Collagen I $\alpha 1$, collagen VI $\alpha 2$, collagen I $\alpha 2$ and collagen IV $\alpha 1/5$ were identified as the most abundant collagens; matrillin 1, vitronectin, fibrillin 1 and laminin $\gamma 1$ as the most abundant ECM glycoproteins; and osteoglycan, perlecan, lumican and prolargin as the most abundant ECM proteoglycans (Fig. 1, first column) [30].

In a parallel study, Li and Tremmel et al. utilized human pancreas tissues to explore the matrisome composition throughout human development [31]. This study identified 185 ECM proteins in the human pancreas throughout development and quantified 117 ECM proteins present at all developmental stages studied (fetal, juvenile, young adult, older adult). The 117 quantified ECM proteins includes 61 core ECM (35 ECM glycoproteins, 19 collagen chains, and 7 proteoglycans) and 56 ECM-associated proteins (28 ECM regulators, 20 ECM-affiliated proteins and 8

Human Pancreas
Asthana et al.

Rank	Protein
1	COL1A1
2	COL1A2
3	COL3A1
4	COL6A3
5	COL6A1
6	COL4A1
7	COL4A2
8	COL6A2
9	HSPG2
10	LUM
11	COL5A2
12	COL5A1
13	FBN1
14	DCN
15	LAMC1
16	OGN
17	NID1
18	LTBP2
19	COL18A1
20	FGB
21	LAMA5
22	LAMB1
23	PRELP
24	FGG
25	VTN
26	TINAGL1
27	LAMA2
28	FGA
29	DPT
30	BGN
31	NID2
32	COL16A1
33	COL14A1
34	ASPN
35	EMILIN1
36	AGRN
37	COL7A1

Human Pancreas
Li and Tremmel et al.

Rank	Protein	Rank	Protein
1	COL1A1	40	OGN
2	COL3A1	41	BGN
3	COL1A2	42	COL14A1
4	COL6A3	43	FRAS1
5	HSPG2	44	POSTN
6	LAMA2	45	THBS1
7	FBN1	46	VWA1
8	LAMA5	47	LAMA4
9	LAMC1	48	IGFBP2
10	COL6A1	49	COL16A1
11	COL6A2	50	COL2A1
12	LAMB1	51	FGL1
13	LAMB2	52	PCOLCE
14	FN1	53	IGFBP7
15	COL4A1	54	COL11A1
16	DCN	55	FBLN1
17	FGG	56	FBLN2
18	COL4A2	57	VWF
19	NID1	58	FBN2
20	FGB	59	MFGE8
21	NID2	60	COL4A3
22	EMILIN1	61	TNXB
23	COL5A2		
24	LUM		
25	FGA		
26	COL5A1		
27	COL18A1		
28	ASPN		
29	PRELP		
30	AGRN		
31	COL15A1		
32	LAMC3		
33	MFAP2		
34	COL12A1		
35	TGFBI		
36	TINAGL1		
37	COL5A3		
38	VTN		
39	CRELD2		

Porcine Pancreas
Gaetani et al.

Rank	Protein
1	COL1A1
2	COL1A2
3	COL1A1
4	COL6A3
5	NID1
6	COL6A1
7	LAMC1
8	COL6A2
9	HSPG2
10	DCN
11	LUM
12	COL4A1/5
13	COL14A1
14	NID1/2
15	FN1
16	COL4A1
17	FBLN5
18	COL5A1
19	OGN
20	COL4A2
21	PRELP
22	DPT
23	COL5A2
24	LAMB1
25	LAMA2
26	FBN2
27	BGN
28	LAMA4
29	ASPN
30	COL18A1
31	COL12A1
32	LAMB2
33	EMILIN1
34	POSTN
35	EFEMP1
36	FBN1
37	AGRN
38	EFEMP2
39	TNXB

Mouse Islets
Naba et al.

Rank	Protein	Rank	Protein	Rank	Protein
1	COL1A1	40	COL7A1	79	FBLN5
2	COL1A1	41	LAMA5	80	COL10A1
3	COL3A1	42	DPT	81	PXDN
4	COL1A2	43	POSTN	82	COL28A1
5	COL2A1	44	COL15A1	83	CRELD1
6	COL5A2	45	SLIT2	84	ECM1
7	COL5A1	46	PRELP	85	LTBP4
8	LAMA1	47	COL4A5	86	MATN2
9	FBN1	48	ASPN	87	THBS1
10	LAMC1	49	SSPO	88	COL24A1
11	LAMB1	50	HMCN1	89	SPOCK2
12	LAMC1	51	COL14A1	90	EFEMP1
13	FBN1	52	OGN	91	TNC
14	DMBT1	53	ELN	92	MMRN1
15	COL6A3	54	COL27A1	93	COL9A2
16	COL6A3	55	LUM	94	LTBP2
17	FGB	56	LAMA4	95	IGFBP7
18	COL11A1	57	VWF	96	ADIPOQ
19	FN1	58	COL16A1	97	AGRN
20	FGA	59	COL12A1	98	COL9A2
21	COL11A2	60	COL12A1	99	HMCN2
22	COL6A2	61	NID2	100	MFAP1A/ MFAP1B
23	COL6A1	62	COL18A1		
24	NID1	63	EMILIN1	101	LTBP1
25	FGG	64	CRELD2	102	FBLN2
26	COL5A3	65	TGFBI	103	TNXB
27	COL4A2	66	VTN		
28	COL4A1	67	PRG2		
29	COL6A6	68	COL4A3		
30	FBN2	69	MFAP2		
31	COL8A1	70	MFAP5		
32	DCN	71	COL4A6		
33	LAMB2	72	LAMA3		
34	COL4A4	73	TINAGL1		
35	VWA5A	74	COL22A1		
36	HSPG2	75	SPARC		
37	LAMA2	76	RELN		
38	BGN	77	TNC		
39	COL6A5	78	VWA1		

Key
Collagens
Proteoglycans
Glycoproteins

Fig. 1 Core matrisome data from studies of whole pancreas and isolated islets. (Data presented are derived from quantitative proteomics of native (non-decellularized) human pancreas in Asthana et al. and Li/Tremmel et al., native porcine pancreas in Gaetani et al., and isolated mouse islets in Naba et al. [30, 32, 36, 38])

secreted factors). In the adult human pancreas, collagen I $\alpha1$, collagen III $\alpha1$, collagen I $\alpha2$ and collagen VI $\alpha3$ were identified as the most abundant collagens; laminins $\alpha2$, $\alpha5$, $\beta1$ and $\gamma1$ and fibrillin 1 as the most abundant ECM glycoproteins; and perlecan, decorin, lumican and prolargin as the most abundant ECM proteoglycans (Fig. 1, second column) [31].

Together, Li and Tremmel et al. [31] and Asthana et al. [30] provide a framework for reconstructing the natural ECM composition of the human pancreas in future tissue engineering applications. Similar studies have been published quantifying the matrisome of adult mouse and porcine pancreata [32–35]. The 2018 publication by Gaetani et al. includes comparable quantitative proteomic data of the native adult porcine pancreas matrisome. Their study quantified 38 core ECM proteins (18 ECM glycoproteins, 13 collagen chains, and 7 proteoglycans). Collagen I $\alpha1$, collagen I $\alpha2$, collagen VI $\alpha3$ and collagen VI $\alpha1$ were identified as the most abundant collagens; laminins $\gamma1$, $\beta1$ $\alpha2$, and $\alpha4$ as the most abundant ECM glycoproteins; and perlecan, decorin, lumican and osteoglycin as the most abundant ECM proteoglycans (Fig. 1, third column). Together, it is clear that Collagens I, II, III, IV, V, and VI are identified in the pancreas across species, with Collagen I, III, and VI being the most abundant in the majority of studies. Laminin α chains 1, 2, 3, 4 and 5, β chains 1 and 2, and γ chains 1 and 3 isoforms have been identified in the pancreas across studies and across species. Other ECM glycoproteins such as fibrillin 1 and 2 and nidogen 1 and 2 have consistently been observed in the pancreas, with fibrillin 1 often being the most abundance ECM glycoprotein in the pancreas. Proteoglycans such as perlecan (also called heparan sulfate proteoglycan 2, HSPG2), decorin and biglycan have also been identified in the pancreas in several studies, with perlecan often being the most abundant [30, 33–37].

2.1 Comparing Whole Pancreas and Islet-Specific ECM

While the aforementioned studies used whole pancreas tissue lysates for proteomic analyses, it is important to understand not only the composition of the whole pancreas matrisome, but also the islet-specific ECM composition. The first proteomic study of islet-specific ECM was published by Naba et al. in 2017 [38]. This study utilized a mild collagenase digestion method to isolate mouse islets to preserve the integrity of the islet ECM. 120 ECM proteins were identified and quantified in normal adult mouse islets, including 71 core matrisome proteins (35 ECM glycoproteins, 28 collagen chains, and 8 proteoglycans) and 49 ECM-associated proteins (28 ECM regulators, 15 ECM-affiliated proteins and 6 secreted factors). Collagen I $\alpha 1$, collagen III $\alpha 1$, collagen I $\alpha 2$ and collagen 2 $\alpha 1$ as the most abundant collagens; laminins $\alpha 1$, $\beta 1$ and $\gamma 1$ and fibrillin 1 as the most abundant ECM glycoproteins; and decorin, perlecan, biglycan and prolargin as the most abundant ECM proteoglycans (Fig. 1, fourth column). The study also compared the matrisome of healthy adult islets to stages in the progression of insulinoma: islets collected at hyperplastic, angiogenic and insulinoma stages. They found 35 ECM proteins to change significantly throughout the progression of insulinoma formation, and consider these ECM proteins to be potential novel targets for anti-angiogenic therapies. Overall, this was the first study to quantitatively assess islet ECM composition at the proteomic level, and provides a database for which to better understand and mimic the islet matrisome. The islet isolation protocol used in this study enabled the ECM of mouse islets to stay intact throughout isolation, but it is unclear whether this method is translatable to the much larger human pancreas, or whether the protocol truly isolated pure mouse islet ECM without damage. The matrisome of human islets has not yet been quantitatively assessed in this way using mass spectrometry.

It is interesting that many of the ECM proteins identified in mouse islets by Naba et al. are also among the most abundant ECM proteins in the human pancreas. This indicates that although there are known differences between the ECM of the acinar and islet compartments of the pancreas, there may also be similarities in some highly abundant ECM components. Li and Tremmel et al. compared expression of selected ECM proteins among the islet and acinar compartments of the pancreas and found that many of the most abundant ECM proteins had relatively equal expression in both regions. In particular, the most abundant ECM components of the whole pancreas – collagen I $\alpha1$, collagen III $\alpha1$ – had almost equal distribution throughout the islet and acinar compartments, with an average adult islet:acinar abundance ratio of 1.2 and 1.1 respectively. However, some less abundant ECM components were found to significantly differ between compartments, such as OGN, FBN2, and EMILIN1 which are all enriched in islets compared to the acinar tissue, and FN1 which was absent in adult human islets, but present in the acinar and connective tissues of the pancreas [31]. Despite these salient findings, it is generally difficult to compare data across studies to consistently identify ECM components that are present and enriched in islets, because the current literature is composed of data derived from different species, using different methods of quantitative mass spectrometry and different equipment. To our knowledge, a comprehensive study quantifying the islet and acinar-specific ECM within a single species at the proteomic level has not yet been published.

Additionally, unlike pancreatic exocrine cells, islet endocrine cells do not themselves express ECM proteins [39]. Therefore, the majority of the ECM within islets is thought to be deposited by mesenchymal stromal cells, endothelial cells, and vascular mural cells that surround and penetrate the islets [39–42]. While endothelial cells are known to deposit basement membrane (mainly collagen IV and laminins $\alpha4$ and $\alpha5$) [39–41], islets also contain many other ECM components. Recently, mouse pancreatic pericytes have been shown to express high levels of basement membrane proteins such as collagen IV, laminins $\alpha2$ and $\alpha4$, but also interstitial ECM proteins found in islets, such as collagens I, III, V and VI and the proteoglycans decorin, biglycan, lumican, osteoglycin and perlecan [40].

For tissue engineering applications aimed at using decellularized pancreas to reconstruct the islet microenvironment, it is important for investigators to understand how the acinar and islet ECM may differ. Such strategies would utilize whole pancreas tissue rather than islets, and therefore the majority of the decellularized ECM would be derived from the exocrine fraction of the pancreas, which comprises more than 95% of the tissue. Therefore, abundant ECM components, such as collagens, that are evenly distributed throughout the pancreas would be present in relevant concentrations to mimic the islet environment. However, minor ECM components that may be present in islets, but not acinar tissue, may need to be added into the decellularized material to achieve islet-like concentrations. More detailed quantitative studies of the islet matrisome will help to guide tissue engineering efforts in the future. Decellularization strategies for endocrine pancreas tissue engineering will be discussed in detail in the next section of this chapter.

Matrisomic studies are useful resources for understanding the composition of healthy pancreas and islet tissues, and may guide future study of the role ECM plays in normal islet function. However, in order to translate these data into useful information for the generation of a scaffold to support islets in culture or transplantation, more work is necessary. First, more information is needed on the differences between the human exocrine pancreas and islet matrisome composition and that of other species. A closer look at how the human islet matrisome compares to that of the non-endocrine pancreas is also essential in order to create a proper environment to support human islets after isolation. It would be impractical to incorporate hundreds of ECM components into a mimetic ECM scaffold, therefore more in-depth work to define how individual ECM components contribute to islet health and function will help to tailor an ECM scaffold that is optimized for islet function without being overly complex.

2.2 ECM in Pancreas and Islet Development

In the interest of understanding the role that ECM has in islet health and function, it is important to acknowledge that ECM is not a static component of the tissue microenvironment. In fact, the matrisome dynamically changes with age and is altered in many disease states, affecting the fate of the resident cells [43, 44]. Many studies have examined aspects of pancreas and islet ECM composition in the context of different developmental and disease states, and it has been postulated that ECM may affect islet maturation and function [6, 39, 40, 45]. It is also well established that ECM is altered in pancreatic diseases, such as pancreatitis and pancreatic cancer [46, 47], and the islet matrisome also appears to change with T1D and T2D [48, 49]. In order to utilize ECM components as diagnostic markers for disease, or to incorporate ECM into scaffolds to support islets in regenerative medicine-based diabetes therapies, we must first understand how the matrisome of the pancreas and islets are altered by disease or change throughout normal development.

In this section we will focus on what is currently known about changes in islet or pancreas ECM with development, from fetal to late adult stages. Early studies to explore the dynamic changes in ECM that occur during development utilized immunofluorescent staining of specific ECM components in tissue samples across specified age ranges to understand how individual ECM components changed with age. Rodent pancreas morphogenesis studies identified that during early pancreas development, cell interactions with the surrounding basement membrane control epithelial branching [50, 51] which precedes the formation of exocrine and endocrine progenitor cells. Changes in collagen I, collagen III and fibronectin expression and organization accompany epithelial branching [52]. Few studies have compared the ECM of human fetal and adult islets, but have found that many islet ECM components significantly change with development [36, 53–55]. Focusing on laminin proteins and detecting expression through immunostaining, Otonkoski et al. found similar expression profiles of most laminin proteins between human fetal and adult islets, except for laminin $\alpha 1$, which was expressed in fetal islets but not in adult islets, and laminin $\beta 2$, which was expressed in adult but not in fetal islets [54]. Riopel et al. focused on differences in collagen gene expression and found that collagens I-VI, IX, XII, XV, XVI, XVIII and XXI had increased expression in fetal islets compared to adult [55]. The first proteomic studies of human fetal and adult pancreas ECM offer a more complete picture of compositional changes that occur from fetal to adulthood, but lack the depth of spatial mapping that other studies have explored. Li and Tremmel et al. explored how the human pancreas ECM changes across four distinct stages: fetal (16–18 gestational weeks), juvenile (age 5–16), younger (age 21–29) and older adult (age 50–61). The study found that 72% (84 proteins) of the 117 quantified ECM proteins had significant changes in protein abundance among these four developmental age groups. Of the core ECM proteins, this included 7 of 19 (37%) collagens, 26 of 35 (74%) of ECM glycoproteins, and 4 of 7 (57%) of proteoglycans. This means that a high fraction of structural ECM proteins change abundance throughout development. Furthermore, the localization of a subset of proteins was quantified in this study. Of 16 core ECM proteins analyzed through immunofluorescent staining, 14 of them (88%) were found to have significant changes in enrichment between the islet and acinar compartments of the pancreas among the four developmental time points [36]. These studies provide clear evidence that human islet ECM is highly dynamic from the developing fetus through adulthood, which suggests that ECM-based methods for supporting stem cell-derived islets through staged differentiation may benefit from employing strategies to mimic these developmental changes in ECM over time in culture. Studies supporting stem cell-derived islets with ECM scaffolds will be discussed later in this chapter.

Several studies have also focused on differences in islet ECM between younger and older adult pancreas donors, with particular interest in how ECM may affect the efficiency of islet isolation between these populations [4, 36, 37, 56, 57]. It is a well-known phenomenon that compared to older adult donors, it is much more difficult to isolate pure, high-quality islets from young adult donors. Two studies by Spiers et al. employing a variety of methodologies including differential enzymatic digestion, immunofluorescent staining, and Raman microspectroscopy (RMS) have established that differences in the ECM environment between donors of younger age (<30 years) and older age (>55 years) significantly affects the digestion efficiency of pancreas and islet tissue with the standard cocktail of collagenases and proteases [56, 57]. Although RMS can identify differences in ECM signal, this methodology is unable to identify specific ECM proteins, and therefore the exact ECM molecules which contribute to this functional property of the tissue could not be specifically defined. Meier et al. went on to show that adult human islet preparations with ≥60% collagen digestion resulted in a higher chance of successful isolation, with a higher number of isolated islets and a much higher chance of being utilized for clinical transplant. In their study, they found that there was not a significant difference in total organ collagen content by donor age, but there was a significant correlation of increased collagen VI abundance with older donor age [4].

Interestingly, Li and Tremmel et al. also found that at the whole-proteome level, there were no differences in the relative abundance of any ECM proteins between younger and older adults [36]. However, by immunfluorescence staining, certain ECM proteins were found to accumulate in and around the islets in older adulthood compared to younger adults. ECM proteins that were found to be enriched in/around older adult islets compared to younger adults include collagen I, collagen V, collagen XII and both laminin $\alpha4$ and $\alpha5$. Collagen VI was found to be enriched (compared to the acinar regions) around both young and older adult human islets [36]. Intriguingly, another study of human pancreas ECM by Berman et al. found that in higher BMI donors, the relative ratio of collagens I and IV were higher, while collagen VI was lower compared to normal BMI donors [37]. This is interesting because donors with higher BMI are also preferred for human islet isolation. Together, these studies suggest that a change in the composition of collagens specifically around the islets impacts islet isolation efficiency. The collagenase blends typically used in human islet isolation are less effective at digesting collagen VI compared to collagens I-V [28, 58], suggesting a mechanism through which islet isolation could be affected by local islet collagen content. It is possible that younger adult islets are surrounded by an ECM membrane that is more highly enriched in collagen VI compared to older adults, which would correlate with lower digestion efficiency. Future studies that quantify the human islet-specific matrisome more thoroughly may help to conclusively identify how these ECM protein ratios may change with aging, and potentially other donor variables, and whether or not these changes coincide with differences in islet health, function or isolation efficiency.

What is currently known about ECM proteins that are present in and around pancreatic islets, how the ECM composition changes throughout development, and which cell types contribute to the deposition of ECM is summarized in Fig. 2.

2.3 Differences in Islet ECM Across Species

Available data about pancreas and islet ECM are derived from many different species, with an abundant amount of data coming from model organisms such as mouse, rat, pig or dog. While some studies have aimed to assess differences in ECM expression across species, little has been published to definitively show whether there are relevant differences in ECM composition between species [6, 59, 60]. However, it is known that physically there are differences in the basement membrane structures of mouse islets compared to human islets. Specifically, rodent islets have a single-layered basement membrane between the endocrine cells and the vasculature, while human islets have a double walled basement membrane, with one membrane facing the endocrine cells and another facing the vasculature [54, 61]. In humans, these two basement membranes have been shown to express different ECM profiles by immunohistochemistry, but it hasn't been well established how the two layers of membrane form, and which cells secrete ECM to construct the inner membrane in the absence of ECM secretion by the endocrine cells themselves. Given that there is a structural difference in the ECM of islets of different species, it could be assumed that ECM composition and signaling may differ as well. Further research is required to establish how these differences manifest in the identity and function of islets of different species. For the purposes of engineering stem cell-derived islets for developmental modeling and for clinical application, it would be ideal and potentially essential to create a human-like ECM environment for the cells. Combining SC-islets with vascular cells may contribute to recreating the structural and molecular composition that matches the native human islet [62].

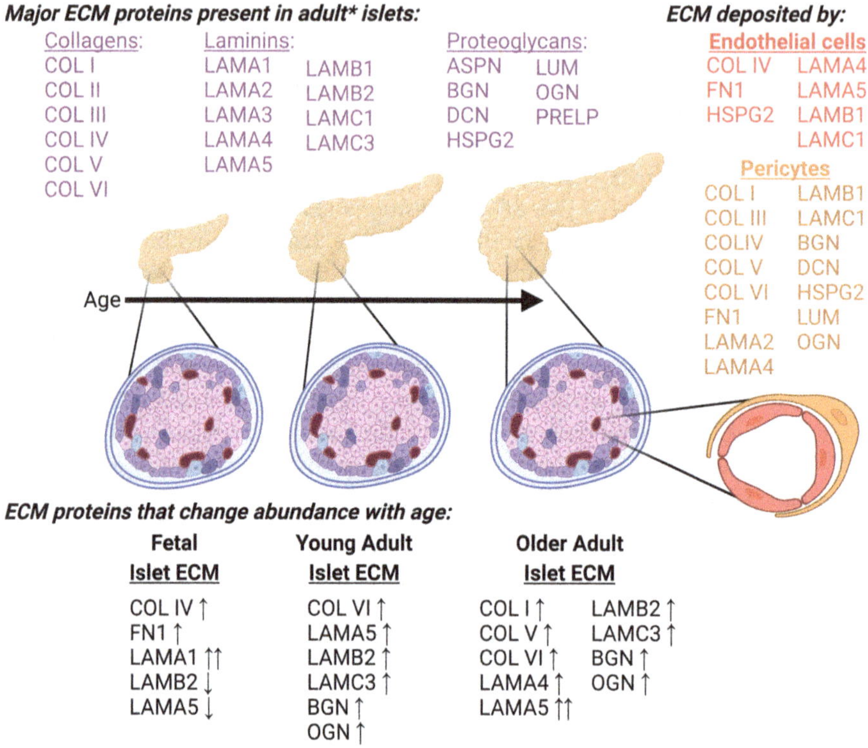

Fig. 2 Islet ECM in development and aging
ECM proteins present in pancreatic islets [4, 30, 32, 36, 38, 54]. Several ECM proteins have been found to be differentially expressed between fetal and adult islets, and additionally between younger and older adult donors. Arrows indicate the relative direction in which ECM proteins change in one age group compared to others, with multiple arrows indicating a more significant change. The cells known to make ECM in and around islets are pericytes and endothelial cells, which secrete different ECM proteins [40].
*Data has been derived from different species, including mouse, pig and human. Data is interpreted from immunofluorescent staining of islet ECM as well as from proteomic profiling of whole pancreas and isolated islets. Created with BioRender.com

2.4 *Roles of Specific ECM Proteins in Islet Health and Function*

In contrast to studies identifying the ECM proteins present in the islet microenvironment, there have also been studies that have examined islet-ECM interactions through the role of individual ECM proteins in islet culture (reviewed in Stendahl et al. 2009 and Sakata et al. 2020) [6, 63]. Studies have compared selected ECM proteins in beta cell and islet co-culture and concluded that either collagen I [64–68], collagen III [69], collagen IV [66, 67, 70–76], collagen V [77], or collagen VI [9, 78] significantly improved beta cell survival and function. However, other studies found that collagen I [15, 79, 80] or collagen IV [15, 17, 80] had negative effects on beta cell survival and insulin production/secretion. Still other studies focused on the roles laminins [71, 81, 82], fibronectin [71, 83], and decorin [84] have on islet survival and function, and found these to be important components of the islet ECM environment.

Some studies have altered the ratio of ECM components in culture, and found that different ratios of ECM proteins had beneficial effects on beta cell survival or function. Weber et al. found that several ECM proteins (collagens I and IV, laminin, fibrinogen, fibronectin, and vitronectin) individually improved survival of the immortalized beta cell line MIN6 and reduced apoptosis compared to 3-D culture without ECM, however a ratio of 25:75 (laminin:collagen IV) was found to stimulate the highest insulin secretion profiles [85]. Another approach by Santini-Gonzalez et al. demonstrated the utility of mixed ECM components through a strategy of thin-coating the outside surface of primary and SC-islets with basement membrane composed of laminin, collagen IV and nidogen and found an improvement in glucose-responsive function over 7 days of culture [26]. Narayanan et al. studied the ECM secreted by a beta cell line, RIN5F, and determined that the ratio of collagen IV:laminin:fibronectin in the secreted ECM was approximately 1:3:3. The authors went on to test various combinations of the three ECM components and found that indeed, SC-islets cultured on the different substrates had the highest beta cell gene expression with a 1:3:3 ratio, mimicking the natural ECM ratio of the RIN5F cells in their study [86].

In addition to studies that have altered the islet ECM environment in culture after islet isolation, other studies have manipulated ECM *in vivo* to study the effect on islet development and function. In one study, global knockout of the α3 chain of collagen V in mice resulted in reduced islet mass and blunted glucose-stimulated insulin secreting function [87]. Another study found that global osteoglycin knockout in mice resulted in impaired glucose tolerance and reduced beta cell mass [88]. Both of these studies were interested in assessing the global effects of loss of these ECM components, and by chance observed changes in islet function. Global knockout of many ECM proteins causes embryonic lethality or serious developmental complications. Because beta cells do not produce the ECM around them, conditional beta cell-specific knockouts of ECM proteins are not effective methods of modifying the islet microenvironment. These two factors have thus far prevented the use of genetic knockout methods for broad scale screening of the effects that ECM proteins have in islet function *in vivo*.

Taken together, one can conclude that ECM interactions in general are important and beneficial to islet survival and function, but it is yet unclear how individual ECM components contribute to these advantages in culture. These studies have generated data about islet-ECM interactions utilizing incomplete ECM scaffolds to support islets. While the data derived therein may be meaningful and informative to understand islet biology, the combination of these approaches with recent pancreas matrisome proteomic data will elevate these studies to better model the islet ECM environment. Together, scaffold design can be based on the ratios of major ECM components in the native islet environment and alterations to these ratios will help to dissect the role that individual components may have in islet health and function.

3 Islet ECM Biomaterials

3.1 Decellularization

ECM can be isolated from tissues through a process called decellularization (decell), in which freeze/thaw cycles, detergents, hyper-/hypotonic solutions, or enzymatic digestions aid in cell lysis and subsequent washing removes cellular material, yielding native ECM. In this way, ECM has been extracted from animal and human organs, and has been proposed for use in regenerative medicine for a variety of applications [89–91]. One decell approach, commonly termed perfusion decellularization, is to deliver detergents through the native vasculature or ductal system to decellularize the tissue with the goal of preserving the macro- and microstructures of an organ [89]. However, seeding cells or islets into whole perfusion-decelled organs has proved challenging [92–97]. Another decell strategy is to solubilize ECM via an acidic pepsin digestion. The digested ECM can be neutralized and warmed to 37 °C to form a hydrogel, or lyophilized to form a sponge [98]. These scaffolds require digestion and reformation of collagen fibrils [99], reorganizing the microstructure and potentially disrupting ECM protein functional domains. On the other hand, a hydrogel can be customized and more easily incorporated into cell culture systems and transplantation strategies. The collagenous nature of native ECM hydrogel is advantageous for *in vivo* applications, where the neutralized digest can be injected as a liquid and forms a gel at physiologic temperatures. Due to the potential challenges of seeding intact 3-D scaffolds, a hydrogel's ease of use has practical advantages in tissue culture and transplantation platforms. General decellularization strategies for extracting pancreatic ECM for use in islet culture are outlined in Fig. 3.

Several studies have been published detailing methods to decell the pancreas of many species including mouse, rat, goat, pig and human, generating pancreatic ECM (P-ECM) [31–35, 37, 92–94, 96, 100–109]. These studies have provided clear evidence that pancreatic decell is achievable, resulting in matrix scaffolds that are free of donor DNA and cellular debris, are not cytotoxic to cultured cells, and are hypoimmunogenic when transplanted. Many initial pancreas decell methods utilized a perfusion approach, via vasculature or the pancreatic duct. Napierala et al. compared perfusion decell of the porcine pancreas through the portal vein, aorta, or pancreatic duct, and found no significant difference in decell efficiency among the three routes [94]. More recent protocols have focused on using pieces of pancreas tissue or homogenized pancreas for decell, followed by the formation of a hydrogel from the resulting ECM. Detergents used have included Triton X-100, sodium dodecyl sulfate (SDS), and sodium deoxycholate, with several recent studies using detergent-free methods. Some protocols include pre-treatment steps to prepare the tissue for decell, including freeze/thaw cycles [31, 34, 101–103, 108, 109], hypo-/hypertonic solution changes [32, 34, 103, 104], or Trypsin treatment [93, 104, 109]. Other protocols have also included protease inhibitors throughout the decell process to better preserve the protein content, or have incorporated nuclease treatments to

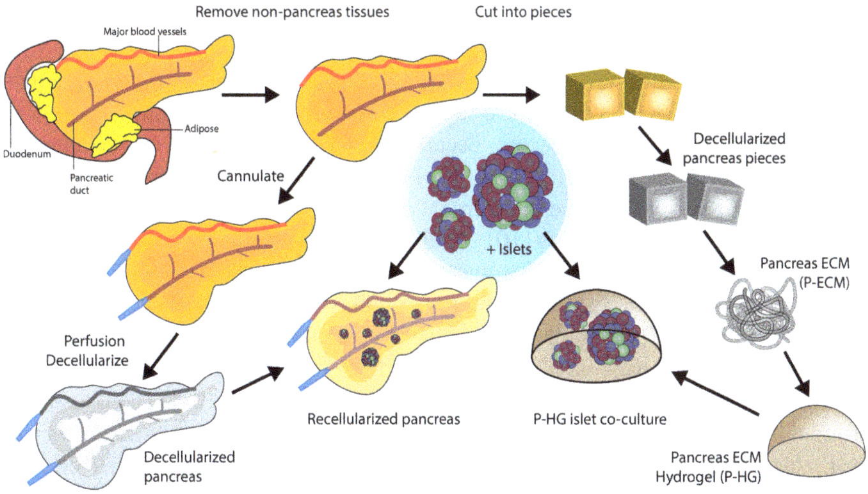

Fig. 3 Schematic representation of pancreas decellularization methods. The pancreas can be decellularized by perfusing detergents through the vasculature or pancreatic duct, keeping the tissue ultrastructure intact (left side). An alternative approach is tissue disruption by cutting or homogenizing, followed by cell lysis, extraction of the pancreas ECM (P-ECM) and reconstitution of the ECM as a hydrogel (P-HG). Either of these scaffolds could be combined with isolated islets or other cell types

more completely remove the native cellular DNA [32]. Gaetani et al. compared the efficacy and impact of different detergent decell protocols on porcine pancreas and assessed the ability of the ECM to form a hydrogel and support cell viability [32]. This study found that Triton-X treatment was a superior method to SDS treatment, as SDS had detrimental effects on gelation of the resulting ECM and SDS-treated scaffolds had a higher toxicity to cells cultured within them. The study went on to confirm that the addition of protease-inhibitors to all stages of the decell protocol improved the retention of ECM proteoglycans and glycoproteins compared to decell performed without protease inhibitors [32]. Overall, the length of detergent treatment appears to be dependent on species and organ size, with smaller organs (mouse, rat) requiring between 3–5 h in detergent, and larger organs (pig, human) needing between 8 h to 3 days [33, 92–94, 96, 102, 104–107, 110]. The detergent used and time exposed to detergent may also affect the microstructure and function of retained matrix molecules [90].

To avoid the potential detrimental effects that exposure to detergents may have on ECM proteins, two studies have developed detergent-free decell protocols for the pancreas [34, 103]. Both studies utilize freezing and thawing the tissue, followed by alternating treatment with hypo- and hypertonic solutions to lyse cells and then wash away the cellular debris. Through proteomic studies of the decellularized pancreas, Hu et al. demonstrate that their detergent-free method enabled identification of a higher number of ECM proteins compared to a detergent-based protocol [34]. On the other hand, Tamburrini et al. demonstrate the ability of detergent-free decellularization to generate a biologically relevant material capable of supporting islet culture [103].

While some studies have aimed to keep the ECM ultrastructure of the pancreas intact throughout decell, others have focused on solubilizing the resulting ECM to create a hydrogel. In the latter case, decell protocols have focused on breaking up the tissue prior to decell [31, 32, 34, 35, 101–103, 108]. Smaller pieces of tissue require shorter periods of time in detergent, and tissue dissociation or homogenization also helps to remove lipids [101, 102]. Additional lipid removal steps, such as treatment with lipases or organic solvents, have been shown to be helpful or even necessary to ensure successful gelation of decellularized pancreas ECM hydrogel [31, 102, 108]. Only four studies have achieved spontaneous formation of hydrogel from decellularized pancreas ECM [31, 32, 102, 108], while others have mixed pancreas ECM with other materials to facilitate gelation [103, 104]. The benefits of using a hydrogel material compared to intact decellularized tissues is that gels can be molded more easily around the cells for culture, and can also be customized with the addition of other soluble or insoluble factors to support cells. Jiang et al. derived hydrogel from porcine lung, bladder and pancreas, and showed that each gel had different mechanical and viscoelastic properties, as well as different ECM composition [108]. Therefore, hydrogel made from a specific organ of interest can provide both the physiological ECM cues and mechanical properties to benefit the matched cell type.

Because there are so many published methods for decellularizing the pancreas, it is important to consider how different methods may differentially alter the retention of ECM components. Protein composition has been assessed through a variety of methods including immunofluorescent staining and microarrays, but to quantitatively assess protein profiles following decell treatment, several groups have used in-depth proteomic analyses. The first study to utilize mass spectrometry-based proteomics on decellularized mouse pancreas, Goh et al., identified 30 ECM-associated proteins in the P-ECM including 17 collagens and 7 laminin proteins [33]. Since then, several studies have published limited proteomic data on decellularized pancreas, generally indicating that P-ECM is abundant in collagens I, III and VI [34, 35, 108]. With the improvement of ECM mass spectrometry technology, three subsequent studies have quantified a larger number of ECM proteins and compared the ECM protein abundance before and after decell [30, 32, 102]. Despite using different decell methods, all studies found that following decellularization the relative abundance of collagens as a percentage of the total ECM increased, while the amount of ECM glycoproteins and proteoglycans is relatively diminished. Gaetani et al. studied how different treatments used for decell affect the protein composition and found that the addition of protease inhibitors and the use of Triton-X (compared with SDS) increased the retention of these less abundant matrisome components, but still resulted in a general enrichment of collagens. Further, Asthana et al. studied how ECM components are differently retained through the solubilization process to make a functional material for cell culture. Their study also showed that collagens were enriched while glycoproteins and proteoglycans were reduced with decell, and this trend was further exaggerated through the solubilization process to make gel

[30]. This is not surprising, as many decell studies have reported similar collagen enrichment in isolated ECM, which may be due to the more insoluble nature of collagens in general as well as the fact that collagen content is the driving component of these materials to form hydrogels [111].

The effect that P-ECM co-culture has on cells and islets has been broadly investigated using a variety of diverse approaches and generating varying results. Cytocompatibility studies testing toxicity of the P-ECM materials on cell survival have indicated that the decelled matrix is not toxic to cells [32, 33, 92, 93, 101, 102, 107]. Initial studies utilized immortalized cells, like the MIN6 and INS-1 beta cell lines to demonstrate cell survival and maintenance of insulin expression and secretion after co-culture with P-ECM [33, 109]. Primary islets are a better model of how these cells respond to environmental cues, and therefore several groups have utilized rat, pig and human islets to assess how P-ECM affects cultured islets. Two studies have found that islets cultured on or within P-ECM survived and functioned equally as well as islets in the absence of matrix over a 4–5 day period [93, 107]. Another group showed that islets within the 3-D scaffolds could function, however this was without a comparison to control islets [94]. DeCarlo et al. performed a longer duration study in which rat islets cultured with rat P-ECM maintained insulin secretion for up to 6 weeks, while naked islets began declining after 2 weeks [105].

Three recent studies have cultured human islets with P-ECM and all determined that P-ECM co-culture improved islet survival and function compared to islets cultured in suspension [31, 103, 108]. Tremmel et al. and Jiang et al. studied the effect of pancreas-specific ECM compared to other ECM gels (kidney-derived, bladder-derived, purified collagen I, alginate) on human islet culture [31, 108]. Both studies found that the pancreas-derived hydrogel had improved islet survival and function. Jiang et al. found that pancreas, lung and bladder-derived hydrogels had significantly different mechanical properties and the ECM protein composition of the three scaffolds were quite different. Nevertheless, in culture with human and rat islets, the bladder and pancreas-derived gels both improved islet function compared to islets cultured in suspension, with no significant differences in performance between the two different gel sources [108]. However, Tremmel et al. found that islet co-culture with various different hydrogels had different effects on islet glucose-stimulated insulin secretion, with the pancreas-derived hydrogel performing the best [31]. This study suggests that there is a beneficial effect of using tissue-matched ECM in islet culture, but further work is required to substantiate these findings. Interestingly, both studies also found that hydrogel co-culture had a beneficial effect on resident islet endothelial cells [31, 108]. Tremmel et al. further found that the arrangement of human islet endocrine cells, the islet architecture, was significantly altered after islet isolation, but was better preserved when the islets were cultured in the P-ECM hydrogel [31]. It was also shown that islet mitochondrial function was improved in P-ECM hydrogel co-culture compared to islets in suspension [31]. These studies indicate that P-ECM improves islet survival and function following isolation, and there may be several biological pathways and cell interactions underlying the improved islet performance.

Because of the promise that tissue-specific ECM may have to help direct the differentiation of stem cells toward a terminal cell fate [112], some studies have attempted to characterize the effect that P-ECM has on hESC differentiation toward an islet-like fate. Narayanan et al. published interesting results using decelled RIN5F (rat insulinoma) beta cell cultures. hESCs plated onto acellular RIN5F ECM and treated with RIN5F conditioned medium over 14 days were shown to differentiate into glucose-responsive insulin producing SC-islets, remarkably without requiring a staged differentiation protocol or through added growth factors [86]. This study highlights the potential P-ECM may have on islets and SC-islets. Similarly, Chaimov et al. found that hepatocytes that were transdifferentiated to beta cells on P-ECM had more than a 4-fold increase in insulin secretion [104]. Sackett et al. showed that hESC-derived pancreas progenitor (PP) cells cultured in P-ECM hydrogel maintained their fate and signature gene expression, but did not assess how P-ECM affected these cells in longer-term culture [102]. Using mouse SC-derived islets, Wan et al. found that SC-islets cultured in 3-D decellularized rat P-ECM had a 2-fold increase in insulin gene expression compared to those in standard 2-D culture, but also did not assess the effect the scaffold had on SC-islet function [96]. Together, these studies suggest that P-ECM may hold potential to support SC-islet differentiation and culture. Yet, there is still too little evidence to indicate whether P-ECM could affect SC-islets in a way that would help overcome deficiencies currently facing the field, such as poor expression of beta cell maturation markers, low levels of total insulin content and deficiencies in insulin secretion compared to primary islets [113].

Very few studies have explored the effects that P-ECM may have on transplanted islets. De Carlo et al. found that rat islets and rat P-ECM transplanted subcutaneously (SQ) within a polyethylene glycol (PEG) hydrogel reversed hyperglycemia for up to 41 days in diabetic rats, while islets with P-ECM but no PEG only reversed hyperglycemia for a few days before returning to a diabetic state [105]. Chaimov et al. also transplanted islets and P-ECM hydrogel SQ within alginate capsules into diabetic mice. They observed a temporary drop in blood glucose, but not a long-term reversal of diabetes [104]. Two studies have transplanted human P-ECM hydrogel into mice with humanized immune systems and found the material to be hypoimmunogenic, however the effect of the matrix on transplanted cells was not evaluated [31, 102]. In a more striking study that has demonstrated the value of co-transplantation with ECM, Yu et al. utilized an “islet viability matrix” (IVM) composed mainly of purified collagen 1 to support SQ islet engraftment [114]. The SQ space is a potential transplant site for islets because: i) the SQ site can be easily accessed during transplantation, ii) islets implanted subcutaneously can be monitored in a non-invasive way, and iii) the site can accommodate a large volume of cells. The SQ space has also been a difficult site to transplant cells within due to low levels of vascularization and high levels of cell death after transplantation [115], requiring higher numbers of islets to reverse diabetes than is required at other well-vascularized sites such as the kidney capsule. Yu et al. transplanted as few as 200 human islets along with the IVM material, successfully lowering the blood glucose of a diabetic mouse, while the same number of islets were ineffective at reversing diabetes when transplanted alone, without IVM [114]. These are promising steps toward creating a strategy for SQ islet transplantation, and P-ECM, or a synthetic composite biomimetic ECM, could play a role in enhancing the microenvironment of this space. More studies are required to confirm the potential benefits of P-ECM on transplanted islets or SC-islets, and to better understand the underlying mechanisms.

3.2 Non-Pancreas-Derived ECM-Based Scaffolds

Prior to the development of decellularization methods and knowledge of the complexity of the ECM, investigators sought to evaluate the impact of individual or cocktails of ECM molecules on islet function. Many studies exploring islet-ECM interactions were previously summarized [6, 116]. For example, Nagata and colleagues observed that collagen I gel or a mixture of collagens I and IV improved islet survival and insulin secretion of isolated rat islets maintained in culture for up to 11 days [67, 117]. As another example, Lucas-Clerc et al. found that isolated human islets embedded in collagen matrix retained their spherical structure and high secretory capacity which was preserved for up to 8 weeks. Collagen-embedded human islets performed significantly better than unembedded islets in culture whose insulin secretion rapidly declined over 2 weeks [118]. Many other studies demonstrate potential benefits of specific individual ECM components on a variety of islets from different species and purified beta cells touting improvements in cell survival, insulin secretion, proliferation, structure and transplantation success [6, 114]. This literature is largely empiric however, only describing phenomena but without dissecting mechanisms. Furthermore, in many cases the matrix components used were poorly defined and/or impure. In other studies, negative effects of matrix components were observed [17].

More recently, building on increasing knowledge of the islet matrisome, investigators are testing composite ECM component-based scaffolds for islet coating and embedding. For example, Santini-Gonzalez and colleagues coated isolated rat and human islets with a solution of laminin, nidogen and collagen IV [26]. In addition to showing the coating was dependent on the presence of Ca^{+2}/Mg^{+2}, they showed modest improvements of human islets dynamic insulin secretory function after 7 days in culture. They also showed successful coating of SC-islets, but studies evaluating possible functional or phenotypic changes were not presented. Additional studies are clearly needed in the context of studying the effects of various combinations of matrix components; but, such studies are hampered by the costs and complexity of empirically testing an array of different ECM components at different concentrations.

ECM component-based scaffolds may be applied to SC-islets and progenitor cells as these components have been described as key determinants of islet differentiation based on defined effects on human fetal pancreas tissue [80, 119–121], as well as of post-transplant engraftment based on the mechanism of preventing apoptosis [114]. Strikingly, very little progress has been reported to date in this context [122]. Applying individual matrix components to 2-D SC-islet cultures, Singh et al. studied the role of basement membrane (BM) proteins on SC-islets structure and function and showed that differentiated human stem cell-derived islet spheroids express BM proteins in a non-polarized web-like structure. However, culturing differentiated SC-islets onto BM proteins laminin 511, collagen IV, or fibronection individually results in polarization of key proteins such as Liprin $\alpha 1$, E-cadherin and ZO1 to the vascular BM, which correlated with improvements in static insulin secretion, as measured by stimulation index, which was shown to be primarily due to reduced basal insulin secretion [25]. Augsornworawat et al. showed that non-growth factor reduced Matrigel aided the formation of islet organoids from hESC-derived beta cells and HUVECs [123].

Other approaches that have been used to support islets include scaffolds derived from tissues other than the pancreas. Due to the extensive vascular network of the lungs, perfusion decellularized rat lung scaffolds were used by Citro et al. and repopulated with endothelial cells to form a vascular network in the decelled organ prior to seeding with islets. After seeding islets into the vascularized endocrine organ, the system was perfusable both *in vitro* and *in vivo* [124]. This is a promising step toward creating a transplantable endocrine organ. Similarly, several studies have used decellularized liver scaffolds to support islets and stem cell-derived islets in culture, although no studies have conclusively demonstrated an advantage of using a liver-derived scaffold over a pancreas-derived scaffold, nor have any compared the composition of the two materials side by side [125–127].

3.3 Non-ECM-Based Polymer Scaffolds

Many natural and synthetic polymer scaffolds for islet transplantation have been proposed and recently reviewed [116, 128, 129]. Suitable synthetic polymers include polydimethylsiloxane (PDMS), polytetrafluoroethylene (PTFE), and poly (D, L-lactide-co-e-caprolactone (PDLLCL), polyvinyl alcohol (PVA), poly(lactide-co-glycolide) (PLG), and polycaprolactone (PCL) whereas poly(ethylene oxide terephthalate)/polybutylene terephthalate (PEOT/PBT) and polysulfone may be trigger danger signals. Natural polymers include fibrin and silk. These materials can serve as a biodegradable matrix and are often fabricated into discrete shapes that facilitate handling and implantation of the islet graft during the transplant procedure. By immobilizing and localizing the graft it may facilitate subsequent removal if needed. Polymer matrices may also stimulate angiogenesis and therefore support engraftment after transplantation, thereby leading to survival of a greater proportion of implanted islets. It may also serve as a platform for sequestering supportive cells (e.g. endothelial cells, mesenchymal stem cells, etc.), molecules (e.g. hepatocyte growth factor or angiogenic factors, etc.) or for local delivery of immunosuppressive medications (i.e. prednisone, Cyclosporine A). Most research in this area has focused on rodent and human islets, but more recently porcine islets, as xenografts, or SC-islets, have been studied in combination with these biodegradable polymers [130–133].

4 Applications of ECM in Regenerative Medicine Therapies for Diabetes

The tissue engineering field as it applies to islets for treating diabetes envisions a combination of cells and materials to overcome current limitations to islet function, survival and ultimately transplantation success of various sources of beta cells. Components of an ideal regenerative medicine-based therapy for diabetes should include: i) a renewable/unlimited cell source, such as porcine islets or differentiated human pluripotent stem cells, ii) potential for adding supportive cells to enhance the survival and functionality of insulin-producing cells, iii) scaffolds, matrices, or hydrogels to support cell health and identity, and iv) a mechanism of overcoming challenges such as hypoxia and immune infiltration at the site of implantation. Just as individual ECM components and non-ECM polymers may be employed for fabricating scaffolds for this purpose, fully intact ECM achieved through decellularization or ever more sophisticated combinations of ECM components have the potential to provide adjunctive and advantageous properties to tissue engineered constructs. As we learn more about the tissue-specific properties of pancreatic islet ECM, and which components are essential for optimized constructs, combinatorial biomimetic hydrogel may be produced to serve as scaffolds for a variety of applications (Fig. 4). Natural ECM or biomimetic ECM hydrogels could be lyophilized and fabricated into scaffolds to support islets in culture or after transplantation. ECM could be used to coat or encapsulate individual islets, or could be used to embed multiple islets in bulk. Additionally, a variety of cell types such as endothelial cells, mesenchymal cells, etc. could be incorporated into multicellular endocrine constructs to form 3-D biomimetic models of the islet environment. ECM hydrogels may provide a niche microenvironment to enhance vascularization after transplantation and prevent cell death through anoikis. The scaffolds could also be modified to provide further protection, such as being loaded with oxygen-generating materials to prevent hypoxia in the initial days following transplantation. ECM hydrogels may also provide a favorable environment for islets in engineered surroundings that isolate them from their normal microenvironment, such as islets encapsulated in immunoprotective polymers, housed within immunoprotective devices, or residing within microfluidic devices for the purpose of *in vitro* functional studies. Though knowledge is just emerging about the many favorable properties of ECM hydrogels, it is clear that given the many potential applications in islet biology and regenerative medicine therapies for diabetes, this field of research will remain robust.

In combination with primary islets or stem cell-derived islets, scaffolds that mimic the natural islet ECM may be useful to support cells both *in vitro* and *in vivo*. Such scaffolds could coat or encapsulate individual islets or multiple islets can be embedded in bulk to support the cells in culture or following transplantation. These scaffolds may be useful substrates to which elements may be added to promote vascularization or reduce hypoxia of a transplanted graft. ECM scaffolds also enable 3-D culture of islets or SC-islets and present the opportunity to model the natural islet microenvironment including ECM signaling as well as various native tissue

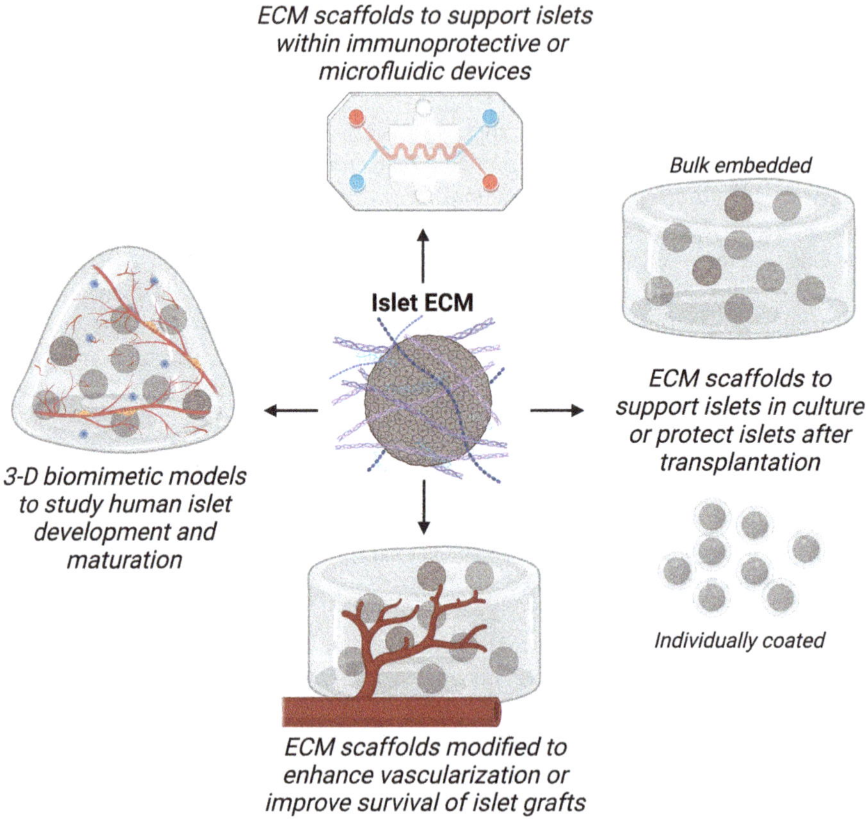

Fig. 4 Applications of ECM scaffolds in islet engineering

components and cell types; these 3-D biomimetic systems could be used to model human development and better understand islet function, maturation, and cell-cell interactions. Similarly, ECM can be incorporated into the design of immunoprotective or microfluidic devices to better support islet health and function within these systems. Created with BioRender.com

References

1. Frantz, C., K.M. Stewart, and V.M. Weaver, *The extracellular matrix at a glance.* J Cell Sci, 2010. **123**(Pt 24): p. 4195–200.
2. Wang, R.N. and L. Rosenberg, *Maintenance of beta-cell function and survival following islet isolation requires re-establishment of the islet-matrix relationship.* J Endocrinol, 1999. **163**(2): p. 181–90.
3. Cross, S.E., et al., *Key Matrix Proteins Within the Pancreatic Islet Basement Membrane Are Differentially Digested During Human Islet Isolation.* Am J Transplant, 2017. **17**(2): p. 451–461.

4. Meier, R.P.H., et al., *Pancreas collagen digestion during islet of Langerhans isolation-a prospective study.* Transpl Int, 2020. **33**(11): p. 1516–1528.
5. Prince, E. and E. Kumacheva, *Design and applications of man-made biomimetic fibrillar hydrogels.* Nature Reviews Materials, 2019. **4**(2): p. 99–115.
6. Stendahl, J.C., D.B. Kaufman, and S.I. Stupp, *Extracellular matrix in pancreatic islets: relevance to scaffold design and transplantation.* Cell Transplant, 2009. **18**(1): p. 1–12.
7. Townsend, S.E. and M. Gannon, *Extracellular Matrix-Associated Factors Play Critical Roles in Regulating Pancreatic beta-Cell Proliferation and Survival.* Endocrinology, 2019. **160**(8): p. 1885–1894.
8. Aamodt, K.I. and A.C. Powers, *Signals in the pancreatic islet microenvironment influence beta-cell proliferation.* Diabetes Obes Metab, 2017. **19 Suppl 1**: p. 124–136.
9. Llacua, L.A., M.M. Faas, and P. de Vos, *Extracellular matrix molecules and their potential contribution to the function of transplanted pancreatic islets.* Diabetologia, 2018. **61**(6): p. 1261–1272.
10. Naba, A., et al., *The extracellular matrix: Tools and insights for the "omics" era.* Matrix Biol, 2016. **49**: p. 10–24.
11. Kular, J.K., S. Basu, and R.I. Sharma, *The extracellular matrix: Structure, composition, age-related differences, tools for analysis and applications for tissue engineering.* J Tissue Eng, 2014. **5**: p. 2041731414557112.
12. Ricard-Blum, S., *The collagen family.* Cold Spring Harb Perspect Biol, 2011. **3**(1): p. a004978.
13. Aumailley, M., *The laminin family.* Cell Adh Migr, 2013. **7**(1): p. 48–55.
14. Fraser, J.R., T.C. Laurent, and U.B. Laurent, *Hyaluronan: its nature, distribution, functions and turnover.* J Intern Med, 1997. **242**(1): p. 27–33.
15. Daoud, J., et al., *The effect of extracellular matrix components on the preservation of human islet function in vitro.* Biomaterials, 2010. **31**(7): p. 1676–82.
16. Daoud, J.T., et al., *Long-term in vitro human pancreatic islet culture using three-dimensional microfabricated scaffolds.* Biomaterials, 2011. **32**(6): p. 1536–42.
17. Kaido, T., et al., *Impact of defined matrix interactions on insulin production by cultured human beta-cells: effect on insulin content, secretion, and gene transcription.* Diabetes, 2006. **55**(10): p. 2723–9.
18. Juang, J.H., et al., *3-D Imaging Reveals Participation of Donor Islet Schwann Cells and Pericytes in Islet Transplantation and Graft Neurovascular Regeneration.* EBioMedicine, 2015. **2**(2): p. 109–19.
19. Nilsson, J., et al., *Recruited fibroblasts reconstitute the peri-islet membrane: a longitudinal imaging study of human islet grafting and revascularisation.* Diabetologia, 2020. **63**(1): p. 137–148.
20. Augsornworawat, P., et al., *Single-Cell Transcriptome Profiling Reveals beta Cell Maturation in Stem Cell-Derived Islets after Transplantation.* Cell Rep, 2020. **32**(8): p. 108067.
21. Maxwell, K.G., et al., *Differential Function and Maturation of Human Stem Cell-Derived Islets After Transplantation.* Stem Cells Transl Med, 2022. **11**(3): p. 322–331.
22. Balboa, D., et al., *Functional, metabolic and transcriptional maturation of human pancreatic islets derived from stem cells.* Nat Biotechnol, 2022.
23. Cottle, L., et al., *Structural and functional polarisation of human pancreatic beta cells in islets from organ donors with and without type 2 diabetes.* Diabetologia, 2021. **64**(3): p. 618–629.
24. Gan, W.J., et al., *Cell polarity defines three distinct domains in pancreatic beta-cells.* J Cell Sci, 2017. **130**(1): p. 143–151.
25. Singh, R., et al., *Enhanced structure and function of human pluripotent stem cell-derived beta-cells cultured on extracellular matrix.* Stem Cells Transl Med, 2020.
26. Santini-Gonzalez, J., et al., *In vitro generation of peri-islet basement membrane-like structures.* Biomaterials, 2021. **273**: p. 120808.
27. Yamada, K.M. and K. Clark, *Cell biology: survival in three dimensions.* Nature, 2002. **419**(6909): p. 790–1.

28. Hughes, S.J., et al., *Characterisation of collagen VI within the islet-exocrine interface of the human pancreas: implications for clinical islet isolation?* Transplantation, 2006. **81**(3): p. 423–6.
29. Wood, G.C., *The formation of fibrils from collagen solutions. 2. A mechanism of collagen-fibril formation.* Biochem J, 1960. **75**: p. 598–605.
30. Asthana, A., et al., *Comprehensive characterization of the human pancreatic proteome for bioengineering applications.* Biomaterials, 2021. **270**: p. 120613.
31. Tremmel, D.M., et al., *A human pancreatic ECM hydrogel optimized for 3-D modeling of the islet microenvironment.* Scientific Reports, 2022. **12**(1).
32. Gaetani, R., et al., *Evaluation of Different Decellularization Protocols on the Generation of Pancreas-Derived Hydrogels.* Tissue Eng Part C Methods, 2018. **24**(12): p. 697-708.
33. Goh, S.K., et al., *Perfusion-decellularized pancreas as a natural 3D scaffold for pancreatic tissue and whole organ engineering.* Biomaterials, 2013. **34**(28): p. 6760–72.
34. Hu, M., et al., *Proteomic and Bioinformatic Analysis of Decellularized Pancreatic Extracellular Matrices.* Molecules, 2021. **26**(21).
35. Vigier, S., et al., *Composition and organization of the pancreatic extracellular matrix by combined methods of immunohistochemistry, proteomics and scanning electron microscopy.* Curr Res Transl Med, 2017. **65**(1): p. 31–39.
36. Li, Z., et al., *Proteome-wide and matrisome-specific alterations during human pancreas development and maturation.* Nat Commun, 2021. **12**(1): p. 1020.
37. Berman, A., et al., *The Influence of the Flow of Detergent and Donor Characteristics on the Extracellular Matrix Composition After Human Pancreas Decellularization.* Transplant Proc, 2020. **52**(7): p. 2043–2049.
38. Naba, A., et al., *Quantitative proteomic profiling of the extracellular matrix of pancreatic islets during the angiogenic switch and insulinoma progression.* Sci Rep, 2017. **7**: p. 40495.
39. Nikolova, G., et al., *The vascular basement membrane: a niche for insulin gene expression and Beta cell proliferation.* Dev Cell, 2006. **10**(3): p. 397–405.
40. Sakhneny, L., A. Epshtein, and L. Landsman, *Pericytes contribute to the islet basement membranes to promote beta-cell gene expression.* Sci Rep, 2021. **11**(1): p. 2378.
41. Hogan, M.F. and R.L. Hull, *The islet endothelial cell: a novel contributor to beta cell secretory dysfunction in diabetes.* Diabetologia, 2017. **60**(6): p. 952–959.
42. Arzouni, A.A., et al., *Mesenchymal stromal cells improve human islet function through released products and extracellular matrix.* Clin Sci (Lond), 2017. **131**(23): p. 2835–2845.
43. Rozario, T. and D.W. DeSimone, *The extracellular matrix in development and morphogenesis: a dynamic view.* Dev Biol, 2010. **341**(1): p. 126–40.
44. Bonnans, C., J. Chou, and Z. Werb, *Remodelling the extracellular matrix in development and disease.* Nat Rev Mol Cell Biol, 2014. **15**(12): p. 786–801.
45. Takahashi, I., et al., *Important role of heparan sulfate in postnatal islet growth and insulin secretion.* Biochem Biophys Res Commun, 2009. **383**(1): p. 113–8.
46. Ferrara, B., et al., *The Extracellular Matrix in Pancreatic Cancer: Description of a Complex Network and Promising Therapeutic Options.* Cancers (Basel), 2021. **13**(17).
47. Gress, T.M., et al., *Role of extracellular matrix in pancreatic diseases.* Digestion, 1998. **59**(6): p. 625–37.
48. Korpos, E., et al., *The peri-islet basement membrane, a barrier to infiltrating leukocytes in type 1 diabetes in mouse and human.* Diabetes, 2013. **62**(2): p. 531–42.
49. Wigger, L., et al., *Multi-omics profiling of living human pancreatic islet donors reveals heterogeneous beta cell trajectories towards type 2 diabetes.* Nat Metab, 2021. **3**(7): p. 1017–1031.
50. Cozzitorto, C. and F.M. Spagnoli, *Pancreas organogenesis: The interplay between surrounding microenvironment(s) and epithelium-intrinsic factors.* Curr Top Dev Biol, 2019. **132**: p. 221–256.
51. Shih, H.P., et al., *ECM Signaling Regulates Collective Cellular Dynamics to Control Pancreas Branching Morphogenesis.* Cell Rep, 2016. **14**(2): p. 169–79.
52. Hisaoka, M., J. Haratake, and H. Hashimoto, *Pancreatic morphogenesis and extracellular matrix organization during rat development.* Differentiation, 1993. **53**(3): p. 163–72.

53. Ma, F., et al., *In Depth Quantification of Extracellular Matrix Proteins from Human Pancreas.* J Proteome Res, 2019. **18**(8): p. 3156–3165.
54. Otonkoski, T., et al., *Unique basement membrane structure of human pancreatic islets: implications for beta-cell growth and differentiation.* Diabetes Obes Metab, 2008. **10 Suppl 4**: p. 119–27.
55. Riopel, M., et al., *Ultrastructural and immunohistochemical analysis of the 8–20 week human fetal pancreas.* Islets, 2014. **6**(4): p. e982949.
56. Spiers, R.M., et al., *Development of a Simple In Vitro Assay to Assess Digestion of the Extracellular Matrix of the Human Pancreas by Collagenase Enzyme Blends.* Cell Transplant, 2018. **27**(7): p. 1039–1046.
57. Spiers, R.M., et al., *Donor age significantly influences the Raman spectroscopic biomolecular fingerprint of human pancreatic extracellular matrix proteins following collagenase-based digestion.* Acta Biomater, 2019. **99**: p. 269–283.
58. Heller-Harrison, R.A. and W.G. Carter, *Pepsin-generated type VI collagen is a degradation product of GP140.* J Biol Chem, 1984. **259**(11): p. 6858–64.
59. van Deijnen, J.H., et al., *Significance of the peri-insular extracellular matrix for islet isolation from the pancreas of rat, dog, pig, and man.* Cell Tissue Res, 1992. **267**(1): p. 139–46.
60. Van Deijnen, J.H., et al., *Distribution of collagens type I, type III and type V in the pancreas of rat, dog, pig and man.* Cell Tissue Res, 1994. **277**(1): p. 115–21.
61. Virtanen, I., et al., *Blood vessels of human islets of Langerhans are surrounded by a double basement membrane.* Diabetologia, 2008. **51**(7): p. 1181–91.
62. Pignatelli, C., et al., *Bioengineering the Vascularized Endocrine Pancreas: A Fine-Tuned Interplay Between Vascularization, Extracellular-Matrix-Based Scaffold Architecture, and Insulin-Producing Cells.* Transpl Int, 2022. **35**: p. 10555.
63. Sakata, N., G. Yoshimatsu, and S. Koadama, *The Roles of Collagen in Islet Transplantation.* OBM Transplantation, 2020. **4**(4).
64. Forget, A., et al., *IGF-2 coated porous collagen microwells for the culture of pancreatic islets.* J Mater Chem B, 2017. **5**(2): p. 220–225.
65. Jalili, R.B., et al., *Fibroblast populated collagen matrix promotes islet survival and reduces the number of islets required for diabetes reversal.* J Cell Physiol, 2011. **226**(7): p. 1813–9.
66. Krishnamurthy, M., et al., *Expression and function of alphabeta1 integrins in pancretic beta (INS-1) cells.* J Cell Commun Signal, 2008. **2**(3-4): p. 67–79.
67. Nagata, N., et al., *Evaluation of insulin secretion of isolated rat islets cultured in extracellular matrix.* Cell Transplant, 2001. **10**(4-5): p. 447-51.
68. Stephens, C.H., et al., *In situ type I oligomeric collagen macroencapsulation promotes islet longevity and function in vitro and in vivo.* Am J Physiol Endocrinol Metab, 2018. **315**(4): p. E650–E661.
69. Olaniru, O.E., et al., *The adhesion receptor GPR56 is activated by extracellular matrix collagen III to improve beta-cell function.* Cell Mol Life Sci, 2018. **75**(21): p. 4007–4019.
70. Davis, N.E., et al., *Enhanced function of pancreatic islets co-encapsulated with ECM proteins and mesenchymal stromal cells in a silk hydrogel.* Biomaterials, 2012. **33**(28): p. 6691–7.
71. Hadavi, E., et al., *Microwell Scaffolds Using Collagen-IV and Laminin-111 Lead to Improved Insulin Secretion of Human Islets.* Tissue Eng Part C Methods, 2019. **25**(2): p. 71–81.
72. Liu, J., et al., *Functionalized self-assembling peptide improves INS-1 beta-cell function and proliferation via the integrin/FAK/ERK/cyclin pathway.* Int J Nanomedicine, 2015. **10**: p. 3519–31.
73. Llacua, A., et al., *Extracellular matrix components supporting human islet function in alginate-based immunoprotective microcapsules for treatment of diabetes.* J Biomed Mater Res A, 2016. **104**(7): p. 1788–96.
74. Pinkse, G.G., et al., *Integrin signaling via RGD peptides and anti-beta1 antibodies confers resistance to apoptosis in islets of Langerhans.* Diabetes, 2006. **55**(2): p. 312–7.
75. Salvay, D.M., et al., *Extracellular matrix protein-coated scaffolds promote the reversal of diabetes after extrahepatic islet transplantation.* Transplantation, 2008. **85**(10): p. 1456–64.

76. Yap, W.T., et al., *Collagen IV-modified scaffolds improve islet survival and function and reduce time to euglycemia.* Tissue Eng Part A, 2013. **19**(21-22): p. 2361–72.
77. Bi, H., K. Ye, and S. Jin, *Proteomic analysis of decellularized pancreatic matrix identifies collagen V as a critical regulator for islet organogenesis from human pluripotent stem cells.* Biomaterials, 2020. **233**: p. 119673.
78. Wang, H., et al., *Culture on a native bone marrow-derived extracellular matrix restores the pancreatic islet basement membrane, preserves islet function, and attenuates islet immunogenicity.* FASEB J, 2020. **34**(6): p. 8044–8056.
79. Edamura, K., et al., *Effect of adhesion or collagen molecules on cell attachment, insulin secretion, and glucose responsiveness in the cultured adult porcine endocrine pancreas: a preliminary study.* Cell Transplant, 2003. **12**(4): p. 439–46.
80. Jiang, F.X., et al., *Laminin-1 promotes differentiation of fetal mouse pancreatic beta-cells.* Diabetes, 1999. **48**(4): p. 722–30.
81. Sigmundsson, K., et al., *Culturing functional pancreatic islets on alpha5-laminins and curative transplantation to diabetic mice.* Matrix Biol, 2018. **70**: p. 5–19.
82. Sojoodi, M., et al., *Enhanced maintenance of rat islets of Langerhans on laminin-coated electrospun nanofibrillar matrix in vitro.* Cell Biol Int, 2013. **37**(4): p. 370–9.
83. Nakashima, Y., et al., *Adhesion characteristics of porcine pancreatic islets and exocrine tissue to coating materials.* Islets, 2018. **10**(3): p. e1460294.
84. Urbanczyk, M., et al., *Decorin improves human pancreatic beta-cell function and regulates ECM expression in vitro.* Matrix Biol, 2022.
85. Weber, L.M., K.N. Hayda, and K.S. Anseth, *Cell-matrix interactions improve beta-cell survival and insulin secretion in three-dimensional culture.* Tissue Eng Part A, 2008. **14**(12): p. 1959–68.
86. Narayanan, K., et al., *Extracellular matrix-mediated differentiation of human embryonic stem cells: differentiation to insulin-secreting beta cells.* Tissue Eng Part A, 2014. **20**(1-2): p. 424–33.
87. Huang, G., et al., *alpha3(V) collagen is critical for glucose homeostasis in mice due to effects in pancreatic islets and peripheral tissues.* J Clin Invest, 2011. **121**(2): p. 769–83.
88. Lee, N.J., et al., *Osteoglycin, a novel coordinator of bone and glucose homeostasis.* Mol Metab, 2018. **13**: p. 30–44.
89. Crapo, P.M., T.W. Gilbert, and S.F. Badylak, *An overview of tissue and whole organ decellularization processes.* Biomaterials, 2011. **32**(12): p. 3233–43.
90. Gilbert, T.W., T.L. Sellaro, and S.F. Badylak, *Decellularization of tissues and organs.* Biomaterials, 2006. **27**(19): p. 3675–83.
91. Tapias, L.F. and H.C. Ott, *Decellularized scaffolds as a platform for bioengineered organs.* Curr Opin Organ Transplant, 2014. **19**(2): p. 145–52.
92. Elebring, E., et al., *Cold-perfusion decellularization of whole-organ porcine pancreas supports human fetal pancreatic cell attachment and expression of endocrine and exocrine markers.* J Tissue Eng, 2017. **8**: p. 2041731417738145.
93. Katsuki, Y., et al., *Endocrine pancreas engineered using porcine islets and partial pancreatic scaffolds.* Pancreatology, 2016. **16**(5): p. 922–30.
94. Napierala, H., et al., *Engineering an endocrine Neo-Pancreas by repopulation of a decellularized rat pancreas with islets of Langerhans.* Sci Rep, 2017. **7**: p. 41777.
95. Uygun, B.E., et al., *Organ reengineering through development of a transplantable recellularized liver graft using decellularized liver matrix.* Nat Med, 2010. **16**(7): p. 814–20.
96. Wan, J., et al., *Culture of iPSCs Derived Pancreatic beta-Like Cells In Vitro Using Decellularized Pancreatic Scaffolds: A Preliminary Trial.* Biomed Res Int, 2017. **2017**: p. 4276928.
97. Batchelder, C.A., M.L. Martinez, and A.F. Tarantal, *Natural Scaffolds for Renal Differentiation of Human Embryonic Stem Cells for Kidney Tissue Engineering.* PLoS One, 2015. **10**(12): p. e0143849.
98. Freytes, D.O., et al., *Preparation and rheological characterization of a gel form of the porcine urinary bladder matrix.* Biomaterials, 2008. **29**(11): p. 1630–7.

99. Soffer-Tsur, N., D. Peer, and T. Dvir, *ECM-based macroporous sponges release essential factors to support the growth of hematopoietic cells.* J Control Release, 2017. **257**: p. 84–90.
100. Singh, G., et al., *Establishment of decellularized extracellular matrix scaffold derived from caprine pancreas as a novel alternative template over porcine pancreatic scaffold for prospective biomedical application.* FASEB J, 2022. **36**(10): p. e22574.
101. Klak, M., et al., *Impact of Porcine Pancreas Decellularization Conditions on the Quality of Obtained dECM.* Int J Mol Sci, 2021. **22**(13).
102. Sackett, S.D., et al., *Extracellular matrix scaffold and hydrogel derived from decellularized and delipidized human pancreas.* Sci Rep, 2018. **8**(1): p. 10452.
103. Tamburrini, R., et al., *Detergent-Free Decellularization of the Human Pancreas for Soluble Extracellular Matrix (ECM) Production.* J Vis Exp, 2020(163).
104. Chaimov, D., et al., *Innovative encapsulation platform based on pancreatic extracellular matrix achieve substantial insulin delivery.* J Control Release, 2017. **257**: p. 91–101.
105. De Carlo, E., et al., *Pancreatic acellular matrix supports islet survival and function in a synthetic tubular device: in vitro and in vivo studies.* Int J Mol Med, 2010. **25**(2): p. 195–202.
106. Mirmalek-Sani, S.H., et al., *Porcine pancreas extracellular matrix as a platform for endocrine pancreas bioengineering.* Biomaterials, 2013. **34**(22): p. 5488–95.
107. Peloso, A., et al., *The Human Pancreas as a Source of Protolerogenic Extracellular Matrix Scaffold for a New-generation Bioartificial Endocrine Pancreas.* Ann Surg, 2016. **264**(1): p. 169–79.
108. Jiang, K., et al., *3-D physiomimetic extracellular matrix hydrogels provide a supportive microenvironment for rodent and human islet culture.* Biomaterials, 2019. **198**: p. 37–48.
109. Yu, H., et al., *The rat pancreatic body tail as a source of a novel extracellular matrix scaffold for endocrine pancreas bioengineering.* J Biol Eng, 2018. **12**: p. 6.
110. Wu, D., et al., *3D Culture of MIN-6 Cells on Decellularized Pancreatic Scaffold: In Vitro and In Vivo Study.* Biomed Res Int, 2015. **2015**: p. 432645.
111. Parenteau-Bareil, R., R. Gauvin, and F. Berthod, *Collagen-Based Biomaterials for Tissue Engineering Applications.* Materials, 2010. **3**(3): p. 1863–1887.
112. Hoshiba, T., et al., *Decellularized Extracellular Matrix as an In Vitro Model to Study the Comprehensive Roles of the ECM in Stem Cell Differentiation.* Stem Cells Int, 2016. **2016**: p. 6397820.
113. Tremmel, D.M., et al., *Mimicking nature-made beta cells: recent advances towards stem cell-derived islets.* Curr Opin Organ Transplant, 2019. **24**(5): p. 574–581.
114. Yu, M., et al., *Islet transplantation in the subcutaneous space achieves long-term euglycaemia in preclinical models of type 1 diabetes.* Nat Metab, 2020. **2**(10): p. 1013–1020.
115. Pepper, A.R., et al., *Revascularization of transplanted pancreatic islets and role of the transplantation site.* Clin Dev Immunol, 2013. **2013**: p. 352315.
116. Tran, R., C. Moraes, and C.A. Hoesli, *Developmentally-Inspired Biomimetic Culture Models to Produce Functional Islet-Like Cells From Pluripotent Precursors.* Front Bioeng Biotechnol, 2020. **8**: p. 583970.
117. Nagata, N.A., K. Inoue, and Y. Tabata, *Co-culture of extracellular matrix suppresses the cell death of rat pancreatic islets.* J Biomater Sci Polym Ed, 2002. **13**(5): p. 579–90.
118. Lucas-Clerc, C., et al., *Long-term culture of human pancreatic islets in an extracellular matrix: morphological and metabolic effects.* Mol Cell Endocrinol, 1993. **94**(1): p. 9–20.
119. Cirulli, V., et al., *Expression and function of alpha(v)beta(3) and alpha(v)beta(5) integrins in the developing pancreas: roles in the adhesion and migration of putative endocrine progenitor cells.* J Cell Biol, 2000. **150**(6): p. 1445–60.
120. Crisera, C.A., et al., *Expression and role of laminin-1 in mouse pancreatic organogenesis.* Diabetes, 2000. **49**(6): p. 936–44.
121. Oberg-Welsh, C., *Long-term culture in matrigel enhances the insulin secretion of fetal porcine islet-like cell clusters in vitro.* Pancreas, 2001. **22**(2): p. 157–63.
122. Tuch, B.E., S.Y. Gao, and J.G. Lees, *Scaffolds for islets and stem cells differentiated into insulin-secreting cells.* Front Biosci (Landmark Ed), 2014. **19**(1): p. 126–38.

123. Augsornworawat, P., et al., *A hydrogel platform for in vitro three dimensional assembly of human stem cell-derived islet cells and endothelial cells.* Acta Biomater, 2019. **97**: p. 272–280.
124. Citro, A., et al., *Biofabrication of a vascularized islet organ for type 1 diabetes.* Biomaterials, 2019. **199**: p. 40–51.
125. Demko, P., et al., *Perfusion-Based Recellularization of Rat Livers with Islets of Langerhans.* Journal of Medical and Biological Engineering, 2022. **42**(2): p. 271–280.
126. Nakayama-Iwatsuki, K., et al., *Acellular matrix derived from rat liver improves the functionality of rat pancreatic islets before or after vitrification.* Cryobiology, 2021. **100**: p. 90–95.
127. Zhou, P., et al., *The dynamic three-dimensional culture of islet-like clusters in decellularized liver scaffolds.* Cell Tissue Res, 2016. **365**(1): p. 157–71.
128. Abadpour, S., et al., *Tissue Engineering Strategies for Improving Beta Cell Transplantation Outcome.* Current Transplantation Reports, 2021. **8**(3): p. 205–219.
129. Smink, A.M., et al., *Polymer scaffolds for pancreatic islet transplantation - Progress and challenges.* Am J Transplant, 2018. **18**(9): p. 2113–2119.
130. Clough, D.W., et al., *Integration of Islet/Beta-Cell Transplants with Host Tissue Using Biomaterial Platforms.* Endocrinology, 2020. **161**(11).
131. Gibly, R.F., et al., *Extrahepatic islet transplantation with microporous polymer scaffolds in syngeneic mouse and allogeneic porcine models.* Biomaterials, 2011. **32**(36): p. 9677–84.
132. Kuppan, P., et al., *Bioabsorption of Subcutaneous Nanofibrous Scaffolds Influences the Engraftment and Function of Neonatal Porcine Islets.* Polymers (Basel), 2022. **14**(6).
133. Youngblood, R.L., et al., *Microporous scaffolds support assembly and differentiation of pancreatic progenitors into beta-cell clusters.* Acta Biomater, 2019. **96**: p. 111–122.

Bioactive Materials for Use in Stem Cell Therapies for the Treatment of Type 1 Diabetes

Jonathan Hinchliffe and Ipsita Roy

1 Immunoisolative Biomaterial Technologies for Type 1 Diabetes

1.1 Classification of Immunoisolating Bioartificial Pancreas

The concept of immunoisolation attempts to address both the immune and morbidity-associated limitations of transplantation, as well as patient compliance and insulin dependence issues inherent in exogenous insulin therapies. Zielinski and Lysaght found that immunoisolation is based upon the idea that if cells are fully enclosed within a selectively permeable, non-immunogenic membrane they are effectively "invisible" to the local immune system [4]. Oxygen and metabolic waste can freely pass between the enclosed cells through the membrane that acts as a physical barrier to immunoglobulin and cell-mediated detection and destruction, whilst allowing for the release of potentially beneficial cell secretions such as testosterone and cortisol. A bioartificial pancreas (also referred to as a semi-artificial pancreas) refines this concept by using a synthetic or bio-based membrane (the "artificial" component") that immunoisolates islets of langerhans for the purpose replicating the endocrine function of the pancreas [17].

Orive et al. found that 17 different cell lineages had been immunoisolated using polymer membranes, citing applications from hormone replacement therapy to research into rare genetic conditions [5]. The main advantage of immunoisolation is the ability to use genotypically distinct cells in the body without fear of an immune response. Not only does this allow the clinician access to previously incompatible tissue for use in the patient but also opens the avenue of xenotransplantation without

J. Hinchliffe · I. Roy (✉)
Department of Materials Science and Engineering, University of Sheffield, Sheffield, UK
e-mail: i.roy@sheffield.ac.uk

L. Piemonti et al. (eds.), *Pluripotent Stem Cell Therapy for Diabetes*,
https://doi.org/10.1007/978-3-031-41943-0_11

genetic modification to display human antigens, making this kind of therapy more financially accessible to health services [6]. Indeed, Sun et al. describe immunoisolation and xenotransplantation using porcine islets in mice as far back as 1993 [7]. Moreover, given proper growth conditions, islets have been shown to be durable *in vivo*. De Vos et al. demonstrated both capsule integrity and islet survival up to 2 years and 200 days, after transplantation, respectively, reducing the need for continual doses of insulin in contemporary diabetes sufferers [8]. Moreover, the relative immune-system inertness of microencapsulated cells reduces the need for immune-suppressant drugs, reducing the risk of infections and potentially cancers [9, 10]

1.2 Immunoisolation Characterization

Given the potential of IRT using immunoisolative devices, several groups have, over the years, attempted to immunoisolate cells contained within the islets of Langerhans. As a result, IRT devices vary in both materials choice and manufacturing options, as well as mechanical, chemical and biological properties, leading to multiple classifications. Rainbow and Lysaght characterized immunoisolatory devices according to the matrix in which the cells are kept, whereas Nafea et al. divided immunoisolative devices according to placement within the human body [11, 12]. Early immunoisolation devices used intravascular devices (IVD) to deliver cell secretions (Fig. 1a). These devices consisted of a hollow tube directing blood flow past a graft containing injected cells that were continually washed in buffer to avoid adhesion.

The cells are then immunoisolated by a semi-permeable membrane. Blood flow past the membrane provides the cells with nutrients and metabolic waste removal whilst the cells are able to secrete their contents directly into the bloodstream proportional to blood glucose concentration, in a process known as glucose-stimulated insulin secretion (GSIS) [13, 14]. Whilst the stability of the capsule and effective diffusion are good advantages for these devices, great care must be taken to minimize the haemostatic properties of these devices. Indeed, early devices were found to have caused both coagulation and haemorrhage *in vivo* [14, 15]. As a result, researchers began exploration of extravascular devices, which instead relied on oxygen and nutrient transport from the surrounding vascularized tissues following implantation, often in the peritoneal space [16]. These extravascular devices include large populations of cells (usually more than 1 million Islet equivalent), fully encapsulated in a bio-inert polymer such as alginate for immune protection (Fig. 1b) [17, 18].

Whilst extravascular devices do not share the thrombogenic properties of IVD's, the reduced proximity to blood vessels requires the *de novo* vascularization of the construct. Until that time encapsulated cells may have to exist in a hypoxic environment, reducing cell viability, survival and secretory function. Given diffusion has a limited effective range of about 150 μm, the large size of macro-encapsulated constructs exacerbates this issue [19, 20], meaning that areas of larger constructs may

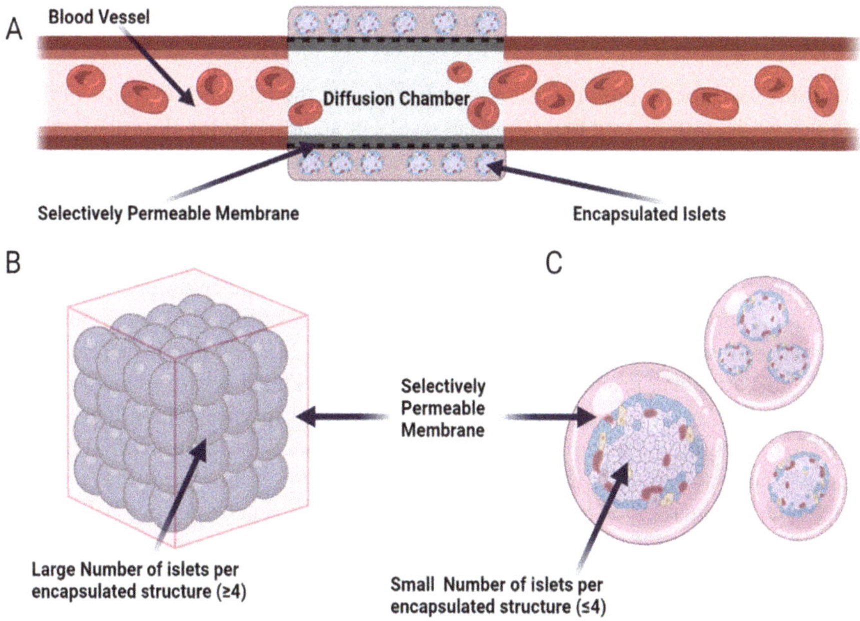

Fig. 1 The three main types of currently used encapsulation devices. (**a**) Intravascular devices operated by connecting a hollow tube to existing blood supply. Oxygen, nutrients, insulin, and metabolic waste products are exchanged between encapsulated islets via a semi-permeable membrane. (**b**) Extravascular devices that do not require a blood supply and operate via diffusion. Traditional extravascular devices use a single construct encapsulating large numbers of islets (**c**) Microencapsulation that produces microcapsules containing relatively small numbers of islets

become "dead zones", where metabolic waste-build up and an anoxic environment threatens the viability of the whole construct [20, 21]. Finally, it has been shown that high concentrations of insulin impede the insulin-secreting capacity of beta cells. This sets a limit on the insulin release capacity of individual constructs and limits their effectiveness in maintaining normoglycaemia *in vivo* [22]. To address some of the limitations of intravascular and extravascular macro-encapsulated devices, recent developments have centred on the use of microencapsulated structures (Fig. 1c). With structures less than 1 mm in diameter, microencapsulated cells display several advantages over macro-encapsulations. Nafea et al. explain that reduced encapsulation size increases the surface area to volume ratio, allowing for greater diffusion [12]. Furthermore, the lower cell numbers and greater diffusion capacity reduces hypoxia in the construct, increasing per-cell insulin production and increasing cell viability. However, microencapsulated structures are not perfect; their small size makes retrieval of faulty devices much harder than macro-encapsulation devices [21].

1.3 Materials Used for Immunoisolation

Materials used for immunoisolation need to enhance the inherent strengths of these devices whilst minimizing their weaknesses. Li et al. describe the desirable properties of successful bioartificial organs, including biological properties like non-thrombogenicity and complement activation resistance, mechanical properties such as matching the tissue strength, crush resistance, ease of manufacture and hydrophilicity [23]. A material common to most classes of immunoisolation devices have been the hydrogels. Hoffman et al. describe hydrogels as a hydrophilic polymer that, upon inundation with water becomes a swollen cross-linked network polymer, able to hold multiple times their dry weight in water [24]. Hydrogels display some of the most attractive properties for immunoisolation. The ability to use a multitude of polymers in hydrogels allows for a wide range of properties, which may be tailored to the required application. Aziz et al. demonstrated the antimicrobial properties of chitosan-based hydrogels, whilst Henderson et al. used the inherent mechanical properties of ionic cross-linked polymers to generate extremely high strength hydrogels [25, 26]. More promising is the ability to immobilize enzymes such as catalase inside the polymer [27], presenting pathways for modulating cell behaviour. Hydrogels are currently used in both macro- and microencapsulation, with Harrington restoring normoglycaemia in Non-Obese Diabetic (NOD) mice using Hyaluronic Acid/Polyethylene Glycol hydrogels and Qi displaying the use of a thin hydrogel sheet for the macroencapsulation of islets [28, 29].

Immunoisolation is not limited to relatively weak, water-swollen polymers. Evidence suggests that strong electrospun polymer fibres constructed from natural polymers such as silk fibroin, polysulfones and polytetraflouroethylene have also been used for immunoisolation [30, 31]. Electrospinning is a methodology, whereupon a polymer solution is charged and injected through a needle, to produce incredibly thin fibres of polymer [32]. Krishnan et al. showed that the generation of dual-hollow electrospun polymer membranes enabled oxygen and nutrient transfer whilst limiting access to the Major Histocompatibility Complex (MHC) and other immune factors [33], regarding the inner fibres as purely “islet hosting” components whereas the outer layer was envisaged for future, vasculature promoting peptide functionalization, increasing long-term oxygen diffusion into the membrane. Finally, metals have been trialled for use in immunoisolation. Zhan et al. sintered titanium to produce encapsulating structures with sintered micropores. The team found that oxygen and small molecules could pass through this membrane whilst selectively isolating Natural Killer (NK) cells. Titanium was used for dual micro-macro encapsulations of cells along with polymer membranes with greater oxygen permittivity like chitosan, protected by the relatively strong titanium stress shield [34].

2 The Need for Bioactivity in Immuno-isolative Insulin Therapies

2.1 Early Research and Successes

Immunoisolation owes its literature base to mouse model experiments in the early 1930s by Bisceglie in 1933. Encapsulated mouse tumour cells were injected into guinea pigs that survived for 12 days. Bisceglie, therefore, concluded that nutrients and waste were transported through the polymer, in both directions [35]. Significantly, the polymer membrane was not attacked by the endogenous immune system. These first attempts were repeated when Chang et al. used interfacial polycondensation techniques to produce micrometre-scale semi-permeable microcapsules of polymer, laying the groundwork for modern cell encapsulation techniques [36]. Darquy and Reach were amongst the first to produce micro-encapsulated constructs containing pancreatic beta cells, secreting insulin in response to elevated glucose levels, without eliciting an immune response from conditioned rabbit serum [37]. They were indeed the pioneers in the development of the architecture of micro-encapsulated structures, still in use today [21, 37]. The greatest leap yet occurred when Shu and He [38] used the alginate encapsulation and peritoneum implantation technique to maintain normoglycaemia in diabetic rats for almost a year (360 days), laying the foundation for work aiming to produce a permanent, biological, immune-resistant structure to correct and preserve normoglycaemic function in diabetic patients.

2.2 Failures and limitations of Immunoisolation for IRT

One of the disadvantages of early immunoisolation solutions for IRT was in the chemistry of the device. The Lum and Sum paper had used the polycondensation technique to generate their microencapsulated structures, combining cationic poly(lysine) with polyanionic gels. Whilst efficient, De Vos et al. noted that 1-step reactions requiring only 2 reactants set limitations on the mechanical, biological and chemical properties of the construct, as modulating one component directly affected the other [39]. This issue was addressed by Wang et al. who used polyelectrolyte complexation to produce multi-component microencapsulated systems, allowing more flexibility in the microencapsulate design [40].

Another indicator of failure in microencapsulated systems is fibrotic tissue formation. Fibrotic tissue is broadly analogous to normal tissue repair, with the exception of delayed myofibroblast senescence and subsequent overproduction of collagenous extracellular matrix (Fig. 2) [41, 42]. Whilst production of fibrotic tissue is more benign than the potential immune response generated by foreign, unencapsulated cells, fibrotic membranes coat the semi-permeable encapsulated membrane, reducing oxygen, waste, and insulin transport, limiting hyperglycaemic

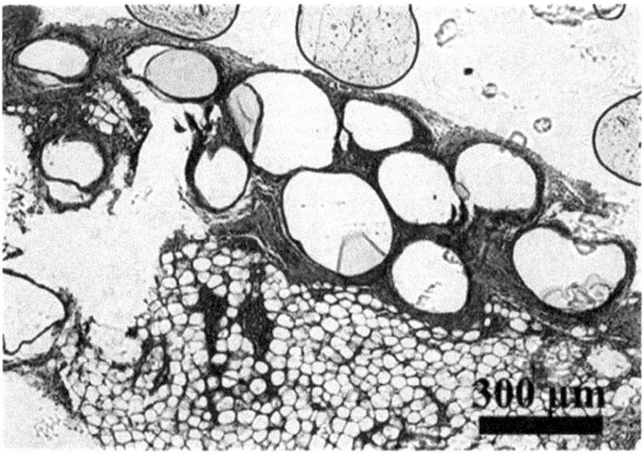

Fig. 2 Fibrotic collagen overgrowth presenting on subcutaneously implanted alginate microcapsules. (Reproduced with permission from Mohammadi et al., 2018 [42])

response [41, 42]. The severity of this issue was compounded by De Vos et al. finding that reductions in beta cell activity occurred not just because of fibrotic tissue formation, but by an imbalance in islet cell division and necrosis [43]. It is hypothesized that hypoxia of islet cells due to suboptimal oxygen permittivity results in the reduction of cell number and, therefore, physiological function [44]. Moreover, the physiological effectiveness of microencapsulated islets in many of the earlier experiments may not be real. Clayton et al. noted that many of the earlier studies into IRT using microencapsulation used streptozotocin (STZ) as the inducer of diabetes in test mice. Studies have shown STZ to be impermanent as a diabetes-inducing agent. Indeed, Hartmann et al. showed that 70% of STZ diabetic mice had recovered by 15 weeks [45, 46]. Despite this, STZ T1D model continues to be used in research due to its improved condition stability compared to other diabetogenic agents such as alloxan. Additionally, STZ is less expensive than viral, or hormone-induced diabetes and does not require the use of mammals such as dog or baboon, which may inflict unnecessary stress and suffering [47, 48].

2.3 *Classical Immunoisolation in the Clinic*

Clinically, microencapsulation for IRT has produced mixed results. A review by Buder et al. describes some of the notable successes and failures encountered in contemporary islet replacement clinical trials [49]. The first clinical trial in a human diabetic model was relatively successful. Soon-Shiong et al. demonstrated the discontinued use of exogenous insulin injections in a 38-year-old diabetic patient following 2 doses of alginate-encapsulated cadaveric islet cells in their 1994 paper [50]. A further study undertaken in four patients by Valdes-Gonzalez et al. was able to significantly reduce patient insulin consumption whilst detecting insulin *in sera* in 50% and 75% of the patients following xenotransplantation of porcine Langerhans and Sertoli cells, respectively [51]. One of the supposed main advantages of immune-isolative microencapsulation for IRT is the proposed "invisibility" to the immune system allowing for unimpeded function over many years. Calafiore et al. demonstrated this clinically, finding not only that microencapsulated islet produced no immune sensitization or adverse immune response on implantation, but that common immune-response markers such as anti MHC I and II as well as islet cell antibodies were notably absent five years after transplantation [52, 53].

Whilst these studies show a good research base for the improvement of biological islet replacement devices, limitations in the studies are clear. Buder et al. noted that the patient described by Shoon-Shiong et al. received medication to prevent immune rejection from a kidney transplant undertaken prior to the clinical trial, artificially reducing the immune response to the implanted microencapsulated islets. Although the attempt made by Valdes-Gonzales et al. was initially successful in reducing exogenous insulin dependency, all four patients were back to prescribed insulin doses by 7 years. Calafiore and Basta noted that descriptions of these trials cannot truly be considered "breakthroughs" in IRT because in none of the trials could complete discontinuation of exogenous insulin be implemented, except for short periods of time for any patient. This indicated that although the IRT was partially effective, the effect was too small to completely maintain normoglycaemia, perhaps due to the *in vitro* issues previously discussed such as hypoxia, poor revascularization, and fibrosis. Moreover, due to the highly experimental nature of these techniques many of these clinical trials are performed on a small number of patients, rendering other underlying conditions, age, and sex to become significant factors in these trials. Buder et al., Calafiore and Basta found only one study using more than 4 patients [49–51]. Although the *in vitro, in vivo* and clinical trials of immunoisolation have shown promising results there are still some major hurdles to overcome to use these techniques as effective insulin therapies.

3 Part 2: Construction of Bioactive Systems for the Treatment of T1D

Initially it may seem that bioactivity would be a disadvantage to immunoisolative technologies using stem cells, especially given the risk of cell recognition and destruction. However, the above disadvantages of current immunoisolative systems have necessitated the incorporation of bioactive components to initiate differentiation of beta cell-like stem cells, as well as preserve their viability long term and maximize their ability to regulate blood glucose concentrations. Whilst many current methods of incorporating bioactivity have been limited to beta cell lines and whole explanted islets, these remain important first steps in allowing for a functional stem-cell based system.

3.1 Bioactive Materials for Oxygen Delivery and Supply

Whilst external artificial pancreas systems have traditionally consisted of insulin pumps [54], recently more advanced systems have utilized living beta cells. However, previously described limitations like hypoxia have led to the development of systems that have both an internal (cell containing) and an external (power and oxygen production) component. Indeed, these systems are currently the most clinically relevant of the implantable devices that act as bioartificial pancreas. This includes the Beta-Air BAP, which consists of a porous, PTFE (polytetraflouroethylene)/alginate composite shell housing live beta cell-containing islets (Fig. 3) [55]. The innovation of this design lies with the transcutaneous ports, which allows for the injection of exogenous oxygen to perfuse the tissues as well as the excellent

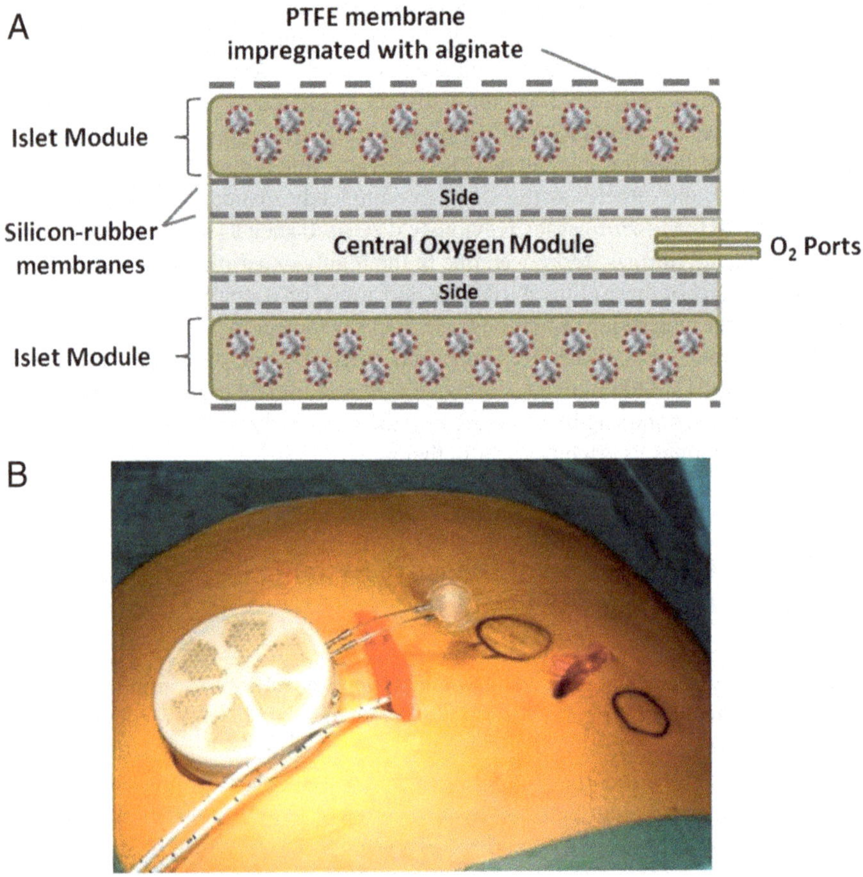

Fig. 3 The implantable bioartificial pancreas device (**a**) Cross-sectional diagram showing the system architecture. (**b**) The "β-Air I" being prepared for the process of implantation. (Reproduced with Permission from Ludwig et al. (2013))

oxygen permeability of the PFTE/alginate composite material. The oxygen supply lies outside of the organism, supplied by oxygen lines that connect to the BAP, implanted using minimally invasive surgery. Phase 1 human clinical trials have already taken place using this system, with patients having up to 2 beta air systems implanted (up to 180 kIEQ) [55, 56]. Whilst the system was removed without evidence of systemic rejection, evidence of fibrosis was found in and around the implant, with extremely limited systemic C-peptide detection (and thus glycaemic control) detected, indicating the need for further optimization to the design, with some research indicating the lack of revascularization. Several other researchers have also taken this approach, such as Procyon Ltd., which is currently developing a miniaturized system utilizing a wearable electrochemical oxygen production system mounted on the wrist and designed to be used alongside a Theracyte™ system, a device that, unlike the beta-air system, can be vascularized pre-implantation [57–59].

Another semi-external system has been described by Tokito et al. where a honeycomb well structure constructed of PDMS is placed inside the organism, with oxygen passage through a transcutaneous, highly microporous PDMS interface [60]. Tokito et al. described improved proliferation and insulin secreting function of INS-1E subclonal rat cell line, with fourfold increase in dsDNA (double-stranded DNA) concentration of the construct over 11 days, *in vitro,* whilst greatly increasing the cell's sensitivity to both increase and decrease in serum glucose concentration. Likewise, the preclinical device known as ceMED (Fig. 4), developed by Yang et al. in collaboration with MIT and Harvard Universities contains two separate chambers, one which collects nutrients through a 10 um semi-permeable and angiogenic membrane. The resulting fluid is moved by an electronic implatable pumping system into the cell containing component, protected against immune cells and moieties by a 200 nm semipermeable (PTFE) barrier [61]. The continuous flow was hypothesized to allow much greater islet encapsulation volumes whilst preserving islet function, which was confirmed in both *in vitro* and *in vivo* analysis, using immunocompetent Lewis rats. Flow through the device not only increased islet viability but also glucose sensitivity, with the flow *in vivo* group displaying 60% less serum glucose (an indicator of insulin secretion) than the groups with no flow. Additionally, the flow groups preserved islet morphology better than the no flow group in both implanted MIN-6 islet cells lines and stem cell-derived beta cells. Whilst the test setup was semi-internal (with power provided by an external large pump), future optimizations could include a fully internal design utilizing battery power, or wireless charging [62].

Lazard et al. also reference the ability for algae to generate an oxygen producing effect *in situ* for a semi-external bioartificial pancreas [63]. Bloch et al. found that under light immersion, co-encapsulates of algae and islets generated a significantly greater islet viability and insulin response to elevated glucose concentration than under dark conditions, as well as under normoxic perfusion protocols [64, 65]. Whilst co-encapsulating and implanting another species into the human environment would certainly have both technical and regulatory challenges, it is not beyond the realm of scientific possibility to imagine a system whereupon a subcutaneous

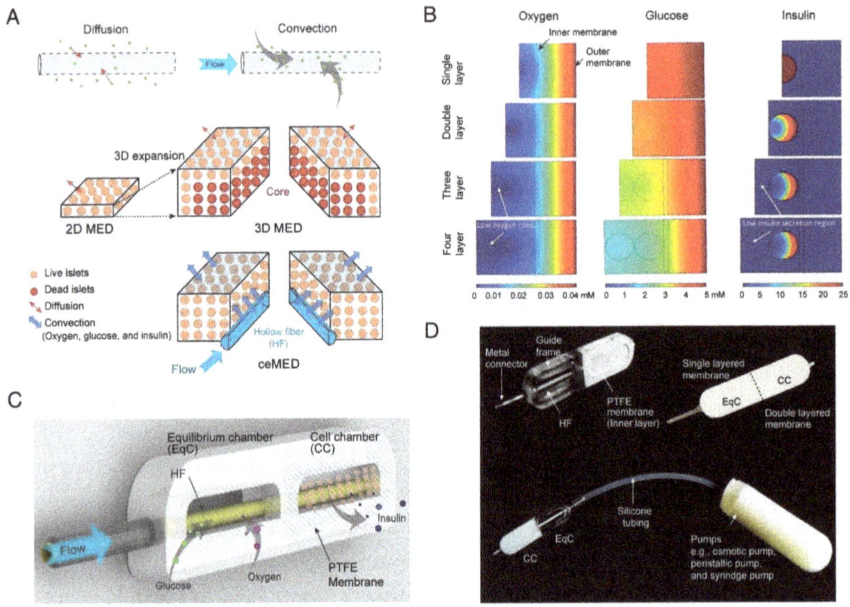

Fig. 4 The ceMED system (2021) (**a**) The expansion of 2D cell culture to 3D, necessitates the use of active perfusion to reduce the formation of a necrotic core (**b**) Simulations of Oxygen, Insulin and Glucose concentrations within 1,2-,3- and 4-layer encapsulated islet systems, with white arrows indicating areas of hypoxia. (**c**) Schematic of the ceMED system, featuring an equilibrium chamber, a cell chamber, and a flow chamber, surrounded by a PTFE membrane. (**d**) The fully assembled system. (Reproduced with permission from Yang et al. (2021))

co-encapsulate of algal and pancreatic beta cells is illuminated from an external source. This would be especially feasibly given that 600–700 nm (red) light is both penetrative to human skin and a photosynthetically active wavelength to genera of algae such as *Chlorella* [66–68].

3.2 *Internal Systems*

Whilst currently not as clinically significant as the semi-internal systems being developed and tested in humans, much work into bioartificial pancreas that are able to be fully implantable without the need for external viability-enhancing systems have been attempted. Whilst many of these systems also focus on mechanically poor encapsulation materials, great strides have been taken to alleviate the traditional drawbacks of materials such as the hydrogels.

3.3 Multi-material Scaffolds as Multimodal Bioactive Agents

Bioactivity is defined by Karas et al. as "The ability of a material to induce specific effects in a biological system after exposure to the said material; the effects include tissue uptake, metabolic or a physiological response" [69]. Analogous to how the biocompatibility of a particular material does not imply compatibility with every tissue type and biological application, bioactivity is not simply just an intrinsic property of a material, and therefore must be contextualized with its effects on both cell/tissue types and for a particular application. In effect then, no material is truly "bioactive" or "bioinert" but is bioactive in a particular way (e.g. revascularization) with a particular tissue type for a specific intended physiological outcome. It may seem initially paradoxical then, that many of the materials used in bioartificial pancreas research are labelled bioinert. Literature describes bioartificial pancreas constructed from polylactic/glycolic acid (PLA/PLGA), polydimethylsiloxane (PDMS), polytetraflouroethylene (PTFE) and polyvinyl alcohol (PVA) for the mechanical stress shield component and bioinert hydrogels such as Gellan Gum (GG), sodium alginate and agarose for the immunoisolation component [70–76]. Whilst these materials have previously been labelled bioinert, the lack of cell binding ligands on hydrogels like agarose and alginate, promotes the maintenance of the spherical islet shape required for maximal insulin secretion whilst reducing the binding ability of immune cells attempting to access the interior [77, 78]. Despite this, given the drawbacks of contemporary bioartificial pancreas (namely revascularization, antimicrobial efficacy, hypoxia, and fibrosis) (Fig. 5) it may be necessary to modify existing materials to improve their bioactive effects on islet cells for insulin replacement therapies, for which materials science has proven effective.

Whilst previous work have made clear some of the failings of past and contemporary immunoisolation attempts, headway has been made to address some of the natural limitations of using mechanically poor materials such as sodium alginate. Since many of the strengthening materials used to improve hydrogel mechanical properties tend to elicit deleterious effects on other properties, such as oxygen permeability, many researchers have concluded that the best way to manage implanted islets is to surround them in a flexible material that can act as a stress shield. This has multiple advantages.

A multimaterial bioartificial pancreas localizes the islets within a single construct, which not only allows for much easier removal in the case of implant failure, but also may potentially enhance the paracrine effect on tissues adjacent to the implant (given islets release endothelial proliferative factors such as VEGF upon hypoxia), which will potentially hasten the revascularization process, reducing the formation of islet necrotic cores. Additionally, an elastomeric scaffold could be functionalized with anti-fibrotic, immunosuppressing or oxygen-producing factors, which can further improve islet viability [79, 80]. Finally, the advent of 3D printing has allowed for submicron geometries, which were previously unavailable. This allows for the previously mentioned islet localization whilst maximizing the advantages of free-floating microencapsulated islets. Indeed, a review by Xu et al. lists

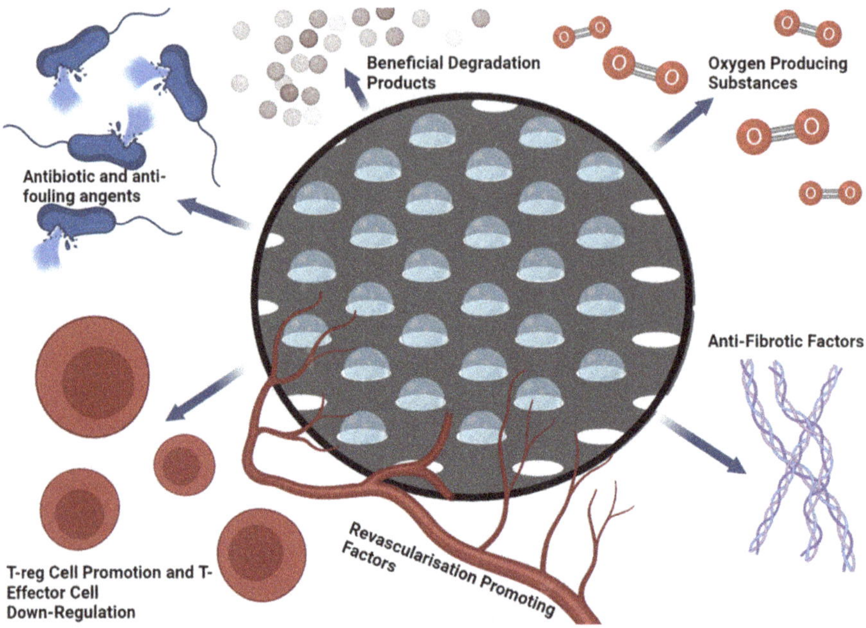

Fig. 5 A schematic demonstrating how the development of a multi-material bioartificial pancreas allows increased strength, but also for the incorporation of multimodal bioactive properties into both the strengthening matrix and the islet encapsulation material

some of the polymers that have been used as the mechanically stabilizing component of a 3D printed bioartificial pancreas, including polylactic acid (PLA), polylactic-co-glycolic acid (PLGA), polyurethane (PU), and polycaprolactone (PCL) [81]. These upcoming sections will detail the progress being made to construct a successful bioactive multimaterial bioartificial pancreas and the features they could incorporate using the above advantages to defeat some of the limitations of contemporary whole stem cell islet encapsulation.

3.4 Oxygen Production

The ideal bioartificial pancreas would be able to endogenously produce oxygen, either by functionalization of the immunoisolating matrix, or by incorporation of micro/nanoparticles of oxygen generating molecules (Fig. 6). These often come in the form of solid peroxides such as calcium peroxide and sodium percarbonate (Eq. 1a–c), Thi et al. [82] demonstrated the use of calcium peroxide, with their successful formation of multifunctional, oxygen producing hydrogels combining it with alginate. In addition, Ke et al. demonstrated the use of sodium percarbonate microcapsules as oxygen producing agents [83]. This approach has also been trialled in the context of a multi-material bioartificial pancreas. Coronel et al.

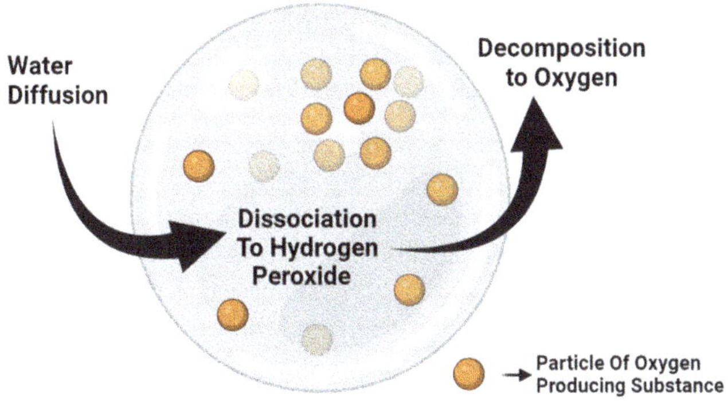

Fig. 6 Release pathways of commonly used oxygen-producing substances. (**a**) Water diffusion into an oxygen-producing scaffold induces the dissociation of the encapsulated particle to generate hydrogen peroxide, which then decomposes to water and oxygen. Commonly used oxygen-producing substances include (**b**) calcium peroxide and (**c**) sodium percarbonate

encapsulated calcium peroxide/PDMS composite structures in immunoisolated islet-laden agarose gels, and found that even under extremely hypoxic conditions, their scaffold maintained enhanced insulin secretion compared to the control [84, 85]. Other proposed methods of producing *in situ* oxygen is through photosynthesis. Bloch et al. showed that illuminated islets co-encapsulated with algae elicited an insulin production response in a light dose-dependent relationship. The multi-material bioartificial pancreas could produce a much greater range of supporting multimodal functions and drug release characteristics to preserve stem cell viability than hydrogels alone [64, 65].

A clear similarity between the different classes of semi-internal devices is an emphasis on oxygen production. However, the maintenance of a large, transcutaneous wound invites both poor patient compliance and pathogenic infection [86]. The Theracyte™ system is an entirely implantable 20 mm pouch designed to contain immunogenic material. In its bioartificial pancreas configuration, the Theracyte system contains multiple layers of various porosities to encourage mechanical stability (in the case of the woven polyester outer layer), revascularization (in the case of the 5 μm middle layer) and immune protection (in the case of the 500 nm inner layer) [57–59]. So far, literature has described Theracyte™ operational use with human, Rhesus monkey and pig islets, with implantation recorded in diabetic mice and non-human primates with partial pancreatectomy-induced diabetes, though human recipients for the treatment of various conditions are recorded in literature [87, 88].

$$CaO_2 + 2\mathrm{H_2O} \rightarrow Ca(OH)_2 + \mathrm{H_2O_2}. \tag{1a}$$

$$2[Na_2CO_3].[3\mathrm{H_2O_2}] + 2\mathrm{H_2O} \rightarrow Na_2CO_{3(aq)} + 3\mathrm{H_2O_2} \tag{1b}$$

$$2\mathrm{H_2O_2} \rightarrow \mathrm{H_2O} + \mathrm{O_2} \tag{1c}$$

Equation 1: (**a**) Dissociation of calcium peroxide to calcium hydroxide and hydrogen peroxide; (**b**) Dissociation of sodium percarbonate to sodium carbonate and hydrogen peroxide (**c**) Decomposition of hydrogen peroxide to water and oxygen

Whilst multiple studies have since confirmed medium-term normoglycaemia following implant deployment in mice and long-term survival and function of islets in Rhesus monkey, some studies have highlighted concern over the prevalence of fibrotic tissue formation post-implantation, which was shown to greatly reduce hormone in circulation (though the highlighting study was using the system as a therapy for a different endocrine condition) [89, 90]. Moreover, Theracyte™ has reduced effectiveness at preventing CD4+ mediated islet cell destruction, with Loudovaris and Charlton finding that CD8+ and B cell depleted SCID mice implanted with Theracyte™ analogous cell delivery systems destroyed the cellular component of the implant within 3 weeks [91]. *In situ* oxygen production of multimaterial bioartificial pancreas has been trialled in both semi-internal (e.g. beta-air) and fully internal systems (though to date clinical data is sparse). Pedraza et al. constructed composite discs containing PDMS and oxygen generating calcium peroxide, which was able to produce sustained oxygen release over 40 days, preventing hypoxia-related death in the islets and preserving cell viability of MIN-6 beta cell lines over 25 days in hypoxic cell culture conditions, whilst retaining comparable GSIS capability [92]. Coronel et al. also produced similar results with their PDMS-based OxySite biomaterial (also a PDMS/calcium peroxide composite) [84, 85], displaying significant increase in implanted islet viability in a rat model following OxySite implantation. This was also true of the design created by Liu et al. where coaxial bioprinting was used to generate struts with an outer macroporous layer and an islet-containing inner layer. Whilst Liu et al. recognized the 1 mm thickness of the outer layer as a disadvantage for the design with regard to oxygen and nutrient diffusion, both cell survival and physiological GSIS activity were both shown experimentally, indicating that improvements in coaxial bioprinting technologies may produce coaxial structures with acceptable diffusion gaps for hypoxia avoidance [93].

3D bioprinting can also be used to confer bioactivity to immunoisolated beta-cell containing structures. Where before bulk hydrogels could not avoid cell hypoxia and necrotic core formation, bioprinting allows for the manufacture of 3D structures at resolutions below effective oxygen diffusion distances. Marchioli et al. used a mixture of gelatin and alginate to generate bioplotted 3D constructs, which were able to maintain the islet cell line and whole islet cell viability for 14 days *in vitro*, whilst Hwang et al. used a multimaterial scaffold of PCL and bioprinted alginate to maintain islet stem cell function, differentiation, and viability over 2 weeks *in vivo* [94, 95].

4 Revascularization

Whilst oxygen production is a more immediate solution to the issues surrounding hypoxia in the stem-cell based islet replacement therapy, prompt revascularization is key to facilitate long-term oxygen delivery using the body's own mechanisms.

Some research has focussed on the modulation of islets in a construct to better tolerate the hypoxic storm. Vaithilingam and Tuch, described other methods including "heat shock" protocols, elevating islet incubation temperature to induce production of heat shock proteins, which have a protective effect against the cytotoxic reactive oxygen species that impair insulin response through cell damage and death [96]. Also, Bellman et al. describe the protective effect of heat shock in islet cells, finding that although nitric oxide mediated cell death was not impaired, NAD^+ production mediated cell death was greatly reduced following culture at 43 °C, reducing islet cell death and cytotoxic stress [97].

However, Vaithilingam/Tuch et al. noted that attempts to reduce hypoxic stress do not address the cause of hypoxia, and as such are only useful for the time immediately after device implantation. Most attempts to reduce long-term hypoxia focus on the improved vascularization of the microencapsulated structures. Phelps et al. incorporated VEGF into hydrogels containing islet cells, finding that microencapsulated structures revascularized within 4 weeks of implantation into rats, whilst producing insulin [98]. The encapsulation of mixed cell cultures containing mesenchymal endothelium progenitors has also been suggested to improve vascularization. Steiner et al. significantly improved the vascularization of bioartificial pancreas utilizing a wraparound arterial loop in the constructs containing microencapsulated MSC's whilst Brissova et al. combined endothelial and islet cells and found it to be an important contributor to vascularization in mice [99, 100]. Whilst the method of immunoisolation and preservation of islet function is important, the location of the implant must also be considered. Highly vascularized regions such as the omentum are often considered as a suitable implantation site. Whilst previous efforts at immunoisolation surrounding highly vascularized regions involved placing the islets directly in the vasculature (intrahepatic and intravascular islet transplantation), recent efforts have placed bioartificial pancreas constructs in contact with vasculature to preserve oxygen uptake and revascularization.

Han et al. described a Biovascular Pancreas (BVP). The construct was made from a decellularized graft and used two layers, the innermost of which was macroporous and is designed to interface with the blood vessel wall. This in turn is connected to a layer of cell-laden hydrogel, which provided biological protection from the immune system. Han et al. found that their BVP produced a 3× greater response to elevated glucose concentrations after 200 min than the control. Subsequent *in vivo* analysis in nude rats found that the device displayed extensive microvascular infiltration from the host tissue although the insulin secretion response only produced near-normoglycaemic conditions for the 90-day test period [101]. Shaheen et al. developed an intravascular bioartificial pancreas (IBAP), with the aim of the construct to connect directly with the existing vasculature [102] (Fig. 7). This was achieved by generating superporous agarose scaffolds with microscale convective pores through which blood flow could be achieved and the immune protective nanopores indicative of classical agarose. The IBAP displayed short-term biphasic GSIS, as well as excellent biocompatibility with human islet cells. However, currently no *in vivo* work using this architecture has been completed. Examples of functionalization of both the encapsulation material and mechanical stabilization structure have

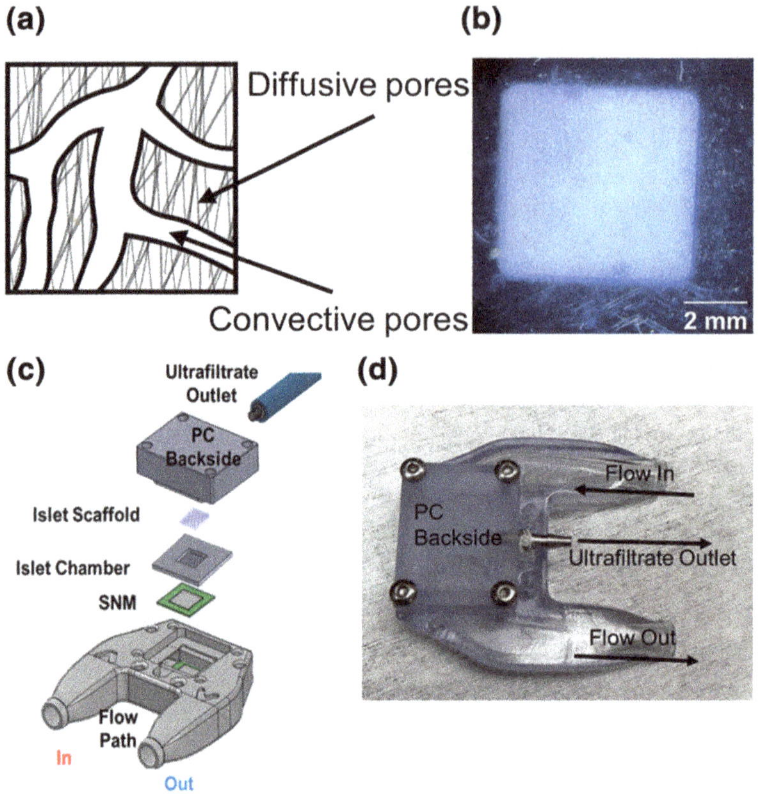

Fig. 7 Superporous agarose (SPA) scaffold and intravascular bioartificial pancreas "iBAP" device assembly. (**a**) Schematic cross-sectional view depicting the convective and diffusive pores in the agarose scaffold. (**b**) SPA scaffold within the islet chamber housing. (**c**) Enlarged view of the components of the iBAP: flow path, silicon nanopore membrane (SNM), cell scaffold within the cell chamber, and polycarbonate (PC) backside and ultrafiltrate outlet. (**d**) Assembled iBAP used for in vitro testing. (Reused with permission from Shaheen et al. (2021))

been described in literature. Gebe et al. described a circular bioartificial pancreas, with capsules of cell-containing alginate surrounding a central VEGF-containing capsule, held in place by a microporous polyvinyl alcohol (PVA) mechanically resistant sponge to allow for VEGF uptake and delivery across the implantation surface [103] (Fig. 8). This design, when implanted into streptozotocin-induced diabetic rats, preserved normoglycaemia for 116 days, up to the implant removal point. Moreover, the VEGF+ group reduced time to normoglycaemia and encouraged revascularization around the implant site under histology.

Alternative methods of constructing a spongey matrix for similar revascularizations systems have included 3D printing, where the ability to make small sub-micron wells provides areas suitable for drug loading. Marchioli et al. utilized rings of heparinized PCL with cell-laden alginate printed into the resultant wells to further improve the immobilization of VEGFA, recognizing the construct-enhanced

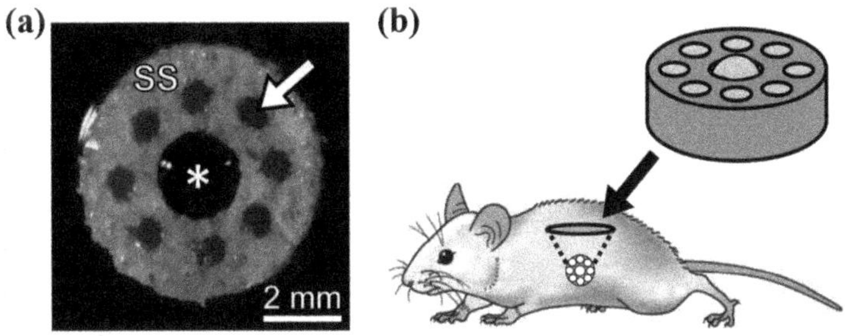

Fig. 8 Diagram of an islet containing scaffold for mice (Gebe et al., 2018). The implant consists of (**a**) a polyvinyl alcohol (PVA) sponge scaffold (SS) with a single VEGF-laden alginate sphere at its centre and surrounded by concentric wells of islet-containing collagen gel. The resultant structure was implanted into (**b**) a subcutaneous pocket of a mouse. (Reproduced with permission from Gebe et al. (2018) [103])

angiogenesis in an *ex ovo* chick model and retaining the ability of encapsulated islets to initiate GSIS *in vitro* (Fig. 9) [104]. Kasojo et al. used heparinization to functionalize a poly(lactide-*co*-caprolactone) pouch with VEGF, providing a controlled release of the hormone over 30 days [105]. Kasojo et al. also implanted their design into Lewis rats, finding that heparinized scaffolds had significantly larger vascular tissue infiltration and were able to achieve at least 50 days of islet viability post-implantation, although large pockets of post-implantation fibrosis around the implant were discovered) [105].

5 Insulin Production Maximization

To maximize the efficiency of insulin replacement devices, the insulin-releasing ability of the devices must be improved. One of the most important factors for islet cell survival and insulin release has been replicating the endogenous pancreatic tissue ECM whilst still preserving immunoisolation. Indeed, Wang and Rosenburg suggested the disruption of the islet-ECM interface may reduce insulin production and even lead to cell death [106]. Hence some studies have attempted to preserve this interface by the co-encapsulation of ECM proteins. Weber and Anseth were able to increase PEG-encapsulated islet insulin production fourfold following co-encapsulation with collagen type IV and laminin. This proposal achieved more recognition in 2018, when Llacua et al. showed that the addition of the same proteins increased cell survival via protection against cytokine activity [107, 108]. This study may also explain why similar biological polymers, such as alginate and ECM-derived polymers like hyaluronic acid can provide better islet survival and function compared to synthetic polymers [109]. Microcapsule sphere morphology may be equally important as biochemical content for insulin production. A team led by Tomei et al. developed a microfluidic device that allows for thin, controlled thickness multilayers on microencapsulated structures [110]. This greatly reduced the

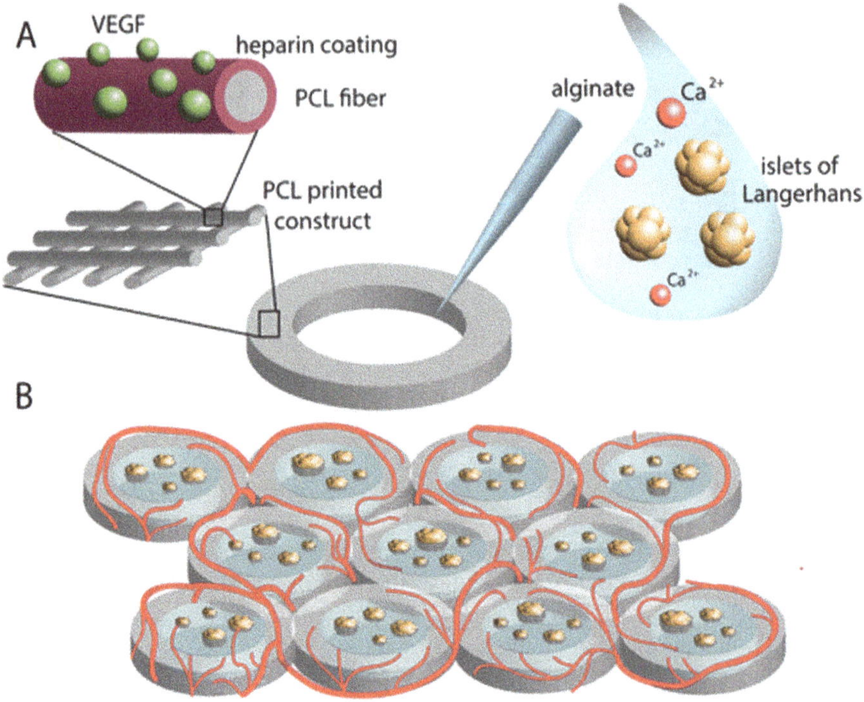

Fig. 9 Schematic of Marchioli et al.'s multimaterial pancreas. (**a**) PCL was 3D printed into ring structures and functionalized with heparin. This was then used to strongly bind VEGF (**b**) Alginate containing islets of Langerhans were then plotted into the rings, which can be printed next to each other for improved space efficiency. (Reused with permission from Marchioli et al. [104])

transplant volume whilst preserving transplanted cell numbers, allowing for a greater insulin secreting effect *in vivo*. This technique was further improved by Stock et al., who encapsulated a new stem-cell derived islet source described in Pagliuca et al. in 5% poly(ethylene glycol)-maleimide (PEG-MAL), with the latter described similar islet cell function compared to human islet cells, resulting in the production of physiologically significant quantities of insulin [111, 112]. These studies confirmed that not only can the scalable technology of islet progenitor differentiation perform like human islets, but that microencapsulation modifications can improve insulin secretion of existing islets to physiological levels for the maintenance of normoglycaemia.

6 Reduction of Fibrotic Tissue Production

Reducing fibrosis was found to be another major issue limiting the *in vivo* effectiveness of implanted microencapsulated islet cells. Once again, the ability to incorporate drugs into hydrogels has led to the reduction of fibrotic tissue formation. Dang et al. used bioluminescent imaging to demonstrate the efficacy of curcumin and dexamethasone in reducing ROS production and inflammatory enzyme activity in mice [113]. Furthermore, the study also co-encapsulated curcumin into alginate microencapsules containing islet cells using electrostatic droplet generation; both increased the control of glucose whilst reducing the growth of fibrotic tissue, though it was noted that the hydrophobic nature of the drugs led to irregularity in the subsequent capsule morphology, setting a limit on the amount of drug that could be encapsulated. Other techniques targeted the cells that mediate the fibrotic response. These included macrophages, where Sung et al. successfully demonstrated the liposome encapsulation of clodronate, a selective inhibitor of macrophages, inducing systemic macrophage depletion and reducing TGF-β1 mRNA expression in a rat kidney fibrosis model [114]. Attempts were made to reduce the differentiation of fibroblasts into myofibroblasts through the use of transforming growth factor (TGF) inhibitors; however, the latter has not currently been co-encapsulated with beta cells or islet secreting analogues [114, 115]. Further, the use of co-encapsulated mesenchymal stem cells (MSCs) has been proposed for the reduction of hypoxia in encapsulated islets. Meier et al. found that not only did MSC conditioned medium induce the release of anti-inflammatory cytokines, but also reduced collagen type I mRNA in mice following PEG microencapsulation of murine islets and implantation in mice. More promisingly, Vaithilingam et al. co-encapsulated MSCs with mice islet cells, improving islet viability from 71.4% to 100% *in vivo* with significantly less fibrotic growth [116, 117]. Another method of reducing the immune response of immunoisolated islets is to modify the shape. Whilst Tomei et al. used layer thickness to increase physiological activity of the islets, Veiseh et al. found that increasing alginate encapsulation size from 0.9 to 1.9mm significantly reduced both fibrotic tissue growth and α-SMA (a marker for fibrotic fibroblast to myofibroblast transdifferentiation) production *in vitro* [110, 118]. These studies indicate that despite progress still being at the pre-clinical stage, there are multiple modalities for reducing fibrosis and immune response for IRT. Whilst beta cell like stem cells may not be directly responsible for reducing fibrotic tissue on their own, it is possible to use stem cell-based therapies to further reduce fibrotic tissue formation. It is already known that certain mesenchymal stem cells derived from bone and adipose tissue improve wound healing and reduce scar tissue formation *in vivo.* For example, Li et al. found that bone mesenchymal stem cells (BMSC) expressing TGF-β3, (higher quantities of which compared to TGF-β1 and 2 is implicated in more organized collagen formation and improved wound healing in foetal tissue) significantly reduced collagen I to III ratio whilst significantly reducing wound depth over a period of 90 days [119]. Whilst coculture of adipose and bone cells with beta cells has been described in literature [120–122], bioactive materials incorporating TGF-β3 have been previously constructed. Mahmoudi et al. used a microfluidic system to produce TGF-β3-loaded alginate nanogel spheres to improve the chondrogenesis of hMSCs *in vitro* whilst Saygili et al. used a similar system to induce improved cartilage repair in a rat model using alginate-polyacrylamide nanoparticles [123].

7 Maintenance of Stem Cell Populations *In Vivo*

While many of the studies described in this chapter mention the maintenance of beta cells and analogues in the eventual *in vivo* environment, the use of beta cell-like stem cells requires the replenishment of cells lost to the immune system, senescence, and degradation of part of a prospective stem cell-based insulin secreting device. This provides an advantage over traditional whole islet-based systems, where the system of beta cell replacement was from the slow replication of existing cells and avoided techniques to induce beta cell proliferation, which carries the risks of neo-plasticity [124–126]. Therefore, it is imperative that stem cell-based systems have the capability both to maintain viability (by the above bioactive modalities) and enable controlled differentiation of beta-like stem cells for continued insulin production.

The use of bioactive compounds to modulate the oxygen content of constructs to enhance insulin secretion and avoid the formation of a necrotic core has been previously described in this book chapter. Similarly, research has shown that modulation of the oxygen environment can positively influence the differentiation of cells to insulin secreting types. Fraker et al. demonstrated the use of perflourocarbon (PFC) perfusion-based oxygen delivery to enhance the differentiation of beta cells from pancreatic buds, with the higher oxygen concentrations favouring endocrine cells over exocrine cells [127]. Later, Cechin et al. showed that oxygen modulation using the same PFC system could produce beta cells from human embryonic stem cells [128]. Future work may focus on the use of oxygen producing biomaterials for more effective stem cell differentiation to insulin secreting cell types.

Soluble factors affecting vascularization of implants may also affect the ability to influence *in situ* differentiation of islet progenitors. Nikolova et al. demonstrated that VEGFA-deficient mouse islets not only experienced reduction in insulin expression, but insulin-containing intracellular secretory granule production [129]. Whilst it was found that VEGFA-initiated endothelial cell proliferation and conversely impaired alpha and beta cell proliferation [132], Guney et al. found that synergistic expression of connective tissue growth factor (CTGF) in both beta and endothelial cells increased both vascularization and beta cell proliferation *in vitro* [130]. Given CTGF's ability to induce proliferation in immature insulin positive cells [131] and the literature precedent of incorporation of CTGF into electrospun scaffolds [132], it may be possible to generate a multimodal revascularization and differentiation platform in a bioactive biomaterial construct.

In addition to the soluble factor binding mentioned previously in this section, additional differentiation signalling for stem cells can be obtained by the surrounding matrix environment the cells adhere to. The alginates commonly used in immunoisolative tissue engineering (with their generally recognized bio-inertness) provide an opportunity to create bioactive hydrogels capable of inducing *in situ* beta cell differentiation from stem cells. The lack of cell binding ligands in alginate promotes the maintenance of spherical islet morphology [75]. The specificity in composition (non and collagenous proteins, polysaccharides, proteoglycans) of native tissue ECM can allow tuneable biomaterials allowing for differentiation for multiple cell lineages, depending on the particular application. In the case of pancreatic tissue engineering ECM has been used to enhance the differentiation of precursors to an insulin secreting state. Narayanan et al. and Singh et al. both demonstrated hESC differentiation to beta cells when grown on native pancreatic ECM [133, 134]. The ability to incorporate ECM in alginate scaffolds has previously allowed for research on multiple tissues including cardiac, hepatic, and renal tissue [135–137]. Islet differentiation through encapsulation in ECM-hydrogel scaffolds has been shown in literature. Formo et al. successfully differentiated hESCs in ECM-containing alginate for 46 days without the formation of a necrotic core. However, it was noted that this setup favoured non-beta endocrine cells proliferation [138]. Wang et al. optimized this by using pancreas-ECM/Alginate hydrogels to improve the differentiation efficiency of mouse iPSCs *in vivo* [139]. Damodaran and Vermette confirmed the recellularization and return to glucose reactivity of decellularized pancreas tissue with INS-1 pancreatic beta cell lines over 120 days, showing whole decellularized tissue as a potential bioactive scaffold for insulin-secreting cells, whilst Citro et al. demonstrated the use of endothelialized lung tissue as a repository for islets, showing physiological glucose reactivity in diabetic rat models over 30 days [140, 141].

This may be due to the mechanical properties and peptide-based signals of the resultant hydrogel. Candiello et al. found that ESCs grown on alginate substrates modified to match native pancreatic tissue mechanical properties elicited significantly greater endodermal islet progenitor traits in a dose-dependent fashion, whilst Richardson et al. found that hESC-pancreatic polypeptide cell (PP) differentiation was optimized in alginates tuned to 4-7kPa stiffness. It can therefore be seen that, in addition to functionalization of previously "bioinert" materials, bioactive molecules and bioinert substrates can be combined and tuned to both support and activate beta cell progenitors *in vitro* and *in vivo* [142, 143].

8 Conclusion

The notion of immunoisolation as a tool for insulin replacement therapy has been proposed as an experimental concept since the 1980s. Immune-compatible oxygen and nutrient-permeable polymer microencapsulation of pancreatic islets, islet cell lines and stem cells have been able to reduce exogenous insulin requirement *in vitro, in vivo* and in limited clinical trials. However, limitations to the long-term effectiveness of these constructs have been fibrotic overgrowth, hypoxia, limited cell viability, stem cell neoplasticity and the impaired production of insulin in the body. Great steps, however, have been taken to address these issues, combining improved vascularization, encapsulation geometry, bioactive materials manufacture, cell viability and function for a long-lasting non-immunogenic bioartificial pancreas. With the first clinical trials featuring these improved IRT devices underway, immune-isolation of beta cell-like stem cell precursors may become the gold standard for long-term normoglycaemia maintenance in diabetic patients in the near future.

References

1. Standl, E., Khunti, K., Hansen, T.B. and Schnell, O., 2019. The global epidemics of diabetes in the 21st century: Current situation and perspectives. European journal of preventive cardiology, 26(2_suppl), pp. 7–14.
2. International Diabetes Federation. IDF Diabetes Atlas, 8th edn. 2017. http://www.diabetesatlas.org (accessed 23 January 2023).
3. Correia, C.R., Nadine, S. and Mano, J.F., 2020. Cell encapsulation systems toward modular tissue regeneration: from immunoisolation to multifunctional devices. Advanced Functional Materials, 30(26), p.1908061.
4. B.A. Zielinski, M.J. Lysaght, Chapter Twenty-Eight - Immunoisolation, R. Lanza, R. Langer, J. Vacanti, Principles of Tissue Engineering (Third Edition), Academic Press, 2007, Pages 399–404, ISBN 9780123706157.
5. Orive, G., Hernández, R.M., Gascón, A.R., Calafiore, R., Chang, T.M.S., De Vos, P., Hortelano, G., Hunkeler, D., Lacík, I. and Pedraz, J.L., 2004. History, challenges and perspectives of cell microencapsulation. TRENDS in Biotechnology, 22(2), pp. 87–92.
6. Matsumoto, S. and Shimoda, M., 2020. Current situation of clinical islet transplantation from allogeneic toward xenogeneic. Journal of Diabetes.
7. Sun, Y.L., Ma, X., Zhou, D., Vacek, I. and Sun, A.M., 1993. Porcine pancreatic islets: isolation, microencapsulation, and xenotransplantation. Artificial organs, 17(8), pp. 727–733.
8. de Vos, P., van Hoogmoed, C.G., van Zanten, J., Netter, S., Strubbe, J.H. and Busscher, H.J., 2003. Long-term biocompatibility, chemistry, and function of microencapsulated pancreatic islets. Biomaterials, 24(2), pp. 305–312.
9. Webb, S.A. and Seppelt, I.M., 2009. Pandemic (H1N1) 2009 influenza ('swine flu') in Australian and New Zealand intensive care. Critical care and resuscitation, 11(3), p. 170.
10. Kawakami, Y., Ohta, S., Sayem, M.A., Tsukamoto, N. and Yaguchi, T., 2020. Immune-resistant mechanisms in cancer immunotherapy. International Journal of Clinical Oncology, pp. 1–8.w
11. R.S. Rainbow, M.J. Lysaght, Chapter II.6.15 – Immunoisolation, B.D. Ratner, A.S. Hoffman, F.J. Schoen, J.E. Lemons, Biomaterials Science (Third Edition), Academic Press, 2013, pp. 1306–1316, ISBN 9780123746269, https://doi.org/10.1016/B978-0-08-087780-8.00123-6.
12. Nafea, E.H., Marson, A., Poole-Warren, L.A. and Martens, P.J., 2011. Immunoisolating semi-permeable membranes for cell encapsulation: focus on hydrogels. Journal of controlled release, 154(2), pp. 110–122.
13. Hu, S. and De Vos, P., 2019. Polymeric approaches to reduce tissue responses against devices applied for islet-cell encapsulation. Frontiers in bioengineering and biotechnology, 7, p. 134.
14. Ludwig, B., 2020. Islet immunoisolation by macroencapsulation. In Transplantation, Bioengineering, and Regeneration of the Endocrine Pancreas (pp. 751–759). Academic Press.
15. Chinnuswami, R., Hussain, A., Loganathan, G., Narayanan, S., Porter, G.D. and Balamurugan, A.N., 2020. Porcine islet cell xenotransplantation. Xenotransplantation-Comprehensive Study, 26, pp. 183–188.
16. Kendall Jr, W.F. and Opara, E.C., 2020. Islet immunoisolation by microencapsulation. In Transplantation, Bioengineering, and Regeneration of the Endocrine Pancreas (pp. 761–786). Academic Press.
17. de Vos, P., 2020. Bioartificial pancreas: challenges and progress. Principles of Tissue Engineering, pp. 665–679.
18. Zhu, Y., Stark, C.J., Madira, S., Ethiraj, S., Venkatesh, A., Anilkumar, S., Jung, J., Lee, S., Wu, C.A., Walsh, S.K. and Stankovich, G.A., 2022. Three-Dimensional Bioprinting with Alginate by Freeform Reversible Embedding of Suspended Hydrogels with Tunable Physical Properties and Cell Proliferation. Bioengineering, 9(12), p. 807.
19. Wu, S., Wang, L., Fang, Y., Huang, H., You, X. and Wu, J., 2021. Advances in encapsulation and delivery strategies for islet transplantation. Advanced Healthcare Materials, 10(20), p. 2100965.

20. Cao, R., Avgoustiniatos, E., Papas, K., de Vos, P. and Lakey, J.R., 2020. Mathematical predictions of oxygen availability in micro-and macro-encapsulated human and porcine pancreatic islets. Journal of Biomedical Materials Research Part B: Applied Biomaterials, 108(2), pp. 343–352.
21. Lum, Z.P., Krestow, M., Tai, I.T., Vacek, I.V.A.N. and Sun, A.M., 1992. Xenografts of rat islets into diabetic mice. An evaluation of new smaller capsules. Transplantation, 53(6), pp. 1180–1183.
22. Bergen, J.F., Mason, N.S., Scharp, D.W. and Sparks, R.E., 1978. Insulin inhibition of islets in transplantation chambers. Artif Organs, 172, pp. 144–50.
23. Li, R.H., 1998. Materials for immunoisolated cell transplantation. Advanced drug delivery reviews, 33(1–2), pp. 87–109.
24. Hoffman, A.S., 2012. Hydrogels for biomedical applications. Advanced drug delivery reviews, 64, pp. 18–23.
25. Aziz, M.A., Cabral, J.D., Brooks, H.J., Moratti, S.C. and Hanton, L.R., 2012. Antimicrobial properties of a chitosan dextran-based hydrogel for surgical use. Antimicrobial agents and chemotherapy, 56(1), pp. 280–287.
26. Henderson, K.J., Zhou, T.C., Otim, K.J. and Shull, K.R., 2010. Ionically cross-linked triblock copolymer hydrogels with high strength. Macromolecules, 43(14), pp. 6193–6201.
27. Kang, S.I. and Bae, Y.H., 2003. A sulfonamide based glucose-responsive hydrogel with covalently immobilized glucose oxidase and catalase. Journal of Controlled Release, 86(1), pp. 115–121.
28. Harrington, S., Ott, L., Karanu, F., Ramachandran, K. and Stehno-Bittel, L., 2020. A Versatile Microencapsulation Platform for Hyaluronic Acid and Polyethylene Glycol. Tissue Engineering Part A.
29. Qi, M., Gu, Y., Sakata, N., Kim, D., Shirouzu, Y., Yamamoto, C., Hiura, A., Sumi, S. and Inoue, K., 2004. PVA hydrogel sheet macroencapsulation for the bioartificial pancreas. Biomaterials, 25(27), pp. 5885–5892.
30. Zhu, Y., Wang, D., Yao, X., Wang, M., Zhao, Y., Lu, Y., Wang, Z. and Guo, Y., 2021. Biomimetic hybrid scaffold of electrospun silk fibroin and pancreatic decellularized extracellular matrix for islet survival. Journal of Biomaterials Science, Polymer Edition, 32(2), pp. 151–165.
31. Ruhela, A., Kasinathan, G.N., Rath, S.N., Sasikala, M. and Sharma, C.S., 2021. Electrospun freestanding hydrophobic fabric as a potential polymer semi-permeable membrane for islet encapsulation. Materials Science and Engineering: C, 118, p. 111409.
32. Xie, X., Chen, Y., Wang, X., Xu, X., Shen, Y., Aldalbahi, A., Fetz, A.E., Bowlin, G.L., El-Newehy, M. and Mo, X., 2020. Electrospinning nanofiber scaffolds for soft and hard tissue regeneration. Journal of Materials Science & Technology, 59, pp. 243–261.
33. Krishnan, L., Clayton, L.R., Boland, E.D., Reed, R.M., Hoying, J.B. and Williams, S.K., 2011, November. Cellular immunoisolation for islet transplantation by a novel dual porosity electrospun membrane. In Transplantation proceedings (Vol. 43, No. 9, pp. 3256–3261). Elsevier.
34. ZHAN, M.J., LI, G., WEI, Q., CUI, H.L. and LIN, L., 2008. Feasibility of Immunoisolation Based on Porous Titanium. Nanotechnology and Precision Engineering, (5), p.5.
35. Bisceglie, V., 1933. Ueber die antineoplastische Immunitat; Ueber die Wachstumsfaehigkeit der heterologen Geschwuelste in erwachsenen Tieren nach Einpflanzung in Kollodiumsaeckchen. Ztschr Krebsforsch, 40, p. 141.
36. Chang, T.M., 1964. Semipermeable microcapsules. Science, 146(3643), pp. 524–525.
37. Darquy, S.A. and Reach, G., 1985. Immunoisolation of pancreatic B cells by microencapsulation. Diabetologia, 28(10), pp. 776–780.
38. Shu, C. and Jundept, H.E., 1995. The long-term observation on transplantation of encapsulated rat islets into diabetic mice. Chinese Journal of Endocrinology and Metabolism, (02).
39. De Vos, P., Wolters, G.H.J., Fritschy, W.M. and Van Schilfgaarde, R., 1993. Obstacles in the application of microencapsulation in islet transplantation. The International journal of artificial organs, 16(4), pp. 205–212.

40. Wang, T., Lacík, I., Brissová, M., Anilkumar, A.V., Prokop, A., Hunkeler, D., Green, R., Shahrokhi, K. and Powers, A.C., 1997. An encapsulation system for the immunoisolation of pancreatic islets. Nature biotechnology, 15(4), pp. 358–362.
41. David, A., Day, J. and Shikanov, A., 2016. Immunoisolation to prevent tissue graft rejection: Current knowledge and future use. Experimental Biology and Medicine, 241(9), pp. 955–961.
42. Mohammadi, M.R., Rodrigez, S., Cao, R., Alexander, M. and Lakey, J.R., 2018. Immune response to subcutaneous implants of alginate microcapsules. Materials Today: Proceedings, 5(7), pp. 15580–15585.
43. De Vos, P., Van Straaten, J.F., Nieuwenhuizen, A.G., de Groot, M., Ploeg, R.J., De Haan, B.J. and Van Schilfgaarde, R., 1999. Why do microencapsulated islet grafts fail in the absence of fibrotic overgrowth? Diabetes, 48(7), pp. 1381–1388
44. Komatsu, H., Kandeel, F. and Mullen, Y., 2018. Impact of oxygen on pancreatic islet survival. Pancreas, 47(5), p. 533.
45. Clayton, H.A., James, R.F.L. and London, N.J.M., 1993. Islet microencapsulation: a review. Acta diabetologica, 30(4), pp. 181–189.
46. Hartmann, K., Besch, W. and Zühlke, H., 1989. Spontaneous recovery of streptozotocin diabetes in mice. Experimental and Clinical Endocrinology & Diabetes, 93(02/03), pp. 225–230.
47. Goyal, S.N., Reddy, N.M., Patil, K.R., Nakhate, K.T., Ojha, S., Patil, C.R. and Agrawal, Y.O., 2016. Challenges and issues with streptozotocin-induced diabetes – a clinically relevant animal model to understand the diabetes pathogenesis and evaluate therapeutics. Chemico-biological interactions, 244, pp. 49–63.
48. Etuk, E.U., 2010. Animals models for studying diabetes mellitus. Agric Biol JN Am, 1(2), pp. 130–134.
49. Buder, B., Alexander, M., Krishnan, R., Chapman, D.W. and Lakey, J.R., 2013. Encapsulated islet transplantation: strategies and clinical trials. Immune network, 13(6), pp. 235–239.
50. Soon-Shiong, P., Heintz, R.E., Merideth, N., Yao, Q.X., Yao, Z., Zheng, T.I.A.N.L.I., Murphy, M., Moloney, M.K., Schmehl, M. and Harris, M., 1994. Insulin independence in a type 1 diabetic patient after encapsulated islet transplantation. Lancet (London, England), 343(8903), pp. 950–951.
51. Valdés-González, R.A., Dorantes, L.M., Garibay, G.N., Bracho-Blanchet, E., Mendez, A.J., Dávila-Pérez, R., Elliott, R.B., Terán, L. and White, D.J., 2005. Xenotransplantation of porcine neonatal islets of Langerhans and Sertoli cells: a 4-year study. European Journal of Endocrinology, 153(3), pp. 419–427.
52. Calafiore, R., Basta, G., Luca, G., Lemmi, A., Montanucci, M.P., Calabrese, G., Racanicchi, L., Mancuso, F. and Brunetti, P., 2006. Microencapsulated pancreatic islet allografts into nonimmunosuppressed patients with type 1 diabetes: first two cases. Diabetes care, 29(1), pp. 137–138.
53. Basta, G., Montanucci, P., Luca, G., Boselli, C., Noya, G., Barbaro, B., Qi, M., Kinzer, K.P., Oberholzer, J. and Calafiore, R., 2011. Long-term metabolic and immunological follow-up of nonimmunosuppressed patients with type 1 diabetes treated with microencapsulated islet allografts: four cases. Diabetes care, 34(11), pp. 2406–2409.
54. Alsaleh, F.M., Smith, F.J., Keady, S. and Taylor, K.M., 2010. Insulin pumps: from inception to the present and toward the future. Journal of clinical pharmacy and therapeutics, 35(2), pp. 127–138.
55. Ludwig, B., Reichel, A., Steffen, A., Zimerman, B., Schally, A.V., Block, N.L., Colton, C.K., Ludwig, S., Kersting, S., Bonifacio, E. and Solimena, M., 2013. Transplantation of human islets without immunosuppression. Proceedings of the National Academy of Sciences, 110(47), pp. 19054–19058.
56. Carlsson, P.O., Espes, D., Sedigh, A., Rotem, A., Zimerman, B., Grinberg, H., Goldman, T., Barkai, U., Avni, Y., Westermark, G.T. and Carlbom, L., 2018. Transplantation of macroencapsulated human islets within the bioartificial pancreas βAir to patients with type 1 diabetes mellitus. American Journal of Transplantation, 18(7), pp. 1735–1744.
57. Kumagai-Braesch, M., Jacobson, S., Mori, H., Jia, X., Takahashi, T., Wernerson, A., Flodström-Tullberg, M. and Tibell, A., 2013. The TheraCyte™ device protects against islet allograft rejection in immunized hosts. Cell transplantation, 22(7), pp. 1137–1146.

58. Boettler, T., Schneider, D., Cheng, Y., Kadoya, K., Brandon, E.P., Martinson, L. and Von Herrath, M., 2016. Pancreatic tissue transplanted in TheraCyte™ encapsulation devices is protected and prevents hyperglycemia in a mouse model of immune-mediated diabetes. Cell transplantation, 25(3), pp. 609–614.
59. Mridha, A.R., Dargaville, T.R., Dalton, P.D., Carroll, L., Morris, M.B., Vaithilingam, V. and Tuch, B.E., 2022. Prevascularized retrievable hybrid implant to enhance function of subcutaneous encapsulated islets. Tissue Engineering Part A, 28(5–6), pp. 212–224.
60. Tokito, F., Shinohara, M., Maruyama, M., Inamura, K., Nishikawa, M. and Sakai, Y., 2021. High density culture of pancreatic islet-like 3D tissue organized in oxygen-permeable porous scaffolds with external oxygen supply. Journal of bioscience and bioengineering, 131(5), pp. 543–548.
61. Yang, K., O'Cearbhaill, E.D., Liu, S.S., Zhou, A., Chitnis, G.D., Hamilos, A.E., Xu, J., Verma, M.K., Giraldo, J.A., Kudo, Y. and Lee, E.A., 2021. A therapeutic convection–enhanced macroencapsulation device for enhancing β cell viability and insulin secretion. Proceedings of the National Academy of Sciences, 118(37), p.e2101258118.
62. Bock, D.C., Marschilok, A.C., Takeuchi, K.J. and Takeuchi, E.S., 2012. Batteries used to power implantable biomedical devices. Electrochimica acta, 84, pp. 155–164.
63. Fan, X., Shangguan, L., Howard, R., Zhang, Y., Peng, Y., Xiong, J., Ma, Y. and Li, X.Y., 2020, April. Towards flexible wireless charging for medical implants using distributed antenna system. In Proceedings of the 26th annual international conference on mobile computing and networking (pp. 1–15).
64. Bloch, K., Papismedov, E., Yavriyants, K., Vorobeychik, M., Beer, S. and Vardi, P., 2006. Photosynthetic oxygen generator for bioartificial pancreas. Tissue engineering, 12(2), pp. 337–344.
65. Bloch, K., Papismedov, E., Yavriyants, K., Vorobeychik, M., Beer, S. and Vardi, P., 2006. Immobilized microalgal cells as an oxygen supply system for encapsulated pancreatic islets: a feasibility study. Artificial organs, 30(9), pp. 715–718.
66. Singh, S.P. and Singh, P., 2015. Effect of temperature and light on the growth of algae species: A review. Renewable and sustainable energy reviews, 50, pp. 431–444.
67. Metsoviti, M.N., Papapolymerou, G., Karapanagiotidis, I.T. and Katsoulas, N., 2019. Effect of light intensity and quality on growth rate and composition of Chlorella vulgaris. Plants, 9(1), p.31.
68. Mustafa, F.H. and Jaafar, M.S., 2013. Comparison of wavelength-dependent penetration depths of lasers in different types of skin in photodynamic therapy. Indian Journal of Physics, 87, pp. 203–209.
69. Karaś, M., Jakubczyk, A., Szymanowska, U., Złotek, U. and Zielińska, E., 2017. Digestion and bioavailability of bioactive phytochemicals. International Journal of Food Science & Technology, 52(2), pp. 291–305.
70. Pinto, E., Zhang, B., Song, S., Bodor, N., Buchwald, P. and Hochhaus, G., 2010. Feasibility of localized immunosuppression: 2. PLA microspheres for the sustained local delivery of a soft immunosuppressant. Die Pharmazie-An International Journal of Pharmaceutical Sciences, 65(6), pp. 429–435.
71. Kang, J., Erdodi, G., Kennedy, J.P., Chou, H., Lu, L. and Grundfest-Broniatowski, S., 2010. Toward a bioartificial pancreas: diffusion of insulin and IgG across immunoprotective membranes with controlled hydrophilic channel diameters. Macromolecular bioscience, 10(4), pp. 369–377.
72. Orive, G., Emerich, D., Khademhosseini, A., Matsumoto, S., Hernández, R.M., Pedraz, J.L., Desai, T., Calafiore, R. and de Vos, P., 2018. Engineering a clinically translatable bioartificial pancreas to treat type I diabetes. Trends in biotechnology, 36(4), pp. 445–456.
73. Yoshimatsu, G., Sakata, N., Tsuchiya, H., Ishida, M., Motoi, F., Egawa, S., Sumi, S., Goto, M. and Unno, M., 2013, June. Development of polyvinyl alcohol bioartificial pancreas with rat islets and mesenchymal stem cells. In Transplantation Proceedings (Vol. 45, No. 5, pp. 1875-1880). Elsevier.

74. Vieira, S., da Silva Morais, A., Garet, E., Silva-Correia, J., Reis, R.L., Gonzalez-Fernandez, A. and Oliveira, J.M., 2021. Methacrylated gellan gum/poly-l-lysine polyelectrolyte complex beads for cell-based therapies. ACS Biomaterials Science & Engineering, 7(10), pp. 4898–4913.
75. Leung, A., Nielsen, L.K., Trau, M. and Timmins, N.E., 2010. Tissue transplantation by stealth—coherent alginate microcapsules for immunoisolation. Biochemical Engineering Journal, 48(3), pp. 337–347.
76. Kobayashi, T., Aomatsu, Y., Iwata, H., Kin, T., Kanehiro, H., Hisanaga, M., Ko, S., Nagao, M. and Nakajima, Y., 2003. Indefinite islet protection from autoimmune destruction in nonobese diabetic mice by agarose microencapsulation without immunosuppression1. Transplantation, 75(5), pp. 619–625.
77. Reaven, E.P., Gold, G., Walker, W. and Reaven, G.M., 1981. Effect of variations in islet size and shape on glucose-stimulated insulin secretion. Hormone and Metabolic Research, 13(12), pp. 673–674.
78. Hu, T. and Lo, A.C., 2021. Collagen–Alginate Composite Hydrogel: Application in Tissue Engineering and Biomedical Sciences. Polymers, 13(11), p.1852.
79. Kim, J., Kang, K., Drogemuller, C.J., Wallace, G.G. and Coates, P.T., 2019. Bioprinting an artificial pancreas for type 1 diabetes. Current diabetes reports, 19, pp. 1–10.
80. Berney, T., Wassmer, C.H., Lebreton, F., Bellofatto, K., Fonseca, L.M., Bignard, J., Hanna, R., Peloso, A. and Berishvili, E., 2022. From islet of Langerhans transplantation to the bioartificial pancreas. La Presse médicale, 51(4), p.104139.
81. Xu, Y., Song, D. and Wang, X., 2022. 3D Bioprinting for Pancreas Engineering/Manufacturing. Polymers, 14(23), p. 5143.
82. Phuong Le Thi, Yunki Lee, Dieu Linh Tran, Thai Thanh Hoang Thi, Kyung Min Park, Ki Dong Park. Calcium peroxide-mediated in situ formation of multifunctional hydrogels with enhanced mesenchymal stem cell behaviors and antibacterial properties. Journal of Materials Chemistry B 2020, 8 (48) , 11033–11043. https://doi.org/10.1039/D0TB02119A
83. Ke, D., Kengla, C., Lee, S.J., Yoo, J.J., Zhu, X. and Murphy, S.V., 2022. Release Kinetics and In Vitro Characterization of Sodium Percarbonate and Calcium Peroxide to Oxygenate Bioprinted Tissue Models. International Journal of Molecular Sciences, 23(12), p. 6842.
84. Coronel, M.M., Liang, J.P., Li, Y. and Stabler, C.L., 2019. Oxygen generating biomaterial improves the function and efficacy of beta cells within a macroencapsulation device. Biomaterials, 210, pp. 1–11.
85. Coronel, M.M., Geusz, R. and Stabler, C.L., 2017. Mitigating hypoxic stress on pancreatic islets via in situ oxygen generating biomaterial. Biomaterials, 129, pp. 139–151.
86. Nair, N., Guedes, M.S., Werts, C. and Gomes-Solecki, M., 2020. The route of infection with Leptospira interrogans serovar Copenhageni affects the kinetics of bacterial dissemination and kidney colonization. PLoS neglected tropical diseases, 14(1), p.e0007950.
87. Boettler, T., Schneider, D., Cheng, Y., Kadoya, K., Brandon, E.P., Martinson, L. and Von Herrath, M., 2016. Pancreatic tissue transplanted in TheraCyte™ encapsulation devices is protected and prevents hyperglycemia in a mouse model of immune-mediated diabetes. Cell transplantation, 25(3), pp. 609–614.
88. Hwang, P.T., Shah, D.K., Garcia, J.A., Bae, C.Y., Lim, D.J., Huiszoon, R.C., Alexander, G.C. and Jun, H.W., 2016. Progress and challenges of the bioartificial pancreas. Nano Convergence, 3(1), pp. 1–11.
89. Kumagai-Braesch, M., Jacobson, S., Mori, H., Jia, X., Takahashi, T., Wernerson, A., Flodström-Tullberg, M. and Tibell, A., 2013. The TheraCyte™ device protects against islet allograft rejection in immunized hosts. Cell transplantation, 22(7), pp. 1137–1146.
90. McKenzie, A.W., Georgiou, H.M., Zhan, Y., Brady, J.L. and Lew, A.M., 2001. Protection of xenografts by a combination of immunoisolation and a single dose of anti-CD4 antibody. Cell Transplantation, 10(2), pp. 183–193.
91. Loudovaris, T., Mandel, T.E. and Charlton, B., 1996. CD4+ T cell mediated destruction of xenografts within cell-impermeable membranes in the absence of CD8+ T cells and B cells1. Transplantation, 61(12), pp. 1678–1684.

92. Pedraza, E., Coronel, M.M., Fraker, C.A., Ricordi, C. and Stabler, C.L., 2012. Preventing hypoxia-induced cell death in beta cells and islets via hydrolytically activated, oxygen-generating biomaterials. Proceedings of the National Academy of Sciences, 109(11), pp. 4245–4250.
93. Liu, X., Carter, S.S.D., Renes, M.J., Kim, J., Rojas-Canales, D.M., Penko, D., Angus, C., Beirne, S., Drogemuller, C.J., Yue, Z. and Coates, P.T., 2019. Development of a coaxial 3D printing platform for biofabrication of implantable islet-containing constructs. Advanced Healthcare Materials, 8(7), p. 1801181.
94. Marchioli, G., van Gurp, L., Van Krieken, P.P., Stamatialis, D., Engelse, M., Van Blitterswijk, C.A., Karperien, M.B.J., de Koning, E., Alblas, J., Moroni, L. and van Apeldoorn, A.A., 2015. Fabrication of three-dimensional bioplotted hydrogel scaffolds for islets of Langerhans transplantation. Biofabrication, 7(2), p.025009.
95. Hwang, D.G., Jo, Y., Kim, M., Yong, U., Cho, S., Choi, Y.M., Kim, J. and Jang, J., 2021. A 3D bioprinted hybrid encapsulation system for delivery of human pluripotent stem cell-derived pancreatic islet-like aggregates. Biofabrication, 14(1), p.014101.
96. Vaithilingam, V. and Tuch, B.E., 2011. Islet transplantation and encapsulation: an update on recent developments. The review of diabetic studies: RDS, 8(1), p. 51.
97. Bellmann, K., Wenz, A., Radons, J., Burkart, V., Kleemann, R. and Kolb, H., 1995. Heat shock induces resistance in rat pancreatic islet cells against nitric oxide, oxygen radicals and streptozotocin toxicity in vitro. The Journal of clinical investigation, 95(6), pp. 2840–2845.
98. Phelps, E.A., Templeman, K.L., Thulé, P.M. and García, A.J., 2015. Engineered VEGF-releasing PEG–MAL hydrogel for pancreatic islet vascularization. Drug delivery and translational research, 5(2), pp. 125–136.
99. Steiner, D., Lingens, L., Fischer, L., Köhn, K., Detsch, R., Boccaccini, A.R., Fey, T., Greil, P., Weis, C., Beier, J.P. and Horch, R.E., 2018. Encapsulation of mesenchymal stem cells improves vascularization of alginate-based scaffolds. Tissue Engineering Part A, 24(17-18), pp. 1320–1331.
100. Brissova, M., Fowler, M., Wiebe, P., Shostak, A., Shiota, M., Radhika, A., Lin, P.C., Gannon, M. and Powers, A.C., 2004. Intraislet endothelial cells contribute to revascularization of transplanted pancreatic islets. Diabetes, 53(5), pp. 1318–1325.
101. Han, E.X., 2021. Development of a Bioartificial Vascular Pancreas for Type 1 Diabetes Treatment (Doctoral dissertation, Yale University).
102. Shaheen, R., Gurlin, R.E., Gologorsky, R., Blaha, C., Munnangi, P., Santandreu, A., Torres, A., Carnese, P., Nair, G.G., Szot, G. and Fissell, W.H., 2021. Superporous agarose scaffolds for encapsulation of adult human islets and human stem-cell-derived β cells for intravascular bioartificial pancreas applications. Journal of Biomedical Materials Research Part A, 109(12), pp. 2438–2448.
103. Gebe, J.A., Preisinger, A., Gooden, M.D., D'Amico, L.A. and Vernon, R.B., 2018. Local, controlled release in vivo of vascular endothelial growth factor within a subcutaneous scaffolded islet implant reduces early islet necrosis and improves performance of the graft. Cell Transplantation, 27(3), pp. 531–541.
104. Marchioli, G., Luca, A.D., de Koning, E., Engelse, M., Van Blitterswijk, C.A., Karperien, M., Van Apeldoorn, A.A. and Moroni, L., 2016. Hybrid polycaprolactone/alginate scaffolds functionalized with VEGF to promote de novo vessel formation for the transplantation of islets of Langerhans. Advanced healthcare materials, 5(13), pp. 1606–1616.
105. Kasoju, N., Pátíková, A., Wawrzynska, E., Vojtíšková, A., Sedlačík, T., Kumorek, M., Pop-Georgievski, O., Sticová, E., Kříž, J. and Kubies, D., 2020. Bioengineering a pre-vascularized pouch for subsequent islet transplantation using VEGF-loaded polylactide capsules. Biomaterials science, 8(2), pp. 631–647.
106. Wang, R.N. and Rosenberg, L., 1999. Maintenance of beta-cell function and survival following islet isolation requires re-establishment of the islet-matrix relationship. Journal of Endocrinology, 163(2), pp. 181–190.
107. Weber, L.M. and Anseth, K.S., 2008. Hydrogel encapsulation environments functionalized with extracellular matrix interactions increase islet insulin secretion. Matrix Biology, 27(8), pp. 667–673.

108. Llacua, L.A., de Haan, B.J. and de Vos, P., 2018. Laminin and collagen IV inclusion in immunoisolating microcapsules reduces cytokine-mediated cell death in human pancreatic islets. Journal of tissue engineering and regenerative medicine, 12(2), pp. 460–467.
109. Velten, F., Laue, C. and Schrezenmeir, J., 1999. The effect of alginate and hyaluronate on the viability and function of immunoisolated neonatal rat islets. Biomaterials, 20(22), pp. 2161–2167.
110. Tomei, A.A., Manzoli, V., Fraker, C.A., Giraldo, J., Velluto, D., Najjar, M., Pileggi, A., Molano, R.D., Ricordi, C., Stabler, C.L. and Hubbell, J.A., 2014. Device design and materials optimization of conformal coating for islets of Langerhans. Proceedings of the National Academy of Sciences, 111(29), pp. 10514–10519.
111. Stock, A.A., Manzoli, V., De Toni, T., Abreu, M.M., Poh, Y.C., Ye, L., Roose, A., Pagliuca, F.W., Thanos, C., Ricordi, C. and Tomei, A.A., 2020. Conformal Coating of Stem Cell-Derived Islets for β Cell Replacement in Type 1 Diabetes. Stem Cell Reports, 14(1), pp. 91–104.
112. Pagliuca, F.W., Millman, J.R., Gürtler, M., Segel, M., Van Dervort, A., Ryu, J.H., Peterson, Q.P., Greiner, D. and Melton, D.A., 2014. Generation of functional human pancreatic β cells in vitro. Cell, 159(2), pp. 428–439.
113. Dang, T.T., Thai, A.V., Cohen, J., Slosberg, J.E., Siniakowicz, K., Doloff, J.C., Ma, M., Hollister-Lock, J., Tang, K.M., Gu, Z. and Cheng, H., 2013. Enhanced function of immunoisolated islets in diabetes therapy by co-encapsulation with an anti-inflammatory drug. Biomaterials, 34(23), pp. 5792–5801.
114. Sung, S.A., Jo, S.K., Cho, W.Y., Won, N.H. and Kim, H.K., 2007. Reduction of renal fibrosis as a result of liposome encapsulated clodronate induced macrophage depletion after unilateral ureteral obstruction in rats. Nephron Experimental Nephrology, 105(1), pp. e1–e9.
115. Okuno, M., Akita, K., Moriwaki, H., Kawada, N., Ikeda, K., Kaneda, K., Suzuki, Y. and Kojima, S., 2001. Prevention of rat hepatic fibrosis by the protease inhibitor, camostat mesilate, via reduced generation of active TGF-β. Gastroenterology, 120(7), pp. 1784–1800.
116. Meier, R.P., Mahou, R., Morel, P., Meyer, J., Montanari, E., Muller, Y.D., Christofilopoulos, P., Wandrey, C., Gonelle-Gispert, C. and Bühler, L.H., 2015. Microencapsulated human mesenchymal stem cells decrease liver fibrosis in mice. Journal of hepatology, 62(3), pp. 634–641.
117. Vaithilingam, V., Evans, M.D., Lewy, D.M., Bean, P.A., Bal, S. and Tuch, B.E., 2017. Co-encapsulation and co-transplantation of mesenchymal stem cells reduces pericapsular fibrosis and improves encapsulated islet survival and function when allografted. Scientific reports, 7(1), pp. 1–13.
118. Veiseh, O., Doloff, J.C., Ma, M., Vegas, A.J., Tam, H.H., Bader, A.R., Li, J., Langan, E., Wyckoff, J., Loo, W.S. and Jhunjhunwala, S., 2015. Size-and shape-dependent foreign body immune response to materials implanted in rodents and non-human primates. Nature materials, 14(6), pp. 643–651.
119. Li, M., Qiu, L., Hu, W., Deng, X., Xu, H., Cao, Y., Xiao, Z., Peng, L., Johnson, S., Alexey, L. and Kingston, P.A., 2018. Genetically-modified bone mesenchymal stem cells with TGF-β3 improve wound healing and reduce scar tissue formation in a rabbit model. Experimental Cell Research, 367(1), pp. 24–29.
120. Yun, I.S., Jeon, Y.R., Lee, W.J., Lee, J.W., Rah, D.K., Tark, K.C. and Lew, D.H., 2012. Effect of human adipose derived stem cells on scar formation and remodeling in a pig model: a pilot study. Dermatologic surgery, 38(10), pp. 1678–1688.
121. Gamble, A., Pawlick, R., Pepper, A.R., Bruni, A., Adesida, A., Senior, P.A., Korbutt, G.S. and Shapiro, A.J., 2018. Improved islet recovery and efficacy through co-culture and co-transplantation of islets with human adipose-derived mesenchymal stem cells. PLoS One, 13(11), p.e0206449.
122. Karaoz, E., Genc, Z.S., Demircan, P.Ç., Aksoy, A. and Duruksu, G., 2010. Protection of rat pancreatic islet function and viability by coculture with rat bone marrow-derived mesenchymal stem cells. Cell death & disease, 1(4), pp. e36–e36.
123. Mahmoudi, Z., Mohammadnejad, J., Bazaz, S.R., Mehrizi, A.A., Saidijam, M., Dinarvand, R., Warkiani, M.E. and Soleimani, M., 2020. Promoted chondrogenesis of hMCSs with

controlled release of TGF-β3 via microfluidics synthesized alginate nanogels. Carbohydrate Polymers, 229, p.115551.
124. Brennand, K. and Melton, D., 2009. Slow and steady is the key to β-cell replication. Journal of cellular and molecular medicine, 13(3), pp. 472–487.
125. Zhang, Z., Yuan, X. M., Li, L. H., & Xie, F. P. (2001). Transdifferentiation in neoplastic development and its pathological implication. Histology and histopathology, 16(4), 1249–1262.
126. Lu, J., Jaafer, R., Bonnavion, R., Bertolino, P. and Zhang, C.X., 2014. Transdifferentiation of pancreatic α-cells into insulin-secreting cells: From experimental models to underlying mechanisms. World journal of diabetes, 5(6), p. 847.
127. Fraker, C.A., Álvarez, S., Papadopoulos, P., Giraldo, J., Gu, W., Ricordi, C., Inverardi, L. and Domínguez-Bendala, J., 2007. Enhanced oxygenation promotes β-cell differentiation in vitro. Stem Cells, 25(12), pp. 3155–3164
128. Cechin, S., Álvarez-Cubela, S., Giraldo, J.A., Molano, R.D., Villate, S., Ricordi, C., Pileggi, A., Inverardi, L., Fraker, C.A. and Domínguez-Bendala, J., 2014. Influence of in vitro and in vivo oxygen modulation on β cell differentiation from human embryonic stem cells. Stem cells translational medicine, 3(3), pp. 277–289.
129. Nikolova, G., Jabs, N., Konstantinova, I., Domogatskaya, A., Tryggvason, K., Sorokin, L., Fässler, R., Gu, G., Gerber, H.P., Ferrara, N. and Melton, D.A., 2006. The vascular basement membrane: a niche for insulin gene expression and β cell proliferation. Developmental cell, 10(3), pp. 397–405.
130. Guney, M.A., Petersen, C.P., Boustani, A., Duncan, M.R., Gunasekaran, U., Menon, R., Warfield, C., Grotendorst, G.R., Means, A.L., Economides, A.N. and Gannon, M., 2011. Connective tissue growth factor acts within both endothelial cells and β cells to promote proliferation of developing β cells. Proceedings of the National Academy of Sciences, 108(37), pp. 15242–15247.
131. Xu, R., Zhang, Z., Toftdal, M.S., Møller, A.C., Dagnaes-Hansen, F., Dong, M., Thomsen, J.S., Brüel, A. and Chen, M., 2019. Synchronous delivery of hydroxyapatite and connective tissue growth factor derived osteoinductive peptide enhanced osteogenesis. Journal of Controlled Release, 301, pp. 129–139.
132. Li, M., Qiu, L., Hu, W., Deng, X., Xu, H., Cao, Y., Xiao, Z., Peng, L., Johnson, S., Alexey, L. and Kingston, P.A., 2018. Genetically-modified bone mesenchymal stem cells with TGF-β3 improve wound healing and reduce scar tissue formation in a rabbit model. Experimental Cell Research, 367(1), pp. 24–29.
133. Narayanan, K., Lim, V.Y., Shen, J., Tan, Z.W., Rajendran, D., Luo, S.C., Gao, S., Wan, A.C. and Ying, J.Y., 2014. Extracellular matrix-mediated differentiation of human embryonic stem cells: differentiation to insulin-secreting beta cells. Tissue Engineering Part A, 20(1-2), pp. 424–433.
134. Singh, R., Cottle, L., Loudovaris, T., Xiao, D., Yang, P., Thomas, H.E., Kebede, M.A. and Thorn, P., 2021. Enhanced structure and function of human pluripotent stem cell-derived beta-cells cultured on extracellular matrix. Stem Cells Translational Medicine, 10(3), pp. 492–505.
135. Curley, C.J., Dolan, E.B., Otten, M., Hinderer, S., Duffy, G.P. and Murphy, B.P., 2019. An injectable alginate/extra cellular matrix (ECM) hydrogel towards acellular treatment of heart failure. Drug Delivery and Translational Research, 9, pp. 1–13.
136. Zhou, C., Zhou, L., Liu, J., Xu, L., Xu, Z., Chen, Z., Ge, Y., Zhao, F., Wu, R., Wang, X. and Jiang, N., 2020. Kidney extracellular matrix hydrogel enhances therapeutic potential of adipose-derived mesenchymal stem cells for renal ischemia reperfusion injury. Acta biomaterialia, 115, pp. 250–263.
137. Kaur, S., Tripathi, D.M., Venugopal, J.R. and Ramakrishna, S., 2020. Advances in biomaterials for hepatic tissue engineering. Current Opinion in Biomedical Engineering, 13, pp. 190–196.
138. Formo, K., Cho, C.H.H., Vallier, L. and Strand, B.L., 2015. Culture of hESC-derived pancreatic progenitors in alginate-based scaffolds. Journal of Biomedical Materials Research Part A, 103(12), pp. 3717–3726.

139. Wang, D., Zhu, Y., Huang, Y., Zhu, J., Zhu, B., Zhao, Y., Lu, Y., Wang, Z. and Guo, Y., 2021. Pancreatic extracellular matrix/alginate hydrogels provide a supportive microenvironment for insulin-producing cells. ACS Biomaterials Science & Engineering, 7(8), pp. 3793–3805.
140. Guruswamy Damodaran, R. and Vermette, P., 2018. Decellularized pancreas as a native extracellular matrix scaffold for pancreatic islet seeding and culture. Journal of tissue engineering and regenerative medicine, 12(5), pp. 1230–1237.
141. Citro, A., Moser, P.T., Dugnani, E., Rajab, T.K., Ren, X., Evangelista-Leite, D., Charest, J.M., Peloso, A., Podesser, B.K., Manenti, F. and Pellegrini, S., 2019. Biofabrication of a vascularized islet organ for type 1 diabetes. Biomaterials, 199, pp. 40–51.
142. Candiello, J., Singh, S.S., Task, K., Kumta, P.N. and Banerjee, I., 2013. Early differentiation patterning of mouse embryonic stem cells in response to variations in alginate substrate stiffness. Journal of biological engineering, 7(1), pp. 1–14.
143. Richardson, T., Barner, S., Candiello, J., Kumta, P.N. and Banerjee, I., 2016. Capsule stiffness regulates the efficiency of pancreatic differentiation of human embryonic stem cells. Acta biomaterialia, 35, pp. 153–165.

Islet Macroencapsulation: Strategies to Boost Islet Graft Oxygenation

Barbara Ludwig, Carolin Heller, Victoria Sarangova, and Petra B. Welzel

1 Introduction

Biological beta cell replacement represents a valid therapeutic approach to attain the desirable metabolic control in patients with insulin-deficient diabetes [10, 58, 59, 69, 88, 112]. Improvements in islet isolation and culture protocols as well as tailored immunosuppressive and anti-inflammatory regimen have positively affected clinical outcome [89]. However, unsatisfactory long-term graft survival due to

All authors contributed equally to this manuscript.

B. Ludwig (✉)
Department of Medicine III, University Hospital Carl Gustav Carus, Technische Universität Dresden, Dresden, Germany

Paul Langerhans Institute Dresden (PLID) of the Helmholtz Center Munich at the University Hospital Carl Gustav Carus, Technische Universität Dresden, Dresden, Germany

German Center for Diabetes Research (DZD e. V.), Neuherberg, Germany

CRTD/DFG-Center for Regenerative Therapies, Technische Universität Dresden, Dresden, Germany
e-mail: Barbara.ludwig@uniklinikum-dresden.de

C. Heller
Department of Medicine III, University Hospital Carl Gustav Carus, Technische Universität Dresden, Dresden, Germany

Paul Langerhans Institute Dresden (PLID) of the Helmholtz Center Munich at the University Hospital Carl Gustav Carus, Technische Universität Dresden, Dresden, Germany

German Center for Diabetes Research (DZD e. V.), Neuherberg, Germany

Leibniz-Institut für Polymerforschung Dresden e.V., Max Bergmann Center of Biomaterials Dresden, Dresden, Germany

L. Piemonti et al. (eds.), *Pluripotent Stem Cell Therapy for Diabetes*,
https://doi.org/10.1007/978-3-031-41943-0_12

allo- and autoimmune reactions and engraftment inefficiency prevent sustained therapeutic effects [118, 122, 124]. Partial or total loss of the islet grafts occurs in more than 50% of patients already after 1 year, particularly within the early post-transplantation phase and mainly attributed to hypoxia, inflammation, and immune rejection [99, 118, 122, 124]. In addition, the morbidity associated with systemic immunosuppression including elevated risk for opportunistic infections and cancer significantly limits the indication for islet transplantation to a subset of critically vulnerable patients [33, 68, 118, 124]. Moreover, islet transplantation is restricted by a major donor shortage, since often more than one human pancreas is needed to achieve a sufficient islet mass [11, 145].

The encapsulation of cells is a concept with a long history that was introduced decades ago [119]. The fundamental strategy of encapsulation is to create a selectively permeable barrier that allows the exchange of oxygen, nutrients and effector molecules, while preventing the migration of immune cells and cytotoxic molecules that could potentially harm the transplanted cells. The recent developments in beta cell replacement, mainly the availability of alternative cell sources, such as insulin-producing cells derived from pluripotent stem cells or from genetically modified xenogeneic donors, generated a renewed attention for macroencapsulation strategies as a safeguard to control the potential risks of these cells such as dedifferentiation of stem cell progenitors or virus transmission from xenogeneic islets [1, 39, 40]. A variety of systems have been reported that address the critical constraint of oxygen supply in cell-based implants by direct oxygenation strategies, fostering revascularization, and enhancing diffusion properties. In native islets, the interaction with the signaling molecules of the extracellular matrix (ECM) helps to define islet morphology, physico-mechanical properties and function [81]. Therefore, recapitulation of the islet native niche has significant importance in encapsulation strategies and may reduce inflammatory reactions and promote long-term efficiency of cell-based implants. Moreover, the implementation of these methodologies, smart device surface design, potentially in combination with local immunosuppression and immunomodulation might allow for transplantation in the absence of systemic immunosuppression and thereby significantly widen the cohort of transplant candidates [12, 40, 83].

Over the last decades, macroencapsulation devices have been proven successful in insulin-deficient animal models by restoring insulin independence and normoglycemia [39]. However, the translation to clinical application has been challenging. Various device systems are being explored in phase I/II clinical trials [16, 21,

V. Sarangova
Department of Medicine III, University Hospital Carl Gustav Carus, Technische Universität Dresden, Dresden, Germany

Leibniz-Institut für Polymerforschung Dresden e.V., Max Bergmann Center of Biomaterials Dresden, Dresden, Germany

P. B. Welzel (✉)
Leibniz-Institut für Polymerforschung Dresden e.V., Max Bergmann Center of Biomaterials Dresden, Dresden, Germany
e-mail: welzel@ipfdd.de

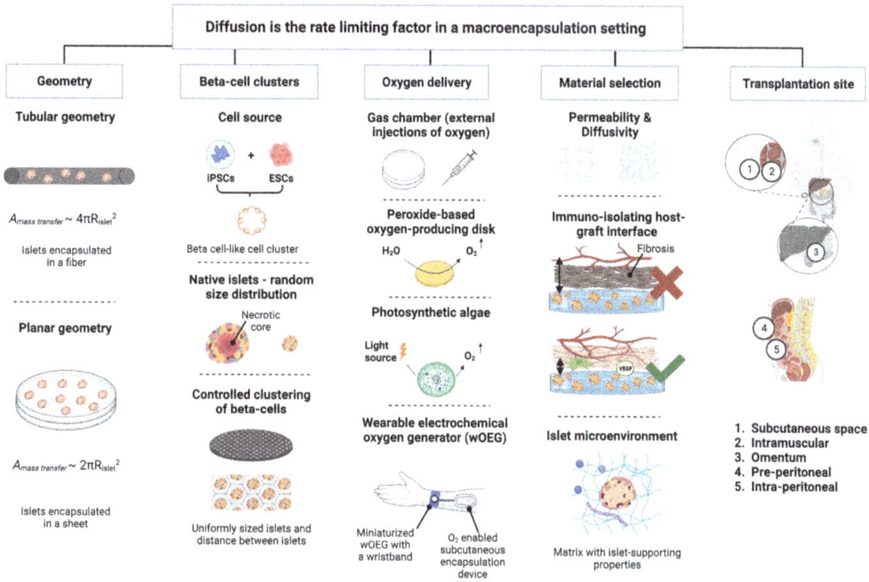

Fig. 1 Interrelated design considerations for continuous and sufficient oxygen supply in macroencapsulation approaches. In an ideal encapsulation system, the device geometry, cell cluster characteristics and arrangement consider diffusion characteristics, implement strategies for bridging the initial hypoxic post-transplantation phase and foster long-term vascularization

83, 115] but still suffer from inadequate transport of effector molecules to and from the implant and the serious issue of insufficient oxygenation particularly in the period between implantation and vascularization of the graft is not resolved [67, 78].

This chapter aims to discuss key aspects of macroencapsulation strategies (Fig. 1), their significant potential for advancing novel transplantation approaches and the natural limitations as diffusion-based systems. A particular focus is on oxygenation of the islet graft, the most critical issue among the numerous challenges of a device for achieving the level of a clinical therapy.

2 Diffusion Is the Limiting Factor in a Macroencapsulation Setting

The distance of the encapsulated cells to the next blood capillary determines the diffusion time of oxygen, glucose, hormones and signalling molecules to reach these cells or to be released from the implant, respectively. This distance has to be covered in different compartments with possibly different properties. Furthermore, passive transport of the relevant molecules is dependent on their concentration gradients between the surrounding of the device and the location of the cells inside the device. Diffusion is not a directed process, but depends on the random movement of

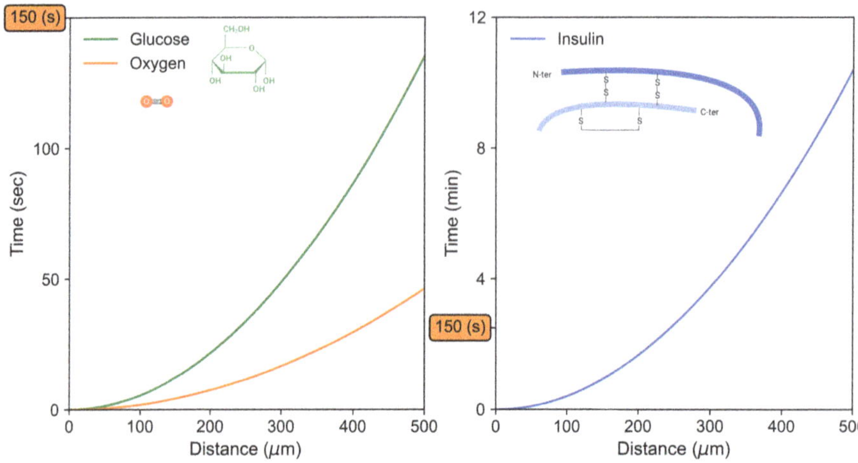

Fig. 2 Exponential increase of molecular diffusion time according to prolonged distance of glucose and oxygen (left) and insulin (right). Diffusion time is calculated based on $t \sim x^2 / 2D$ with t as time, x as distance, and D as diffusion coefficient for the respective molecules in water at 37 °C. The diffusivity of insulin in alginate beads is reported to be 48% lower [77, 138]

molecules along the concentration gradient from high to low concentration. The time of diffusion for a given molecule grows exponentially with the distance it has to overcome (Fig. 2).

Consequently, the distances that need to be covered by the relevant molecules to diffuse to or away from the islets must be kept at a minimum. On the outside of the device, capillaries have to grow as close as possible to the device surface without interfering with fibrotic tissue (Sect. 6). Within the device, the islets must be placed intimate to the host-graft interface. Moreover, islet diameter and inter-islet spacing are critical due to competition in oxygen consumption (Sect. 4).

3 Macroscale Device Design

Macroscopic systems for beta cell replacement combine the total cell volume in a single, defined container/scaffold that ideally can be easily transplanted and retrieved. There are two conceptual strategies, open and closed systems. An open device allows blood vessels to enter the device and directly interact with the implanted cells. The direct vascularization of the implanted cells is intended to allow for robust and consistent engraftment but necessitates the use of immune suppression therapy because the implanted cells are not hidden from the immune system. A closed device is designed to prevent immune cells from directly contacting the implanted cells and provoking an immune response. The interexchange of nutrients, oxygen, and hormones is exclusively depending on diffusion.

Generally, there are two main design concepts regarding the geometry of islet macroencapsulation devices: a planar device geometry (designed as a planar diffusion chamber or simple slab) or a tubular geometry in which the islets are integrated. Most device types are designed for extravascular implantation. These devices are associated with only minor implantation complications, but most preclinical and clinical trials were compromised by the insufficient oxygen and nutrients diffusion into the core of the device and to the cells [5–7]. However, in these early studies important measures for further encapsulation development such as biocompatibility, host cell membrane overgrowth, and prevention of allograft rejection were established. Some approaches were inspired by the development of hollow fibre technology for renal dialysis. Islets were integrated in hollow fibres and used as diffusion devices [63, 92, 148]. With this strategy, first non-curative preclinical studies in rodents were successfully performed and could demonstrate the recovery of viable and functional human islets after several weeks of implantation [85, 86]. In theory, a tubular geometry offers superior characteristics in terms of mass transfer across the membrane due to the high surface area-to-volume ratio, but poses practical issues. The main limitation of this clever technology is the low packing density of the device types that prevents transplantation of potentially curative islet doses. A fibre containing a relevant islet mass would need to be 12.8 m long to fit a curative islet mass, unlikely conceivable in a clinical setting [46]. An advantage of solely hydrogel-based devices are their biocompatibility and facile mass transfer, but they typically exhibit low mechanical strength [79]. Yet innovative concepts for strengthening hydrogel fibres for macroencapsulation are being explored [79, 143]. An example is the TRAFFIC encapsulation device as described by An et al., a thread-reinforced alginate fibre, which showed great potential in maintaining cell viability and reversing diabetes in small and large animal models [8].

According to novel computational tools to evaluate and optimize bioartificial pancreas device geometry, the most widely used macroencapsulation device technologies of the robust planar geometry type are suitable and adaptable, provoking less necrosis and functional cell loss and, therefore, reduced cell dose [46]. They consist of encapsulated islets immobilized in flat slabs and/or a sealed flat chamber. This configuration is believed to provide superior stability and in addition allow for facile retrieval. However, these three-dimensional (3D) structures typically lead to insufficient oxygenation, in particular of larger islets clusters, as not enough oxygen can be delivered to oxygenate the core of the clusters [22]. Various strategies address this issue by accelerating vascularization such as incorporating specific design features, local pro-angiogenic agent delivery, and pre-vascularization (Sect. 6.1). Still, these approaches require days to weeks to generate a functional vascular network or repeated surgery [117].

Another alternative approach are intravascular devices with great advantages regarding oxygen and nutrient supply. They consist of hollow fibres or tubes filled with cell clusters and fixed to the recipient's vascular system. The "Hockey Puck" device perfused with arterial blood was demonstrated successful in curative large animal models [70–72]. However, disastrous bleeding complications in the recipient animals (disconnection of the carotid artery cannulae to the device) stopped this and other approaches from further development and clinical exploration. Given the recent technical improvements, intravascular devices might justify a second chance.

There have been several approaches that address the diffusion of molecules within a cell-based implant. In bioprinting approaches islets have been embedded in alginate and printed into scaffolds with macropores. This arrangement increases the available surface area for diffusion [44, 45]. The SONIC device developed at Cornell University mimics the trachea of insects for the distribution of oxygen. A skeletal structure made of a poly(vinylidene fluoride-*co*-hexafluoropropylene) (PVDF-HFP) polymer creates a network with a high oxygen diffusivity. This skeleton distributes oxygen quickly to the islets in the interstitial lumina. However, these approaches do not include a component for mechanical strength and do pose risks for mechanical stress damage to the hydrogel material.

A further approach is a convection-based encapsulation device developed at Massachusetts Institute of Technology [147]. This device setup is built around a fibre with a continuous flow of liquid. The flow carries away molecules via convection, which increases the concentration gradient and thereby reduces diffusion times within the device.

In order to strengthen the numerous research approaches, computational modelling is a desirable method. Computational platforms allow to investigate the therapeutic capacity of bioartificial pancreas devices and specifically address the constraints of mass transfer within. Most recently, Ernst et al. presented a computational tool for investigating the therapeutic potential of generic and user-programmable cell-based implants, accounting for relevant stochastic properties including size distribution and random localization of the cells within the device [46]. Such open-source tools might tremendously enhance efficiency and prevent repeated failure in this competitive field.

4 Cluster Size, Arrangement and Cell Source

Acknowledging that diffusion properties, oxygen supply and mass transport capacity are major determinants for a successful encapsulation concept, special attention must be paid to both the individual properties of islet clusters and their arrangement to each other. In addition, the particular characteristics and needs of different islet sources in terms of oxygen demand and secretion dynamics must be considered leading to specifically tailored conceptual and design approaches for an optimal encapsulation setting.

4.1 Significance of Homogenous and Small Cell Clusters

In a macroencapsulation setting and independent of the islet source, it is essential to use intact cell clusters with high purity and ideally minimal size distribution in order to reduce oxygen demand, maximize applicable density and obtain optimal diffusion characteristics. In fact, there are multiple indications that smaller pancreatic islets, species-independent, are superior to larger ones in terms of survival and insulin secretion due to reduced susceptibility to ischemia [60, 75]. Using theoretical modelling to identify the optimal size for islets, based on oxygen consumption rate and nutrient diffusion throughout the islets, resulted in a diameter of 100 μm [60]. In particular, within a macroencapsulation device, where oxygen consumption rate and diffusion characteristics are essential for function, the generation of islets with optimal dimensions may have significant beneficial effects on islet survival after transplantation. Moreover, immunogenicity in the transplant setting seems to be attenuated in reaggregated islet clusters [126]. It is hypothesized that hypoxia and core necrosis of islets result in the release of danger-associated molecular patterns (DAMPs) and subsequent damage to neighboring islets from DAMP-induced immune activation [37]. For native islets, the strategy of islet dissociation and reaggregation to form size-defined homogeneous islet preparations represents a powerful tool to advance islet encapsulation (Ludwig/Heller et al., "unpublished"). In the context of stem cell-derived islet clusters, it is an obvious advantage and compelling opportunity to control for islet size using microwell plate technologies already in the late stages of differentiation. The advantage of homogenous islet preparations is maximally exploited by placing them in defined arrangement to each other. Conversely, uniform clusters are also a mandatory prerequisite for the production of defined micropatterns of islets within a device (Fig. 3) (Sect. 4.2).

Commercially manufactured microwell-culture plates for the production of cell clusters are available, though the geometry (e.g. well diameter and distance between wells) is fixed and they are not designed to be placed inside a macroencapsulation device. Generally, they only enable a low packaging density of islets with microwells of about 400–500 μm in diameter.

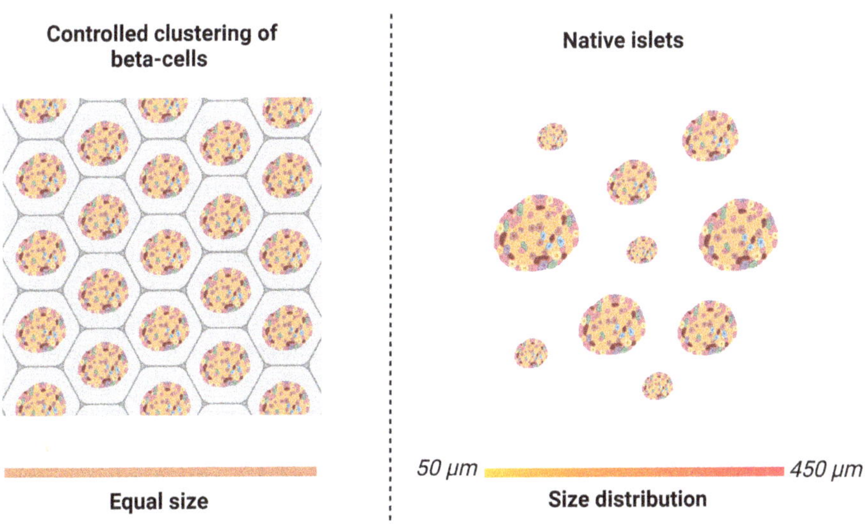

Fig. 3 Significance of homogenous cell clusters. In an encapsulation setting, cell clusters of defined and optimal size are a crucial prerequisite for their optimal arrangement and consequently oxygen supply within the device

For the fabrication of tailored micropatterns, different techniques from photo and soft lithography are available. These enable the production of high-resolution structures in large quantities. To create three-dimensional microstructures, replica moulding, embossing and solvent-assisted micromoulding techniques can be used [109]. Different factors must be considered for the production of microwell-plates that should be used within an encapsulation device. (1) The method of microwell array production must be scalable up to the clinical needs. The production method therefore needs to be fast and reproducible and the material must be easily available in large quantities. (2) The structures must be sterilizable during the production process. (3) The material used must be non-toxic to the cells, durable, favourably be soft and of low swelling characteristics to avoid a change of dimensions of the wells.

Various materials have been used for the creation of user-defined microwells. For example, poly(ethylene glycol) (PEG) has been cross-linked on PDMS stamps to produce 75 μm small microwells for retinal organoid culture [38]. Furthermore, a commonly used material for islet cell clustering is agarose. It has the advantage that it is a soft material as well and relatively easily moldable by a short heating process [60, 91]. PDMS has been used as a basis for microwell culture as well. For example, Tu et al. cultured pseudo islets in concave PDMS microwells [137] or Rojas-Canales et al. used a microwell-based platform fabricated from PDMS for shipment of islets [116].

4.2 Significance of Islet Arrangement: Microscale Device Design

Enduring oxygen supply at the site of implant is essential to mitigate hypoxia-induced cell death and dysfunction, particularly for highly metabolically active cells such as pancreatic islet clusters. Besides the structure and configuration of individual islet clusters, their arrangement within the islet module is thereby of crucial relevance.

Albeit previously often neglected, the microarchitecture within the implant to control cell arrangement is of major significance. One of the critical issues in many encapsulation devices is the random arrangement of cell clusters of variable size in varying distance to the device surface with high impact on oxygen supply, but also on metabolic responsiveness [46, 127]. Important design criteria for an optimized device should therefore focus on minimizing the distance of the islet clusters to the device surface and thus to the surrounding vasculature (Fig. 4) and on arranging the clusters as dense as possible in a regular manner without preventing sufficient oxygen supply due to competition.

A preferential solution is the regular arrangement of the islets in a monolayer, supported by e.g. a hydrogel array of microwells (Fig. 4). In a monolayer arrangement, the mass transport of oxygen and nutrients towards the islets takes place from the area above and/or below them. The oxygen gradient establishes from the implant

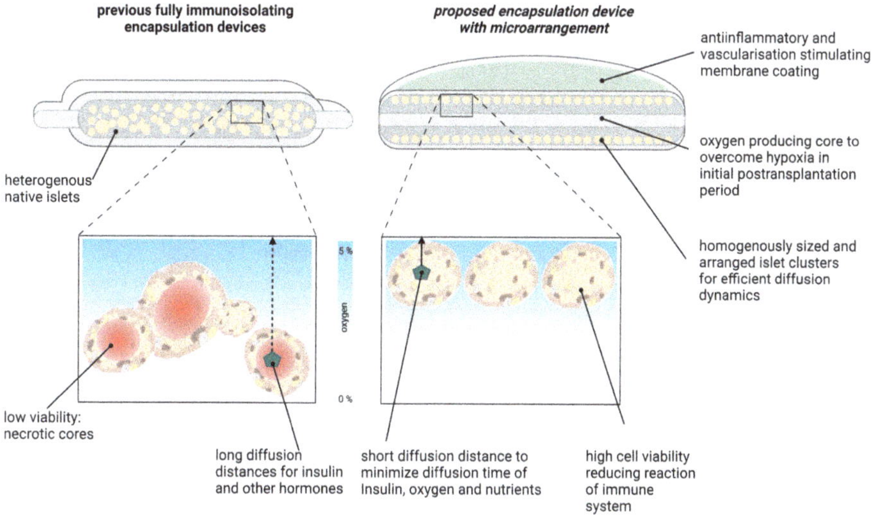

Fig. 4 Microarchitecture and arrangement of islet clusters within a macroencapsulation device as major prerequisite for islet survival and device efficiency

surface, which is the oxygen source towards the oxygen consuming cells, which are the oxygen sink. The surface area above an islet therefore determines the area from which oxygen can diffuse towards the islet (oxygen diffusive area). Accordingly, the larger the distance between two neighbouring islets, the larger is the area from which oxygen and nutrients diffuse towards them without being consumed by the neighbouring islet. Therefore, for the arrangement of islets in a monolayer within a device of a suitable size for clinical application, an optimum between large oxygen diffusion surface area and high islet number per area must be established (Fig. 5).

Some customized strategies for micropatterned device systems have been presented and proven successfully in preclinical settings [20, 127, 128].

In view of clinical translation, inevitable scalability, reproducibility, and material stability are essential and yet challenging requirements.

4.3 *Significance of the Islet Source*

Among the various considerations to enhance diffusion properties, optimize oxygen supply and promote mass transport, the actual islet source is of critical relevance and tailored approaches are mandatory. Stem cell-sourced implants pose additional safety concerns due to their potential for non-terminal differentiation; thus, the use of macroencapsulation devices made of customized functional biomaterials aims at ensuring site retention and graft retrieval. Moreover, stem cell-derived insulin-producing cells have been shown to undergo phenotypic maturation when

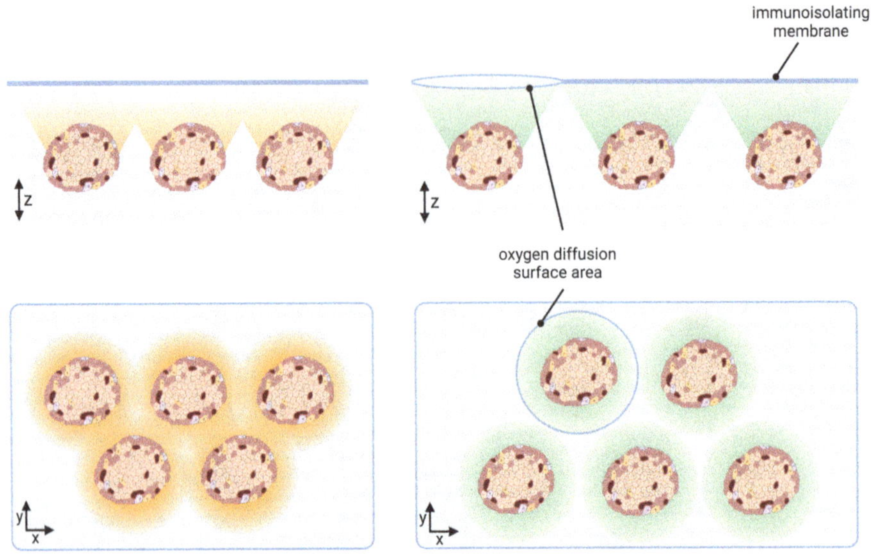

Fig. 5 Oxygen diffusion surface area of islets. The oxygen diffusion surface area is dependent on the packing density of islets in the *x*- and *y*-dimension. Close packaging leads to competition for oxygen (left, orange), while wider spacing leaves enough oxygen to single islets (right, green)

transplanted in vivo [135]. In addition to a maturing cell cargo, also the peri-implant environment is dynamic ranging from implant vascularization to fibrotic encapsulation, thereby critically determining fate and functionality of the transplanted cells (Sect. 6). The oxygen demand depends and varies strongly among different cell sources as well. Ernst and co-workers reported measurements of maximum basal oxygen consumption rates vary from 133.6 nmol O_2 min^{-1} mg DNA^{-1} in human islets to 225.9 nmol O_2 min^{-1} mg DNA^{-1} in stem cell-derived beta cells to 346.6 nmol O_2 min^{-1} mg DNA^{-1} in rat islets [46]. The oxygen demand of stem cell-derived beta cells is 69% higher than that of human islets. This high heterogeneity emphasizes the relevance of appropriate experimental design during the developmental process of encapsulation strategies and the necessity of customized device characteristics based on the intended cell source.

There are also specific approaches to improve islet hypoxia resistance through genetic modification and co-implantation of "helper cells". In particular, overexpression of genes involved in promoting cell survival and adaptation to low oxygen levels, such as hypoxia-inducible factor-1 alpha (HIF-1α) and erythropoietin (EPO), as well as co-implantation of human amniotic epithelial cells (hAECs) has shown promise. For example, Fenjves et al. used adeno-associated virus (AAV) to overexpress EPO in human islets, leading to increased islet viability and function in hypoxic conditions [51]. Cheng et al. showed that overexpression of the hypoxia-inducible factor 1-alpha (HIF-1α) in islets improved their survival and function under hypoxic conditions [26]. Another study explored the capacity of hAECs to enhance the viability and efficacy of transplanted islets in a murine diabetes model.

The findings showed that co-implantation of hAECs with islets significantly improved islet survival and vascularization, leading to better glycemia and insulin production [73]. Overall, these studies suggest that genetic modification of islets to improve hypoxia resistance and co-implantation with "helper-cells" may be a promising approach particularly for islets intended for encapsulation.

5 Oxygen Delivery Methods

The high oxygen demand of the islet graft is one of the main limitations of islet macroencapsulation approaches. Due to the inevitable delay of angiogenesis in the early post-transplantation period, encapsulated islets face a challenge to sustain their viability and functionality potency under hypoxic conditions [29, 128] (Fig. 6). Excessive efforts have been made to introduce different strategies in the past decades to approach the problem of hypoxia [50, 103]. An uninterrupted continuous sustained delivery of oxygen is necessary for the viability of islets in the post-transplantation period (Fig. 1).

Common strategies for increased oxygenation of the cell graft include gas reservoirs, oxygen-transporting materials (perfluorocarbon- and haemoglobin-based materials), electrochemical oxygen generators, or oxygen-generating biomaterials [103]. The gas reservoir approach is appealing due to its pragmatism but challenged by clinical scalability and the daily expenditure on the patient side [21, 82, 83]. Fully implantable electrochemical oxygen generators are an attractive alternative that have been introduced decades ago [146]. This system avoids the issue of refilling and improves long-term functionality of a cell-based implant. The concept is based on continuous production of oxygen from water electrolysis [103]. This

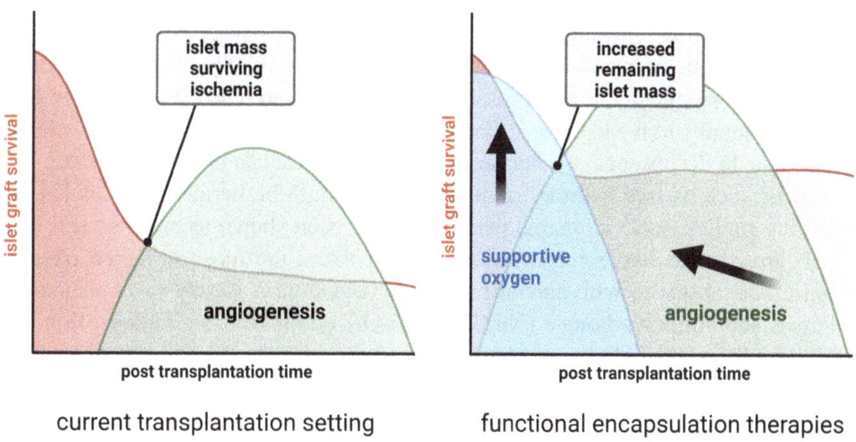

Fig. 6 Islet graft survival is depending on oxygen. In many current transplant settings a large amount of the islet mass is lost due to the lack of oxygen. Oxygen in the early post-transplantation phase can support the islets until sufficient revascularization took place to fully supply the graft

battery-powered and water-refillable technology could be integrated in macroencapsulation devices as currently explored by Procyon Technologies LLC in collaboration with Novo Nordisk (Procyon Technologies LLC [108]). In the current version the patient has to carry a wristband, which may affect flexibility [32]. A similar approach is taken by Giner Life Sciences for production and supplementation of oxygen to cell-based implants via electrochemical oxygen generators (Giner Life Sciences [55]). Other approaches explore the use of algae for local photosynthesis [25, 48, 62]. However, these technologies have certain downsides, such as the limited transplantation site of the device to skin or subcutaneous tissues, difficulties in achieving regulatory approval due to complex biological techniques and procedures, and low viability of the islet graft [48].

5.1 Oxygen-Generating Materials

In situ oxygen-generating materials (OGM) exhibit various advantages of flexible size and minimal patient management. Typical OGMs are liquid peroxides (hydrogen peroxide) and solid inorganic peroxides or percarbonates (calcium peroxide, magnesium peroxide, and sodium percarbonate). When interacting with water, OGMs can produce and deliver oxygen (O_2) by means of a chemical reaction. For instance, calcium peroxide forms calcium hydroxide and hydrogen peroxide, which decomposes into oxygen and water. The rate of oxygen release depends on a set of different environmental factors such as temperature, pH, purity and solubility of the peroxide, and the presence of catalysts. Notably, the rate of oxygen release might be significantly affected by the formation of metal hydroxides, which increases the pH of the microenvironment. Therefore, it is necessary to use additional buffer systems in order to maintain the pH balance of the system [54].

Liquid hydrogen peroxide (H_2O_2) as used in various biomedical applications provides high solubility in water, which contributes to faster oxygen release [120]. The main mediator in the reaction of hydrolytic decomposition of hydrogen peroxide is catalase, an enzyme that is also present in human liver and blood [96]. Notably, this enzyme contains iron and organic functional groups facilitating the oxygen-release properties. In the absence of catalase, hydrogen peroxide can produce cytotoxic by-products, such as free hydroxyl radicals, which might be harmful to cells [2]. In previous studies, solid inorganic peroxides have been shown to promote cell survival, growth and tissue regeneration. Among them, the most sustained oxygen release was observed with calcium peroxide (CaO_2) in a variety of applications, whereas sodium percarbonate ($Na_2CO_3 \cdot 1.5H_2O_2$) established the higher solubility and more biocompatible byproducts. Magnesium peroxide (MgO_2) was characterized by the slowest oxygen formation, due to its relatively low solubility coefficient [30].

Nevertheless, the main disadvantage of using OGMs is the relatively fast oxygen production at the initial stages, in particular for beta cells that are sensitive to high oxygen tension and free radical stress. In order to prevent initial burst oxygen release mediated by the direct contact of an OGM with water [17], the use of carrier materials is required. Integrating OGMs into scaffolds, nanofibers, or microspheres enables defined and sustained oxygen release to the cells [96, 101, 142].

Polymeric materials such as polylactic acid (PLA), poly(lactic-co-glycolic acid) (PLGA), ε-polycaprolactone (PCL), polyurethane (PU), and poly(*N*-vinylpyrrolidone) (PVP) were used as such carrier materials to embed OGMs in order to supply high oxygen demanding cell transplants or to support ischemic tissues as for instance summarized by Ashammakhi et al. [9]. The combination of hydrogels such as gellan gum, alginate, and gelatin methacryloyl (GelMA) polymer networks with calcium peroxide also yielded composite materials capable of generating oxygen for several days in hypoxic conditions [4, 90, 93, 98]. Colombani et al. proposed a hyaluronic acid-based cryogel system as a potential carrier material for calcium peroxide [28]. The material that was additionally conjugated to catalase to boost local oxygen release while minimizing cytotoxicity was tested with melanoma cells; however, the system might be adapted for application in islet macroencapsulation strategies as well.

To bridge the oxygen demand of islet clusters protected in a macroencapsulation device in the early post-transplantation period, the use of biodegradable and hydrophilic polymers for OGMs embedding may not be sufficient due to the often still rather fast oxygen release. To address this, non-degradable hydrophobic polymers like PDMS have been suggested as carrier materials to provide implant stability. In 2012, Pedraza et al. demonstrated in vitro that the oxygenation of islets could be maintained for more than 3 weeks under hypoxic conditions by using a disk-shaped PDMS scaffold mixed with calcium peroxide to generate oxygen [105] (Fig. 7).

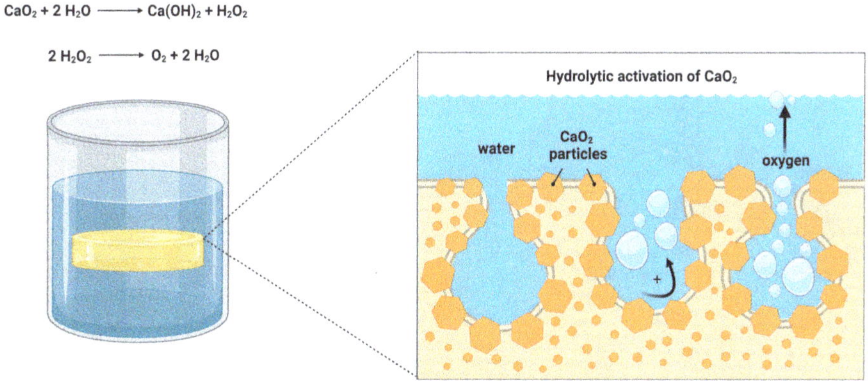

Fig. 7 Oxygen generation from a CaO_2-based composite material. Water diffuses into the carrier material resulting in hydrolysis of CaO_2 with subsequent local release of oxygen for cellular consumption. (Modified from Coronel et al. [30])

5.2 Applications in Islet Transplantation

Various strategies for delivering oxygen to macroencapsulated islets have been reported and some have shown promising results. The "β Air" device from Beta-O_2 Technologies Ltd. is a prominent example of an encapsulation device with a refillable intracapsular supportive oxygen chamber. This extravascular planar device consists of immunoisolating membranes and the alginate-embedded islet graft and additionally a gas chamber, in which oxygen periodically is injected using an extracorporeal pump [21, 82, 83]. The schematic overview of the device is depicted in Fig. 8.

To address the issue of external oxygen supply, several studies have developed internal oxygen-generating biomaterials. As already mentioned in Sect. 5.1, peroxides embedded in polymers are quite attractive. For instance, Lee et al. showed that calcium peroxide-PDMS scaffolds can provide sustained oxygen release for more than 24 hours in vitro [74]. Similarly, Coronel et al. established an oxygen generating device (OxySite), which provided higher transplantation efficiency and strong pancreatic intra-islet vascularization under hypoxic conditions. The study demonstrated a positive effect of OxySite on islet cell viability, supporting aerobic metabolism, and reducing the generation of inflammatory cytokines [29]. Another potential oxygen-generating source for islet cell transplantation was described by McQuilling et al. The established devices represent sodium percarbonate and calcium peroxide particles encapsulated in alginate, accordingly. Both oxygen-generating compounds showed prolonged oxygen release in the early post-transplantation period, which lasted for around 7 days, demonstrating their potential to improve islet cell viability. It has been shown that the use of higher concentrations of sodium percarbonate led

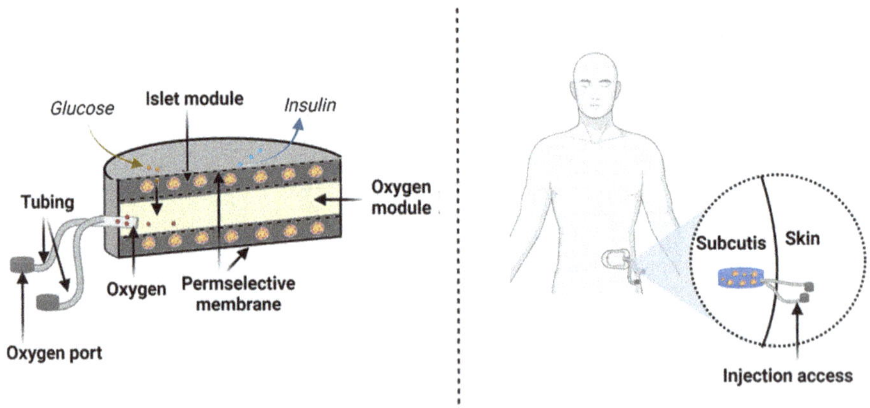

Fig. 8 Oxygen supply in the "β Air" device. The core of the device contains a gas chamber refillable with oxygen via subcutaneous access ports. Two compartments housing the pancreatic islets immobilized in alginate are surrounding the gas reservoir. The plastic housing of the chamber gives mechanical stability while an alginate coated hydrophilic poly(tetrafluoroethylene) (PTFE) membrane sets up the immunoisolating barrier

to damage of islet cells due to formation of free radicals, probably induced by the higher sodium percarbonate solubility compared to calcium peroxide [90].

An in situ gelling hydrogel system designed to release oxygen, consisting of thiolated hyaluronic acid, 8-arm poly(ethylene glycol) acrylate, and calcium peroxide, was found to provide oxygen for a minimum of 30 hours. The system also maintained the viability and health of encapsulated insulin-secreting insulinoma INS-1E cell clusters for a period of 3 days of culture, demonstrating its potential for injection-based cell administration [133].

6 Material Selection

Important design parameters that can influence the delivery of nutrients and oxygen to the islets as well are the physical (mechanical and microstructural) and biochemical properties of the materials used for macroencapsulation [15, 32, 56, 102, 110] (Fig. 1). These are on the one hand side the embedding materials in contact with the transplanted islets (intracapsular islet microenvironment) and on the other hand side the materials in contact with the surrounding tissue (device surface) – that often also give structural support and/or act as permselective element (immunoisolating host-graft interface). Depending on the device design, the material(s) used inside the capsule and at the capsule surface are equal or different in biochemical and/or physical properties.

The macroencapsulation device as a foreign material elicits non-specific inflammatory responses at the host-graft interface with activation of the complement system and fibrous encapsulation of the device. Thus, the delivery of oxygen and nutrients to the islets is hindered or completely inhibited. The induced stress might increase the additional secretion of cytokines and release of DAMPs, resulting in a vicious cycle precipitating a broad inflammatory process [41, 104].

Recent advances in biomaterial research and technologies were applied to address these problems. Material research in nano- and microencapsulation approaches supported and inspired the improvement of macroencapsulation devices. However, there is still a large need to design novel customized devices and membranes that recapitulate key properties of the natural pancreatic environment and promote islet cell engraftment and long-term survival and function in an extrahepatic site. Ideal materials should provide mechanical protection and a supportive extracellular environment as a desirable niche for the islets, be biocompatible and additionally allow for tailored biofunctionalization.

6.1 The Immunoisolating Host-Graft Interface

Although islet transplantation has demonstrated its potential in treating type 1 diabetes, the chronic need for immunosuppression to establish graft acceptance is a major limitation. Transplantation of allogeneic, xenogeneic or stem cell-derived beta cell grafts induces complex interactions between the foreign cells and the recipient that involve non-immunological factors as well as innate and adaptive immunity. These reactions can cause rapid or gradual functional graft loss. It is assumed that the rejection of unencapsulated allogeneic islets is primarily driven by direct T-cell recognition of foreign antigens, which requires direct contact of cytotoxic T-cells and the islet graft. Therefore, macroencapsulation of islets may enable the transplantation of pancreatic islets in the absence of systemic immunosuppression by protecting the graft through a mechanical barrier [18]. The "β Air" device is an example: it has shown to enable transplantation of porcine islets in diabetic primates without any immunosuppression [82]. Lots on functional beta cell immunoprotection can be learned from encapsulated xenogeneic islets. Devices and especially their functional membranes that have been able to protect xenogeneic islets from the immune system can be assumed to also protect stem cell-derived islets.

In the context of macroencapsulation, immune mediators such as antibodies, complement, and cytokines that are potentially harmful to transplanted cells are of particular importance since the immune barrier is based upon size exclusion membranes. Some of the referred molecules are similar in size to insulin or significantly smaller (e.g. nitric oxide). Furthermore, pancreatic islets are particularly susceptible to inflammatory processes. Although, within a macroencapsulation device the islets are shielded from circulating blood that typically triggers rapid thrombo-inflammatory reactions (Instant Blood-Mediated Inflammatory Reaction, IBMIR) in intraportal islet transplantation and dramatically affects the survival and function [99], inflammatory mediators such as IL-1ß, produced as a result of infiltrating immune cells or beta cells themselves, may significantly influence the short- and long-term fate of the islets [41]. Permselective elements may have the potential to protect the islets to a certain extent, but size exclusion alone cannot prevent the entry of all immune mediators. The combination of macroencapsulation and local immunosuppression may open up a novel and potent strategy to even better protect the transplanted cells.

A highly significant factor for a functional macroencapsulation device with persistent proper diffusion characteristics for oxygen, nutrients and hormones is a successful tissue integration, preventing fibrosis and enhancing vascularization. Therefore, besides suitable perm-selectivity and diffusion characteristics, the surface of the encapsulation device also needs to reduce the inflammatory response towards the foreign-body material and direct neovascularization towards the host-graft interface. The foreign-body response results in the deposition of a dense fibrotic layer around the device that delays or prevents the diffusion of oxygen and functional metabolites between the outside and inside of the device (Fig. 9). Thus, keys to the successful transplantation of the encapsulated beta cells are the biocompatibility and the proangiogenic properties of the immunoisolating element or the device surface in general.

Consequently, the selection of the materials utilized for the outermost surface of the macroencapsulation device plays an important role. For immunoisolation, membranes composed of hydrophilic expanded poly(tetrafluoroethylene) (ePTFE) are

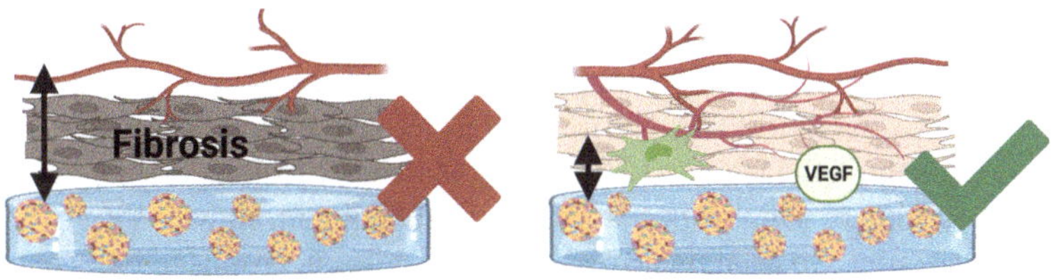

Fig. 9 Diffusion pathway of oxygen, nutrients and effector molecules in the z-dimension. The inflammatory response to the foreign surface induces a fibrotic layer around the encapsulation device. This creates an additional barrier that molecules have to pass

frequently used in islet transplantation. These membranes, for instance combined with a PCL membrane or an alginate hydrogel, prevented the penetration of immune cells, providing protection of the islet graft [24, 83]. Another approach used poly(ether sulfone)/poly(*N*-vinylpyrrolidone) (PES/PVP) blends to fabricate microporous membranes with integrated molded microwells [127]. In addition, polymers have been applied to locally release signaling molecules in a controlled manner, which prevent or minimize the immune system's attack on the transplanted beta cells, as well as promote the survival and function of transplanted beta cells [3, 15, 32, 107, 110].

Goswami et al. highlighted several types of immunomodulatory factors that have shown promise in preclinical studies, such as the incorporation of the chemokine CXCL12 with alginate hydrogel encapsulation of stem cell-derived islets [3] and the use of the small molecule GW2580 [113], contribute to reducing the foreign body response to the implant [56].

Coating of hydrogel fibers or concentric multiphasic constructs for cell encapsulation with zwitterionic polymer materials resulted in superior biocompatibility and reduced fibrosis [76, 79, 80]. For example, the SHIELD device is based on a zwitterionic alginate hydrogel, which helped to alleviate fibrotic reactions and confer long-term functionality to the device [80]. Similarly, Ma et al. introduced a biocompatible nanoporous device made of zwitterionic polyurethane to encapsulate pancreatic islets for treating type 1 diabetes. The device protects islets form the immune system, allows the exchange of nutrients and oxygen, and has shown promising outcomes in diabetic mice for up to 12 weeks [79].

Besides preventing a fibrotic response, it is further of interest to enhance vascularization at the host-graft interface to deliver sufficient oxygen to the graft. There are several approaches based on biochemical properties, but also just on physical properties, such as for example structural properties of the applied materials. For instance, the TheraCyte™ system is composed of two membranes made of PTFE – a cell impermeable membrane (400 nm pore size) and a membrane with larger pores (5 μm) to promote angiogenesis. This double layer approach seeks to reduce diffusion time delays by development of a new vasculature within the large pores, while the smaller pores immunoprotect the encapsulated islets [121]. Other approaches have used membranes decorated with intermittent lines on the device surface for alignment of endothelial cells for improved vascularization [129]. Proangiogenic factors that have been investigated for their potential to support beta cell replacement therapies include vascular endothelial growth factor (VEGF) [53, 110], basic fibroblast growth factor (bFGF) [110], platelet-derived growth factor (PDGF) [19], and insulin-like growth factor-1 (IGF-1) [61]. Marchioli et al. reported a two-phasic scaffold consisting of a highly porous PCL-ring functionalized with VEGF to facilitate vascularization and an alginate hydrogel core with encapsulated islets to protect them from the host immune system [87].

In own studies, macroporous device membranes were impregnated with multifunctional biohybrid hydrogels [52, 150] that enable release of pro-angiogenic growth factors (e.g. VEGF) in a controlled manner to improve vascularization in the early post-transplant phase (Ludwig/Heller et al., "unpublished"). This is an example, how tissue integration can be specifically supported by one of the materials building up the permselective immunoisolating element. Furthermore, proangiogenic helper cells, such as mesenchymal stroma cells, endothelial cells, stellate cells, or amniotic epithelial cells, have been co-encapsulated with beta cells for local modulation of the graft environment and for a better tissue integration and vascularization.

The closer in proximity to the host-graft interface blood capillaries are formed the less time molecules, especially oxygen, need to diffuse towards the encapsulated cells. Anticipating these mechanisms and the various influencing factors such as the device design and size, improved biocompatibility has been demonstrated in several recent studies [139, 140].

6.2 *Islet Microenvironment*

Beyond immunoisolation, biomaterials have a high potential to act as an artificial cellular microenvironment for the transplanted cell clusters that locally provide them with mechanical support, but also with ECM proteins, relevant peptide sequences, or signaling molecules [31, 35, 66, 130], similar as in their natural in vivo environment. In their physiological niche, islets are embedded within an ECM providing the cells with structural, physical and chemical cues. Hydrogels have been shown to be advantageous to mimic various natural tissue microenvironments [42, 149]. These highly cross-linked hydrophilic polymer networks can absorb large amounts of water, which allows diffusion and transport of solute molecules (e.g., nutrients, hormones, oxygen) within the network. This transport is highly dependent on the mesh size of the polymeric network similar as described for the permselective element. Consequently, a hydrogel matrix surrounding the islets within a macroencapsulation device (intracapsular islet microenvironment) needs to be optimized with regard to its mesh size as well in order to ensure sufficient transport. Tailored mechanical properties and biomolecular functionalization can further enhance functional long-term survival of the graft.

The most widely studied materials for encapsulation of islets, including micro-encapsulation approaches, are alginate-based hydrogels. However, these materials might not be the most suitable ones. Stability in vivo, including permselectivity to cytokines, antibodies, and cells, has been long argued [13, 94, 100]. Other natural polymers were employed for islet embedding in pre-clinical trials, such as cellulose, agarose, collagen, chitosan, and gelatin; however, none of them yield fully satisfying outcomes with respect to therapeutic potency [47]. In particular, incorporating ECM components (such as fibronectin, laminin and collagen) into the islet-surrounding matrix has been shown to enhance the function and long-term survival of transplanted islets [81]. Moreover, for instance human pancreatic artery-derived ECM has been reported to inhibit the expansion of human naive CD4 T-cells while supporting various pancreatic cell types [106]. Another study [64] compared in situ encapsulation of human and rodent islets within ECM-based hydrogels derived from porcine bladder and pancreas, which showed favorable functional stability and retained islet-resident ECs. However, the hydrogels gradually lost some of their components, had poor mechanical properties, degraded quickly in vivo, and did not sustain vascularization, resulting in graft loss [64, 107]. One possible approach to overcome these drawbacks in mimicking the islet microarchitecture is to combine natural and synthetic hydrogel precursors within one biohybrid material [19, 34, 95]. Fully synthetic polymers offer some advantages like for instance high purity and low batch-to-batch variations as well as enhanced mechanical properties of the cross-linked hydrogel network [144]. In particular, fully synthetic hydrogels based on PEG were widely used [149]. For instance, native islets were embedded in a slab made of a PEG-based hydrogel for transplantation into immunocompetent mice. The introduction of microchannels to enhance the transport characteristics resulted in a better functionality of the devices [114]. This and other studies show that in addition to the biochemical material properties, introduction of microstructures or macroporosity into the material and/or minimizing material thickness utilizing various methods and technologies improve diffusion characteristics and increase mechanical stability [23, 132].

A different approach to re-establish islet microenvironment is derived from advancements in tissue engineering and the utilization of decellularized organs as a support system for cell therapies. Various studies demonstrated preservation of the entire organization of the ECM and microarchitecture, with the aim of replicating the complexity of the native endocrine microenvironment [27, 57, 107].

7 Transplantation Site

After all design considerations, the establishment of an optimal transplantation site is an imperative and yet challenging requisite for cell survival and device efficacy. Ideal oxygen, nutrient, and hormone diffusion conditions, physiological route of hormone delivery, minimal surrounding tissue remodelling induction and clinically applicable scale are critical determinants. Various potential implantation sites have been discussed in the literature for macroencapsulation devices. Most trials are performed in the subcutaneous space [12, 97, 136, 141] followed by the omentum [14], intramuscular [49], intra-peritoneal [68, 84], and pre-peritoneal [82, 83]. The subcutaneous space is easily accessible for implantation and retrieval and even biopsy. However, the subcutaneous space is a bradytrophic tissue with poor vascularization and while various rodent models of subcutaneous macro-device implantation were successful, they (partially) failed after upscaling to large animal models or in clinical application [12, 21, 83, 134]. The rare vascularization is causing poor oxygen supply to the encapsulated islets and prohibits a relevant route for insulin diffusion. Moreover, the route of insulin drainage is less physiological compared to the portal drainage as reached by an omental and intraperitoneal site. The latter can accommodate large volumes; however, the intraperitoneal access can result in complications associated with the surgical procedure. The intramuscular site is generally attractive due to the high level of vascularization and the affinity to neovascularization and the possible minimally invasive implantation method [43, 131]. Another approach is the pre-vascularization of the transplantation site by local administration of vascular growth or trophic factors, or by the implantation of foreign material to induce strong neovascularization prior to implanting the cell containing devices several weeks later [36, 65]. No consensus has been reached upon an optimal implantation site. Depending on device geometry, surface characteristics and cell volume, individual tailored sites might be appropriate to minimally affect biomechanics of the device, minimize the trauma to the surrounding tissue and thereby mitigate foreign body response and inflammation inducing fibrosis of the implant and thereby impairing oxygen and molecular transport.

8 Perspectives

Intrahepatic islet transplantation has successfully restored glycemic regulation in patients with type 1 diabetes [123]; however, the pool of eligible recipients of this procedure is limited, in part, because patients must endure chronic immunosuppression to prevent the immune destruction of the cells. Islet encapsulation is an expanding field that attempts to overcome these hurdles by providing the means to transplant islets without immunosuppression, creating a beneficial microenvironment for long-term function, and may enable the safe usage of alternative sources for insulin-producing cells such as stem cell-derived insulin-producing cell clusters. Numerous small and some large animal models have demonstrated the promise of encapsulated islet transplantation. The clinical experience with encapsulated islets is still limited. The key issues that need to be addressed are definition of the optimal capsule configuration to effectively deliver oxygen and nutrients to the islet graft while integrating a sufficient islet mass, the use of anti-inflammatory agents to prohibit foreign body response and promote sustained islet function, and eliminate the need for immunosuppression. A major challenge is further to reach consensus about appropriate animal models. It is a matter of common knowledge that rodent models although indispensable cannot predict success in large animal models or man. Therefore, clinically relevant large animal models are of great relevance in order to adequately evaluate the clinical potency of encapsulation systems and eventually satisfy regulatory agencies in order to proceed to clinical studies. Moreover, particularly in large animal models and eventually in clinical pilot trials, more attention must be paid to metabolic monitoring and kinetic measurements in order to appropriately assess the physiological insulin secretory and regulating capacity of transplanted insulin-producing cells.

The great interest in islet encapsulation is due to the exploitation of alternative cell sources. A variety of studies have suggested the potential of encapsulated islet xenotransplantation to treat diabetes. Furthermore, in recent years effective methods have been established for genetic modification of pigs. Somatic cell nuclear transfer using genetically modified donor cells and the new prospects opened through CRISPR/Cas technology facilitates the generation of tailored pig models for xenotransplantation. However, despite all advances in this field, there is a serious reservation against xenotransplantation, which is mainly associated with safety concerns. Thus, macroencapsulation may offer a safe strategy that will lead to a clinically viable therapy that is both effective and safe.

Stem cell-derived islet cells are certainly the most attractive alternative, virtually unlimited source for transplantation. Despite the recent milestone advances [111, 125], two principle issues have not been resolved yet. Beta cells generated from human embryonic stem cells or induced pluripotent stem cells may show persistence of undifferentiated cells with the potential competence to proliferate by undergoing malignant transformation. Further, transplanted beta cells may be recognized by the abundant T-cells activated against beta cell-antigens, which developed in autoimmune type 1 diabetes, and subsequently be attacked and destroyed. However, encapsulation techniques may allow safe enclosing of stem cell-derived insulin-producing cells and tumor formation and immune attacks can be prevented.

Computational modelling, combinatorial methods and high-throughput screening approaches might have significant impact on current and future developments to foster targeted modifications, avoid repeated failures, and finally facilitate the identification of optimal biomaterials for the specific cell environment and the device-host interface.

Based on the recent tremendous advancements, it is reasonable to envision that within the next decade the cure of diabetes will rely on the conjoined application of cell-based treatments and biomaterial science. The latter may provide important contributes and fulfill the pretension of a simple and safe therapy as a functional cure for diabetes.

Acknowledgements All figures were (partially) created using BioRender.com. This work was in part supported by the Deutsche Forschungsgemeinschaft (DFG, German Research Foundation) (GRK 2251 and TRR 127) and by the Horizon 2020 program of the European Union (FTI; 950762).

Literature

1. Abadpour, S., Wang, C., Niemi, E. M. & Scholz, H. 2021. Tissue Engineering Strategies for Improving Beta Cell Transplantation Outcome. *Current Transplantation Reports,* 8, 205–219.
2. Abdi, S. I., Ng, S. M. & Lim, J. O. 2011. An enzyme-modulated oxygen-producing micro-system for regenerative therapeutics. *Int J Pharm,* 409, 203–5.
3. Alagpulinsa, D. A., Cao, J. J. L., Driscoll, R. K., Sîrbulescu, R. F., Penson, M. F. E., Sremac, M., Engquist, E. N., Brauns, T. A., Markmann, J. F., Melton, D. A. & Poznansky, M. C. 2019. Alginate-microencapsulation of human stem cell-derived β cells with CXCL12 prolongs their survival and function in immunocompetent mice without systemic immunosuppression. *Am J Transplant,* 19, 1930–1940.
4. Alemdar, N., Leijten, J., Camci-Unal, G., Hjortnaes, J., Ribas, J., Paul, A., Mostafalu, P., Gaharwar, A. K., Qiu, Y., Sonkusale, S., Liao, R. & Khademhosseini, A. 2017. Oxygen-Generating Photo-Cross-Linkable Hydrogels Support Cardiac Progenitor Cell Survival by Reducing Hypoxia-Induced Necrosis. *ACS Biomater Sci Eng,* 3, 1964–1971.
5. Algire, G. H. 1957a. Diffusion-chamber techniques for studies of cellular immunity. *Ann N Y Acad Sci,* 69, 663–7.
6. Algire, G. H. 1957b. Summary of studies of transplantation of homologous tissues. *Fed Proc,* 16, 601–2.
7. Algire, G. H., Borders, M. L. & Evans, V. J. 1958. Studies of heterografts in diffusion chambers in mice. *J Natl Cancer Inst,* 20, 1187–1201.
8. An, D., Chiu, A., Flanders, J. A., Song, W., Shou, D., Lu, Y. C., Grunnet, L. G., Winkel, L., Ingvorsen, C., Christophersen, N. S., Fels, J. J., Sand, F. W., Ji, Y., Qi, L., Pardo, Y., Luo, D., Silberstein, M., Fan, J. & Ma, M. 2018. Designing a retrievable and scalable cell encapsulation device for potential treatment of type 1 diabetes. *Proc Natl Acad Sci U S A,* 115, E263–e272.
9. Ashammakhi, N., Darabi, M. A., Kehr, N. S., Erdem, A., Hu, S. K., Dokmeci, M. R., Nasr, A. S. & Khademhosseini, A. 2020. Advances in Controlled Oxygen Generating Biomaterials for Tissue Engineering and Regenerative Therapy. *Biomacromolecules,* 21, 56–72.
10. Baidal, D. A., Ballou, C. M., Rickels, M. R., Berney, T., Pattou, F., Payne, E. H., Barton, F. B. & Alejandro, R. 2023. Predictive Value of C-Peptide Measures for Clinical Outcomes of β-Cell Replacement Therapy in Type 1 Diabetes: Report From the Collaborative Islet Transplant Registry (CITR). *Diabetes Care.*
11. Balamurugan, A. N., Naziruddin, B., Lockridge, A., Tiwari, M., Loganathan, G., Takita, M., Matsumoto, S., Papas, K., Trieger, M., Rainis, H., Kin, T., Kay, T. W., Wease, S., Messinger, S., Ricordi, C., Alejandro, R., Markmann, J., Kerr-Conti, J., Rickels, M. R., Liu, C., Zhang, X., Witkowski, P., Posselt, A., Maffi, P., Secchi, A., Berney, T., O'Connell, P. J., Hering, B. J. & Barton, F. B. 2014. Islet product characteristics and factors related to successful human islet transplantation from the Collaborative Islet Transplant Registry (CITR) 1999–2010. *Am J Transplant,* 14, 2595–606.
12. Barkai, U., Rotem, A. & de Vos, P. 2016. Survival of encapsulated islets: More than a membrane story. *World J Transplant,* 6, 69–90.
13. Basta, G. & Calafiore, R. 2011. Immunoisolation of pancreatic islet grafts with no recipient's immunosuppression: actual and future perspectives. *Curr Diab Rep,* 11, 384–91.
14. Berman, D. M., O'Neil, J. J., Coffey, L. C., Chaffanjon, P. C., Kenyon, N. M., Ruiz, P., Jr., Pileggi, A., Ricordi, C. & Kenyon, N. S. 2009. Long-term survival of nonhuman primate islets implanted in an omental pouch on a biodegradable scaffold. *Am J Transplant,* 9, 91–104.
15. Berney, T., Wassmer, C. H., Lebreton, F., Bellofatto, K., Fonseca, L. M., Bignard, J., Hanna, R., Peloso, A. & Berishvili, E. 2022. From islet of Langerhans transplantation to the bioartificial pancreas. *Presse Med,* 51, 104139.
16. Boettler, T., Schneider, D., Cheng, Y., Kadoya, K., Brandon, E. P., Martinson, L. & von Herrath, M. 2016. Pancreatic Tissue Transplanted in TheraCyte Encapsulation Devices Is

Protected and Prevents Hyperglycemia in a Mouse Model of Immune-Mediated Diabetes. *Cell Transplant,* 25, 609–14.
17. Borg, D. J. & Bonifacio, E. 2011. The use of biomaterials in islet transplantation. *Curr Diab Rep,* 11, 434–44.
18. Bose, S., Volpatti, L. R., Thiono, D., Yesilyurt, V., McGladrigan, C., Tang, Y., Facklam, A., Wang, A., Jhunjhunwala, S., Veiseh, O., Hollister-Lock, J., Bhattacharya, C., Weir, G. C., Greiner, D. L., Langer, R. & Anderson, D. G. 2020. A retrievable implant for the long-term encapsulation and survival of therapeutic xenogeneic cells. *Nat Biomed Eng,* 4, 814–826.
19. Brady, A. C., Martino, M. M., Pedraza, E., Sukert, S., Pileggi, A., Ricordi, C., Hubbell, J. A. & Stabler, C. L. 2013. Proangiogenic hydrogels within macroporous scaffolds enhance islet engraftment in an extrahepatic site. *Tissue Eng Part A,* 19, 2544–52.
20. Buitinga, M., Truckenmüller, R., Engelse, M. A., Moroni, L., Ten Hoopen, H. W., van Blitterswijk, C. A., de Koning, E. J., van Apeldoorn, A. A. & Karperien, M. 2013. Microwell scaffolds for the extrahepatic transplantation of islets of Langerhans. *PLoS One,* 8, e64772.
21. Carlsson, P. O., Espes, D., Sedigh, A., Rotem, A., Zimerman, B., Grinberg, H., Goldman, T., Barkai, U., Avni, Y., Westermark, G. T., Carlbom, L., Ahlström, H., Eriksson, O., Olerud, J. & Korsgren, O. 2018. Transplantation of macroencapsulated human islets within the bioartificial pancreas βAir to patients with type 1 diabetes mellitus. *Am J Transplant,* 18, 1735–1744.
22. Carmeliet, P. & Jain, R. K. 2000. Angiogenesis in cancer and other diseases. *Nature,* 407, 249–57.
23. Chabert, M. & Viovy, J. L. 2008. Microfluidic high-throughput encapsulation and hydrodynamic self-sorting of single cells. *Proc Natl Acad Sci U S A,* 105, 3191–6.
24. Chang, R., Faleo, G., Russ, H. A., Parent, A. V., Elledge, S. K., Bernards, D. A., Allen, J. L., Villanueva, K., Hebrok, M., Tang, Q. & Desai, T. A. 2017. Nanoporous Immunoprotective Device for Stem-Cell-Derived β-Cell Replacement Therapy. *ACS Nano,* 11, 7747–7757.
25. Chávez, M. N., Moellhoff, N., Schenck, T. L., Egaña, J. T. & Nickelsen, J. 2020. Photosymbiosis for Biomedical Applications. *Frontiers in Bioengineering and Biotechnology,* 8.
26. Chou, F. C., Huang, S. H. & Sytwu, H. K. 2012. Genetically engineered islets and alternative sources of insulin-producing cells for treating autoimmune diabetes: quo vadis? *Int J Endocrinol,* 2012, 296485.
27. Citro, A., Neroni, A., Pignatelli, C., Campo, F., Policardi, M., Monieri, M., Pellegrini, S., Dugnani, E., Manenti, F., Maffia, M. C., Valla, L., Kemter, E., Marzinotto, I., Olgasi, C., Cucci, A., Follenzi, A., Lampasona, V., Wolf, E. & Piemonti, L. 2023. Directed self-assembly of a xenogeneic vascularized endocrine pancreas for type 1 diabetes. *Nat Commun,* 14, 878.
28. Colombani, T., Eggermont, L., Hatfield, S., Rezaeeyazdi, M., Memic, A., Sitkovsky, M. & Bencherif, S. 2020. *Oxygen-generating cryogels restore T cell-mediated cytotoxicity in hypoxic tumors.*
29. Coronel, M. M., Geusz, R. & Stabler, C. L. 2017. Mitigating hypoxic stress on pancreatic islets via in situ oxygen generating biomaterial. *Biomaterials,* 129, 139–151.
30. Coronel, M. M., Liang, J. P., Li, Y. & Stabler, C. L. 2019. Oxygen generating biomaterial improves the function and efficacy of beta cells within a macroencapsulation device. *Biomaterials,* 210, 1–11.
31. Coronel, M. M. & Stabler, C. L. 2013. Engineering a local microenvironment for pancreatic islet replacement. *Curr Opin Biotechnol,* 24, 900–8.
32. Dang, H. P., Chen, H., Dargaville, T. R. & Tuch, B. E. 2022. Cell delivery systems: Toward the next generation of cell therapies for type 1 diabetes. *J Cell Mol Med,* 26, 4756–4767.
33. Dantal, J. & Soulillou, J. P. 2005. Immunosuppressive drugs and the risk of cancer after organ transplantation. *N Engl J Med,* 352, 1371–3.
34. Daoud, J. T., Petropavlovskaia, M. S., Patapas, J. M., Degrandpre, C. E., Diraddo, R. W., Rosenberg, L. & Tabrizian, M. 2011. Long-term in vitro human pancreatic islet culture using three-dimensional microfabricated scaffolds. *Biomaterials,* 32, 1536–42.
35. Davis, N. E., Beenken-Rothkopf, L. N., Mirsoian, A., Kojic, N., Kaplan, D. L., Barron, A. E. & Fontaine, M. J. 2012. Enhanced function of pancreatic islets co-encapsulated with ECM proteins and mesenchymal stromal cells in a silk hydrogel. *Biomaterials,* 33, 6691–7.

36. De Vos, P., Hillebrands, J. L., De Haan, B. J., Strubbe, J. H. & Van Schilfgaarde, R. 1997. Efficacy of a prevascularized expanded polytetrafluoroethylene solid support system as a transplantation site for pancreatic islets. *Transplantation,* 63, 824–30.
37. de Vos, P., Lazarjani, H. A., Poncelet, D. & Faas, M. M. 2014. Polymers in cell encapsulation from an enveloped cell perspective. *Adv Drug Deliv Rev,* 67–68, 15–34.
38. Decembrini, S., Hoehnel, S., Brandenberg, N., Arsenijevic, Y. & Lutolf, M. P. 2020. Hydrogel-based milliwell arrays for standardized and scalable retinal organoid cultures. *Sci Rep,* 10, 10275.
39. Desai, T. & Shea, L. D. 2017. Advances in islet encapsulation technologies. *Nat Rev Drug Discov,* 16, 338–350.
40. Desai, T. A. & Tang, Q. 2018. Islet encapsulation therapy – racing towards the finish line? *Nat Rev Endocrinol,* 14, 630–632.
41. Donath, M. Y., Boni-Schnetzler, M., Ellingsgaard, H., Halban, P. A. & Ehses, J. A. 2010. Cytokine production by islets in health and diabetes: cellular origin, regulation and function. *Trends Endocrinol Metab,* 21, 261–7.
42. Drury, J. L. & Mooney, D. J. 2003. Hydrogels for tissue engineering: scaffold design variables and applications. *Biomaterials,* 24, 4337–51.
43. Duffy, G. P., Robinson, S. T., O'Connor, R., Wylie, R., Mauerhofer, C., Bellavia, G., Straino, S., Cianfarani, F., Mendez, K., Beatty, R., Levey, R., O'Sullivan, J., Mcdonough, L., Kelly, H., Roche, E. T. & Dolan, E. B. 2020. Implantable Therapeutic Reservoir Systems for Diverse Clinical Applications in Large Animal Models. *Adv Healthc Mater,* 9, e2000305.
44. Duin, S., Bhandarkar, S., Lehmann, S., Kemter, E., Wolf, E., Gelinsky, M., Ludwig, B. & Lode, A. 2022. Viability and Functionality of Neonatal Porcine Islet-like Cell Clusters Bioprinted in Alginate-Based Bioinks. *Biomedicines,* 10.
45. Duin, S., Schütz, K., Ahlfeld, T., Lehmann, S., Lode, A., Ludwig, B. & Gelinsky, M. 2019. 3D Bioprinting of Functional Islets of Langerhans in an Alginate/Methylcellulose Hydrogel Blend. *Adv Healthc Mater,* 8, e1801631.
46. Ernst, A. U., Wang, L. H., Worland, S. C., Marfil-Garza, B. A., Wang, X., Liu, W., Chiu, A., Kin, T., O'Gorman, D., Steinschneider, S., Datta, A. K., Papas, K. K., James Shapiro, A. M. & Ma, M. 2022. A predictive computational platform for optimizing the design of bioartificial pancreas devices. *Nat Commun,* 13, 6031.
47. Espona-Noguera, A., Ciriza, J., Cañibano-Hernández, A., Orive, G., Hernández, R. M. M., Saenz Del Burgo, L. & Pedraz, J. L. 2019. Review of Advanced Hydrogel-Based Cell Encapsulation Systems for Insulin Delivery in Type 1 Diabetes Mellitus. *Pharmaceutics,* 11.
48. Evron, Y., Zimermann, B., Ludwig, B., Barkai, U., Colton, C., Weir, G., Arieli, B., Maimon, S., Shalev, N., Yavriyants, K., Goldman, T., Gendler, Z., Eizen, L., Vardi, P., Bloch, K., Barthel, A., Bornstein, S. & Rotem, A. 2014. Oxygen Supply by Photosynthesis to an Implantable Islet Cell Device. *Hormone and Metabolic Research,* 47.
49. Farina, M., Alexander, J. F., Thekkedath, U., Ferrari, M. & Grattoni, A. 2019. Cell encapsulation: Overcoming barriers in cell transplantation in diabetes and beyond. *Adv Drug Deliv Rev,* 139, 92–115.
50. Farris, A. L., Rindone, A. N. & Grayson, W. L. 2016. Oxygen Delivering Biomaterials for Tissue Engineering. *J Mater Chem B,* 4, 3422–3432.
51. Fenjves, E. S., Ochoa, M. S., Gay-Rabinstein, C., Molano, R. D., Pileggi, A., Mendez, A. J., Inverardi, L. & Ricordi, C. 2004. Adenoviral gene transfer of erythropoietin confers cytoprotection to isolated pancreatic islets. *Transplantation,* 77, 13–8.
52. Freudenberg, U., Liang, Y., Kiick, K. L. & Werner, C. 2016. Glycosaminoglycan-Based Biohybrid Hydrogels: A Sweet and Smart Choice for Multifunctional Biomaterials. *Adv Mater,* 28, 8861–8891.
53. Gebe, J. A., Preisinger, A., Gooden, M. D., D'Amico, L. A. & Vernon, R. B. 2018. Local, Controlled Release In Vivo of Vascular Endothelial Growth Factor Within a Subcutaneous Scaffolded Islet Implant Reduces Early Islet Necrosis and Improves Performance of the Graft. *Cell Transplant,* 27, 531–541.

54. Gholipourmalekabadi, M., Zhao, S., Harrison, B. S., Mozafari, M. & Seifalian, A. M. 2016. Oxygen-Generating Biomaterials: A New, Viable Paradigm for Tissue Engineering? *Trends Biotechnol,* 34, 1010–1021.
55. Giner Life Sciences, A. F. G. L. S. accessed February 2023. https://www.ginerinc.com/life-sciences-applications [Online]. [Accessed].
56. Goswami, D., Domingo-Lopez, D. A., Ward, N. A., Millman, J. R., Duffy, G. P., Dolan, E. B. & Roche, E. T. 2021. Design Considerations for Macroencapsulation Devices for Stem Cell Derived Islets for the Treatment of Type 1 Diabetes. *Adv Sci (Weinh),* 8, e2100820.
57. Guruswamy Damodaran, R. & Vermette, P. 2018. Decellularized pancreas as a native extracellular matrix scaffold for pancreatic islet seeding and culture. *J Tissue Eng Regen Med,* 12, 1230–1237.
58. Hering, B. J., Ballou, C. M., Bellin, M. D., Payne, E. H., Kandeel, F., Witkowski, P., Alejandro, R., Rickels, M. R. & Barton, F. B. 2023. Factors associated with favourable 5 year outcomes in islet transplant alone recipients with type 1 diabetes complicated by severe hypoglycaemia in the Collaborative Islet Transplant Registry. *Diabetologia,* 66, 163–173.
59. Hering, B. J., Clarke, W. R., Bridges, N. D., Eggerman, T. L., Alejandro, R., Bellin, M. D., Chaloner, K., Czarniecki, C. W., Goldstein, J. S., Hunsicker, L. G., Kaufman, D. B., Korsgren, O., Larsen, C. P., Luo, X., Markmann, J. F., Naji, A., Oberholzer, J., Posselt, A. M., Rickels, M. R., Ricordi, C., Robien, M. A., Senior, P. A., Shapiro, A. M., Stock, P. G. & Turgeon, N. A. 2016. Phase 3 Trial of Transplantation of Human Islets in Type 1 Diabetes Complicated by Severe Hypoglycemia. *Diabetes Care,* 39, 1230–40.
60. Hilderink, J., Spijker, S., Carlotti, F., Lange, L., Engelse, M., van Blitterswijk, C., de Koning, E., Karperien, M. & van Apeldoorn, A. 2015. Controlled aggregation of primary human pancreatic islet cells leads to glucose-responsive pseudoislets comparable to native islets. *J Cell Mol Med,* 19, 1836–46.
61. Hlavaty, K. A., Gibly, R. F., Zhang, X., Rives, C. B., Graham, J. G., Lowe, W. L., Jr., Luo, X. & Shea, L. D. 2014. Enhancing human islet transplantation by localized release of trophic factors from PLG scaffolds. *Am J Transplant,* 14, 1523–32.
62. Hopfner, U., Schenck, T. L., Chávez, M. N., Machens, H. G., Bohne, A. V., Nickelsen, J., Giunta, R. E. & Egaña, J. T. 2014. Development of photosynthetic biomaterials for in vitro tissue engineering. *Acta Biomater,* 10, 2712–7.
63. Hymer, W. C., Wilbur, D. L., Page, R., Hibbard, E., Kelsey, R. C. & Hatfield, J. M. 1981. Pituitary hollow fiber units in vivo and in vitro. *Neuroendocrinology,* 32, 339–49.
64. Jiang, K., Chaimov, D., Patel, S. N., Liang, J.-P., Wiggins, S. C., Samojlik, M. M., Rubiano, A., Simmons, C. S., & Stabler, C. L. 2019. 3-D physiomimetic extracellular matrix hydrogels provide a supportive microenvironment for rodent and human islet culture. *Biomaterials, 198*, 37–48. https://doi.org/10.1016/j.biomaterials.2018.08.057
65. Juang, J. H., Bonner-Weir, S., Ogawa, Y., Vacanti, J. P. & Weir, G. C. 1996. Outcome of subcutaneous islet transplantation improved by polymer device. *Transplantation,* 61, 1557–61.
66. Kerby, A., Jones, E. S., Jones, P. M. & King, A. J. 2013. Co-transplantation of islets with mesenchymal stem cells in microcapsules demonstrates graft outcome can be improved in an isolated-graft model of islet transplantation in mice. *Cytotherapy,* 15, 192–200.
67. Komatsu, H., Cook, C., Wang, C. H., Medrano, L., Lin, H., Kandeel, F., Tai, Y. C. & Mullen, Y. 2017. Oxygen environment and islet size are the primary limiting factors of isolated pancreatic islet survival. *PLoS One,* 12, e0183780.
68. Krishnan, R., Alexander, M., Robles, L., Foster, C. E., 3rd & Lakey, J. R. 2014. Islet and stem cell encapsulation for clinical transplantation. *Rev Diabet Stud,* 11, 84–101.
69. Lablanche, S., Vantyghem, M. C., Kessler, L., Wojtusciszyn, A., Borot, S., Thivolet, C., Girerd, S., Bosco, D., Bosson, J. L., Colin, C., Tetaz, R., Logerot, S., Kerr-Conte, J., Renard, E., Penfornis, A., Morelon, E., Buron, F., Skaare, K., Grguric, G., Camillo-Brault, C., Egelhofer, H., Benomar, K., Badet, L., Berney, T., Pattou, F. & Benhamou, P. Y. 2018. Islet transplantation versus insulin therapy in patients with type 1 diabetes with severe hypoglycaemia or poorly controlled glycaemia after kidney transplantation (TRIMECO): a multicentre, randomised controlled trial. *Lancet Diabetes Endocrinol,* 6, 527–537.

70. Lanza, R. P., Beyer, A. M., Staruk, J. E. & Chick, W. L. 1993. Biohybrid artificial pancreas. Long-term function of discordant islet xenografts in streptozotocin diabetic rats. *Transplantation,* 56, 1067–72.
71. Lanza, R. P., Borland, K. M., Lodge, P., Carretta, M., Sullivan, S. J., Muller, T. E., Solomon, B. A., Maki, T., Monaco, A. P. & Chick, W. L. 1992a. Treatment of severely diabetic pancreatectomized dogs using a diffusion-based hybrid pancreas. *Diabetes,* 41, 886–9.
72. Lanza, R. P., Borland, K. M., Staruk, J. E., Appel, M. C., Solomon, B. A. & Chick, W. L. 1992b. Transplantation of encapsulated canine islets into spontaneously diabetic BB/Wor rats without immunosuppression. *Endocrinology,* 131, 637–42.
73. Lebreton, F., Bellofatto, K., Wassmer, C. H., Perez, L., Lavallard, V., Parnaud, G., Cottet-Dumoulin, D., Kerr-Conte, J., Pattou, F., Bosco, D., Othenin-Girard, V., Martinez de Tejada, B. & Berishvili, E. 2020. Shielding islets with human amniotic epithelial cells enhances islet engraftment and revascularization in a murine diabetes model. *Am J Transplant,* 20, 1551–1561.
74. Lee, E. M., Jung, J. I., Alam, Z., Yi, H. G., Kim, H., Choi, J. W., Hurh, S., Kim, Y. J., Jeong, J. C., Yang, J., Oh, K. H., Kim, H. C., Lee, B. C., Choi, I., Cho, D. W. & Ahn, C. 2018. Effect of an oxygen-generating scaffold on the viability and insulin secretion function of porcine neonatal pancreatic cell clusters. *Xenotransplantation,* 25, e12378.
75. Lehmann, R., Zuellig, R. A., Kugelmeier, P., Baenninger, P. B., Moritz, W., Perren, A., Clavien, P. A., Weber, M. & Spinas, G. A. 2007. Superiority of small islets in human islet transplantation. *Diabetes,* 56, 594–603.
76. Li, B., Yuan, Z., Jain, P., Hung, H. C., He, Y., Lin, X., McMullen, P. & Jiang, S. 2020. De novo design of functional zwitterionic biomimetic material for immunomodulation. *Sci Adv,* 6, eaba0754.
77. Li, R. H., Altreuter, D. H. & Gentile, F. T. 1996. Transport characterization of hydrogel matrices for cell encapsulation. *Biotechnol Bioeng,* 50, 365–73.
78. Liang, J. P., Accolla, R. P., Soundirarajan, M., Emerson, A., Coronel, M. M. & Stabler, C. L. 2021. Engineering a macroporous oxygen-generating scaffold for enhancing islet cell transplantation within an extrahepatic site. *Acta Biomater,* 130, 268–280.
79. Liu, Q., Wang, X., Chiu, A., Liu, W., Fuchs, S., Wang, B., Wang, L. H., Flanders, J., Zhang, Y., Wang, K., Melero-Martin, J. M. & Ma, M. 2021. A Zwitterionic Polyurethane Nanoporous Device with Low Foreign-Body Response for Islet Encapsulation. *Adv Mater,* 33, e2102852.
80. Liu, W., Flanders, J. A., Wang, L. H., Liu, Q., Bowers, D. T., Wang, K., Chiu, A., Wang, X., Ernst, A. U., Shariati, K., Caserto, J. S., Parker, B., Gao, D., Plesser, M. D., Grunnet, L. G., Rescan, C., Pimentel Carletto, R., Winkel, L., Melero-Martin, J. M. & Ma, M. 2022. A Safe, Fibrosis-Mitigating, and Scalable Encapsulation Device Supports Long-Term Function of Insulin-Producing Cells. *Small,* 18, e2104899.
81. Llacua, A., Faas, M. & de Vos, P. 2018. Extracellular matrix molecules and their potential contribution to the function of transplanted pancreatic islets. *Diabetologia,* 61.
82. Ludwig, B., Ludwig, S., Steffen, A., Knauf, Y., Zimerman, B., Heinke, S., Lehmann, S., Schubert, U., Schmid, J., Bleyer, M., Schonmann, U., Colton, C. K., Bonifacio, E., Solimena, M., Reichel, A., Schally, A. V., Rotem, A., Barkai, U., Grinberg-Rashi, H., Kaup, F. J., Avni, Y., Jones, P. & Bornstein, S. R. 2017. Favorable outcome of experimental islet xenotransplantation without immunosuppression in a nonhuman primate model of diabetes. *Proc Natl Acad Sci U S A,* 114, 11745–11750.
83. Ludwig, B., Reichel, A., Steffen, A., Zimerman, B., Schally, A. V., Block, N. L., Colton, C. K., Ludwig, S., Kersting, S., Bonifacio, E., Solimena, M., Gendler, Z., Rotem, A., Barkai, U. & Bornstein, S. R. 2013. Transplantation of human islets without immunosuppression. *Proc Natl Acad Sci U S A,* 110, 19054–8.
84. Magisson, J., Sassi, A., Xhema, D., Kobalyan, A., Gianello, P., Mourer, B., Tran, N., Burcez, C. T., Bou Aoun, R. & Sigrist, S. 2020. Safety and function of a new pre-vascularized bioartificial pancreas in an allogeneic rat model. *J Tissue Eng,* 11, 2041731420924818.
85. Maki, T., Mullon, C. J., Solomon, B. A. & Monaco, A. P. 1995. Novel delivery of pancreatic islet cells to treat insulin-dependent diabetes mellitus. *Clin Pharmacokinet,* 28, 471–82.

86. Maki, T., Ubhi, C. S., Sanchez-Farpon, H., Sullivan, S. J., Borland, K., Muller, T. E., Solomon, B. A., Chick, W. L. & Monaco, A. P. 1991. Successful treatment of diabetes with the biohybrid artificial pancreas in dogs. *Transplantation,* 51, 43–51.
87. Marchioli, G., Luca, A. D., De Koning, E., Engelse, M., Van Blitterswijk, C. A., Karperien, M., Van Apeldoorn, A. A. & Moroni, L. 2016. Hybrid Polycaprolactone/Alginate Scaffolds Functionalized with VEGF to Promote de Novo Vessel Formation for the Transplantation of Islets of Langerhans. *Adv Healthc Mater,* 5, 1606–16.
88. Markmann, J. F., Rickels, M. R., Eggerman, T. L., Bridges, N. D., Lafontant, D. E., Qidwai, J., Foster, E., Clarke, W. R., Kamoun, M., Alejandro, R., Bellin, M. D., Chaloner, K., Czarniecki, C. W., Goldstein, J. S., Hering, B. J., Hunsicker, L. G., Kaufman, D. B., Korsgren, O., Larsen, C. P., Luo, X., Naji, A., Oberholzer, J., Posselt, A. M., Ricordi, C., Senior, P. A., Shapiro, A. M. J., Stock, P. G. & Turgeon, N. A. 2021. Phase 3 trial of human islet-after-kidney transplantation in type 1 diabetes. *Am J Transplant,* 21, 1477–1492.
89. Matsumoto, S., Takita, M., Chaussabel, D., Noguchi, H., Shimoda, M., Sugimoto, K., Itoh, T., Chujo, D., Sorelle, J., Onaca, N., Naziruddin, B. & Levy, M. F. 2011. Improving efficacy of clinical islet transplantation with iodixanol-based islet purification, thymoglobulin induction, and blockage of IL-1β and TNF-α. *Cell Transplant,* 20, 1641–7.
90. Mcquilling, J. P., Sittadjody, S., Pendergraft, S., Farney, A. C. & Opara, E. C. 2017. Applications of particulate oxygen-generating substances (POGS) in the bioartificial pancreas. *Biomater Sci,* 5, 2437–2447.
91. Mirab, F., Kang, Y. J. & Majd, S. 2019. Preparation and characterization of size-controlled glioma spheroids using agarose hydrogel microwells. *PLoS One,* 14, e0211078.
92. Monaco, A. P., Maki, T., Ozato, H., Carretta, M., Sullivan, S. J., Borland, K. M., Mahoney, M. D., Chick, W. L., Muller, T. E., Wolfrum, J. & et al. 1991. Transplantation of islet allografts and xenografts in totally pancreatectomized diabetic dogs using the hybrid artificial pancreas. *Ann Surg,* 214, 339–60; discussion 361–2.
93. Montesdeoca, C. Y. C., Afewerki, S., Stocco, T. D., Corat, M. A. F., de Paula, M. M. M., Marciano, F. R. & Lobo, A. O. 2020. Oxygen-generating smart hydrogels supporting chondrocytes survival in oxygen-free environments. *Colloids Surf B Biointerfaces,* 194, 111192.
94. Morch, Y. A., Donati, I., Strand, B. L. & Skjak-Braek, G. 2006. Effect of Ca2+, Ba2+, and Sr2+ on alginate microbeads. *Biomacromolecules,* 7, 1471–80.
95. Muthyala, S., Bhonde, R. R. & Nair, P. D. 2010. Cytocompatibility studies of mouse pancreatic islets on gelatin--PVP semi IPN scaffolds in vitro: potential implication towards pancreatic tissue engineering. *Islets,* 2, 357–66.
96. Nandi, A., Yan, L. J., Jana, C. K. & Das, N. 2019. Role of Catalase in Oxidative Stress- and Age-Associated Degenerative Diseases. *Oxid Med Cell Longev,* 2019, 9613090.
97. Neufeld, T., Ludwig, B., Barkai, U., Weir, G. C., Colton, C. K., Evron, Y., Balyura, M., Yavriyants, K., Zimermann, B., Azarov, D., Maimon, S., Shabtay, N., Rozenshtein, T., Lorber, D., Steffen, A., Willenz, U., Bloch, K., Vardi, P., Taube, R., de Vos, P., Lewis, E. C., Bornstein, S. R. & Rotem, A. 2013. The efficacy of an immunoisolating membrane system for islet xenotransplantation in minipigs. *PLoS One,* 8, e70150.
98. Newland, B., Baeger, M., Eigel, D., Newland, H. & Werner, C. 2017. Oxygen-Producing Gellan Gum Hydrogels for Dual Delivery of Either Oxygen or Peroxide with Doxorubicin. *ACS Biomater Sci Eng,* 3, 787–792.
99. Nilsson, B., Ekdahl, K. N. & Korsgren, O. 2011. Control of instant blood-mediated inflammatory reaction to improve islets of Langerhans engraftment. *Curr Opin Organ Transplant,* 16, 620–6.
100. O'Sullivan, E. S., Vegas, A., Anderson, D. G. & Weir, G. C. 2011. Islets transplanted in immunoisolation devices: a review of the progress and the challenges that remain. *Endocr Rev,* 32, 827–44.
101. Oh, S. H., Ward, C. L., Atala, A., Yoo, J. J. & Harrison, B. S. 2009. Oxygen generating scaffolds for enhancing engineered tissue survival. *Biomaterials,* 30, 757–62.
102. Opara, A., Jost, A., Dagogo-Jack, S. & Opara, E. C. 2021. Islet cell encapsulation – Application in diabetes treatment. *Exp Biol Med (Maywood),* 246, 2570–2578.

103. Papas, K. K., De Leon, H., Suszynski, T. M. & Johnson, R. C. 2019. Oxygenation strategies for encapsulated islet and beta cell transplants. *Adv Drug Deliv Rev,* 139**,** 139–156.
104. Paredes-Juarez, G. A., Sahasrabudhe, N. M., Tjoelker, R. S., de Haan, B. J., Engelse, M. A., de Koning, E. J., Faas, M. M. & de Vos, P. 2015. DAMP production by human islets under low oxygen and nutrients in the presence or absence of an immunoisolating-capsule and necrostatin-1. *Sci Rep,* 5**,** 14623.
105. Pedraza, E., Coronel, M. M., Fraker, C. A., Ricordi, C. & Stabler, C. L. 2012. Preventing hypoxia-induced cell death in beta cells and islets via hydrolytically activated, oxygen-generating biomaterials. *Proc Natl Acad Sci U S A,* 109**,** 4245–50.
106. Peloso, A., Urbani, L., Cravedi, P., Katari, R., Maghsoudlou, P., Fallas, M. E. A., Sordi, V., Citro, A., Purroy, C., Niu, G., McQuilling, J. P., Sittadjody, S., Farney, A. C., Iskandar, S. S., Zambon, J. P., Rogers, J., Stratta, R. J., Opara, E. C., Piemonti, L., Furdui, C. M., Soker, S., De Coppi, P., & Orlando G. 2016. The human pancreas as a source of protolerogenic extracellular matrix scaffold for a new-generation bioartificial endocrine pancreas. *Annals of Surgery, 264*(1), 169–79. https://doi.org/10.1097/SLA.0000000000001364
107. Pignatelli, C., Campo, F., Neroni, A., Piemonti, L. & Citro, A. 2022. Bioengineering the Vascularized Endocrine Pancreas: A Fine-Tuned Interplay Between Vascularization, Extracellular-Matrix-Based Scaffold Architecture, and Insulin-Producing Cells. *Transplant International,* 35.
108. Procyon Technologies LLC and Novo Nordisk A/S to Collaborate on the Development of a Stem-Cell Based Therapy for Type 1 Diabetes. accessed February 2023. https://www.prnewswire.com/news-releases/procyon-technologies-llc-and-novo-nordisk-as-to-collaborate-on-the-development-of-a-stem-cell-based-therapy-for-type-1-diabetes-301188131.html [Online]. [Accessed].
109. Qin, D., Xia, Y. & Whitesides, G. M. 2010. Soft lithography for micro- and nanoscale patterning. *Nat Protoc,* 5**,** 491–502.
110. Quizon, M. J. & García, A. J. 2022. Engineering β Cell Replacement Therapies for Type 1 Diabetes: Biomaterial Advances and Considerations for Macroscale Constructs. *Annual Review of Pathology: Mechanisms of Disease,* 17**,** 485–513.
111. Ramzy, A., Thompson, D. M., Ward-Hartstonge, K. A., Ivison, S., Cook, L., Garcia, R. V., Loyal, J., Kim, P. T. W., Warnock, G. L., Levings, M. K. & Kieffer, T. J. 2021. Implanted pluripotent stem-cell-derived pancreatic endoderm cells secrete glucose-responsive C-peptide in patients with type 1 diabetes. *Cell Stem Cell,* 28**,** 2047–2061.e5.
112. Reid, L., Baxter, F. & Forbes, S. 2021. Effects of islet transplantation on microvascular and macrovascular complications in type 1 diabetes. *Diabet Med,* 38**,** e14570.
113. Ricard, N., Zhang, J., Zhuang, Z. W. & Simons, M. 2019. Isoform-Specific Roles of ERK1 and ERK2 in Arteriogenesis. *Cells,* 9.
114. Rios, P. D., Zhang, X., Luo, X. & Shea, L. D. 2016. Mold-casted non-degradable, islet macro-encapsulating hydrogel devices for restoration of normoglycemia in diabetic mice. *Biotechnol Bioeng,* 113**,** 2485–95.
115. Robert, T., De Mesmaeker, I., Stangé, G. M., Suenens, K. G., Ling, Z., Kroon, E. J. & Pipeleers, D. G. 2018. Functional Beta Cell Mass from Device-Encapsulated hESC-Derived Pancreatic Endoderm Achieving Metabolic Control. *Stem Cell Reports,* 10**,** 739–750.
116. Rojas-Canales, D. M., Waibel, M., Forget, A., Penko, D., Nitschke, J., Harding, F. J., Delalat, B., Blencowe, A., Loudovaris, T., Grey, S. T., Thomas, H. E., Kay, T. W. H., Drogemuller, C. J., Voelcker, N. H. & Coates, P. T. 2018. Oxygen-permeable microwell device maintains islet mass and integrity during shipping. *Endocr Connect,* 7**,** 490–503.
117. Rouwkema, J., Rivron, N. C. & van Blitterswijk, C. A. 2008. Vascularization in tissue engineering. *Trends Biotechnol,* 26**,** 434–41.
118. Ryan, E. A., Paty, B. W., Senior, P. A., Bigam, D., Alfadhli, E., Kneteman, N. M., Lakey, J. R. & Shapiro, A. M. 2005. Five-year follow-up after clinical islet transplantation. *Diabetes,* 54**,** 2060–9.
119. Scharp, D. W., Mason, N. S. & Sparks, R. E. 1984. Islet immuno-isolation: the use of hybrid artificial organs to prevent islet tissue rejection. *World J Surg,* 8**,** 221–9.

120. Schmidtke, T., White, D. & Woolard, C. 1999. Oxygen release kinetics from solid phase oxygen in Arctic Alaska. *J Hazard Mater,* 64, 157–65.
121. Schweicher, J., Nyitray, C. & Desai, T. A. 2014. Membranes to achieve immunoprotection of transplanted islets. *Front Biosci (Landmark Ed),* 19, 49–76.
122. Shapiro, A. M., Lakey, J. R., Ryan, E. A., Korbutt, G. S., Toth, E., Warnock, G. L., Kneteman, N. M. & Rajotte, R. V. 2000. Islet transplantation in seven patients with type 1 diabetes mellitus using a glucocorticoid-free immunosuppressive regimen. *N Engl J Med,* 343, 230–8.
123. Shapiro, A. M., Pokrywczynska, M. & Ricordi, C. 2017. Clinical pancreatic islet transplantation. *Nat Rev Endocrinol,* 13, 268–277.
124. Shapiro, A. M., Ricordi, C., Hering, B. J., Auchincloss, H., Lindblad, R., Robertson, R. P., Secchi, A., Brendel, M. D., Berney, T., Brennan, D. C., Cagliero, E., Alejandro, R., Ryan, E. A., Dimercurio, B., Morel, P., Polonsky, K. S., Reems, J. A., Bretzel, R. G., Bertuzzi, F., Froud, T., Kandaswamy, R., Sutherland, D. E., Eisenbarth, G., Segal, M., Preiksaitis, J., Korbutt, G. S., Barton, F. B., Viviano, L., Seyfert-Margolis, V., Bluestone, J. & Lakey, J. R. 2006. International trial of the Edmonton protocol for islet transplantation. *N Engl J Med,* 355, 1318–30.
125. Shapiro, A. M. J., Thompson, D., Donner, T. W., Bellin, M. D., Hsueh, W., Pettus, J., Wilensky, J., Daniels, M., Wang, R. M., Brandon, E. P., Jaiman, M. S., Kroon, E. J., D'Amour, K. A. & Foyt, H. L. 2021. Insulin expression and C-peptide in type 1 diabetes subjects implanted with stem cell-derived pancreatic endoderm cells in an encapsulation device. *Cell Rep Med,* 2, 100466.
126. Shizuru, J., Trager, D. & Merrell, R. C. 1985. Structure, function, and immune properties of reassociated islet cells. *Diabetes,* 34, 898–903.
127. Skrzypek, K., Groot Nibbelink, M., van Lente, J., Buitinga, M., Engelse, M. A., de Koning, E. J. P., Karperien, M., van Apeldoorn, A. & Stamatialis, D. 2017. Pancreatic islet macroencapsulation using microwell porous membranes. *Sci Rep,* 7, 9186.
128. Skrzypek, K., Nibbelink, M. G., Karbaat, L. P., Karperien, M., van Apeldoorn, A. & Stamatialis, D. 2018. An important step towards a prevascularized islet macroencapsulation device-effect of micropatterned membranes on development of endothelial cell network. *J Mater Sci Mater Med,* 29, 91.
129. Smink, A. M., Skrzypek, K., Liefers-Visser, J. A. L., Kuwabara, R., de Haan, B. J., de Vos, P. & Stamatialis, D. 2021. In vivovascularization and islet function in a microwell device for pancreatic islet transplantation. *Biomed Mater,* 16.
130. Stendahl, J. C., Wang, L. J., Chow, L. W., Kaufman, D. B. & Stupp, S. I. 2008. Growth factor delivery from self-assembling nanofibers to facilitate islet transplantation. *Transplantation,* 86, 478–81.
131. Svensson, J., Lau, J., Sandberg, M. & Carlsson, P. O. 2011. High vascular density and oxygenation of pancreatic islets transplanted in clusters into striated muscle. *Cell Transplant,* 20, 783–8.
132. Teramura, Y. & Iwata, H. 2010. Bioartificial pancreas microencapsulation and conformal coating of islet of Langerhans. *Adv Drug Deliv Rev,* 62, 827–40.
133. Toftdal, M. S., Taebnia, N., Kadumudi, F. B., Andresen, T. L., Frogne, T., Winkel, L., Grunnet, L. G. & Dolatshahi-Pirouz, A. 2021. Oxygen releasing hydrogels for beta cell assisted therapy. *Int J Pharm,* 602, 120595.
134. Tomei, A. A., Villa, C. & Ricordi, C. 2015. Development of an encapsulated stem cell-based therapy for diabetes. *Expert Opin Biol Ther,* 15, 1321–36.
135. Tremmel, D. M., Mitchell, S. A., Sackett, S. D. & Odorico, J. S. 2019. Mimicking nature-made beta cells: recent advances towards stem cell-derived islets. *Curr Opin Organ Transplant,* 24, 574–581.
136. Trivedi, N., Steil, G. M., Colton, C. K., Bonner-Weir, S. & Weir, G. C. 2000. Improved vascularization of planar membrane diffusion devices following continuous infusion of vascular endothelial growth factor. *Cell Transplant,* 9, 115–24.
137. Tu, T. Y., Wang, Z., Bai, J., Sun, W., Peng, W. K., Huang, R. Y., Thiery, J. P. & Kamm, R. D. 2014. Rapid prototyping of concave microwells for the formation of 3D multicellular cancer aggregates for drug screening. *Adv Healthc Mater,* 3, 609–16.

138. Tziampazis, E. & Sambanis, A. 1995. Tissue engineering of a bioartificial pancreas: modeling the cell environment and device function. *Biotechnol Prog,* 11, 115–26.
139. Vegas, A. J., Veiseh, O., Doloff, J. C., Ma, M., Tam, H. H., Bratlie, K., Li, J., Bader, A. R., Langan, E., Olejnik, K., Fenton, P., Kang, J. W., Hollister-Locke, J., Bochenek, M. A., Chiu, A., Siebert, S., Tang, K., Jhunjhunwala, S., Aresta-Dasilva, S., Dholakia, N., Thakrar, R., Vietti, T., Chen, M., Cohen, J., Siniakowicz, K., Qi, M., Mcgarrigle, J., Graham, A. C., Lyle, S., Harlan, D. M., Greiner, D. L., Oberholzer, J., Weir, G. C., Langer, R. & Anderson, D. G. 2016. Combinatorial hydrogel library enables identification of materials that mitigate the foreign body response in primates. *Nat Biotechnol,* 34, 345–52.
140. Veiseh, O., Doloff, J. C., Ma, M., Vegas, A. J., Tam, H. H., Bader, A. R., Li, J., Langan, E., Wyckoff, J., Loo, W. S., Jhunjhunwala, S., Chiu, A., Siebert, S., Tang, K., Hollister-Lock, J., Aresta-Dasilva, S., Bochenek, M., Mendoza-Elias, J., Wang, Y., Qi, M., Lavin, D. M., Chen, M., Dholakia, N., Thakrar, R., Lacik, I., Weir, G. C., Oberholzer, J., Greiner, D. L., Langer, R. & Anderson, D. G. 2015. Size- and shape-dependent foreign body immune response to materials implanted in rodents and non-human primates. *Nat Mater,* 14, 643–51.
141. Vlahos, A. E., Talior-Volodarsky, I., Kinney, S. M. & Sefton, M. V. 2021. A scalable deviceless biomaterial approach for subcutaneous islet transplantation. *Biomaterials,* 269, 120499.
142. Wang, J., Zhu, Y., Bawa, H. K., Ng, G., Wu, Y., Libera, M., van der Mei, H. C., Busscher, H. J. & Yu, X. 2011. Oxygen-generating nanofiber cell scaffolds with antimicrobial properties. *ACS Appl Mater Interfaces,* 3, 67–73.
143. Wang, X., Maxwell, K. G., Wang, K., Bowers, D. T., Flanders, J. A., Liu, W., Wang, L. H., Liu, Q., Liu, C., Naji, A., Wang, Y., Wang, B., Chen, J., Ernst, A. U., Melero-Martin, J. M., Millman, J. R. & Ma, M. 2021. A nanofibrous encapsulation device for safe delivery of insulin-producing cells to treat type 1 diabetes. *Sci Transl Med,* 13.
144. Weber, L. M., Lopez, C. G. & Anseth, K. S. 2009. Effects of PEG hydrogel crosslinking density on protein diffusion and encapsulated islet survival and function. *J Biomed Mater Res A,* 90, 720–9.
145. Weir, G. C., Bonner-Weir, S. & Leahy, J. L. 1990. Islet mass and function in diabetes and transplantation. *Diabetes,* 39, 401–5.
146. Wu, H., Avgoustiniatos, E. S., Swette, L., Bonner-Weir, S., Weir, G. C. & Colton, C. K. 1999. In situ electrochemical oxygen generation with an immunoisolation device. *Ann N Y Acad Sci,* 875, 105–25.
147. Yang, K., O'Cearbhaill, E. D., Liu, S. S., Zhou, A., Chitnis, G. D., Hamilos, A. E., Xu, J., Verma, M. K. S., Giraldo, J. A., Kudo, Y., Lee, E. A., Lee, Y., Pop, R., Langer, R., Melton, D. A., Greiner, D. L. & Karp, J. M. 2021. A therapeutic convection-enhanced macroencapsulation device for enhancing β cell viability and insulin secretion. *Proc Natl Acad Sci U S A,* 118.
148. Zekorn, T., Siebers, U., Filip, L., Mauer, K., Schmitt, U., Bretzel, R. G. & Federlin, K. 1989. Bioartificial pancreas: the use of different hollow fibers as a diffusion chamber. *Transplant Proc,* 21, 2748–50.
149. Zhang, Y., Li, L., Mu, J., Chen, J., Feng, S. & Gao, J. 2020. Implantation of a functional TEMPO-hydrogel induces recovery from rat spinal cord transection through promoting nerve regeneration and protecting bladder tissue. *Biomater Sci,* 8, 1695–1701.
150. Zieris, A., Chwalek, K., Prokoph, S., Levental, K. R., Welzel, P. B., Freudenberg, U. & Werner, C. 2011. Dual independent delivery of pro-angiogenic growth factors from starPEG-heparin hydrogels. *J Control Release,* 156, 28–36.

Part III
Immunoescape

Immunogenicity of Stem Cell Derived Beta Cells

Nicoline H. M. den Hollander and Bart O. Roep

1 Introduction

The field of beta cell replacement therapy has made major leaps during the last decades [1]. Pancreas transplantation already serves as a last resort for T1D patients with end-stage kidney failure and has gained an impressively good prognosis to reverse insulin-dependency durably and even reduce early-stage diabetic complications [2]. Indeed, the clinical success of pancreas allograft transplantation competes with that of kidney. In this light, it is perhaps surprising that the clinical success of transplantation of islets derived from pancreatic donors falls behind. Several features reduce the efficacy of intraportal islet transplantation, including the injection into the blood leading to instant blood-mediated inflammatory reaction (IBMIR) and their destiny in the liver where they will face glucose and lipid toxicity as well as accumulation of drugs [3]. Other implantation sites have been explored but failed to offer suitable alternatives [4, 5]. One important reason may be their immediate exposure to the immune system that harbours islet autoreactive T cells initially responsible for the destruction of endogenous pancreatic beta cells [6]. It is, therefore, important to consider the processes involved in the beta cell destruction process leading to T1D.

N. H. M. den Hollander · B. O. Roep (✉)
Department of Internal Medicine, Section Immunomodulation & Regenerative Medicine, Leiden University Medical Center, Leiden, The Netherlands
e-mail: b.o.roep@lumc.nl

L. Piemonti et al. (eds.), *Pluripotent Stem Cell Therapy for Diabetes*,
https://doi.org/10.1007/978-3-031-41943-0_13

2 The Cause of Beta Cell Loss in T1D

Type 1 diabetes is generally believed to result from the autoimmune-mediated destruction of pancreatic beta cells in individuals with increased genetic risk, in whom immune regulation remained insufficient to counter islet autoimmunity [7]. T cells play a detrimental role in this process. A wide variety of candidate proteins in beta cells has been identified as targets of these islet-specific autoreactive T cells, including (pro)insulin, glutamic acid decarboxylase 65 kDa (GAD65), the insulinoma-antigen 2 (IA-2), zinc transporter 8 (ZnT8), and many others. Both CD4 and CD8 T cells play a role. They dominate the inflammatory lesion in the pancreatic islets (insulitis) and some of these T1D patient-derived T cells even have been shown to selectively destroy beta cells *in vivo* in humanized mouse models, underscoring their pathogenic nature and detrimental role in the beta cell destruction process leading to T1D [8–11]. Impaired immune regulation contributes to the loss of immune tolerance, perhaps not so much as a result of reduced numbers of regulatory T cells (Tregs), but rather due to their impaired capacity to inhibit effector T cells [12, 13]. Indeed, blocking Tregs in cancer patients with so-called checkpoint inhibiting therapeutic monoclonal antibodies unleashed islet autoimmunity and T1D-like disease in addition to successfully activating tumor-specific effector T cells to cope with cancer [14–16]. We, therefore, contend that "benign islet autoimmunity" does not exist and can prove pathogenic (diabetogenic) when conditions change [7].

Other than impaired immune regulation, the balance between immune tolerance and islet autoimmunity may also be disturbed if beta cells change faces [17]. Beta cells are able to produce very high levels of insulin protein in a very short time period in response to high glucose levels. This demanding task may lead to unfolded protein response (UPR) in which the endoplasmic reticulum (ER) upregulates chaperones to cope with unfolded proteins [18]. This phenomenon may also result in increased misfolded proteins that trigger apoptotic pathways via NF-kB, which increases cytokine and chemokine release, notably CXCL10, attracting immune cells [19, 20]. ER stress can also induce neoantigen generation by beta cells and upregulation of HLA Class I that promote beta cell immunogenicity and destruction [18]. One such mechanism involves the activation of the enzyme tissue transglutaminase that can lead to deamidation of proinsulin and create neoepitopes with superior binding affinity to HLA-DQ8cis and -trans (composed of the DQα-chain of DQ2 and the β-chain of DQ8) that impose (very) high genetic risk for T1D [21]. T cell reactivity to this post-translationally modified proinsulin is exclusively detected in T1D patients [21]. Another major perturbation of beta cell homeostasis results from ribosomal infidelity and involves the generation of defective ribosomal protein products resulting from misreads of insulin mRNA (INS-DRiP) [22]. The generation of the neoantigen INS-DRiP by stressed beta cells dramatically increases their immunogenicity. CD8 T cells against DRiP are exclusively detectable in the insulitic lesions of T1D patients and selectively destroy beta cells [9, 22]. Collectively, these novel observations led us to propose that T1D results from mistakes of stressed beta cells, rather than the immune system (Fig. 1) [7]. Instead, the immune system

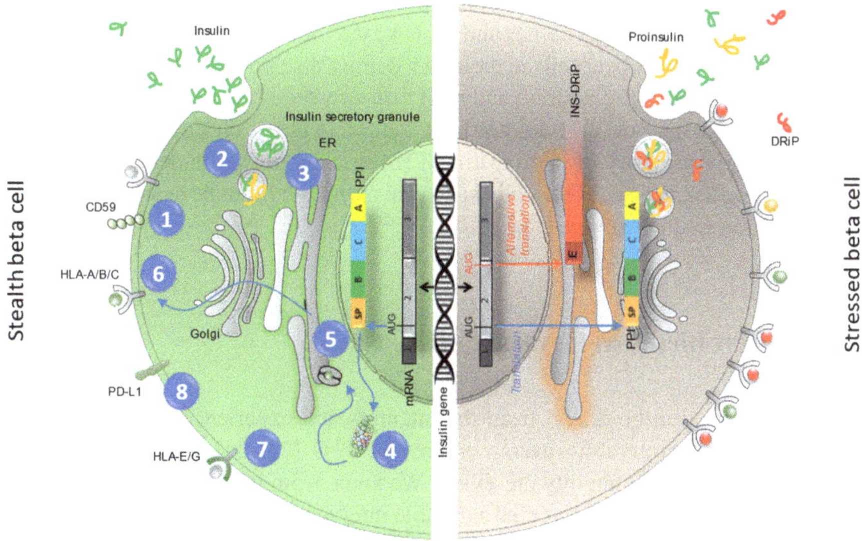

Fig. 1 Turning beta-cells "stealth" to the immune system. A schematic representation of cellular processes within a healthy beta cell (left) and dysregulated processes within a stressed beta cell (right) that increase immunogenicity. The incentive in beta-cell replacement therapy is to create a source of stem cell derived beta cells that is resilient to stress and does not provoke the immune system to react. Stem cell derived beta cells are hypoimmunogenic by nature: **(1)** they express more that CD59 prevents complement-mediated cell lysis, which is downregulated in stressed beta cells; **(2)** they produce less insulin released via secretory granules, which may also contain proinsulin and defective ribosomal insulin product (INS-DRiP) in case of beta cell stress, as these immature granules lack protease activity and cannot catalyse proinsulin into insulin and C-peptide. **(3)** Ribosomes in the endoplasmic reticulum (ER) translate mRNA into preproinsulin. In stressed beta cells, ribosomes start translation at a different start site, leading to an alternative splicing variant (INS-DRiP). Less insulin production imposes less biosynthetic burden and less risk for ribosomal infidelity that would have generated INS-DRiP and other immunogenic neoantigens upon stress**.** Knocking out translation initiation INS-DRiP would prevent the generation of this immunodominant neoantigen and reduce beta-cell immunogenicity. **(4)** The proteasome cleaves preproinsulin into proinsulin peptides, yet, during stress, this proteasome changes into an immunoproteasome; cleaving more rapidly and facilitating transportation into the ER. Inhibiting proteasomal activity will reduce the peptide repertoire expressed by HLA on the surface of beta cells and this reduces beta-cell immunogenicity. **(5)** Transporter associated with antigen processing (TAP), which is upregulated in T1D beta cells, transports proinsulin peptides into the ER. As viruses learned, reducing peptide transport by TAP into the ER will reduce peptide loading of HLA and thus immunogenicity. **(6)** HLA-A/B/C hyperexpression during stress recruits immune cells that recognize neoantigens (e.g., proinsulin, DRiP) and induce T cell-mediated cytotoxicity. Reducing HLA expressing will reduce immunogenicity, but increases the risk for innate immunity by NK T cells. **(7)** HLA-E/G in turn prevents NK cell activation and promote tolerogenesis. **(8)** PD-L1 is expressed on beta cells during inflammation to inhibit T cell-mediated cytotoxicity, but when the beta cell suffers from prolonged stress, PD-L1 is eventually lost; marking the start of beta cell destruction. Genetic engineering of iPSC to express PD-L1 would endorse beta cells derived from these stem cells with immune protection

may act with good intentions, as similar ribosomal infidelity occurs in cancers and infection to expose affected tissues and provoke the immune system to act. This new insight into the role of beta cells in their own demise also points to the need to target beta cells with therapy to destress and revitalize stressed beta cells and may explain why immunotherapy of T1D did not have a durable effect on disease progression [23]. At best, a minority of participating T1D patients showed a delay in progression. As long as (stressed) beta cells keep provoking the immune system, immunotherapy alone will not suffice [7].

3 The Problems in Reversing Disease

Several lessons can be drawn from the attempts and experiences to stop disease progression and preferably reverse or prevent T1D [23, 24]. Immunotherapy is unavoidable if (auto)immunity pre-exists. We know from interventions in T1D that immune suppression has not yet led a durable intervention in T1D progression [23]. In the context of islet transplantation, we also learned that the current choice of immune suppression may prevent allograft rejection but is insufficient to cope with pre-existent islet autoimmunity [25]. Once islet autoimmunity recurs following islet transplantation, it may also precipitate allograft rejection to further limit the impact of beta cell replacement therapy. Another consequence of immune suppression is that it will affect our chances to re-educate the immune system to reinstate immune tolerance, the ultimate challenge in both transplantation and autoimmune diseases. Unfortunately, in case of a donor as source of new beta cells (be it cadaveric pancreas donor, embryo or iPSC), we need to deal with both types of immune reactivity impacting the beta cell graft: allo- and autoimmunity [1, 25]. Endocrine cells also happen to be particularly vulnerable to certain immunosuppressive agents (e.g., corticosteroids) that are actually harming beta cells, while other drugs impair engraftment and angiogenesis – two essential features for beta cells that restore glycaemic control. Furthermore, neoantigens such as the post-transcriptionally modified DRiP and the post-translationally modified deamidated proinsulin elicit a stronger immune response than native autoantigens, because neoantigens do not contribute to thymic education and selection of the immune repertoire [9, 21, 22]. Therefore, they bind HLA with stronger avidity, which induce a strong response that is similar towards viruses or other pathogens. These findings highlight the importance to prevent beta cells from stress – avoiding formation of neoantigens. Reducing beta cell stress may also involve implanting new sources of beta cells. This offers opportunities to treat and edit these new beta cells, in particular if these can be produced from stem cells.

4 Reducing Immunogenicity of Stem Cell-Derived Beta Cells for Clinical Therapy

Decades of efforts to avoid immune attack of beta cells grafts by encapsulation have left us frustrated and discouraged [1]. Even if islets can be put into alginate capsules or behind membranes, they still need to get unimpaired and rapid access to blood for their oxygen, glucose sensing and insulin release. Early efforts to encapsulate embryonic stem cell-derived beta cell progenitors suffered from innate immune reactions against the foreign body implanted, resulting in scarring and fibrosis and affecting the rate of differentiation, maturation and function of beta cells [1, 26] (Table 1). This is a gentle reminder that, if encapsulation is pursued, the features of the membranes to minimize foreign body responses and activation of both the innate

Table 1 Reducing immunogenicity of stem cell derived beta cells

Opportunity		Refs
Their stem cell origin	Less biosynthetic stress	[27]
	Less HLA	[64, 68]
	Less autoantigens	[9, 21, 22]
	Less neoantigens	[27, 28, 33, 48]
	Expression of inhibitory molecules	
Microenvironment	Implantation site	[70]
	Engineering beta cells to produce anti-inflammatory cytokines	[20]
	Eliminate CXCL10 production	[19]
	Avoid type I interferon signaling (TYK2i)	[60–63]
	Beta cell promoting drugs (GLP-1, TXNIPi, EGF)	
	Immune suppressive drugs	
Destress beta cells	Reduce biosynthetic burden	[22]
	Avoid unfolded protein responses	[18, 22]
Antigen expression	Reduce beta cell stress	[18, 22]
	Reduce ribosomal infidelity	[18, 22]
	Prevent neoantigen production	[22]
	Reduce HLA expression	[43]
	Interfere with antigen processing	[43]
	Interfere with epitope transport	[43]
	Mismatch HLA class I with recipient	[35]
Increase their tolerability	Introduce PD-L1	[48, 70]
	Introduce HLA-G/E	[30–34, 41, 42]
	Introduce CD59	[27, 28]
	Knock out CXCL10	[20]
	Knock out CD9	[72]
	Knock out HLA A/B/C, except HLA-A2	[41, 42]

and adaptive immune system, as well as to protect the endocrine content, allow angiogenesis and prevent disease recurrence and rejection. More recent clinical trials assessed perforated capsules: allowing vascularization to reduce hypoxia and allow nutrient supply, but requiring long-term immune suppression, which makes the recipient more susceptible to (potentially dangerous) infections, and allowing escape of grafted (stem) cells that could theoretically activate the immune system or lead to teratoma [1]. Optimizing the chances of their endocrine content is also essential, to create an environment favourable for end-differentiation and function of stem cell-derived endocrine cells in combination with a hypo-inflammatory milieu to discourage the immune system to come in play. To treat or engineer beta cells to foster this incentive, Mother Nature holds several pointers.

5 Lessons from Stem Cells

Stem cell-derived beta cells are already hypoimmunogenic by their nature, at least in the absence of inflammation: they express relatively low levels of HLA in combination with immune inhibitory molecules, such as the complement inhibitor CD59 that is known as a "suicide inhibiting" protein, which prevents complement activation [27, 28]. Thus, CD59 on beta cells promotes their viability. Stem cell-derived beta cells are also less stressed by nature because they produce less insulin. This may be a blessing in disguise that avoids generation of stress-induced neoantigens and production of inflammatory mediators such as IFNα and CXCL10 [19, 20]. Human alpha cells turned into insulin-producing cells are less immunogenic to T1D patient-derived CD8 T-cell clones to DRiP [29]. Lessons from pregnancy – nature's most successful mechanism to induce tolerance between mother and foetus – include the favourable expression of HLA-E and -G in the maternal/foetal interface, as well as production of anti-inflammatory cytokines such as IL-10, and TGFβ, as well as IDO, prostaglandins and EGF [30–33]. The success of tolerance induction between mother and foetus is so strong that even many years after birth, HLA matching of the graft with the recipient is as effective as matching with the recipient's mother [34]. Of course, other mechanisms to induce immune tolerance come in play as well that are not easily translatable to clinical stem cell transplantation, such as the induction of mixed chimerism [35]. Importantly, immune suppressive drugs will impair such desirable tolerance promoting mechanisms. We must also bear in mind that the bar is high in case of transplantation into recipients suffering from an autoimmune disease: it is easier to induce tolerance in the naive than to restore tolerance in the primed [36].

6 Lessons from Viruses

Viruses have developed unparalleled skills to trick the immune system and avoid their clearance from the infected host [37–40]. It is conceivable that some of these can be exploited to endorse stem cell-derived beta cells with the benefits of immune evasion. For instance, cytomegalovirus reduces the expression of HLA molecules, the antennae on the surface of cells that the immune system requires to detect affected tissue [37–40]. It is important, though, to appreciate the level of sophistication that viruses acquired: they do not completely remove HLA, because this would be a signal for natural killer cells in our ready-to-go innate immune system to destroy the cell. Instead, viruses subtly impair antigen processing, thus reducing the peptide repertoire expressed and presented by HLA, as well as interfere with the transport of peptides into the ER. The lesson would, therefore, be to perhaps reduce, but not abolish, HLA expression on the surface of stem cell-derived beta cells. Unfortunately, pharma and academia commonly explore knockout of HLA, perhaps in combination with knock-in of tolerance promoting HLA-E or -G. The results so far have indeed pointed to the subsequent risk of NK cell activation and graft destruction.

HLA-A, HLA-B and HLA-C present their antigen to CD8, which can destroy beta cells. On the other hand, HLA-E promotes tolerance by avoiding NK cell activation. Therefore, knocking out these classical MHC Class I subtypes and preserving HLA-E seems to be a good approach. However, knocking out these classical subtypes also prevents expression of HLA-E during inflammation. Therefore, HLA editing requires precise craftsmanship to specifically induce the desired pathways (immune surveillance) and abolish the others (T cell-mediated cytotoxicity and NK cell activation) [41, 42].

We engineered human β cells to express herpesvirus-encoded immune-evasion proteins, so-called immunevasins [43]. The capacity of immunevasins to protect β cells from autoreactive T-cell killing was evaluated *in vitro* and *in vivo* in humanized mice. Lentiviral vectors were used for efficient genetic modification of primary human β cells without impairing their function (glucose sensing and insulin production). Autoreactive cytotoxic CD8(+) T-cell clones isolated from T1D patients with recent onset diabetes selectively destroyed human β cells, while co expression of the human cytomegalovirus-encoded US2 protein and serine proteinase inhibitor 9 offered highly efficient protection *in vitro* [43]. Moreover, preserved human insulin production and C-peptide secretion was achieved upon co-implantation of these genetically modified pseudo-islets with β-cell-specific cytotoxic T cells into immunodeficient mice. Collectively, these data provide proof of concept that human β cells can be efficiently genetically modified to provide protection from killing mediated by autoreactive T cells and retain their function *in vitro* and *in vivo* and illustrate the opportunities to be guided by ancient immune evasion strategies developed by viruses.

7 Lessons from Cancer

Some cancers, too, take advantage of reducing (but not eliminating) HLA expression [44]. HLA is not only important to induce inflammation but also to induce regulation. Inflammatory cytokines IFN-α and IFN-γ are responsible for upregulation of HLA Class I and programmed cell death ligand-1 (PD-L1). Ligation of PD-L1 to PD-1 inhibits effector T cells [45, 46]. In cancer, anti-PD-1/PD-L1 antibodies have been investigated to reduce immune evasiveness of tumour cells. However, around 1% of cancer patients treated with checkpoint blockade anti-cancer therapy develops T1D [15, 16, 47]. Thus, PD-L1 and HLA overexpression are pathophysiological changes that may actually be protective against T1D progression. Therefore, elimination of HLA during inflammation may also worsen the disease, as beta cells cannot inhibit autoreactive T cells/effector cells through PD-L1. The critical role of PD-L1 ligation in protection from T1D was already discussed. Therapeutic antibodies targeting PD1 and its ligand PD-L1 used in cancer immunotherapy to interfere with immune regulation may lead to autoimmune diseases, including T1D. Interestingly, PD-L1 is expressed in the islets of people with T1D and is upregulated by IFN-α and-γ, possibly to attenuate the autoimmune assault. Human stem cell-derived beta cells engineered to present PD-L1 improved transplant survival in NOD mice carrying human HLA class I despite massive leukocyte infiltrations around the islet graft and showed reduced stimulation of diabetogenic CD8 T cells when compared to unmodified stem cell-derived beta cells [48]. Yet, targeting HLA class I and PD-L1 alone does not sufficiently protect stem cell-derived islet cells from graft rejection. Additional genetic engineering of these cells to secrete IL-10 or TGF-β promoted a tolerogenic local microenvironment by recruiting Tregs to the islet grafts to offer immune protection [49]. Such hypoimmunogenic microenvironments resemble the immune evasive tumor microenvironment, illustrating the opportunities to take lessons from cancer [50].

8 Lessons from T1D Immunopathogenesis and Immunotherapy

It proved very difficult to identify effective immune suppressive therapy to delay, reverse or prevent T1D. The reasons for the partial success or failure of immune intervention therapy in T1D is not fully understood [23]. Nonetheless, our insight into the immunopathogenesis of T1D has changed spectacularly in the last decade, due to access to the pancreatic lesions of T1D patients [8, 51]. First, the degree of beta cell inflammation is much less dramatic than animal models of autoimmune diabetes suggested [8, 51–53]. At the supposed peak of inflammation (around diagnosis), less than 20% of islets show leukocyte infiltration (insulitis), with the average number of infiltrating leukocytes around 15 cells per islet, while 50% of islets or more show insulin content, even in cases where the T1D pancreas donor died from hyperglycemia and insulin deficiency [8]. Curiously, from clinical islet transplantation it is known that 10% of original beta cell mass can suffice to restore glycaemic control [1]. The implications of these findings for T1D patients are profound. First, there is a new sense of urgency that remaining beta cell mass can and should be preserved, even years after diagnosis (currently, immunotherapy is only offered at, or just before, diagnosis of T1D). Second, there is an apparent disconnect between beta cell mass and function: even if no insulin production is detectable, beta cells may still be there, be it dormant (senescent, hibernating), but this cannot yet be determined due to a lack of biomarkers of beta cell mass [55]. The new observation that the rate of inflammation is quite limited, and in few islets only, may imply that systemic, "big hammer" and non-specific immune suppressive therapies may have limited and transient impact [36, 56]. Both the insulitic lesion and the disease process are heterogeneous, as are T1D patients, which adds challenges to intervention therapy and may imply that different therapeutic strategies are required for different patients (precision medicine; Ref. [24, 54]). Restoring tolerance is still in its infancy, but is not impossible, and early steps have been made to re-educate the immune system not to respond to beta cells, for instance by gene therapy and "inverse" vaccination using islet autoantigenic peptides either in solution or loaded onto tolerogenic dendritic cells that prime islet antigen-specific regulatory T cells and suppresses islet autoimmunity [57–59]. Standard of care of T1D patients is still limited to treating its symptoms and complications (insulin shortage, dysglycemia, retinopathy, neuropathy, nephropathy, cardiovascular disease), not its cause (autoimmune destruction of beta cells). To date, disease progression can be slowed down in some, but not in most patients; it cannot be stopped or reversed [23]. While the reason for this problem is still not fully understood, the problem actually may not primarily be immune related, but instead beta cells might lie at the heart of the problem and pathology [7]. Indeed, the pivotal mistake is likely made by stressed beta cells, where the immune system is responding to abnormalities caused by beta cell distress, and with "good intentions", as the immune system has been taught and trained to recognize and respond to such abnormalities that also occur upon infection or tumorigenesis to fight off pathogens and cancer [7]. This still would qualify T1D as an autoimmune disease, but the new insight opens novel opportunities of intervention therapy targeting beta cells rather than, or next to, immunotherapy. As explained earlier, ER stress in response to high insulin demands triggers the synthesis of neoantigens and proinflammatory responses [18, 22, 50]. Therefore, it may prove useful to minimize such stress in stem cell-derived beta cells. This can be overcome by administration of stress-reducing agents such as glucagon-like protein-1 (GLP-1) agonists, thioredoxin-interacting protein (TXNIP) inhibitors and endothelial growth factor (EGF) [60–62]. We contend that the primary incentive in T1D pathogenesis and immunotherapy, as well as beta cell replacement therapy, is to make beta cells "happy" again, under the presumption that the immune tissue is not interested in happy self-tissue [7]. If, however, tissue is stressed by infection, tumorigenesis, metabolism and immune responses, it may cry for "help" by the immune system. Neoantigen expression is a hallmark of "danger" and has all the ingredients to provoke the immune system to act. Avoiding beta cells from eliciting danger signals would, therefore, be a key opportunity to avoid their destruction, either in the pancreas or following beta cell replacement therapy.

9 Lessons from Clinical Islet Transplantation

To reiterate, immune suppressive drugs do not suffice to deal with pre-existent islet autoimmunity and increases risk for infection and cancer [63–68]. If T1D patients elicit T cell autoimmunity to GAD65 and IA-2 up to decades after T1D diagnosis, they never achieve insulin independence following islet transplantation, despite using the current standards of immune suppressive therapy in clinical transplantation [64, 68]. Yet, if T cells of T1D patients neither respond to GAD65 nor IA-2, they have a very favourable prognosis to achieve insulin independence 1 year after islet implantation (>85%). Other favourable prognostic features include low leukocyte kinds and good quality islet grafts [64]. These observations in the clinic underscore the challenge of immune memory and the pre-existent autoimmune condition of the islet recipients [36, 56]. Curiously, mismatching the islet graft for HLA class I may avoid recurrence of CD8 T cell autoimmunity in pre-immunized T1D islet allograft recipients and promote graft survival and function – underscoring that allo-immunity leading to graft rejection may not be a major issue anymore in beta cell replacement therapy, provided that the patient receives immune suppressive medication [69]. Yet, tapering of immune suppressive therapy once the islet graft stops functioning can precipitate comorbidity (e.g., Graves' diseases), emphasizing the desire to develop beta cell replacement strategies not requiring immune suppressive drugs [63, 65]. Identifying a hypo-inflammatory or tolerance promoting implantation site proves difficult [4, 5]. It is, therefore, attractive to consider endorsing the stem cell-derived beta cell graft either acquiring intrinsic immune evasive features (e.g., through genetic engineering of the stem cells) or through creation an artificial niche by building a hypo-inflammatory and tolerance promoting scaffold that offers the protective milieu for stem cell-derived beta cells to thrive and function.

10 Future Perspectives

As we share our scientific knowledge, we are getting closer to creating unlimited sources of beta cells and making stem cell-derived beta cell therapy applicable to a larger group of T1D patients. However, we are not there yet. Each field of T1D research, be it stem cell therapy or immunology, has its own challenges and limitations that delay finding an all-in-one treatment that is effective, harmless and durable. As described earlier, stem cell-derived beta cell therapy becomes more interesting when advancements are made in building immune tolerance – and vice versa. Thus, these strengths should be combined and complement each other to reach the ultimate goal: insulin independence.

Regarding the microenvironment, *in vitro* induced beta cells are not as viable when they are implanted *in vivo*, because of encapsulation difficulties [26]. The need for encapsulation becomes clear when the newly implanted beta cells face IBMIR and allograft rejection or induce graft versus host disease (GvHD) themselves. Yet, encapsulation can also induce foreign body response and completely isolate the implanted cells with non-vascularized, fibrotic tissue, which not only severely impairs their ability to sense glucose fluctuations and release insulin, but also deprives them from nutrients – threatening their viability [26]. New approaches are focusing on micropores within the capsule, but these, in turn, recruited immune cells [70]. Thus, finding the right balance between vascularization and encapsulation is very challenging and it seems necessary to use additional immunosuppressive therapy to prevent allograft rejection. Most immunosuppressive agents, however, need long-term administration to maintain treatment efficacy [63, 65]. Moreover, these agents are often too systemic and non-specific – eliminating a broad range of T cells that are needed to fight chronic infection, in thus far have little impact on pre-existing (auto-)immunity [71].

Hence, our focus should shift towards restoring immune tolerance to islets, for instance by inducing regulatory cells, rather than eliminating effector cells. Ideally, antigen-specific regulatory immune cells are induced that inhibit proliferation of antigen-specific effector cells. For instance, tolerogenic antigen-specific dendritic cells are created from patient-derived monocytes that induce tolerance and improve HbA1c, even in long-standing patients [57, 58]. The downside is the time and effort that is required to treat a single patient, which would not be a realistic solution for global application. Moreover, patients still need insulin treatment to acquire such improvement in HbA1c, which is why this therapy would be even more effective when combined with stem cell-derived beta cell therapy.

It seems likely that shifting the balance towards immune regulation is not sufficient to regenerate insulin production, which is why beta cell immunogenicity itself also needs to be addressed (Fig. 1). Interestingly, some markers were found in stem cell-derived beta cells that actually promote their viability, such as PD-L1 and CD59, which inhibit effector T cells [7, 48] and complement activation [27, 28]. Furthermore, CXCL10 is a marker that attracts immune cells, which could be knocked out to promote beta cell survival [20]. More recently, CD9 was discovered as a marker for beta cell senescence. Beta cells that have low CD9 expression are more responsive to glucose fluctuations and produce less inflammatory cytokines [72]. Another mechanism to reduce immunogenicity is blocking signalling of inflammatory cytokines IFN-α and IFN-γ using JAK inhibitors [73]. However, IFN-γ signalling is needed to clear infections from the gut and protect mucosal barrier. Therefore, TYK2 inhibitors are currently investigated that selectively block TYK2 association with the IFN-α receptor (IFNAR) upon IFN-α stimulation, inhibiting the downstream pathway for antigen processing and MHC Class I upregulation [19]. TYK2 inhibitors almost completely block T cell-mediated cytotoxicity and are, therefore, a promising therapy for prevention of T1D or in combination with stem cell-derived beta cell transplantation [74]. Furthermore, progression has been made in HLA editing. Since knocking out HLA-A/B/C and maintaining HLA-E still induced NK cell activation, because HLA-E needs signalling of HLA-A to be expressed, researchers have now shown that knocking out HLA-A/B/C except for HLA-A2 successfully increases HLA-E expression upon IFN-γ stimulation in a humanized mouse model; maintaining immune surveillance while protecting beta cells from autoreactive T cell activation and NK cell killing [42]. Lastly, stem cell-derived beta cells would benefit from stress-reducing agents such as GLP-1, TXNIP inhibitors and EGF [60–63], as these can enhance glucose-induced insulin release and also prevent immune cell recruitment – relieving beta cell stress and preventing neoantigen production.

References

1. Odorico J, Markmann J, Melton D, et al. Report of the Key Opinion Leaders Meeting on Stem Cell-derived Beta Cells. *Transplantation* 2018; **102**(8): 1223–9.
2. Smets YF, Westendorp RG, van der Pijl JW, et al. Effect of simultaneous pancreas-kidney transplantation on mortality of patients with type-1 diabetes mellitus and end-stage renal failure. *Lancet* 1999; **353**(9168): 1915–9.
3. Harlan DM, Kenyon NS, Korsgren O, Roep BO, Immunology of Diabetes S. Current advances and travails in islet transplantation. *Diabetes* 2009; **58**(10): 2175–84.
4. Van Hulle F, De Groot K, Hilbrands R, et al. Function and composition of pancreatic islet cell implants in omentum of type 1 diabetes patients. *Am J Transplant* 2022; **22**(3): 927–36.
5. Maffi P, Nano R, Monti P, et al. Islet Allotransplantation in the Bone Marrow of Patients With Type 1 Diabetes: A Pilot Randomized Trial. *Transplantation* 2019; **103**(4): 839–51.
6. Moberg L, Johansson H, Lukinius A, et al. Production of tissue factor by pancreatic islet cells as a trigger of detrimental thrombotic reactions in clinical islet transplantation. *Lancet* 2002; **360**(9350): 2039–45.
7. Roep BO, Thomaidou S, van Tienhoven R, Zaldumbide A. Type 1 diabetes mellitus as a disease of the beta cell (do not blame the immune system?). *Nat Rev Endocrinol* 2021; **17**(3): 150–61.
8. Coppieters KT, Dotta F, Amirian N, et al. Demonstration of islet-autoreactive CD8 T cells in insulitic lesions from recent onset and long-term type 1 diabetes patients. *J Exp Med* 2012; **209**(1): 51–60.
9. Anderson AM, Landry LG, Alkanani AA, et al. Human islet T cells are highly reactive to preproinsulin in type 1 diabetes. *Proc Natl Acad Sci U S A* 2021; **118**(41).
10. Tan S, Li Y, Xia J, et al. Type 1 diabetes induction in humanized mice. *Proc Natl Acad Sci U S A* 2017; **114**(41): 10954–9.
11. Michels AW, Landry LG, McDaniel KA, et al. Islet-Derived CD4 T Cells Targeting Proinsulin in Human Autoimmune Diabetes. *Diabetes* 2017; **66**(3): 722–34.
12. Lindley S, Dayan CM, Bishop A, Roep BO, Peakman M, Tree TI. Defective suppressor function in CD4(+)CD25(+) T-cells from patients with type 1 diabetes. *Diabetes* 2005; **54**(1): 92–9.
13. Schneider A, Rieck M, Sanda S, Pihoker C, Greenbaum C, Buckner JH. The effector T cells of diabetic subjects are resistant to regulation via CD4+ FOXP3+ regulatory T cells. *J Immunol* 2008; **181**(10): 7350–5.
14. Perdigoto AL, Quandt Z, Anderson M, Herold KC. Checkpoint inhibitor-induced insulin-dependent diabetes: an emerging syndrome. *Lancet Diabetes Endocrinol* 2019; **7**(6): 421–3.
15. Samoa RA, Lee HS, Kil SH, Roep BO. Anti-PD-1 Therapy-Associated Type 1 Diabetes in a Pediatric Patient With Relapsed Classical Hodgkin Lymphoma. *Diabetes Care* 2020; **43**(9): 2293–5.
16. de Filette JMK, Pen JJ, Decoster L, et al. Immune checkpoint inhibitors and type 1 diabetes mellitus: a case report and systematic review. *Eur J Endocrinol* 2019; **181**(3): 363–74.
17. van Lummel M, Zaldumbide A, Roep BO. Changing faces, unmasking the beta cell: post-translational modification of antigens in type 1 diabetes. *Curr Opin Endocrinol Diabetes Obes* 2013; **20**(4): 299–306.
18. Thomaidou S, Kracht MJL, van der Slik A, et al. beta Cell Stress Shapes CTL Immune Recognition of Preproinsulin Signal Peptide by Posttranscriptional Regulation of Endoplasmic Reticulum Aminopeptidase 1. *Diabetes* 2020; **69**(4): 670–80.

19. Marroqui L, Dos Santos RS, Floyel T, et al. TYK2, a Candidate Gene for Type 1 Diabetes, Modulates Apoptosis and the Innate Immune Response in Human Pancreatic beta Cells. *Diabetes* 2015; **64**(11): 3808–17.
20. Roep BO, Kleijwegt FS, van Halteren AG, et al. Islet inflammation and CXCL10 in recent-onset type 1 diabetes. *Clin Exp Immunol* 2010; **159**(3): 338–43.
21. van Lummel M, Duinkerken G, van Veelen PA, et al. Posttranslational modification of HLA-DQ binding islet autoantigens in type 1 diabetes. *Diabetes* 2014; **63**(1): 237–47.
22. Kracht MJ, van Lummel M, Nikolic T, et al. Autoimmunity against a defective ribosomal insulin gene product in type 1 diabetes. *Nat Med* 2017; **23**(4): 501–7.
23. Atkinson MA, Roep BO, Posgai A, Wheeler DCS, Peakman M. The challenge of modulating beta cell autoimmunity in type 1 diabetes. *Lancet Diabetes Endocrinol* 2019; **7**(1): 52–64.
24. Battaglia M, Ahmed S, Anderson MS, et al. Introducing the Endotype Concept to Address the Challenge of Disease Heterogeneity in Type 1 Diabetes. *Diabetes Care* 2020; **43**(1): 5–12.
25. Roep BO, Stobbe I, Duinkerken G, et al. Auto- and alloimmune reactivity to human islet allografts transplanted into type 1 diabetic patients. *Diabetes* 1999; **48**(3): 484–90.
26. Alagpulinsa DA, Cao JJL, Driscoll RK, et al. Alginate-microencapsulation of human stem cell-derived beta cells with CXCL12 prolongs their survival and function in immunocompetent mice without systemic immunosuppression. *Am J Transplant* 2019; **19**(7): 1930–40.
27. van der Torren CR, Zaldumbide A, Duinkerken G, et al. Immunogenicity of human embryonic stem cell-derived beta cells. *Diabetologia* 2017; **60**(1): 126–33.
28. van der Torren CR, Zaldumbide A, Roelen DL, et al. Innate and adaptive immunity to human beta cell lines: implications for beta cell therapy. *Diabetologia* 2016; **59**(1): 170–5.
29. Furuyama K, Chera S, van Gurp L, et al. Diabetes relief in mice by glucose-sensing insulin-secreting human alpha-cells. *Nature* 2019; **567**(7746): 43–8.
30. Craenmehr MHC, Haasnoot GW, Drabbels JJM, et al. Soluble HLA-G levels in seminal plasma are associated with HLA-G 3′UTR genotypes and haplotypes. *HLA* 2019; **94**(4): 339–46.
31. Craenmehr MHC, Nederlof I, Cao M, et al. Increased HLA-G Expression in Term Placenta of Women with a History of Recurrent Miscarriage Despite Their Genetic Predisposition to Decreased HLA-G Levels. *Int J Mol Sci* 2019; **20**(3).
32. Claas FH, Gijbels Y, van der Velden-de Munck J, van Rood JJ. Induction of B cell unresponsiveness to noninherited maternal HLA antigens during fetal life. *Science* 1988; **241**(4874): 1815–7.
33. Gornalusse GG, Hirata RK, Funk SE, et al. HLA-E-expressing pluripotent stem cells escape allogeneic responses and lysis by NK cells. *Nat Biotechnol* 2017; **35**(8): 765–72.
34. Burlingham WJ, Grailer AP, Heisey DM, et al. The effect of tolerance to noninherited maternal HLA antigens on the survival of renal transplants from sibling donors. *N Engl J Med* 1998; **339**(23): 1657–64.
35. Eikmans M, van Halteren AG, van Besien K, van Rood JJ, Drabbels JJ, Claas FH. Naturally acquired microchimerism: implications for transplantation outcome and novel methodologies for detection. *Chimerism* 2014; **5**(2): 24–39.
36. Roep BO, Wheeler DCS, Peakman M. Antigen-based immune modulation therapy for type 1 diabetes: the era of precision medicine. *Lancet Diabetes Endocrinol* 2019; **7**(1): 65–74.
37. Wiertz EJ, Mukherjee S, Ploegh HL. Viruses use stealth technology to escape from the host immune system. *Mol Med Today* 1997; **3**(3): 116–23.
38. Wiertz EJ, Tortorella D, Bogyo M, et al. Sec61-mediated transfer of a membrane protein from the endoplasmic reticulum to the proteasome for destruction. *Nature* 1996; **384**(6608): 432–8.
39. Jones TR, Wiertz EJ, Sun L, Fish KN, Nelson JA, Ploegh HL. Human cytomegalovirus US3 impairs transport and maturation of major histocompatibility complex class I heavy chains. *Proc Natl Acad Sci U S A* 1996; **93**(21): 11327–33.
40. Wiertz EJ, Jones TR, Sun L, Bogyo M, Geuze HJ, Ploegh HL. The human cytomegalovirus US11 gene product dislocates MHC class I heavy chains from the endoplasmic reticulum to the cytosol. *Cell* 1996; **84**(5): 769–79.
41. Cirulli V, Zalatan J, McMaster M, et al. The class I HLA repertoire of pancreatic islets comprises the nonclassical class Ib antigen HLA-G. Diabetes. 2006;55(5):1214–1222. https://doi.org/10.2337/db05-0731

42. Parent AV, Faleo G, Chavez J, Saxton M, Berrios DI, Kerper NR, Tang Q, Hebrok M. Selective deletion of human leukocyte antigens protects stem cell-derived islets from immune rejection. Cell Rep. 2021 Aug 17;36(7):109538. https://doi.org/10.1016/j.celrep.2021.109538. PMID: 34407395
43. Zaldumbide A, Alkemade G, Carlotti F, et al. Genetically engineered human islets protected from CD8-mediated autoimmune destruction in vivo. *Mol Ther* 2013; **21**(8): 1592–601.
44. Koopman LA, van Der Slik AR, Giphart MJ, Fleuren GJ. Human leukocyte antigen class I gene mutations in cervical cancer. *J Natl Cancer Inst* 1999; **91**(19): 1669–77.
45. Colli ML, Hill JLE, Marroqui L, et al. PDL1 is expressed in the islets of people with type 1 diabetes and is up-regulated by interferons-alpha and-gamma via IRF1 induction. *EBioMedicine* 2018; **36**: 367–75.
46. Perdigoto AL, Deng S, Du KC, et al. Immune cells and their inflammatory mediators modify beta cells and cause checkpoint inhibitor-induced diabetes. *JCI Insight* 2022; **7**(17).
47. Stamatouli AM, Quandt Z, Perdigoto AL, et al. Collateral Damage: Insulin-Dependent Diabetes Induced With Checkpoint Inhibitors. *Diabetes* 2018; **67**(8): 1471–80.
48. Santini-Gonzalez J, Castro-Gutierrez R, Becker MW, Rancourt C, Russ HA, Phelps EA. Human stem cell derived beta like cells engineered to present PD-L1 improve transplant survival in NOD mice carrying human HLA class I. *Front Endocrinol (Lausanne)* 2022; **13**: 989815.
49. Gerace D, Zhou Q, Kenty JH, et al. Engineering human stem cell-derived islets to evade immune rejection and promote localized immune tolerance. *Cell Rep Med* 2023; **4**(1): 100879.
50. Kracht MJL, Zaldumbide A, Roep BO. Neoantigens and Microenvironment in Type 1 Diabetes: Lessons from Antitumor Immunity. *Trends Endocrinol Metab* 2016; **27**(6): 353–62.
51. Campbell-Thompson M, Fu A, Kaddis JS, et al. Insulitis and beta Cell Mass in the Natural History of Type 1 Diabetes. *Diabetes* 2016; **65**(3): 719–31.
52. Leete P, Oram RA, McDonald TJ, et al. Studies of insulin and proinsulin in pancreas and serum support the existence of aetiopathological endotypes of type 1 diabetes associated with age at diagnosis. *Diabetologia* 2020; **63**(6): 1258–67.
53. Leete P, Willcox A, Krogvold L, et al. Differential Insulitic Profiles Determine the Extent of beta Cell Destruction and the Age at Onset of Type 1 Diabetes. *Diabetes* 2016; **65**(5): 1362–9.
54. Arif S, Leete P, Nguyen V, et al. Blood and islet phenotypes indicate immunological heterogeneity in type 1 diabetes. *Diabetes* 2014; **63**(11): 3835–45.
55. Roep BO, Peakman M. Surrogate end points in the design of immunotherapy trials: emerging lessons from type 1 diabetes. *Nat Rev Immunol* 2010; **10**(2): 145–52.
56. Malmegrim KC, de Azevedo JT, Arruda LC, et al. Immunological Balance Is Associated with Clinical Outcome after Autologous Hematopoietic Stem Cell Transplantation in Type 1 Diabetes. *Front Immunol* 2017; **8**: 167.
57. Nikolic T, Suwandi JS, Wesselius J, et al. Tolerogenic dendritic cells pulsed with islet antigen induce long-term reduction in T-cell autoreactivity in type 1 diabetes patients. *Front Immunol* 2022; **13**: 1054968.
58. Nikolic T, Zwaginga JJ, Uitbeijerse BS, et al. Safety and feasibility of intradermal injection with tolerogenic dendritic cells pulsed with proinsulin peptide-for type 1 diabetes. *Lancet Diabetes Endocrinol* 2020; **8**(6): 470–2.
59. Alhadj Ali M, Liu YF, Arif S, et al. Metabolic and immune effects of immunotherapy with proinsulin peptide in human new-onset type 1 diabetes. *Sci Transl Med* 2017; **9**(402).
60. von Herrath M, Bain SC, Bode B, et al. Anti-interleukin-21 antibody and liraglutide for the preservation of beta cell function in adults with recent-onset type 1 diabetes: a randomised, double-blind, placebo-controlled, phase 2 trial. *Lancet Diabetes Endocrinol* 2021; **9**(4): 212–24.
61. Lam TKT, Cherney DZI. Beta cell preservation in patients with type 1 diabetes. *Nat Med* 2018; **24**(8): 1089–90.
62. Ovalle F, Grimes T, Xu G, et al. Verapamil and beta cell function in adults with recent-onset type 1 diabetes. *Nat Med* 2018; **24**(8): 1108–12.
63. Huurman VA, van der Torren CR, Gillard P, et al. Immune responses against islet allografts during tapering of immunosuppression--a pilot study in 5 subjects. *Clin Exp Immunol* 2012; **169**(2): 190–8.

64. Hilbrands R, Huurman VA, Gillard P, et al. Differences in baseline lymphocyte counts and autoreactivity are associated with differences in outcome of islet cell transplantation in type 1 diabetic patients. *Diabetes* 2009; **58**(10): 2267–76.
65. Gillard P, Huurman V, Van der Auwera B, et al. Graves hyperthyroidism after stopping immunosuppressive therapy in type 1 diabetic Islet cell recipients with pretransplant TPO autoantibodies. *Diabetes Care* 2009; **32**(10): 1817–9.
66. Roelen DL, Huurman VA, Hilbrands R, et al. Relevance of cytotoxic alloreactivity under different immunosuppressive regimens in clinical islet cell transplantation. *Clin Exp Immunol* 2009; **156**(1): 141–8.
67. Huurman VA, Velthuis JH, Hilbrands R, et al. Allograft-specific cytokine profiles associate with clinical outcome after islet cell transplantation. *Am J Transplant* 2009; **9**(2): 382–8.
68. Huurman VA, Hilbrands R, Pinkse GG, et al. Cellular islet autoimmunity associates with clinical outcome of islet cell transplantation. *PLoS One* 2008; **3**(6): e2435.
69. Velthuis JH, Unger WW, Abreu JR, et al. Simultaneous detection of circulating autoreactive CD8+ T-cells specific for different islet cell-associated epitopes using combinatorial MHC multimers. *Diabetes* 2010; **59**(7): 1721–30.
70. Yoshihara E, O'Connor C, Gasser E, Wei Z, Oh TG, Tseng TW, et al. Immune-evasive human islet-like organoids ameliorate diabetes. Nature. (2020) 586:606–11. https://doi.org/10.1038/s41586-020-2631-z
71. Srinivas TR, Meier-Kriesche HU. Minimizing immunosuppression, an alternative approach to reducing side effects: objectives and interim result. Clin J Am Soc Nephrol. 2008;3 Suppl 2(Suppl 2):S101–S116. https://doi.org/10.2215/CJN.03510807 Immune suppression infections
72. Li X, Yang KY, Chan VW, et al. Single-Cell RNA-Seq Reveals that CD9 Is a Negative Marker of Glucose-Responsive Pancreatic β-like Cells Derived from Human Pluripotent Stem Cells. Stem Cell Reports. 2020;15(5):1111–1126. https://doi.org/10.1016/j.stemcr.2020.09.009
73. Assadiasl S, Fatahi Y, Mosharmovahed B, Mohebbi B, Nicknam MH. Baricitinib: From Rheumatoid Arthritis to COVID-19. J Clin Pharmacol. 2021;61(10):1274–1285. https://doi.org/10.1002/jcph.1874
74. Chandra V, Ibrahim H, Halliez C, Prasad RB, Vecchio F, Dwivedi OP, Kvist J, Balboa D, Saarimäki-Vire J, Montaser H, Barsby T, Lithovius V, Artner I, Gopalakrishnan S, Groop L, Mallone R, Eizirik DL, Otonkoski T. The type 1 diabetes gene TYK2 regulates β-cell development and its responses to interferon-α. Nat Commun. 2022 Oct 26;13(1):6363. https://doi.org/10.1038/s41467-022-34069-z. PMID: 36289205; PMCID: PMC9606380.

Immune Evasive Stem Cell Islets

Federica Cuozzo, Valeria Sordi, and Lorenzo Piemonti

1 Mechanism of Rejection

A major player in the immune recognition of allogeneic ESC- or iPSC-derived β cells is the mismatch between donor and recipient human leukocyte antigens (HLA). These are encoded by six genes in the major histocompatibility complex (MHC) and result in two classes of highly polymorphic glycoproteins that bind self- and non-self peptides for presentation on antigen-presenting cells (APCs) [49, 61]. Class I is further divided into three major subclasses, HLA-A, HLA-B and HLA-C, and three minor classes named HLA-E, HLA-F and HLA-G [18, 63]. Class I molecules are expressed on the majority of nucleated cells and are responsible for the presentation of intracellular and viral proteins for recognition by $CD8^+$ cytotoxic T lymphocytes (CTLs). Class I molecules are complosed of a 44 kDa α-chain comprising an intracellular region, a transmembrane domain containing the $CD8^+$ docking region and three extracellular immunoglobulin (Ig)-like domains complexed with the light chain β2-microglobulin (B2M) [3, 61]. Class II, including HLA-DR, HLA-DP and HLA-DQ, is expressed in specialized APCs, such as B lymphocytes, dendritic cells and macrophages and it presents pathogens or fragments of extracellular proteins to $CD4^+$ helper T cells [11, 49, 61]. Molecules belonging to HLA class II have a 32 kDa α-chain and a 28 kDa β-chain consisting of an intracellular tail, a $CD4^+$-docking transmembrane region and two extracellular Ig-like domains [3, 61]. Owing to the polymorphic nature of the MHC, it is highly improbable to find fully

F. Cuozzo · V. Sordi
Diabetes Research Institute, IRCCS San Raffaele Hospital, Milan, Italy

L. Piemonti (✉)
Diabetes Research Institute and Clinical Unit of Regenerative Medicine and Transplant, IRCCS San Raffaele Hospital, Milan, Italy

Vita-Salute San Raffaele University, Milan, Italy
e-mail: piemonti.lorenzo@hsr.it

L. Piemonti et al. (eds.), *Pluripotent Stem Cell Therapy for Diabetes*,
https://doi.org/10.1007/978-3-031-41943-0_14

HLA-matched donors, thus it is likely that the transplantation of allografts would elicit the activation of T lymphocytes in the recipient [41, 61]. Upon transplantation, immune rejection of the graft can occur in two different ways, namely the direct and indirect pathways of allograft recognition (Fig. 1) [2, 18, 61]. These pathways are distinguished based on the immune cells triggering the rejection. Direct allorecognition occurs when host CD8$^+$ and CD4$^+$ T cells are activated by the interaction with donor HLA class I and class II complexed with donor peptides on donor APCs [58]. Importantly, this pathway recruits between 1 and 5% of all T cells, accounting for early and acute graft rejection [2]. Conversely, indirect allorecognition is achieved when donor-derived antigens are engulfed and processed in the recipient APCs. The derived peptides are then presented on the recipient APCs, complexed with the recipient HLA class I and II for recognition by CD8$^+$ and CD4$^+$ T cells, respectively [18, 51, 58]. Upon binding the presented peptides through the T cell receptor (TCR), CD4$^+$ T cells interact with co-stimulatory molecules, such as CD80/CD28 and CD40/CD154, and become activated [61]. As a consequence, gene transcription of

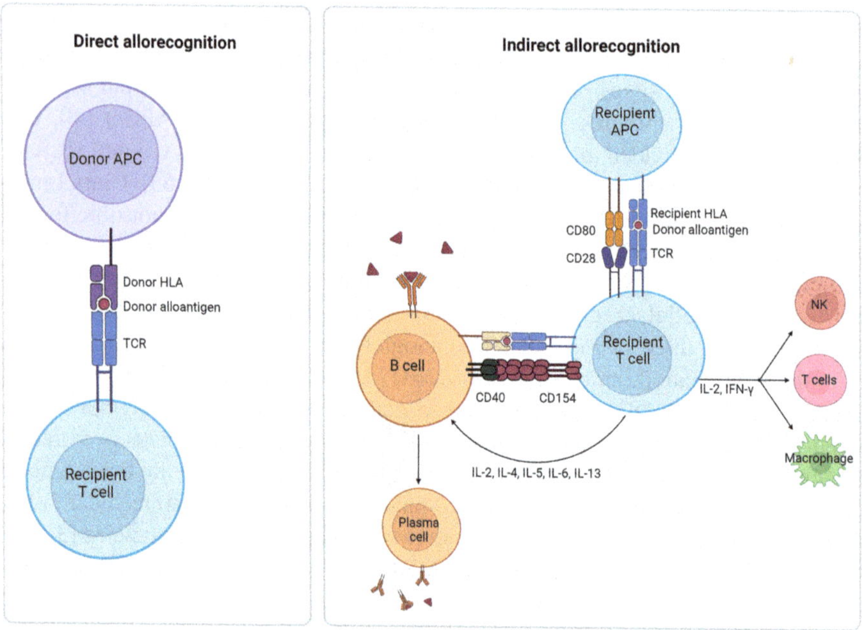

Fig. 1 Direct and indirect pathways of allorecognition. On the left-hand side, the schematic shows the mechanism of direct recognition by recipient T cells of the donor alloantigen cell presented on donor APC. On the right-hand side, the indirect pathways of allorecognition is depicted. Donor alloantigens are presented on the surface of the recipient APC for recognition by the recipient T cells. The latter interacting with CD80/CD28, CD40/CD154 induce the expression of IL-2, IFN-γ, IL-4, IL-5, IL-6, IL-13 leading to the differentiation of B cells into plasma cells and the activation of T cells, NK, and macrophages. APC, antigen-presenting cell; IL-2, interleukin 2; IFN-γ, interferon γ; IL-4, interleukin 4; IL-5, interleukin 5; IL-6, interleukin 6; IL-13, interleukin 13; NK, natural killer

a wide range of cytokines, including interleukin 2 (IL-2), interleukin 4 (IL-4), interleukin 5 (IL-5), interleukin 6 (IL-6), interleukin 13 (IL-13) and interferon γ (IFN-γ), is induced [51, 61]. IL-2 and IFN-γ secreted by activated CD4+ T helper cells contribute to the differentiation of CD8+ T cells into functional CTLs as well as to the activation of non-antigen-specific effectors such as macrophages and natural killer (NK) cells [61]. Moreover, the release of IFN-γ is associated with the upregulation of HLA genes in the allogeneic graft, leading to increased sensitivity of host T cells towards donor HLA [11, 41]. The release of IL-2, IL-4, IL-5, IL-6, and IL-13 by T helper cells is involved in the differentiation of B cells into antibody-secreting plasma cells (Fig. 1) [61]. CTLs are responsible for direct target cell cytotoxic lysis, obtained via cytoskeletal alteration of the CTLs themselves and their targets, resulting in the formation of a supramolecular adhesion complex (SMAC). This is an immunological synapse through which CTL lytic granules pass, inducing the release of apoptotic granzymes and perforins in target cells [60, 61]. Non-antigen-specific effector cells elicit the so-called delayed-type hypersensitivity (DTH) response, which induces allorejection via augmented vascular permeability, endothelial activation and increased leukocyte migration, prompted by the release of pro-inflammatory cytokines and chemokines [61]. B lymphocytes can act as both APCs and antibody-producing plasma cells. The former bind donor proteins on their surface Ig and process them into shorter peptides to present on APC B cells to recruit CD4+ T helper cells. These, through the release of cytokines, can further enhance the proliferation and maturation of B lymphocytes. Thus, the mature plasma cells produce monoclonal antibodies against the original donor protein, which is opsonized and degraded by complement-mediated cell lysis (Fig. 1) [61]. Noticeably, NK play a fundamental role in discriminating self-HLA and destroying donor cells that fail to express sufficient HLA class I molecules [37, 41]. This mechanism is known as missing-self recognition and it allows NK cells to orchestrate allorejection even in the absence of HLA class I molecules [36, 41]. The ability of NK cells to detect the presence or absence of specific HLA class I alleles is the result of a finely tuned balance of inhibitory and activating receptors. Indeed, the inhibitory killer cell immunoglobulin-like receptors (KIR) on NK cells can detect the presence of HLA-A, HLA-B and HLA-C preventing NK-mediated cell lysis. Likewise, the C-type lectin-like receptor NKG2A binds to HLA-E resulting in the inhibition of NK cells [33]. In contrast, receptors such as NKG2D, DNAM-1 and natural cytotoxicity receptors (NCRs), namely NKp30, NKp44 and NKp46 play a potent activating function in NK cells [32, 57]. Under standard conditions, HLA class I molecules expressed on the surface of healthy cells act as ligands, which binging to inhibitory receptors on NK cells elicit immune surveillance and protection. Conversely, when HLA class I molecules are absent, NK cells are subjected to a lower inhibitory stimulus; in fact, they are activated by binding to upregulated activating receptors present on non-self cells [34]. Activated NK cells undergo calcium rises and cytoskeletal changes leading to the formation of an immunological synapse releasing perforin and granzymes [38, 64]. In addition to these cytotoxic molecules, cytokines such as IFN-γ, tumor necrosis factor-α (TNF-α), IL-5, IL-13 interleukin 10 (IL-10) granulocyte macrophage-colony stimulating factor (GM-CSF) and chemokines such as CCL1, CCL2, CCL3, CCL4, CCL5 and CXCL8 are also released [4, 15, 50]. These exert a pro-inflammatory function eventually leading to target cell apoptosis [43].

2 Invisible Cells

In light of the strong immune response deriving from HLA mismatch, the success of ESC- or iPSC-derived β cell transplantation is strictly linked to the use of effective strategies to shield the cells from allorejection. To do this, conventional immunosuppressive therapies with IL-2 receptor monoclonal antibody or ATG, paired with maintenance using sirolimus and tacrolimus, are traditionally employed [1]. However, the prolonged use of immunosuppressants is far less than ideal as it increases the risk of developing teratomas and causes toxicity, which might lead to systemic immunosuppression [30, 40]. To address this, interruption of immunosuppressive therapy is required, which eventually results in graft rejection [41]. Despite increasing efforts to develop immunosuppressive drugs with limited side effects, such as molecular target drugs, low molecular weight compounds and mild immunosuppressive regimens, the continued use of immunosuppressants remains a high-risk option [41, 59]. Thus, the identification of strategies to transplant cells that are "invisible" to the host immune system is gaining increasing attention in the scientific community.

2.1 HLA Matched Cells

One option is to generate a pluripotent stem cell bank expressing a selection of HLA haplotypes matching a large number of patients belonging to different geographical and ethnic groups [59, 62]. To match all or most of the existing haplotypes, access to thousands of ESC or iPSC donors would be required [59, 61, 62]. Indeed, more than 9000 alleles are involved in the expression of HLA haplotypes [59]. It was estimated that to match only the 20 most frequent haplotypes in 50–22% of European Americans and African Americans, 26,000 and 110,000 donors would be needed from each population [21]. The magnitude of such a project suggests the potential for a smaller but highly selected group of donor cells for use in a wide range of graft hosts. Since only one HLA haplotype needs matching to avoid immune rejection, HLA-homozygous cell lines can be used as donors for several recipients [49]. Therefore, the development of a clinical good manufacturing practice (GMP) "haplobank" from a list of HLA homozygous iPSC donors classified based on their optimal utility is desirable. While this does not guarantee the availability of suitable HLA types for all recipients, it enables high percentages of successful transplantations in various populations [59]. Nonetheless, the creation of a national or international haplobank ensuring standard procedures and equity of distribution continues to be an ambitious project. It requires concerted efforts from large groups of collaborators worldwide and the development of more manageable approaches is crucial [21]. In line with this, in a work from 2013 Riolobos and colleagues established a method to generate HLA-homozygous cells from heterozygous ESC donors. Through a two-step gene targeting and mitotic recombination strategy, they were able to derive HLA homozygous ESC to create a small haplobank [49]. However, their practice is hindered by the need for embryonic donors, which is associated with the ethical problem of HLA haplotyping and the selection of sperm and egg donors for ESC production [55]. Although this limitation can be overcome using iPSC derived from HLA-homozygous donors sharing one haplotype with the recipient [49], other approaches have been developed to engineer ESC and iPSC lines to escape immune rejection.

2.2 Abrogation of *B2M*

To date, the most successful approach to reduce immunogenicity is the abrogation of the *B2M* gene (Fig. 2). This encodes for a common subunit of HLA class I molecules and its knockout results in HLA class I negative iPSC. *B2M*-lacking cells have the great potential of functioning as universal donors for the transplantation of cells that do not express HLA class II, such as solid organ cells [49]. Importantly, the generation of a universal donor is also advantageous compared to the creation of a haplobank and especially for patient-specific cell reprogramming and differentiation [49]. Several groups have achieved disruption of *B2M* in diverse ways. For instance, in the aforementioned study, Riolobos et al. also inactivated *B2M* in ESC using *Cre*-recombinase to ablate two adeno-associated virus (AAV)-inserted cassettes into exon 1 of the *B2M* gene [49]. Through this method, they successfully achieved *B2M* silencing as confirmed by flow cytometry, immunoblotting and PCR data. They showed that *B2M*$^{-/-}$ human ESC (hESC) differentiated for 15–18 days elicit reduced allogeneic response in HLA-A*0201-restricted CD8$^+$ T cells (HLA-A*0201 being the most common allele in HLA-A2 subtypes among Caucasians [65] and in peripheral blood mononucleated cells (PBMC) [49]. Similarly, in a later study, the expression of *B2M* was abrogated by targeting exons 2 and 3 of both alleles of the gene in hESC [67]. Indeed, exon 2 and exon 3 were, respectively, replaced with NEO^R and PURO^R cassettes via homologous recombination, obtaining *B2M*$^{-/-}$ hESC with preserved self-renewal capacity, genomic stability and in vitro pluripotency. In this work, hESC were differentiated for 14 days and shown to prevent the expression of HLA class I molecules even after stimulation with IFN-γ. Importantly, gene editing, as described in this work, protected B2M-null cells from CD8$^+$ response, though showing increased susceptibility to NK cell-mediated destruction [67]. Importantly, this study by Wang and colleagues was the first to exploit an in vivo model to show proof-of-concept for *B2M* ablation in hESC. While in all studies aimed at the generation of invisible cells for cell therapy in vitro models of the immune response are used, these methods can sometimes be insufficient

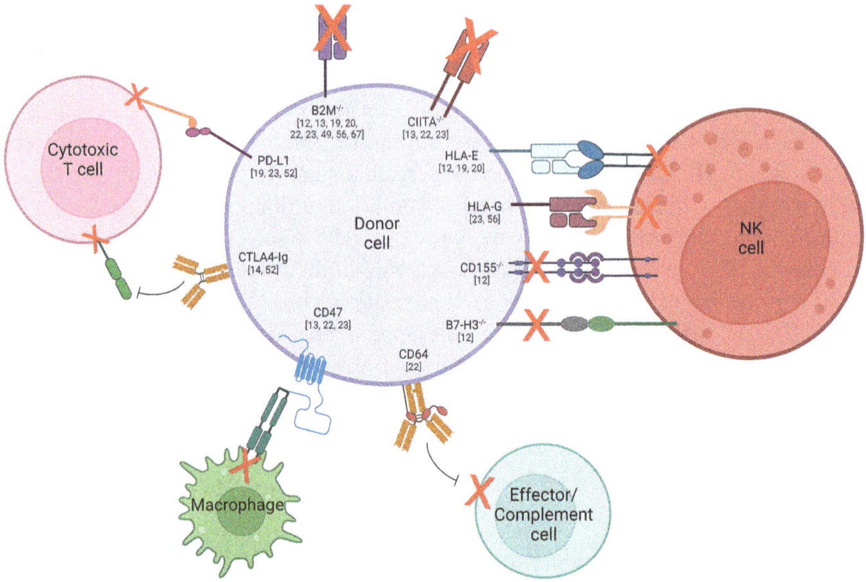

Fig. 2 Generation of invisible cells. Schematic summarizing different approaches to the production of cells eliciting reduced immunogenicity, through the abrogation of B2M, the evasion of NK-mediated response, and the induction of tolerogenic genes. Citations to published studies employing one or more of the described methods are included in the figure. B2M, β2-microglobulin; NK, natural killer; CIITA, class II MHC transactivator; HLA-E, human leukocyte antigen class E; HLA-G, human leukocyte antigen class G; CTLA4-Ig, cytotoxic T lymphocytes antigen 4 immunoglobulin; PD-L1, programmed death 1 ligand

to recapitulate the complexity of in vivo allorejection. To validate the potential of the suggested strategies, it is important to focus on their efficacy in models that more closely resemble the efficacy of such strategies in patients. As previously noted, Wang et al. obtained a *B2M*$^{-/-}$ hESC cell line that could successfully evade CD8$^+$ T cell-mediated immunity in vitro [67]. To test whether these results could be replicated in vivo, immunocompetent and immunodeficient SCID mice were used. These were transplanted with *B2M*$^{-/-}$ hESC and assessed for teratoma formation either in an NK-expressing and NK-depleted background. As anticipated, *B2M*$^{-/-}$ cells did not elicit a CD8$^+$ T cell response in any mouse models or NK cell backgrounds. However, NK cell inhibition was necessary for the HLA class I-null hESC to escape NK cell missing-self response in immunocompetent as well as in SCID mice [67]. Since NK cells play an important function in preventing viral infections and tumorigenesis, such depletion might result in the long-term development of infections and tumors [67]. The development of ever-evolving strategies to knock out *B2M* has facilitated the generation of hypoimmunogenic pluripotent cells retaining their potential to differentiate. However, not only does silencing of the accessory chain B2M prevent the expression of HLA class I, subclasses A, B and C, but it also avoids the expression of HLA-E and HLA-G, which are required for NK cell tolerance [16, 35]. Therefore, *B2M* ablation is often accompanied by additional strategies to ensure evasion from the NK cell cytotoxic response and to increase immune tolerance.

2.3 Evasion of NK Cytotoxicity

In 2017, the same team that initially developed the two cassette insertions followed by *Cre*-recombinase method for *B2M* excision established a system to escape NK cells-mediated lysis. Upon silencing *B2M* via *Cre*-recombinase removing the inserted *HyTK* and *TKNeo* fusion genes, they implemented further AAV vectors to knock in the expression of the minimally polymorphic subclass HLA-E single-chain dimer and trimer in hESC [20]. Of note, HLA-E binds to the common NK cell inhibitory receptor NKG2A, shielding the cells from the missing-self response [25] (Fig. 2). The resulting B2M-null, HLA-E expressing hESC and their 15–20 days differentiated derivatives were protected from $CD8^+$ T cells as well as from NK cell-mediated lysis and anti-HLA-antibodies. Thus, they were able to create a cell line that, albeit expressing HLA-E, did not appear as allogeneic to the recipient and could successfully escape host immunity [20]. To validate their strategy in vivo, Gornalusse and colleagues employed immunodeficient NSG-B2m mice transplanted with allogeneic NKG2A-expressing NK cells or $CD8^+$ T cells [20]. In particular, they used growing luciferase-expressing B2M-null, HLA-E single chain expressing teratomas derived from ESC as a readout of the protective effect of gene editing compared to $B2M^{+/+}$ controls. Confirming their results in vitro, the authors found that $CD8^+$ and NK cells invade and induce necrosis to a significantly higher degree in $B2M^{+/+}$ teratomas than in their engineered counterparts [20]. Therefore, the authors proposed HLA class I-null, HLA-E expressing ESC as suitable universal donors for transplantation of differentiated cells that do not express HLA class II molecules. Importantly, subjects lacking HLA class I molecules might display reduced clearance of infections and malignancies [20]. Hence the induction of HLA-E, which can present peptides for recognition by T cells, could prove beneficial in these contexts [24, 47]. However, it is worth noting that the use of this model is still far from the standards of clinical practice and would require improving its safety through extensive pre-clinical testing [20]. While the approach described above for the knock-in of HLA-E proved effective, it did not address the need for a method to protect donor cells from NK cell response when these lack NKG2A [5, 45]. For instance, other receptors exerting an inhibitory function on NK cells such as LILRB1 and KIR2DL4 display a higher affinity for HLA-G [9, 53, 54]. The latter is expressed during pregnancy when it confers maternal tolerance to fetal tissue and has been reported to have the widest range of target lymphocytes, including $CD8^+$ and $CD4^+$ T cells and APCs [7, 44]. Therefore, in a work from 2020 a system to edit hESC to lack the expression of HLA major subclasses while expressing HLA-G was developed [56] (Fig. 2). Importantly, of the seven existing isoforms of HLA-G, only G1 and G5 bind to B2M with all three globular domains [7, 8]. In this study, CRISPR/Cas9-driven *B2M* knockout was achieved by targeting exons 2 and 3 of

B2M. Subsequently, HLA-G1 and G5-specific gRNA were used to induce the biallelic knock-in of HLA-G in the frame of endogenous *B2M* loci resulting in two B2M-HLA-G fusion proteins [56]. The deriving hESC and the cardiomyocytes differentiated from them were tested for immunogenicity via their effect on proliferation and activation of PBMC and T cell infiltration. While undifferentiated hESC did not elicit an immune response regardless of the gene editing, the differentiated derivatives were protected from T cell immunity when B2M-null. However, the expression of HLA-G1 alone or HLA-G1 and G5 combined was crucial to protect the differentiated cells from NK cell missing-self response. Of note, the best protection from NK cell cytotoxicity was obtained when differentiated hESC expressed both membrane-bound HLA-G1 and soluble HLA-G5 isoforms. This suggested that the secretion of soluble HLA-G5 might be needed to guarantee a low immunogenic environment [56]. To investigate the efficacy of their hypoimmunogenic HLA class I knockout, HLA-G knock-in model in vivo Shi et al. performed a teratoma formation assay [56]. To do this, they injected PBMCs into immunodeficient M-NSG mice transplanted with edited hESC-derived embryoid bodies. In agreement with previous literature, teratomas derived from cells lacking B2M or B2M and HLA-G1 or B2M, HLA-G1 and G5 were all protected from T cell infiltration. On the other hand, *B2M* knockout alone did not provide protection from NK cell missing-self response for which the induced expression of HLA-G in its membrane-bound or soluble form is indispensable [56]. Moreover, the mRNA levels of IL-1, IL-8 and TNF-α were significantly lower in teratomas deriving from HLA-G1- and G5-expressing hESC, indicating for the first time that the HLA-G5 isoform is crucial to maintaining a nonanomalous immune environment [56]. Moving on from hESC, in response to the need for more easily available universal donors, several studies were conducted on iPSC. For instance, in their study, Chimienti et al. conducted an in-depth investigation of the expression of NK-activating ligands in several iPSC lines and derived β cells. The expression patterns of a long list of ligands, including CD155, CD112, MICA, MICB, BAG6/BAT3, B7-H6 and B7-H3 were interrogated during the differentiation of iPSC into β cells. From this, the DNAM1 ligand CD155 and the Nkp30 ligand B7-H3 emerged as the most expressed mRNA and proteins of the batch [12]. Thus, a novel approach to escape NK cell-mediated immunity via the knockout of these NK-activating ligands was proposed. To identify the most effective ligand knockout as well as to assess the possibility of a synergistic effect three engineered iPSC lines were generated. Two of these lines were double knockouts lacking *B2M* and *CD155* or *B7-H3*, and the third line was a triple knockout of *B2M*, *CD155* and *B7-H3* (Fig. 2). Following co-culture with the $CD16^+$ subset of NK cells both $B2M^{-/-}CD155^{-/-}$ and $B2M^{-/-}B7\text{-}H3^{-/-}$ protected iPSC from missing-self response, whereas the concomitant loss of *CD155* and *B7-H3* did not exert any synergistic effect [12]. Conversely, insulin-secreting β cells differentiated from iPSC displayed enhanced survival rates when lacking both *CD155* and *B7-H3* [12]. Strikingly, 10 days upon transplantation $B2M^{-/-}B7\text{-}H3^{-/-}CD155^{-/-}$ iPSC-derived pancreatic progenitors (PP) showed improved survival compared to $B2M^{-/-}$ PP during in vivo testing. As a readout of this, the team measured the bioluminescence of grafts in NSG mice injected with human NK cells 24 hours prior and after transplantation with luciferase-expressing *B2M*$^{-/-}$*B7-H3*$^{-/-}$*CD155*$^{-/-}$, *B2M*$^{-/-}$, *B2M*$^{-/-}$ *HLA-E*$^{+/+}$ and WT cells [12]. Moreover, *B2M*$^{-/-}$*B7-H3*$^{-/-}$*CD155*$^{-/-}$ PP stained positively for the β cell-specific factor Pdx-1 and insulin, whereas *B2M*$^{-/-}$ presented significantly lower levels of both markers. This indicated that PP cells lacking B2M as well as B7-H3 and CD155 but not their B2M-null counterparts were committed to differentiation into β cells, suggesting the importance of protection from the immune system for appropriate PP maturation [12]. Remarkably, the authors developed a way to achieve immune compatibility by identification and knockout of key checkpoints of the NK cell-mediated response, a potentially less invasive technique than previously described knock-in methods [12]. It is important to note that this approach represents a significant paradigm shift in the field of invisible cell generation. Rather than shielding the donor cells from NK cytotoxicity through knock-in of generic protective genes, Chimienti et al. envisioned a way to identify what iPSC and their derivatives express and use for NK cell recognition and target it specifically.

2.4 Tolerogenic Gene Expression

Noticeably, the induction of tolerogenic genes as well as the modulation of NK ligands and receptors paved the way for several applications complementing the efficacy of the *B2M* knockout. As early as 2014, the protective role of the knock-in of PD-L1 and cytotoxic T lymphocyte antigen 4 (CTLA4)-Ig in hESC transplanted into humanized mice was described [52]. PD-L1 binds to the B7 T cell surface molecules family member named programmed death 1 (PD-1), which inhibits T cell immunity [17, 52]. Conversely, CTLA4 binds to other members of the B7 family such as CD80 and CD86 triggering a strong stimulatory pathway in T cells, making CTLA4-Ig useful to prevent T cell activation [52, 66](Fig. 2). Another important regulator often targeted to keep the immune response under control is CD47. This is an Ig-like protein carrying a "don't eat me" signal ensuring the recognition of autologous cells and their protection from phagocytosis. Indeed, CD47 is a ubiquitous membrane protein that binds to its receptor SIRPα on macrophages eliciting an inhibitory function on these [29] (Fig. 2). Moreover, CD47 is marginally involved in other functions such as neutrophil migration, T cell co-stimulation and axon extension [29]. One possible strategy to delete HLA-A, HLA-B and HLA-C while preserving the expression of HLA minor subclasses and inducing tolerogenic gene expression in iPSC was proposed by Han and coworkers in 2019. They employed CRISPR/Cas9 genome-editing technology to specifically target the genes encoding for HLA major subclasses A, B and C and the class II MHC transactivator (CIITA) [23]. CIITA is a non-DNA-binding coactivator and the key regulator of HLA class II expression, hence its excision results in HLA class II-null cells [10, 48] (Fig. 2). Although only polymorphic HLA class I genes were targeted, the expression of HLA minor subclass E was impaired on the APC cell surface [23]. Thus, to shield the iPSC from lysis by NK cells, Han et al. further engineered the cells to express

HLA-G. This ligates several NK inhibitory receptors with high affinity making it more likely to avoid the missing-self response, compared to other HLA minor subclasses [23]. Moreover, the expression of the immunomodulatory factors CD47 and programmed death ligand-1 (PD-L1) was induced (Fig. 2) through integration in the *AAVS1* safe harbor locus to overcome macrophagic response and counter residual T cell-mediated immunity, respectively. Therefore, the team was capable of generating hypoimmunogenic iPSC that maintained their pluripotency and differentiating potential and could efficiently be differentiated into endothelial cells (ECs) and vascular smooth muscle cells (VSMCs). Both the iPSC and their derivatives elicited reduced T cell response. Interestingly, from the comparison between cells where only HLA class I and II were silenced and cells where tolerogenic factors were induced, PD-L1 emerged as a stronger modulator of $CD8^+$ than $CD4^+$ T cells [23]. The overexpression of HLA-G and CD47 successfully protected the cells from NK- and macrophage-mediated response, providing a means by which long-term transplant rejection might be faced. However, partial T cell response was still evoked by all combinations of engineered cells compared to the negative control [23]. In vivo, the T cell recall response of wild-type (WT) versus edited hESC was measured by means of teratoma growth. Upon transplantation with hESC, immunodeficient SCID mice were sensitized through the injection of human $CD8^+$ T cells and assessed for teratoma formation. As expected, the abrogation of HLA class I major subclasses and HLA class II resulted in teratomas of significantly bigger volume than WT. Although tending toward a bigger size, hESC expressing HLA-G, PD-L1 and CD47 did not exert the formation of significantly more voluminous teratomas [23]. Moreover, all engineered hESC showed decreased T cell infiltration compared to WT as confirmed by the limited mRNA expression of CD8 and IL-2 [23]. Of note, the efficacy of the proposed model in shielding edited cells from NK cytotoxicity and macrophagic attack was limited to in vitro experimentation. Both human and mouse iPSC have been engineered to lack HLA class I and II and overexpress the tolerogenic CD47 in a work by Deuse et al. [13] (Fig. 2). Using a strategy similar to that described above, they employed CRISP/Cas9 to silence *B2M* and *CIITA* (or *B2m* and *Ciita* in rodent cells) and to knock-in the *CD47* (or *Cd47*) transgene. Mouse cells retained their pluripotency and differentiation potential, hence could be differentiated into ECs, smooth muscle cells (SMCs) and cardiomyocytes (CMs) [13]. Similarly, upon gene editing, human iPSC derived from $CD34^+$ cord blood maintained a normal karyotype and pluripotency, thus being differentiated into ECs and CMs. In both species, gene engineering did not alter the expression of NK cell ligands [13]. The hypoimmunogenicity of the generated models was tested in vivo, in particular through transplantation of mouse and human cells in allogeneic BALB/c and humanized NSG-SGM3 mice, respectively [13, 69]. Human as well as rodent engineered iPSC and their differentiated derivatives accomplished long-term survival as opposed to their WT counterparts. Although the expression of HLA minor subclasses, such as HLA-E and G was not induced in these models, CD47 overexpression was found to be sufficient to mitigate innate cellular immunity, including NK cell cytotoxicity [13]. In a recently published study by the same authors, the overexpression of the *CD64* transgene was implemented in the previously developed *B2M$^{-/-}$CIITA$^{-/-}$CD47^{+}* model [22]. While a number of approaches have been identified to face cellular immunity, in this work the authors aimed at developing a method to protect from antibody-mediated cellular cytotoxicity (ADCC) and complement-dependent cytotoxicity (CDC). To do this, they developed a system where the overexpression of *CD46* or its truncated form *CD64t* on the transplanted cells provided a high-affinity, competitive receptor for the IgG free fragment crystallizable domain (F_c). In standard conditions, F_c, as well as the antigen-binding fragments (F_{ab}) of the IgG are responsible for the activation of effector cells or complement and epitope binding, respectively, resulting in target cell destruction [6, 22]. Conversely, in hypoimmunogenic iPSC and their derivatives, the addition of CD46 serves to capture IgG F_c, displacing its binding to effector or complement cells. Thus, both human and mouse *B2M$^{-/-}$CIITA$^{-/-}$CD47^{+}CD64^{+}* iPSC successfully evaded innate and adaptive cell killing as well as antibody-mediated immunity in vitro [22] (Fig. 2). Similarly, *B2M$^{-/-}$CIITA$^{-/-}$CD47^{+}CD64^{+}* iPSC displayed increased survival when transplanted into NSG mice injected with allogeneic macrophages or NK cells or humanized NSG-SGM3 sensitized with human NK cells [22]. Importantly, the potential of this model for employability in cell therapy was assessed through the differentiation of the engineered iPSC into cells representative of three relevant disease models. Indeed, CAR T cells, thyroid cells and β cells derived from hypoimmunogenic *CD64t*-overexpressing iPSC were used to establish their efficacy in the treatment of solid tumors, Hashimoto's thyroiditis and T1D, respectively [22]. It was found that CAR T cells transduced to express *CD64t* had as high tumor-killing capacity as regular CAR T cells, but prevented HLA and non-HLA antibody-mediated immunity. Thyroid epithelial cells edited to overexpress anti-thyroid peroxidase (*TPO*) and *CD64t* displayed higher protection from anti-TPO IgG compared to *TPO*-only-overexpressing cells, inferring the importance of CD64 in protecting from autoimmunity. Ultimately, the protective role of *CD64t* enhanced expression was assayed in iPSC-derived β cells. Importantly, *CD64t*-overexpressing β cells retained intact glucose metabolism and insulin secretion. Moreover, they were resistant to HLA antibody killing and had a survival rate comparable to that of engrafted β cells not undergoing antibody challenge [22]. Further feeding the β cell and islet replacement field, Gerace et al. generated immune-evasive iPSC-derived islets by standard silencing of *B2M* followed by induction of the *PD-L1* and *HLA-E* transgenes [19] (Fig. 2). Although all edited cell lines achieved successful differentiation, the authors found that the overexpression of *PD-L1* alone or combined with *B2M* knockout did not protect iPSC-derived β cells from xenorejection upon transplantation in B6/albino mice [19]. As per previous works, HLA class I-null cells were protected from T cell response in vitro as well as upon transplantation in humanized NSG mice. Surprisingly, although in iPSC-derived β cells, the lack of HLA proved to be sufficient to evade NK cell cytotoxicity, the induction of HLA-E did not confer increased protection. Thus, the authors suggested that iPSC-derived islets might inherently possess an NK cell-resistant phenotype characterized by the deficiency of non-HLA NK cell activating ligands and expression of non-HLA inhibitory ligands [19]. Furthermore, the team introduced a new strategy of allograft protection through the induction of localized immune tolerance [19]. Indeed, they edited iPSC to express the tolerogenic cytokines IL-2 mutein (N88D), transforming growth factor β (TGF-β) and IL-10. IL-2 plays a crucial role in the generation and expansion of regulatory T cells (T_{reg}), which are essential to guarantee immunological homeostasis and suppress the self-immune response [25, 46]. The IL-2 mutant form N88D is known to act as a T_{reg}-selective molecule, preferentially expanding T_{reg} over $CD4^+$ and $CD8^+$ T cells due to its reduced affinity for the IL-2Rβγ receptor [46]. Moreover, IL-10 and TGF-β are key cytokines for the production of secondary suppressor cells from T_{regs} [26]. Therefore, the effect of N88D, IL-10 and TGF-β knock-in in hypoimmunogenic iPSC and β cells derived from them was assessed. Importantly, engineered iPSC-derived β cells showed increased glucose-stimulated insulin secretion compared to controls. Moreover, the induction of cytokines resulted in the generation and stabilization of T_{regs} in iPSC-derived β cells, achieving localized immune tolerance and increased survival to PBMCs cytotoxicity [19].

3 Gene Editing of Isolated Islets

Due to the extent of the social burden posed by T1D, the identification of a functional strategy to definitively escape β cell graft rejection is imperative. While great potential lies in the use of stem cell-derived invisible β cells, it is worth noting that scientific efforts are also going towards editing isolated islets to prevent immune rejection. In a study from 2015, El Khatib and colleagues engineered mouse β cells to overexpress a PD-L1-CTLA4Ig fusion protein and IL-10 [14]. Upon transplantation in allogeneic non-obese diabetic (NOD) mice, PD-L1-CTLAIg and IL-10 over-expressing β cells resulted in delayed onset of hyperglycemia and preserved β cell mass. However, they failed to restore healthy blood glucose levels in NOD mice with early-onset diabetes. Moreover, albeit protected from hyperacute allorejection in autoimmune diabetic mice, edited β cells were still susceptible to chronic rejection [14]. Again through the integration of a fusion construct, a study by Paul et al. described the protective function of PD-L1 and indoleamine 2,3-dioxygease (IDO) overexpression in mouse islets [42]. IDO is an enzyme mediating the oxidative metabolism leading to the generation of kynurenine from tryptophan. Kynurenine, in turn, stimulates apoptosis, loss of function, inhibition of TCR-activated T cell cycle and activation of T_{regs}, all resulting in the suppression of adaptive and innate immunity [39]. In this work, the authors generated a chimeric fusion transgene named PIDO, inducing the expression of both PD-L1 and IDO in islets from C57BL6/J mice. Subsequently, they transplanted the edited islets into diabetic BALB/c or NOD mice, which displayed reduced autoimmune destruction of the allografts and restored normoglycemia [42]. Moreover, upon transplantation, graft cells presented with decreased levels of infiltrating pro-inflammatory cells and enhanced levels of T_{regs}. However, transplantation of PIDO-expressing islets did not elicit long-term acquired memory tolerance in the host. Ultimately, the team demonstrated the employability of the PIDO transgene to generate immune evasive porcine islets for transplantation in rodent or canine xenorecipients [42]. Another strategy aimed at engineering isolated islets to improve immune tolerance is based on the induction of a chimeric Fas-ligand (FasL) presenting a streptavidin core (SA-FasL), which can be transiently presented on the surface of biotinylated cells [68, 70]. Activated T cells are characterized by an upregulation of the Fas receptor (CD95), which binding to FasL plays a central role in the process of T cell activation-induced cell death (AICD). The latter is a crucial pathway for the maintenance of immune homeostasis and tolerance to self-antigens [31]. In agreement with this, SA-FasL-engineered islets transplanted in allogeneic mice induced local apoptosis of $CD4^+$ and $CD8^+$ T cells, but not of T_{regs}, leading to indefinite survival of the graft and reversion of diabetes [70]. Following systemic, TGF-β-dependent early phase tolerance, SA-FasL-engineered islets elicited graft-localized tolerance during the maintenance phase. Importantly, their local protective function extended to later transplantation with non-edited islets provided that these were engrafted in the same site where SA-FasL islets had evoked an immune privilege before [68].

Two recent papers from the same group of scientists have made significant advancements in the field of creating invisible cells for β cell replacement. In the first paper, gene editing techniques were applied to human donor islets, modifying them to achieve a hypoimmune HLA class I and class II negative phenotype, along with CD47 overexpression. These edited cells were then dissociated into single cells, modified, and reaggregated to form human HIP pseudoislets (p-islets). It was demonstrated that these human HIP p-islets could survive, engraft, and improve diabetes in immunocompetent, allogeneic, diabetic humanized mice. Furthermore, the HIP p-islet cells exhibited the ability to evade autoimmune destruction in autologous, diabetic humanized autoimmune mice [27, 28].

In the second paper, rhesus macaque HIP cells were engineered by depleting HLA class I and II molecules and overexpressing CD47 ($B2M^{-/-}CIITA^{-/-}CD47^+$), which were then transplanted intramuscularly into four allogeneic rhesus macaques. HIP cells demonstrated unrestricted survival for 16 weeks in fully immunocompetent allogeneic recipients and differentiated into various lineages, whereas allogeneic wild-type cells were strongly rejected. Additionally, human HIP cells were differentiated into insulin producing pancreatic islet cells and survived in immunocompetent, allogeneic diabetic humanized mice for 4 weeks, effectively ameliorating diabetes. Edited primary rhesus macaque islets with HIP modifications were able to survive for 40 weeks in an allogeneic rhesus macaque recipient without the need for immunosuppression, whereas unedited islets were rapidly rejected [27, 28].

These new results involving the transplantation of gene-edited stealth islets, which do not require immunosuppression, offer promising prospects for the broader adoption of this therapy.

Acknowledgements This study was supported by PNC-E.3 Iniziativa LSH-TA, Fondazione Italiana Diabete (FID), SOStegno 70 Insieme ai ragazzi diabetici Associazione Onlus (Project "Beta is better"), and fundraising campaign "Un brutto t1po".

References

1. Barton, F. B., Rickels, M. R., Alejandro, R., Hering, B. J., Wease, S., Naziruddin, B., . . . Shapiro, A. M. (2012). Improvement in outcomes of clinical islet transplantation: 1999–2010. *Diabetes Care, 35*(7), 1436–1445. https://doi.org/10.2337/dc12-0063
2. Benichou, G., & Thomson, A. W. (2009). Direct versus indirect allorecognition pathways: on the right track. *Am J Transplant, 9*(4), 655–656. https://doi.org/10.1111/j.1600-6143.2009.02572.x
3. Bjorkman, P. J., Saper, M. A., Samraoui, B., Bennett, W. S., Strominger, J. L., & Wiley, D. C. (1987). Structure of the human class I histocompatibility antigen, HLA-A2. *Nature, 329*(6139), 506–512. https://doi.org/10.1038/329506a0
4. Bluman, E. M., Bartynski, K. J., Avalos, B. R., & Caligiuri, M. A. (1996). Human natural killer cells produce abundant macrophage inflammatory protein-1 alpha in response to monocyte-derived cytokines. *J Clin Invest, 97*(12), 2722–2727. https://doi.org/10.1172/JCI118726
5. Braud, V. M., Allan, D. S., O'Callaghan, C. A., Soderstrom, K., D'Andrea, A., Ogg, G. S., . . . McMichael, A. J. (1998). HLA-E binds to natural killer cell receptors CD94/NKG2A, B and C. *Nature, 391*(6669), 795–799. https://doi.org/10.1038/35869
6. Bruhns, P., Iannascoli, B., England, P., Mancardi, D. A., Fernandez, N., Jorieux, S., & Daeron, M. (2009). Specificity and affinity of human Fcgamma receptors and their polymorphic variants for human IgG subclasses. *Blood, 113*(16), 3716–3725. https://doi.org/10.1182/blood-2008-09-179754
7. Carosella, E. D., Favier, B., Rouas-Freiss, N., Moreau, P., & Lemaoult, J. (2008). Beyond the increasing complexity of the immunomodulatory HLA-G molecule. *Blood, 111*(10), 4862–4870. https://doi.org/10.1182/blood-2007-12-127662
8. Carosella, E. D., Moreau, P., Le Maoult, J., Le Discorde, M., Dausset, J., & Rouas-Freiss, N. (2003). HLA-G molecules: from maternal-fetal tolerance to tissue acceptance. *Adv Immunol, 81*, 199–252. https://doi.org/10.1016/s0065-2776(03)81006-4
9. Caumartin, J., Favier, B., Daouya, M., Guillard, C., Moreau, P., Carosella, E. D., & LeMaoult, J. (2007). Trogocytosis-based generation of suppressive NK cells. *EMBO J, 26*(5), 1423–1433. https://doi.org/10.1038/sj.emboj.7601570
10. Chen, H., Li, Y., Lin, X., Cui, D., Cui, C., Li, H., & Xiao, L. (2015a). Functional disruption of human leukocyte antigen II in human embryonic stem cell. *Biological Research, 48*(1), 59. https://doi.org/10.1186/s40659-015-0051-6
11. Chen, H. F., Yu, C. Y., Chen, M. J., Chou, S. H., Chiang, M. S., Chou, W. H., . . . Ho, H. N. (2015b). Characteristic expression of major histocompatibility complex and immune privilege genes in human pluripotent stem cells and their derivatives. *Cell Transplant, 24*(5), 845–864. https://doi.org/10.3727/096368913X674639
12. Chimienti, R., Baccega, T., Torchio, S., Manenti, F., Pellegrini, S., Cospito, A., . . . Piemonti, L. (2022). Engineering of immune checkpoints B7-H3 and CD155 enhances immune compatibility of MHC-I(-/-) iPSCs for beta cell replacement. *Cell Rep, 40*(13), 111423. https://doi.org/10.1016/j.celrep.2022.111423
13. Deuse, T., Hu, X., Gravina, A., Wang, D., Tediashvili, G., De, C., . . . Schrepfer, S. (2019). Hypoimmunogenic derivatives of induced pluripotent stem cells evade immune rejection in fully immunocompetent allogeneic recipients. *Nat Biotechnol, 37*(3), 252–258. https://doi.org/10.1038/s41587-019-0016-3
14. El Khatib, M. M., Sakuma, T., Tonne, J. M., Mohamed, M. S., Holditch, S. J., Lu, B., . . . Ikeda, Y. (2015). beta-Cell-targeted blockage of PD1 and CTLA4 pathways prevents development of autoimmune diabetes and acute allogeneic islets rejection. *Gene Ther, 22*(5), 430–438. https://doi.org/10.1038/gt.2015.18
15. Fauriat, C., Long, E. O., Ljunggren, H. G., & Bryceson, Y. T. (2010). Regulation of human NK-cell cytokine and chemokine production by target cell recognition. *Blood, 115*(11), 2167–2176. https://doi.org/10.1182/blood-2009-08-238469

16. Ferreira, L. M. R., Meissner, T. B., Tilburgs, T., & Strominger, J. L. (2017). HLA-G: At the Interface of Maternal-Fetal Tolerance. *Trends Immunol, 38*(4), 272–286. https://doi.org/10.1016/j.it.2017.01.009
17. Fife, B. T., & Bluestone, J. A. (2008). Control of peripheral T-cell tolerance and autoimmunity via the CTLA-4 and PD-1 pathways. *Immunological Reviews, 224*(1), 166–182. https://doi.org/10.1111/j.1600-065X.2008.00662.x
18. Frederiksen, H. R., Doehn, U., Tveden-Nyborg, P., & Freude, K. K. (2020). Non-immunogenic Induced Pluripotent Stem Cells, a Promising Way Forward for Allogenic Transplantations for Neurological Disorders. *Front Genome Ed, 2*, 623717. https://doi.org/10.3389/fgeed.2020.623717
19. Gerace, D., Zhou, Q., Kenty, J. H., Veres, A., Sintov, E., Wang, X., . . . Melton, D. A. (2023). Engineering human stem cell-derived islets to evade immune rejection and promote localized immune tolerance. *Cell Rep Med, 4*(1), 100879. https://doi.org/10.1016/j.xcrm.2022.100879
20. Gornalusse, G. G., Hirata, R. K., Funk, S. E., Riolobos, L., Lopes, V. S., Manske, G., . . . Russell, D. W. (2017). HLA-E-expressing pluripotent stem cells escape allogeneic responses and lysis by NK cells. *Nat Biotechnol, 35*(8), 765–772. https://doi.org/10.1038/nbt.3860
21. Gourraud, P. A., Gilson, L., Girard, M., & Peschanski, M. (2012). The role of human leukocyte antigen matching in the development of multiethnic "haplobank" of induced pluripotent stem cell lines. *Stem Cells, 30*(2), 180–186. https://doi.org/10.1002/stem.772
22. Gravina, A., Tediashvili, G., Rajalingam, R., Quandt, Z., Deisenroth, C., Schrepfer, S., & Deuse, T. (2023). Protection of cell therapeutics from antibody-mediated killing by CD64 overexpression. *Nat Biotechnol.* https://doi.org/10.1038/s41587-022-01540-7
23. Han, X., Wang, M., Duan, S., Franco, P. J., Kenty, J. H., Hedrick, P., . . . Cowan, C. A. (2019). Generation of hypoimmunogenic human pluripotent stem cells. *Proc Natl Acad Sci U S A, 116*(21), 10441–10446. https://doi.org/10.1073/pnas.1902566116
24. Hansen, S. G., Wu, H. L., Burwitz, B. J., Hughes, C. M., Hammond, K. B., Ventura, A. B., . . . Picker, L. J. (2016). Broadly targeted CD8(+) T cell responses restricted by major histocompatibility complex E. *Science, 351*(6274), 714–720. https://doi.org/10.1126/science.aac9475
25. Horowitz, A., Strauss-Albee, D. M., Leipold, M., Kubo, J., Nemat-Gorgani, N., Dogan, O. C., . . . Blish, C. A. (2013). Genetic and environmental determinants of human NK cell diversity revealed by mass cytometry. *Sci Transl Med, 5*(208), 208ra145. https://doi.org/10.1126/scitranslmed.3006702
26. Horwitz, D. A., Zheng, S. G., Wang, J., & Gray, J. D. (2008). Critical role of IL-2 and TGF-beta in generation, function and stabilization of Foxp3+CD4+ Treg. *Eur J Immunol, 38*(4), 912–915. https://doi.org/10.1002/eji.200738109
27. Hu X, Gattis C, Olroyd AG, Friera AM, White K, Young C, Basco R, Lamba M, Wells F, Ankala R, Dowdle WE, Lin A, Egenberger K, Rukstalis JM, Millman JR, Connolly AJ, Deuse T, Schrepfer S. (2023a) Human hypoimmune primary pancreatic islets avoid rejection and autoimmunity and alleviate diabetes in allogeneic humanized mice. Sci Transl Med. 15(691):eadg5794. https://doi.org/10.1126/scitranslmed.adg5794.
28. Hu X, White K, Olroyd AG, DeJesus R, Dominguez AA, Dowdle WE, Friera AM, Young C, Wells F, Chu EY, Ito CE, Krishnapura H, Jain S, Ankala R, McGill TJ, Lin A, Egenberger K, Gagnon A, Michael Rukstalis J, Hogrebe NJ, Gattis C, Basco R, Millman JR, Kievit P, Davis MM, Lanier LL, Connolly AJ, Deuse T, Schrepfer S. (2023b). Hypoimmune induced pluripotent stem cells survive long term in fully immunocompetent, allogeneic rhesus macaques. *Nat Biotechnol*. https://doi.org/10.1038/s41587-023-01784-x.
29. Jaiswal, S., Jamieson, C. H., Pang, W. W., Park, C. Y., Chao, M. P., Majeti, R., . . . Weissman, I. L. (2009). CD47 is upregulated on circulating hematopoietic stem cells and leukemia cells to avoid phagocytosis. *Cell, 138*(2), 271–285. https://doi.org/10.1016/j.cell.2009.05.046
30. Johnson, J. D., Ao, Z., Ao, P., Li, H., Dai, L. J., He, Z., . . . Warnock, G. L. (2009). Different effects of FK506, rapamycin, and mycophenolate mofetil on glucose-stimulated insulin release and apoptosis in human islets. *Cell Transplant, 18*(8), 833–845. https://doi.org/10.3727/096368909X471198

31. Ju, S. T., Panka, D. J., Cui, H., Ettinger, R., el-Khatib, M., Sherr, D. H., . . . Marshak-Rothstein, A. (1995). Fas(CD95)/FasL interactions required for programmed cell death after T-cell activation. *Nature, 373*(6513), 444–448. https://doi.org/10.1038/373444a0
32. Kruse, P. H., Matta, J., Ugolini, S., & Vivier, E. (2014). Natural cytotoxicity receptors and their ligands. *Immunol Cell Biol, 92*(3), 221–229. https://doi.org/10.1038/icb.2013.98
33. Lanier, L. L. (2005). NK cell recognition. *Annu Rev Immunol, 23*, 225–274. https://doi.org/10.1146/annurev.immunol.23.021704.115526
34. Lanier, L. L. (2008). Up on the tightrope: natural killer cell activation and inhibition. *Nat Immunol, 9*(5), 495–502. https://doi.org/10.1038/ni1581
35. Lee, N., Llano, M., Carretero, M., Ishitani, A., Navarro, F., López-Botet, M., & Geraghty, D. E. (1998). HLA-E is a major ligand for the natural killer inhibitory receptor CD94/NKG2A. *Proceedings of the National Academy of Sciences, 95*(9), 5199–5204. https://doi.org/10.1073/pnas.95.9.5199
36. Liao, N. S., Bix, M., Zijlstra, M., Jaenisch, R., & Raulet, D. (1991). MHC class I deficiency: susceptibility to natural killer (NK) cells and impaired NK activity. *Science, 253*(5016), 199–202. https://doi.org/10.1126/science.1853205
37. Ljunggren, H. G., & Karre, K. (1990). In search of the 'missing self': MHC molecules and NK cell recognition. *Immunol Today, 11*(7), 237–244. https://doi.org/10.1016/0167-5699(90)90097-s
38. Martinez-Lostao, L., Anel, A., & Pardo, J. (2015). How Do Cytotoxic Lymphocytes Kill Cancer Cells? *Clin Cancer Res, 21*(22), 5047–5056. https://doi.org/10.1158/1078-0432.CCR-15-0685
39. Munn, D. H., Sharma, M. D., Baban, B., Harding, H. P., Zhang, Y., Ron, D., & Mellor, A. L. (2005). GCN2 kinase in T cells mediates proliferative arrest and anergy induction in response to indoleamine 2,3-dioxygenase. *Immunity, 22*(5), 633–642. https://doi.org/10.1016/j.immuni.2005.03.013
40. Opelz, G., & Dohler, B. (2013). Association between steroid dosage and death with a functioning graft after kidney transplantation. *Am J Transplant, 13*(8), 2096–2105. https://doi.org/10.1111/ajt.12313
41. Otsuka, R., Wada, H., Murata, T., & Seino, K. I. (2020). Immune reaction and regulation in transplantation based on pluripotent stem cell technology. *Inflamm Regen, 40*, 12. https://doi.org/10.1186/s41232-020-00125-8
42. Paul, P. K., Das, R., Drow, T., Nylen, E. A., de Souza, A. H., Wang, Z., . . . Galipeau, J. (2022). Islet allografts expressing a PD-L1 and IDO fusion protein evade immune rejection and reverse preexisting diabetes in immunocompetent mice without systemic immunosuppression. *Am J Transplant, 22*(11), 2571–2585. https://doi.org/10.1111/ajt.17162
43. Paul, S., & Lal, G. (2017). The Molecular Mechanism of Natural Killer Cells Function and Its Importance in Cancer Immunotherapy. *Front Immunol, 8*, 1124. https://doi.org/10.3389/fimmu.2017.01124
44. Pazmany, L., Mandelboim, O., Vales-Gomez, M., Davis, D. M., Reyburn, H. T., & Strominger, J. L. (1996). Protection from natural killer cell-mediated lysis by HLA-G expression on target cells. *Science, 274*(5288), 792–795. https://doi.org/10.1126/science.274.5288.792
45. Pegram, H. J., Andrews, D. M., Smyth, M. J., Darcy, P. K., & Kershaw, M. H. (2011). Activating and inhibitory receptors of natural killer cells. *Immunol Cell Biol, 89*(2), 216–224. https://doi.org/10.1038/icb.2010.78
46. Peterson, L. B., Bell, C. J. M., Howlett, S. K., Pekalski, M. L., Brady, K., Hinton, H., . . . Wicker, L. S. (2018). A long-lived IL-2 mutein that selectively activates and expands regulatory T cells as a therapy for autoimmune disease. *J Autoimmun, 95*, 1–14. https://doi.org/10.1016/j.jaut.2018.10.017
47. Pietra, G., Romagnani, C., Mazzarino, P., Falco, M., Millo, E., Moretta, A., . . . Mingari, M. C. (2003). HLA-E-restricted recognition of cytomegalovirus-derived peptides by human CD8+ cytolytic T lymphocytes. *Proc Natl Acad Sci U S A, 100*(19), 10896–10901. https://doi.org/10.1073/pnas.1834449100

48. Reith, W., LeibundGut-Landmann, S., & Waldburger, J. M. (2005). Regulation of MHC class II gene expression by the class II transactivator. *Nat Rev Immunol, 5*(10), 793–806. https://doi.org/10.1038/nri1708
49. Riolobos, L., Hirata, R. K., Turtle, C. J., Wang, P. R., Gornalusse, G. G., Zavajlevski, M., . . . Russell, D. W. (2013). HLA engineering of human pluripotent stem cells. *Mol Ther, 21*(6), 1232–1241. https://doi.org/10.1038/mt.2013.59
50. Roda, J. M., Parihar, R., Magro, C., Nuovo, G. J., Tridandapani, S., & Carson, W. E., 3rd. (2006). Natural killer cells produce T cell-recruiting chemokines in response to antibody-coated tumor cells. *Cancer Res, 66*(1), 517–526. https://doi.org/10.1158/0008-5472.CAN-05-2429
51. Rogers, N. J. L., R.I. (2002). Allorecognition. *American Journal of Transplantation* (1), 97–102. https://doi.org/10.1034/j.1600-6143.2001.10201.x
52. Rong, Z., Wang, M., Hu, Z., Stradner, M., Zhu, S., Kong, H., . . . Fu, X. (2014). An effective approach to prevent immune rejection of human ESC-derived allografts. *Cell Stem Cell, 14*(1), 121–130. https://doi.org/10.1016/j.stem.2013.11.014
53. Rouas-Freiss, N., Goncalves, R. M., Menier, C., Dausset, J., & Carosella, E. D. (1997a). Direct evidence to support the role of HLA-G in protecting the fetus from maternal uterine natural killer cytolysis. *Proc Natl Acad Sci U S A, 94*(21), 11520–11525. https://doi.org/10.1073/pnas.94.21.11520
54. Rouas-Freiss, N., Marchal, R. E., Kirszenbaum, M., Dausset, J., & Carosella, E. D. (1997b). The alpha1 domain of HLA-G1 and HLA-G2 inhibits cytotoxicity induced by natural killer cells: is HLA-G the public ligand for natural killer cell inhibitory receptors? *Proc Natl Acad Sci U S A, 94*(10), 5249–5254. https://doi.org/10.1073/pnas.94.10.5249
55. Ruth, R. F., Liza, D., Alison, S. B.-H., Dawn Mueller, A., Hilary, B., Dan, W. B., . . . John, D. G. (2003). Public Stem Cell Banks: Considerations of Justice in Stem Cell Research and Therapy. *The Hastings Center Report, 33*(6), 13–27. https://doi.org/10.2307/3527822
56. Shi, L., Li, W., Liu, Y., Chen, Z., Hui, Y., Hao, P., . . . Zhang, X. (2020). Generation of hypo-immunogenic human pluripotent stem cells via expression of membrane-bound and secreted beta2m-HLA-G fusion proteins. *Stem Cells, 38*(11), 1423–1437. https://doi.org/10.1002/stem.3269
57. Shibuya, A., Campbell, D., Hannum, C., Yssel, H., Franz-Bacon, K., McClanahan, T., . . . Phillips, J. H. (1996). DNAM-1, a novel adhesion molecule involved in the cytolytic function of T lymphocytes. *Immunity, 4*(6), 573–581. https://doi.org/10.1016/s1074-7613(00)70060-4
58. Song, G., Wang, S., Barkestani, M. N., Mullan, C., Fan, M., Jiang, B., . . . Jane-Wit, D. (2022). Membrane attack complexes, endothelial cell activation, and direct allorecognition. *Front Immunol, 13*, 1020889. https://doi.org/10.3389/fimmu.2022.1020889
59. Sordi, V., Pellegrini, S., & Piemonti, L. (2017). Immunological Issues After Stem Cell-Based beta Cell Replacement. *Curr Diab Rep, 17*(9), 68. https://doi.org/10.1007/s11892-017-0901-4
60. Stinchcombe, J. C., & Griffiths, G. M. (2007). Secretory mechanisms in cell-mediated cytotoxicity. *Annu Rev Cell Dev Biol, 23*, 495–517. https://doi.org/10.1146/annurev.cellbio.23.090506.123521
61. Taylor, C. J., Bolton, E. M., & Bradley, J. A. (2011). Immunological considerations for embryonic and induced pluripotent stem cell banking. *Philos Trans R Soc Lond B Biol Sci, 366*(1575), 2312–2322. https://doi.org/10.1098/rstb.2011.0030
62. Taylor, C. J., Peacock, S., Chaudhry, A. N., Bradley, J. A., & Bolton, E. M. (2012). Generating an iPSC bank for HLA-matched tissue transplantation based on known donor and recipient HLA types. *Cell Stem Cell, 11*(2), 147–152. https://doi.org/10.1016/j.stem.2012.07.014
63. Ting, J. P., & Trowsdale, J. (2002). Genetic control of MHC class II expression. *Cell, 109 Suppl*, S21–33. https://doi.org/10.1016/s0092-8674(02)00696-7
64. Topham, N. J., & Hewitt, E. W. (2009). Natural killer cell cytotoxicity: how do they pull the trigger? *Immunology, 128*(1), 7–15. https://doi.org/10.1111/j.1365-2567.2009.03123.x
65. Torikai, H., Akatsuka, Y., Miyauchi, H., Terakura, S., Onizuka, M., Tsujimura, K., . . . Takahashi, T. (2007). The HLA-A*0201-restricted minor histocompatibility antigen HA-1H

peptide can also be presented by another HLA-A2 subtype, A*0206. *Bone Marrow Transplant, 40*(2), 165–174. https://doi.org/10.1038/sj.bmt.1705689
66. Walker, L. S., & Abbas, A. K. (2002). The enemy within: keeping self-reactive T cells at bay in the periphery. *Nat Rev Immunol, 2*(1), 11–19. https://doi.org/10.1038/nri701
67. Wang, D., Quan, Y., Yan, Q., Morales, J. E., & Wetsel, R. A. (2015). Targeted Disruption of the beta2-Microglobulin Gene Minimizes the Immunogenicity of Human Embryonic Stem Cells. *Stem Cells Transl Med, 4*(10), 1234–1245. https://doi.org/10.5966/sctm.2015-0049
68. Woodward, K. B., Zhao, H., Shrestha, P., Batra, L., Tan, M., Grimany-Nuno, O., . . . Yolcu, E. S. (2020). Pancreatic islets engineered with a FasL protein induce systemic tolerance at the induction phase that evolves into long-term graft-localized immune privilege. *Am J Transplant, 20*(5), 1285–1295. https://doi.org/10.1111/ajt.15747
69. Wunderlich, M., Chou, F. S., Link, K. A., Mizukawa, B., Perry, R. L., Carroll, M., & Mulloy, J. C. (2010). AML xenograft efficiency is significantly improved in NOD/SCID-IL2RG mice constitutively expressing human SCF, GM-CSF and IL-3. *Leukemia, 24*(10), 1785–1788. https://doi.org/10.1038/leu.2010.158
70. Yolcu, E. S., Zhao, H., Bandura-Morgan, L., Lacelle, C., Woodward, K. B., Askenasy, N., & Shirwan, H. (2011). Pancreatic islets engineered with SA-FasL protein establish robust localized tolerance by inducing regulatory T cells in mice. *J Immunol, 187*(11), 5901–5909. https://doi.org/10.4049/jimmunol.1003266

Islet Immunoengineering

Leonor N. Teles, Chris M. Li, Zachary M. Wilkes, Aaron A. Stock, and Alice A. Tomei

1 Introduction

From 2000 to 2020, 4365 islet allotransplantation procedures were performed in 2170 recipients across Europe, North America, Asia, Oceania, and South America [1]. The islet transplantation procedure replaces the insulin-producing pancreatic β cells that are destroyed in type 1 diabetes (T1D), which provides clear benefits in

Leonor N. Teles and Chris M. Li are co-first authors of this chapter.

L. N. Teles · Z. M. Wilkes
Diabetes Research Institute, University of Miami Miller School of Medicine, Miami, FL, USA

Department of Biomedical Engineering, University of Miami, Miami, FL, USA

C. M. Li
Diabetes Research Institute, University of Miami Miller School of Medicine, Miami, FL, USA

Department of Microbiology and Immunology, University of Miami School of Medicine, Miami, FL, USA

A. A. Stock
Diabetes Research Institute, University of Miami Miller School of Medicine, Miami, FL, USA

A. A. Tomei (✉)
Diabetes Research Institute, University of Miami Miller School of Medicine, Miami, FL, USA

Department of Biomedical Engineering, University of Miami, Miami, FL, USA

Department of Surgery, University of Miami Miller School of Medicine, Miami, FL, USA
e-mail: atomei@miami.edu

L. Piemonti et al. (eds.), *Pluripotent Stem Cell Therapy for Diabetes*,
https://doi.org/10.1007/978-3-031-41943-0_15

patient outcome and quality of life ranging from a reduction in severe hypoglycemic events to restoring full metabolic control and insulin independence [2–5]. Despite these benefits, islet transplantation remains a last-line approach indicated only for patients with the most severe cases of T1D (those experiencing hypoglycemia unawareness, severe hypoglycemic episodes, and who are refractory to insulin therapy), and thus this curative procedure is not widely available to the millions of patients with T1D [6, 7]. Instead, the treatment for T1D is lifelong disease management through exogenous insulin administration [8]. Although such insulin treatment can now be automated [8], islet transplantation represents a curative deviceless approach [9]. To widen the availability of islet transplantation, the fields of regenerative medicine and islet immunoengineering must address key challenges in the accessibility and applicability of this procedure.

The accessibility of the islet transplantation is challenged by the combination of limited islet tissue supply, which currently relies on cadaveric organ donors, and the current site of islet transplantation, the hepatic portal vein. A percutaneous transhepatic approach is used to deliver transplanted islets into the hepatic portal vein due to its ease of access, high safety and low morbidity, and enhanced islet survival [6, 10]. However, islets transplanted in this fashion are immediately attacked by the instant blood-mediated inflammatory reaction (IBMIR), a nonspecific immune response characterized by coagulation and the formation of thrombi, activation of complement, and immune cell infiltration into islets [11, 12]. Due to the IBMIR, most islets are lost in the first few weeks following transplantation [13], requiring patients to undergo multiple islet infusions. Ultimately, one islet transplant recipient requires a volume of tissue equivalent from that of multiple donors, exacerbating the limited islet tissue supply [13].

In tandem, the applicability of the islet transplantation procedure is challenged again by intrahepatic transplant site as well as the need for chronic, systemic post-transplant immunosuppression. Although the percutaneous transhepatic approach makes islet transplantation one of the safest transplantation procedures [6], the islet graft becomes irretrievable in the case of adverse events. This location also requires the systemic use of post-transplant immunosuppressive therapy, which is required to protect islet grafts from both allogeneic rejection and recurrent autoimmunity [14] by the recipient and needed for the lifetime of the graft. However, such immunosuppression also significantly interferes with normal immune function throughout the rest of the body; chronic systemic immunosuppression can not only preclude potential tolerance induction to islet grafts, but also carries significant risk of infection [15–17], malignancy [6, 18, 19], and toxicity [18, 20] to other organs.

Stem cell-derived islets (SC islets) represent a breakthrough technology in regenerative medicine that advances islet transplantation as a cure for T1D. Recently, such islets have successfully differentiated into mature β cells following transplantation into patients and were able to mediate metabolic control comparable to that of primary islet transplants [21]. As a potentially limitless source of islet tissue, these islets can overcome our reliance on scarce islet donors and address the challenge in the accessibility of islet transplantation.

Unfortunately, without further immunoengineering approaches, SC islets will face the same challenges in the applicability of islet transplantation. The ability to retrieve the graft following transplantation becomes even more pertinent with the use of SC islets, which present a risk for tumorigenesis [22–24]. SC islet transplantation will also continue to require the use of chronic, systemic immunosuppression, as these islets will remain susceptible to direct and indirect allorecognition pathways (where the recipient's immune system recognizes the non-self major histocompatibility complexes and foreign alloantigens shed from these islets, respectively [25–27]). Although stem cells are regarded as immune-privileged due to decreased or absent expression of human leukocyte antigens (HLA) and may be able to avoid allorecognition by the recipient, even minimal expression of non-self HLA, epigenetic memory of induced pluripotent stem cells, or differentiation into end-stage β cells renders them vulnerable to rejection [28, 29]. These islets will also remain susceptible to the recurrent autoimmunity against β cell-specific and insulin-related antigens that originally produced the disease [14]. Ultimately, SC islets will increase the accessibility of islet transplantation, but the same challenges will continue to limit its applicability as a cure for T1D.

Thus, the field of islet immunoengineering focuses on developing strategies to increase the applicability of islet transplantation, using ones derived from stem cells or otherwise, to (1) protect islets from recipient immune responses while sparing overall immunity against pathogens and cancer cells, and (2) promote full graft integration into transplant recipients, which is critical to long-term functional restoration of physiological and homeostatic glucose metabolism. Key to all of these immunoengineering approaches is the development of alternative transplantations sites that are confined and well-vascularized [21, 30–32]. Such alternative sites will improve safety and efficacy over the current intrahepatic transplantation site by eliminating the risk for thrombosis, the IBMIR, and enabling retrieval of the graft in the case of adverse events. Furthermore, a confined and well-vascularized site will enable the use of powerful immunoengineering approaches that focus on enhancing islet delivery and support to and within the transplantation site, and achieving non-systemic, targeted, and localized immunosuppressive and immunomodulatory drug delivery. The ultimate goal of these islet immunoengineering approaches is to develop a one-time "fix and forget" treatment that fully integrates the graft into the recipient and restores homeostatic metabolic control, all without the need for chronic, systemic immunosuppression.

This chapter will discuss these immunoengineering approaches, specifically ones that (Fig. 1):

- Optimize biomaterial properties for islet/SC islet delivery platforms to enhance integration and long-term function in extrahepatic confined sites
- Utilize nano-, micro-, and macroscale biomaterials for targeted and/or local immunosuppressive and immunomodulatory drug delivery within the islet/SC islet transplant site

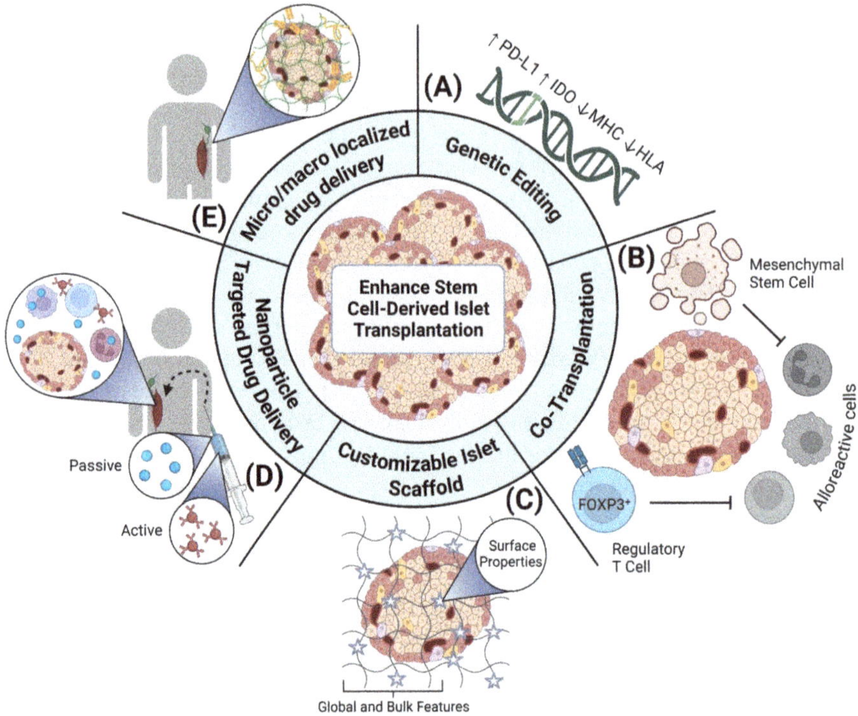

Fig. 1 Immunoengineering approaches to increase the applicability of stem cell-derived islet transplantation as a cure for type 1 diabetes. (**a**) Stem cell-derived islets can be genetically edited to decrease their immunogenicity (decreased major histocompatibility complex (MHC)/human leukocyte antigen (HLA) expression) and increase expression of immunoregulatory molecules (programmed death ligand-1 (PD-L1) or indoleamine 2,3-dioxygenase (IDO)). (**b**) Islets can be co-transplanted alongside immunomodulatory cells such as regulatory T cells and mesenchymal stem cells, which can regulate recipient alloimmunity and autoimmunity, and protect transplanted islets. (**c**) Islets can be delivered within highly customizable biomaterial scaffolds, whose surface, global, and bulk properties can be tuned to modulate immunity and promote islet engraftment and long-term functionality. (**d**) Immunosuppressive and immunomodulatory drugs can be encapsulated in nanoparticles and administered intravenously. Nanoparticles can then target the islet transplantation site, either passively or actively when constructed with targeting molecules, and deliver drugs locally while avoiding systemic adverse effects. (**e**) Immunosuppressive and immunomodulatory drugs can be loaded into micro- and macroscopic biomaterials, such as hydrogels or solid scaffolds, and implanted alongside islets. Following implantation, biomaterials promote drug release locally with highly customizable kinetics while avoiding systemic adverse effects

- Genetically modify SC islets to reduce their immunogenicity and enable expression of immune checkpoint inhibitor molecules
- Co-transplant engineered immunomodulatory and -regulatory cells alongside islets/SC islets.

2 Immune Challenges Against Implanted Materials and Transplanted Islets

Regardless of the source, islet transplantation induces unique and potent host immune responses that eventually lead to graft failure, requiring the use of chronic and systemic immunosuppression. This section summarizes the temporal sequence of host immune responses to islet transplant grafts, including the instant blood-mediated inflammatory reaction and alloimmunity (Fig. 2). Additionally, some islet immunoengineering approaches aim to control these immune responses locally by implanting immunomodulatory biomaterials alongside islets. While biomaterials provide unique means to modulate drug solubility, release kinetics, and biodistribution that are highly tunable based on the biomaterial formulation, biomaterial implants are also faced by an acute and chronic foreign body reaction, which adds another layer of immunological complexity. However, the customizability of biomaterials also enables them to minimize the foreign body reaction. Setting aside the potential for localized and sustained release of immunomodulatory drugs by biomaterials (described further in Sect. 4), the chemical and physical composition of the biomaterials themselves can influence immune cell behavior and serve as an important foundation for designing biomaterials.

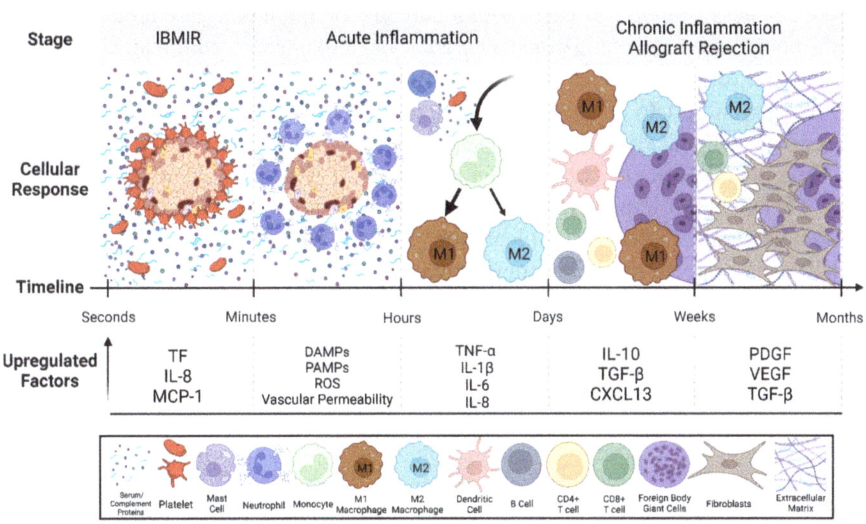

Fig. 2 Summary of the foreign body reaction to biomaterials, the immune reaction to islet transplantation, and the instant blood-mediated inflammatory reaction (IBMIR) in the hepatic portal vein site used in islet transplantation. TF tissue factor, IL interleukin, MCP monocyte chemotactic protein, DAMPs damage-associated molecular patterns, PAMPs pathogen-associated molecular patterns, ROS reactive oxygen species, TNF tumor necrosis factor, TGF transforming growth factor, CXCL C-X-C motif chemokine ligand, PDGF platelet-derived growth factor, VEGF vascular endothelial growth factor

Overall, despite the challenges of foreign body responses to biomaterials, each of the immunological challenges above can be met using the versatility of biomaterials, which can be designed to target any stage, cell, or pathway of the immune response to transplanted islets, shift them to desired outcomes, and improve the long-term survival of islet allografts.

2.1 The Instant Blood-Mediated Inflammatory Reaction

Currently, islet transplantations are performed using a percutaneous transhepatic approach to access the hepatic portal vein and deliver islets throughout the liver's microvasculature. This procedure and location have key advantages in its ease of access and safety (i.e., by interventional radiology, making this one of the safest transplantation procedures), as well as enhanced islet survival [6, 10]. Unfortunately, a key drawback to the intrahepatic site is the instant blood-mediated inflammatory reaction (IBMIR), a complex nonspecific response of the innate immune system to the graft that occurs immediately upon transplantation and concurrently with the early stages of the foreign body response [12]. Due to the procedure of isolating islets from donor pancreata, islets are covered in tissue factor, a potent trigger of coagulation, and cytokines such as interleukin-8 (IL-8) and monocyte chemotactic protein-1 (MCP-1, also known as CCL2) [33]. Thus, the IBMIR is triggered by the direct contact of implanted islets with a patient's blood, leading to rapid activation of the coagulation and complement cascades and leukocyte recruitment. Ultimately, around half of islets are lost in the first few minutes following transplantation, and the rest are subject to allorejection and recurrence of autoimmunity requiring the use of systemic and chronic immunosuppression [13, 34].

Thus, the IBMIR, in combination with pre-existing autoimmunity in patients with T1D and their inevitable development of alloimmunity, poses major immunological challenges in islet transplantation. To meet the challenge presented by the IBMIR, intensive research into transplantation sites alternative to the liver, such as subcutaneous sites, is underway [21, 31, 32]. Establishing and utilizing a confined extrahepatic transplantation site will not only circumvent the IBMIR but will also localize islets and enable the co-delivery of immunomodulatory cells, drugs, and biomaterials to prevent allorejection and recurrence of autoimmunity.

2.2 The Foreign Body Response Against Implanted Materials

When a material is implanted into the body, an unavoidable immune reaction known as the foreign body reaction (FBR) parallels the normal wound healing response. This pro-inflammatory immune reaction attempts to eliminate the foreign material through degradation and/or phagocytosis. Failing this, the trajectory of the wound healing response shifts from regeneration to repair, chiefly characterized by fibroblasts surrounding the implant and creating a fibrotic barrier to isolate it from the rest of the body [35]. This section will describe the cellular and molecular processes that characterize the multi-phase host response to implanted biomaterials that leads to the FBR. The FBR is inevitable in the context of biomaterials used for islet transplantation and, if unaddressed, leads to limited graft function or loss. The FBR poses a challenge for long-term implantation because it can [36]:

- Hamper nutrient delivery in cell-based implants and waste removal
- Limit the performance and durability of the implant
- Impair the implant's ability to integrate into the body and interact with the target tissue
- Hinder biosensing and biosecretory functions of the delivered cells
- Induce pain and discomfort
- Ultimately lead to failure and the need for follow-up transplants and procedures.

In the case of islet transplantation, the FBR itself can cause transplant rejection, but recurrent autoimmunity and the IBMIR play yet another complex and detrimental role in the integration process [13]. This intricate interplay is also described in this chapter. Because the FBR cannot be avoided, attempts to attenuate or skew this response to implanted biomaterials and protect the transplant and surrounding tissue are highly desirable [36]. Such attempts include developing bio-inert and/or immunomodulatory materials, designing materials with properties that prevent fibrosis and promote a healing response, and devising drug delivery platforms for local, long-lasting mitigation of the FBR.

The FBR begins immediately after a material is implanted and proceeds like the response to any injury, where damage signals trigger an acute inflammation stage, which facilitates attempts to remove invading threats [35]. The implantation itself leads to tissue damage and blood extravasation, which then triggers an instantaneous infiltration of inflammatory mediator cells and blood proteins, including albumin, fibrinogen, fibronectin and vitronectin [35]. Following this, the Vroman effect takes place, which describes the dynamic blood serum protein adsorption-desorption to a material surface and is highly specific to the material [37]. Protein adsorption to the implanted material occurs nonspecifically and creates a provisional matrix. This initial layer then directs the immune system towards the implantation site, which enables immune cells to gather and interact with the foreign material [35]. Integrin signals mediate activation and adhesion of host immune cells to the implanted biomaterial, now covered in blood serum proteins [38]. Occurring in parallel with the Vroman effect, leukocytes detect damage-associated molecular patterns (DAMPs) released during the transplantation process and implantation of the biomaterial, which initiate the inflammatory response [38].

Key cells in this initial acute inflammatory reaction to the implanted material are neutrophils, regarded as the first line of defense against pathogens, which rapidly respond and migrate to any injured tissue [39]. Neutrophils possess a unique ability to sense chemical gradients, like the platelet complement proteins in the blood serum that have deposited onto the surface of the transplant, allowing them to polarize and rapidly migrate towards implanted material within minutes [40]. Upon arriving to the site of transplantation, neutrophils attach to the protein layer, coat the implant, and release pro-inflammatory factors such as proteolytic enzymes and reactive oxygen species [41]. Additionally, these factors also enhance the recruitment and effector functions of other immune cells [41]. Mast cells also quickly become active at the site of inflammation, and alongside neutrophils, they undergo degranulation, expelling membrane-bound secretory granules that further promote a pro-inflammatory environment and recruit other cells [42]. The released substances and oxidative environment can significantly damage the implant in both its biomaterial structure and cellular cargo [43, 44]. Furthermore, neutrophils secrete neutrophil extracellular traps, extracellular fibers that further contribute to material biofouling and inflammation [45, 46].

All the chemical signals released thus far trigger vascular permeabilization and attracts the next contributors to the FBR: monocytes [38, 47]. Monocytes are macrophage precursors that, upon arriving at the site of inflammation, differentiate into macrophages. At this early phase of inflammation, macrophages adopt an M1 polarization, a phenotype characterized by secretion of pro-inflammatory cytokines like tumor necrosis factor-α (TNFα) and interleukins (IL) IL-1β, IL-6, and IL-8 [38, 47–49]. Macrophages remain at the transplant site for multiple days to continuously phagocytose the foreign material. Because material implants are often too large to be effectively phagocytosed, macrophages will continue to secrete molecules that break down the foreign material and then engulf the resulting pieces [50]. With prolonged exposure to this hostile, macrophage-mediated inflammatory environment, the surface of biomaterials can degrade and crack [35]. The persistence of the implant and continued pro-inflammatory signaling causes macrophages to undergo "frustrated phagocytosis," fusion, and formation of foreign body giant cells (FBGC) [51, 52]. With their unique multinucleated morphology, FBGCs are a hallmark of the FBR. At this stage, days after transplantation, the unresolved immune response shifts to the chronic inflammatory stage.

Chronic inflammation precludes resolution of the immune response, full regeneration of the implant site, and integration and biocompatibility of the biomaterial. Instead, the chronic inflammatory phase is characterized by repair, with deposition of connective tissue and fibrosis around the implant [35, 39]. Antigen-presenting dendritic cells are key players in activating macrophages and T cells and in mediating the shift from chronic inflammation to non-regenerative repair [52]. Additionally, macrophage polarization transitions away from the pro-inflammatory M1 phenotype toward an anti-inflammatory M2 phenotype. Although the M2 phenotype is an important mediator of wound healing, this phenotype also mediates fibrotic repair in the context of chronic inflammation [53]. Therefore, there is a fine balance between pro-regenerative and pro-fibrotic M2-like macrophages, and this balance may shift depending on biomaterial properties and released substances upon implantation [53]. M2 macrophages secrete cytokines including IL-10, TGF-β, and platelet-derived growth factor (PDGF), which, along with vascular endothelial growth factor (VEGF) secreted from FBGC and myeloid cells, establish a pro-fibrotic milieu [36]. This environment supports the recruitment of fibroblasts to the area, their differentiation into myofibroblasts, and boosts their production and deposition of extracellular matrix at the implant site [54]. All these combined cellular responses lead to fibrotic capsule thickening over months until complete isolation of the implant from surrounding tissue is achieved, forming the FBR capsule [35].

The inevitable FBR capsule poses one of the biggest challenges in material implantation. If the capsule continuously grows, it can form a physical barrier between the material and the surrounding tissue [35]. If the material implant is intended to deliver a special therapeutic agent in response to biostimulation by the host as in the case of biomaterials for islet delivery, the FBR capsule could impede that dynamic bidirectional process [35]. However, if an implant is designed to release local, large amounts of therapeutic agents (drug depot), the material may be less impaired by the FBR [55]. The FBR is a well-described cellular mechanism, but the exact material properties that trigger the immune response after it is implanted are not well understood [36]. The fibrotic encapsulation of the biomaterial implant can be influenced by the choice of biomaterial type, size, transplant site, and other factors; therefore, the FBR cannot be avoided but may be managed through careful material design. Section 3 of this chapter details how materials may be optimized in the field of immunoengineering to modulate local host immune responses and improve the functionality of biomaterial-containing islet grafts.

2.3 Alloimmunity and Islet Transplant Rejection

Upon the transplantation of an allogeneic cellular graft, alloimmunity occurs; these immune responses of the transplant recipient's immune system are specific to non-self antigens (allo- or iso-antigens) but from the same species (human to human) [13]. Alloimmunity is explained in chapter "An Ethical Perspective on the Social Value of Cell-Based Technologies in Type 1 Diabetes", but a summary is presented next.

Allogeneic rejection occurs across three timeframes: hyperacute, acute, and chronic [56]. Hyperacute rejection happens within minutes and mainly results from pre-existing anti-donor tissue antibodies that activate the complement pathway and cause graft thrombosis [56]. Although hyperacute rejection primarily occurs due to donor–recipient major histocompatibility complex (MHC) mismatching, islet recipients can develop broad MHC sensitivity from receiving mixed-donor batches of islets across multiple infusions [6]. On the other hand, acute and chronic rejection are mediated by cellular and humoral immunity occurring from weeks to months and months to years, respectively [56].

There are *three* stages of alloimmunity driving graft rejection: (*i*) recognition, (*ii*) clonal expansion and effector cell maturation, and (*iii*) graft destruction [57]. The main triggers of this immune response against the allograft are the donor MHC molecules on the surface of transplanted islets, which are often not fully matched, even in the case of stem cell-derived islets [57, 58]. Host innate immune cells including natural killer cells and macrophages can identify non-self MHC class I or alloantibody tags and then clear the foreign material [58]. Recognition by adaptive immunity relies on antigen presentation occurring through the *direct* and the *indirect* pathways [25]. In the *direct* pathway, recipient T cells alloreactive to the graft can be activated by donor antigen-presenting cells (APCs) transplanted within the graft, which present antigens in intact donor MHC molecules [25]. The *indirect* pathway is mediated by recipient APCs, which process donor-derived antigens and present them in recipient MHC molecules to recipient alloreactive T cells [26]. '*Semi-direct*' pathways have also been described, in which recipient APCs obtain donor MHC via cell-to-cell contact and trigger host T-cell responses, leading to chronic rejection [26]. Following these allorecognition pathways, alloreactive $CD8^+$ T cells and B cells, with the help of alloreactive $CD4^+$ T cells, undergo clonal expansion, effector cell maturation, and mediate acute and chronic rejection of allografts.

3 Biomaterial Properties and Modulation of the Immune Response

Islet transplantation outcomes can be improved with the use of biomaterials, which possess intrinsic properties that influence the implant's ability to heal, function, and modulate immune cells [39]. Biomaterials are advantageous platforms for cell transplant because they can improve cell survival, fate, and integration by fine-tuning material properties and incorporating bioactive cues. Biomaterials have a robust effect on macrophages and angiogenic responses, which, in turn, largely determines graft function [59]. However, biomaterials are inherently invasive when implanted into the body and trigger the FBR described previously. Resolution of this reaction and integration into the host's body is required for long-term islet engraftment and function. A key determinant of this outcome is macrophage phenotype, specifically the balance between M1 and M2 macrophage phenotypes [60]. Biomaterials can shift these phenotypes one way or the other, and the properties of the biomaterial or its functionalization for the presentation and/or controlled release of immunomodulatory molecules can be tuned to produce a desired outcome. This section focuses on the effect of biomaterials' physical properties on inflammatory responses, which can be tuned to improve transplantation, summarized in Table 1. While this section strives to evaluate the individual effect of each physical property, the net immunomodulatory effect of different biomaterial preparations will vary.

3.1 Types of Biomaterials

The type of biomaterial used in transplantation determines macrophage adhesion and activation, which defines the subsequent activation and adhesion of lymphocytes. Biomaterials are generally divided into metallic and non-metallic materials. Those that are non-metallic can be further classified as organic (natural and synthetic polymers) or inorganic (ceramic materials) [39]. Islet transplantation most commonly utilizes synthetic and natural polymers, making them key materials, which are described further in chapters "Extracellular Matrix to Support Beta Cell Health and Function", "Bioactive Materials for Use in Stem Cell Therapies for the Treatment of Type 1 Diabetes", and "Considerations Pertaining to Implant Sites for Cell-Based Insulin Replacement Therapies". Synthetic polymers have tunable biomechanical and biodegradability properties, which can be modulated to control subsequent immune responses. These types of biomaterials are usually inert, easily reproducible, have long shelf lives and tunable degradation rates, and can be fabricated into complex shapes. However, synthetic biomaterials offer poor cell adhesion unless they are modified and trigger stronger host innate immune responses compared to natural polymers [61]. Natural polymers better resemble native extracellular matrix (ECM), generally making them more biocompatible, conducive to cell adhesion sites, and grants bioactive properties. However, they have higher sterilization costs, lot-to-lot variability, low mechanical properties (increased failure rates), and may induce unpredictable immune responses because of batch-to-batch variabilities. Their degradation rates also prevent their use in immunoisolation platforms.

In islet immunoengineering, natural materials like chitosan, alginate, collagen and hyaluronic acid (HA) and synthetic materials like poly lactic-*co*-glycolic acid (PLGA), poly(ethylene glycol) (PEG), poly (D,L-lactide-*co*-ϵ-caprolactone) (PDLLCL), and methacrylic acid-based materials have been widely used for local immunomodulation (Table 1) [62].

3.2 Global Features of Biomaterials

3.2.1 Shape and Size

Biomaterial shape and size play a crucial role in the healing process by determining cells and proteins adhesion to the material; immune cell migration, proliferation, differentiation, cytokine secretion; and macrophage fusion [36]. Implant shape and the presence of surface protrusions or dentations determine the availability of attachment sites, affecting how macrophages uptake the biomaterial and ultimately influencing the FBR [63]. Generally, the presence of vertices facilitates cell adhesion and reaction, making spherical-shaped implants the least immunogenic [64]. Indeed, spherical implants have been used in islet transplantation and produced better outcomes compared to non-spherical implants [64]. It is also important to note that related shapes such as elliptical or cylindrical biomaterials are more rapidly internalized by macrophages compared to spherical shapes, and that elongated worm-shaped biomaterials trigger little to no phagocytosis [50]. Regarding size, biomaterials with diameters larger than 1.5 mm are more biocompatible and trigger less FBR and fibrosis, due to the difficulty for macrophages to adhere to, phagocytose, and degrade larger materials [62]. While overall shape and size are important factors in determining foreign body reactions, the local geometry (surface topography) of biomaterials is another important consideration.

Table 1 The influence of biomaterial properties on the innate and adaptive immune system

Property		Example	Target immunity/cell type	Biological effect	References
Biomaterial type	Natural	Chitosan	Neutrophils	↑ Migration and IL-8	[71]
			Dendritic cells	↑ Maturation (compared to alginate and HA)	[72]
			Natural killer cells	↓ Metabolism ↑ Number	[73]
		HA	Dendritic cells	↓ Activation (compared to PLGA and chitosan)	[72]
			Macrophages	↓ M1 markers (IL-6, IFN-γ, and MCP-1 activation) ↑ M2 markers (IL-10)	[74]
		Other biomimetic extracellular matrix components	Macrophages	↓ M1 markers ↑ M2 makers (proregenerative CD206)	[71]
		Alginate	Dendritic cells	↑ Adhesion (compared to HA) ↓ Activation (compared to PLGA and chitosan)	[72]
			Adaptive immunity	↓ B and T cell activation markers (compared to PLGA and chitosan)	[72]
	Synthetic	PLGA	Dendritic cells	↑ Maturation (compared to alginate and HA)	[72]
			Mast cells	↑ Recruitment ↑ Degranulation	[75]
			Dendritic cells	↑ Maturation (compared to alginate and HA)	[72]
			Macrophages	↓ Activation ↓ Pro-inflammatory cytokines	[76]
		PEG	Neutrophils	↑ Oxidative stress enzyme release ↓ Matrix metalloproteinase 9 (tissue remodeling activator)	[77]
			Innate immunity	↓ Neutrophil and protein adhesion	[66]

Global feature	Shape	Spherical		Long-term in vivo implantation	↓ Foreign body reaction and fibrosis	[64]
		Rods, amorphous		Macrophages and monocytes	↑ Recognition and phagocytosis	[78]
	Size	>1.5 mm		Long-term in vivo implantation	↓ Foreign body reaction and fibrosis	[64]
	Porosity	<30μm		Macrophages	↑ Pro-inflammatory phenotype	[65]
		Pores that block direct cell contact		Adaptive immunity	Direct antigen recognition pathway blocking	[13]
Surface properties	Surface chemistry	Amino (positive charge)		Macrophages	↑ Anti-inflammatory phenotype	[39, 67]
		Hydroxyl (neutral charge)		Innate immunity	↑ Integrin binding and cell adhesion of all chemical groups	
		Carboxyl (negative charge)		Long-term in vivo implantation	↓ Fibrotic capsule thickness	
				Macrophages	↑ Pro-inflammatory phenotype	
	Hydrophobicity	Hydrophilic		Macrophages	↓ M1 markers (TNF-α, IL-1β, MCP-1) ↓ foreign-body giant cells	[79]
				Dendritic cells	↓ Maturation	[80]
				Monocyte	↓ Adhesion	[81]
	Fouling	Non-fouling		Innate immunity	↓ Monocyte and protein adhesion	[82]
				Macrophages	↓ Pro-inflammatory phenotype	[66]
	Topography	Roughness	Smooth	Neutrophils	↓ Neutrophil recruitment and IL-1 production	[39]
				Macrophages	↓ M1 markers (TNF-α) ↓ foreign-body giant cells	[64]
		Patterning	Micro-patterning	Macrophages	↑ M2 makers (IL-4, IL-13) (with parallel gratings for cell elongation)	[83]
					↑TNF-α and vascular endothelial growth factor in patterns with larger widths	[84]
					↓ Monocyte fusion ↑ M1/M2 hybrid phenotype	[84]

(continued)

Table 1 (Continued)

Property		Example	Target immunity/cell type	Biological effect	References
Bulk properties	Elasticity	Lower	Macrophages	↓ Adherence	[62]
				↑ Pro-inflammatory phenotype (most materials)	[65, 68]
		Higher		↑ Pro-inflammatory phenotype ↓ Phagocytosis (collagen scaffolds)	[85]
			Dendritic cells	↑ Activation (PLGA)	[86]
	Molecular weight	Larger HA chains	Macrophages	↑ Anti-inflammatory phenotype	[87]
			T cells	↓ Activation	[88, 89]
			Dendritic cells	↓ Activation	[90]

HA hyaluronic Acid, *PLGA* poly(lactic-*co*-glycolic acid), *PEG* poly(ethylene glycol), *IL* interleukin, *IFN* interferon, *MCP* monocyte chemotactic protein, *TNF* tumor necrosis factor

3.2.2 Porosity

Many biomaterial implants have pores to enhance nutrient diffusion and cell integration. From an immunological perspective, pore size can modulate macrophages toward pro- or anti-inflammatory phenotypes, though contradictory patterns have been reported. It has been found that smaller pores (around 30μm) resulted in stronger pro-inflammatory macrophage phenotypes compared to larger pores (around 80μm) [65]. At the same time, it has also been found that smaller pore size constructs (around 40μm) induced more anti-inflammatory macrophage phenotypes compared to larger pores (100μm) [63]. This discrepancy can likely be attributed to the difficulty in isolating the effect of single biomaterial properties and that a combination of features such as biomaterial type, pore size, and mechanical properties can induce drastically different immune reactions or acute versus chronic host reactions.

3.3 Surface Properties of Biomaterials

3.3.1 Surface Chemistry, Hydrophobicity, and Fouling

Surface properties include surface chemistry, charge, and hydrophobicity, which are interlinked by biomaterial chemical moieties. For example, the immune system can specifically recognize hydrophobic regions in biological molecules as danger-associated molecular patterns and eliminate them through degradation and phagocytosis or contain them through FBGC formation [39, 66]. Hydrophobicity is not the sole factor determining biomaterial immunogenicity, and hydrophilicity and overall adhesiveness also play important roles. Biomaterials can be designed to have poor adhesive surfaces to prevent platelets and complement from adhering to the biomaterial surface, thus limiting the IBMIR, and macrophages and lymphocytes adhere to and are activated by hydrophilic and anionic surfaces [39]. On the other hand, hydrophilic and neutrally charged surfaces are associated with $CD4^+$ T lymphocyte interactions, and hydrophobic surfaces with $CD8^+$ T lymphocyte adhesion. Chemical groups that have been used for biomaterial functionalization include amino (-NH2), carboxyl (-COOH), hydroxyl (-OH), and methyl (-CH3) groups, each with different charges and hydrophobicities [39, 67]. In short, biomaterials with amino and hydroxyl groups trigger the highest inflammatory cell infiltration and result in thicker fibrotic capsules, while negatively charged carboxylated surfaces hamper inflammatory reactions and abrogate macrophage-mediated inflammation while promoting regulatory T cells (Tregs) [67]. However, there have been contradictory results on the influence of hydrophobic, neutral -CH3 moieties on fibrotic capsule thickness, cell adhesion, and inflammatory cell migration. Therefore, while chemical moieties alone can influence the immune response, further insight is needed to understand their role in combination with other biomaterial properties (Table 1).

3.3.2 Surface Topography

By acting as the interface with immune cells, the surface topography of implanted materials can also impact the immune response. At the nano- and microscale, the roughness of a material affects protein adsorption and cellular adherence, migration, proliferation, and differentiation. It has been widely shown that smoother surfaces impede protein adsorption and lead to milder immune responses to biomaterial implants [62]. Smoother surfaces pose a lower chance of protein adsorption in 'bumps' on the material surface, which otherwise would aid in recruiting cells and trigger an immune-inflammatory cascade. Furthermore, rougher surfaces are considered to trigger more rapid production of reactive oxygen species compared to smooth materials, which could exacerbate inflammatory responses to the biomaterial implant [36]. This finding extends to surfaces with patterns featuring various shapes and dimensions, such as higher macrophage attachment on micropillars with 5μm diameters [63]. In materials with limited cell adhesion surfaces, cells attach in between such micropatterns [63]. On the other hand, on highly adhesive surfaces, macrophages adsorb micropillars via phagocytosis. Finally, when implanting fibrous materials, fiber alignment can affect immune reaction, where aligned fiber topographies trigger less fibrotic capsule formation and cell infiltration than randomly aligned fibers [66].

3.4 Bulk Properties of Biomaterials

3.4.1 Elasticity

The bulk elasticity of a material describes the extent a material compresses under a given amount of external pressure, and this value can modulate the foreign body reaction. Generally, materials with lower moduli of elasticity around 5 kPa or less (that is, softer materials), more strongly induce inflammatory phenotypes in macrophages compared to those with moduli above 40 kPa (stiffer materials). These effects have been demonstrated both in vitro and in vivo [65, 68]. In animal models at 3 days post-implantation, stiffer scaffolds integrate into the body with better outcomes compared to softer scaffolds, which have increased fibrotic encapsulation and inflammation. However, in the case of collagen scaffolds, which are commonly used in T1D for prevention of wound complications, the results are the opposite: a higher modulus induced pro-inflammatory macrophage phenotypes and reduced phagocytosis [63].

3.4.2 Molecular Weight

Some biomaterials can have varying molecular weights, which can also impact the immune response. One such biomaterial is hyaluronic acid (HA), and it has been shown that lower molecular weight HA triggers dendritic and T cell activation and cytokine secretion through Toll-like receptor (TLR)-2 and TLR-4 signaling [69]. In the same line, macrophages develop a pro-inflammatory M1 phenotype in response to lower molecular weight HA and but an M2 phenotype develops with higher molecular weight HA [69]. Therefore, larger HA chains have more anti-inflammatory properties. This pattern is important for biomaterials with more easily tunable molecular weights such as PEG, HA, PLGA and PDLLCL, but may be more difficult to implement and evaluate in other natural biomaterials like chitosan that require complex hydrolysis reactions to change the molecular weight [70].

3.5 Concluding Remarks

The field of islet immunoengineering focuses on utilizing biomaterials to modulate local immune responses and enhance functional outcomes of transplanted islets. Such biomaterial-mediated improvements may include optimizing biomaterial properties to decrease (1) direct and/or indirect immune responses responsible for islet rejection and recurrence of autoimmunity as well as (2) foreign body responses harmful immune responses while (3) promoting integration of biomaterial scaffolds for islet delivery. This section summarized how different biomaterial properties may be tuned to develop islet encapsulation platforms and islet delivery scaffolds that can improve transplant function and long-term graft acceptance.

4 Biomaterial Drug Delivery Strategies to Eliminate the Need for Chronic Systemic Immunosuppression After Islet Transplantation

SC islets will provide a limitless source of islet tissue and overcome our current and long-standing reliance on donor-derived tissue. However, their transplantation will continue to be challenged by allogeneic rejection and recurrence of autoimmunity in transplant recipients [14]. Currently, systemic immunosuppression is used to prevent such rejection and autoimmunity, but it must be taken continuously to maintain the graft and is associated with significant risk of infection [15–17], malignancy [6, 18, 19], suppression of immune tolerance [91], and drug-related toxicity [18, 20]. Furthermore, these regimens have an annual cost equal to 20% of the one-time cost of allogeneic islet transplant (including islet sourcing, screening, and medical procedure) [92]. The need for chronic immunosuppression has restricted islet transplantation only to patients with the most severe cases of T1D and limits the procedure's benefit to quality of life as patients effectively trade one lifelong condition for another [4, 93, 94]. Ultimately, while the supply of donor-derived tissue limits the availability of pancreatic islet transplantation, the need for immunosuppression limits its applicability as a cure for T1D. Thus, islet transplantation without the need for chronic systemic immunosuppression represents a major goal of the field.

The first allogeneic transplantations of donor-derived islets occurred in 1980, and these early recipients underwent immunosuppressive treatment with intravenous anti-lymphocyte globulin, oral azathioprine and/or cyclophosphamide, and oral prednisone [95, 96]. Respectively, this regimen is intended to deplete, inhibit the proliferation of, and inactivate leukocytes at large, which includes those that would mediate rejection. This broad and indiscriminate inhibition of the immune system characterizes *immunosuppression* and results in the adverse effects mentioned above. Aside from these immunological risks, immunosuppressive drugs carry inherent toxicity and/or side effects. In particular, prednisone and related steroidal immunosuppressants are well known for inducing insulin resistance and directly counteract the efficacy of islet transplantation. Thus, in 2000, a steroid-free immunosuppressive regimen known as the Edmonton protocol was developed, and patients achieved insulin independence up to 5 years after allogeneic transplant [97–99]. Since then, various combinations of immunosuppressing drugs before and after islet transplant have fallen in and out of favor, but the drugs themselves and the method of delivery (i.e. systemic administration) have remained largely unchanged [19].

New avenues to prevent rejection while eliminating the need for chronic systemic immunosuppression can be found in novel approaches to immunomodulation and localized drug delivery to the transplantation site. In contrast to immunosuppression, *immunomodulation* aims to influence and direct the immune response to a desired outcome rather than wholly suppressing it. Immunomodulatory drugs do not have clear distinctions from immunosuppressive drugs, but generally tend to affect immune cell signaling, come in the form of a peptide or antibody, and exert their effects extracellularly [100]. By this definition, immunomodulatory drugs can include interleukin cytokines, interleukin receptor antagonists, chemokines, and molecules affecting immune checkpoints. Notably, while immunomodulation may produce desirable effects at the transplantation site, their systemic administration can still disrupt normal immune function in other locations of the body. Thus, both immunosuppressive and immunomodulatory regimens stand to benefit from improved localized drug delivery that exerts beneficial effects at the transplantation site while avoiding off-target or toxic systemic effects (Fig. 3) [101].

4.1 Drug Delivery via Biomaterials Improves Bioavailability and Stability

Inducing therapeutic effect within the graft while avoiding off-target and systemic effects can be achieved when drug is delivered only to the area of islet transplantation, an approach that relies on biomaterials. Biomaterial encapsulation of drugs not only enables graft targeting and local release (described in the next section) but also benefits their pharmacokinetic and pharmacodynamic properties. Specifically, biomaterials can increase the solubility and stability of hydrophobic and hydrophilic drugs while also providing controlled and sustained delivery for longer periods of time than unformulated drugs [102]. Table 2 depicts examples of biomaterial

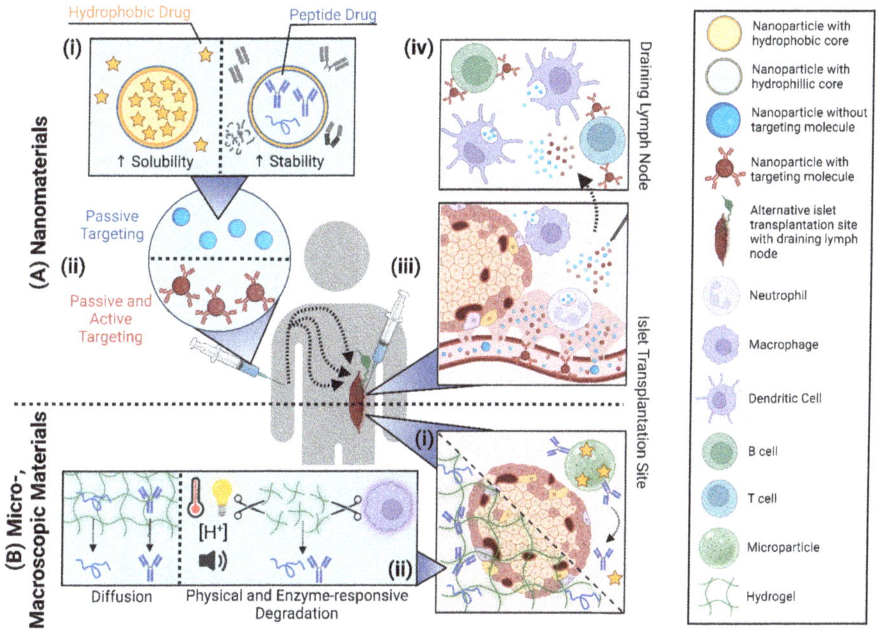

Fig. 3 Biomaterial-mediated methods for targeted, localized, and sustained drug delivery to islet transplant sites alternative to the liver. (**a**) Advantages and biological considerations of drug delivery using *nanoscale materials*. (i) Nanoparticles can be configured to have hydrophobic cores (left, yellow) or hydrophilic cores (right, blue). Encapsulating hydrophobic and peptide drugs within biomaterials can increase their solubility and stability, respectively. (ii) Nanoparticles can be additionally modified with targeting molecules to increase drug delivery to specific cells of interest (active targeting). With or without active targeting, (iii) intravenously injected nanoparticles accumulate in the islet transplant site due to the enhanced permeation and retention effect, secondary to permeabilized local vasculature due to local inflammation (passive targeting). Alternatively, nanoparticles can be injected directly into accessible graft sites (e.g. subcutaneously). Following either method of administration, nanoparticles within the transplant site can be uptaken or diffuse directly into local cells, delivering drug payloads intracellularly, and/or mediate sustained release of drug over time. (iv) Nanoparticles within the graft site can transit to the nearest draining lymph node in a cell-free state or after being uptaken by immune cells migrating from the transplantation site. There, they can also release drug or be uptaken by local cells of the lymph node. (**b**) Advantages and biological considerations of drug delivery using *micro- and macroscopic materials*. (i), top right: Islets can be co-transplanted with microparticles large enough to resist phagocytosis, which then form a local drug depot and exhibit sustained drug release over time. (i), bottom left: Islets can be co-transplanted alongside or within drug-loaded hydrogels and solid scaffolds (e.g., PDMS or PLGA), also forming local drug depots and mediating sustained release of drug over time. (ii) Hydrogels/scaffolds can passively release drugs through mechanisms such as diffusion, or they can be engineered to become responsive to physical stimuli (high temperature, light, pH, and ultrasound) or enzymatic degradation

Table 2 Examples of biomaterial delivery of immunosuppressive and immunomodulatory drugs and release kinetics

	Drug	Mechanism of action	Solubility in water (g/L)	Delivery method	Sustained release	References
Immunosuppressive	Rapamycin (Sirolimus)	Mammalian target of rapamycin (mTOR) inhibition	Insoluble; logP = 4.80	PCL microparticle	Constant through day 10	[104]
				PLGA microparticle	Delayed release starting day 10 through day 30	[104]
				Polymeric bile-acid nanoparticle	Up to 20 days, in vivo	[134]
	Cyclosporin A	Inhibition of IL-2 production and lymphocyte signaling	Insoluble; logP: 2.92; 3.30	PEG-PPS nanomicelle	First order through day 15	[107]
				PEG-OES nanofibril	First order through day 4	[107]
	Dexamethasone	Multiple genomic effects	logP = 1.83	PLGA micelle	Biphasic. Majority released day 1, through day 10–30	[109]
				PDMS scaffold	Biphasic. Majority released day 10, through day 30	[110]
				PLGA microparticle	50% released at 10 days	[135]
Immunomodulatory	IL-1 receptor antagonist	Binds to IL-1 receptor 1	Soluble	Pluronic F127 hydrogel	Through 12 hours	[113]
	CTLA4-Ig	Binds to CD80, CD86	Soluble	Acellular dermal matrix	Immobile	[136]
	Fas ligand	Pro-apoptotic signaling	Soluble	PEG microgel	Microgels present through day 21 in vivo	[112, 137]
	CCL1	Treg bioactive factor	Soluble	Alginate hydrogel	Not quantified	[114]
	IL-2	Treg bioactive factor	Soluble	Alginate hydrogel	Not quantified	[114]
	IL-33	Expansion of ST2+ Tregs	Soluble	PLGA scaffold	Biphasic. Peak release at day 1, through day 7	[115]
	CXCL12	T cell chemorepellant	Soluble	Incorporated into alginate microcapsules	Burst release through day 1, stable release from day 4 through day 25	[116]
	CCL21	Naïve and regulatory T cell homing	Soluble	PEG hydrogel	Biphasic. Release depends on cross-linker used. Peak release at 6 h, through day 5; Peak release at day 7, through day 28	[111]

PCL polycaprolactone, *PLGA* poly(lactic-*co*-glycolic) acid, *PEG-PPS* poly(ethylene glycol)-poly(propylene sulfide), *PEG-OES* poly(ethylene glycol)-oligo(ethylene sulfide), *PDMS* polydimethylsiloxane, *IL* interleukin, *PEG* poly(ethylene glycol), *Treg* regulatory T cell delivery of immunosuppressive drugs currently used in clinic and experimental immunomodulatory drugs in the context of islet transplantation, where these patterns can be observed.

Key immunosuppressive drugs such as rapamycin (sirolimus), calcineurin inhibitors, and glucocorticoids have low solubility in water, which decreases their bioavailability, requires higher dosage, and narrows the therapeutic window [103]. Furthermore, these drugs must be solubilized in non-inert carriers such as Cremophor EL, which has been associated with hypersensitivity reactions, dyslipidemia, and neuropathy [104–106]. Instead, biomaterial particles carrying these drugs are configured to contain a hydrophobic core resembling micelles, which can dramatically increase their solubility in water. For example, cyclosporin A has a solubility of 0.012 g/L in water and loading into amphiphilic self-assembling block copolymer-based nanomicelles increases it more than 100-fold to 1.88 g/L [107]. Thus, packaging immunosuppressive drugs in biomaterials increases their solubility, decreases the dose needed, and eliminates the need for current carriers (Fig. 3).

Immunosuppressive and immunomodulatory drugs can present challenges with stability and serum half-life, which can be merely minutes to hours for protein and peptide drugs [108]. Biomaterials can offer a solution by enabling sustained release of drug over time (Table 2). Indeed, materials containing these agents have demonstrated sustained release by continuously eluting drugs for up to a month (it is important to note though that these experiments are commonly performed in vitro and release kinetics may change in vivo). One consideration for sustained drug release lies within the release kinetics of biomaterials. It is a common finding for drugs to be released in a biphasic/burst pattern, where most of the loaded drug is released relatively early after delivery, followed by diminished but measurable release thereafter. This consideration can be addressed through the diversity and customizability of biomaterial vehicles. Without specific drug-linking approaches, drug release from biomaterials is determined by diffusion through and/or decomposition of the material via degradation, deformation, or swelling; factors that can all be tuned according to formulation and chemical composition of the biomaterial [108]. For example, particle size being equal, rapamycin can have a constant sustained release from microparticles ending at day 10 or a delayed release only starting at day 10, depending on whether the particle is composed of polycaprolactone (PCL) or poly(lactic-*co*-glycolic acid) (PLGA) [104]. Cyclosporin A can have exponential, first-order release kinetics from poly(ethylene glycol) (PEG)-based nanoparticles lasting through 4 or 15 days depending on whether the particle is micelle- or rod-shaped [107]. Dexamethasone release from microscale PLGA micelles peaks at 1 day, while release from a polydimethylsiloxane (PDMS) microporous scaffold peaks at 10 days [109, 110]. Finally, CCL21 chemokine loaded into a PEG hydrogel can have peak release at 6 hours or 7 days [111], depending on the drug-biomaterial crosslinking chemistry, which determines the gel's degradation characteristics. In all, the chemical composition and structure of particles and hydrogels can not only be optimized to carry a particular drug, but to also deliver it in a desired timeframe. A combination of delivery approaches for immunosuppressive and immunomodulatory could be tuned to release drug targeting specific phases of an immune response, such as an early innate foreign body reaction or delayed adaptive allogeneic rejection.

4.2 Drug Delivery via Biomaterials Enables Islet Graft Targeting and Local Release

Aside from benefits of biomaterial-mediated drug to bioavailability and sustained release, islet transplantation synergizes with the capability of biomaterials to deliver drug directly and specifically to the transplantation site. This is because islets have a key advantage over solid organs (heart, kidney, lung, etc.) in that micrometer-scale islets or islet progenitors freely suspended in small volumes (from 300μL to 10 mL) can be transplanted in various places throughout the body [21, 31, 98]. A well-defined, vascularized islet transplantation site is uniquely compatible and synergizes with localized drug delivery by biomaterials, an approach that can achieve therapeutic concentrations of drug to prevent islet rejection while avoiding systemic distribution of immunosuppressive and immunomodulatory drugs.

The sizes of biomaterials for local drug delivery can range from nanometers to centimeters. Their size and method of administration determines distribution through the body and interaction with the immune system (Table 3). Generally, nanoparticles can be suitable for intravenous administration and localization (both passively and actively, discussed further in the next section) to a transplantation site, with subsequent uptake by cells leading to intracellular drug delivery. On the other hand, microparticles and macroscopic scaffolds can be co-implanted alongside islets, remain in the transplantation site, and act as a drug depot that releases over time to the local extracellular space.

Microparticles larger than 10μm and macroscopic gels and scaffolds (on the order of millimeters and larger) are suitable for direct implantation alongside transplanted islets. In this setting, these biomaterials can present or elute drug directly in the transplantation site and immediately achieve local drug delivery. Implantation of these biomaterials will likely be compatible with future methods and locations of islet transplantation, as microparticles and macroscopic hydrogels can be surgically implanted alongside islets or injected into the graft site [108, 112]. In experimental islet transplantation, microparticles have been used to deliver hydrophobic immunosuppressants such as rapamycin/sirolimus and dexamethasone while macroscopic hydrogels have been used to deliver hydrophilic immunomodulators (Table 2). Due to their size and in contrast to nanoparticles, microparticles and macroscopic hydrogels are resistant to phagocytosis and thus release drug extracellularly, a delivery method compatible with intracellular-acting immunosuppressants and required for extracellular-acting immunomodulators. Simultaneously, the presence of these materials will create a constant interface for foreign body immune reactions,

Table 3 Size distribution and biological considerations of drug-delivering biomaterials

	Size	Biological consideration	Application	References
Nanoparticles (nm)	5–10	Particles can pass through the glomerulus and abundant small pores in normal tissue endothelium	Cyclosporin A nanofiber (5 nm diameter, 1um length)	[107]
	20	Lower size limit for effective APC antigen processing and cross-presentation	Cyclosporin A nanomicelle	[107, 122]
	40	Optimal size for transit through the lymphatic system		[122]
	100	Upper size limit for nanoparticles to target cell organelles		[118, 120]
		IV: upper size limit to pass through endothelial barriers (blood-brain, cardiac endothelium), can be engulfed by mononuclear phagocytic cells in liver and spleen		
	175	IV: upper size limit to pass through liver endothelial cell fenestrations		[118]
	200	SC: Upper size limit for diffusion through endothelial cell junctions and cell-free passage to draining LN. Above this, nanoparticles are phagocytosed and transported to draining LNs with 24 h delay IV: Upper size limit for longer circulation time due to low uptake by reticuloendothelial system Upper size limit for non-phagocytic clathrin-mediated endocytosis		[118, 121, 122]
	345		Rapamycin bile-acid-polymer nanoparticle	[134]
	400	Suggested upper size limit for nanoparticles to possess EPR capability		[118]
	500	Upper size limit for nanoparticle penetration through cell membranes. Non-phagocytic caveolae-mediated endocytosis can occur		[118, 121]

(continued)

Table 3 (Continued)

	Size	Biological consideration	Application	References
Microparticles (μm)	1	Lower size limit for non-phagocytic macropinosome uptake		[121]
	2–3	Optimal size for APC phagocytosis		[123, 124]
	5	Upper size limit for particles to circulate through capillaries and non-phagocytic macropinosome uptake		[121, 123]
	10	Upper size limit for phagocytosis. Local administration results in formation of drug depot	CCL1 and IL-2-supplemented alginate microencapsulation IL-1Ra microparticles Dexamethasone PLGA micelles	[108, 109, 114, 123, 138]
	20		Dexamethasone PLGA microparticle	[135]
	100		Rapamycin microparticles	[104]
	200		Streptavidin-Fas ligand microgel	[112, 137]
	300		Alginate microencapsulation supplemented with CXCL12	[108]
Macrogels, polymeric scaffolds (mm)	1.5	For implanted biomaterials larger than 1.5 mm, decreased foreign body reaction and fibrosis		[64]
			0.5 mL subcutaneous CTLA4Ig hydrogel	[138]
			1 × 5mm CTLA4-loaded matrices	[136]
			Dexamethasone in PDMS scaffold	[110]
			IL-1Ra Pluronic F127 hydrogel	[113]
			IL-33 PLGA scaffold	[115]
			CCL21 PEG hydrogel	[111]

APC antigen-presenting cell, *IV* intravenous administration route, *SC* subcutaneous administration route, *LN* lymph node, *EPR* enhanced permeation and retention, *IL* interleukin, *IL-1Ra* interleukin-1 receptor antagonist, *PLGA* poly(lactic-*co*-glycolic) acid, *PDMS* polydimethylsiloxane, PEG poly(ethylene glycol) leading to the consideration that these materials should be degradable once the drug payload has been depleted [108, 111]. Peptide/protein drugs have additional considerations when it comes to loading into a hydrogel or scaffold. In the simplest approach, peptides and proteins can be passively incorporated by mixing into a hydrogel as it forms, provided that the fabrication method does not involve conditions that would disrupt their structure and function. Drugs loaded in this fashion leave the hydrogel through diffusion or degradation and deformation of the hydrogel [108]. This approach has been used in the context of diabetes to deliver interleukin-1 receptor antagonist [113], interleukin-2 and -33 [114, 115], and CXCL12 [116]. Peptides and proteins can also be actively bound to hydrogels. Multiple reaction pathways can achieve direct covalent conjugation [108] but may alter protein function. CCL21 has been covalently tethered to hydrogels without affecting its function [111], while Fas ligand has been fused to streptavidin and was then tethered to biotinylated microparticles [112]. Finally, drug-carrying nanoparticles can be tethered to hydrogels, which can increase drug stability, achieve local and sustained delivery of nanoparticles without relying on systemic circulation, and avoid deactivating modifications in the case of peptide drugs [111].

Nanoparticles are those on the scale of nanometers, making them amenable to intravenous injection (Fig. 3 and Table 3). This delivery route can be useful for localized post-transplantation drug delivery after larger, co-implanted microparticles and macroscopic hydrogels have exhausted their drug supply, thereby offering a way to "refill" local drug depots [117]. For intravenous injection, multiple size thresholds govern nanoparticle distribution throughout the body. At the systems level, nanoparticles with diameters up to 400 nm are more relevant for intravenous injection as they can accumulate in areas with leaky vasculature through a phenomenon known as the enhanced permeation and retention (EPR) effect [118, 119]. However, it is also important to note that the EPR effect does not prevent nanoparticle distribution to other areas of the body such as the liver and spleen, where mononuclear phagocytic cells uptake and clear the particles [120]. As size decreases below 200 nm, nanoparticles will diffuse through or between endothelial cells more easily and exhibit longer circulation times due to lower uptake by the reticuloendothelial system [118, 120, 121]. The 200 nm threshold is also important for local administration routes (subcutaneous, intramuscular, and mucosal). Particles delivered in these fashions that are smaller than this threshold are able to transit to the draining lymph node in a cell-free state, whereas larger ones up to 4μm must first be phagocytosed by local antigen-presenting cells and then be carried to the draining lymph node [122–124]. Nanoparticle size also influences uptake and intracellular distribution. Unless specifically modified, immune cells may more readily uptake nanoparticles through phagocytosis due to extensive opsonization by immunoglobulins and complement during circulation compared to non-phagocytic cells [121]. However, non-phagocytic endocytosis pathways such as clathrin- or caveolae-mediated endocytosis can allow non-phagocytic cells to uptake nanoparticles [121]. Regarding intracellular targeting, phagocytosis and clathrin-mediated endocytosis result in particle delivery to degradative lysosomes, whereas caveolae-mediated endocytosis can bypass this pathway and be more suitable for delivery of sensitive cargo (peptides, nucleic acids) [121]. Finally, due to these mechanisms, nanoparticles and smaller microparticles are well-suited to deliver drugs with intracellular mechanisms of action, which typically includes non-biological immunosuppressive drugs and excludes biological immunomodulatory drugs.

4.3 Customization of Biomaterials to Further Enhance Targeted and Local Drug Delivery

Biomaterials can be modified aside from chemical composition to further enhance drug delivery to sites of islet transplantation. Namely, circulating nanoparticles can be modified with targeting molecules to improve delivery to specific cells of interest [102], and hydrogels/solid scaffolds can be designed to degrade in the presence of specific environmental factors (stimulus-responsive activation) [125, 126].

Modification of circulating drug-carrying nanoparticles with receptor-specific ligands can potentially enable active targeting in addition to the passive EPR effect nanoparticles [102, 127]. In cancer, nanoparticles have been modified with monoclonal antibodies against cell-specific surface proteins, ligands of receptors expressed by target cells, glycoproteins, RNA aptamers, and various small peptides to improve cancer cell targeting [102]. Because various approaches can be used to achieve nanoparticle modification, such as the avidin–biotin interaction [127] or oligodeoxynucleotide binding [117], these cancer-targeting approaches can be adapted to islet-targeting approaches by simply changing the targeting moiety. Indeed, human β cell-specific RNA aptamers [128] and glucagon-like peptide-1 (GLP-1) analogs targeting the GLP-1 receptor highly expressed on β cells [129] could be used for this purpose.

Both particle and hydrogel biomaterials can be modified to release drug payloads upon encountering specific environmental conditions, resulting in stimulus-responsive activation. Some drug-loaded biomaterials can be activated by physical stimuli such as ultrasound [130], pH [131], temperature, light, pressure, and electric fields [132]. Of these, the externally applied stimuli triggering drug release from a co-implanted hydrogel would be compatible with the ongoing research in accessible, subcutaneous alternative islet transplantation sites mentioned previously. Biomaterials can also be enzyme-responsive and activatable by biological stimuli. In cancer, hydrogels have been designed to degrade in the presence of matrix metalloproteinases and hyaluronidases, enzymes that are both upregulated in the tumor microenvironment [125, 126]. In islet transplantation, hydrolytically degradable hydrogels have been used [111, 133], and further understanding of the immune response against transplanted islets (with and without biomaterials) could inform the design of future enzyme-responsive hydrogels for local drug delivery.

4.4 Concluding Remarks

Stem cell-derived islet technology will represent a breakthrough in the field of islet transplantation by providing a limitless source of donor tissue and vastly increase the accessibility of islet transplantation. Unfortunately, even with these islets, recipients will still require immunosuppression to prevent alloimmune rejection and recurrent autoimmunity against transplanted islets. This need for chronic systemic immunosuppression following islet transplantation has and will continue to limit the applicability of islet transplantation. Strategies for targeted, localized delivery of current immunosuppressive and future immunomodulatory drugs using biomaterials have strong potential to eliminate chronic systemic immunosuppressive and break through this second barrier. Biomaterials are highly customizable in their chemical composition, size, structure, and bioactive profile, allowing them to optimally package, localize, and sustainably release drugs in sites of islet transplantation. Delivering currently used immunosuppressive drugs with biomaterials can widen their therapeutic window by lowering their needed dose, avoiding systemic toxicity, and maintaining therapeutic efficacy at the transplantation site. When combined with biomaterial delivery of immunomodulatory drugs in development, tolerance to the islet graft could also be induced, thereby achieving a one-time "fix and forget" restoration of metabolic homeostasis in patients with T1D.

5 Approaches for Gene Editing of Stem Cell-Derived Islets

Gene editing technologies function as an invaluable tool in the field of immunoengineering, allowing for persistent and accurate modifications to a cell's phenotype. The past use of genetic engineering platforms has primarily focused on altering nucleotide sequences in a cell's genome to knock in advantageous genes or knock out deleterious genes. Recently, genetic engineering technologies have grown to include strategies to alter gene expression via epigenetic regulatory mechanisms. Of importance in the immunoengineering of SC islets, human leukocyte antigen (HLA) loci can be excised from the genome, decreasing the immunogenicity of the SC islets, which results in decreased allogenic immune responses. Additionally, immune checkpoint ligands, such as PD-L1, can be overexpressed to allow for SC islets to directly modulate the allogenic immune response, similar to how tumors generate a locally suppressive microenvironment.

5.1 Gene Editing via CRISPR-Cas9

The landmark discovery of clustered regularly interspaced short palindromic repeat (CRISPR) genomic regions followed by CRISPR associated genes (Cas) in bacteria has yielded a robust and widely applicable toolbox for genomic engineering [139]. Originally, the CRISPR-Cas9 system was developed as a platform for nucleotide sequence modifications through either insertions or deletions (indels) at specific genome positions [139, 140]. The basic CRISPR-Cas9 gene editing system consists of two primary components, the Cas9 endonuclease and a single guide RNA (sgRNA) [140]. The Cas9 protein has both an RNA and DNA interacting domain, where the RNA binding domain binds the sgRNA and the DNA domain initiates a double-strand break in the cell's DNA [140, 141]. An sgRNA is composed of two distinct RNA sequences on the same RNA molecule. The first is complementary to the targeted DNA sequence where a cut is desired and the second is recognized by the Cas9 for recruitment to the site [140, 141].

In brief, CRISPR-Cas9 employs a genetic construct coding for both the Cas9 and sgRNA that is introduced to targeted cells' genome via lentiviral plasmid transduction or transient transfection [139]. Upon transcription of the sgRNA and translation of the Cas9, an sgRNA-Cas9 complex is formed, which then targets the genomic region of interest, resulting in a double-strand break [140]. Following a double-strand break, endogenous DNA repair is initiated via either non-homologous end joining (NHEJ) or homology-directed repair (HDR) pathways [139, 140]. NHEJ is an error-prone mechanism that can lead to indels at the break site, which can ultimately result in a *knockout* of the target [140]. Contrary to NHEJ, the HDR pathways use a deliverable repair template with the desired *knock-in* sequence flanked by sequences homologous to those on either side of the double-strand break [140].

More recently, there have been advancements in CRISPR-Cas fusion systems for targeted epigenetic modification, direct activation, or suppression [142, 143]. Of importance to this work was the development of null- Cas9 (dCas9), which is a non-catalytically active form of the Cas9 enzyme [142]. In these fusion proteins, the dCas9 domain acts as a homing molecule to the site for modification, while an active protein can add (e.g., DNA methyltransferases) or remove (e.g., TET enzymes) epigenetic modulatory groups from the site [142, 144–148]. Alternatively, direct activation and suppression can be achieved by fusing proteins with effector, activator (VP64), or repressor (KRAB) domains [142, 149–151].

A strategy using dCas9 systems is the select upregulation of immune checkpoint ligands, primarily Programmed Death Ligand 1 (PD-L1), which acts as an inhibitory signal to T cells. The trifusion protein dCas9-VPR contains three transactivation domains, which upregulate transcription by recruiting the molecular machinery needed for transcription to occur [142, 143]. In this way, dCas9-VPR with an sgRNA targeting the PD-L1 promoter can be co-delivered to stem cell-derived beta cells and induce overexpression of PD-L1, thus modulating β cell–T cell interactions.

5.2 Human Leukocyte Antigen Knockout

Human leukocyte antigens (HLAs), the major histocompatibility complex molecules present in humans, are proteins that bind to peptide antigens and display them to various immune cells [152]. When this interaction occurs, the immune cell may or may not elicit an effector function depending on the identity of the peptide and the HLA subclass. There are two HLA subtypes—class I and class II. HLA class II molecules are expressed by professional antigen-presenting cells (APCs) such as B cells, dendritic cells (DCs), and macrophages. APCs internalize soluble self and non-self antigens, process them into short polypeptides, load them on HLA class II, and display them on the cell surface where they may be surveilled by $CD4^+$ T helper (T_h) cells. In contrast, HLA class I molecules are expressed by all nucleated cells to present peptides originating from within the cell to $CD8^+$ T cells. These peptides of intracellular origin may derive from normally translated self-peptides, or from those generated via viral infection, and thereby enable the immune system to distinguish healthy cells that should be ignored from infected ones that must be eliminated.

Within the family of HLA class I, there are classical and non-classical subtypes. The classical subtypes are HLA-A, HLA-B, and HLA-C, which are primarily responsible for the display of intracellular peptides to the immune system for reasons outlined above. The non-classical subtypes are HLA-E, HLA-F, and HLA-G, which through a variety of mechanisms prevent the activation of natural killer (NK), T, and B cells in sites like the maternal-fetal interface [153].

In the context of cell transplantation, because HLA-A/B/C exhibit a high degree of polymorphism across their gene pool and also present self-peptides, they are considered to be the primary drivers of immunological allogeneic rejection [152]. As such, a strategy to reduce the allorecognition of transplanted SC islets by the host immune system is to prevent their HLA class I expression. Without HLA class I, the host adaptive immune system, starting with $CD8^+$ T cell surveillance, is unable to detect allogeneic SC islets and destroy them. However, a lack of HLA expression is associated with an abnormal cell state, and so these cells activate and are destroyed by NK cells. Thus, a strategy to achieve adaptive immune evasion while preventing NK cell activation is to delete the classical HLA class I genes while preserving the non-classical ones.

These genetic modifications can be effectively achieved via CRISPR-Cas9 editing at the pluripotent stem cell (PSC) level before pancreatic endoderm differentiation [154]. The combination of HLA I genes that are deleted or preserved can vary and lead to different outcomes in the context of allorecognition. One configuration preserved the HLA-E/F/G genes and deleted nearly all HLA-A/B/C genes, except for the HLA-A2 allele. This is because for HLA-E to exact its tolerogenic effect on NK cells, it needs to present the signal peptide of HLA-A2 [154].

This HLA class I re-configuration did not affect the ability of PSCs to differentiate into SC islet cells, with final preparations of wild type (WT) and gene edited cells yielding final preparations with similar fractions of mature SC β cells characterized by NKX6.1 and C-peptide expression of about 34%. Upon in vitro incubation with NK cells, the HLA-A2/E/F/G SC islets cells elicited reduced NK cell activation as compared to WT, as measured by CD107a expression of less than 10% versus 15% by WT. Further, in humanized NSG mice, transplanted HLA-A2/E/F/G SC islets cells persisted in the spleen at about 75% at 8 weeks after transplant, versus only about 30% by WT SC islets [154]. Taken together, these results demonstrate the ability of selective HLA class I deletion to limit allorecognition of SC islets and, therefore, is a promising strategy for SC islet transplantation with reduced and potentially no immunosuppression.

5.3 PD-L1 Expression

Many lessons from tumor biology, particularly those involving immune tolerance and evasion, can be adapted to the immunoprotection of transplanted SC islets [155]. Programmed death ligand-1 (PD-L1) is a protein, often expressed by many types of cancers, that binds to PD-1, a receptor on T and B cells and monocytes. This binding interaction leads to a regulatory signaling pathway, which arrests the proliferation of antigen-specific T cells and preserves regulatory T cells (Tregs) [156]. In one configuration, SC islets engineered to express PD-L1 are achieved via lentiviral transfection of human induced pluripotent stem cells (iPS) cells where the lentivirus is used to insert the PD-L1 gene along with an upstream UCN3 promoter so that PD-L1 is expressed by SC β and α cells [157]. When this genetic modification is achieved, iPS may then undergo standard SC islet differentiation protocol to generate SC islets with similar oxygen consumption rates (OCRs), a measure of metabolic activity, to primary human islets. Further, the PD-L1 expressing SC islets can secrete C-peptide, and therefore, insulin, in response to a glucose challenge indicating that such genetic modification does not affect their primary function [157].

Upon xenotransplantation under the kidney capsule of immunocompetent diabetic C57BL/6 mice and transplantation in humanized NSG mice, PD-L1 expressing SC islets normalized blood glucose for up to 50 and 27 days, respectively [157]. This contrasts with WT SC islets, which were rejected by day 10 and 7 post-transplant, respectively. In the humanized NSG model, recipients of PD-L1 SC islets exhibited significant quantities of serum human C-peptide, as opposed to WT, which were C-peptide negative by day 4 after transplant. Graft analysis after explant showed the persistence of PD-L1 expressing SC islets as measured by GFP expression, while little to no GFP positive cells were found in grafts receiving WT SC islets. Finally, PD-L1 SC islet grafts contained a threefold lower fraction of leukocytes (by CD45 expression) than WT SC islet grafts. These observations provide strong evidence that PD-L1 expression by SC islets can transiently prevent immune rejection but may require other interventions for durable tolerance.

5.4 Indoleamime-2,3-Dioxygenase Expression

Indoleamine-2,3-dioxygenase (IDO) is an enzyme expressed by professional APCs, such as B cells, DCs, and macrophages, which regulates immunity by altering immunometabolism [158]. Specifically, this enzyme catalyzes the oxidation of the amino acid tryptophan to N-formylkynurenine (KYN). The action of IDO alters the immune response in two ways: the effect of local tryptophan depletion and the action of the KYN product [159]. T cells require tryptophan to proliferate, and so a lack of available tryptophan leads to cell cycle arrest and T cell anergy. KYN itself acts to create a more tolerogenic microenvironment by inducing new Tregs and increasing the suppressor function of already existing Tregs.

Given these properties, IDO can play an important role in the immunoprotection of SC islet grafts. This can be achieved by engineering SC islets or accessory cells to constitutively express IDO. In one configuration, primary islets were transplanted with fibroblasts transfected via lentiviral vector to express IDO [160]. Coculture studies showed that IDO-expressing fibroblasts do not impact the viability or function of β cells. In an allogeneic islet transplant model, diabetic C57BL/6 mice receiving islets and IDO-fibroblasts in a scaffold experienced normalized blood glucose levels for about 50 days, which contrasts with those that received islets and normal fibroblasts, where normoglycemia did not persist past 10 days. Graft histology revealed insulin positive islets and Foxp3$^+$ Tregs at significantly higher quantities in grafts of islets and IDO-fibroblasts as compared to the control. There were also lower fractions of proliferating CD4 and CD8 T cells in grafts of islets and IDO-fibroblasts as compared to control. These findings highlight the potential of IDO as a potent immunomodulator for the protection of SC islet allografts.

6 Islet Co-Transplantation with Immunomodulatory Cells

Stem cell-derived islets can be co-transplanted with immunomodulatory cells to improve graft function and viability. While there are many potential cell therapies for co-transplantation currently under investigation, this section focuses on mesenchymal stem cells and regulatory T cells, some of the most-researched cell therapies for co-transplant strategies to delay allorejection in SC islet transplant. Mesenchymal stem cells are studied in a wide variety of contexts for their regenerative and immunomodulatory properties, the latter of which holds promise in modulating allogeneic immune responses against transplanted islets. Likewise, regulatory T cells can support islets by creating a tolerogenic environment through the secretion of anti-inflammatory cytokines and suppression of allo- and autoreactive T cells.

6.1 Co-Transplantation with Mesenchymal Stem Cells

Mesenchymal stem cells (MSCs) are pluripotent cells under extensive study as therapies for a wide variety of diseases. However, their clinical value arises not only from their stem/progenitor capabilities but also from their immunomodulatory properties [28]. MSCs can regulate both innate and adaptive immune cells through direct cell–cell contact such as intracellular and vascular cell adhesion molecules, galectin 1, programmed death ligands 1 and 2, mitochondrial transfer, and through the MSC secretome (such as transforming growth factor-β1 (TGF-β1), tumor necrosis factor-α (TNF-α), prostaglandin E2 (PGE2), interferon-γ (IFN-γ), hepatocyte (HGF) and fibroblast (FGF) growth factors, indoleamine 2,3-dioxygenase (IDO), and pro-angiogenic factors) encapsulated in MSC-secreted extracellular vesicles [161, 162].

These immunomodulatory pathways have been shown in preclinical studies of syngeneic and allogeneic MSC-islet co-transplantations to improve revascularization [163], prolong graft survival [164, 165], improve insulin secretory function, reduce fibrosis [166], and suppress host immune response to the graft [162, 167, 168]. MSCs have also been studied in the context of islet encapsulation in alginate and were found to improve the functional outcome of islet allotransplants both when co-encapsulated with or co-transplanted alongside islets [169]. Finally, MSC-derived exosomes alone have been shown to reduce hypoxia-induced apoptosis in transplanted β cells [170].

The clinical utility of MSCs faces challenges in standardizing their source, preconditioning protocols, and cryopreservation. However, their cell-free extracellular vesicles retain immunomodulatory and regenerative properties and may represent a more clinically practical application as drug-carrying particles [161, 171, 172]. Alternatively, selected bioactive factors found within MSC-secreted vesicles can be encapsulated in synthetic vesicles for local and/or targeted delivery to islet grafts.

6.2 Co-Transplantation with Regulatory T Cells

Regulatory T cells (Tregs) are another subset of cells that could be co-transplanted with stem cell-derived islets to modulate local immune responses to the graft. Tregs are known for their role in maintaining tolerance by halting the activation, proliferation, maturation, and effector functions of many immune cells, including NK cells, $CD4^+$ and $CD8^+$ T cells, B cells, and dendritic cells [173]. They do so by secreting regulatory cytokines such as IL-10, transforming growth factor β (TGF-β), IDO, and/or by cell-cell contact mechanisms [174]. With these tolerogenic abilities, Treg therapy is currently being explored in the prevention and treatment of early onset T1D but also has the potential to benefit allogeneic islet transplantation [175–177].

Tregs can benefit allogeneic islet transplantation by inducing peripheral tolerance to transplanted islet alloantigens, thereby improving islet survival and eliminating the need for chronic immunosuppression [62]. This has indeed been demonstrated in preclinical settings with polyclonal Treg co-transplantation that secrete TGF-β and induce tolerance to alloantigens [178]. Finally, SC islets stand to address the limited availability of donor islet tissue, and the potential benefit of Treg co-transplantation with these islets has yet to be explored. Importantly, preclinical and clinical studies with recipient-derived Tregs in solid organ transplantation have shown that peripheral T cell tolerance induction is not necessarily durable. Therefore, methods to enhance Treg function is an active research area [173].

Avenues to enhance Treg-mediated tolerance to islet grafts lie in developing donor-specific Tregs, either by culturing isolated Tregs with donor-derived cells, developing Tregs with engineered T cell receptors (TCRs), or generating Tregs with chimeric antigen receptors (CARs) (Fig. 4) [179]. These approaches have the common goal of inducing Treg recruitment and suppression that is specifically targeted to islet grafts. Because Treg activation and suppression are antigen-dependent,

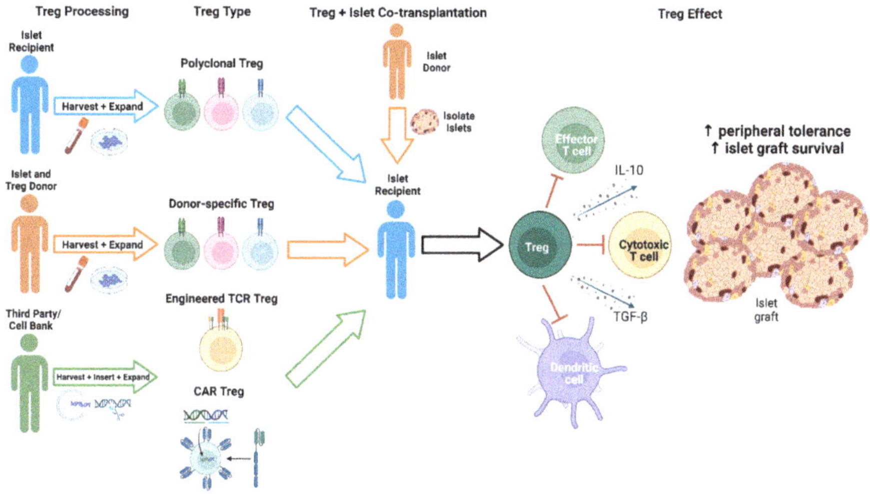

Fig. 4 Avenues for Treg co-transplantation with islets and their ability to induce peripheral tolerance via immune cell regulation and cytokine secretion. Tregs may be adoptively transferred after transplantation and engineered to target the transplant site or co-transplanted with islets during the implantation procedure. Treg regulatory T cell, TCR T cell receptor, IL-10 interleukin-10, TGF-β transforming growth factor β

donor-derived Tregs with MHCs matching those of donor-derived tissue allows these Tregs to target the transplanted islet tissue (donor-specific Tregs) [180]. Here, a restricted TCR repertoire is expanded using the donor's antigen-presenting cells, and these Tregs have been shown to traffic to the graft site [181]. Cultured, antigen-specific Tregs are often generated by stimulating isolated $CD4^{+}CD25^{+}$ (and sometimes $CD127^{+}$) Tregs with donor-derived peripheral blood mononuclear cells [182], CD40L-activated donor B cells [183], and/or allogeneic mature or monocyte-derived dendritic cells [184, 185], thus yielding donor-specific Tregs. However, this approach has been challenged as the induced Tregs are often contaminated with effector T cells during the in vitro expansion, which requires additional tissue collection from donors, limiting the potential applicability of this procedure [180]. Researchers have, therefore, engineered Tregs with a predetermined antigen specificity by transfecting viral vectors encoding specific TCRs [186]. Unfortunately, this approach is also hindered by the need for donor and recipient cells for Treg expansion and the risks associated with using viral vectors.

CAR Tregs have the potential to address the challenges in Treg therapy mentioned above and are under investigation [187]. CARs replace the TCR with a hybrid receptor containing a single-chain variable fragment specific for a selected antigen (i.e., the binding moiety of a monoclonal antibody) and an extracellular hinge region while also incorporating intracellular signaling domains that initiate T cell activation pathways [186]. The exact structure and mechanism of the CAR's intracellular domains have evolved over time to facilitate CAR T cell activation independent from APC co-stimulation. For example, first-generation CARs have the CD3 signaling domain common to any T lymphocyte, second-generation CARs build in a CD28-derived co-stimulatory region for T cell activation, and third-generation CARs additionally add a 4-1BB-derived co-stimulatory region [188]. Ultimately, CAR Tregs have selectable specificity and function independent of antigen presentation and co-stimulatory activation, eliminating the need to expand antigen-specific Tregs or identify relevant alloantigens [189]. Since CAR Tregs no longer have a traditional TCR, they could be engineered to be specific to selected islet-specific targets, such as the GLP-1 receptor [129, 190]. CAR Tregs are generally more effective in suppression than polyclonal Tregs due to their ability to accumulate at target tissues expressing the selected antigen [186]. This can enable targeted immunosuppression with fewer side effects, though some can persist such as cytokine "storms" and neuronal cytotoxicity seen in anti-tumor CAR T cells [188]. For islet transplantation, the CAR can be designed to induce Treg accumulation and suppression at the graft site, and these CAR Tregs could be directly co-transplanted alongside islets [191]. These Tregs could then secrete anti-inflammatory cytokines like IL-10 and IL-35, TGF-β, which also modulate antigen-presenting cells [62]. Given that affinity may be a critical factor in therapeutic benefit, studies comparing the affinity of a wild-type TCR to that of a CAR are ongoing. Thus, there is no clear answer regarding the effectiveness of CAR Tregs at suppressing immune responses compared to wild-type or engineered Tregs, especially in the context of islet transplantation. As more researchers explore stem cell-derived islets, co-transplantation with engineered cells remains a promising approach in islet immunoengineering to reduce the need for immunosuppression and support long-term islet function.

References

1. Berney T, Andres A, Bellin MD, de Koning EJP, Johnson PRV, Kay TWH, et al. A Worldwide Survey of Activities and Practices in Clinical Islet of Langerhans Transplantation. Transpl Int. 2022;35:10507.
2. Vantyghem MC, Chetboun M, Gmyr V, Jannin A, Espiard S, Le Mapihan K, et al. Ten-Year Outcome of Islet Alone or Islet After Kidney Transplantation in Type 1 Diabetes: A Prospective Parallel-Arm Cohort Study. Diabetes Care. 2019;42(11):2042–9.
3. Lablanche S, Vantyghem MC, Kessler L, Wojtusciszyn A, Borot S, Thivolet C, et al. Islet transplantation versus insulin therapy in patients with type 1 diabetes with severe hypoglycaemia or poorly controlled glycaemia after kidney transplantation (TRIMECO): a multicentre, randomised controlled trial. Lancet Diabetes Endocrinol. 2018;6(7):527–37.
4. Barton FB, Rickels MR, Alejandro R, Hering BJ, Wease S, Naziruddin B, et al. Improvement in outcomes of clinical islet transplantation: 1999–2010. Diabetes Care. 2012;35(7):1436–45.
5. Markmann JF, Rickels MR, Eggerman TL, Bridges ND, Lafontant DE, Qidwai J, et al. Phase 3 trial of human islet-after-kidney transplantation in type 1 diabetes. Am J Transplant. 2021;21(4):1477–92.
6. Shapiro AM, Pokrywczynska M, Ricordi C. Clinical pancreatic islet transplantation. Nat Rev Endocrinol. 2017;13(5):268–77.
7. Mahase E. Type 1 diabetes: Global prevalence is set to double by 2040, study estimates. BMJ. 2022;378:o2289.

8. Subramanian S, Baidal D. The Management of Type 1 Diabetes. In: Feingold KR, Anawalt B, Boyce A, Chrousos G, de Herder WW, Dhatariya K, et al., editors. Endotext. South Dartmouth (MA); 2000.
9. Senior P, Lam A, Farnsworth K, Perkins B, Rabasa-Lhoret R. Assessment of Risks and Benefits of Beta Cell Replacement Versus Automated Insulin Delivery Systems for Type 1 Diabetes. Curr Diab Rep. 2020;20(10):52.
10. Nilsson B, Ekdahl KN, Korsgren O. Control of instant blood-mediated inflammatory reaction to improve islets of Langerhans engraftment. Curr Opin Organ Transplant. 2011;16(6):620-6.
11. Kanak MA, Takita M, Kunnathodi F, Lawrence MC, Levy MF, Naziruddin B. Inflammatory response in islet transplantation. Int J Endocrinol. 2014;2014:451035.
12. Jeyagaran A, Lu CE, Zbinden A, Birkenfeld AL, Brucker SY, Layland SL. Type 1 diabetes and engineering enhanced islet transplantation. Adv Drug Deliv Rev. 2022;189:114481.
13. Samojlik MM, Stabler CL. Designing biomaterials for the modulation of allogeneic and autoimmune responses to cellular implants in Type 1 Diabetes. Acta Biomater. 2021;133:87–101.
14. Burke GW, 3rd, Vendrame F, Virdi SK, Ciancio G, Chen L, Ruiz P, et al. Lessons From Pancreas Transplantation in Type 1 Diabetes: Recurrence of Islet Autoimmunity. Curr Diab Rep. 2015;15(12):121.
15. Vincenti F, Rostaing L, Grinyo J, Rice K, Steinberg S, Gaite L, et al. Belatacept and Long-Term Outcomes in Kidney Transplantation. N Engl J Med. 2016;374(4):333–43.
16. Fishman JA. Opportunistic infections--coming to the limits of immunosuppression? Cold Spring Harb Perspect Med. 2013;3(10):a015669.
17. Roberts MB, Fishman JA. Immunosuppressive Agents and Infectious Risk in Transplantation: Managing the "Net State of Immunosuppression". Clin Infect Dis. 2021;73(7):e1302–e17.
18. Katabathina V, Menias CO, Pickhardt P, Lubner M, Prasad SR. Complications of Immunosuppressive Therapy in Solid Organ Transplantation. Radiol Clin North Am. 2016;54(2):303–19.
19. Registry CIT. Scientific Summary of the Collaborative Islet Transplant Registry (CITR) 2020 Eleventh Allograft Data Report. 2022.
20. Ingelfinger JR, Schwartz RS. Immunosuppression--the promise of specificity. N Engl J Med. 2005;353(8):836–9.
21. Ramzy A, Thompson DM, Ward-Hartstonge KA, Ivison S, Cook L, Garcia RV, et al. Implanted pluripotent stem-cell-derived pancreatic endoderm cells secrete glucose-responsive C-peptide in patients with type 1 diabetes. Cell Stem Cell. 2021;28(12):2047–61 e5.
22. Ben-David U, Benvenisty N. The tumorigenicity of human embryonic and induced pluripotent stem cells. Nat Rev Cancer. 2011;11(4):268–77.
23. Gutierrez-Aranda I, Ramos-Mejia V, Bueno C, Munoz-Lopez M, Real PJ, Macia A, et al. Human induced pluripotent stem cells develop teratoma more efficiently and faster than human embryonic stem cells regardless the site of injection. Stem Cells. 2010;28(9):1568–70.
24. Han L, He H, Yang Y, Meng Q, Ye F, Chen G, et al. Distinctive Clinical and Pathologic Features of Immature Teratomas Arising from Induced Pluripotent Stem Cell-Derived Beta Cell Injection in a Diabetes Patient. Stem Cells Dev. 2022;31(5-6):97–101.
25. Seetharam A, Tiriveedhi V, Mohanakumar T. Alloimmunity and autoimmunity in chronic rejection. Curr Opin Organ Transplant. 2010;15(4):531–6.
26. Ali JM, Bolton EM, Bradley JA, Pettigrew GJ. Allorecognition pathways in transplant rejection and tolerance. Transplantation. 2013;96(8):681–8.
27. Siu JHY, Surendrakumar V, Richards JA, Pettigrew GJ. T cell Allorecognition Pathways in Solid Organ Transplantation. Front Immunol. 2018;9:2548.
28. Ankrum JA, Ong JF, Karp JM. Mesenchymal stem cells: immune evasive, not immune privileged. Nat Biotechnol. 2014;32(3):252–60.
29. Ubiali F, Nava S, Nessi V, Frigerio S, Parati E, Bernasconi P, et al. Allorecognition of human neural stem cells by peripheral blood lymphocytes despite low expression of MHC molecules: role of TGF-beta in modulating proliferation. Int Immunol. 2007;19(9):1063–74.

30. Pepper AR, Gala-Lopez B, Pawlick R, Merani S, Kin T, Shapiro AM. A prevascularized subcutaneous device-less site for islet and cellular transplantation. Nat Biotechnol. 2015;33(5):518–23.
31. Shapiro AMJ, Thompson D, Donner TW, Bellin MD, Hsueh W, Pettus J, et al. Insulin expression and C-peptide in type 1 diabetes subjects implanted with stem cell-derived pancreatic endoderm cells in an encapsulation device. Cell Rep Med. 2021;2(12):100466.
32. Tomei AA. Engineering Confined and Prevascularized Sites for Islet Transplantation. Transplantation. 2018;102(11):1793–4.
33. Delaune V, Berney T, Lacotte S, Toso C. Intraportal islet transplantation: the impact of the liver microenvironment. Transpl Int. 2017;30(3):227–38.
34. Eich T, Eriksson O, Lundgren T, Nordic Network for Clinical Islet T. Visualization of early engraftment in clinical islet transplantation by positron-emission tomography. N Engl J Med. 2007;356(26):2754–5.
35. Carnicer-Lombarte A, Chen ST, Malliaras GG, Barone DG. Foreign Body Reaction to Implanted Biomaterials and Its Impact in Nerve Neuroprosthetics. Front Bioeng Biotechnol. 2021;9:622524.
36. Veiseh O, Vegas AJ. Domesticating the foreign body response: Recent advances and applications. Adv Drug Deliv Rev. 2019;144:148–61.
37. Jung SY, Lim SM, Albertorio F, Kim G, Gurau MC, Yang RD, et al. The Vroman effect: a molecular level description of fibrinogen displacement. J Am Chem Soc. 2003;125(42):12782–6.
38. Anderson JM, Rodriguez A, Chang DT. Foreign body reaction to biomaterials. Semin Immunol. 2008;20(2):86–100.
39. Mariani E, Lisignoli G, Borzi RM, Pulsatelli L. Biomaterials: Foreign Bodies or Tuners for the Immune Response? Int J Mol Sci. 2019;20(3).
40. Rosales C. Neutrophil: A Cell with Many Roles in Inflammation or Several Cell Types? Front Physiol. 2018;9:113.
41. Selders GS, Fetz AE, Radic MZ, Bowlin GL. An overview of the role of neutrophils in innate immunity, inflammation and host-biomaterial integration. Regen Biomater. 2017;4(1):55–68.
42. Krystel-Whittemore M, Dileepan KN, Wood JG. Mast Cell: A Multi-Functional Master Cell. Front Immunol. 2015;6:620.
43. Wiggins MJ, Wilkoff B, Anderson JM, Hiltner A. Biodegradation of polyether polyurethane inner insulation in bipolar pacemaker leads. J Biomed Mater Res. 2001;58(3):302–7.
44. Labow RS, Meek E, Santerre JP. Hydrolytic degradation of poly(carbonate)-urethanes by monocyte-derived macrophages. Biomaterials. 2001;22(22):3025–33.
45. Kaplan MJ, Radic M. Neutrophil extracellular traps: double-edged swords of innate immunity. J Immunol. 2012;189(6):2689–95.
46. Jhunjhunwala S, Aresta-DaSilva S, Tang K, Alvarez D, Webber MJ, Tang BC, et al. Neutrophil Responses to Sterile Implant Materials. PLoS One. 2015;10(9):e0137550.
47. Yamashiro S, Kamohara H, Wang JM, Yang D, Gong WH, Yoshimura T. Phenotypic and functional change of cytokine-activated neutrophils: inflammatory neutrophils are heterogeneous and enhance adaptive immune responses. J Leukoc Biol. 2001;69(5):698–704.
48. Franz S, Rammelt S, Scharnweber D, Simon JC. Immune responses to implants - a review of the implications for the design of immunomodulatory biomaterials. Biomaterials. 2011;32(28):6692–709.
49. Mesure L, De Visscher G, Vranken I, Lebacq A, Flameng W. Gene expression study of monocytes/macrophages during early foreign body reaction and identification of potential precursors of myofibroblasts. PLoS One. 2010;5(9):e12949.
50. Sheikh Z, Brooks PJ, Barzilay O, Fine N, Glogauer M. Macrophages, Foreign Body Giant Cells and Their Response to Implantable Biomaterials. Materials (Basel). 2015;8(9):5671–701.
51. Ahmadzadeh K, Vanoppen M, Rose CD, Matthys P, Wouters CH. Multinucleated Giant Cells: Current Insights in Phenotype, Biological Activities, and Mechanism of Formation. Front Cell Dev Biol. 2022;10:873226.

52. Vishwakarma A, Bhise NS, Evangelista MB, Rouwkema J, Dokmeci MR, Ghaemmaghami AM, et al. Engineering Immunomodulatory Biomaterials To Tune the Inflammatory Response. Trends Biotechnol. 2016;34(6):470–82.
53. Julier Z, Park AJ, Briquez PS, Martino MM. Promoting tissue regeneration by modulating the immune system. Acta Biomater. 2017;53:13–28.
54. Noskovicova N, Hinz B, Pakshir P. Implant Fibrosis and the Underappreciated Role of Myofibroblasts in the Foreign Body Reaction. Cells. 2021;10(7).
55. Fayzullin A, Bakulina A, Mikaelyan K, Shekhter A, Guller A. Implantable Drug Delivery Systems and Foreign Body Reaction: Traversing the Current Clinical Landscape. Bioengineering (Basel). 2021;8(12).
56. Moreau A, Varey E, Anegon I, Cuturi MC. Effector mechanisms of rejection. Cold Spring Harb Perspect Med. 2013;3(11).
57. Guo Y, Sun T, Gu Y, Liu X, Ke Q, Wei X, et al. Rational Synthesis of Chabazite (CHA) Zeolites with Controlled Si/Al Ratio and Their CO2 /CH4 /N2 Adsorptive Separation Performances. Chem Asian J. 2018;13(21):3222–30.
58. Taylor CJ, Bolton EM, Bradley JA. Immunological considerations for embryonic and induced pluripotent stem cell banking. Philos Trans R Soc Lond B Biol Sci. 2011;366(1575):2312–22.
59. Mitrousis N, Fokina A, Shoichet MS. Biomaterials for cell transplantation. Nature Reviews Materials. 2018;3(11):441–56.
60. Nair A, Tang L. Influence of scaffold design on host immune and stem cell responses. Semin Immunol. 2017;29:62–71.
61. Vasconcelos DP, Aguas AP, Barbosa MA, Pelegrin P, Barbosa JN. The inflammasome in host response to biomaterials: Bridging inflammation and tissue regeneration. Acta Biomater. 2019;83:1–12.
62. Wang X, Brown NK, Wang B, Shariati K, Wang K, Fuchs S, et al. Local Immunomodulatory Strategies to Prevent Allo-Rejection in Transplantation of Insulin-Producing Cells. Adv Sci (Weinh). 2021;8(17):e2003708.
63. Antmen E, Vrana NE, Hasirci V. The role of biomaterials and scaffolds in immune responses in regenerative medicine: macrophage phenotype modulation by biomaterial properties and scaffold architectures. Biomater Sci. 2021;9(24):8090–110.
64. Veiseh O, Doloff JC, Ma M, Vegas AJ, Tam HH, Bader AR, et al. Size- and shape-dependent foreign body immune response to materials implanted in rodents and non-human primates. Nat Mater. 2015;14(6):643–51.
65. Jiang S, Lyu C, Zhao P, Li W, Kong W, Huang C, et al. Cryoprotectant enables structural control of porous scaffolds for exploration of cellular mechano-responsiveness in 3D. Nat Commun. 2019;10(1):3491.
66. Boehler RM, Graham JG, Shea LD. Tissue engineering tools for modulation of the immune response. Biotechniques. 2011;51(4):239–40, 42, 44 passim.
67. Thevenot P, Hu W, Tang L. Surface chemistry influences implant biocompatibility. Curr Top Med Chem. 2008;8(4):270–80.
68. Camarero-Espinosa S, Carlos-Oliveira M, Liu H, Mano JF, Bouvy N, Moroni L. 3D Printed Dual-Porosity Scaffolds: The Combined Effect of Stiffness and Porosity in the Modulation of Macrophage Polarization. Adv Healthc Mater. 2022;11(1):e2101415.
69. Andorko JI, Jewell CM. Designing biomaterials with immunomodulatory properties for tissue engineering and regenerative medicine. Bioeng Transl Med. 2017;2(2):139–55.
70. Goncalves C, Ferreira N, Lourenco L. Production of Low Molecular Weight Chitosan and Chitooligosaccharides (COS): A Review. Polymers (Basel). 2021;13(15).
71. Sadtler K, Wolf MT, Ganguly S, Moad CA, Chung L, Majumdar S, et al. Divergent immune responses to synthetic and biological scaffolds. Biomaterials. 2019;192:405–15.
72. Babensee JE, Paranjpe A. Differential levels of dendritic cell maturation on different biomaterials used in combination products. J Biomed Mater Res A. 2005;74(4):503–10.
73. Caires HR, Esteves T, Quelhas P, Barbosa MA, Navarro M, Almeida CR. Macrophage interactions with polylactic acid and chitosan scaffolds lead to improved recruitment of human mesenchymal stem/stromal cells: a comprehensive study with different immune cells. J R Soc Interface. 2016;13(122).

74. Kajahn J, Franz S, Rueckert E, Forstreuter I, Hintze V, Moeller S, et al. Artificial extracellular matrices composed of collagen I and high sulfated hyaluronan modulate monocyte to macrophage differentiation under conditions of sterile inflammation. Biomatter. 2012;2(4):226–36.
75. Thevenot PT, Nair AM, Shen J, Lotfi P, Ko CY, Tang L. The effect of incorporation of SDF-1alpha into PLGA scaffolds on stem cell recruitment and the inflammatory response. Biomaterials. 2010;31(14):3997–4008.
76. Scislowska-Czarnecka A, Szmigiel D, Genet M, Dupont-Gillain C, Pamula E, Kolaczkowska E. Oxygen plasma surface modification augments poly(L-lactide-co-glycolide) cytocompatibility toward osteoblasts and minimizes immune activation of macrophages. J Biomed Mater Res A. 2015;103(12):3965–77.
77. Cohen HC, Lieberthal TJ, Kao WJ. Poly(ethylene glycol)-containing hydrogels promote the release of primary granules from human blood-derived polymorphonuclear leukocytes. J Biomed Mater Res A. 2014;102(12):4252–61.
78. Champion JA, Katare YK, Mitragotri S. Particle shape: a new design parameter for micro- and nanoscale drug delivery carriers. J Control Release. 2007;121(1–2):3–9.
79. Li G, Yang P, Guo X, Huang N, Shen R. An in vitro evaluation of inflammation response of titanium functionalized with heparin/fibronectin complex. Cytokine. 2011;56(2):208–17.
80. Kou PM, Babensee JE. Macrophage and dendritic cell phenotypic diversity in the context of biomaterials. J Biomed Mater Res A. 2011;96(1):239–60.
81. Hezi-Yamit A, Sullivan C, Wong J, David L, Chen M, Cheng P, et al. Impact of polymer hydrophilicity on biocompatibility: implication for DES polymer design. J Biomed Mater Res A. 2009;90(1):133–41.
82. Bridges AW, Singh N, Burns KL, Babensee JE, Andrew Lyon L, Garcia AJ. Reduced acute inflammatory responses to microgel conformal coatings. Biomaterials. 2008;29(35):4605–15.
83. McWhorter FY, Wang T, Nguyen P, Chung T, Liu WF. Modulation of macrophage phenotype by cell shape. Proc Natl Acad Sci U S A. 2013;110(43):17253–8.
84. Chen S, Jones JA, Xu Y, Low HY, Anderson JM, Leong KW. Characterization of topographical effects on macrophage behavior in a foreign body response model. Biomaterials. 2010;31(13):3479–91.
85. Sridharan R, Cavanagh B, Cameron AR, Kelly DJ, O'Brien FJ. Material stiffness influences the polarization state, function and migration mode of macrophages. Acta Biomater. 2019;89:47–59.
86. Kim J, Li WA, Sands W, Mooney DJ. Effect of pore structure of macroporous poly(lactide-co-glycolide) scaffolds on the in vivo enrichment of dendritic cells. ACS Appl Mater Interfaces. 2014;6(11):8505–12.
87. Sokolowska M, Chen LY, Eberlein M, Martinez-Anton A, Liu Y, Alsaaty S, et al. Low molecular weight hyaluronan activates cytosolic phospholipase A2alpha and eicosanoid production in monocytes and macrophages. J Biol Chem. 2014;289(7):4470–88.
88. Scheibner KA, Lutz MA, Boodoo S, Fenton MJ, Powell JD, Horton MR. Hyaluronan fragments act as an endogenous danger signal by engaging TLR2. J Immunol. 2006;177(2):1272–81.
89. Tesar BM, Jiang D, Liang J, Palmer SM, Noble PW, Goldstein DR. The role of hyaluronan degradation products as innate alloimmune agonists. Am J Transplant. 2006;6(11):2622–35.
90. Termeer CC, Hennies J, Voith U, Ahrens T, Weiss JM, Prehm P, et al. Oligosaccharides of hyaluronan are potent activators of dendritic cells. J Immunol. 2000;165(4):1863–70.
91. Glatigny S, Hollbacher B, Motley SJ, Tan C, Hundhausen C, Buckner JH, et al. Abatacept Targets T Follicular Helper and Regulatory T Cells, Disrupting Molecular Pathways That Regulate Their Proliferation and Maintenance. J Immunol. 2019;202(5):1373–82.
92. Beckwith J, Nyman JA, Flanagan B, Schrover R, Schuurman HJ. A health economic analysis of clinical islet transplantation. Clin Transplant. 2012;26(1):23–33.
93. Gil APP, Lunardi AC, Santana FR, Bergamim J, Sarmento LA, Cristelli MP, et al. Impact of Renal Transplantation and Immunosuppressive Therapy on Muscle Strength, Functional Capacity, and Quality of Life: A Longitudinal Study. Transplant Proc. 2020;52(5):1279–83.

94. Sommerer C, Zeier M. Reporting Quality-of-Life Outcomes in Clinical Trials of Immunosuppressive Therapy in Kidney Transplantation. Am J Kidney Dis. 2016;67(5):722–3.
95. Sutherland DE, Matas AJ, Goetz FC, Najarian JS. Transplantation of dispersed pancreatic islet tissue in humans: autografts and allografts. Diabetes. 1980;29 Suppl 1:31–44.
96. Largiader F, Kolb E, Binswanger U. A long-term functioning human pancreatic islet allotransplant. Transplantation. 1980;29(1):76–7.
97. Shapiro AM, Ricordi C, Hering BJ, Auchincloss H, Lindblad R, Robertson RP, et al. International trial of the Edmonton protocol for islet transplantation. N Engl J Med. 2006;355(13):1318–30.
98. Shapiro AM, Lakey JR, Ryan EA, Korbutt GS, Toth E, Warnock GL, et al. Islet transplantation in seven patients with type 1 diabetes mellitus using a glucocorticoid-free immunosuppressive regimen. N Engl J Med. 2000;343(4):230–8.
99. Hirshberg B, Rother KI, Digon BJ, 3rd, Lee J, Gaglia JL, Hines K, et al. Benefits and risks of solitary islet transplantation for type 1 diabetes using steroid-sparing immunosuppression: the National Institutes of Health experience. Diabetes Care. 2003;26(12):3288–95.
100. Isaacs JD, Burmester GR. Smart battles: immunosuppression versus immunomodulation in the inflammatory RMDs. Ann Rheum Dis. 2020;79(8):991–3.
101. Bentley ER, Little SR. Local delivery strategies to restore immune homeostasis in the context of inflammation. Adv Drug Deliv Rev. 2021;178:113971.
102. Sharma N, Bietar K, Stochaj U. Targeting nanoparticles to malignant tumors. Biochim Biophys Acta Rev Cancer. 2022;1877(3):188703.
103. Ponticelli C, Glassock RJ. Prevention of complications from use of conventional immunosuppressants: a critical review. J Nephrol. 2019;32(6):851–70.
104. Fan Y, Zheng X, Ali Y, Berggren PO, Loo SCJ. Local release of rapamycin by microparticles delays islet rejection within the anterior chamber of the eye. Sci Rep. 2019;9(1):3918.
105. Gelderblom H, Verweij J, Nooter K, Sparreboom A. Cremophor EL: the drawbacks and advantages of vehicle selection for drug formulation. Eur J Cancer. 2001;37(13):1590–8.
106. Park K. Facing the truth about nanotechnology in drug delivery. ACS Nano. 2013;7(9):7442–7.
107. Velluto D, Bojadzic D, De Toni T, Buchwald P, Tomei AA. Drug-Integrating Amphiphilic Nanomaterial Assemblies: 1. Spatiotemporal control of cyclosporine delivery and activity using nanomicelles and nanofibrils. J Control Release. 2021;329:955–70.
108. Li J, Mooney DJ. Designing hydrogels for controlled drug delivery. Nat Rev Mater. 2016;1(12).
109. Kuppan P, Kelly S, Polishevska K, Hojanepesov O, Seeberger K, Korbutt GS, et al. Co-localized immune protection using dexamethasone-eluting micelles in a murine islet allograft model. Am J Transplant. 2020;20(3):714–25.
110. Jiang K, Weaver JD, Li Y, Chen X, Liang J, Stabler CL. Local release of dexamethasone from macroporous scaffolds accelerates islet transplant engraftment by promotion of anti-inflammatory M2 macrophages. Biomaterials. 2017;114:71–81.
111. Zisi Tegou F, Velluto D, Stock AA, Fitzgerald SN, Stealey S, Zustiak SP, et al. CCL21 and beta-cell antigen releasing hydrogels as tolerance-inducing therapy in Type I diabetes. J Control Release. 2022;348:499–517.
112. Headen DM, Woodward KB, Coronel MM, Shrestha P, Weaver JD, Zhao H, et al. Local immunomodulation Fas ligand-engineered biomaterials achieves allogeneic islet graft acceptance. Nat Mater. 2018;17(8):732–9.
113. Akash MS, Rehman K, Li N, Gao JQ, Sun H, Chen S. Sustained delivery of IL-1Ra from pluronic F127-based thermosensitive gel prolongs its therapeutic potentials. Pharm Res. 2012;29(12):3475–85.
114. Kim J, Hope CM, Gantumur N, Perkins GB, Stead SO, Yue Z, et al. Encapsulation of Human Natural and Induced Regulatory T-Cells in IL-2 and CCL1 Supplemented Alginate-GelMA Hydrogel for 3D Bioprinting. Advanced Functional Materials. 2020;30(15).

115. Liu JMH, Zhang X, Joe S, Luo X, Shea LD. Evaluation of biomaterial scaffold delivery of IL-33 as a localized immunomodulatory agent to support cell transplantation in adipose tissue. J Immunol Regen Med. 2018;1:1–12.
116. Chen T, Yuan J, Duncanson S, Hibert ML, Kodish BC, Mylavaganam G, et al. Alginate encapsulant incorporating CXCL12 supports long-term allo- and xenoislet transplantation without systemic immune suppression. Am J Transplant. 2015;15(3):618–27.
117. Brudno Y, Silva EA, Kearney CJ, Lewin SA, Miller A, Martinick KD, et al. Refilling drug delivery depots through the blood. Proc Natl Acad Sci U S A. 2014;111(35):12722–7.
118. Bae YH, Park K. Targeted drug delivery to tumors: myths, reality and possibility. J Control Release. 2011;153(3):198–205.
119. Wong AD, Ye M, Ulmschneider MB, Searson PC. Quantitative Analysis of the Enhanced Permeation and Retention (EPR) Effect. PLoS One. 2015;10(5):e0123461.
120. Getts DR, Shea LD, Miller SD, King NJ. Harnessing nanoparticles for immune modulation. Trends Immunol. 2015;36(7):419–27.
121. Hillaireau H, Couvreur P. Nanocarriers' entry into the cell: relevance to drug delivery. Cell Mol Life Sci. 2009;66(17):2873–96.
122. Bachmann MF, Jennings GT. Vaccine delivery: a matter of size, geometry, kinetics and molecular patterns. Nat Rev Immunol. 2010;10(11):787–96.
123. Fisher JD, Acharya AP, Little SR. Micro and nanoparticle drug delivery systems for preventing allotransplant rejection. Clin Immunol. 2015;160(1):24–35.
124. Champion JA, Walker A, Mitragotri S. Role of particle size in phagocytosis of polymeric microspheres. Pharm Res. 2008;25(8):1815–21.
125. Li H, Yu SS, Miteva M, Nelson CE, Werfel T, Giorgio TD, et al. Matrix Metalloproteinase Responsive, Proximity-activated Polymeric Nanoparticles for siRNA Delivery. Adv Funct Mater. 2013;23(24):3040–52.
126. Liwinska W, Waleka-Bagiel E, Stojek Z, Karbarz M, Zabost E. Enzyme-triggered- and tumor-targeted delivery with tunable, methacrylated poly(ethylene glycols) and hyaluronic acid hybrid nanogels. Drug Deliv. 2022;29(1):2561–78.
127. Jain A, Cheng K. The principles and applications of avidin-based nanoparticles in drug delivery and diagnosis. J Control Release. 2017;245:27–40.
128. Van Simaeys D, De La Fuente A, Zilio S, Zoso A, Kuznetsova V, Alcazar O, et al. RNA aptamers specific for transmembrane p24 trafficking protein 6 and Clusterin for the targeted delivery of imaging reagents and RNA therapeutics to human beta cells. Nat Commun. 2022;13(1):1815.
129. Alavi SE, Ebrahimi Shahmabadi H. GLP-1 peptide analogs for targeting pancreatic beta cells. Drug Discov Today. 2021;26(8):1936–43.
130. Yan F, Li L, Deng Z, Jin Q, Chen J, Yang W, et al. Paclitaxel-liposome-microbubble complexes as ultrasound-triggered therapeutic drug delivery carriers. J Control Release. 2013;166(3):246–55.
131. Xie J, Li A, Li J. Advances in pH-Sensitive Polymers for Smart Insulin Delivery. Macromol Rapid Commun. 2017;38(23).
132. He L, Zuo Q, Xie S, Huang Y, Xue W. Intelligent hydrogels for drug delivery system. Recent Pat Drug Deliv Formul. 2011;5(3):265–74.
133. Kinney SM, Ortaleza K, Vlahos AE, Sefton MV. Degradable methacrylic acid-based synthetic hydrogel for subcutaneous islet transplantation. Biomaterials. 2022;281:121342.
134. Lee JS, Han P, Chaudhury R, Khan S, Bickerton S, McHugh MD, et al. Metabolic and immunomodulatory control of type 1 diabetes via orally delivered bile-acid-polymer nanocarriers of insulin or rapamycin. Nat Biomed Eng. 2021;5(9):983–97.
135. Primavera R, Razavi M, Kevadiya BD, Wang J, Vykunta A, Di Mascolo D, et al. Enhancing islet transplantation using a biocompatible collagen-PDMS bioscaffold enriched with dexamethasone-microplates. Biofabrication. 2021;13(3).
136. Zhang W, Gorantla VS, Campbell PG, Li Y, Yang Y, Komatsu C, et al. Biopatterned CTLA4/Fc Matrices Facilitate Local Immunomodulation, Engraftment, and Glucose Homeostasis After Pancreatic Islet Transplantation. Diabetes. 2016;65(12):3660–6.

137. Lei J, Coronel MM, Yolcu ES, Deng H, Grimany-Nuno O, Hunckler MD, et al. FasL microgels induce immune acceptance of islet allografts in nonhuman primates. Sci Adv. 2022;8(19):eabm9881.
138. Hou H, Fu S, Liu C, Chen J, Hsu BR. The graft survival protection of subcutaneous allogeneic islets with hydrogel grafting and encapsulated by CTLA4Ig and IL1ra. Polymer Journal. 2013;46:136–44.
139. Hsu PD, Lander ES, Zhang F. Development and applications of CRISPR-Cas9 for genome engineering. Cell. 2014;157(6):1262–78.
140. Janik E, Niemcewicz M, Ceremuga M, Krzowski L, Saluk-Bijak J, Bijak M. Various Aspects of a Gene Editing System-CRISPR-Cas9. Int J Mol Sci. 2020;21(24).
141. Doudna JA, Charpentier E. Genome editing. The new frontier of genome engineering with CRISPR-Cas9. Science. 2014;346(6213):1258096.
142. Brezgin S, Kostyusheva A, Kostyushev D, Chulanov V. Dead Cas Systems: Types, Principles, and Applications. Int J Mol Sci. 2019;20(23).
143. Chavez A, Scheiman J, Vora S, Pruitt BW, Tuttle M, E PRI, et al. Highly efficient Cas9-mediated transcriptional programming. Nat Methods. 2015;12(4):326–8.
144. Choudhury SR, Cui Y, Lubecka K, Stefanska B, Irudayaraj J. CRISPR-dCas9 mediated TET1 targeting for selective DNA demethylation at BRCA1 promoter. Oncotarget. 2016;7(29):46545–56.
145. Liu XS, Wu H, Ji X, Stelzer Y, Wu X, Czauderna S, et al. Editing DNA Methylation in the Mammalian Genome. Cell. 2016;167(1):233–47 e17.
146. McDonald JI, Celik H, Rois LE, Fishberger G, Fowler T, Rees R, et al. Reprogrammable CRISPR/Cas9-based system for inducing site-specific DNA methylation. Biol Open. 2016;5(6):866–74.
147. Morita S, Noguchi H, Horii T, Nakabayashi K, Kimura M, Okamura K, et al. Targeted DNA demethylation in vivo using dCas9-peptide repeat and scFv-TET1 catalytic domain fusions. Nat Biotechnol. 2016;34(10):1060–5.
148. Vojta A, Dobrinic P, Tadic V, Bockor L, Korac P, Julg B, et al. Repurposing the CRISPR-Cas9 system for targeted DNA methylation. Nucleic Acids Res. 2016;44(12):5615–28.
149. Gilbert LA, Larson MH, Morsut L, Liu Z, Brar GA, Torres SE, et al. CRISPR-mediated modular RNA-guided regulation of transcription in eukaryotes. Cell. 2013;154(2):442–51.
150. Maeder ML, Linder SJ, Cascio VM, Fu Y, Ho QH, Joung JK. CRISPR RNA-guided activation of endogenous human genes. Nat Methods. 2013;10(10):977–9.
151. Mali P, Aach J, Stranges PB, Esvelt KM, Moosburner M, Kosuri S, et al. CAS9 transcriptional activators for target specificity screening and paired nickases for cooperative genome engineering. Nat Biotechnol. 2013;31(9):833–8.
152. Wyatt RC, Lanzoni G, Russell MA, Gerling I, Richardson SJ. What the HLA-I!-Classical and Non-classical HLA Class I and Their Potential Roles in Type 1 Diabetes. Curr Diab Rep. 2019;19(12):159.
153. Hackmon R, Pinnaduwage L, Zhang J, Lye SJ, Geraghty DE, Dunk CE. Definitive class I human leukocyte antigen expression in gestational placentation: HLA-F, HLA-E, HLA-C, and HLA-G in extravillous trophoblast invasion on placentation, pregnancy, and parturition. Am J Reprod Immunol. 2017;77(6).
154. Parent AV, Faleo G, Chavez J, Saxton M, Berrios DI, Kerper NR, et al. Selective deletion of human leukocyte antigens protects stem cell-derived islets from immune rejection. Cell Rep. 2021;36(7):109538.
155. Ai L, Xu A, Xu J. Roles of PD-1/PD-L1 Pathway: Signaling, Cancer, and Beyond. Adv Exp Med Biol. 2020;1248:33-59.
156. Francisco LM, Sage PT, Sharpe AH. The PD-1 pathway in tolerance and autoimmunity. Immunol Rev. 2010;236:219–42.
157. Yoshihara E, O'Connor C, Gasser E, Wei Z, Oh TG, Tseng TW, et al. Immune-evasive human islet-like organoids ameliorate diabetes. Nature. 2020;586(7830):606–11.

158. Munn DH, Mellor AL. Indoleamine 2,3 dioxygenase and metabolic control of immune responses. Trends Immunol. 2013;34(3):137–43.
159. Wu H, Gong J, Liu Y. Indoleamine 2, 3-dioxygenase regulation of immune response (Review). Mol Med Rep. 2018;17(4):4867–73.
160. Hosseini-Tabatabaei A, Jalili RB, Khosravi-Maharlooei M, Hartwell R, Kilani RT, Zhang Y, et al. Immunoprotection and Functional Improvement of Allogeneic Islets in Diabetic Mice, Using a Stable Indoleamine 2,3-Dioxygenase Producing Scaffold. Transplantation. 2015;99(7):1341–8.
161. Song N, Scholtemeijer M, Shah K. Mesenchymal Stem Cell Immunomodulation: Mechanisms and Therapeutic Potential. Trends Pharmacol Sci. 2020;41(9):653–64.
162. Shrestha M, Nguyen TT, Park J, Choi JU, Yook S, Jeong JH. Immunomodulation effect of mesenchymal stem cells in islet transplantation. Biomed Pharmacother. 2021;142:112042.
163. Figliuzzi M, Cornolti R, Perico N, Rota C, Morigi M, Remuzzi G, et al. Bone marrow-derived mesenchymal stem cells improve islet graft function in diabetic rats. Transplant Proc. 2009;41(5):1797-800.
164. Berman DM, Willman MA, Han D, Kleiner G, Kenyon NM, Cabrera O, et al. Mesenchymal stem cells enhance allogeneic islet engraftment in nonhuman primates. Diabetes. 2010;59(10):2558–68.
165. Ding Y, Xu D, Feng G, Bushell A, Muschel RJ, Wood KJ. Mesenchymal stem cells prevent the rejection of fully allogenic islet grafts by the immunosuppressive activity of matrix metalloproteinase-2 and -9. Diabetes. 2009;58(8):1797–806.
166. Fransson M, Brannstrom J, Duprez I, Essand M, Le Blanc K, Korsgren O, et al. Mesenchymal stromal cells support endothelial cell interactions in an intramuscular islet transplantation model. Regen Med Res. 2015;3:1.
167. Wu H, Wen D, Mahato RI. Third-party mesenchymal stem cells improved human islet transplantation in a humanized diabetic mouse model. Mol Ther. 2013;21(9):1778–86.
168. Yoshimatsu G, Sakata N, Tsuchiya H, Minowa T, Takemura T, Morita H, et al. The co-transplantation of bone marrow derived mesenchymal stem cells reduced inflammation in intramuscular islet transplantation. PLoS One. 2015;10(2):e0117561.
169. Vaithilingam V, Evans MDM, Lewy DM, Bean PA, Bal S, Tuch BE. Co-encapsulation and co-transplantation of mesenchymal stem cells reduces pericapsular fibrosis and improves encapsulated islet survival and function when allografted. Sci Rep. 2017;7(1):10059.
170. Chen J, Chen J, Cheng Y, Fu Y, Zhao H, Tang M, et al. Mesenchymal stem cell-derived exosomes protect beta cells against hypoxia-induced apoptosis via miR-21 by alleviating ER stress and inhibiting p38 MAPK phosphorylation. Stem Cell Res Ther. 2020;11(1):97.
171. Hade MD, Suire CN, Suo Z. Mesenchymal Stem Cell-Derived Exosomes: Applications in Regenerative Medicine. Cells. 2021;10(8).
172. Khayambashi P, Iyer J, Pillai S, Upadhyay A, Zhang Y, Tran SD. Hydrogel Encapsulation of Mesenchymal Stem Cells and Their Derived Exosomes for Tissue Engineering. Int J Mol Sci. 2021;22(2).
173. Sojka DK, Huang YH, Fowell DJ. Mechanisms of regulatory T-cell suppression - a diverse arsenal for a moving target. Immunology. 2008;124(1):13–22.
174. Vignali DA, Collison LW, Workman CJ. How regulatory T cells work. Nat Rev Immunol. 2008;8(7):523–32.
175. Volfson-Sedletsky V, Jones At, Hernandez-Escalante J, Dooms H. Emerging Therapeutic Strategies to Restore Regulatory T Cell Control of Islet Autoimmunity in Type 1 Diabetes. Front Immunol. 2021;12:635767.
176. Bluestone JA, Buckner JH, Fitch M, Gitelman SE, Gupta S, Hellerstein MK, et al. Type 1 diabetes immunotherapy using polyclonal regulatory T cells. Sci Transl Med. 2015;7(315):315ra189.
177. Marek-Trzonkowska N, Mysliwiec M, Iwaszkiewicz-Grzes D, Gliwinski M, Derkowska I, Zalinska M, et al. Factors affecting long-term efficacy of T regulatory cell-based therapy in type 1 diabetes. J Transl Med. 2016;14(1):332.

178. Pathak S, Meyer EH. Tregs and Mixed Chimerism as Approaches for Tolerance Induction in Islet Transplantation. Front Immunol. 2020;11:612737.
179. Selck C, Dominguez-Villar M. Antigen-Specific Regulatory T Cell Therapy in Autoimmune Diseases and Transplantation. Front Immunol. 2021;12:661875.
180. Mathew JM, Voss JH, McEwen ST, Konieczna I, Chakraborty A, Huang X, et al. Generation and Characterization of Alloantigen-Specific Regulatory T Cells For Clinical Transplant Tolerance. Sci Rep. 2018;8(1):1136.
181. Cabello-Kindelan C, Mackey S, Sands A, Rodriguez J, Vazquez C, Pugliese A, et al. Immunomodulation Followed by Antigen-Specific Treg Infusion Controls Islet Autoimmunity. Diabetes. 2020;69(2):215–27.
182. Peters JH, Hilbrands LB, Koenen HJ, Joosten I. Ex vivo generation of human alloantigen-specific regulatory T cells from CD4(pos)CD25(high) T cells for immunotherapy. PLoS One. 2008;3(5):e2233.
183. Putnam AL, Safinia N, Medvec A, Laszkowska M, Wray M, Mintz MA, et al. Clinical grade manufacturing of human alloantigen-reactive regulatory T cells for use in transplantation. Am J Transplant. 2013;13(11):3010–20.
184. Cherai M, Hamel Y, Baillou C, Touil S, Guillot-Delost M, Charlotte F, et al. Generation of Human Alloantigen-Specific Regulatory T Cells Under Good Manufacturing Practice-Compliant Conditions for Cell Therapy. Cell Transplant. 2015;24(12):2527–40.
185. Veerapathran A, Pidala J, Beato F, Yu XZ, Anasetti C. Ex vivo expansion of human Tregs specific for alloantigens presented directly or indirectly. Blood. 2011;118(20):5671–80.
186. Zhang Q, Lu W, Liang CL, Chen Y, Liu H, Qiu F, et al. Chimeric Antigen Receptor (CAR) Treg: A Promising Approach to Inducing Immunological Tolerance. Front Immunol. 2018;9:2359.
187. Muller YD, Ferreira LMR, Ronin E, Ho P, Nguyen V, Faleo G, et al. Precision Engineering of an Anti-HLA-A2 Chimeric Antigen Receptor in Regulatory T Cells for Transplant Immune Tolerance. Front Immunol. 2021;12:686439.
188. Hong M, Clubb JD, Chen YY. Engineering CAR-T Cells for Next-Generation Cancer Therapy. Cancer Cell. 2020;38(4):473–88.
189. Feins S, Kong W, Williams EF, Milone MC, Fraietta JA. An introduction to chimeric antigen receptor (CAR) T-cell immunotherapy for human cancer. Am J Hematol. 2019;94(S1):S3–S9.
190. Vinet L, Lamprianou S, Babic A, Lange N, Thorel F, Herrera PL, et al. Targeting GLP-1 receptors for repeated magnetic resonance imaging differentiates graded losses of pancreatic beta cells in mice. Diabetologia. 2015;58(2):304–12.
191. Pierini A, Iliopoulou BP, Peiris H, Perez-Cruz M, Baker J, Hsu K, et al. T cells expressing chimeric antigen receptor promote immune tolerance. JCI Insight. 2017;2(20).

Part IV
Preclinical Model and Translational Approaches

Considerations Pertaining to Implant Sites for Cell-Based Insulin Replacement Therapies

Braulio A. Marfil-Garza, Nerea Cuesta-Gomez, and A. M. James Shapiro

1 Introduction

Type 1 diabetes (T1D) is a major global health issue. According to the Type 1 Diabetes Index Model, in 2021 there were about 8.4 million individuals worldwide living with T1D. More importantly, its prevalence is projected to increase by 60–107% by 2040 [68]. Severe and recurrent hypoglycemia is a major limitation to attain optimal glycemic control [44] and affects up to 38% of people with T1D [2]. Pancreatic islet transplantation (ITx) has become a valuable option to treat recurrent and severe hypoglycemia in people with T1D experiencing otherwise intractable severe and recurrent hypoglycemia [38]. Recent long-term reports highlight its safety in terms of patient survival [107, 112, 122, 201] and procedural complications [78, 122], as well as its efficacy in abrogating severe hypoglycemia while simultaneously improving glycemic control long term [78, 107, 122, 201].

B. A. Marfil-Garza
Alberta Diabetes Institute, Edmonton, AB, Canada

Clinical Islet Transplant Program, University of Alberta, Edmonton, AB, Canada

Tecnologico de Monterrey, Institute for Obesity Research, Hospital Zambrano Hellion, TecSalud, San Pedro Garza Garcia, Mexico

N. Cuesta-Gomez
Alberta Diabetes Institute, Edmonton, AB, Canada

Department of Surgery, University of Alberta, Edmonton, AB, Canada

A. M. J. Shapiro (✉)
Alberta Diabetes Institute, Edmonton, AB, Canada

Clinical Islet Transplant Program, University of Alberta, Edmonton, AB, Canada

Department of Surgery, University of Alberta, Edmonton, AB, Canada
e-mail: jshapiro@ualberta.ca

L. Piemonti et al. (eds.), *Pluripotent Stem Cell Therapy for Diabetes*,
https://doi.org/10.1007/978-3-031-41943-0_16

As the field enters an era of cell-based therapies and regenerative medicine for T1D, successful clinical implementation of these treatments demands thoughtful consideration regarding cell culture processes, physiological development and adaptation, implantation procedures, clinical measures of success, and the overall expected balance of risks and benefits. In this chapter, we focus on discussing implantation sites and techniques, and their potential implications for current and future β-cell replacement therapies. We synthesize key physiological concepts and elaborate upon aspects regarding graft survival and function. Finally, we succinctly review the current clinical evidence concerning intraportal and extrahepatic implantation sites and suggest potential research avenues for future research.

2 Type 1 Diabetes

T1D is an autoimmune disorder with both genetic and environmental contributions. It is characterized by lymphocytic infiltration of the islets (insulitis) followed by loss of β cell mass resulting in symptomatic hyperglycemia that requires exogenous insulin [53]. The natural history of T1D can be classified into three stages, which can last from months to decades [8, 9]. The first stage represents normoglycemic individuals that have developed two or more T1D-associated islet autoantibodies and, subsequently, autoreactive $CD4^+$ and $CD8^+$ T cell responses [148]. Insulin, 65 kDa glutamic acid decarboxylase (GAD65), insulinoma-associated protein 2 (IA-2), and zinc transport 8 (ZnT8) are the four major antigens and the presence of two or more autoantibodies confers a nearly 100% chance of progressing to overt T1D within 20 years [3, 83]. At stage 2, there is evidence of dysglycemia (i.e., intolerance to glucose or impaired fasting glycemia) and at stage 3, overt T1D develops, with the typical presence of symptoms, such as polyuria and polydipsia, weight loss, fatigue, and diabetic ketoacidosis, among others.

Throughout the pre-symptomatic stages, β-cell destruction is mediated principally by T lymphocytes, although B lymphocytes, macrophages, dendritic cells (DCs) and natural killer cells (NK cells) also contribute. Autoimmunity against β-cells begins in the pancreatic lymph node where naive autoreactive T cells that have escaped negative selection in the thymus are activated by β-cell antigens presented by DCs [52, 86]. Upon activation, autoreactive $CD4^+$ and $CD8^+$ T cells travel to the islets and target several proteins expressed by β-cells, promoting in this manner insulitis. Once peripheral insulitis has been established, the cytokines and chemokines secreted upon T cell receptor triggering recruit other effector cells, like macrophages and NK cells to induce β-cell death [33, 71] as well as B cells and DCs, whose primary function is antigen presentation to allow further progression of the disease [18, 32, 128, 180]. Despite Tregs being present within the islets, the ratio of autoreactive T cell to Treg progressively increases throughout insulitis, which is another important component of the pathophysiology of T1D [120]. Upon the destruction of β-cells, autoreactive T cells then acquire a circulating memory T cell phenotype, which might particularly relevant for ITx and stem cell-based replacement therapies [30, 49].

2.1 *Autoimmunity After Islet Transplantation*

Recurrent autoimmunity following ITx due to autoantibodies and circulating autoreactive memory T cells remains a major issue precluding optimal function and long-term graft survival. Since autoimmune recurrence would likely mimic the immune activation and infiltration occurring at T1D onset, autoantibodies were initially considered to be prognostic markers for the detection of recurrent autoimmunity. However, previous studies monitoring autoantibody levels pre- and post-transplantation remain inconclusive, with some reporting a negative correlation between upregulation of islet autoantibodies and long-term islet function/survival [24, 25, 124, 142, 144, 204, 205], while others have found no correlation at all [3, 80, 82, 170].

Despite the unclear role of autoantibodies in autoimmune recurrence following ITx, in the presence of β-cell-derived antigens, autoreactive memory T cells and B cells can quickly expand and give rise to effector T and B cells, resulting in insulitis with mononuclear cell infiltrate and selective β-cell death in the patients that have undergone pancreas transplantation or ITx [80, 82, 149, 169, 187, 192, 193]. It remains unknown whether the implant location could have an effect in delaying immune infiltration; however, due to the high resistance of the liver against immune responses, it is unlikely that any other implantation site could help delay graft rejection by modulating immune infiltration. Conversely, IBMIR is considered one of the causes of islet death following intraportal islet infusion, and hence, finding alternative transplant sites that could potentially minimize initial inflammation, and hence cell death, could potentially delay or prevent autoimmune recurrence, but this remains to be proven clinically [108]. Conversely, transplanting islets or stem cell-derived islet-like clusters (SC-islets) to alternative more ischemic sites could actually amplify local injury and augment both auto- and alloimmune-related destructive events.

3 Islet Biology and Physiology

A comprehensive discussion on islet biology and physiology is beyond the scope of this chapter and the reader is referred to an extensive review of the human pancreatic islet that has been recently accomplished by Walker et al. [213]. Nevertheless, we will touch on key considerations to provide context for the following sections.

The pancreatic islet of Langerhans is composed of 24–60% insulin-producing β-cells, 10–30% glucagon-producing α-cells, ~10% somatostatin-producing δ-cells, ~1% pancreatic polypeptide containing cells, also known as PP cells, and ~1% ghrelin cells [48, 182]. These hormones regulate each other and act together to ensure blood glucose homeostasis. For this reason, diabetes should not be considered a monohormonal disease, which is a key consideration for the design and development of novel stem cell-based therapies.

β-cells contain secretory granules with insulin, which is released in response to high concentrations of glucose and other stimuli upon food ingestion. Briefly, nutrient uptake initiates glycolysis and oxidative metabolism, resulting in ATP production and the subsequent closure of ATP-sensitive K^+ channels (K_{ATP}) in the β-cell membrane. Accumulation of K^+ in the cytosol induces membrane depolarization and the opening of voltage-dependent Ca^{2+} channels, promoting an influx of Ca^{2+} and the release of insulin (detailed mechanisms are revised by Rorsman and Ashcroft [172]). Despite all β-cells containing insulin granules, two populations based on the secretory patterns have been identified: responsive and unresponsive to glucose challenges [72]. α-cells are central in the counterregulatory response to insulin through the secretion of glucagon under low concentrations of glucose, a protective mechanism for hypoglycemia that is commonly impaired in T1D. Hypoglycemia results on low levels of intracellular ATP and the subsequent closure of (K_{ATP}) within α-cells. Reduced influx of K^+ will cause the depolarization of the cell membrane and the influx of Ca^{2+} into the cell, triggering glucagon secretion [131]. Somatostatin is secreted by δ-cells in the presence of nutrients and functions as a negative regulator of insulin, but also glucagon and pancreatic polypeptide secretion from adjacent cells. PP cells secrete pancreatic polypeptide in response to arginine, resulting in glucagon release inhibition under low glucose concentrations and satiety [4, 48]. Lastly, ghrelin is secreted by the ε-cells in the islet during fasting periods and inhibits insulin secretion while also regulating glucagon, somatostatin and pancreatic polypeptide secretion [48, 176].

The complex intra-islet network, beyond paracrine regulation, is also regulated by gastrointestinal hormones, the incretins. These include the glucagon-like peptide-1 (GLP-1) and glucose-dependent insulinotropic polypeptide (GIP) [155]. Upon nutrient intake, GLP-1 and GIP are secreted into the bloodstream, and then bind to G protein-coupled receptors (GLP-1R and GIPR, respectively) on the surface of β-cells, resulting in activation of downstream pathways ultimately potentiating insulin secretion [36, 39, 181]. GIP can also stimulate glucagon secretion under low glucose concentrations to ameliorate hypoglycemia [39].

In the context of ITx, there are other critical factors regulating hormone secretion beyond glycemic levels including the extent of vascularization and the degree of inflammation. Despite accounting for 1–2% of the pancreas mass, islets require 5–20% of the pancreatic blood flow. The process of islet isolation leads to a loss of vascular structures, hence, upon transplantation, islets are completely dependent on passive oxygen diffusion, which leads to central core necrosis and impaired function [102]. Indeed, it has been demonstrated that relatively mild hypoxia (pO_2 of 27 mmHg) decreases insulin release by 50% [54]. While hypoxia affects the function of all islet cells, it predominantly affects β-cells [65]. The partial preservation of α-cells could partially explain why abrogation of hypoglycemia is a more uniform outcome following ITx, although more research is needed to fully elucidate this phenomenon. Beyond hypoxia-mediated cell death, the lack of vascularization also leads to the loss of hemodynamic regulatory mechanisms that are necessary for physiological hormone secretion [87]. Similar impairment of islet secretory function might be related to the loss of innervation during the isolation process and delayed/incomplete reinnervation [73]. Finally, cell death, regardless of the cause, can provoke innate inflammatory responses through the release of damage-associated molecular patterns, which in turn promote leukocyte trafficking and localized cytokine and chemokine release [91], ultimately impairing islet function [206]. Overall, suboptimal oxygenation and vascularization, as well as incomplete innervation of the transplanted islets, results in impaired glucose homeostasis and cell death, which promotes immune infiltration and compromises islet engraftment, survival and function.

4 Intraportal Islet Transplantation

The standard process for ITx involves isolation of the islets from deceased-donor pancreata followed by infusion into the liver intraportal circulation using ultrasound and fluoroscopic guidance (Fig. 1) [123]. While intraportal delivery of islets is safe and the only implantation approach shown to consistently support adequate engraftment and sustained graft survival, it is associated with a ~7% complication rate [122], mostly related to bleeding (including hepatic hematoma, arteriovenous fistula and pseudoaneurysm formation) [114, 211] and thrombosis [92, 185]. The risk of bleeding is directly proportional to the number of islet infusions [211], while thrombosis is positively correlated with packed cell volume [92] (even in autologous ITx [26]). Other complications related to intraportal ITx include transient increase in transaminase levels, which typically resolves between 4 and 8 weeks post-ITx [152] and hepatic steatosis [20, 207], which seems to be correlated with islet mass [207] and currently of uncertain, but probably minimal clinical relevance [85].

Other important limitations with intraportal ITx are related to the innate inflammatory responses occurring immediately and chronically post-transplant, chronic hypoxemia and impaired glucose responsiveness and metabolic function. These events are depicted in Fig. 2. The advantages and disadvantages with this implantation site are delineated in Table 1.

Innate inflammatory reactions post-ITx, collectively referred as instant blood-mediated inflammatory response (IBMIR), are a major factor for islet death during and immediately post-infusion and contribute to a loss of up to 50% of the infused islet mass within 24 hours [58, 59]. Considering that most ITx recipients require two to three islet infusions to achieve optimal outcomes [122, 200], ameliorating or preventing IBMIR remains a research priority. IBMIR is a multi-layered innate immune response triggered by direct exposure of islets (and tissue factor) to the recipients' blood. It is not limited to allo- or xenotransplantation, as evidence in autologous ITx exists [132]. Immediately after infusion, exposure to blood promotes platelet activation, which triggers the secretion of prothrombotic factors as well as pro-inflammatory cytokines that activate the coagulation and complement systems, ultimately leading to cell-mediated injury (i.e., via neutrophils, monocytes and macrophages) [91]. These responses synergize with hypoxia-mediated cell death, activation of inflammatory pathways within islets themselves (e.g., MAP and

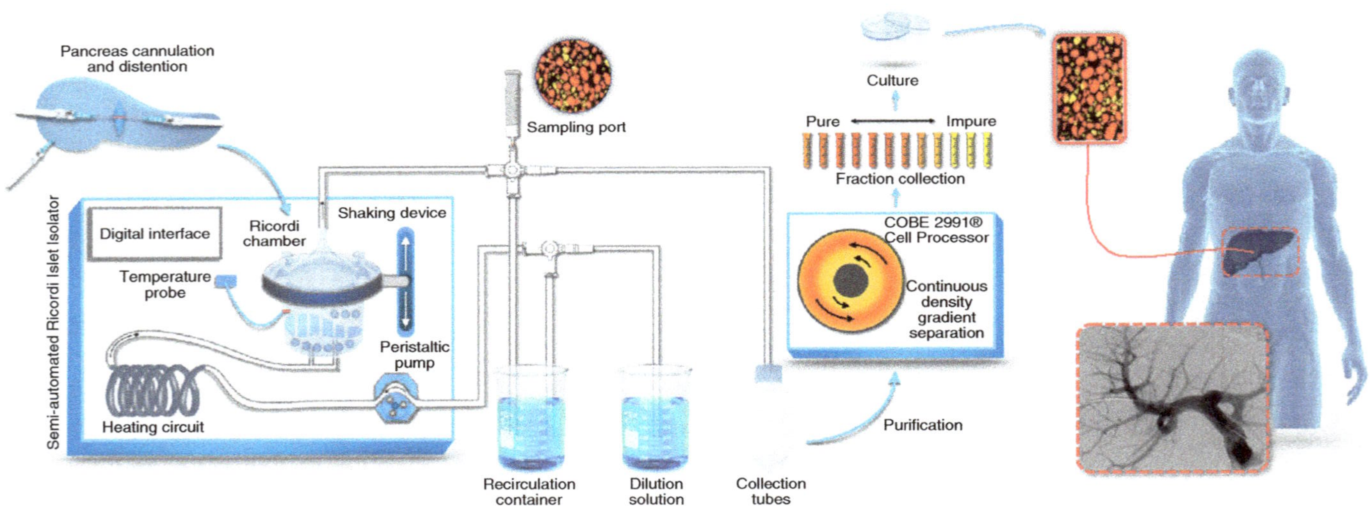

Fig. 1 Pancreatic islet transplantation
After "decontaminating" the pancreas (removing spleen, duodenum and peripancreatic fat), a small incision in the mid-body of the pancreas is done to expose the main pancreatic duct, which is cannulated using two catheters; the outflow tract at the major papilla is sealed using a third catheter. Pancreas distention with cold collagenase infused through both catheters using a perfusion machine (not shown) follows. The pancreas is then sliced and placed in the Ricordi chamber for digestion. Enzymatic digestion is sustained by activating collagenase through controlled warming (~36 °C) using a heating circuit and a temperature probe. Mechanical digestion is facilitated using orbital shaking of the chamber and silicon nitride/metal marbles inside the chamber. The solution is recirculated through the circuit until the pancreas is appropriately digested and islets are visibly dislodged from the exocrine tissue; this is evaluated by taking samples at predefined time points through a sampling port and staining them with dithizone. Once digestion is appropriate, the circuit is "opened" to dilute the solution and stop enzymatic activity. The islets are then collected. The collected tissue is introduced into a cell processor for purification using continuous density gradient centrifugation. Tissue fractions are collected from the cell processor and the highly pure fractions are selected to be cultured and, subsequently, transplanted. Percutaneous and transhepatic cannulation of a peripheral branch of the portal vein is done using ultrasound and fluoroscopy. Islets are loaded into a gravity infusion bag with heparin (70 IU/kg). Sequential portal pressure monitoring is done and the cannulation tract is sealed with a collagen hemostatic powder. (Adapted from Marfil-Garza et al. [123])

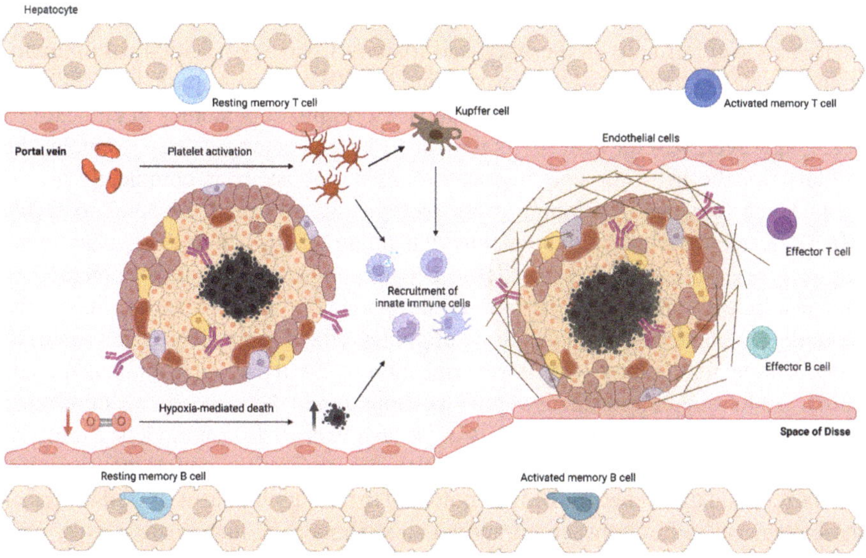

Fig. 2 Schematic representation of the main events occurring upon intraportal islet transplantation
In the presence of human blood, transplanted human islets trigger an instant blood-mediated inflammatory reaction (IBMIR), which consists of the activation of circulating platelets and the subsequent release of prothrombotic factors as well as pro-inflammatory cytokines that activate the complement system and recruit leukocytes. Immune cell infiltration induces the secretion of pro-inflammatory cytokines by Kupffer cells, further exacerbating immune infiltration. After infusion, islets rely on passive diffusion to fulfill their oxygen requirements, which coupled with the low oxygen tension in the liver, results in inadequate revascularization, oxidative stress and hypoxia-mediated core necrosis and cell death. Apoptotic cells release damage-associated molecular patterns, which further recruit and activate leukocytes. Increased cytokine secretion by the innate immune system in combination with the expression of islet antigens in the newly transplanted islets promotes the activation and expansion of autoreactive memory T cells and B cells, giving rise to effector T cells and B cells

Table 1 Advantages and disadvantages with intraportal islet transplantation

Advantages	Disadvantages
Portal delivery of insulin – physiological	Procedural risks (e.g., bleeding, thrombosis)
Minimally invasive	Difficult to monitor
Technically feasible and reproducible	Does not allow graft retrieval
Vast experience	Increased exposure to immunosuppressants
Direct exposure to blood – good oxygenation and nutrient delivery	Direct exposure to blood – strong innate immune responses (IBMIR)
	Hostile immune environment

IBMIR instant blood-mediated inflammatory reaction

JNK kinases, IL-1β and NFAT/NF-*k*B- signaling), local innate immune responses exerted by Kupffer cells in the liver, and toxicity from pharmacological immunosuppression [91]. Several strategies targeting IBMIR, such as heparin [101], TNF-α and IL-1 inhibitors [16, 17, 194] have shown to improve outcomes in clinical ITx [122]; however, several aspects of remain to be characterized. For example, the influence of the liver immune microenvironment has not been completely unraveled, which prevents developing more biologically informed therapies to tackle IBMIR. Evidence from small animal models might be difficult to extrapolate to the clinic, given the differences in islet-to-hepatic sinusoid size ratios compared to humans. Indeed, one reason the murine intraportal ITx models cannot achieve the same degree of success as in the clinic might be related to the degree and extent of embolization that occurs with intraportal infusion in these small animal models [126]. Addressing these unanswered questions could be done by incorporating autologous pig islet transplant models or ex vivo liver perfusion models. Similarly, xenogeneic ITx and SC-islets transplantation demands investigation into potential similarities and difference in IBMIR compared to primary human islets, and whether these responses differ between intraportal and extrahepatic implantation sites.

In parallel to unraveling the pathophysiological mechanisms of IBMIR, novel therapeutic agents proposed in preclinical studies could be considered in clinical trials moving forward. Cibinetide, an agonist of the innate repair receptor (a.k.a. as common β-subunit of the erythropoietin receptor) shows a favorable safety profile in clinical trials [113] and has been recently reported to preserve human islet survival and function in a pro-inflammatory in vitro environment by decreasing platelet consumption, but also in vivo by regulation of macrophage infiltration [222]. Another potential approach to tackle IBMIR is to strive for a more complete inhibition of cytokine responses early post-transplant. To this extent, tocilizumab, a competitive inhibitor of IL-6 has shown to preserve function of human islet in a cytokine-based in vitro pro-inflammatory assays, and displayed equal potency to anakinra, which is currently used in clinical ITx [174]. While promising, anti-inflammatory effects of tocilizumab should be weighed against evidence suggesting delayed revascularization of transplanted islets [127], and a potentially increased rate of opportunistic infections in clinical trials [130]. Tofacitinib, a JAK inhibitor that inhibits the production of macrophage-mediated inflammatory cytokines, has also shown positive results in an allogeneic cynomolgus macaque islet transplant model that used anakinra, adalimumab, thymoglobulin (basiliximab for second infusion) and sirolimus as immunosuppression regime [97]. Another option is to explore classical anti-inflammatory agents, such as dexamethasone. A single, low dose (e.g., 4–6 mg) course of dexamethasone, particularly through localized release (e.g., intraportal infusion), might mitigate IBMIR by promoting macrophage polarization towards an anti-inflammatory phenotype (M2) [89, 106, 146]. These effects may be additive or synergistic with intraportal infusion of anti-inflammatory agents. However, more in vitro and in vivo preclinical studies using human islets are needed to fully support this intervention. Other potential interventions to mitigate IBMIR that can be implemented at the pre-transplant/culture stage include mitomycin C [221], inhibitors of the pro-inflammatory c-Jun NH_2-terminal kinase (JNK) pathway [133], co-culture with mesenchymal stromal cells [100] or pancreatic stellate cells [139], among others. Although potentially effective, in vitro strategies to ameliorate the impact of IBMIR might be difficult to approve in countries such as the U.S.A., where regulatory agencies have raised concerns (perhaps unjustified) [219] on the biological effect of pre-transplant culture time on islets. As mentioned before, these investigations should become a priority as they would also inform other potential explorations involving extrahepatic implantation sites.

5 Physiology After Islet Transplantation

Most studies evaluating physiological and glucose counterregulatory responses following ITx have naturally included recipients of liver intraportal ITx. There is controversy as to whether these responses become normal following ITx. In this section, we will summarize the current evidence. To provide a more comprehensive picture of this important topic, we will include evidence from islet autologous and allogeneic ITx. Some potential implications of the evidence obtained from preclinical research endeavors will also be incorporated to provide some guidance on future avenues to be explored.

There are several physiological aspects that can be evaluated post-ITx and these typically include determination of β-cell mass and insulin secretion responses, as well as counterregulatory responses to hypoglycemia. Determination of β-cell/islet mass and insulin secretion commonly involve glucose-potentiation of insulin/C--peptide release in response to non-glucose secretagogues (e.g., arginine), as well as hyperglycemic clamps [166]. These tests have been shown to correlate with transplanted β-cell mass in studies of segmental and whole pancreas transplantation [40, 168], but also intrahepatic autologous [167, 196] and allogeneic ITx [173]. For evaluation of counterregulatory responses, hyperinsulinemic hypoglycemic and euglycemic clamps have been implemented [158, 163].

Adequate interpretation of β-cell/islet mass determinations in vivo should consider the theoretical islet mass in people without T1D. The typical number of islets in a pancreas of healthy individual ranges from 3.2 to 14.8 million [84, 104]. It appears that islets ≤100 μm in diameter represent most (~80%) of the islets in a human pancreas, and that the number of those with a diameter >23 μm are only about 1.5 million [104]. This has been suggested to translate into ~0.6–1.1 million islet equivalents (IEQs) [98]. Of relevance, an IEQ is defined as an islet with a diameter of 150 μm. Although the IEQ has enabled standardizing islet "doses" across centers, it remains flawed and suffers from significant inter- and intra-rater variability due to its subjectivity [81]. Additionally, small islets are typically lost during islet isolation, but also are underrepresented in IEQ counts. In fact, original estimations of islet mass using the IEQ excluded those islets with <50 μm in diameter. While some corrections have been implemented, overall, it appears that this method overestimates IEQ counts (and islet mass) by 10–30% [99]. The typical islet size index (islet equivalents/islet numbers) is approximately 1.1 [13, 122], which means that the typical transplanted islet is slightly above 150 μm in diameter. Considering that larger islets are more prone to hypoxia-mediated cell death due to the transplanted islet's complete dependence on passive oxygen diffusion [102] efforts to recover smaller islets during isolation remain vital. Regarding transplanted islet mass, this varies significantly and ranges from 6.5 to 15.5 IEQ/kg [42, 61, 107, 111, 122], which amounts to ~0.5–1.1 million IEQs in a standard 70 kg recipient. Since up to 50% of islets are lost during and immediately post-infusion [58, 59], it is anticipated that the surviving transplanted islet cell mass will be lower than a healthy person (~0.2–0.45 million IEQ vs ~0.6–1.1 million IEQ), but could still represent between 18% and 30% of the islet mass found in a non-diabetic pancreas. These theoretical approximations coincide with evidence in the autologous [162, 167] and the allogeneic ITx setting [159, 161] suggesting that long-term optimal function (i.e., insulin independence) may require transplantation and engraftment of an islet mass ≥20–30% of normal. Studies in preclinical T1D also coincide with these estimations [67]. However, it is less clear how much islet mass is required to support fully physiological responses and other beneficial outcomes beyond insulin independence, such as abrogation of hypoglycemia.

Estimations of islet mass by measuring secretory capacity (i.e., insulin responses) are sensitive, and the level of correlation with the transplanted islet mass does suggest they approximate transplanted islet mass rather precisely [166]. However, they rely on some considerations that might not be universally present. The most important is that correlations with transplanted islet mass assume that 100% of the transplanted islets survive and that they are fully functional. Additionally, reports have typically included patients with insulin independence and optimal glycemic control post-transplant (either autologous or allogeneic). This outcome is unfortunately not the norm and thus it is currently unknown whether contemporary evidence is fully applicable to recipients with less favorable outcomes. Considering that these tests are cumbersome and not widely available, other measures to evaluate graft function and to establish potential benchmarks for future clinical endeavors would be valuable. In this regard, in a recent retrospective study, we have shown that fasting C-peptide levels could correctly identify recipients with insulin independence, but also those achieving abrogation of hypoglycemia, with a level of ≥0.12 nmol/L correctly identifying 97.4% of these patients (sensitivity = 97.3, specificity = 100%) [208]. Somewhat similar values from the Collaborative Islet Transplant Registry have been also reported [11]. Overall, these results are informative for future studies, including those evaluating stem cell-based replacement therapies, as well as immunotherapies in people with T1D.

Beyond islet mass, determination of secretory capacity also involves the kinetics and physiological nature of insulin responses from the graft. Of relevance, there is some preclinical evidence that suggests that liver intraportal delivery of islets does not support physiological insulin secretion. Studies from Per-Ola Carlsson's group have shown that, in rats, intraportally delivered islets can only respond appropriately to glucose and arginine when delivered through the hepatic artery, and not the portal vein [110]. These findings can be explained by defects in neovascularization and perfusion, but also suggest that nutrient sensing in the islet occurs only through newly formed blood vessels, particularly those tributaries from the hepatic artery [77, 110]. Other possible contributing factors include the lack of innervation, which has potential impact on regulation of intra-islet blood flow and ultimately glucose sensing and insulin secretion kinetics. Innervation is lost during islet isolation and the timeline and degree of reinnervation remains largely unknown [73]. Another potential confounder is the fact that murine models of liver intraportal ITx might be biased due to the inherent differences in islet-to-hepatic sinusoid ratio, which would be associated with a higher degree of liver damage and potential inflammation upon intraportal islet delivery, as compared to humans. In accordance with data from preclinical studies, there is also controversy with the evidence from clinical reports. For example, an early report by Pyzdrowski et al. showed that following autologous ITx, glucose- and arginine-stimulated responses were normal, although decreased in magnitude [150]. In contrast, Meier et al., using hyperglycemic clamps reported deficient first-phase insulin secretion, as well as blunted glucose-potentiated arginine insulin responses in allogeneic ITx recipients compared to heathy controls [125]. Interestingly, the pulsatile pattern of insulin secretion was restored in ITx recipients; however, its pulse mass and amplitude were around 50% of healthy controls. The authors also performed direct transhepatic catheterization and showed that first-pass insulin extraction was similar in ITx recipients vs controls [125]. More recently, Rickels et al. have shown that acute arginine-alone and glucose-potentiated arginine insulin responses were also blunted in allogeneic ITx recipients compared to non-diabetic subjects. Remarkably, the authors report that a modified induction immunosuppression regimen (antithymocyte globulin and etanercept) substantially improved these blunted responses, even after 1 year post-transplant [159]. Preservation of graft function and these pulsatile patterns of insulin secretion might have added benefits in terms of hepatic glucose production, a central mechanism to prevent hypoglycemia, as previous studies have shown [117, 158]. Finally, responses to incretins, which are vital for physiological insulin secretion, have also been shown to be preserved after clinical ITx, although to a lesser extent compared to both whole pancreas recipients and healthy controls [63, 160, 210]. Together, these findings suggest that, while insulin secretion follows similar kinetics to healthy controls, and thus some degree of synchrony might be reestablished with intraportal ITx, reduced islet mass inherently leads to overall blunted responses.

In terms of the counterregulatory responses after ITx, it appears that similar or even lower islet masses relative to healthy individuals are needed to preserve them. Briefly, it should be highlighted that counterregulatory responses are partly dependent on paracrine signalling within the islet, as appropriate glucagon secretion from α-cells is dependent on insulin signalling from β-cells. The other component of these counterregulatory responses consists of epinephrine, growth hormone and cortisol secretion, as well as changes in the central nervous system to modulate behavior. These coordinated responses lead to increased hepatic glucose production, decreased peripheral glucose utilization, and promotion of food ingestion [177]. Recurrent hypoglycemia impairs all of these responses. In the ITx setting, earlier evidence suggested impaired counterregulatory responses (i.e., glucagon secretion) to hypoglycemia [94, 138]. However, further studies showed an improvement in glucagon responses in ITx recipients when compared to people with T1D on the waitlist for an ITx, although only partially when compared to healthy controls. It has been also demonstrated that these responses are activated at a normal glycemic threshold [138], which would be necessary for protection of hypoglycemia. Additionally, catecholamine responses and autonomic symptoms with hypoglycemia have been shown to be similar between ITx recipients and healthy controls, and substantially improved as compared to T1D subjects, which suggests recovery of central nervous system recognition of hypoglycemia, one of the main underlying phenomena causing and perpetuating hypoglycemia [157].

Altogether, current evidence suggests that ITx recipients have blunted insulin responses as well as counterregulatory responses to hypoglycemia compared to healthy controls, which is likely a result of a low number of engrafted islets, a relatively lower degree of vascularization and a lack of innervation. However, more studies are needed to fully assess the long-term implications of these blunted insulin responses, but also to determine how these responses compare at extrahepatic implantation sites.

6 Extrahepatic Pancreatic Islet Transplantation

Intraportal islet infusion is the clinical gold standard given that it is the only anatomical site proven to consistently support islet engraftment and long-term graft function. However, extrahepatic islet implantation has always been a desired approach. Indeed, the first islet allotransplant, performed by Charles Pybus in Newcastle-upon-Tyne, England in 1914, used the subcutaneous site as the implantation site [183]. To date, extrahepatic ITx in humans has been attempted at the subcutaneous site [10, 64, 115, 116, 209], the intramuscular space [19, 41, 137, 153], the intravascular space [31, 147], the gastric submucosa [209, 220], the omentum and intraperitoneal space [12, 121, 178, 199, 209], the spleen [109, 218], the kidney subcapsular space [90], and the bone marrow [118, 119], with ongoing clinical trials evaluating the anterior chamber of eye (NCT02916680, NCT02846571). In terms of insulin independence, the first case after ITx in a person with T1D, reported by Largiadèr and Kolb, was achieved by intrasplenic implantation [109]. Similarly, other cases of insulin independence have been reported with ITx into the omentum [12, 190], the subcutaneous space [198] and the extrahepatic intravascular space [31, 47, 147]. Unfortunately, there are few direct comparative studies with the intraportal site. Recently, we have described our experience with extrahepatic ITx into the subcutaneous space, the gastric submucosa and the greater omentum and compared it with our 20-year cohort of patients receiving intraportal ITx. In our experience, none of the extrahepatic sites achieved substantial fasting or stimulated C-peptide secretion and all recipients had early graft failure [209]. Importantly, recipients who subsequently underwent intraportal ITx achieved similar outcomes as those undergoing “upfront” intraportal ITx, suggesting that, while these extrahepatic sites were incapable of sustaining engraftment and graft survival, no detrimental effects, including sensitization, were caused by these interventions [209]. A limitation with this report is that it included only nine patients in the extrahepatic ITx cohort group, which highlights the need for larger, perhaps multicenter, studies.

Beyond the desire to avoid procedural risks, there are other potential advantages driving efforts into improving and realizing extrahepatic ITx. It should also be emphasized that a promising area of opportunity remains in terms of dissecting and thoroughly characterizing the physiological differences between each implantation sites. Naturally, a comparison with intraportal ITx would seem to be the most relevant and preclinical studies tackling these important matters should take this into account for experimental design and execution. Table 2 provides a preliminary assessment of some potential physiological, immunological and technical considerations to guide preclinical and clinical endeavors exploring extrahepatic ITx. In the next section, key aspects of extrahepatic implantation sites are synthesized and discussed. Of note, the clinical characteristics of recipients of extrahepatic ITx might not be similar to those receiving intraportal infusions. Thus, comparisons in terms of outcomes should proceed with caution.

6.1 Intraperitoneal Space and Greater Omentum

The intraperitoneal space and greater omentum have accrued a considerable amount of evidence from preclinical studies, but also clinical trials of encapsulated islet transplantation, being the most studied site for this purpose [121]. It has shown to support insulin independence for at least 1 year, albeit in a single patient [12]. Additionally, there is evidence in the autologous ITx setting that patients having combined intraportal and intraperitoneal infusions (due to large tissue volumes) have better [70], or at least indistinguishable [167], responses to those having intraportal infusions exclusively, suggesting that islets placed in the peritoneal cavity can survive and function.

Nevertheless, it seems that the peritoneal space and/or greater omentum requires more refinements as it does not consistently support islet engraftment and long-term function as compared to the intraportal liver circulation [209]. Despite these results, there are myriad reasons to support endeavors to further optimize and use this anatomical location for ITx. This site is partly favored due to its safety profile, because it is not technically challenging and since it can accommodate large volumes of tissue. The latter is of utmost importance for encapsulated cell therapies, including SC-islets. From a physiological standpoint, this anatomical location theoretically enables physiological insulin secretion due to its communications with the portal circulation [189].

Despite its favorable profile, this site has some caveats. First, the final location of islets transplanted into the peritoneum is unknown. Evidence in non-human primates suggests that these migrate to dependent areas, such as the pelvis [22], which

Table 2 Considerations of selected implantation sites for pancreatic islet transplantation

Factor	Liver intraportal	Spleen	Intraperitoneal/ omental pouch	Gastric submucosa	Kidney subcapsular	Subcutaneous	Intramuscular	Bone marrow
Clinical evidence	++++	+	+++	+	+	+	+	+
Experimental evidence	+++	+	+++	+	++++	+++	++	++
Vascularization	⟵ High → ⟵ Venous blood →		Minimal Passive diffusion → needs neovasc.	Moderate Passive diffusion → needs neovasc.	Moderate Passive diffusion → needs neovasc.	Minimal Passive diffusion → needs neovasc.	Minimal Passive diffusion → needs neovasc.	High Venous blood
Oxygenation (pO_2)[a]	~28–55 mmHg		~18–38 mmHg	~57.6–61.2 mmHg	~42–72 mmHg	~28–48 mmHg	~25–38 mmHg	~49–52 mmHg
Insulin delivery and metabolism	⟵⟵⟵⟵⟵⟵ Intraportal (physiological) ⟶⟶⟶⟶⟶				Systemic	Systemic	Systemic	Systemic
Innate immunity	Higher risk of IBMIR Kupffer cells	Higher risk of IBMIR	Higher risk of IBMIR	Lower risk of IBMIR	Lower risk of IBMIR	Lower risk of IBMIR	Lower risk of IBMIR	Higher risk of IBMIR
Procedural risks	Moderate (interventional radiology)		Moderate (laparoscopic surgery)	Minimal (upper gastrointestinal endoscopy)	High (laparoscopic/ open surgery)	Minimal (surgical dissection)	Minimal (interventional radiology vs surgical dissection)	Minimal (intraosseous infusion, iliac crest)
Potential for encapsulated ITx	+	+	+++	++	+	+++	+++	+
Potential for monitoring/ biopsy	⟵⟵⟵⟵⟵⟵ Minimal ⟶⟶⟶⟶⟶			Moderate	Moderate	High	High	Moderate
Potential for retrieval	+	+	+	+	++	+++	+++	+

Adapted from Barrett et al. [14] and Solass et al. [189]

IP intraperitoneal, *Neovasc* neovascularization, *IBMIR* instant blood-mediated inflammatory response

[a]Different animal models are included

would promote aggregation, impair vascularization and oxygen diffusion, and ultimately trigger/exacerbate hypoxia-mediated cell death. Implantation into defined compartments, such as the *bursa omentalis* could partly address this issue [22, 151] while contributing to facilitating graft recovery (proven at least for encapsulated islets). Moreover, it has been shown to have more favorable immunological conditions to support long-term graft survival [15, 62]. However, these iterations demand a more invasive and technically challenging implantation procedure. Second, the intraperitoneal space is relatively hypoxic as compared to the native pancreas and other implantation sites. The visceral peritoneum receives an absolute blood flow that is ~60 times less than the pancreatic islets, ~40 times less than the kidneys, ~6–8 times less than the liver, similar to the greater omentum and skin, and ~4 times higher than the striated muscle [14, 189]. Third, although recapitulation of physiological insulin secretion due to portal vein drainage is anticipated, this is still controversial, as there is evidence demonstrating more variability of insulin absorption and kinetics as compared to systemic (intravenous) insulin administration [165]. Similarly, there is evidence suggesting higher and earlier peak systemic insulin concentrations acutely and a more transient suppression of hepatic glucose production and lipolysis as compared to subcutaneous insulin administration [93]. A recent systematic review does suggest improved overall glucose control with continuous intraperitoneal insulin infusion compared to continuous subcutaneous insulin infusion (mean difference in HbA1c levels of −0.61 [−0.94 to −0.28]), although some study heterogeneity exists [55]. Of relevance, a clinical report has shown that intraperitoneal administration of insulin compared unfavorably to intraportal ITx in terms of glycemic control [202]. Overall, there are some compelling arguments to continue exploring the intraperitoneal space and greater omentum as a potential home for ITx, but also for SC-islets. These efforts should be accompanied by a comprehensive evaluation of the physiological responses post-transplant and appropriate comparisons with the intraportal site.

6.2 Spleen

It is anticipated that ITx into the spleen or splenic circulation would enable a similar degree of engraftment as the intraportal circulation. Additionally, insulin responses would, in theory, be physiological as this organ also has drainage into the portal vein. Importantly, the splenic circulation was the first implantation site that allowed insulin independence in a human subject [109]. However, while the spleen could be a suitable alternative to the intraportal circulation, it is precisely these similarities that introduce a certain level of futility to further attempting ITx into this site. It is also more technically challenging in its current form, and it has been shown to be associated with a higher incidence of thrombosis and splenic rupture [56, 218]. Indeed, splenic infarction and rupture is a potential complication of this technique, as the human spleen lacks expansive and muscular auto-transfusion responses found in large animals such as the dog. Admittedly, this evidence comes from earlier studies including relatively large tissue volumes, hence it cannot be extrapolated to the present day given the substantial advances in islet isolation protocols. However, it seems that the spleen would offer no metabolic or functional advantages to the standard intraportal liver circulation, with a potentially higher risk profile, providing weak arguments to proceed with further clinical trials using this site.

6.3 Kidney Capsule

The kidney subcapsular space is the most effective site to reverse diabetes in preclinical models especially in mice, even when directly compared to the liver intraportal circulation, the intramuscular and the peritoneal space/omentum [96, 175]. Early clinical reports suggested similar efficacy to intraportal transplantation [90]; however, since protocols have improved disproportionately using the latter site, these early results cannot be extrapolated to the present. Notably, Groth and colleagues carried out renal subcapsular porcine fetal islet xenotransplants in two T1D recipient undergoing living-donor kidney transplantation, and showed proof of concept that pig islets could survive at the kidney subcapsular space as evidenced by positive insulin and glucagon staining in percutaneously obtained graft biopsies. Unfortunately, the degree of immune responses led to rejection despite strong immunosuppression, which ultimately precluded long-term graft survival in these patients [69]. With the success of clinical intraportal ITx, the kidney subcapsular approach has been neglected. Nevertheless, there are several considerations that merit discussion. First, the human kidney capsule is believed to be more rigid compared to mice, and thus less able to accommodate sufficient packed cell volumes. These notions were put forward when islet isolation protocols yielded large impure preparations [188], and recent evidence with non-human primate models shows that the kidney subcapsular space can accommodate clinically relevant islet masses [145]. In our center, the median packed cell volume is 3 mL (interquartile range 2.5–4 mL) (personal communication), which might be adequately fitted into the human kidney subcapsular space, even if it requires two or three different locations within the same kidney. Second, implantation into this site would require more invasive surgical procedures. While this is true, surgical innovations might allow safer implantation techniques. Conversely, this site could allow better access to non-invasive and invasive graft monitoring compared to the standard liver intraportal circulation, as has been shown by Groth et al. [69]. Third, early inflammatory responses, such as IBMIR, could be reduced as direct exposure to blood would be more limited. This could transform ITx by decreasing the number of infusions required to achieve durable clinical success. A concern for hypoxia would be marginally justified given the wealth of data supporting the kidney subcapsular space as the optimal implantation site in preclinical models; however, more information in humans is certainly required. Fourth, subcapsular islet transplantation may protect the islets from high intraportal concentrations of immunosuppressants (compared to systemic venous levels) that could, in theory, impair β-cell function [51, 184]. While

ITx into the kidney subcapsular space might require refinement of current surgical procedures to justify its implementation as an alternative to the liver intraportal site, characterizing the properties that explain its success in preclinical models should be a priority.

6.4 Subcutaneous Space

The subcutaneous (SC) space remains theoretically attractive as an implantation site due to the possibility of minimally invasive implantation procedures (potentially ambulatory), its capacity to accommodate large tissue volumes, its accessibility for graft monitoring and retrieval (if required), and its limited direct contact with blood (thus minimizing IBMIR). However, this latter aspect, its relatively avascular nature, limits nutrient and oxygen delivery, which severely impacts engraftment, survival and function. Preclinical and clinical endeavors are ongoing to optimize this implantation site for ITx, including prevascularization strategies, exogenous oxygen delivery and/or oxygen-releasing biomaterials [66]. Importantly, the advent of potentially limitless and immunoevasive cell sources through SC-islets could open the door to a true cure for T1D. Even if graft attrition ensues due to non-immune causes (e.g., β-cell exhaustion), the limitless possibility of SC-islets and the potential for retrievability with subcutaneous implantation could allow for replenishment without any major inconveniences.

Prevascularization strategies have been attempted in the clinic using subcutaneously implanted polymer chambered medical devices, but with only limited cell survival histologically and with no detectable circulating C-peptide [64, 209]. A more recent clinical report evaluating an iteration of these devices showed some clinical impact in terms of glycemic control, although only when supplementary intraportal islet infusions were also administered, confounding interpretation of the contribution of the subcutaneous islet component [10]. Similarly, a recent report on macroencapsulated transplantation of stem cell-derived pancreatic endocrine progenitors (PEC-Direct) into the SC space shows that, while cells can differentiate and survive at this site, engraftment is minimal and long-term function is marginal [154, 186]. This latter approach, however, did not use any techniques to enable prevascularization. While potentially promising, a major limitation of these macroencapsulation systems is that they still required immunosuppression and were not able to fully prevent hypoxia-mediated cell death. Of relevance, while many studies have reported lower oxygen tension levels at the SC space, there is substantial variability in measuring techniques and conditions (e.g., temperature and probe location) [21, 27]. This highlights a need to conduct studies to precisely assess tissue oxygenation at the SC site. The introduction of novel prevascularization strategies [141] also provides a unique opportunity to evaluate the changing dynamics of tissue oxygenation and should also be pursued in the future. In addition, it is currently unclear whether SC-islets are sensitive to hypoxia in the same manner as primary human islets, which constitutes another area of opportunity for researchers in the field.

Non-physiological insulin kinetics (i.e., systemic insulin delivery) is another obvious concern and a potential roadblock to advance the SC space as a new home for islets. As previously mentioned, the intraperitoneal space constitutes a more physiological delivery route and compares favorably to the SC route in terms of acute insulin responses [55]. Moreover, delayed insulin responses with subcutaneously implanted grafts would only be exacerbated by current encapsulation techniques and biomaterials [29, 103]. This is a problem shared with other sites with systemic insulin delivery, and that even the most advanced exogenous insulin delivery systems have not been able to circumvent. Chronic localized changes, such as fibrosis and scarring at the implantation site (including the SC space) could further impact insulin secretion kinetics. Finally, regarding immunogenicity, there is preliminary evidence that the SC space shows a greater immunoresponsiveness to alginate capsules compared to the intraportal and subcapsular kidney space [6]. Whether this could be related to the immunological milieu per se or to a larger extent of hypoxia-mediated cell death remains to be assessed and represents an important area of opportunity. Overall, the SC space remains a highly desirable implantation site, but whether it can be optimized to enable successful and durable engraftment, as well as physiological insulin responses, remains to be determined.

6.5 Intramuscular Space

The intramuscular (IM) space shares many advantages and disadvantages with the SC space (Table 2). Similarly, there is scarce clinical evidence. While a case report on autologous ITx into the brachioradialis muscle of a 7-year-old patient showed proof of concept for graft survival and improved glycemic control 2 years after transplant [153], a report including allogeneic ITx demonstrated less encouraging results. This small case series of four patients undergoing allogeneic ITx into the brachioradialis muscle demonstrated safety in the procedures, albeit only transient function with no impact on glycemic control or insulin requirements [19]. Particular to the IM site is the fact that it possesses characteristics that support metabolic demands of the parathyroid glands [79], which has prompted some efforts to translate this outcomes to ITx. Some of the pro-angiogenic and pro-survival effect driving the success of IM parathyroid gland implantation seem to be driven by resident $CD34^{+}$ cells, and preclinical evidence shows that co-transplantation of parathyroid gland tissue (including these cells) can support engraftment of mouse and human islets, as well as stem cell-derived insulin-producing cells [216]. In fact, an ongoing clinical trial, the Pancreatic Islets and Parathyroid Gland Co-transplantation for Treatment of Diabetes in the Intra-Muscular Site (PARADIGM, NCT03977662), seeks to build on this preliminary experience. Naturally, results with this trial will be extremely information and naturally would also pave the way to efforts into other implantation sites, such as the SC space.

6.6 Gastric Submucosa

One of the major advantages with the gastric submucosal site is it technical appeal, as it could be readily performed endoscopically. Preclinical studies including rats [50] and pigs [57] provided strong evidence to pursue pilot clinical trials. In this regard, a two-case series of autologous ITx suggested that islets can survive long-term (follow-up for 3 years) at this site, and function sufficiently to facilitate diabetes management (decrease insulin requirements) or maintain insulin independence [220]. Unfortunately, our recent experience suggest that this is not the case for allogeneic ITx [209]. It appears that, while technically appealing, the gastric submucosal site still requires additional strategies to improve engraftment and long-term function.

6.7 Bone Marrow

The intraosseous implantation of islets represents an attractive alternative to the intraportal circulation. Preclinical evidence showed promising results in models including mice [34] and Rhesus monkeys [214], even when compared to the intraportal circulation. These motivated a clinical trial in autologous ITx from the Milan group, where four patients with contraindications for intraportal islet delivery had their islets implanted into the iliac crest. Here, Maffi et al. report overall safety with the procedure (including no hematological or osseous abnormalities), but also successful engraftment and adequate insulin secretory responses, as well as evidence of graft survival for at least 12-months post/transplant [118]. With this, the same group carried a pilot randomized clinical trial comparing the liver intraportal (n = 6) vs intraosseous (n = 3) site for ITx under full immunosuppression with basiliximab plus anakinra and etanercept, and maintenance with tacrolimus and sirolimus [119]. In this report, the authors again document a safe profile with this technique. However, while initial engraftment was observed in all three patients with intraosseous ITx, graft survival was short-lived and lasted a median of 122 days, with no beneficial impacts in glycemic control throughout follow-up. In contrast, recipients of liver intraportal ITx maintained function over time (median graft survival of 624 days) and experienced positive outcomes in terms of improved glycemic control and decreased insulin requirements [119]. Importantly, $CD8^+$ and $CD4^+$ effector memory T cells, as well as markers of lymphocyte activation were significantly higher in the intraosseous recipients compared to liver intraportal. Similarly, the risk of increasing autoantibody titers was more pronounced in recipients of intraosseous ITx vs liver intraportal, and this increase also developed faster [119]. While this study provides a comprehensive assessment of this technique and its immunological responses, many mechanistic aspects explaining graft loss remain to be determined. Overall, the technical feasibility, safety and reproducibility of intraosseous ITx provide strong arguments to pursue endeavors to further characterize this site in terms of it immunological milieu (and provide more direct comparisons with the liver intraportal site) and its physiological performance, and to continue exploring its role as a potential new home for ITx and other stem cell-based β-cell replacement therapies.

6.8 Novel Extrahepatic Implantation Sites

Several immune privileged sites have been attempted in preclinical models; however, evidence is lacking in humans. Two ongoing pilot clinical trials spearheaded by the Diabetes Research Institute in Miami (NCT02846571) and the University Hospital in Basel, Switzerland (NCT02916680) are evaluating islet transplantation into the anterior chamber of the eye of severely visual impaired recipients. The Miami group plans to enroll people with T1D with or without a prior kidney transplant, while the Basel group plans to enroll people with T1D that are already on immunosuppression (e.g., post-kidney transplant). While it is expected that lower islet masses could be transplanted into this site, the vascularized and immunoprivileged nature of this site has been shown to enable diabetes reversal at lower islet masses, as compared to other sites [129, 191], and it could ultimately derive in improved glycemic control or abrogation of severe hypoglycemia, albeit insulin independence might be substantially harder to achieve. Moreover, this site has proven to allow in vivo monitoring of vascularization and islet macro architecture, which certainly represents an intriguing and potentially practice-changing concept. Nevertheless, more studies are required, particularly those focusing on characterizing its immunoprivileged nature, given that there is evidence that alloimmune responses and autoimmune recurrence can occur at this site [1, 129].

Other immune privileged sites include the thymus, the brain, and the testis. These anatomical locations have mainly been explored in preclinical trials with intriguing results. The thymus role in lymphocyte "negative selection" made it an attractive candidate for extrahepatic ITx and has been studied since the 1990s [35]. While successful in rodent, canine and porcine models, its mechanisms for immunoprivilege remain unknown and its potential for clinical translation unexplored. An early case report by Arias-Diaz et al. attempted combined intraportal and intrathymic ITx and showed overall safety and good outcomes up to 14 months post-transplant [5]. However, results obtained in this case might have been confounded by the large transplanted islet mass (1.9 million IEQ from seven donors) and the fact that 94% of that mass was transplanted into the portal vein. Altogether, the immunological contribution of intrathymic ITx in this report was not assessed and remains largely unknown [5]. The brain has been used as a site for ITx in various animal models. Insulin secreted from the transplanted islet can cross the blood-barrier and research endeavors have reported diabetes reversal without immunosuppression [136]. An additional consideration for this site pertains to the role of insulin resistance on neurodegenerative disorders such as Alzheimer's and Parkinson's disease. Operative risks might be prohibitive with this approach, however. Finally, the testes are considered immunoprivileged, mainly thanks to the presence of Sertoli cells. Beyond implantation of islets at this site, Sertoli cells have been co-transplanted with the purpose of inducing immunological tolerance, which has been achieved in some studies [195]. Overall, current research with ITx into immunoprivileged sites retains value from a proof-of-concept standpoint, albeit its clinical relevance and potential for translation remains limited.

7 Special Considerations

Renewable in vitro produced and/or expanded pluripotent stem cells (PSCs) offer a potentially unlimited source for the mass production of SC-islets. A surge of bioengineering and cellular approaches have been explored to protect primary human or pig islets, as well as SC-islets, from the immune system and prevent, or delay, autoimmune recurrence [88]. A summary of considerations around these technologies, but also regarding these novel tissue sources for transplantation, is presented in this section.

7.1 Stem Cell Replacement Therapies

Following a six-stage differentiation protocol that mimics embryonic pancreatic development, pluripotent stem cells (PSCs) can be differentiated in vitro into pancreatic progenitors, which can be further differentiated in vitro and in vivo into SC-islets with functional insulin, glucagon and somatostatin secreting cells [105, 156, 186]. PSCs are first differentiated into definitive endoderm (stage-1), followed by differentiation into primitive gut tube (stage-2) and posterior foregut (stage-3) before developing as pancreatic progenitors (stage-4). At this stage, cells can be transplanted to further mature in vivo, or can be differentiated in vitro into pancreatic endocrine precursors (stage-5), and lastly into SC-islets (stage-6) [46].

Currently, several clinical trials are evaluating the safety and efficacy of SC-islets. ViaCyte Inc. (acquired by Vertex Pharma, closing pending) has conducted clinical trials to test the safety and tolerability of macroencapsulated embryonic stem cell (ESC)-derived pancreatic progenitor cells (PEC) in subjects with T1D (NCT02939118, NCT03162926 and NCT03163511). One year of data from the first cohorts of T1D patients that received subcutaneous implantation of PEC-Direct cells combined with an immunosuppressive regimen (clinical trial NCT03163511) showed that the devices explanted had histological evidence of engraftment and were positive for insulin expression. Additionally, recipients showed human C-peptide in peripheral blood, which correlated with improvements (albeit marginal) in glycemic control [154, 186]. Recently, CRISPR Therapeutics and ViaCyte have initiated phase 1 clinical trials to evaluate the safety, tolerability, and immune evasion potential of gene-edited PEC-01 cells (NCT05210530, NCT05565248) in people with T1D without the use of any immunosuppression. Similarly, Vertex Pharmaceutical Inc. has initiated clinical trials to test the functionality of implanted SC-islets in a non-retrievable intraportal site with full dose immunosuppression, which resulted in insulin independence [164]. In contrast, Sigilon Therapeutics has developed a gel-based microencapsulation method, which can accommodate 30,000 stem cell-derived β-cells per microcapsule [203].

In vitro generation of SC-islets provides a unique opportunity to deliver adequate islet masses (i.e., ~20,000 IEQ/kg) to ensure the survival of sufficient islet masses despite IBMIR (~30% of healthy controls to achieve insulin independence). Access to an unlimited number of SC-islets would also allow for implantation on more inhospitable environments, like the SC space, with its already mentioned advantages. Furthermore, in vitro differentiation of PSCs enables the generation of insulin-secreting organoids, where endothelial and/or epithelial cells could be incorporated with SC-islets for better engraftment and improved vascularization [217, 223]. Naturally, production of patient-specific SC-islets via the differentiation of autologous induced PSCs could enable immunosuppression-free transplantation, thus preventing all adverse effects related to induction and maintenance immunosuppression. However, strategies to protect the SC-islets from autoimmune recurrence would still likely be required.

7.2 Approaches to Improve Islet Survival and Engraftment

As mentioned in previous sections, suboptimal engraftment and cell death remain major roadblocks to achieve long-term graft survival following ITx. Several strategies to protect islets have been explored, including cellular encapsulation, immunoregulatory cell therapies, and stem cell engineering. Herein, we highlight some key considerations for future endeavors.

Micro- and macroencapsulation of islets using biocompatible materials seeks to prevent the interaction between the graft and immune effector cells. However, proteins released by encapsulated islets can still activate the innate immune responses and ultimately lead to graft loss [103]. In addition, encapsulation intrinsically impairs the diffusion of nutrients and oxygen due to the added physical barrier, which combined with the fibrotic response and foreign body reaction elicited by most current biomaterials, further limits oxygen availability [212]. The lack of active glucose and insulin transport mechanisms in encapsulation devices additionally contributes to delayed glucose sensing and hormone secretion [76, 103]. Oxygen restrictions also affect the metabolic function of β-cells as well as their survival, resulting in the release of damage-associated molecular patterns followed, which culminates in immune infiltration and cell death. In situ oxygen generation has demonstrated decreased hypoxia and immune infiltration while improving graft outcomes in mice [43, 134, 140, 215], but further research in more stringent large animal models and testing in primary human islets as well as on SC-islets is required.

An alternative strategy consists of immunoregulation through co-transplantation with mesenchymal stromal cells (MSCs) or Tregs. The array of cytokines and chemokines secreted by MSCs fosters a strong localized immunomodulatory and anti-inflammatory milieu [45]. Co-transplantation with islets has been shown to modulate host immune responses to the graft but also improve revascularization, promoting an overall improved outcome in terms of glycemic control in animal models [23, 75, 95, 135, 224]. It is likely that in the clinic, co-transplantation of MSC with islets (or SC-islets) into the liver intraportal circulation would result in the migration of MSCs into the lung microcirculation, which would have undesired repercussions. Thus, these cells would have to be tested in extrahepatic sites, with or without added cell encapsulation technologies to improve safety. Alternatively, the immunomodulatory molecules and trophic factors secreted by MSCs could be exploited in co-culture systems, or even through the use of MSC-derived exosomes, which has been suggested recently [37]. For an extensive review in this topic the reader is referred to Arzouni et al. [7]. Ex vivo expanded autologous Tregs have also been shown to promote graft survival and increase acute rejection-free survival in the context of liver and kidney transplantation [74, 179, 197]. These approaches will likely require the introduction of more specific cellular products, such as those exploring engineered T cell receptors and chimeric antigen receptors, which could simultaneously tackle alloimmune, but also autoimmune responses. Considerations regarding the timing of infusion, as well as the potential crosstalk with current immunosuppressive regimens should also be discussed and evaluated [60].

Ultimately, stem cell engineering has the potential to eliminate immune-mediated graft loss post-transplant. Overexpression of immunomodulatory molecules such as CD274 and/or CTLA-4 on SC-islets could prevent immune recognition and improve long-term graft survival without any need for immunosuppression [171, 223]. Alternatively, removal of islet antigens known to be targeted by autoantibodies (e.g., IA, GAD65, IA-2 and ZnT8) would delay the activation and amplification of memory T cell and B cell responses and ultimately, T1D recurrence. Furthermore, elimination of HLA class I and HLA class II molecules with expression of immunologically inert chimeric HLA proteins could enable the generation of a universal PSC line that would potentially avoid immune recognition. However, graft rejection also arises due to non-HLA mediated allograft immune responses, which might introduce an extra layer of complexity [143, 225]. In addition, stem cell engineering for the overexpression of genes promoting angiogenesis, including VEGF, could potentially improve the revascularization and engraftment of SC-islets [28], but further research is required. For these reasons, clinical outcomes of CRISPRCas9 gene-edited PEC-01 cells (VCTX210A, NCT05210530) without immunosuppression will provide valuable information on the impact of these gene-editing approaches and pave the way for future research efforts in this area.

8 Conclusions and Final Recommendations

The era of stem cell-based regenerative cell therapies in diabetes has been inaugurated. Myriad proof-of-concept studies in the preclinical arena have now translated into the clinic, with many more endeavors underway. Realizing a cell-based treatment and potential cure for diabetes introduces unique opportunities, but also challenges. It allows tailoring of specific cell products to the patients' needs and context to maximize efficiency, enables engineering cells with certain characteristics that would enhance their performance, but also opens the possibility of exploring novel implantation techniques and anatomical locations that would ensure safety. Increasing knowledge in stem cell biology and ongoing advances in gene editing technologies will certainly address the issue of tailoring cell products and optimizing their performance. However, much more research is needed to characterize the safety, technical feasibility, physiological implications and long-term safety of different implantation sites for ITx and stem cell-based therapies in diabetes. We propose the following considerations to guide future endeavors exploring these aspects (Table 3). It is our belief that this proposed framework could enable a more efficient and stringent approach to ongoing and future research efforts in the field. This

Table 3 Considerations for future research exploring implantation sites for islet transplantation and novel cell-based therapies in diabetes

Aspects	Considerations
Technical	Invasiveness Procedural complexity Logistical considerations (i.e., need for site preparation or prevascularization) Potential for post-transplant monitoring/biopsy Potential for retrieval, if required Potential for retransplantation or replenishment
Physiological	Potential to support early and long-term islet engraftment Potential to provide adequate oxygenation and vascularization Potential for appropriate glucose sensing Route of insulin delivery (portal vs systemic) Potential to support physiological insulin responses
Immunological	Potential to promote/ameliorate IBMIR Potential to provide any degree of immunoprotection Potential to support macro- or microencapsulated cell products Interactions with current immunosuppressive regimens (healing, cicatrization, fibrosis, etc.)
Previous evidence	Biological plausibility Potential for clinical translation Strength and rigor of the evidence (including direct comparisons with the standard intraportal site) The type of animal models that have been tested The potential physiological differences between the animal models and humans

Adapted from Merani et al. [126]
IBMIR instant blood-mediated inflammatory reaction

would ensure safe and effective clinical translation. Altogether, the future looks promising. The proof of concept that cellular therapies can provide substantial benefits for selected people with T1D has already been established. Novel stem cell-based therapies have the potential to advance this into a true cure. Valuable cells are being generated daily all over the world and they are rapidly making their way into the clinic. It is our turn to find them the best home.

Acknowledgements Braulio A. Marfil–Garza was supported by a scholarship from the Alberta Transplant Institute and is currently the Director of Pancreatic Islet Transplantation at TecSalud - Tecnologico de Monterrey. Nerea Cuesta-Gomez is currently supported by the Natural Sciences and Engineering Research Council – Collaborative Research and Training Experience (NSERC-CREATE) program and Juvenile Diabetes Research Foundation Canada. AM James Shapiro is supported through a Canada Research Chair (Tier 1) in Regenerative Medicine and Transplant Surgery, and through grant support from the Juvenile Diabetes Research Foundation, Diabetes Canada, The Canadian Donation and Transplant Research Project, the Diabetes Research Institute Foundation of Canada, the Alberta Diabetes Foundation, and the Canadian Stem Cell Network.

References

1. Abdulreda, M. H., Faleo, G., Molano, R. D., Lopez-Cabezas, M., Molina, J., Tan, Y., Echeverria, O. A., Zahr-Akrawi, E., Rodriguez-Diaz, R., Edlund, P. K., Leibiger, I., Bayer, A. L., Perez, V., Ricordi, C., Caicedo, A., Pileggi, A. & Berggren, P. O. 2011. High-resolution, noninvasive longitudinal live imaging of immune responses. *Proc Natl Acad Sci U S A,* 108, 12863–8.
2. Alwafi, H., Alsharif, A. A., Wei, L., Langan, D., Naser, A. Y., Mongkhon, P., Bell, J. S., Ilomaki, J., Al Metwazi, M. S., Man, K. K. C., Fang, G. & Wong, I. C. K. 2020. Incidence and prevalence of hypoglycaemia in type 1 and type 2 diabetes individuals: A systematic review and meta-analysis. *Diabetes Res Clin Pract,* 170, 108522.
3. Anteby, R., Lucander, A., Bachul, P. J., Pyda, J., Grybowski, D., Basto, L., Generette, G. S., Perea, L., Golab, K., Wang, L. J., Tibudan, M., Thomas, C., Fung, J. & Witkowski, P. 2021. Evaluating the Prognostic Value of Islet Autoantibody Monitoring in Islet Transplant Recipients with Long-Standing Type 1 Diabetes Mellitus. *J Clin Med,* 10.
4. Aragón, F., Karaca, M., Novials, A., Maldonado, R., Maechler, P. & Rubí, B. 2015. Pancreatic polypeptide regulates glucagon release through PPYR1 receptors expressed in mouse and human alpha-cells. *Biochim Biophys Acta,* 1850, 343–51.
5. Arias-Diaz, J., Vara, E., Balibrea, J. L., Garcia, C., Maranes, A., Diaz, A., Lezana, A. H., Gallego, J., Rodriguez, R., Maranes, J. P. & Barrientos, A. 1996. CT-guided fine-needle approach for intrathymic islet transplantation in a diabetic patient. *Pancreas,* 12, 100–2.
6. Arifin, D. R., Valdeig, S., Anders, R. A., Bulte, J. W. & Weiss, C. R. 2016. Magnetoencapsulated human islets xenotransplanted into swine: a comparison of different transplantation sites. *Xenotransplantation,* 23, 211–21.
7. Arzouni, A. A., Vargas-Seymour, A., Nardi, N., King, A. J. F. & Jones, P. M. 2018. Using Mesenchymal Stromal Cells in Islet Transplantation. *Stem Cells Transl Med,* 7, 559–563.
8. Atkinson, M. A. & Eisenbarth, G. S. 2001. Type 1 diabetes: new perspectives on disease pathogenesis and treatment. *Lancet,* 358, 221–9.
9. Atkinson, M. A., Eisenbarth, G. S. & Michels, A. W. 2014. Type 1 diabetes. *Lancet,* 383, 69–82.

10. Bachul, P., Generette, G., Perez-Gutierrez, A., Borek, P., Wang, L., Golab, K., Basto, L., Perea, L., Tibudan, M., Juengel, B., Kumar, J., Thomas, C., Philipson, L., Fung, J. & Witkowski, P. 307.5 Modified approach allowed for improved islet allotransplantation into pre-vascularized Sernova Cell PouchTM device - preliminary results of the phase I/II clinical trial at University of Chicago. IPITA Virtual Congress 2021, 2021 Virtual.
11. Baidal, D., Rickels, M., Ballou, C., Payne, E., Barton, F. & Alejandro, R. 2021. Predictive Value of C-peptide Measures for Clinical Outcomes of Islet Transplantation in Type 1 Diabetes: A Report From the Collaborative Islet Transplant Registry (CITR). *Transplantation,* 105**,** S4.
12. Baidal, D. A., Ricordi, C., Berman, D. M., Alvarez, A., Padilla, N., Ciancio, G., Linetsky, E., Pileggi, A. & Alejandro, R. 2017. Bioengineering of an Intraabdominal Endocrine Pancreas. *N Engl J Med,* 376**,** 1887–1889.
13. Balamurugan, A. N., Naziruddin, B., Lockridge, A., Tiwari, M., Loganathan, G., Takita, M., Matsumoto, S., Papas, K., Trieger, M., Rainis, H., Kin, T., Kay, T. W., Wease, S., Messinger, S., Ricordi, C., Alejandro, R., Markmann, J., Kerr-Conti, J., Rickels, M. R., Liu, C., Zhang, X., Witkowski, P., Posselt, A., Maffi, P., Secchi, A., Berney, T., O'Connell, P. J., Hering, B. J. & Barton, F. B. 2014. Islet product characteristics and factors related to successful human islet transplantation from the Collaborative Islet Transplant Registry (CITR) 1999-2010. *Am J Transplant,* 14**,** 2595–606.
14. Barrett, K. E., Barman, S. M., Boitano, S. & Brooks, H. L. 2018. Editors. *Ganong's Review of Medical Physiology, 25e.* New York, NY: McGraw-Hill Education.
15. Bartholomeus, K., Jacobs-Tulleneers-Thevissen, D., Shouyue, S., Suenens, K., In't Veld, P. A., Pipeleers-Marichal, M., Pipeleers, D. G. & Hellemans, K. 2013. Omentum is better site than kidney capsule for growth, differentiation, and vascularization of immature porcine beta-cell implants in immunodeficient rats. *Transplantation,* 96**,** 1026–33.
16. Barton, F. B., Rickels, M. R., Alejandro, R., Hering, B. J., Wease, S., Naziruddin, B., Oberholzer, J., Odorico, J. S., Garfinkel, M. R., Levy, M., Pattou, F., Berney, T., Secchi, A., Messinger, S., Senior, P. A., Maffi, P., Posselt, A., Stock, P. G., Kaufman, D. B., Luo, X., Kandeel, F., Cagliero, E., Turgeon, N. A., Witkowski, P., Naji, A., O'Connell, P. J., Greenbaum, C., Kudva, Y. C., Brayman, K. L., Aull, M. J., Larsen, C., Kay, T. W., Fernandez, L. A., Vantyghem, M. C., Bellin, M. & Shapiro, A. M. 2012. Improvement in outcomes of clinical islet transplantation: 1999-2010. *Diabetes Care,* 35**,** 1436–45.
17. Bellin, M. D., Barton, F. B., Heitman, A., Harmon, J. V., Kandaswamy, R., Balamurugan, A. N., Sutherland, D. E., Alejandro, R. & Hering, B. J. 2012. Potent induction immunotherapy promotes long-term insulin independence after islet transplantation in type 1 diabetes. *Am J Transplant,* 12**,** 1576–83.
18. Belz, G. T., Behrens, G. M., Smith, C. M., Miller, J. F., Jones, C., Lejon, K., Fathman, C. G., Mueller, S. N., Shortman, K., Carbone, F. R. & Heath, W. R. 2002. The CD8alpha(+) dendritic cell is responsible for inducing peripheral self-tolerance to tissue-associated antigens. *J Exp Med,* 196**,** 1099–104.
19. Bertuzzi, F., Colussi, G., Lauterio, A. & De Carlis, L. 2018. Intramuscular islet allotransplantation in type 1 diabetes mellitus. *Eur Rev Med Pharmacol Sci,* 22**,** 1731–1736.
20. Bhargava, R., Senior, P. A., Ackerman, T. E., Ryan, E. A., Paty, B. W., Lakey, J. R. & Shapiro, A. M. 2004. Prevalence of hepatic steatosis after islet transplantation and its relation to graft function. *Diabetes,* 53**,** 1311–7.
21. Blake, D. F., Young, D. A. & Brown, L. H. 2018. Transcutaneous oximetry: variability in normal values for the upper and lower limb. *Diving Hyperb Med,* 48**,** 2–9.
22. Bochenek, M. A., Veiseh, O., Vegas, A. J., Mcgarrigle, J. J., Qi, M., Marchese, E., Omami, M., Doloff, J. C., Mendoza-Elias, J., Nourmohammadzadeh, M., Khan, A., Yeh, C. C., Xing, Y., Isa, D., Ghani, S., Li, J., Landry, C., Bader, A. R., Olejnik, K., Chen, M., Hollister-Lock, J., Wang, Y., Greiner, D. L., Weir, G. C., Strand, B. L., Rokstad, A. M. A., Lacik, I., Langer, R., Anderson, D. G. & Oberholzer, J. 2018. Alginate encapsulation as long-term immune protection of allogeneic pancreatic islet cells transplanted into the omental bursa of macaques. *Nat Biomed Eng,* 2**,** 810–821.

23. Borg, D. J., Weigelt, M., Wilhelm, C., Gerlach, M., Bickle, M., Speier, S., Bonifacio, E. & Hommel, A. 2014. Mesenchymal stromal cells improve transplanted islet survival and islet function in a syngeneic mouse model. *Diabetologia,* 57, 522–31.
24. Bosi, E., Bottazzo, G. F., Secchi, A., Pozza, G., Shattock, M., Saunders, A., Gelet, A., Touraine, J. L., Traeger, J. & Dubernard, J. M. 1989. Islet cell autoimmunity in type I diabetic patients after HLA-mismatched pancreas transplantation. *Diabetes,* 38 Suppl 1, 82–4.
25. Bosi, E., Braghi, S., Maffi, P., Scirpoli, M., Bertuzzi, F., Pozza, G., Secchi, A. & Bonifacio, E. 2001. Autoantibody response to islet transplantation in type 1 diabetes. *Diabetes,* 50, 2464–71.
26. Boucher, A. A., Wastvedt, S., Hodges, J. S., Beilman, G. J., Kirchner, V. A., Pruett, T. L., Hering, B. J., Schwarzenberg, S. J., Downs, E., Freeman, M., Trikudanathan, G., Chinnakotla, S. & Bellin, M. D. 2021. Portal Vein Thrombosis May Be More Strongly Associated With Islet Infusion Than Extreme Thrombocytosis After Total Pancreatectomy With Islet Autotransplantation. *Transplantation,* 105, 2499–2506.
27. Boyko, E. J., Ahroni, J. H. & Stensel, V. L. 2001. Tissue oxygenation and skin blood flow in the diabetic foot: responses to cutaneous warming. *Foot Ankle Int,* 22, 711–4.
28. Brissova, M., Shostak, A., Shiota, M., Wiebe, P. O., Poffenberger, G., Kantz, J., Chen, Z., Carr, C., Jerome, W. G., Chen, J., Baldwin, H. S., Nicholson, W., Bader, D. M., Jetton, T., Gannon, M. & Powers, A. C. 2006. Pancreatic islet production of vascular endothelial growth factor--a is essential for islet vascularization, revascularization, and function. *Diabetes,* 55, 2974–85.
29. Buchwald, P., Tamayo-Garcia, A., Manzoli, V., Tomei, A. A. & Stabler, C. L. 2018. Glucose-stimulated insulin release: Parallel perifusion studies of free and hydrogel encapsulated human pancreatic islets. *Biotechnol Bioeng,* 115, 232–245.
30. Bulek, A. M., Cole, D. K., Skowera, A., Dolton, G., Gras, S., Madura, F., Fuller, A., Miles, J. J., Gostick, E., Price, D. A., Drijfhout, J. W., Knight, R. R., Huang, G. C., Lissin, N., Molloy, P. E., Wooldridge, L., Jakobsen, B. K., Rossjohn, J., Peakman, M., Rizkallah, P. J. & Sewell, A. K. 2012. Structural basis for the killing of human beta cells by CD8(+) T cells in type 1 diabetes. *Nat Immunol,* 13, 283–9.
31. Calafiore, R. 1992. Transplantation of microencapsulated pancreatic human islets for therapy of diabetes mellitus. A preliminary report. *ASAIO J,* 38, 34–7.
32. Calderon, B., Suri, A., Miller, M. J. & Unanue, E. R. 2008. Dendritic cells in islets of Langerhans constitutively present beta cell-derived peptides bound to their class II MHC molecules. *Proc Natl Acad Sci U S A,* 105, 6121–6.
33. Calderon, B., Suri, A. & Unanue, E. R. 2006. In CD4+ T-cell-induced diabetes, macrophages are the final effector cells that mediate islet beta-cell killing: studies from an acute model. *Am J Pathol,* 169, 2137–47.
34. Cantarelli, E., Melzi, R., Mercalli, A., Sordi, V., Ferrari, G., Lederer, C. W., Mrak, E., Rubinacci, A., Ponzoni, M., Sitia, G., Guidotti, L. G., Bonifacio, E. & Piemonti, L. 2009. Bone marrow as an alternative site for islet transplantation. *Blood,* 114, 4566–74.
35. Cantarelli, E. & Piemonti, L. 2011. Alternative transplantation sites for pancreatic islet grafts. *Curr Diab Rep,* 11, 364–74.
36. Carlessi, R., Chen, Y., Rowlands, J., Cruzat, V. F., Keane, K. N., Egan, L., Mamotte, C., Stokes, R., Gunton, J. E., Bittencourt, P. I. H. & Newsholme, P. 2017. GLP-1 receptor signalling promotes β-cell glucose metabolism via mTOR-dependent HIF-1α activation. *Sci Rep,* 7, 2661.
37. Chen, J., Cheng, Y., Fu, Y., Zhao, H., Tang, M., Lin, N., Shi, X., Lei, Y., Wang, S., Huang, L., Wu, W. & Tan, J. 2020. Mesenchymal stem cell-derived exosomes protect beta cells against hypoxia-induced apoptosis via miR-21 by alleviating ER stress and inhibiting p38 MAPK phosphorylation. *Stem Cell Res Ther,* 11, 97.
38. Choudhary, P., Rickels, M. R., Senior, P. A., Vantyghem, M. C., Maffi, P., Kay, T. W., Keymeulen, B., Inagaki, N., Saudek, F., Lehmann, R. & Hering, B. J. 2015. Evidence-informed clinical practice recommendations for treatment of type 1 diabetes complicated by problematic hypoglycemia. *Diabetes Care,* 38, 1016–29.

39. Christensen, M., Vedtofte, L., Holst, J. J., Vilsbøll, T. & Knop, F. K. 2011. Glucose-dependent insulinotropic polypeptide: a bifunctional glucose-dependent regulator of glucagon and insulin secretion in humans. *Diabetes,* 60, 3103–9.
40. Christiansen, E., Tibell, A., Volund, A., Rasmussen, K., Groth, C. G., Holst, J. J., Pedersen, O., Christensen, N. J. & Madsbad, S. 1996. Pancreatic endocrine function in recipients of segmental and whole pancreas transplantation. *J Clin Endocrinol Metab,* 81, 3972–9.
41. Christoffersson, G., Henriksnas, J., Johansson, L., Rolny, C., Ahlstrom, H., Caballero-Corbalan, J., Segersvard, R., Permert, J., Korsgren, O., Carlsson, P. O. & Phillipson, M. 2010. Clinical and experimental pancreatic islet transplantation to striated muscle: establishment of a vascular system similar to that in native islets. *Diabetes,* 59, 2569–78.
42. CITR 2017. Tenth Annual Report. *In:* Center, C. C. (ed.). Collaborative Islet Transplant Registry.
43. Coronel, M. M., Liang, J. P., Li, Y. & Stabler, C. L. 2019. Oxygen generating biomaterial improves the function and efficacy of beta cells within a macroencapsulation device. *Biomaterials,* 210, 1–11.
44. Cryer, P. E. 2014. Glycemic goals in diabetes: trade-off between glycemic control and iatrogenic hypoglycemia. *Diabetes,* 63, 2188–95.
45. Cuesta-Gomez, N., Graham, G. J. & Campbell, J. D. M. 2021. Chemokines and their receptors: predictors of the therapeutic potential of mesenchymal stromal cells. *J Transl Med,* 19, 156.
46. Cuesta-Gomez, N., Verhoeff, K., Jasra, I. T., Pawlick, R., Dadheech, N. & Shapiro, A. M. J. 2022. Characterization of stem-cell-derived islets during differentiation and after implantation. *Cell Rep,* 40, 111238.
47. Cugnenc, P. H., Bethoux, J. P., Altman, J. J., Bismuth, H., Wind, P., Drevillon, C., Tessier, C., Moulonguet, L. & Chretien, Y. 1990. [Implantation of pancreatic islets in arteriolar epiploic flap. Preliminary note on 3 cases]. *Chirurgie,* 116, 268–74.
48. Da Silva Xavier, G. 2018. The Cells of the Islets of Langerhans. *J Clin Med,* 7.
49. Danke, N. A., Yang, J., Greenbaum, C. & Kwok, W. W. 2005. Comparative study of GAD65-specific CD4+ T cells in healthy and type 1 diabetic subjects. *J Autoimmun,* 25, 303–11.
50. De Mesquita, G. H. A., Jardim, Y. J., Iuamoto, L. R., Suguita, F. Y., Essu, F. F., Oliveira, L. T., Meyer, A., Crescenzi, A., Rocha-Santos, V., Galvao, F. H. F., Andraus, W., Chaib, E. & D'Albuquerque, L. A. C. 2018. Gastric fundus submucosa as a site for islets transplantation: An experimental study. *Ann Med Surg (Lond),* 28, 1–5.
51. Desai, N. M., Goss, J. A., Deng, S., Wolf, B. A., Markmann, E., Palanjian, M., Shock, A. P., Feliciano, S., Brunicardi, F. C., Barker, C. F., Naji, A. & Markmann, J. F. 2003. Elevated portal vein drug levels of sirolimus and tacrolimus in islet transplant recipients: local immunosuppression or islet toxicity? *Transplantation,* 76, 1623–5.
52. Dilorenzo, T. P. 2011. Multiple antigens versus single major antigen in type 1 diabetes: arguing for multiple antigens. *Diabetes Metab Res Rev,* 27, 778–83.
53. Dimeglio, L. A., Evans-Molina, C. & Oram, R. A. 2018. Type 1 diabetes. *Lancet,* 391, 2449–2462.
54. Dionne, K. E., Colton, C. K. & Yarmush, M. L. 1993. Effect of hypoxia on insulin secretion by isolated rat and canine islets of Langerhans. *Diabetes,* 42, 12–21.
55. Dirnena-Fusini, I., Am, M. K., Fougner, A. L., Carlsen, S. M. & Christiansen, S. C. 2021. Physiological effects of intraperitoneal versus subcutaneous insulin infusion in patients with diabetes mellitus type 1: A systematic review and meta-analysis. *PLoS One,* 16, e0249611.
56. Du Toit, D. F., Reece-Smith, H., Mcshane, P., Denton, T. & Morris, P. J. 1982. Effect of cyclosporin A on allotransplanted pancreatic fragments to the spleen of totally pancreatectomized dogs. *Transplantation,* 33, 302–7.
57. Echeverri, G. J., Mcgrath, K., Bottino, R., Hara, H., Dons, E. M., Van Der Windt, D. J., Ekser, B., Casu, A., Houser, S., Ezzelarab, M., Wagner, R., Trucco, M., Lakkis, F. G. & Cooper, D. K. 2009. Endoscopic gastric submucosal transplantation of islets (ENDO-STI): technique and initial results in diabetic pigs. *Am J Transplant,* 9, 2485–96.

58. Eich, T., Eriksson, O., Lundgren, T. & Nordic Network for Clinical Islet, T. 2007. Visualization of early engraftment in clinical islet transplantation by positron-emission tomography. *N Engl J Med,* 356, 2754–5.
59. Eriksson, O., Eich, T., Sundin, A., Tibell, A., Tufveson, G., Andersson, H., Felldin, M., Foss, A., Kyllönen, L., Langstrom, B., Nilsson, B., Korsgren, O. & Lundgren, T. 2009. Positron emission tomography in clinical islet transplantation. *Am J Transplant,* 9, 2816–24.
60. Ferreira, L. M. R., Muller, Y. D., Bluestone, J. A. & Tang, Q. 2019. Next-generation regulatory T cell therapy. *Nat Rev Drug Discov,* 18, 749–769.
61. Frank, A., Deng, S., Huang, X., Velidedeoglu, E., Bae, Y. S., Liu, C., Abt, P., Stephenson, R., Mohiuddin, M., Thambipillai, T., Markmann, E., Palanjian, M., Sellers, M., Naji, A., Barker, C. F. & Markmann, J. F. 2004. Transplantation for type I diabetes: comparison of vascularized whole-organ pancreas with isolated pancreatic islets. *Ann Surg,* 240, 631–40; discussion 640–3.
62. Fukuda, S., Yabe, S. G., Nishida, J., Takeda, F., Nashiro, K. & Okochi, H. 2019. The intraperitoneal space is more favorable than the subcutaneous one for transplanting alginate fiber containing iPS-derived islet-like cells. *Regen Ther,* 11, 65–72.
63. Fung, M., Thompson, D., Shapiro, R. J., Warnock, G. L., Andersen, D. K., Elahi, D. & Meneilly, G. S. 2006. Effect of glucagon-like peptide-1 (7-37) on beta-cell function after islet transplantation in type 1 diabetes. *Diabetes Res Clin Pract,* 74, 189–93.
64. Gala-Lopez, B., Pepper, A. R., Dinyari, P., Malcolm, A., Kin, T., Pawlick, R., Senior, P. A. & Shapiro, A. M. 2016. Subcutaneous clinical islet transplanta on in a prevascularized subcutaneous pouch – preliminary experience. *CellR4,* 4, e2132.
65. Gerber, P. A. & Rutter, G. A. 2017. The Role of Oxidative Stress and Hypoxia in Pancreatic Beta-Cell Dysfunction in Diabetes Mellitus. *Antioxid Redox Signal,* 26, 501–518.
66. Goswami, D., Domingo-Lopez, D. A., Ward, N. A., Millman, J. R., Duffy, G. P., Dolan, E. B. & Roche, E. T. 2021. Design Considerations for Macroencapsulation Devices for Stem Cell Derived Islets for the Treatment of Type 1 Diabetes. *Adv Sci (Weinh),* 8, e2100820.
67. Greenbaum, C. J., Prigeon, R. L. & D'Alessio, D. A. 2002. Impaired beta-cell function, incretin effect, and glucagon suppression in patients with type 1 diabetes who have normal fasting glucose. *Diabetes,* 51, 951–7.
68. Gregory, G. A., Robinson, T. I. G., Linklater, S. E., Wang, F., Colagiuri, S., De Beaufort, C., Donaghue, K. C., International Diabetes Federation Diabetes Atlas Type 1 Diabetes in Adults Special Interest, G., Magliano, D. J., Maniam, J., Orchard, T. J., Rai, P. & Ogle, G. D. 2022. Global incidence, prevalence, and mortality of type 1 diabetes in 2021 with projection to 2040: a modelling study. *Lancet Diabetes Endocrinol,* 10, 741–760.
69. Groth, C. G. 1995. Transplantation of porcine fetal pancreas to diabetic patients. *Lancet,* 345, 735.
70. Gupta, V., Wahoff, D. C., Rooney, D. P., Poitout, V., Sutherland, D. E., Kendall, D. M. & Robertson, R. P. 1997. The defective glucagon response from transplanted intrahepatic pancreatic islets during hypoglycemia is transplantation site-determined. *Diabetes,* 46, 28–33.
71. Gur, C., Porgador, A., Elboim, M., Gazit, R., Mizrahi, S., Stern-Ginossar, N., Achdout, H., Ghadially, H., Dor, Y., Nir, T., Doviner, V., Hershkovitz, O., Mendelson, M., Naparstek, Y. & Mandelboim, O. 2010. The activating receptor NKp46 is essential for the development of type 1 diabetes. *Nat Immunol,* 11, 121–8.
72. Gutierrez, G. D., Gromada, J. & Sussel, L. 2017. Heterogeneity of the Pancreatic Beta Cell. *Front Genet,* 8, 22.
73. Hampton, R. F., Jimenez-Gonzalez, M. & Stanley, S. A. 2022. Unravelling innervation of pancreatic islets. *Diabetologia,* 65, 1069–1084.
74. Harden, P. N., Game, D. S., Sawitzki, B., Van Der Net, J. B., Hester, J., Bushell, A., Issa, F., Brook, M. O., Alzhrani, A., Schlickeiser, S., Scotta, C., Petchey, W., Streitz, M., Blancho, G., Tang, Q., Markmann, J., Lechler, R. I., Roberts, I. S. D., Friend, P. J., Hilton, R., Geissler, E. K., Wood, K. J. & Lombardi, G. 2021. Feasibility, long-term safety, and immune monitoring of regulatory T cell therapy in living donor kidney transplant recipients. *Am J Transplant,* 21, 1603–1611.

75. Hayward, J. A., Ellis, C. E., Seeberger, K., Lee, T., Salama, B., Mulet-Sierra, A., Kuppan, P., Adesida, A. & Korbutt, G. S. 2017. Cotransplantation of Mesenchymal Stem Cells With Neonatal Porcine Islets Improve Graft Function in Diabetic Mice. *Diabetes,* 66, 1312–1321.
76. Hellman, B., Salehi, A., Gylfe, E., Dansk, H. & Grapengiesser, E. 2009. Glucose generates coincident insulin and somatostatin pulses and antisynchronous glucagon pulses from human pancreatic islets. *Endocrinology,* 150, 5334–40.
77. Henriksnas, J., Lau, J., Zang, G., Berggren, P. O., Kohler, M. & Carlsson, P. O. 2012. Markedly decreased blood perfusion of pancreatic islets transplanted intraportally into the liver: disruption of islet integrity necessary for islet revascularization. *Diabetes,* 61, 665–73.
78. Hering, B. J., Ballou, C. M., Bellin, M. D., Payne, E. H., Kandeel, F., Witkowski, P., Alejandro, R., Rickels, M. R. & Barton, F. B. 2022. Factors associated with favourable 5 year outcomes in islet transplant alone recipients with type 1 diabetes complicated by severe hypoglycaemia in the Collaborative Islet Transplant Registry. *Diabetologia.*
79. Hicks, G., George, R. & Sywak, M. 2017. Short and long-term impact of parathyroid autotransplantation on parathyroid function after total thyroidectomy. *Gland Surg,* 6, S75–S85.
80. Hilbrands, R., Huurman, V. A., Gillard, P., Velthuis, J. H., De Waele, M., Mathieu, C., Kaufman, L., Pipeleers-Marichal, M., Ling, Z., Movahedi, B., Jacobs-Tulleneers-Thevissen, D., Monbaliu, D., Ysebaert, D., Gorus, F. K., Roep, B. O., Pipeleers, D. G. & Keymeulen, B. 2009. Differences in baseline lymphocyte counts and autoreactivity are associated with differences in outcome of islet cell transplantation in type 1 diabetic patients. *Diabetes,* 58, 2267–76.
81. Huang, H. H., Harrington, S. & Stehno-Bittel, L. 2018. The Flaws and Future of Islet Volume Measurements. *Cell Transplant,* 27, 1017–1026.
82. Huurman, V. A., Hilbrands, R., Pinkse, G. G., Gillard, P., Duinkerken, G., Van De Linde, P., Van Der Meer-Prins, P. M., Versteeg-Van Der Voort Maarschalk, M. F., Verbeeck, K., Alizadeh, B. Z., Mathieu, C., Gorus, F. K., Roelen, D. L., Claas, F. H., Keymeulen, B., Pipeleers, D. G. & Roep, B. O. 2008. Cellular islet autoimmunity associates with clinical outcome of islet cell transplantation. *PLoS One,* 3, e2435.
83. Insel, R. A., Dunne, J. L., Atkinson, M. A., Chiang, J. L., Dabelea, D., Gottlieb, P. A., Greenbaum, C. J., Herold, K. C., Krischer, J. P., Lernmark, Å., Ratner, R. E., Rewers, M. J., Schatz, D. A., Skyler, J. S., Sosenko, J. M. & Ziegler, A. G. 2015. Staging presymptomatic type 1 diabetes: a scientific statement of JDRF, the Endocrine Society, and the American Diabetes Association. *Diabetes Care,* 38, 1964–74.
84. Ionescu-Tirgoviste, C., Gagniuc, P. A., Gubceac, E., Mardare, L., Popescu, I., Dima, S. & Militaru, M. 2015. A 3D map of the islet routes throughout the healthy human pancreas. *Sci Rep,* 5, 14634.
85. Jackson, S., Mager, D. R., Bhargava, R., Ackerman, T., Imes, S., Hubert, G., Koh, A., Shapiro, A. M. & Senior, P. A. 2013. Long-term follow-up of hepatic ultrasound findings in subjects with magnetic resonance imaging defined hepatic steatosis following clinical islet transplantation: a case-control study. *Islets,* 5, 16–21.
86. Jaeckel, E., Lipes, M. A. & Von Boehmer, H. 2004. Recessive tolerance to preproinsulin 2 reduces but does not abolish type 1 diabetes. *Nat Immunol,* 5, 1028–35.
87. Jansson, L. & Carlsson, P. O. 2019. Pancreatic Blood Flow with Special Emphasis on Blood Perfusion of the Islets of Langerhans. *Compr Physiol,* 9, 799–837.
88. Jeyagaran, A., Lu, C. E., Zbinden, A., Birkenfeld, A. L., Brucker, S. Y. & Layland, S. L. 2022. Type 1 diabetes and engineering enhanced islet transplantation. *Adv Drug Deliv Rev,* 189, 114481.
89. Jiang, K., Weaver, J. D., Li, Y., Chen, X., Liang, J. & Stabler, C. L. 2017. Local release of dexamethasone from macroporous scaffolds accelerates islet transplant engraftment by promotion of anti-inflammatory M2 macrophages. *Biomaterials,* 114, 71–81.
90. Jindal, R. M., Sidner, R. A., Mcdaniel, H. B., Johnson, M. S. & Fineberg, S. E. 1998. Intraportal vs kidney subcapsular site for human pancreatic islet transplantation. *Transplant Proc,* 30, 398–9.

91. Kanak, M. A., Takita, M., Kunnathodi, F., Lawrence, M. C., Levy, M. F. & Naziruddin, B. 2014. Inflammatory response in islet transplantation. *Int J Endocrinol,* 2014, 451035.
92. Kawahara, T., Kin, T., Kashkoush, S., Gala-Lopez, B., Bigam, D. L., Kneteman, N. M., Koh, A., Senior, P. A. & Shapiro, A. M. 2011. Portal vein thrombosis is a potentially preventable complication in clinical islet transplantation. *Am J Transplant,* 11, 2700–7.
93. Kelley, D. E., Henry, R. R. & Edelman, S. V. 1996. Acute effects of intraperitoneal versus subcutaneous insulin delivery on glucose homeostasis in patients with NIDDM. Veterans Affairs Implantable Insulin Pump Study Group. *Diabetes Care,* 19, 1237–42.
94. Kendall, D. M., Teuscher, A. U. & Robertson, R. P. 1997. Defective glucagon secretion during sustained hypoglycemia following successful islet allo- and autotransplantation in humans. *Diabetes,* 46, 23–7.
95. Kerby, A., Jones, E. S., Jones, P. M. & King, A. J. 2013. Co-transplantation of islets with mesenchymal stem cells in microcapsules demonstrates graft outcome can be improved in an isolated-graft model of islet transplantation in mice. *Cytotherapy,* 15, 192–200.
96. Kim, H. I., Yu, J. E., Park, C. G. & Kim, S. J. 2010. Comparison of four pancreatic islet implantation sites. *J Korean Med Sci,* 25, 203–10.
97. Kim, J. M., Shin, J. S., Min, B. H., Kang, S. J., Yoon, I. H., Chung, H., Kim, J., Hwang, E. S., Ha, J. & Park, C. G. 2019. JAK3 inhibitor-based immunosuppression in allogeneic islet transplantation in cynomolgus monkeys. *Islets,* 11, 119–128.
98. Kin, T. 2010. Islet isolation for clinical transplantation. *Adv Exp Med Biol,* 654, 683–710.
99. Kissler, H. J., Niland, J. C., Olack, B., Ricordi, C., Hering, B. J., Naji, A., Kandeel, F., Oberholzer, J., Fernandez, L., Contreras, J., Stiller, T., Sowinski, J. & Kaufman, D. B. 2010. Validation of methodologies for quantifying isolated human islets: an Islet Cell Resources study. *Clin Transplant,* 24, 236–42.
100. Koehler, N., Buhler, L., Egger, B. & Gonelle-Gispert, C. 2022. Multipotent Mesenchymal Stromal Cells Interact and Support Islet of Langerhans Viability and Function. *Front Endocrinol (Lausanne),* 13, 822191.
101. Koh, A., Senior, P., Salam, A., Kin, T., Imes, S., Dinyari, P., Malcolm, A., Toso, C., Nilsson, B., Korsgren, O. & Shapiro, A. M. 2010. Insulin-heparin infusions peritransplant substantially improve single-donor clinical islet transplant success. *Transplantation,* 89, 465–71.
102. Komatsu, H., Kandeel, F. & Mullen, Y. 2018. Impact of Oxygen on Pancreatic Islet Survival. *Pancreas,* 47, 533–543.
103. Korsgren, O. 2017. Islet Encapsulation: Physiological Possibilities and Limitations. *Diabetes,* 66, 1748–1754.
104. Korsgren, O., Nilsson, B., Berne, C., Felldin, M., Foss, A., Kallen, R., Lundgren, T., Salmela, K., Tibell, A. & Tufveson, G. 2005. Current status of clinical islet transplantation. *Transplantation,* 79, 1289–93.
105. Kroon, E., Martinson, L. A., Kadoya, K., Bang, A. G., Kelly, O. G., Eliazer, S., Young, H., Richardson, M., Smart, N. G., Cunningham, J., Agulnick, A. D., D'Amour, K. A., Carpenter, M. K. & Baetge, E. E. 2008. Pancreatic endoderm derived from human embryonic stem cells generates glucose-responsive insulin-secreting cells in vivo. *Nat Biotechnol,* 26, 443–52.
106. Kuppan, P., Kelly, S., Polishevska, K., Hojanepesov, O., Seeberger, K., Korbutt, G. S. & Pepper, A. R. 2020. Co-localized immune protection using dexamethasone-eluting micelles in a murine islet allograft model. *Am J Transplant,* 20, 714–725.
107. Lablanche, S., Borot, S., Wojtusciszyn, A., Skaare, K., Penfornis, A., Malvezzi, P., Badet, L., Thivolet, C., Morelon, E., Buron, F., Renard, E., Tauveron, I., Villard, O., Munch, M., Sommacal, S., Clouaire, L., Jacquet, M., Gonsaud, L., Camillo-Brault, C., Colin, C., Bosson, J. L., Bosco, D., Berney, T., Kessler, L., Benhamou, P. Y. & Network, G. 2021. Ten-year outcomes of islet transplantation in patients with type 1 diabetes: Data from the Swiss-French GRAGIL network. *Am J Transplant,* 21, 3725–3733.
108. Lang, K. S., Georgiev, P., Recher, M., Navarini, A. A., Bergthaler, A., Heikenwalder, M., Harris, N. L., Junt, T., Odermatt, B., Clavien, P. A., Pircher, H., Akira, S., Hengartner, H. & Zinkernagel, R. M. 2006. Immunoprivileged status of the liver is controlled by Toll-like receptor 3 signaling. *J Clin Invest,* 116, 2456–63.

109. Largiader, F., Kolb, E. & Binswanger, U. 1980. A long-term functioning human pancreatic islet allotransplant. *Transplantation,* 29, 76–7.
110. Lau, J., Jansson, L. & Carlsson, P. O. 2006. Islets transplanted intraportally into the liver are stimulated to insulin and glucagon release exclusively through the hepatic artery. *Am J Transplant,* 6, 967–75.
111. Lemos, J. R. N., Baidal, D. A., Poggioli, R., Fuenmayor, V., Chavez, C., Alvarez, A., Linetsky, E., Mauvais-Jarvis, F., Ricordi, C. & Alejandro, R. 2021. Prolonged islet allograft function is associated with female sex in patients after islet transplantation. *J Clin Endocrinol Metab,* 107, e973–e979.
112. Lemos, J. R. N., Baidal, D. A., Ricordi, C., Fuenmayor, V., Alvarez, A. & Alejandro, R. 2021. Survival After Islet Transplantation in Subjects With Type 1 Diabetes: Twenty-Year Follow-Up. *Diabetes Care,* 44, e67–e68.
113. Lois, N., Gardner, E., Mcfarland, M., Armstrong, D., Mcnally, C., Lavery, N. J., Campbell, C., Kirk, R. I., Bajorunas, D., Dunne, A., Cerami, A. & Brines, M. 2020. A Phase 2 Clinical Trial on the Use of Cibinetide for the Treatment of Diabetic Macular Edema. *J Clin Med,* 9.
114. Low, G., Hussein, N., Owen, R. J., Toso, C., Patel, V. H., Bhargava, R. & Shapiro, A. M. 2010. Role of imaging in clinical islet transplantation. *Radiographics,* 30, 353–66.
115. Ludwig, B., Ludwig, S., Steffen, A., Knauf, Y., Zimerman, B., Heinke, S., Lehmann, S., Schubert, U., Schmid, J., Bleyer, M., Schonmann, U., Colton, C. K., Bonifacio, E., Solimena, M., Reichel, A., Schally, A. V., Rotem, A., Barkai, U., Grinberg-Rashi, H., Kaup, F. J., Avni, Y., Jones, P. & Bornstein, S. R. 2017. Favorable outcome of experimental islet xenotransplantation without immunosuppression in a nonhuman primate model of diabetes. *Proc Natl Acad Sci U S A,* 114, 11745–11750.
116. Ludwig, B., Reichel, A., Steffen, A., Zimerman, B., Schally, A. V., Block, N. L., Colton, C. K., Ludwig, S., Kersting, S., Bonifacio, E., Solimena, M., Gendler, Z., Rotem, A., Barkai, U. & Bornstein, S. R. 2013. Transplantation of human islets without immunosuppression. *Proc Natl Acad Sci U S A,* 110, 19054–8.
117. Luzi, L., Hering, B. J., Socci, C., Raptis, G., Battezzati, A., Terruzzi, I., Falqui, L., Brandhorst, H., Brandhorst, D., Regalia, E., Brambilla, E., Secchi, A., Perseghin, G., Maffi, P., Bianchi, E., Mazzaferro, V., Gennari, L., Di Carlo, V., Federlin, K., Pozza, G. & Bretzel, R. G. 1996. Metabolic effects of successful intraportal islet transplantation in insulin-dependent diabetes mellitus. *J Clin Invest,* 97, 2611–8.
118. Maffi, P., Balzano, G., Ponzoni, M., Nano, R., Sordi, V., Melzi, R., Mercalli, A., Scavini, M., Esposito, A., Peccatori, J., Cantarelli, E., Messina, C., Bernardi, M., Del Maschio, A., Staudacher, C., Doglioni, C., Ciceri, F., Secchi, A. & Piemonti, L. 2013. Autologous pancreatic islet transplantation in human bone marrow. *Diabetes,* 62, 3523–31.
119. Maffi, P., Nano, R., Monti, P., Melzi, R., Sordi, V., Mercalli, A., Pellegrini, S., Ponzoni, M., Peccatori, J., Messina, C., Nocco, A., Cardillo, M., Scavini, M., Magistretti, P., Doglioni, C., Ciceri, F., Bloem, S. J., Roep, B. O., Secchi, A. & Piemonti, L. 2019. Islet Allotransplantation in the Bone Marrow of Patients With Type 1 Diabetes: A Pilot Randomized Trial. *Transplantation,* 103, 839–851.
120. Marfil-Garza, B. A., Hefler, J., Bermudez De Leon, M., Pawlick, R., Dadheech, N. & Shapiro, A. M. J. 2021. Progress in Translational Regulatory T Cell Therapies for Type 1 Diabetes and Islet Transplantation. *Endocr Rev,* 42, 198–218.
121. Marfil-Garza, B. A., Polishevska, K., Pepper, A. R. & Korbutt, G. S. 2020. Current State and Evidence of Cellular Encapsulation Strategies in Type 1 Diabetes. *Compr Physiol,* 10, 839–878.
122. Marfil-Garza, B. A., Imes, S., Verhoeff, K., Hefler, J., Lam, A., Dajani, K., Anderson, B., O'Gorman, D., Kin, T., Bigam, D., Senior, P. A., Shapiro, A. M. J. 2022. Pancreatic islet transplantation in type 1 diabetes: 20-year experience from a single-centre cohort in Canada. *Lancet Diabetes Endocrinol,* 10, 519–532.
123. Marfil-Garza, B. A., Shapiro, A. M. J. & Kin, T. 2021. Clinical Islet Transplantation: Current Progress and New Frontiers. *J Hepatobiliary Pancreat Sci.*

124. Martins, L. S. 2014. Autoimmune diabetes recurrence should be routinely monitored after pancreas transplantation. *World J Transplant,* 4, 183–7.
125. Meier, J. J., Hong-Mcatee, I., Galasso, R., Veldhuis, J. D., Moran, A., Hering, B. J. & Butler, P. C. 2006. Intrahepatic transplanted islets in humans secrete insulin in a coordinate pulsatile manner directly into the liver. *Diabetes,* 55, 2324–32.
126. Merani, S., Toso, C., Emamaullee, J. & Shapiro, A. M. 2008. Optimal implantation site for pancreatic islet transplantation. *Br J Surg,* 95, 1449–61.
127. Min, B. H., Shin, J. S., Kim, J. M., Kang, S. J., Kim, H. J., Yoon, I. H., Park, S. K., Choi, J. W., Lee, M. S. & Park, C. G. 2018. Delayed revascularization of islets after transplantation by IL-6 blockade in pig to non-human primate islet xenotransplantation model. *Xenotransplantation,* 25.
128. Mohan, J. F., Levisetti, M. G., Calderon, B., Herzog, J. W., Petzold, S. J. & Unanue, E. R. 2010. Unique autoreactive T cells recognize insulin peptides generated within the islets of Langerhans in autoimmune diabetes. *Nat Immunol,* 11, 350–4.
129. Mojibian, M., Harder, B., Hurlburt, A., Bruin, J. E., Asadi, A. & Kieffer, T. J. 2013. Implanted islets in the anterior chamber of the eye are prone to autoimmune attack in a mouse model of diabetes. *Diabetologia,* 56, 2213–21.
130. Mosharmovahed, B., Fatahi, Y., Mohebbi, B., Ghorbanian, S. A. & Assadiasl, S. 2020. Tocilizumab in transplantation. *Eur J Clin Pharmacol,* 76, 765–773.
131. Müller, T. D., Finan, B., Clemmensen, C., Dimarchi, R. D. & Tschöp, M. H. 2017. The New Biology and Pharmacology of Glucagon. *Physiol Rev,* 97, 721–766.
132. Naziruddin, B., Iwahashi, S., Kanak, M. A., Takita, M., Itoh, T. & Levy, M. F. 2014. Evidence for instant blood-mediated inflammatory reaction in clinical autologous islet transplantation. *Am J Transplant,* 14, 428–37.
133. Noguchi, H., Miyagi-Shiohira, C., Nakashima, Y., Ebi, N., Hamada, E., Tamaki, Y., Kuwae, K., Kobayashi, N., Saitoh, I. & Watanabe, M. 2018. Modified cell-permeable JNK inhibitors efficiently prevents islet apoptosis and improves the outcome of islet transplantation. *Sci Rep,* 8, 11082.
134. Oh, S. H., Ward, C. L., Atala, A., Yoo, J. J. & Harrison, B. S. 2009. Oxygen generating scaffolds for enhancing engineered tissue survival. *Biomaterials,* 30, 757–62.
135. Park, K. S., Kim, Y. S., Kim, J. H., Choi, B., Kim, S. H., Tan, A. H., Lee, M. S., Lee, M. K., Kwon, C. H., Joh, J. W., Kim, S. J. & Kim, K. W. 2010. Trophic molecules derived from human mesenchymal stem cells enhance survival, function, and angiogenesis of isolated islets after transplantation. *Transplantation,* 89, 509–17.
136. Park, S. & Lee, D. 2017. The anterior chamber of the eye as a site for pancreatic islet transplantation. *Journal of Industrial and Engineering Chemistry*, 29–35.
137. Pattou, F., Kerr-Conte, J. & Wild, D. 2010. GLP-1-receptor scanning for imaging of human beta cells transplanted in muscle. *N Engl J Med,* 363, 1289–90.
138. Paty, B. W., Ryan, E. A., Shapiro, A. M., Lakey, J. R. & Robertson, R. P. 2002. Intrahepatic islet transplantation in type 1 diabetic patients does not restore hypoglycemic hormonal counterregulation or symptom recognition after insulin independence. *Diabetes,* 51, 3428–34.
139. Paul, P. K., Das, R., Drow, T. J., De Souza, A. H., Balamurugan, A. N., Belt Davis, D. & Galipeau, J. 2022. Pancreatic Stellate Cells Prolong Ex Vivo Islet Viability and Function and Improve Engraftment. *Stem Cells Transl Med.*
140. Pedraza, E., Coronel, M. M., Fraker, C. A., Ricordi, C. & Stabler, C. L. 2012. Preventing hypoxia-induced cell death in beta cells and islets via hydrolytically activated, oxygen-generating biomaterials. *Proc Natl Acad Sci U S A,* 109, 4245–50.
141. Pepper, A. R., Gala-Lopez, B., Pawlick, R., Merani, S., Kin, T. & Shapiro, A. M. 2015. A prevascularized subcutaneous device-less site for islet and cellular transplantation. *Nat Biotechnol,* 33, 518–23.
142. Pestana, N., Malheiro, J., Silva, F., Silva, A., Ribeiro, C., Pedroso, S., Almeida, M., Dias, L., Henriques, A. C. & Martins, S. 2020. Impact of Pancreatic Autoantibodies in Pancreas Graft Survival After Pancreas-Kidney Transplantation. *Transplant Proc,* 52, 1370–1375.

143. Petersdorf, E. W. 2004. HLA matching in allogeneic stem cell transplantation. *Curr Opin Hematol,* 11, 386–91.
144. Piemonti, L., Everly, M. J., Maffi, P., Scavini, M., Poli, F., Nano, R., Cardillo, M., Melzi, R., Mercalli, A., Sordi, V., Lampasona, V., Espadas De Arias, A., Scalamogna, M., Bosi, E., Bonifacio, E., Secchi, A. & Terasaki, P. I. 2013. Alloantibody and autoantibody monitoring predicts islet transplantation outcome in human type 1 diabetes. *Diabetes,* 62, 1656–64.
145. Pomposelli, T., Schuetz, C., Wang, P. & Yamada, K. 2021. A Strategy to Simultaneously Cure Type 1 Diabetes and Diabetic Nephropathy by Transplant of Composite Islet-Kidney Grafts. *Front Endocrinol (Lausanne),* 12, 632605.
146. Primavera, R., Razavi, M., Kevadiya, B. D., Wang, J., Vykunta, A., Di Mascolo, D., Decuzzi, P. & Thakor, A. S. 2021. Enhancing islet transplantation using a biocompatible collagen-PDMS bioscaffold enriched with dexamethasone-microplates. *Biofabrication,* 13.
147. Prochorov, A. V., Tretjak, S. I., Goranov, V. A., Glinnik, A. A. & Goltsev, M. V. 2008. Treatment of insulin dependent diabetes mellitus with intravascular transplantation of pancreatic islet cells without immunosuppressive therapy. *Adv Med Sci,* 53, 240–4.
148. Pugliese, A. 2017. Autoreactive T cells in type 1 diabetes. *J Clin Invest,* 127, 2881–2891.
149. Pugliese, A., Reijonen, H. K., Nepom, J. & Burke, G. W. 2011. Recurrence of autoimmunity in pancreas transplant patients: research update. *Diabetes Manag (Lond),* 1, 229–238.
150. Pyzdrowski, K. L., Kendall, D. M., Halter, J. B., Nakhleh, R. E., Sutherland, D. E. & Robertson, R. P. 1992. Preserved insulin secretion and insulin independence in recipients of islet autografts. *N Engl J Med,* 327, 220–6.
151. Qi, M., Lacik, I., Kollarikova, G., Strand, B. L., Formo, K., Wang, Y., Marchese, E., Mendoza-Elias, J. E., Kinzer, K. P., Gatti, F., Paushter, D., Patel, S. & Oberholzer, J. 2011. A recommended laparoscopic procedure for implantation of microcapsules in the peritoneal cavity of non-human primates. *J Surg Res,* 168, e117–23.
152. Rafael, E., Ryan, E. A., Paty, B. W., Oberholzer, J., Imes, S., Senior, P., Mcdonald, C., Lakey, J. R. & Shapiro, A. M. 2003. Changes in liver enzymes after clinical islet transplantation. *Transplantation,* 76, 1280–4.
153. Rafael, E., Tibell, A., Ryden, M., Lundgren, T., Savendahl, L., Borgstrom, B., Arnelo, U., Isaksson, B., Nilsson, B., Korsgren, O. & Permert, J. 2008. Intramuscular autotransplantation of pancreatic islets in a 7-year-old child: a 2-year follow-up. *Am J Transplant,* 8, 458–62.
154. Ramzy, A., Thompson, D. M., Ward-Hartstonge, K. A., Ivison, S., Cook, L., Garcia, R. V., Loyal, J., Kim, P. T. W., Warnock, G. L., Levings, M. K. & Kieffer, T. J. 2021. Implanted pluripotent stem-cell-derived pancreatic endoderm cells secrete glucose-responsive C-peptide in patients with type 1 diabetes. *Cell Stem Cell,* 28, 2047–2061 e5.
155. Rehfeld, J. F. 2018. The Origin and Understanding of the Incretin Concept. *Front Endocrinol (Lausanne),* 9, 387.
156. Rezania, A., Bruin, J. E., Arora, P., Rubin, A., Batushansky, I., Asadi, A., O'Dwyer, S., Quiskamp, N., Mojibian, M., Albrecht, T., Yang, Y. H., Johnson, J. D. & Kieffer, T. J. 2014. Reversal of diabetes with insulin-producing cells derived in vitro from human pluripotent stem cells. *Nat Biotechnol,* 32, 1121–33.
157. Rickels, M. R. 2012. Recovery of endocrine function after islet and pancreas transplantation. *Curr Diab Rep,* 12, 587–96.
158. Rickels, M. R., Fuller, C., Dalton-Bakes, C., Markmann, E., Palanjian, M., Cullison, K., Tiao, J., Kapoor, S., Liu, C., Naji, A. & Teff, K. L. 2015. Restoration of Glucose Counterregulation by Islet Transplantation in Long-standing Type 1 Diabetes. *Diabetes,* 64, 1713–8.
159. Rickels, M. R., Liu, C., Shlansky-Goldberg, R. D., Soleimanpour, S. A., Vivek, K., Kamoun, M., Min, Z., Markmann, E., Palangian, M., Dalton-Bakes, C., Fuller, C., Chiou, A. J., Barker, C. F., Luning Prak, E. T. & Naji, A. 2013. Improvement in beta-cell secretory capacity after human islet transplantation according to the CIT07 protocol. *Diabetes,* 62, 2890–7.
160. Rickels, M. R., Mueller, R., Markmann, J. F. & Naji, A. 2009. Effect of glucagon-like peptide-1 on beta- and alpha-cell function in isolated islet and whole pancreas transplant recipients. *J Clin Endocrinol Metab,* 94, 181–9.

161. Rickels, M. R., Mueller, R., Teff, K. L. & Naji, A. 2010. beta-Cell secretory capacity and demand in recipients of islet, pancreas, and kidney transplants. *J Clin Endocrinol Metab,* 95, 1238–46.
162. Rickels, M. R. & Robertson, R. P. 2019. Pancreatic Islet Transplantation in Humans: Recent Progress and Future Directions. *Endocr Rev,* 40, 631–668.
163. Rickels, M. R., Schutta, M. H., Mueller, R., Kapoor, S., Markmann, J. F., Naji, A. & Teff, K. L. 2007. Glycemic thresholds for activation of counterregulatory hormone and symptom responses in islet transplant recipients. *J Clin Endocrinol Metab,* 92, 873–9.
164. Ricordi, C., Naji, A., Rickels, M. R., Bruinsma, B., Marigowda, G., Ross, L., Wang, C., Pagliuca, F. W., Sanna, B., Kean, L. S., Peters, A., Witkowski, P. & Markmann, J. F. 2022. Stem cell-derived, fully differentiated islet cells for type 1 diabetes. *58th European Association for the Study of Diabetes Meeting.* Stockholm, Sweden.
165. Robert, J. J., Chauvet, D., Darmaun, D. & Leblanc, H. 1993. Hepatic glucose production during intraperitoneal and intravenous closed-loop insulin regulation of blood glucose in type 1 (insulin-dependent) diabetic patients. *Diabetologia,* 36, 1185–90.
166. Robertson, R. P. 2007. Estimation of beta-cell mass by metabolic tests: necessary, but how sufficient? *Diabetes,* 56, 2420–4.
167. Robertson, R. P., Bogachus, L. D., Oseid, E., Parazzoli, S., Patti, M. E., Rickels, M. R., Schuetz, C., Dunn, T., Pruett, T., Balamurugan, A. N., Sutherland, D. E., Beilman, G. & Bellin, M. D. 2015. Assessment of beta-cell mass and alpha- and beta-cell survival and function by arginine stimulation in human autologous islet recipients. *Diabetes,* 64, 565–72.
168. Robertson, R. P., Sutherland, D. E. & Lanz, K. J. 1999. Normoglycemia and preserved insulin secretory reserve in diabetic patients 10-18 years after pancreas transplantation. *Diabetes,* 48, 1737–40.
169. Roep, B. O. 2008. Islet autoreactive CD8 T-cells in type 1 diabetes: licensed to kill? *Diabetes,* 57, 1156.
170. Roep, B. O., Stobbe, I., Duinkerken, G., Van Rood, J. J., Lernmark, A., Keymeulen, B., Pipeleers, D., Claas, F. H. & De Vries, R. R. 1999. Auto- and alloimmune reactivity to human islet allografts transplanted into type 1 diabetic patients. *Diabetes,* 48, 484–90.
171. Rong, Z., Wang, M., Hu, Z., Stradner, M., Zhu, S., Kong, H., Yi, H., Goldrath, A., Yang, Y. G., Xu, Y. & Fu, X. 2014. An effective approach to prevent immune rejection of human ESC-derived allografts. *Cell Stem Cell,* 14, 121–30.
172. Rorsman, P. & Ashcroft, F. M. 2018. Pancreatic β-Cell Electrical Activity and Insulin Secretion: Of Mice and Men. *Physiol Rev,* 98, 117–214.
173. Ryan, E. A., Lakey, J. R., Paty, B. W., Imes, S., Korbutt, G. S., Kneteman, N. M., Bigam, D., Rajotte, R. V. & Shapiro, A. M. 2002. Successful islet transplantation: continued insulin reserve provides long-term glycemic control. *Diabetes,* 51, 2148–57.
174. Sahraoui, A., Kloster-Jensen, K., Ueland, T., Korsgren, O., Foss, A. & Scholz, H. 2014. Anakinra and tocilizumab enhance survival and function of human islets during culture: implications for clinical islet transplantation. *Cell Transplant,* 23, 1199–211.
175. Sakata, N., Tan, A., Chan, N., Obenaus, A., Mace, J., Peverini, R., Sowers, L., Chinnock, R. & Hathout, E. 2009. Efficacy comparison between intraportal and subcapsular islet transplants in a murine diabetic model. *Transplant Proc,* 41, 346–9.
176. Salehi, A., Dornonville De La Cour, C., Håkanson, R. & Lundquist, I. 2004. Effects of ghrelin on insulin and glucagon secretion: a study of isolated pancreatic islets and intact mice. *Regul Pept,* 118, 143–50.
177. Sanchez-Rangel, E., Deajon-Jackson, J. & Hwang, J. J. 2022. Pathophysiology and management of hypoglycemia in diabetes. *Ann N Y Acad Sci.*
178. Saudek, F., Hladikova, Z., Hagerf, B., Nemetova, L., Girman, P., Kriz, J., Marada, T., Habart, D., Berkova, Z., Leontovyc, I. & Fronek, J. 2022. Transplantation of Pancreatic Islets Into the Omentum Using a Biocompatible Plasma-Thrombin Gel: First Experience at the Institute for Clinical and Experimental Medicine in Prague. *Transplant Proc,* 54, 806–810.

179. Sawitzki, B., Harden, P. N., Reinke, P., Moreau, A., Hutchinson, J. A., Game, D. S., Tang, Q., Guinan, E. C., Battaglia, M., Burlingham, W. J., Roberts, I. S. D., Streitz, M., Josien, R., Boger, C. A., Scotta, C., Markmann, J. F., Hester, J. L., Juerchott, K., Braudeau, C., James, B., Contreras-Ruiz, L., Van Der Net, J. B., Bergler, T., Caldara, R., Petchey, W., Edinger, M., Dupas, N., Kapinsky, M., Mutzbauer, I., Otto, N. M., Ollinger, R., Hernandez-Fuentes, M. P., Issa, F., Ahrens, N., Meyenberg, C., Karitzky, S., Kunzendorf, U., Knechtle, S. J., Grinyo, J., Morris, P. J., Brent, L., Bushell, A., Turka, L. A., Bluestone, J. A., Lechler, R. I., Schlitt, H. J., Cuturi, M. C., Schlickeiser, S., Friend, P. J., Miloud, T., Scheffold, A., Secchi, A., Crisalli, K., Kang, S. M., Hilton, R., Banas, B., Blancho, G., Volk, H. D., Lombardi, G., Wood, K. J. & Geissler, E. K. 2020. Regulatory cell therapy in kidney transplantation (The ONE Study): a harmonised design and analysis of seven non-randomised, single-arm, phase 1/2A trials. *Lancet,* 395, 1627–1639.
180. Saxena, V., Ondr, J. K., Magnusen, A. F., Munn, D. H. & Katz, J. D. 2007. The countervailing actions of myeloid and plasmacytoid dendritic cells control autoimmune diabetes in the nonobese diabetic mouse. *J Immunol,* 179, 5041–53.
181. Seino, S. & Shibasaki, T. 2005. PKA-dependent and PKA-independent pathways for cAMP-regulated exocytosis. *Physiol Rev,* 85, 1303–42.
182. Shapiro, A. M., Lakey, J. R., Ryan, E. A., Korbutt, G. S., Toth, E., Warnock, G. L., Kneteman, N. M. & Rajotte, R. V. 2000. Islet transplantation in seven patients with type 1 diabetes mellitus using a glucocorticoid-free immunosuppressive regimen. *N Engl J Med,* 343, 230–8.
183. Shapiro, A. M. J. 2007. A Historical Perspective on Experimental and Clinical Islet Transplantation. *In:* Shapiro, A. M. J. & Shaw, A. M. (eds.) *Islet Transplantation and Beta-cell Replacement Therapy.* 1st edition ed. New York, NY: Informa Healthcare.
184. Shapiro, A. M. J., Gallant, H. L., Hao, E. G., Lakey, J. R., Mccready, T., Rajotte, R. V., Yatscoff, R. W. & Kneteman, N. M. 2005. The portal immunosuppressive storm: relevance to islet transplantation? *Ther Drug Monit,* 27, 35–7.
185. Shapiro, A. M. J., Lakey, J. R., Rajotte, R. V., Warnock, G. L., Friedlich, M. S., Jewell, L. D. & Kneteman, N. M. 1995. Portal vein thrombosis after transplantation of partially purified pancreatic islets in a combined human liver/islet allograft. *Transplantation,* 59, 1060–3.
186. Shapiro, A. M. J., Thompson, D., Donner, T. W., Bellin, M. D., Hsueh, W., Pettus, J., Wilensky, J., Daniels, M., Wang, R. M., Brandon, E. P., Jaiman, M. S., Kroon, E. J., D'Amour, K. A. & Foyt, H. L. 2021. Insulin expression and C-peptide in type 1 diabetes subjects implanted with stem cell-derived pancreatic endoderm cells in an encapsulation device. *Cell Rep Med,* 2, 100466.
187. Sibley, R. K. & Sutherland, D. E. 1987. Pancreas transplantation. An immunohistologic and histopathologic examination of 100 grafts. *Am J Pathol,* 128, 151–70.
188. Smood, B., Bottino, R., Hara, H. & Cooper, D. K. C. 2019. Is the renal subcapsular space the preferred site for clinical porcine islet xenotransplantation? Review article. *Int J Surg,* 69, 100–107.
189. Solass, W., Horvath, P., Struller, F., Konigsrainer, I., Beckert, S., Konigsrainer, A., Weinreich, F. J. & Schenk, M. 2016. Functional vascular anatomy of the peritoneum in health and disease. *Pleura Peritoneum,* 1, 145–158.
190. Soon-Shiong, P., Heintz, R. E., Merideth, N., Yao, Q. X., Yao, Z., Zheng, T., Murphy, M., Moloney, M. K., Schmehl, M., Harris, M., et al. 1994. Insulin independence in a type 1 diabetic patient after encapsulated islet transplantation. *Lancet,* 343, 950–1.
191. Speier, S., Nyqvist, D., Cabrera, O., Yu, J., Molano, R. D., Pileggi, A., Moede, T., Kohler, M., Wilbertz, J., Leibiger, B., Ricordi, C., Leibiger, I. B., Caicedo, A. & Berggren, P. O. 2008. Noninvasive in vivo imaging of pancreatic islet cell biology. *Nat Med,* 14, 574–8.
192. Stegall, M. D., Lafferty, K. J., Kam, I. & Gill, R. G. 1996. Evidence of recurrent autoimmunity in human allogeneic islet transplantation. *Transplantation,* 61, 1272–4.
193. Sutherland, D. E., Sibley, R., Xu, X. Z., Michael, A., Srikanta, A. M., Taub, F., Najarian, J. & Goetz, F. C. 1984. Twin-to-twin pancreas transplantation: reversal and reenactment of the pathogenesis of type I diabetes. *Trans Assoc Am Physicians,* 97, 80–7.

194. Szempruch, K. R., Banerjee, O., Mccall, R. C. & Desai, C. S. 2019. Use of anti-inflammatory agents in clinical islet cell transplants: A qualitative systematic analysis. *Islets,* 11, 65–75.
195. Takemoto, N., Liu, X., Takii, K., Teramura, Y. & Iwata, H. 2014. Transplantation of co-aggregates of Sertoli cells and islet cells into liver without immunosuppression. *Transplantation,* 97, 287–93.
196. Teuscher, A. U., Kendall, D. M., Smets, Y. F., Leone, J. P., Sutherland, D. E. & Robertson, R. P. 1998. Successful islet autotransplantation in humans: functional insulin secretory reserve as an estimate of surviving islet cell mass. *Diabetes,* 47, 324–30.
197. Todo, S., Yamashita, K., Goto, R., Zaitsu, M., Nagatsu, A., Oura, T., Watanabe, M., Aoyagi, T., Suzuki, T., Shimamura, T., Kamiyama, T., Sato, N., Sugita, J., Hatanaka, K., Bashuda, H., Habu, S., Demetris, A. J. & Okumura, K. 2016. A pilot study of operational tolerance with a regulatory T-cell-based cell therapy in living donor liver transplantation. *Hepatology,* 64, 632–43.
198. Valdes-Gonzalez, R. A., Dorantes, L. M., Garibay, G. N., Bracho-Blanchet, E., Mendez, A. J., Davila-Perez, R., Elliott, R. B., Teran, L. & White, D. J. 2005. Xenotransplantation of porcine neonatal islets of Langerhans and Sertoli cells: a 4-year study. *Eur J Endocrinol,* 153, 419–27.
199. Van Hulle, F., De Groot, K., Hilbrands, R., Van De Velde, U., Suenens, K., Stange, G., De Mesmaeker, I., De Paep, D. L., Ling, Z., Roep, B., Gillard, P., Pipeleers, D., Keymeulen, B. & Jacobs-Tulleneers-Thevissen, D. 2022. Function and composition of pancreatic islet cell implants in omentum of type 1 diabetes patients. *Am J Transplant,* 22, 927–936.
200. Vantyghem, M.-C., Chetboun, M., Gmyr, V., Jannin, A., Espiard, S., Le Mapihan, K., Raverdy, V., Delalleau, N., Machuron, F., Hubert, T., Frimat, M., Van Belle, E., Hazzan, M., Pigny, P., Noel, C., Caiazzo, R., Kerr-Conte, J. & Pattou, F. 2019. Ten-Year Outcome of Islet Alone or Islet After Kidney Transplantation in Type 1 Diabetes: A Prospective Parallel-Arm Cohort Study. *Diabetes Care,* dc190401.
201. Vantyghem, M. C., Chetboun, M., Gmyr, V., Jannin, A., Espiard, S., Le Mapihan, K., Raverdy, V., Delalleau, N., Machuron, F., Hubert, T., Frimat, M., Van Belle, E., Hazzan, M., Pigny, P., Noel, C., Caiazzo, R., Kerr-Conte, J. & Pattou, F. 2019. Ten-Year Outcome of Islet Alone or Islet After Kidney Transplantation in Type 1 Diabetes: A Prospective Parallel-Arm Cohort Study. *Diabetes Care,* 42, 2042–2049.
202. Vantyghem, M. C., Marcelli-Tourvieille, S., Fermon, C., Duhamel, A., Raverdy, V., Arnalsteen, L., Kerr-Conte, J., Noel, C., Fontaine, P. & Pattou, F. 2009. Intraperitoneal insulin infusion versus islet transplantation: comparative study in patients with type 1 diabetes. *Transplantation,* 87, 66–71.
203. Vegas, A. J., Veiseh, O., Gurtler, M., Millman, J. R., Pagliuca, F. W., Bader, A. R., Doloff, J. C., Li, J., Chen, M., Olejnik, K., Tam, H. H., Jhunjhunwala, S., Langan, E., Aresta-Dasilva, S., Gandham, S., Mcgarrigle, J. J., Bochenek, M. A., Hollister-Lock, J., Oberholzer, J., Greiner, D. L., Weir, G. C., Melton, D. A., Langer, R. & Anderson, D. G. 2016. Long-term glycemic control using polymer-encapsulated human stem cell-derived beta cells in immune-competent mice. *Nat Med,* 22, 306–11.
204. Vendrame, F., Hopfner, Y. Y., Diamantopoulos, S., Virdi, S. K., Allende, G., Snowhite, I. V., Reijonen, H. K., Chen, L., Ruiz, P., Ciancio, G., Hutton, J. C., Messinger, S., Burke, G. W. & Pugliese, A. 2016. Risk Factors for Type 1 Diabetes Recurrence in Immunosuppressed Recipients of Simultaneous Pancreas-Kidney Transplants. *Am J Transplant,* 16, 235–45.
205. Vendrame, F., Pileggi, A., Laughlin, E., Allende, G., Martin-Pagola, A., Molano, R. D., Diamantopoulos, S., Standifer, N., Geubtner, K., Falk, B. A., Ichii, H., Takahashi, H., Snowhite, I., Chen, Z., Mendez, A., Chen, L., Sageshima, J., Ruiz, P., Ciancio, G., Ricordi, C., Reijonen, H., Nepom, G. T., Burke, G. W. & Pugliese, A. 2010. Recurrence of type 1 diabetes after simultaneous pancreas-kidney transplantation, despite immunosuppression, is associated with autoantibodies and pathogenic autoreactive CD4 T-cells. *Diabetes,* 59, 947–57.
206. Vénéreau, E., Ceriotti, C. & Bianchi, M. E. 2015. DAMPs from Cell Death to New Life. *Front Immunol,* 6, 422.

207. Venturini, M., Maffi, P., Querques, G., Agostini, G., Piemonti, L., Sironi, S., De Cobelli, F., Fiorina, P., Secchi, A. & Del Maschio, A. 2015. Hepatic steatosis after islet transplantation: Can ultrasound predict the clinical outcome? A longitudinal study in 108 patients. *Pharmacol Res,* 98, 52–9.
208. Verhoeff, K., Marfil-Garza, B. A., Dajani, K., Bigam, D. L., Anderson, B., Kin, T., Lam, A., O'Gorman, D., Senior, P. A. & Shapiro, A. M. J. 2022. C-peptide Targets and Patient-centered Outcomes of Relevance to Cellular Transplantation for Diabetes. *Transplantation.*
209. Verhoeff, K., Marfil-Garza, B. A., Sandha, G., Cooper, D., Dajani, K., Bigam, D. L., Anderson, B., Kin, T., Lam, A., O'Gorman, D., Senior, P. A., Ricordi, C. & Shapiro, A. M. J. 2022. Outcomes Following Extrahepatic and Intraportal Pancreatic Islet Transplantation: A Comparative Cohort Study. *Transplantation,* 106, 2224–2231.
210. Vethakkan, S. R., Walters, J. M., Gooley, J. L., Boston, R. C., Kay, T. W., Goodman, D. J., Jenkins, A. J. & Ward, G. M. 2014. The incretin response after successful islet transplantation. *Transplantation,* 97, e9–11.
211. Villiger, P., Ryan, E. A., Owen, R., O'Kelly, K., Oberholzer, J., Al Saif, F., Kin, T., Wang, H., Larsen, I., Blitz, S. L., Menon, V., Senior, P., Bigam, D. L., Paty, B., Kneteman, N. M., Lakey, J. R. & Shapiro, A. M. 2005. Prevention of bleeding after islet transplantation: lessons learned from a multivariate analysis of 132 cases at a single institution. *Am J Transplant,* 5, 2992–8.
212. Vlahos, A. E. & Sefton, M. V. 2018. Muted fibrosis from protected islets. *Nat Biomed Eng,* 2, 791–792.
213. Walker, J. T., Saunders, D. C., Brissova, M. & Powers, A. C. 2021. The Human Islet: Mini-Organ With Mega-Impact. *Endocr Rev,* 42, 605–657.
214. Wang, C., Du, X., He, S., Yuan, Y., Han, P., Wang, D., Chen, Y., Liu, J., Tian, B., Yang, G., Yi, S., Gao, F., Zhong, Z., Li, H., Cheng, J. & Lu, Y. 2017. A preclinical evaluation of alternative site for islet allotransplantation. *PLoS One,* 12, e0174505.
215. Wang, L. H., Ernst, A. U., Flanders, J. A., Liu, W., Wang, X., Datta, A. K., Epel, B., Kotecha, M., Papas, K. K. & Ma, M. 2021. An inverse-breathing encapsulation system for cell delivery. *Sci Adv,* 7.
216. Ward, C., Faleo, G., Duh, Q., Hebrok, M., Chang, W., Stock, P. & Tang, Q. 2018. Reversal of Diabetes and Preservation of Pancreatic Islet Grafts in the Extrahepatic Space with Novel Parathyroid Gland Co-Transplantation. *Transplantation,* 102, S368–S369.
217. Wassmer, C. H., Lebreton, F., Bellofatto, K., Perez, L., Cottet-Dumoulin, D., Andres, A., Bosco, D., Berney, T., Othenin-Girard, V., Martinez De Tejada, B., Cohen, M., Olgasi, C., Follenzi, A., Berishvili, E. & Consortium, V. 2021. Bio-Engineering of Pre-Vascularized Islet Organoids for the Treatment of Type 1 Diabetes. *Transpl Int,* 35, 10214.
218. White, S. A., London, N. J., Johnson, P. R., Davies, J. E., Pollard, C., Contractor, H. H., Hughes, D. P., Robertson, G. S., Musto, P. P. & Dennison, A. R. 2000. The risks of total pancreatectomy and splenic islet autotransplantation. *Cell Transplant,* 9, 19–24.
219. Witkowski, P., Odorico, J., Pyda, J., Anteby, R., Stratta, R. J., Schrope, B. A., Hardy, M. A., Buse, J., Leventhal, J. R., Cui, W., Hussein, S., Niederhaus, S., Gaglia, J., Desai, C. S., Wijkstrom, M., Kandeel, F., Bachul, P. J., Becker, Y. T., Wang, L. J., Robertson, R. P., Olaitan, O. K., Kozlowski, T., Abrams, P. L., Josephson, M. A., Andreoni, K. A., Harland, R. C., Kandaswamy, R., Posselt, A. M., Szot, G. L., Ricordi, C. & On Behalf of the Islets for US, C. 2021. Arguments against the Requirement of a Biological License Application for Human Pancreatic Islets: The Position Statement of the Islets for US Collaborative Presented during the FDA Advisory Committee Meeting. *J Clin Med,* 10.
220. Wszola, M., Berman, A., Gorski, L., Ostaszewska, A., Serwanska-Swietek, M., Krajewska, M., Lipinska, A., Chmura, A. & Kwiatkowski, A. 2018. Endoscopic Islet Autotransplantation Into Gastric Submucosa-1000-Day Follow-up of Patients. *Transplant Proc,* 50, 2119–2123.
221. Yamane, K., Anazawa, T., Tada, S., Fujimoto, N., Inoguchi, K., Emoto, N., Nagai, K., Masui, T., Okajima, H., Takaori, K., Sumi, S. & Uemoto, S. 2020. Mitomycin C treatment improves pancreatic islet graft longevity in intraportal islet transplantation by suppressing proinflammatory response. *Sci Rep,* 10, 12086.

222. Yao, M., Domogatskaya, A., Agren, N., Watanabe, M., Tokodai, K., Brines, M., Cerami, A., Ericzon, B. G., Kumagai-Braesch, M. & Lundgren, T. 2021. Cibinetide Protects Isolated Human Islets in a Stressful Environment and Improves Engraftment in the Perspective of Intra Portal Islet Transplantation. *Cell Transplant,* 30, 9636897211039739.
223. Yoshihara, E., O'Connor, C., Gasser, E., Wei, Z., Oh, T. G., Tseng, T. W., Wang, D., Cayabyab, F., Dai, Y., Yu, R. T., Liddle, C., Atkins, A. R., Downes, M. & Evans, R. M. 2020. Immune-evasive human islet-like organoids ameliorate diabetes. *Nature,* 586, 606–611.
224. Yoshimatsu, G., Sakata, N., Tsuchiya, H., Minowa, T., Takemura, T., Morita, H., Hata, T., Fukase, M., Aoki, T., Ishida, M., Motoi, F., Naitoh, T., Katayose, Y., Egawa, S. & Unno, M. 2015. The co-transplantation of bone marrow derived mesenchymal stem cells reduced inflammation in intramuscular islet transplantation. *PLoS One,* 10, e0117561.
225. Zhang, Q. & Reed, E. F. 2016. The importance of non-HLA antibodies in transplantation. *Nat Rev Nephrol,* 12, 484–95.

Genetic Safety Switches for Pluripotent Stem Cell-Derived Therapies for Diabetes

Dena E. Cohen and Jon S. Odorico

Abbreviations

CID	Chemical inducer of dimerization
GCV	Ganciclovir
GSIS	Glucose-stimulated insulin secretion
hESC	Human embryonic stem cell
hiPSC	Human-induced pluripotent stem cell
hPSC	Human pluripotent stem cell
HSV-TK	Herpes simplex virus thymidine kinase
iC9	Inducible caspase 9
NTR	Nitroreductase

1 Introduction

Since the advent of human pluripotent stem cell (hPSC) technology, it has been recognized that the transplantation of patients with hPSC-derived products may place recipients at risk for tumor formation. This potential risk extends to hPSC-derived pancreatic endocrine cell populations, where residual replicative cells with tumorigenic potential may persist despite the many weeks of directed

D. E. Cohen
Regenerative Medical Solutions, Inc, Madison, WI, USA

J. S. Odorico (✉)
Regenerative Medical Solutions, Inc, Madison, WI, USA

Division of Transplantation, Department of Surgery, University of Wisconsin School of Medicine and Public Health, Madison, WI, USA
e-mail: jon@surgery.wisc.edu

L. Piemonti et al. (eds.), *Pluripotent Stem Cell Therapy for Diabetes*,
https://doi.org/10.1007/978-3-031-41943-0_17

differentiation that are typically required to produce insulin-expressing cells [1, 2]. Although such cells represent only a very small proportion of the intended therapeutic cell population, the risks imposed by their presence cannot be discounted. The number of pluripotent cells necessary to achieve teratoma formation is quite small—on the order of hundreds or thousands of cells may be sufficient [3, 4]. When considering an estimated therapeutic dose of ~10^9 hPSC-derived pancreatic cells, analogous to the amount used for cadaveric donor islet transplants [5, 6], it becomes clear that even a tiny proportion of residual pluripotent cells or other highly replicative cells may pose a non-negligible risk.

Over the past several decades, the use of genetically encoded safety switches has emerged as an attractive option for enhancing the safety profile of hPSC-derived therapies. The overarching principal of all such strategies is the introduction of a "suicide gene" into the cell population prior to transplantation. The suicide gene renders cells that express it sensitive to a particular small molecule, the application of which induces apoptosis. Because normal human cells do not encode the suicide gene, treatment with the relevant small molecules will leave the recipient patients' own cells unharmed. Fortuitously for the field of regenerative medicine, such systems have been the subject of extensive development and testing in the field of hematology and oncology, where they have been investigated for use in gene-directed enzyme prodrug therapy [7, 8], and as a means to eliminate transplanted T cells in the event of graft versus host disease, or for use in chimeric antigen receptor T cell therapy [9]. Clinical trials of suicide gene-mediated approaches in these contexts have been ongoing for over 25 years, providing a rich knowledge base with respect to their safety and efficacy.

Design of a suicide gene-based approach for enhancing the safety of a transplanted cell products relies upon the thoughtful selection of three main elements (Fig. 1): a suicide gene/drug pair, a promoter to drive the cell type specificity of suicide gene expression, and a gene delivery system to introduce the suicide gene/promoter construct into the desired cell population. This chapter provides an overview of the main options currently available for each of these three key components, and the considerations that might favor one approach over another. Finally, two recent examples of the application of these technologies to hPSC-derived islet-like cells are summarized. For purposes of this discussion, human embryonic stem cells (hESCs) and human-induced pluripotent stem cells (hiPSCs) are generally not distinguished, because suicide gene constructs appear to behave similarly in both cell types.

2 Suicide Genes

Suicide genes encode proteins that are not otherwise expressed in human cells, and that can induce death of the expressing cell upon exposure to a small molecule. Often, suicide genes are derived from viral or bacterial enzyme sequences, conferring upon the expressing cell the ability to convert a non-toxic pro-drug to a toxic

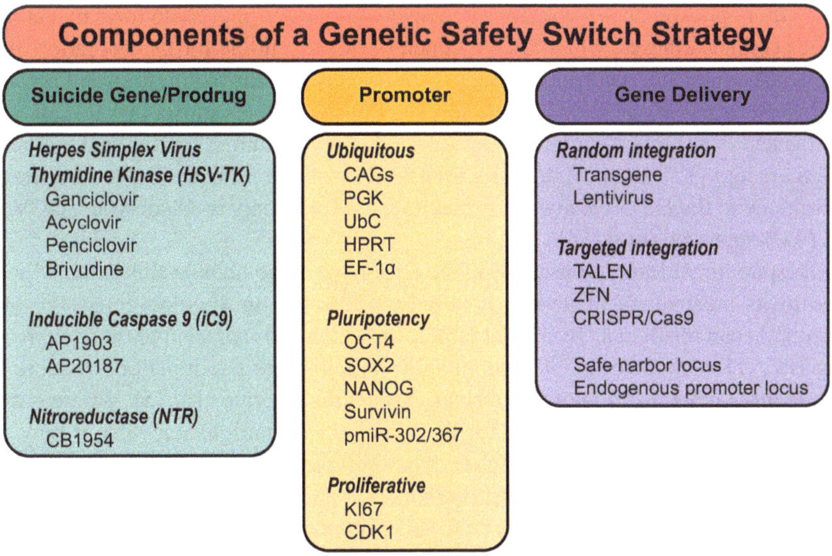

Fig. 1 The three key components of a genetic safety switch strategy are displayed. For any given strategy, a suicide gene/prodrug pair must be selected (left). Next, a promoter must be selected to drive expression of the suicide gene in cells of the desired type (center). Finally, an approach for delivering the suicide gene construct to the genome of hPSC must be chosen (right). In the event that targeting using gene editing tools is selected, a further determination must be made as to whether to target a safe harbor locus or the endogenous locus of the selected gene promoter (lower right)

product. Other suicide genes are derived from human protein sequences that are engineered to endow them with novel functionality. Here, we briefly describe three well-known suicide genes, all of which have been studied in the context of hPSC and their derivatives. Development of suicide gene technology is an area of very active research, and numerous suicide genes have been developed that have yet to be evaluated in hPSC. Likewise, efforts to improve the properties of the suicide genes described here through modification of the protein or small molecule components are ongoing. This section is intended to provide the reader with an understanding of the basic principles that govern selection of a suicide gene for a desired application; principles that will continue to apply even as the technology moves forward.

2.1 HSV-TK

Perhaps the most well-known suicide gene is the herpes simplex virus thymidine kinase (HSV-TK) gene, which catalyzes monophosphorylation of the non-toxic pro-drug ganciclovir (GCV) or other similar nucleoside analogs.

GCV-monophosphate is subsequently converted to the triphosphate form by endogenous cellular kinases, which then competes with thymidine for incorporation during DNA synthesis. Incorporation of GCV into DNA leads to single strand breaks, cell cycle arrest at the G2/M transition, and subsequent apoptosis [10, 11]. It was first tested as a safety switch for a cell-based therapy in human T cell transfer almost 30 years ago [12]. In addition to its long history of use as a suicide gene, a major advantage of the HSV-TK system is that its most commonly used substrate, GCV, is an FDA-approved drug [13].

Despite its widespread use, HSV-TK does have some notable limitations. First, due to its mechanism of action, it is only efficacious in dividing cells [10, 11]. Further, it can result in a "bystander effect," in which cells adjacent to those expressing HSV-TK may take up GCV-monophosphate through gap junctions, leading to the death of cells that do not themselves express the enzyme [14, 15]. Whereas this effect has been capitalized upon for the treatment of malignancies [8], it may be undesirable in the context of hPSC-derived therapies, where ablation of undesired cells may inadvertently lead to the elimination of adjacent therapeutic cells within a graft. The HSV-TK protein is known to be immunogenic in humans, and its immunogenicity does not appear to be dependent upon prior HSV infection [16]. However, this may be less of a concern for situations in which immunosuppression is envisioned as part of the strategy for delivering the hPSC-based therapy.

High-level expression of HSV-TK in hPSC may be toxic even in the absence of GCV treatment [17, 18]. HSV-TK expression appears to result in accumulation of thymidine monophosphate in hPSC, leading to imbalances the nucleotide pool available for DNA synthesis [17]. This may in turn lead to unwanted apoptosis via a mechanism similar to a "thymidine block," where cell cycle progression is deliberately halted via treatment with excess thymidine [19]. The potential cytotoxicity of HSV-TK should, therefore, be considered for any strategy in which the gene will be expressed in hPSC. The cytotoxicity may make it difficult to isolate or expand clones of hPSC harboring an HSV-TK based safety switch, or may drive mutations in or silencing of the HSV-TK gene in surviving cells.

High concentrations of GCV have likewise been shown to cause nonspecific toxicity to hPSC in the absence of HSV-TK expression [20–22]. Other prodrugs such as penciclovir [22] or brivudine [21] appear to be less intrinsically toxic to hPSC, but similarly effective at inducing cell death in an HSV-TK-dependent manner. Although less well-studied than GCV, these alternate prodrugs may offer advantages when using HSV-TK as a suicide gene in hPSC. Together, these limitations highlight the importance of carefully titrating HSV-TK gene expression, thoughtful selection of the most appropriate prodrug, and design of a prodrug dosing regimen to leverage this suicide gene/prodrug pair in hPSC.

2.2 NTR

The enzyme nitroreductase (NTR), derived from *E. coli*, has also been extensively explored as a suicide gene in T cell transfer, although it has not been extensively tested in the context of hPSC, with only one published example to date [23]. This enzyme acts on the prodrug CB1954, creating an alkylating agent capable of cross-linking DNA, thereby inducing apoptosis. Importantly, the apoptosis induced by the NTR/CB1954 system is cell cycle-independent, so it may be used to eliminate undesired non-dividing cells, a potential advantage over the HSV-TK/GCV system [24].

As with HSV-TK, a possible limitation of NTR is its immunogenicity as a bacterial gene product [25]. The toxic metabolites of CB1954 are extremely cell-permeable, resulting in potent bystander effects [26], and the drug, which is not FDA-approved, has been associated with dose-dependent hepatotoxicity in vivo [27]. Next-generation small molecules are currently been developed that might make NTR-based approaches more clinically tractable [28]. Further study of this suicide gene in hPSC and their derivatives is required to more fully understand its utility in this context.

2.3 Inducible Caspase 9

Inducible caspase 9 (iC9) induces cell death using a mechanism that is quite distinct from the other two suicide gene systems described above. When endogenous caspase 9 is activated by signals such as cytochrome c release from damaged mitochondria, it dimerizes and triggers a signaling cascade leading to apoptosis [29]. iC9 is a fusion protein engineered to take advantage of this property. It is comprised of a modified caspases 9 sequence that lacks the endogenous dimerization domain, fused to an engineered version of the FK506 binding protein 12 (*FKBP12*) dimerization domain [30]. Dimerization of iC9 can be triggered by treatment with small molecules such as AP1903 or AP20187, referred to as a chemical inducers of dimerization (CID), leading to rapid apoptosis of cells expressing the fusion protein.

Advantages of this system include the use of a non-immunogenic fusion protein based on human protein sequences [31], the absence of a bystander effect due to the unique mechanism of action of AP1903, the extremely rapid kinetics of elimination of cells expressing iC9 [32, 33], and the demonstrated safety of AP1903 in patients [34]. Due to its unique mechanism of action, iC9 works in both dividing and non-dividing cells. A disadvantage of the iC9 system is that upregulation of apoptotic pathway components such as *BCL2* may provide a novel mechanism whereby hPSC may become resistant to its effects [35]. Some non-specific toxicity of AP20187 on hPSC has also been reported [36].

3 Gene Promoters

A critical decision when developing a suicide gene safety switch is the selection of the promoter used to drive expression of the suicide gene in the desired cell population. The promoter used must be selective and specific, to avoid unintended elimination of desired cells. Further, the promoter must be strong enough to drive robust expression of the suicide gene product, in order to ensure rapid and efficient elimination of the targeted cells, but not so strong as to produce cytotoxic effects.

Once a promoter has been selected, assuming it is derived from a human gene, a further consideration is whether to target the suicide gene to the endogenous locus of the desired promoter, or to instead use a smaller promoter fragment to drive expression from a different genomic integration site. The use of the endogenous promoter via a knock-in approach may offer advantages in terms of specificity, because in addition to the gene promoter, expression of the suicide gene will be governed by all relevant regulatory elements, plus the appropriate chromatin context. Potential disadvantages of the use of the endogenous promoter include the need to use targeted genome editing to place the suicide gene at the desired site, possible disruption of the endogenous gene product, and the risk of lower expression levels of the product of the suicide gene if tools such as the viral 2A self-cleaving peptide or an internal ribosome entry site (IRES) are employed to facilitate expression of the suicide gene product off the same transcript as the endogenous gene product [37].

Another option is to use a smaller promoter fragment derived from the desired gene as a component of the suicide gene construct. An advantage of this approach is that such constructs are amenable to placement either in a safe harbor locus within the genome, or to random integration into the genome via transfection or viral transduction. The endogenous genetic locus is thus not manipulated, which may be important if it encodes a key gene involved in specifying either pluripotency or a particular subset of the therapeutic cell population. On the other hand, use of a smaller promoter fragment may result in reduced specificity of expression pattern due to the absence of distal regulatory elements or the lack of the appropriate chromatin context. Transgenic constructs integrated into the genome randomly may also be more susceptible to silencing or induce unexpected functional mutations. Regardless of whether or not the endogenous locus is used, the expression pattern of the suicide gene must be carefully checked against the expression pattern of the endogenous gene to ensure adequate fidelity.

The majority of published studies regarding the use of suicide gene technology in hPSC focus on the use of one of three broad categories of promoters: ubiquitous promoters, which will result in expression of the suicide gene in all cells of the graft; pluripotency promoters, which restrict expression to undifferentiated hPSC; or proliferative cell promoters, which restrict expression to actively dividing cells.

3.1 Ubiquitous Promoters

A ubiquitous promoter is one that is expressed in hPSC and all of their potential derivatives. Such promoters are typically derived from so-called “housekeeping” genes, which encode essential metabolic enzymes or structural components of mammalian cells. The use of such a promoter in the context of a genetic safety switch implies that all the cells in a population of hPSC-derived cells will be ablated upon drug treatment, not just proliferative cells or specific undesired cell types.

Numerous well-characterized ubiquitous promoters have been tested in the context of genetic safety switches for hPSC, including chicken beta actin, rabbit ß-globin promoter (CAGs), phosphoglycerate kinase (PGK), ubiquitin (UbC), hypoxanthine phosphoribosyltransferase 1 (HPRT1), and elongation factor 1α (EF-1α). These promoters are almost exclusively used in the context of transgenic constructs, instead of the endogenous locus, due to their well-characterized nature and compact size. All of these promoters have been used successfully for the creation of genetic safety switches in hPSC [17, 20–22, 38–47], with no clear rationale for selecting one above the others.

A few studies have attempted to directly compare different ubiquitous promoters for use in driving suicide gene expression. In one side-by-side comparison, the UbC and EF-1α performed similarly in undifferentiated hPSC in vitro and following transplantation of hPSC-derived neural progenitor cells in vivo, with both promoters driving adequate expression of iC9 resulting in equal and near-complete graft ablation following drug administration [39]. In a different study, the EF-1α promoter was directly compared to CAGs, as well as the endogenous PPP1R12C promoter at the AAVS1 safe-harbor locus, which is also a ubiquitous promoter [48]. This study found that the PPP1R12C promoter was too weak to drive sufficient transgene expression for efficient cell elimination, and that the EF-1α promoter underwent silencing in iPSC, consistent with prior reports. Of the three promoters tested, only the CAGs promoter was sufficient to allow for induction of apoptosis in hPSC in vitro and elimination of teratomas in vivo. However, many other groups have reported success with the EF-1α promoter, and thus there may be context- or cell-line-specific effects with respect to transgene expression level and promoter silencing.

3.2 Pluripotency Promoters

Because the most often cited concern with respect to hPSC-derived therapies is teratoma formation, many groups have opted to use promoters that drive suicide gene expression only in pluripotent cells. The promoters of canonical pluripotency genes including *OCT4* [41, 49, 50], *SOX2* [51], and *NANOG* [46, 52] have all been tested in this context. Others have tried promoters for survivin (*BIRC5*, [20]), and the hPSC-restricted microRNA gene pmiR-302/367 [36]. An advantage of such an approach is that use of this type of promoter should leave desired cell types in an hPSC-derived graft intact, while eliminating hPSC with tumorigenic potential. Consistent with this idea, suicide genes driven by pluripotency promoters can efficiently and rapidly induce cell death in cultured hPSC [53].

These findings have been extended in vivo, where multiple groups have demonstrated that treatment of mice with the appropriate small molecules proximal to cell transplantation can preclude the formation of teratomas [20]. However, if a teratoma is allowed sufficient time to form prior to drug administration, the differentiated cells of the tumor, which no longer express pluripotency genes, are resistant to induction of cell death and the tumor may be unaffected by drug treatment [41, 49]. Thus, unlike with a ubiquitous promoter, established teratomas may not be eliminated via drug treatment in the context of a pluripotency promoter.

An important consideration when selecting a pluripotency promoter for use in genetic safety switches is that not all pluripotency gene promoters are in fact restricted to pluripotent cells. For example, *OCT4* is reactivated in some types of adult stem cells [54], and *SOX2* is reactivated in both endodermal [55] and ectodermal [56] derivatives. Thus, when selecting a pluripotency promoter for a genetic safety switch strategy, it is important to bear in mind the extent to which a selected pluripotency promoter may be reactivated in therapeutic cells within a graft. The use of a pluripotency promoter also precludes elimination of undesired but non-pluripotent cells from a graft.

3.3 Proliferative Cell Promoters

Whereas use of a pluripotency promoter restricts expression of a suicide gene to one type of potentially tumorigenic cell, use of a promoter that drives gene expression in all proliferative cells should enable selective ablation of any actively dividing (and therefore, potentially tumorigenic) cell in an hPSC-derived population. The promoters with this property that have been tested thus far are *KI67* [22, 57] and *CDK1* [58, 59]. Interestingly, these promoters have thus far only been investigated for use with the HSV-TK suicide gene, whose activity is restricted to dividing cells via its mechanism of action (see above), thus resulting in a double layer of specificity of the genetic switch for dividing cells. All of the tested constructs appear to behave similarly, with efficient elimination of undifferentiated hPSC in vitro, and the ability to block tumor growth in vivo without ablating non-dividing hPSC-derived cells when drug treatment was applied proximal to cell transplantation [22, 57–59]. Given that many therapeutic cell types, including pancreatic beta cells, are post-mitotic, such an approach may offer an attractive mechanism to reduce tumor risk without the need to completely destroy engrafted cells. This may be advantageous in instances such as islet-like grafts, where the engrafted cell population requires a prolonged incubation in vivo to achieve full functionality.

4 Gene Delivery Approaches

Following the selection of a suicide gene and a promoter to drive its expression, the final step in designing a genetically encoded safety switch is the selection of a method for delivering the construct to the desired hPSC cell line. To date, random integration via lentiviral transduction has been the most widely used approach, likely due to its comparative ease of use, particularly when genome editing techniques for hPSC were still being developed. However, as genome editing techniques have become much more readily accessible to researchers over the past decade, they have become more popular. Targeted insertion of safety switches may ultimately represent the more clinically translatable approach since the genetic modifications present in the resultant cells are deliberate, clonal, and can be precisely characterized.

4.1 Random Integration

In the simplest scenario, safety switches may be allowed to randomly integrate into the genome of hPSC. Thus far, this has most commonly been implemented using lentiviral transduction [17, 18, 20, 22, 36, 38, 39, 41, 44, 45, 57], although transfection of linearized plasmids [40] and transposon-based insertion have also been attempted [21]. Focusing on lentiviral transduction, advantages of this approach include the relative technical simplicity of viral production, the efficiency of the transduction of hPSC, and strong transgene expression even when the viral copy number per cell is low. For example, expression of a suicide gene was found to be more robust following lentiviral delivery than upon targeted integration of the identical construct into the AAVS1 safe harbor locus [41].

Lentiviral constructs are limited to an insert size of approximately 10 kb, which must be kept in mind when designing genetic safety switches to be delivered with this approach. Furthermore, the virus integrates randomly into the human genome, and may do so multiple times within a single cell, leading to insertional mutagenesis with potentially deleterious effects. It is difficult to establish clonal cell populations following delivery of a construct via lentivirus [41], and thus cell populations produced by this approach are genetically heterogeneous in nature; indeed, many groups do not attempt to isolate clonal lines from transduced hPSC. Despite these disadvantages, the rapidity with which lentiviral constructs may be built, delivered, and tested, has made them an attractive platform for testing genetic safety switches [20].

4.2 Targeted Integration

Historically, targeted integration of transgenes in hPSC was limited by the extremely low frequency of homologous recombination in these cells. However, the advent of technologies to introduce precisely targeted single- or double-strand breaks into the genome has dramatically enhanced the ability to introduce modifications at desired genomic positions [60]. Early technologies such as TALEN and zinc finger nucleases (ZFN) have been to a large extent supplanted by the advent of CRISPR-Cas9 [61], but all three of these technologies have been applied in the context of creating genomic safety switches.

The major advantage of using a targeted gene editing strategy to introduce a genetic safety switch is the specificity of genome engineering, allowing for the placement of the switch at a precise position in the genome. This enables the use of safe-harbor loci such as AAVS1 [62], where suicide gene constructs may be integrated without deleterious effects on cell function [42, 43, 48]. It also enables the placement of suicide genes such that their expression is driven by the endogenous promoter of a desired gene [17, 53, 58, 63], enabling faithful recapitulation of the desired expression pattern. Thanks to the popularity of targeted genome editing, tools to target many commonly used loci have already been developed, enabling their rapid adoption for use in the creation of genetic safety switches.

A major disadvantage of targeted genome editing is the slow nature of the process: it can be very time-consuming to select, expand, and characterize targeted hPSC lines [64]. Further, although technological advances continue to improve the specificity of genome editing tools such as CRISPR-Cas9, their use may still introduce off-target mutations into the genome. Computational tools can be used to identify the most likely off-target sites for evaluation by genomic sequencing, although such targeted sequencing cannot rule out the introduction of mutations elsewhere in the genome.

5 Escape

No genetic safety switch can be completely foolproof. There are multiple mechanisms whereby cells harboring such switches may escape their influence, particularly when cell populations are placed under selection pressure by addition of the small molecules that trigger cell death. An early study found that hPSC expressing HSV-TK under a constitutive promoter developed resistance to GCV at a frequency of $10^{-6} - 10^{-7}$ [40]. Mechanisms for resistance include accumulation of mutations in the suicide gene [45], or silencing of the suicide gene construct via DNA methylation [35, 48]. In the case of iC9, up-regulation of the endogenous BCL2 gene has also been observed as a resistance mechanism [35, 53]. Suicide gene constructs may also be removed from the genome via loss of heterozygosity, although this risk can be mitigated through targeted, homozygous insertion of the genetic safety switch to produce cell lines in which the risk of drug resistance is extremely low [58]. For any given genetic safety switch strategy, it will, therefore, be important to perform empirical testing to determine the rate of emergence of resistant clones upon drug selection. The use of multiple switches that rely on distinct small molecules may be another approach to mitigate the risk of cells escaping selection.

6 Use of Suicide Gene Technology in Islet-Like Cell Products

The evidence base for use of suicide genes in the context of the production of pancreatic endocrine cells from hPSC remains sparse, with only two published studies as of this writing. In one of these studies [53], the CRISPR-Cas9 nickase system was used to target a 2A self-cleaving peptide followed by the coding sequence for iC9 to the endogenous *SOX2* locus in H1 hESC. *SOX2* is widely utilized as a pluripotency marker and is abundantly expressed in undifferentiated hESC [65]. The authors confirmed that apoptosis could be efficiently induced in correctly targeted, undifferentiated cells (referred to as H1-iC9) by application of AP1903.

To test the utility and specificity of their suicide gene construct in the context of differentiated cell products, the authors next differentiated H1-iC9 into hematopoietic cells and neuronal cells using standard protocols. As expected, the endogenous SOX2 gene was strongly down-regulated during hematopoietic differentiation, and hematopoietic cells generated from H1-iC9 cells were resistant to AP1903-induced apoptosis. The picture during neuronal differentiation is more complicated, because SOX2 is a known marker of neural progenitor cells (NPC; [56]). When H1-iC9 cells were differentiated toward the neuronal lineage, a mixture of SOX2+ NPC and SOX2−/TuJ1+ neuron-like cells was produced. As anticipated, application of AP1903 selectively ablated the SOX2+ NPC, whereas the SOX2−/TuJ1+ neuron-like cells were spared.

Finally, the authors differentiated H1-iC9 cells toward the pancreatic lineage, using a previously published protocol [66]. Differentiation of H1-iC9 produced the expected populations of PDX1+/NKX6–1+ progenitors at Stage 4, and INS+/NKX6–1+ beta-like cells at Stage 7. Both of these cell types were resistant to AP1903-mediated apoptosis. Importantly, SOX2+/FOXA2+ cells were also observed at Stage 7, consistent with the role of SOX2 in the formation of anterior foregut endoderm [55]. Surprisingly, this sub-population of cells was resistant to AP1903-mediated apoptosis despite detection of the iC9 protein by immunofluorescence. The mechanism for this resistance remains unclear. However, mixing studies in which pluripotent H1-iC9 cells were spiked into Stage 7 populations confirmed the ability of AP1903 to efficiently eliminate the pluripotent cells, indicating that the drug resistance observed in the SOX2+/FOXA2+ population was not simply an effect of the culture conditions.

In a second example of the use of suicide gene technology in pancreatic cell populations, investigators sought to establish a genetically modified cell line that would enable ablation of residual pluripotent cells and the elimination of any insulin negative cells remaining at the conclusion of a pancreatic differentiation protocol [23]. To accomplish this goal, two constructs were electroporated into H1 hESC. The first construct consisted of a loxP-flanked transgene comprising a constitutive CMV promoter driving expression of the NTR suicide gene. The second construct contained a transgene consisting of the HSV-TK suicide gene driven by the human telomerase reverse transcriptase (hTERT) promoter, followed by the coding sequence for Cre recombinase driven by the human insulin promoter. Therefore, pluripotent cells, which express hTERT but not insulin, should be sensitive to both GCV and CB1954. By contrast, in insulin-expressing cells, the hTERT promoter should be silent, while the human insulin promoter is active. Thus, the HSV-TK gene should be silent, while the NTR gene will be excised by Cre recombinase, rendering insulin-expressing cells resistant to both drugs.

Transgenic hESC (referred to as λH1) were confirmed to be sensitive to both GCV and CB1954, both in vitro and following transplantation in vivo. The cells were then subjected to a beta cell differentiation protocol, using the parental H1 line as a comparator. Both cell lines differentiated with equal efficiency, producing approximately 20–30% PDX1+/NKX6−1+/INS+ cells at the end of the protocol. Following completion of the differentiations, a subset of cultures were treated with GCV and CB1954 for a further 7 days, whereas control cultures received no drug treatment. In differentiated λH1, the proportion of PDX1+/NKX6−1+/INS+ cells increased to nearly 80% following drug treatment, whereas no change was observed in either untreated cells or drug-treated differentiated H1 cells. PDX1−/NKX6−1− cells were efficiently eliminated from the population of differentiated transgenic cells following drug treatment. In contrast, PDX1+/NKX6.1+/INS− cells were resistant to GCV/CB1954 treatment. Based on mRNA expression data, it appears possible that insulin promoter activity precedes expression of the insulin protein in pancreatic progenitors, resulting in premature excision of the NTR transgene.

Both in vitro and in vivo assays were used to assess the functionality of beta-like cells derived from λH1. As might be expected based on the enrichment for beta-like cells following drug treatment of differentiated λH1, the selected populations displayed a higher insulin content per ug DNA compared to similarly treated H1 controls. Static and dynamic glucose-stimulated insulin secretion (GSIS) assays revealed enhanced glucose responsiveness in drug-selected cells. In vivo functionality of the cells was tested using three experimental conditions. Immunocompromised mice transplanted with either fully differentiated H1 cells, fully differentiated λH1 cells pre-treated with CB1954 and GCV in vitro, and fully differentiated transgenic cells in which recipient mice were treated with CB1954 and GCV in vivo beginning 3 weeks after transplantation. All 3 groups evinced equivalent glucose-responsive secretion of human c-peptide at 6 and 8 weeks after surgery. However, necropsy revealed OCT4+ cells and teratoma formation in mice transplanted with fully differentiated H1 cells, whereas these features were absent from grafts of the transgenic cells, whether drug treatment occurred in vitro or in vivo.

These two studies demonstrate the feasibility of using suicide gene technology to enhance the purity and safety of stem cell-derived pancreatic cell populations. None of the three drugs tested appeared to have nonspecific beta cell toxicity, which is encouraging for future applications of these technologies. Together, the studies underscore the critical importance of promoter selection when using suicide gene approaches. The first study demonstrated re-activation of the canonical pluripotency gene *SOX2* in later stages of differentiation in the pancreatic and neuronal lineages, whereas the second study revealed activation of the insulin gene promoter days or weeks ahead of the production of detectable insulin protein. Future investigations of the use of suicide genes in pancreatic cell populations will need to explore the use of gene promoters that are tailored to the desired use case, be that elimination of pluripotent or replicative cells, or the selection of desired cell types at the conclusion of a differentiation protocol. In order to maximize efficacy and specificity, each promoter/suicide gene construct will need to be carefully tested in the cell line and differentiation protocol for which its ultimate clinical application is intended. All pancreatic cell products harboring suicide gene constructs will also need to undergo more thorough in vivo testing to confirm both beta-like cell functionality and suicide gene utility in diabetic animal models.

7 Conclusions

Over the past two decades, extensive research has led to exciting progress in the use of suicide genes as safety switches for future hPSC-derived therapies. An ever-expanding tool kit of gene delivery technologies, genome editing strategies, suicide genes, and promoters with a range of specificities has set the stage for the development of strategies to enrich desired therapeutic cell populations while reducing the risk for tumorigenicity following transplantation. Although testing of these technologies in the context of pancreatic endocrine cells is still in its early stages, encouraging preliminary findings support a bright future for the use of these approaches to enhance the safety profile of hPSC-based therapies for diabetes.

References

1. Nair GG, Liu JS, Russ HA, Tran S, Saxton MS, Chen R, et al. Recapitulating endocrine cell clustering in culture promotes maturation of human stem-cell-derived β cells. Nat Cell Biol. 2019;21(2):263–74.
2. Veres A, Faust AL, Bushnell HL, Engquist EN, Kenty JH, Harb G, et al. Charting cellular identity during human in vitro β-cell differentiation. Nature. 2019;569(7756):368–73.
3. Hentze H, Soong PL, Wang ST, Phillips BW, Putti TC, Dunn NR. Teratoma formation by human embryonic stem cells: evaluation of essential parameters for future safety studies. Stem Cell Res. 2009;2(3):198–210.
4. Lee AS, Tang C, Cao F, Xie X, van der Bogt K, Hwang A, et al. Effects of cell number on teratoma formation by human embryonic stem cells. Cell Cycle. 2009;8(16):2608–12.
5. Markmann JF, Deng S, Huang X, Desai NM, Velidedeoglu EH, Lui C, et al. Insulin independence following isolated islet transplantation and single islet infusions. Ann Surg. 2003;237(6):741–9; discussion 9-50.
6. Ryan EA, Lakey JR, Paty BW, Imes S, Korbutt GS, Kneteman NM, et al. Successful islet transplantation: continued insulin reserve provides long-term glycemic control. Diabetes. 2002;51(7):2148–57.
7. Malekshah OM, Chen X, Nomani A, Sarkar S, Hatefi A. Enzyme/Prodrug Systems for Cancer Gene Therapy. Curr Pharmacol Rep. 2016;2(6):299–308.
8. Zhang J, Kale V, Chen M. Gene-directed enzyme prodrug therapy. AAPS J. 2015;17(1):102–10.
9. Brandt LJB, Barnkob MB, Michaels YS, Heiselberg J, Barington T. Emerging Approaches for Regulation and Control of CAR T Cells: A Mini Review. Front Immunol. 2020;11:326.

10. Halloran PJ, Fenton RG. Irreversible G2-M arrest and cytoskeletal reorganization induced by cytotoxic nucleoside analogues. Cancer Res. 1998;58(17):3855–65.
11. Rubsam LZ, Davidson BL, Shewach DS. Superior cytotoxicity with ganciclovir compared with acyclovir and 1-beta-D-arabinofuranosylthymine in herpes simplex virus-thymidine kinase-expressing cells: a novel paradigm for cell killing. Cancer Res. 1998;58(17):3873–82.
12. Bonini C, Ferrari G, Verzeletti S, Servida P, Zappone E, Ruggieri L, et al. HSV-TK gene transfer into donor lymphocytes for control of allogeneic graft-versus-leukemia. Science. 1997;276(5319):1719–24.
13. Baker R. FDA approves oral ganciclovir as first drug to prevent CMV disease. Food and Drug Administration. BETA. 1995:8.
14. Bi WL, Parysek LM, Warnick R, Stambrook PJ. In vitro evidence that metabolic cooperation is responsible for the bystander effect observed with HSV tk retroviral gene therapy. Hum Gene Ther. 1993;4(6):725–31.
15. Fick J, Barker FG, Dazin P, Westphale EM, Beyer EC, Israel MA. The extent of heterocellular communication mediated by gap junctions is predictive of bystander tumor cytotoxicity in vitro. Proc Natl Acad Sci U S A. 1995;92(24):11071–5.
16. Traversari C, Marktel S, Magnani Z, Mangia P, Russo V, Ciceri F, et al. The potential immunogenicity of the TK suicide gene does not prevent full clinical benefit associated with the use of TK-transduced donor lymphocytes in HSCT for hematologic malignancies. Blood. 2007;109(11):4708–15.
17. Iwasawa C, Tamura R, Sugiura Y, Suzuki S, Kuzumaki N, Narita M, et al. Increased Cytotoxicity of Herpes Simplex Virus Thymidine Kinase Expression in Human Induced Pluripotent Stem Cells. Int J Mol Sci. 2019;20(4).
18. Kojima K, Miyoshi H, Nagoshi N, Kohyama J, Itakura G, Kawabata S, et al. Selective Ablation of Tumorigenic Cells Following Human Induced Pluripotent Stem Cell-Derived Neural Stem/Progenitor Cell Transplantation in Spinal Cord Injury. Stem Cells Transl Med. 2019;8(3):260–70.
19. Bjursell G, Reichard P. Effects of thymidine on deoxyribonucleoside triphosphate pools and deoxyribonucleic acid synthesis in Chinese hamster ovary cells. J Biol Chem. 1973;248(11):3904–9.
20. Ide K, Mitsui K, Irie R, Matsushita Y, Ijichi N, Toyodome S, et al. A Novel Construction of Lentiviral Vectors for Eliminating Tumorigenic Cells from Pluripotent Stem Cells. Stem Cells. 2018;36(2):230–9.
21. Lou Z, Post A, Rodgers CE, Chamankhah M, Hong J, Ahuja CS, et al. Neural Progenitor Cells Expressing Herpes Simplex Virus-Thymidine Kinase for Ablation Have Differential Chemosensitivity to Brivudine and Ganciclovir. Front Cell Neurosci. 2021;15:638021.
22. Locatelli M, Delhaes F, Cherpin O, Black ME, Carnesecchi S, Preynat-Seauve O, et al. Optimization of Thymidine Kinase-Based Safety Switch for Neural Cell Therapy. Cells. 2022;11(3).
23. Qadir MMF, Álvarez-Cubela S, Belle K, Sapir T, Messaggio F, Johnson KB, et al. A Double Fail-Safe Approach to Prevent Tumorigenesis and Select Pancreatic β Cells from Human Embryonic Stem Cells. Stem Cell Reports. 2019;12(3):611–23.
24. Bridgewater JA, Springer CJ, Knox RJ, Minton NP, Michael NP, Collins MK. Expression of the bacterial nitroreductase enzyme in mammalian cells renders them selectively sensitive to killing by the prodrug CB1954. Eur J Cancer. 1995;31A(13–14):2362–70.
25. Patel P, Young JG, Mautner V, Ashdown D, Bonney S, Pineda RG, et al. A phase I/II clinical trial in localized prostate cancer of an adenovirus expressing nitroreductase with CB1954 [correction of CB1984]. Mol Ther. 2009;17(7):1292–9.
26. Bridgewater JA, Knox RJ, Pitts JD, Collins MK, Springer CJ. The bystander effect of the nitroreductase/CB1954 enzyme/prodrug system is due to a cell-permeable metabolite. Hum Gene Ther. 1997;8(6):709–17.
27. Chung-Faye G, Palmer D, Anderson D, Clark J, Downes M, Baddeley J, et al. Virus-directed, enzyme prodrug therapy with nitroimidazole reductase: a phase I and pharmacokinetic study of its prodrug, CB1954. Clin Cancer Res. 2001;7(9):2662–8.

28. Sharrock AV, McManaway SP, Rich MH, Mumm JS, Hermans IF, Tercel M, et al. Engineering the. Front Pharmacol. 2021;12:701456.
29. Skulachev VP. Cytochrome c in the apoptotic and antioxidant cascades. FEBS Lett. 1998;423(3):275–80.
30. Straathof KC, Pulè MA, Yotnda P, Dotti G, Vanin EF, Brenner MK, et al. An inducible caspase 9 safety switch for T-cell therapy. Blood. 2005;105(11):4247–54.
31. Arber C, Abhyankar H, Heslop HE, Brenner MK, Liu H, Dotti G, et al. The immunogenicity of virus-derived 2A sequences in immunocompetent individuals. Gene Ther. 2013;20(9):958–62.
32. Di Stasi A, Tey SK, Dotti G, Fujita Y, Kennedy-Nasser A, Martinez C, et al. Inducible apoptosis as a safety switch for adoptive cell therapy. N Engl J Med. 2011;365(18):1673–83.
33. Zhou X, Dotti G, Krance RA, Martinez CA, Naik S, Kamble RT, et al. Inducible caspase-9 suicide gene controls adverse effects from alloreplete T cells after haploidentical stem cell transplantation. Blood. 2015;125(26):4103–13.
34. Iuliucci JD, Oliver SD, Morley S, Ward C, Ward J, Dalgarno D, et al. Intravenous safety and pharmacokinetics of a novel dimerizer drug, AP1903, in healthy volunteers. J Clin Pharmacol. 2001;41(8):870–9.
35. Yagyu S, Hoyos V, Del Bufalo F, Brenner MK. Multiple mechanisms determine the sensitivity of human-induced pluripotent stem cells to the inducible caspase-9 safety switch. Mol Ther Methods Clin Dev. 2016;3:16003.
36. Bedel A, Beliveau F, Lamrissi-Garcia I, Rousseau B, Moranvillier I, Rucheton B, et al. Preventing Pluripotent Cell Teratoma in Regenerative Medicine Applied to Hematology Disorders. Stem Cells Transl Med. 2017;6(2):382–93.
37. Wang X, Marchisio MA. Synthetic polycistronic sequences in eukaryotes. Synth Syst Biotechnol. 2021;6(4):254–61.
38. Yagyu S, Hoyos V, Del Bufalo F, Brenner MK. An Inducible Caspase-9 Suicide Gene to Improve the Safety of Therapy Using Human Induced Pluripotent Stem Cells. Mol Ther. 2015;23(9):1475–85.
39. Itakura G, Kawabata S, Ando M, Nishiyama Y, Sugai K, Ozaki M, et al. Fail-Safe System against Potential Tumorigenicity after Transplantation of iPSC Derivatives. Stem Cell Reports. 2017;8(3):673–84.
40. Schuldiner M, Itskovitz-Eldor J, Benvenisty N. Selective ablation of human embryonic stem cells expressing a "suicide" gene. Stem Cells. 2003;21(3):257–65.
41. Dahlke J, Schott JW, Vollmer Barbosa P, Klatt D, Selich A, Lachmann N, et al. Efficient Genetic Safety Switches for Future Application of iPSC-Derived Cell Transplants. J Pers Med. 2021;11(6).
42. Lipus A, Janosz E, Ackermann M, Hetzel M, Dahlke J, Buchegger T, et al. Targeted Integration of Inducible Caspase-9 in Human iPSCs Allows Efficient. Int J Mol Sci. 2020;21(7).
43. Kimura Y, Shofuda T, Higuchi Y, Nagamori I, Oda M, Nakamori M, et al. Human Genomic Safe Harbors and the Suicide Gene-Based Safeguard System for iPSC-Based Cell Therapy. Stem Cells Transl Med. 2019;8(7):627–38.
44. Sułkowski M, Konieczny P, Chlebanowska P, Majka M. Introduction of Exogenous HSV-TK Suicide Gene Increases Safety of Keratinocyte-Derived Induced Pluripotent Stem Cells by Providing Genetic "Emergency Exit" Switch. Int J Mol Sci. 2018;19(1).
45. Kotini AG, de Stanchina E, Themeli M, Sadelain M, Papapetrou EP. Escape Mutations, Ganciclovir Resistance, and Teratoma Formation in Human iPSCs Expressing an HSVtk Suicide Gene. Mol Ther Nucleic Acids. 2016;5(2):e284.
46. Cheng F, Ke Q, Chen F, Cai B, Gao Y, Ye C, et al. Protecting against wayward human induced pluripotent stem cells with a suicide gene. Biomaterials. 2012;33(11):3195–204.
47. Anderson D, Self T, Mellor IR, Goh G, Hill SJ, Denning C. Transgenic enrichment of cardiomyocytes from human embryonic stem cells. Mol Ther. 2007;15(11):2027–36.
48. Shi ZD, Tchao J, Wu L, Carman AJ. Precision installation of a highly efficient suicide gene safety switch in human induced pluripotent stem cells. Stem Cells Transl Med. 2020;9(11):1378–88.

49. Liu Y, Yang Y, Suo Y, Li C, Chen M, Zheng S, et al. Inducible caspase-9 suicide gene under control of endogenous oct4 to safeguard mouse and human pluripotent stem cell therapy. Mol Ther Methods Clin Dev. 2022;24:332–41.
50. Ou W, Li P, Reiser J. Targeting of herpes simplex virus 1 thymidine kinase gene sequences into the OCT4 locus of human induced pluripotent stem cells. PLoS One. 2013;8(11):e81131.
51. Wu C, Hong SG, Winkler T, Spencer DM, Jares A, Ichwan B, et al. Development of an inducible caspase-9 safety switch for pluripotent stem cell-based therapies. Mol Ther Methods Clin Dev. 2014;1:14053.
52. Rong Z, Fu X, Wang M, Xu Y. A scalable approach to prevent teratoma formation of human embryonic stem cells. J Biol Chem. 2012;287(39):32338–45.
53. Wu Y, Chang T, Long Y, Huang H, Kandeel F, Yee JK. Using Gene Editing to Establish a Safeguard System for Pluripotent Stem-Cell-Based Therapies. iScience. 2019;22:409–22.
54. Liedtke S, Stephan M, Kögler G. Oct4 expression revisited: potential pitfalls for data misinterpretation in stem cell research. Biol Chem. 2008;389(7):845–50.
55. Green MD, Chen A, Nostro MC, d'Souza SL, Schaniel C, Lemischka IR, et al. Generation of anterior foregut endoderm from human embryonic and induced pluripotent stem cells. Nat Biotechnol. 2011;29(3):267–72.
56. Graham V, Khudyakov J, Ellis P, Pevny L. SOX2 functions to maintain neural progenitor identity. Neuron. 2003;39(5):749–65.
57. Tieng V, Cherpin O, Gutzwiller E, Zambon AC, Delgado C, Salmon P, et al. Elimination of proliferating cells from CNS grafts using a Ki67 promoter-driven thymidine kinase. Mol Ther Methods Clin Dev. 2016;6:16069.
58. Liang Q, Monetti C, Shutova MV, Neely EJ, Hacibekiroglu S, Yang H, et al. Linking a cell-division gene and a suicide gene to define and improve cell therapy safety. Nature. 2018;563(7733):701–4.
59. de Luzy IR, Law KCL, Moriarty N, Hunt CPJ, Durnall JC, Thompson LH, et al. Human stem cells harboring a suicide gene improve the safety and standardisation of neural transplants in Parkinsonian rats. Nat Commun. 2021;12(1):3275.
60. Montoliu L. Transgenesis and Genome Engineering: A Historical Review. Methods Mol Biol. 2023;2631:1–32.
61. Hendriks D, Clevers H, Artegiani B. CRISPR-Cas Tools and Their Application in Genetic Engineering of Human Stem Cells and Organoids. Cell Stem Cell. 2020;27(5):705–31.
62. Sadelain M, Papapetrou EP, Bushman FD. Safe harbours for the integration of new DNA in the human genome. Nat Rev Cancer. 2011;12(1):51–8.
63. Liu S, Zhang Q, Shy AN, Yi M, He H, Lu S, et al. Enzymatically Forming Intranuclear Peptide Assemblies for Selectively Killing Human Induced Pluripotent Stem Cells. J Am Chem Soc. 2021;143(38):15852–62.
64. Chang YJ, Cui X, Levi SR, Jenny LA, Tsang SH. CRISPR Manipulations in Stem Cell Lines. Methods Mol Biol. 2023;2560:249–56.
65. Yu J, Vodyanik MA, Smuga-Otto K, Antosiewicz-Bourget J, Frane JL, Tian S, et al. Induced pluripotent stem cell lines derived from human somatic cells. Science. 2007;318(5858):1917–20.
66. Rezania A, Bruin JE, Arora P, Rubin A, Batushansky I, Asadi A, et al. Reversal of diabetes with insulin-producing cells derived in vitro from human pluripotent stem cells. Nat Biotechnol. 2014;32(11):1121–33.

Safety Issues Related to Pluripotent Stem Cell-Based Therapies: Tumour Risk

Sanne Hillenius, Joaquin Montilla-Rojo, Thomas F. Eleveld, Daniela C. F. Salvatori, and Leendert H. J. Looijenga

1 Introduction

Pluripotent stem cells (PSCs) have gained significant scientific attention, aiding in understanding human development and diseases while also providing a tool to explore the molecular mechanisms of diseases and new possibilities for therapeutic applications. Over the last four decades significant strides have been made in PSC research, starting from the first successful isolation and prolonged culture of mouse embryonic stem cells (mESCs) isolated from the inner cell mass of pre-implantation embryos [1, 2]. The ability to culture ESCs from other species soon followed, including human ESCs (hESCs) [3–7]. However, ethical concerns regarding the use of hESCs for research as well as the need for patient-specific cell therapies called for an additional source for human PSCs (hPSCs). This led to the pivotal scientific achievement of generating induced pluripotent stem cells (iPSCs) through the reprogramming of terminally differentiated adult fibroblasts by initially using a defined cocktail of the pluripotency factors *OCT3/4*, *SOX2*, *KLF4,* and *c-MYC* [8–10]. The seminal studies leading to this achievement were led by Prof. Dr. Shinya Yamanaka who was later awarded the Nobel Prize for Physiology and Medicine in 2012 [11]. Since then, PSC-based in vitro models have been investigated for a legion of diseases with clinical trials ensuing in less than a decade [12–18].

Sanne Hillenius and Joaquin Montilla-Rojo contributed equally.

S. Hillenius · T. F. Eleveld · L. H. J. Looijenga (✉)
Princess Máxima Center for Pediatric Oncology, Utrecht, The Netherlands
e-mail: L.Looijenga@prinsesmaximacentrum.nl

J. Montilla-Rojo · D. C. F. Salvatori
Anatomy and Physiology, Department Clinical Sciences, Faculty of Veterinary Medicine, Utrecht University, 3584 CL Utrecht, The Netherlands

L. Piemonti et al. (eds.), *Pluripotent Stem Cell Therapy for Diabetes*,
https://doi.org/10.1007/978-3-031-41943-0_18

hPSC-based clinical trials have also been initiated for diabetes mellitus type 1 and 2 (T1D and T2D, respectively) due to great efforts in the generation of pancreatic cell lineages from hPSCs [19–29] and improvements in the bioengineering of cell delivery systems [30–32].

Despite promising (pre-)clinical results, safety concerns of hPSC-derived therapies in general have been raised regarding potential tumour formation risk [33–35]. The intrinsic pluripotent and self-renewing nature of hPSCs inherently serves as a double-edged sword. It provides the opportunity to indefinitely produce patient-specific cell types derived from all three embryonic germ layers (endoderm, mesoderm and ectoderm) via directed differentiation and maturation. However, if residual undifferentiated cells remain in the final cell product these may demonstrate uncontrolled proliferation and (partial) derailed differentiation, potentially leading to benign or malignant tumour formation when transplanted into patients. These concerns are not unfounded as several case studies have reported tumour development in patients after receiving hPSC-based therapies [36, 37]. Notably, Han and colleagues reported the development of a malignant teratoma (a subtype of germ cell tumours, hereinafter referred to as GCTs) containing residual undifferentiated cells in a patient with T2D who had been transplanted with autologous hPSC-derived pancreatic cells as part of a clinical study conducted in China [37]. Fortunately, to date, tumour formation has not been reported in the aforementioned hPSC-based T1D clinical trials [32, 38–40]. However, several of these trials have used cells that resemble a progenitor-like state known as pancreatic endoderm cells (PECs) and early pre-clinical studies generating PEC with subsequent transplantation in mice demonstrated that 15–46% of mice developed tumours, indicating that their use in clinical trials requires cautious monitoring [20, 41–45]. Additionally, improved differentiation protocols have shown further maturation of PEC into stem cell islets (SCIs) containing multiple pancreatic cell types including β-like cells and these SCIs demonstrated additional maturation in vivo post-transplantation in mice, resembling adult human islets [21, 22, 25–27, 29, 45–51].

In addition to the risk of residual undifferentiated cells, the presence of any off-target cell type post-differentiation and -maturation may also pose a significant risk. For example, several studies on SCI differentiation also identified an off-target cell population post-differentiation that resembles intestinal enterochromaffin cells (SC-EC cells). While the cause and effect of this population is largely unknown, its identification begs the question whether other (potentially risky) off-target cell populations may also arise during SCI differentiation that have not yet been identified. Despite efforts to enrich for the final SCI population for mature pancreatic islet cell types through biomarker identification and cell sorting techniques, pure populations containing solely the target cell types have not yet been achieved [25, 41, 46, 49, 52–55]. While specific examples of research on hPSC-based therapies for T1D and T2D are mentioned here, these potential risks concern hPSC-based therapies in general [35]. Of note, while in vitro differentiation of hPSCs towards the final desired cell type is crucial, proper subsequent maturation towards the terminally differentiated cell types of interest (either in vitro or after administration to the patient) is equally important for the efficiency and safety of hPSC-based therapies, thus constituting another aspect that requires thorough investigation. Therefore, from hereon, when referring to differentiation in this chapter, we also refer to subsequent maturation processes.

Notably, the use of hPSCs for regenerative medicine carries another risk. In contrary to PSCs generally being characterised by a low mutation rate compared to somatic cells, accumulating data clearly underlines that hPSCs acquire recurrent (epi)genetic aberrations in culture, many of which are similar to those found in several types of cancers [34, 35, 56–84]. These aberrations can range from whole chromosomal aneuploidy and smaller chromosomal structural rearrangements to (potentially) deleterious single nucleotide mutations. Although the current understanding of the phenotypic changes associated with these aberrations is still in its infancy, specifically regarding their potential for (benign or malignant) tumour formation, the parallels between hPSCs and multiple cancer types present a major hurdle for the clinical application. In fact, in 2015, the first hPSC-based clinical trial (RIKEN) was halted due to the detection of de novo acquired mutations in the final cell product, leading to concerns regarding the safety of their use [85]. Evidently, hPSC-based therapies require standardised and continuous characterisation both during preclinical development and clinical application to ensure their safety. This chapter will, therefore, highlight the known (epi)genetic aberrations observed in hPSCs, their similarities to cancer and will describe techniques that can potentially aid in ensuring the safety of hPSC-derived therapies.

2 Recurrent (Epi)Genetic Aberrations in Human Pluripotent Stem Cells

Different recurrent (epi)genetic aberrations present in hPSCs have been described. These may originate from the source from which the PSCs were derived, i.e., the developing embryos in the case of hESCs or donor somatic cells in the case of hiPSCs [5, 8, 63, 65, 66, 81, 86]. Moreover, specifically in hiPSCs the acquisition of (epi)genetic aberrations has also been linked to the process of cellular reprogramming, which leads to profound changes in the (epi)genetic expression profile, highlighting the necessity to establish proper quality controls and functional analyses after this stage in hPSC culture [63, 64, 66, 86–88]. Additionally, many aberrations found in hPSCs have been linked to (prolonged) in vitro culture [34, 35, 59, 70, 75, 89–96]. This phenomenon is known as culture adaptation, where cells acquire (epi) genetic aberrations due to the selective pressures induced by culture conditions (which remain poorly understood), leading to a potential growth advantage, resistance to apoptosis and a subsequent progressive dominance of these variant cells over their unaltered counterparts. This may, in part, be driven by the increased proliferation capacity of hPSCs, owing to an enhanced cell cycle progression rate compared to somatic cells which may be accompanied by increased DNA replication stress and subsequent damage [35, 97–99]. hPSCs have been reported to have a low threshold for apoptosis in response to DNA damage and selection against this trait may exacerbate culture adaptation and lead to preferential overgrowth of these variant cells [35]. Generally, the recurrent genetic aberrations may include (sub) chromosomal amplifications, deletions and rearrangements as well as other types of mutations that affect key cell cycle-related genes and deregulate normal cell division, and differentiation. Notably, several of these aberrations resemble those found in human malignancies (Table 1 and Fig. 1) and have been linked to a greater resilience to culture conditions due to a higher cloning efficiency, growth factor independence, impaired differentiation potential and the overall acquisition of cancer traits in vitro [62, 70, 74, 91, 94, 100, 101]. Additionally, multiple studies have reported single nucleotide mutations in known cancer-related genes in hPSCs several of which are highlighted further below [34, 70, 72, 75, 76, 78, 80].

2.1 Highlighting the Most Recurrent Chromosomal Aberrations in hPSCs

While aberrations affecting almost all chromosomes have been described several have been shown to occur more frequently, suggesting a selective advantage to their alteration which may be attributed to the alteration of specific genes within these chromosomes, known as driver genes [33–35]. However, the selection dynamics for these chromosomal aberrations and the identification of driver genes remain largely elusive. Although several differences have been observed both in hESCs and hiPSCs, the most frequent chromosomal aberrations are the complete gain of chromosomes 12 and X, and the partial gains of chromosomes 1, 12, 17, 20 and X (Table 1), which are also highly identifiable in multiple cancer types (Fig. 1), including malignant GCTs (Box 1) [33–35, 56–61, 63, 65–69, 72, 73, 75, 81, 83, 84, 102–119]. For example, in 2020, Assou and colleagues reported that the total recurrence of the abnormalities found in five hotspots located in chromosomes 20, 12, 17, X and 1 (ordered based on level of frequency) accounted for 58.7% of all the aberrations in hPSCs in culture [77]. More specifically, the most frequently amplified subchromosomal regions are 20q11, 12p11-pter, 17q11-q25, 1q21-q25, Xp11–22 and the full trisomy of chromosome 12 and chromosome X (the latter specifically in female cell lines). Although duplications of whole chromosomes and partial amplification of chromosomal regions are the most prevalent, partial or complete loss of chromosomes has also been reported, mainly affecting chromosomes 1, 3, 4, 8 and 18 [66, 67, 74].

Box 1: Pluripotency and Malignancy: Lessons Learned from Germ Cell Tumours

Historically, the concept of pluripotent stem cells precedes the discovery of ESCs and iPSCs and was inaugurated by research on human and mouse (gonadal) germ cell tumours [149–156]. In human pathology, GCTs represent a heterogeneous collection of solid tumours that arise from the deregulation of the cells of the early embryo and germ line, including ESCs and primordial germ cells (PGCs) as well as its progeny that give rise to the reproductive cells of both sexes [125, 156]. These tumours can arise in multiple sites along the midline of the body (owing in part to the migratory nature of PGCs), including the gonads. An interesting subtype of GCTs are the teratomas, which in principle represent benign tumour masses containing differentiated and fully matured cell types derived from the three embryonic germ layers [125, 156]. Although their clinical presentation has bewildered many for centuries, their further characterisation was made possible through the identification of the 129t mouse strain in 1954 which presents a high incidence of spontaneous testicular teratomas as well as teratocarcinomas (which are the malignant counterpart of teratomas) within the seminiferous tubules of the developing mouse gonad as early as 13 days of embryonic development [150, 151, 157, 158]. Teratocarcinomas were subsequently shown to contain undifferentiated embryonal carcinoma (EC) cells that, when reinjected into mice, generated new teratocarcinomas, thus demonstrating their malignancy as well as the single cell origin of these tumours [150, 151, 159]. Moreover, these resulting secondary tumours also contained fully differentiated and matured tissues from all three germ layers, indicating that EC cells are pluripotent. Additional mouse studies led to the suggestion that teratomas and EC cells in teratocarcinomas arise from PGCs as the ectopic transplantation of the genital ridges of mice also gave rise to similar tumours [159]. Besides similarities to PGCs, studies performed in the 1970s showed that teratomas and EC cells are closely related to the inner cell mass of a mouse blastocyst-stage embryos, with an important finding that mouse EC cells can contribute to embryonic development [160–162], providing another close link between EC cells and the pluripotent cells of early embryonic development which are now known to be the ESCs. It was also the subsequent success in culturing EC cells and establishing human EC cell lines as well as the identification of specific membrane biomarkers [155, 163–167] that ultimately led to the successful isolation and propagation of mESCs [1, 2], further paving the way to the field of pluripotent stem cell research as we know it today. In fact, currently, EC cells (which are also now known to express the pluripotency markers OCT3/4 and SOX2) are considered the malignant counterpart of ESCs [152, 153, 155, 156, 168]. Notably, the close connection between hPSCs, PGCs, and GCTs is also reflected in the similarities observed in recurrent chromosomal aberrations [56, 59, 90, 125, 156]. Aneuploidy of chromosome 12, specifically isochromosome 12p (i12p), is one of the most characteristic aberrations in malignant GCTs; however, aberrations in chromosome 1, 17, 20 and X have also been reported [56, 59, 90, 125, 130, 156, 169, 170]. This suggests that hPSCs and deregulated PGCs may have similar selection mechanisms that lead to these similarities in chromosomal aberrations. Evidently, the unique and historical connections between hPSCs, PGCs and GCTs implicate the relevance of both PGC and GCT research in furthering our understanding of the cause and effect of the (epi)genetic instability observed in hPSCs and its implications in the safety of PSC-based therapies.

Table 1 Most recurrent (sub)chromosomal alterations observed in hPSCs

Chromosomes affected by recurrent genetic aberrations in hPSCs	Type of aberration	Some reported phenotypes linked to aberrations	Putative driver genes in the affected regions and their function		Cancer types commonly displaying similar aberrations	References
1	Trisomy c.1 (gain of the full chromosome)		**1. *POU5F1P4***	*POU5F1* pseudogene 4, regulates stemness maintenance by regulating *POU5F1* expression (key pluripotency factor)	Breast cancer [120, 121]	[59, 61, 62, 66–69, 72–74, 77, 81, 83, 110, 124]
	Gain of entire q-arm		**2. *CD30***	Member of the TNF-receptor superfamily and regulator of apoptosis, it mediates the activation of the NF-κβ pathway regulating cell proliferation and survival	Multiple myeloma [122]	
	Structural rearrangement within the q-arm (q21–25; q31–36)		**3. *MDM4***	Encodes a protein binding the tumour suppressor *TP53*, thus inhibiting its activity and consequently suppressing apoptosis	Cervical cancer [123]	

12	Trisomy c.12 (gain of entire chromosome)	In the undifferentiated state, trisomy 12 cells exhibit increased proliferation and teratocarcinoma formation; Variants with a 12p gain also show a reduced differentiation capacity	**4. *SOX5***	Promotes cell growth and migration though modulation of DNMT/p21 pathway, as well as regulating chondrogenesis and cartilage differentiation in early embryos	Malignant testicular germ cell tumours [125]	[59, 63, 117, 125, 126, 128–130]
	Gain of entire p-arm		**5. *KRAS***	Regulator of cell division through the modulation of RAS/MAPK pathway	Chronic lymphocytic leukemia [126]	
	Partial gain of p-arm (p11-pter)		**6. *GDF3***	Inhibitor of BMP signalling, regulates stemness maintenance and differentiation	Colorectal tumours [127]	
	Isochromosome c.12p		**7. *DPPA3***	Related to pluripotency and stemness maintenance through chromatin condensation		
			8. *NANOG*	Pluripotency and stemness maintenance master controller		

(continued)

Table 1 (continued)

Chromosomes affected by recurrent genetic aberrations in hPSCs	Type of aberration	Some reported phenotypes linked to aberrations	Putative driver genes in the affected regions and their function		Cancer types commonly displaying similar aberrations	References
17	Trisomy c.17 (gain of entire chromosome)	Variants with a chromosome 17q gain exhibit selective growth advantage, supercompetitive phenotype in high cell density cultures and altered differentiation patterns	**9. *TP53***	Pivotal in regulation of cell cycle progression, controls growth arrest, DNA repair and apoptosis through the transactivation of its target genes	Breast cancer [131]	[56, 59, 96, 132, 133]
	Gain of entire q-arm		**10. *MAPK7***	Member of the MAPK family, involved in regulation of cell proliferation, differentiation, transcription regulation and development		
	Partial gain of q-arm (q11–25)		**11. *GRB7***	Mediates matrix metalloprotease 9 (MMP9) and EFG-mediated ERK activation, regulating cell migration and invasion		
			12. *CASC3*	Core component of the exon junction complex, promoting transcription-wide activation and nonsense-mediated decay		
			13. *SURVIVIN (BIRC5)*	Regulates cell division and inhibition of apoptosis by blocking caspase activation		

18	Interstitial deletion (18q)		**14.** ***BCL2***	Apoptosis regulator	Colorectal cancer [134, 135]	[69, 73, 83, 112, 118, 136]
			15. ***NEDD4L***	Key regulator of protein lysosomal degradation through ubiquitination, it is involved in DNA repair, autophagy and cell growth	Esophageal squamous cell carcinoma ([69, 73, 83, 112, 118, 136])	
					Prostate cancer [137]	

(continued)

Table 1 (continued)

Chromosomes affected by recurrent genetic aberrations in hPSCs	Type of aberration	Some reported phenotypes linked to aberrations	Putative driver genes in the affected regions and their function		Cancer types commonly displaying similar aberrations	References
20	Trisomy c.20 (gain of entire chromosome)	Selective growth advantage in undifferentiated state reduced genetic stability and altered differentiation patterns	**16.** ***BCL2L1***	Member of the BCL2 family of regulators, plays a crucial role in anti-apoptotic cellular pathways	Breast cancer [138]	[60, 67, 68, 70, 77, 79, 81, 91, 100, 111, 118, 142–144]
	Isochromosome c.20q		**17.** ***DNMT3B***	DNA methylation regulator, plays a major role in gene silencing and activation during cancer progression	Cervical cancer [139]	
	Interstitial duplication (20q11)		**18.** ***ID1***	Regulator of cellular growth and senescence via NANOG	Colorectal cancer [140, 141]	
			19. ***TPX2***	Regulator of microtubule assembly during M phase of mitosis		
			20. ***HCK***	Tyrosine kinase specially expressed in hematopoietic lineages supporting proliferation and survival by physical association with oncogenic fusion proteins		

X	Trisomy c. X in female lines (gain of entire chromosome)		**21.** ***ELK1***	Transcription activator regulating DNA binding with key repressive and activating roles in pluripotency maintenance	Male breast cancer [145]	[72, 77, 112]
	Partial gain q-arm (q11–22)		**22.** ***ARAF***	Regulator of MEK/ERK pathway, ultimately promotes cell cycle progression and proliferation	Prostate cancer [146]	

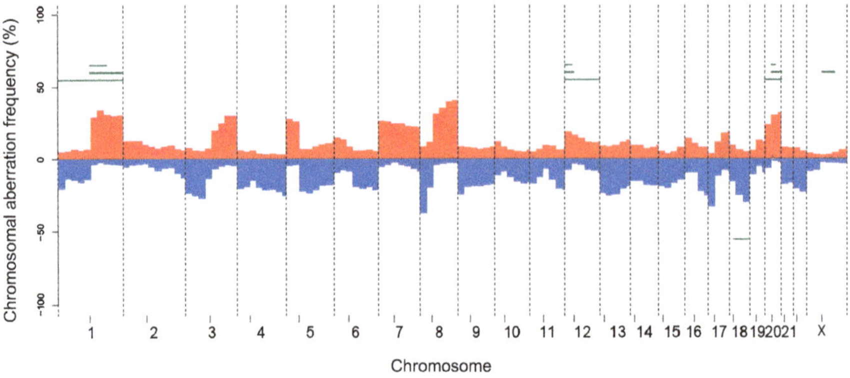

Fig. 1 Chromosomal aberrations in cancer versus hPSCs. The data shown here is based upon data generated by the TCGA Research Network (https://www.cancer.gov/ccg/research/genome-sequencing/tcga) obtained through cBioPortal (https://www.cBioPortal.org [147, 148] and represents the copy number changes (i.e. chromosomal aberrations) observed in all included epithelial cancer types. Top and bottom vertical bars represent chromosomal amplifications (in red) and deletions (in blue), respectively. The X-axis shows all chromosomes. Each chromosome is divided into bins spanning 25 Mb and the small demarcation above each chromosome number indicates the centromere (dividing the p- and q-arm). The Y-axis represents the frequency (in percentage) of either the amplification or deletion of the chromosomal bin. Horizontal bars at the top (gain) and bottom (loss) indicate common chromosomal aberrations found in hPSCs in chromosome 1, 12, 17, 20 and X, ranging from whole chromosomal aneuploidy to (partial) chromosome arm aberrations

The most frequently identified chromosomal aberration in hPSCs is the sub-chromosomal amplification of 20q11, more specifically 20q11.21 [60, 67, 68, 70, 72, 77, 81, 100, 118, 142, 143], with multiple studies reporting >20% occurrence in 125 screened hESC lines [67] and 18% among 34 hiPSC lines [68]. Among the protein-coding genes located in this locus, the most commonly upregulated in abnormal hPSC cultures is *BCL2L1* (also known as *BCL-XL*), a gene belonging to the *BCL2* family playing a pivotal role in anti-apoptotic cellular pathways [70, 100, 142]. Reportedly, hPSC lines containing the 20q11 amplicon present higher population doubling rates than control lines, which is mostly attributed to the overexpression of *BCL2L1* [70, 95]. Notably, *BCL2L1* has been identified as a key factor promoting adenoma-to-carcinoma progression in human colorectal cancer [171]. An additional gene located in this frequently amplified locus is *DNMT3b,* a pluripotency-associated gene that regulates the maintenance of DNA methylation, which was shown to play a role in gene silencing during cancer progression and to have altered expression levels in hPSCs with acquired 20q11 amplification [68, 118, 172]. Moreover, two other protein-coding genes located in this locus, *ID1* and *TPX2*, have been associated with cancer progression in salivary gland carcinoma and hepatocellular carcinoma, respectively, and are also overexpressed in hPSCs with a 20q11 amplification [70, 95, 100, 173–175]. Notably, *TPX2* was shown to lead to induction of *BCL2L1* expression, resulting in increased cell survival [95]. This suggests not only a locational but also a functional link between these two genes. The overexpression of these known proto-oncogenes in 20q11 in both cancer and hPSCs suggests their potential implication in malignant transformation and their candidacy as genetic markers for potential malignant transformation in hPSCs.

Additionally, trisomy of chromosome 12 is the most commonly observed whole chromosome aneuploidy in hPSCs, followed by smaller partial gains within the chromosome, including isochromosome 12p which is highly characteristic of GCTs [56, 59, 63, 66, 69, 74, 86, 90, 125, 169]. Although not fully elucidated, several studies have suggested that aberrations in chromosome 12 may be mediated through several putative driver genes, including *NANOG*, *GDF3*, *KRAS* and *DPPA3* [35, 59, 63, 91, 117, 128, 129, 176–178]. In these studies, cells harbouring chromosome 12 aneuploidy showed overexpression of these genes which regulate cell cycle progression and are suspected to confer a growth advantage over unaltered cells. Keller and colleagues further demonstrated that *NANOG* and *GDF3* play a role in delaying differentiation, resulting in residual undifferentiated cells in culture after directed differentiation efforts [178]. Moreover, upon injection into mouse models, hPSCs harbouring aneuploidies in chromosome 12 resulted in tumours presenting with increased expression of these and other cancer driver genes while displaying malignancy-related histopathological traits compared to their diploid counterparts [128]. Others have not observed similar outcomes in vivo [179], thus suggesting the need for further investigation towards the effects in vivo. Nevertheless, the association between specific aberrations in chromosome 12 and malignant behaviour of hPSCs suggest they may enhance tumour formation upon their in vivo injection, thus highlighting its relevance in the context of hPSC-based therapy safety.

Furthermore, partial or complete gains of chromosome 17 have also been frequently reported in hESCs and to a lesser extent in hiPSCs [56, 59, 63, 66, 68, 69, 72, 74, 77, 112, 132]. It has been hypothesised that the gain of extra copies of genes encoding proteins with pivotal roles in cell proliferation, migration and apoptosis specially in the 17q11–25 region such as *MAPK7*, *GRB7*, *CASC3*, *STAT3* and *BIRC5* may drive the impact of this recurrent chromosomal abnormality [56, 59, 68, 132]. Overexpression of these genes has been previously correlated to cancer progression and poor prognosis, such as in breast cancer [56, 59, 68, 133, 180]. In contrast to these gains, specific losses in chromosome 17 (specifically 17p13.1) and single nucleotide mutations in specific genes have also been reported in both hESCs and hiPSCs, including *TP53* [73, 75, 76, 78, 81]. TP53 is a famous tumour suppressor gene known as the 'guardian of the genome' due to its fundamental role in regulating a multitude of processes such as the cell cycle and DNA repair pathways [181–184]. *TP53* mutations occur in over 50% of cancers, many of which result in a loss of function or have a dominant negative effect [184–187]. Similar mutations have now also been reported in hPSCs and mutated cells were shown to have a selective advantage in culture [75, 76, 78], suggesting a strong selective pressure to inactivate the proper functioning of the *TP53* gene. Another candidate driver gene located in chromosome 17 is *BIRC5*, an inhibitor of apoptosis that is highly expressed in neuroblastomas and was shown to play a role in cancer stem cells and teratoma formation [188, 189].

Chromosomal structural rearrangements in the q-arm of chromosome 1 (ranging from 1q25–46) and less frequently chromosome 1 trisomy have often been reported in hPSCs [59, 61, 62, 67–69, 72, 74, 77, 81, 83, 110]. Little is known about putative driver genes, except for *MDM4*, an important regulator of TP53, which was suggested by Merkle and colleagues due its location in chromosome 1q and showed recurrent duplication in the analysed hPSCs [81]. Similarly, several reports have identified both trisomy and smaller rearrangements in chromosome X [57, 59, 67, 74, 112] and some putative driver genes have been reported, including *ELK1* and *ARAF* [59]. Overall, aberrations in chromosome 1 and X are not as well characterised as those in chromosomes 12, 17 and 20, highlighting the need to further understand and validate their effects and their relevance in hPSC safety.

2.2 Recurrent Small Genetic Aberrations in hPSCs

In addition to chromosomal aberrations, recurrent smaller genetic aberrations in hPSCs (including single nucleotide mutations) are being identified at a rapid rate, mostly aided by the advent of next-generation sequencing techniques [64, 78, 80, 81, 190]. While the aforementioned mutations in *TP53* remain to be the most prevalent among hPSCs [75, 76, 191], additional mutations in genes linked to cancer have been identified. For example, using whole exome sequencing (WES) over 100 mutations were identified in 22 hiPSC lines in 2011 [64], half of which are known cancer mutations based on data from the COSMIC cancer database (https://cancer.sanger.ac.uk/cosmic) [192]. Many of these mutations were shown to be present in the original fibroblasts used to generate the various hiPSC lines; however, several were de novo mutations associated with reprogramming and subsequent culture. Moreover, an RNA-sequencing study performed a decade later also found cancer-related mutations within four genes (*TP53, EGFR, PATZ1*, and *CDK12*) in 56 out of 172 samples of two famous hESC lines (H1 and H9) obtained from different research labs worldwide [78]. Similarly, a study that employed whole genome sequencing (WGS) and single nucleotide polymorphism (SNP) genotyping of 143 different hESC lines identified cancer-related mutations in 10 genes (involved in mostly the *TP53* pathway and DNA damage response pathways) among 15 hESC lines [81]. Finally, a study analysing WGS and WES data from over 600 hiPSC lines identified recurrent inactivating mutations in *BCOR* acquired in vitro, which resulted in global transcriptome changes and a hampered differentiation capacity compared to their wild-type counterparts [190]. *BCOR* is part of the polycomb repressive complex 1, involved in the epigenetic modification of histones and has been shown to be involved in maintaining pluripotency [190, 193, 194] and various types of *BCOR* mutations have been associated with multiple cancers [193]. Of note, BCOR is located on p-arm of the X chromosome and the inactivating nature of its mutations found in hPSCs seemingly contrasts the aforementioned (sub)chromosomal gains recurrently observed in hPSCs, again highlighting the need for further investigation of aberrations observed in this chromosome [190, 193], Nevertheless, recurrently acquired *TP53* and *BCOR* mutations in hPSCs along with other altered cancer-related genes represent potential risks associated with the use of hPSC-based therapies.

2.3 Epigenetic Aberrations in hPSCs

Apart from genetic abnormalities, aberrations and overall instability in the epigenetic constitution of hPSCs have also been observed and linked to culture adaptations [34, 67, 73, 82, 83, 89, 195–198]. In general, hPSCs greatly rely on epigenetic and transcriptional regulatory machinery to control the expression of pluripotency factors [199–201]. Although hESCs and hiPSCs share numerous characteristics in their epigenetic profiles, hiPSCs display altered DNA methylation and histone modifications related to an incomplete erasure of the epigenetic signature during the reprogramming process [202, 203]. Additionally, loss of epigenetic stability has also been observed as a result of in vitro adaptation (e.g., methylation blueprint changes). Several of these changes that occur both during reprogramming and culture are similar to those associated with malignant transformation [204–207]. Overall, these alterations could hamper hPSC differentiation potential and efficiency [82, 208–211] and impact large regions of the chromosome, potentially leading to proto-oncogene activation and/or tumour suppressor gene inactivation, mimicking cancer events and thus jeopardising hPSC-derived therapies [212–218]. Although the full impact of epigenetic aberrations varies depending on the regions affected, it promotes genetic instability, potentially leading to further genetic aberrations [206]. These epigenetic aberrations are also relevant in the context of X chromosome inactivation (XCI). In female cells, XCI is a highly regulated process in vivo that facilitates an equilibrated dosage of X-linked gene expression achieved through epigenetic imprinting (i.e., silencing) of one X chromosome [219–222]. Of note, XCI also has also been found to occur in male malignant GCTs due to the presence of super numerical X chromosomes, even resulting in the identification of XIST (an important regulator of XCI) as a potential blood-based molecular biomarker for these tumours [223–225]. In contrast, hPSCs in culture have been shown to reactivate their X chromosome either as a consequence of the reprogramming process or due to culture adaptation [59, 212, 220, 226–229]. This may lead to overexpression of genes that ultimately confer a growth advantage, potentially calling the safety of such lines for clinical application into question. Although much is still unknown about the effects of these epigenetic aberrations in hPSCs, especially regarding their potential implication in malignant transformation, they are gaining significant attention, further aided by the rise of high-throughput techniques [82, 199, 206].

2.4 The Effects of Different Reprogramming and Culture Techniques on hPSC (Epi)Genetic Integrity

While outside the scope of this current chapter, it is worth noting that increasing evidence demonstrates that different reprogramming techniques (in the case of hiPSCs) and different hPSC culture conditions differentially affect hPSC genetic integrity and function. As mentioned before, de novo genetic aberrations in hiPSCs have been associated with reprogramming and several studies reported differences between different reprogramming techniques (e.g., integrative versus non-integrative); however, this remains disputed and requires further investigation [63–66, 77, 86–88, 230–234]. Moreover, the degree of recurrent genetic aberrations in hPSCs appears to differ depending on the employed culture method. For example, hPSCs cultured on feeder cells (mostly mouse embryonic fibroblasts) versus compound-defined feeder-free conditions as well as the method used to passage hPSCs (enzymatic versus mechanical) appear to have different effects on their genetic integrity [59, 73, 83]. Garitaonandia and colleagues compared the occurrence of genetic aberrations in hESCs and hiPSCs cultured under multiple conditions and demonstrated a higher incidence of chromosomal duplications (mostly in chromosome 20q and 12) and deletions (mostly in chromosome 17q) in cells that were passaged enzymatically and cultured under feeder-free conditions [73]. Moreover, lower hPSC mutation rates and increased reprogramming efficiency have been observed when hPSCs are both reprogrammed and cultured under hypoxic rather than normoxic conditions [108, 115, 235–239]. A recent study also reported a lower fraction of residual undifferentiated cells post-differentiation and a lower frequency of teratoma formation in mice when PSCs were cultured and differentiated to cardiomyocyte-like cells under hypoxic conditions [240]. The list of parameters involved in different reprogramming and culture methods that may differentially affect the (epi)genetic integrity of PSCs may be virtually limitless and further comparative studies are required to understand their (epi)genetic and phenotypic effects in hPSCs. Such studies will aid in identifying the optimal conditions for hPSC culture that maintain their (epi)genetic integrity. Moreover, the implementation of standardised and GMP-grade culture conditions will improve the quality of hPSCs in culture and likely reduce the degree of (epi)genetic instability [191, 241].

3 Techniques to Assess the Safety of Pluripotent Stem Cells for Regenerative Medicine

As previously mentioned, ensuring the safety of hPSC-based therapies may require a lack of both cancer-driving (epi)genetic aberrations as well as residual undifferentiated and off-target cells. Realistically, a complete lack of (epi)genetic aberrations may be difficult to achieve during all stages of hPSC derivation, culture, and differentiation. Therefore, an important feat is not only the standardised monitoring of their occurrence but also the understanding of their specific impact and risk for unfavourable outcomes which will require a complimentary repertoire of at least (epi)genetic, transcriptomic and proteomic analyses. In line with this, the International Society for Stem Cell Research (ISSCR) recently published a set of extensive guidelines regarding best-practices in hPSC culture for (pre)clinical research and application [191]. These guidelines emphasise the need for standardised and routine screening of hPSCs for acquired genetic aberrations and the best methods to employ for this purpose, providing an important stepping stone for all researchers in the field of regenerative medicine. Nevertheless, it is critical to further investigate the relevance of and develop strategies for these recurrent aberrations in terms of tumour formation which will aid in overcoming this specific hurdle in the use of hPSC-based therapies in the clinic. In addition, the efficiency of directed differentiation protocols and techniques aimed at preventing or removing residual undifferentiated and other off-target cells will need to be further developed. This section provides an overview of the techniques that have been and will continue to be valuable for answering these questions and to aid in ensuring the safety of hPSC-based therapies (Fig. 2).

3.1 Assessing Pluripotency and Malignancy: The Teratoma Assays and (Partial) Alternatives

An assay that has long been considered the golden standard in assessing the potency of putative hPSCs is the in vivo teratoma assay, which was developed based on early studies of teratomas in mice (see Box 1) [242–244]. This assay consists of the injection of a cell population into immunodeficient mice. There, supported by the presence of an intricate mixture of factors both from the local microenvironment as well as those circulating within the host organism, the cells may form tumours within several weeks [245]. These tumours are then removed and subjected to histological analysis for morphological features and lineage-specific marker expression to assess their composition [246]. If the tumours contain cell lineages derived from all embryonic germ layers (as seen in teratomas) the injected cells are regarded as developmentally pluripotent [107, 153, 155, 245, 247]. This assay, therefore, enables the assessment of the potency of the cells and may potentially enable the identification of residual undifferentiated cells in the final cell product. Moreover, in contrast to many of the techniques outlined in this section, the teratoma assay is especially unique as it enables the direct assessment of intrinsic cell malignancy. According to the International Stem cell Banking Initiative [243], *"teratomas comprising non-proliferating somatic tissue may be further labelled as 'benign', 'mature' or 'fully differentiated', whereas teratomas composed of immature, proliferating foetal-like tissues may be labelled 'immature'"*. This is directly parallel to what is observed in GCTs, where the presence of immature tissues such as EC cells (considered the malignant counterpart of ESCs [168]; see Box 1) or extraembryonic tissues like yolk sac elements and primitive neuroectoderm is considered a sign of malignancy [170]. Due to the similarities between spontaneous and experimental teratomas, the same criteria are applied to determine malignancy of the injected cells in the teratoma assay (see Box 1) [167, 191].

Although the teratoma assay remains highly popular worldwide, it is increasingly scrutinised due to several key limitations of its use. For example, from an ethical perspective, studies employing the teratoma assay often require large numbers of animals and a lengthy study duration (usually several months), which potentially leads to animal suffering [244, 248–250]. Moreover, while this assay has been employed for decades, it has never been fully standardised. Parameters influencing the comparison of the outcomes among studies, like number and preparation of transplanted cells, mouse strain, site of transplantation, length of monitoring time

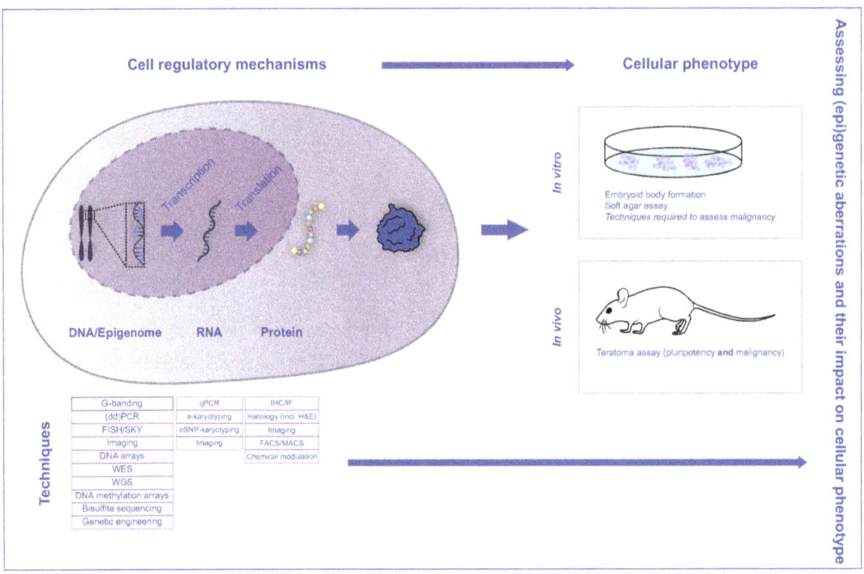

Fig. 2 Available techniques to assess and modulate the cell regulatory mechanisms and the consequent phenotypes that govern hPSCs and their derivatives. Many techniques are available that can aid in further interrogating and improving the safety of hPSC-based therapies for regenerative medicine. These include techniques that analyse hPSCs and their derivatives for their (epi)genetic integrity as well as (aberrant) gene expression and function at multiple levels (epigenome, DNA, RNA and protein). Additional techniques also allow for the modulation of hPSCs and their derivatives. These also aid in identifying undesired cell populations such as residual undifferentiated cells and off-target populations that arise during directed differentiation. Notably, only the in vivo teratoma assay can assess cell malignancy and in vitro techniques need to be developed to serve as a replacement (see italic text in the box listing the in vitro techniques). (dd) PCR, (droplet digital) polymerase chain reaction; FISH, fluorescent in situ hybridisation; SKY, spectral karyotyping; SNP, single nucleotide polymorphism; aCGH, comparative genomic hybridisation array; WES, whole exome sequencing; WGS, whole genome sequencing; qPCR, quantitative PCR; IHC/IF, immunohistochemistry/immunofluorescence; H&E, haematoxylin and eosin staining; FACS/MACS, fluorescence–/magnetic activated cell sorting

for tumour formation, and analytical methodology of the resulting tumour, are usually poorly specified and often incomplete despite several calls for standardisation [242, 250–252]. Additionally, despite the relevance of the assay in the assessment of a cell product's safety, little is still known about the underlying mechanisms behind tumour formation and what defines malignant potential, as the effects of the in vivo niche supporting tumour growth are poorly understood [253–255]. Because of these limitations, the field aims to move away from the use of this in vivo *assay*, as advised in the aforementioned ISSCR guidelines which state that the teratoma assay is only to be used when no other in vitro methods are available to study cell characteristics [191]. Thus, efforts to find alternative in vitro methods for assessing malignancy potential and the safety of hPSCs are actively ongoing, focusing on a better understanding of the underlying mechanisms behind tumour formation without the need of in vivo models.

One approach to achieve this is the use of embryoid bodies (EBs) as a model to determine the differentiation capacity of hPSCs in vitro [191, 256]. During EB culture, cells form three-dimensional aggregates and, in lieu of defined chemical factors that induce lineage-specific differentiation, cells are permitted to differentiate unbiasedly, resulting in the emergence of structures resembling early embryoid differentiation [244, 257]. If the EBs contain structures from all three germ layers (commonly determined by histological analysis using lineage-specific markers) then the original cells are considered pluripotent [191]. This assay may, therefore, be useful in determining whether a final cell product contains residual undifferentiated cells. Unfortunately, while many different methodologies have been described for improved EB formation [258, 259], to our knowledge, none can fully mimic in vivo (malignant) tumour formation highlighting the continued reliance on the in vivo teratoma assay. This could in part be caused by the failure of EB models to properly mimic a microenvironment similar to the in vivo teratoma assay, including interactions with other cells and the extracellular matrix, which may highly affect hPSC behaviour and their subsequent potential for tumour development. Thus, further research into adapting the EB formation techniques to enable the assessment of cell malignancy will be crucial in assessing the safety of hPSC-based therapies.

In vitro methods aiming to detect intrinsic malignant potential have also been explored based on our knowledge of common phenotypic traits of transformed cells. This includes the assessment of the invasive capacity of the cells of interest. Through assays like the soft agar assay, the ability of the cells to grow in an anchorage-independent growth can be tested, a characteristic considered a hallmark of malignant transformation [260–262]. This is supported by a study reporting that this assay can distinguish wild-type hiPSCs and established human ovarian teratocarcinoma lines [263]. Nevertheless, this assay focuses on one of the multiple hallmarks that characterise cancer development, potentially excluding others such as induction of angiogenesis, evasion of growth suppressors and overall (epi)genetic instability [264]. In addition, only invasive transformed cells can grow in these anchorage-free conditions, thus being an invalid assay for detecting benign residual undifferentiated cells from a cell pool [260]. This emphasises the limitations of using assays that assess single phenotypic characteristics and calls for more comprehensive in vitro analyses.

3.2 Additional Techniques to Assess and Modulate Pluripotency, Aberrant Pluripotent Cells and Undesired Cell Populations

While the abovementioned techniques remain relevant for the assessment of cell pluripotency and malignancy and are subject to development, they do not yet fully capture the complexity and depth of the aberrations observed in hPSCs nor their cause and effect, especially regarding the risk of tumour formation. These also give limited information about the presence and makeup of undesired cell populations post-differentiation (i.e. residual undifferentiated cells and other off-target populations). However, in the past two decades, an extensive arsenal of techniques has been developed for the detailed characterisation of cells in vitro which can further aid in assessing the safety of hPSCs and their derivatives. These techniques continue to develop and are accompanied by both specific strengths and limitations, the latter of which includes variable sensitivities (i.e. regarding detection of aberrations at low allelic frequencies), differences in resolution, throughput, cost and time consumption (aspects that are extensively discussed elsewhere: [33, 35, 59, 191, 265, 266]). Here, we discuss these techniques based on their applicability in assessing (epi)genetic integrity and the identification of undesired cell populations after directed cell differentiation.

Histological techniques have long been valuable in assessing protein expression and cellular organisation. This includes the macroscopic analysis of tissues (e.g. of tumours obtained in the teratoma assay or EB formation assay) most often performed after staining with haematoxylin and eosin (H&E) and subsequent analysis by an experienced (veterinary) pathologist [167]. Moreover, immunohistochemistry remains relevant for identifying cell types based on cell type specific markers. Regarding hPSC-based therapies, the guidelines published by the ISSCR this year have listed several important pluripotency and lineage-specific markers [191]. The use of these cell-specific markers will continue to be crucial to assess hPSCs at all stages of differentiation and for the identification of residual undifferentiated or other off-target cell populations. Moreover, advances in imaging techniques and artificial intelligence have proven and will continue to be valuable in tracking the hPSC behaviour during reprogramming, culture and differentiation [94, 267–270]. Nonetheless, histological assessments and imaging techniques will need to be complemented by other techniques to distinguish (epi)genetically aberrant cells and their functional effects.

Techniques assessing the transcriptome (i.e. gene expression) can also be informative in identifying hPSC genetic aberrations. Low-throughput, high-resolution techniques such as quantitative PCR (qPCR) can evaluate the relative expression of specific target genes, which has also been adapted to identify specific chromosomal aberrations [70, 73, 74, 191, 266, 271, 272]. Moreover, as mentioned before, high-resolution and high-throughput transcriptomic sequencing techniques such as (single cell) RNA sequencing can identify small genetic aberrations at a staggering resolution, including cancer-related mutations [75, 76, 78, 112]. Notably, these sequencing techniques also allow for the clustering of similar cell populations based on their gene expression profile, thus allowing for the detection of the desired and potentially off-target cell populations post-differentiation. This is exemplified by the identification of the on-target pancreatic cell types observed during SCI differentiation and off-target populations such as the SC-EC cells [25, 32] Moreover, adaptations of RNA-sequencing such as eSNP-karyotyping enable the assessment of chromosomal aberrations [33, 63, 71, 273]. Nevertheless, these transcriptomic techniques do not allow for functional analyses and, therefore, do not provide information about the potential malignancy of cells.

Furthermore, the initial identification of recurrent genetic aberrations in hPSCs was made possible through the use of a classic karyotyping technique known as Giemsa banding (G-banding) [56, 58, 63, 73, 83, 274, 275]. This allows for the visualisation of large (sub)chromosomal abnormalities, including losses, gains and structural rearrangements [275–277]. Although informative, the low resolution of karyotyping only permits the visualisation of large chromosomic aberrations (>5 Mb), thus overlooking those of smaller sizes that may still have large effects on cellular behaviour [33, 72, 74, 106]. Moreover, this technique does not allow for the analysis of an entire hPSC batch indicating that its sensitivity is limited to the number of metaphases analysed and, therefore, cannot fully exclude the presence of aberrant cells [35, 74]. Also, karyotyping only provides information about the genome and therefore does not give insight into gene expression nor cell function, including cell (pluri)potency or malignancy. Additional cytogenetic and genetic analyses with improved resolutions compared to traditional G-banding have been employed to identify hPSC aberrations. These include droplet digital PCR (ddPCR), fluorescent in situ hybridisation (FISH) and a multi-colour FISH-based technique known as spectral karyotyping (SKY) which enables the simultaneous visualisation of all chromosome pairs in a cell [33, 35, 58, 74, 77, 278, 279]. Moreover, high-throughput, high-resolution techniques to assess the genome have frequently been employed such as DNA arrays, including SNP arrays and comparative genomic hybridisation arrays (aCGH) [56, 57, 61, 63, 65–68, 72, 83, 280]. More recently, WGS and WES techniques offer an even higher resolution that have further illuminated hPSC genetic aberrations, including the identification of recurrent cancer-related mutations [64, 75, 81, 190]. Regarding the epigenome, several of the aforementioned techniques can also be employed to analyse the expression and mutational status of epigenetic regulators [82]. However, additional techniques that can assess aberrant chromatin accessibility, DNA methylation and XCI are also available, including DNA methylation arrays and bisulfite sequencing [65, 73, 82, 198, 281]. While these (epi)genetic techniques remain highly valuable, they do not identify specific cell types, nor do they provide functional information (e.g., regarding malignant potential).

Notably, the (epi)genetic and transcriptomic sequencing studies that identified cancer-related mutations in hPSCs often employed data from comprehensive databases containing (epi)genetic and transcriptomic sequencing data from a large range of cancer types such as COSMIC (https://cancer.sanger.ac.uk/cosmic; [192]) and The Cancer Genome Atlas program (TCGA) (https://www.cancer.gov/ccg/research/genome-sequencing/tcga) [75, 78, 80, 81, 190]. Future sequencing studies will greatly benefit from the collaborative development of comprehensive databases with similar data for a large number of hPSC lines. An example of such an initiative is the Human-Induced Pluripotent Stem Cell Initiative (HipSci, https://www.hipsci.org/) [112, 284] which has catalogued near 1000 hiPSCc lines including extensive information about their (epi)genome, transcriptome, and proteome. Further extension of such resources (also for hESCs) will undoubtedly aid the research on the (epi)genetic integrity of hPSCs and the safety of their use in the clinic.

Commercial bioinformatic tools based on gene expression data have also been valuable in assessing cell pluripotency, including PluriTest® [285] and Scorecard [286, 287] which have been validated by the International Stem Cell Initiative (ISCI) [244]. PluriTest®, which is based on microarray data, compares the transcriptome of a test cell line to a large number of reference lines that have been validated for their pluripotency. ScoreCard is based on qPCR data and also assesses pluripotency by comparing a test cell line to a panel of known pluripotency markers. In addition, ScoreCard also assesses the differentiation capacity of a test cell line post differentiation (e.g., using EB formation assays) through comparison with a large gene panel of known lineage-specific markers [286, 287]. As pluripotency can be evaluated, the presence of residual undifferentiated cells within the final cell product may be also identified. Nonetheless, these techniques lack the resolution to distinguish hPSCs with intrinsic malignant potential [179], nor do they give information about their mutational status.

3.3 Interventional Techniques to Eliminate Off-Target Cell Populations

Monitoring cells for their (epi)genetic status and controlling their subsequent phenotype prior to their use for clinical applications remains a difficult task and is further complicated by a currently limited understanding of what defines malignant potential in hPSCs. Therefore, to complimentarily aid in safeguarding the use of hPSCs for clinical applications, techniques are also being developed that aim to enrich for the desired cell population(s) post-differentiation [34, 45]. An example of this is the use of mechanical manipulations, which is especially relevant in the context of SCI differentiation where the single-cell dissociation and subsequent (cryopreservation and) reaggregation of endodermal cells efficiently decreases the number of off-target non-endocrine cells [25, 45, 288]. Additionally, the use of fluorescence- and magnetic-activated cell sorting techniques (FACS and MACS, respectively) has proven effective in eliminating off-target cell populations in vitro using lineage-specific markers [289, 290]. These techniques are also complemented by data obtained from high-throughput sequencing techniques, expediting the identification of cell-type specific markers and thus the distinction between different cell populations, with increasing accuracy. For example, regarding SCI differentiation, several studies reported that isolation for GP2a + enriched for pancreatic progenitor cells and subsequent claims were made that no teratomas formed after injection of these cells into mice [46, 53, 55]. Similarly, others have demonstrated that CD49a + isolation during SCI differentiation enriches for stem-cell β cells (responsible for insulin production) while also eliminating SC-EC cells [25].

Chemical modulation during cell culture has also proven to be an effective strategy to prevent the presence of off-target cell populations. For instance, PluriSIn-1 (and oleate synthesis inhibitor) and quecertin (also known as YM155, an inhibitor of the previously mentioned Survivin protein) were reported to eliminate residual undifferentiated cells from culture and subsequently prevented teratoma formation compared to control conditions [291–293]. Moreover, it was recently reported that the addition of exogenous nucleosides to hPSC culture aided in relieving replication stress and reduced the level of acquired mutations in vitro, highlighting that chemical modulation of hPSC culture also aids in maintaining their (epi)genetic integrity [294]. Finally, genetic engineering of hPSC lines such as the introduction of suicide genes have also gained significant attention [295]. A recent example of this is the integration of the inducible Caspase 9 gene (iCas9) under the promoter of an endogenously expressed gene specific to hPSCs, which has been shown to ablate residual hPSCs upon induction with corresponding small molecules [296–299]. Notably, while their use is highly promising, the risk of off-target effects of both chemical modulation and genetic engineering practices continues to warrant caution and further investigation, especially due to the ultimate clinical application of the modulated cell products.

It has become apparent that while many techniques are available that provide insight into the risks of hPSCs-based therapies and how to tackle them, their specific drawbacks preclude any singular technique from providing an all-encompassing solution. Therefore, a standardised collection of techniques that is performed routinely is required to monitor hPSC (epi)genetic integrity and the nature of their derivatives during differentiation and to assess their (potentially malignant) effects. This is exemplified by the aforementioned ISSCR guidelines published this year [191]; however, ongoing development of techniques (especially those assessing the phenotypic consequences in terms of malignancy) and structural implementation of such guidelines remains an important goal to ultimately develop safe hPSC-based therapies.

4 Conclusion

The risk of tumour formation in hPSC-based therapies remains a major hurdle in the field of regenerative medicine. One of the main contributing factors to this risk is the propensity of hPSCs to acquire recurrent (epi)genetic aberrations that range from complete aneuploidy to single nucleotide mutations with many resembling those found in cancers. The last two decades have shown that chromosomal aberrations (mostly amplifications) occur most frequently in chromosomes 1, 12, 17, and X, in addition to several small mutations affecting single genes such as *TP53* and *BCOR*. Recurrent epigenetic aberrations have also been observed. Moreover, the risk of undesired cell populations in the final cell population after directed differentiation further complicate the safe use of hPSC-based therapies. This primarily regards residual undifferentiated (pluripotent) cells, a phenomenon that is strongly paralleled by EC cells that are responsible for malignant tumours observed in mouse experiments (teratocarcinomas) and in GCT patients. Additionally, off-target populations post-differentiation such as the SC-EC population during SCI differentiation are becoming increasingly recognised yet their consequences are still largely unknown. Therefore, significant efforts are now focused on mitigating the risk of tumour formation in hPSC-based therapies by tackling these two aspects.

Critical to this field of research is the use of appropriate techniques. While traditional karyotyping (G-banding) has been fundamental in the first identification of recurrent large chromosomal aberrations, other techniques have furthered the characterisation of the (epi)genetic integrity of hPSCs with increasing accuracy, throughput and resolution. These include techniques such as DNA arrays (e.g., aCGH and SNP-arrays), FISH-based techniques, PCR-based techniques (e.g., ddPCR and qPCR), and histological analyses combined with advanced imaging techniques. Most recently, high-throughput sequencing techniques such as WES, WGS, (single-cell) RNA-sequencing and bisulfite sequencing have opened new avenues to characterise hPSCs at previously unimaginable proportions, strongly accelerating this field of research. Moreover, in an effort to eliminate unwanted cells, multiple techniques are being employed to enrich for the on-target cell population. These include mechanical modulations (e.g., cell sorting techniques and physical modifications during cell culture), chemical modulations (e.g., the use of inhibitory small molecules), and genetic engineering (e.g., introduction of suicide genes). Further development and optimisation of these techniques will be crucial to ensure the purity of the final cell product. Finally, the functional assessment of hPSCs, mainly regarding pluripotency and potential malignancy, has heavily relied on the in vivo teratoma assay for many decades. Despite its technical uniqueness in assessing cell malignancy, several key limitations (e.g., ethical concerns and lack of standardisation) call for an alternative approach. While in vitro techniques such as the EB formation assay can currently assess pluripotency, comprehensive in vitro techniques to reliably assess cell malignancy are still lacking. Further research into developing such techniques is vital to both facilitate detailed characterisations of the functional effect of recurrent (epi)genetic aberrations and to ensure the elimination off-target and residual undifferentiated cells.

Thus, while limited as single techniques, a standardised combination of multiple techniques will unquestionably advance research towards unravelling the cause and effect of the (epi)genetic aberrations observed in hPSCs, reducing potentially unwanted cell populations, and defining the relevant and predictive parameters of cell malignancy. Moreover, the transition from guidelines (such as the ISSCR 2023 guidelines) towards legislation will also provide an important steppingstone towards the universal standardisation of hPSC assessment. Ultimately, these collective efforts will pave the way towards safe hPSC-based therapies.

References

1. Evans MJ, Kaufman MH. Establishment in culture of pluripotential cells from mouse embryos. Nature. 1981;292(5819):154–6.
2. Martin GR. Isolation of a pluripotent cell line from early mouse embryos cultured in medium conditioned by teratocarcinoma stem cells. Proc Natl Acad Sci U S A. 1981;78(12):7634–8.
3. Thomson JA, Kalishman J, Golos TG, Durning M, Harris CP, Becker RA, et al. Isolation of a primate embryonic stem cell line. Proc Natl Acad Sci U S A. 1995;92(17):7844–8.
4. Thomson JA, Kalishman J, Golos TG, Durning M, Harris CP, Hearn JP. Pluripotent cell lines derived from common marmoset (Callithrix jacchus) blastocysts. Biol Reprod. 1996;55(2):254–9.
5. Thomson JA, Itskovitz-Eldor J, Shapiro SS, Waknitz MA, Swiergiel JJ, Marshall VS, et al. Embryonic stem cell lines derived from human blastocysts. Science. 1998;282(5391):1145–7.
6. Buehr M, Meek S, Blair K, Yang J, Ure J, Silva J, et al. Capture of authentic embryonic stem cells from rat blastocysts. Cell. 2008;135(7):1287–98.
7. Li P, Tong C, Mehrian-Shai R, Jia L, Wu N, Yan Y, et al. Germline competent embryonic stem cells derived from rat blastocysts. Cell. 2008;135(7):1299–310.
8. Takahashi K, Yamanaka S. Induction of pluripotent stem cells from mouse embryonic and adult fibroblast cultures by defined factors. Cell. 2006;126(4):663–76.
9. Takahashi K, Tanabe K, Ohnuki M, Narita M, Ichisaka T, Tomoda K, et al. Induction of pluripotent stem cells from adult human fibroblasts by defined factors. Cell. 2007;131(5):861–72.
10. Yu J, Vodyanik MA, Smuga-Otto K, Antosiewicz-Bourget J, Frane JL, Tian S, et al. Induced pluripotent stem cell lines derived from human somatic cells. Science. 2007;318(5858):1917–20.
11. The Nobel Prize in Physiology or Medicine 2012.2012 07 August 2023.
12. Trounson A, DeWitt ND. Pluripotent stem cells progressing to the clinic. Nat Rev Mol Cell Biol. 2016;17(3):194–200.
13. Shi Y, Inoue H, Wu JC, Yamanaka S. Induced pluripotent stem cell technology: a decade of progress. Nat Rev Drug Discov. 2017;16(2):115–30.
14. Eguizabal C, Aran B, Chuva de Sousa Lopes SM, Geens M, Heindryckx B, Panula S, et al. Two decades of embryonic stem cells: a historical overview. Hum Reprod Open. 2019;2019(1):hoy024.
15. Wiegand C, Banerjee I. Recent advances in the applications of iPSC technology. Curr Opin Biotechnol. 2019;60:250–8.
16. Yamanaka S. Pluripotent Stem Cell-Based Cell Therapy-Promise and Challenges. Cell Stem Cell. 2020;27(4):523–31.
17. Ji Y, Hu C, Chen Z, Li Y, Dai J, Zhang J, et al. Clinical trials of stem cell-based therapies for pediatric diseases: a comprehensive analysis of trials registered on ClinicalTrials.gov and the ICTRP portal site. Stem Cell Res Ther. 2022;13(1):307.

18. Kim JY, Nam Y, Rim YA, Ju JH. Review of the Current Trends in Clinical Trials Involving Induced Pluripotent Stem Cells. Stem Cell Rev Rep. 2022;18(1):142–54.
19. D'Amour KA, Bang AG, Eliazer S, Kelly OG, Agulnick AD, Smart NG, et al. Production of pancreatic hormone-expressing endocrine cells from human embryonic stem cells. Nat Biotechnol. 2006;24(11):1392–401.
20. Kroon E, Martinson LA, Kadoya K, Bang AG, Kelly OG, Eliazer S, et al. Pancreatic endoderm derived from human embryonic stem cells generates glucose-responsive insulin-secreting cells in vivo. Nat Biotechnol. 2008;26(4):443–52.
21. Rezania A, Bruin JE, Arora P, Rubin A, Batushansky I, Asadi A, et al. Reversal of diabetes with insulin-producing cells derived in vitro from human pluripotent stem cells. Nat Biotechnol. 2014;32(11):1121–33.
22. Pagliuca FW, Millman JR, Gurtler M, Segel M, Van Dervort A, Ryu JH, et al. Generation of functional human pancreatic beta cells in vitro. Cell. 2014;159(2):428–39.
23. Russ HA, Parent AV, Ringler JJ, Hennings TG, Nair GG, Shveygert M, et al. Controlled induction of human pancreatic progenitors produces functional beta-like cells in vitro. EMBO J. 2015;34(13):1759–72.
24. Nostro MC, Sarangi F, Yang C, Holland A, Elefanty AG, Stanley EG, et al. Efficient generation of NKX6-1+ pancreatic progenitors from multiple human pluripotent stem cell lines. Stem Cell Reports. 2015;4(4):591–604.
25. Veres A, Faust AL, Bushnell HL, Engquist EN, Kenty JH, Harb G, et al. Charting cellular identity during human in vitro beta-cell differentiation. Nature. 2019;569(7756):368–73.
26. Velazco-Cruz L, Song J, Maxwell KG, Goedegebuure MM, Augsornworawat P, Hogrebe NJ, et al. Acquisition of Dynamic Function in Human Stem Cell-Derived beta Cells. Stem Cell Reports. 2019;12(2):351–65.
27. Nair GG, Liu JS, Russ HA, Tran S, Saxton MS, Chen R, et al. Recapitulating endocrine cell clustering in culture promotes maturation of human stem-cell-derived beta cells. Nat Cell Biol. 2019;21(2):263–74.
28. Barsby T, Ibrahim H, Lithovius V, Montaser H, Balboa D, Vahakangas E, et al. Differentiating functional human islet-like aggregates from pluripotent stem cells. STAR Protoc. 2022;3(4):101711.
29. Balboa D, Barsby T, Lithovius V, Saarimaki-Vire J, Omar-Hmeadi M, Dyachok O, et al. Functional, metabolic and transcriptional maturation of human pancreatic islets derived from stem cells. Nat Biotechnol. 2022;40(7):1042–55.
30. Siehler J, Blochinger AK, Meier M, Lickert H. Engineering islets from stem cells for advanced therapies of diabetes. Nat Rev Drug Discov. 2021;20(12):920–40.
31. Dang HP, Chen H, Dargaville TR, Tuch BE. Cell delivery systems: Toward the next generation of cell therapies for type 1 diabetes. J Cell Mol Med. 2022;26(18):4756–67.
32. Hogrebe NJ, Ishahak M, Millman JR. Developments in stem cell-derived islet replacement therapy for treating type 1 diabetes. Cell Stem Cell. 2023;30(5):530–48.
33. Andrews PW, Barbaric I, Benvenisty N, Draper JS, Ludwig T, Merkle FT, et al. The consequences of recurrent genetic and epigenetic variants in human pluripotent stem cells. Cell Stem Cell. 2022;29(12):1624–36.
34. Lezmi E, Benvenisty N. The Tumorigenic Potential of Human Pluripotent Stem Cells. Stem Cells Transl Med. 2022;11(8):791–6.
35. Halliwell J, Barbaric I, Andrews PW. Acquired genetic changes in human pluripotent stem cells: origins and consequences. Nat Rev Mol Cell Biol. 2020;21(12):715–28.
36. Amariglio N, Hirshberg A, Scheithauer BW, Cohen Y, Loewenthal R, Trakhtenbrot L, et al. Donor-derived brain tumor following neural stem cell transplantation in an ataxia telangiectasia patient. PLoS Med. 2009;6(2):e1000029.
37. Han L, He H, Yang Y, Meng Q, Ye F, Chen G, et al. Distinctive Clinical and Pathologic Features of Immature Teratomas Arising from Induced Pluripotent Stem Cell-Derived Beta Cell Injection in a Diabetes Patient. Stem Cells Dev. 2022;31(5–6):97–101.

38. Carlsson PO, Espes D, Sedigh A, Rotem A, Zimerman B, Grinberg H, et al. Transplantation of macroencapsulated human islets within the bioartificial pancreas betaAir to patients with type 1 diabetes mellitus. Am J Transplant. 2018;18(7):1735–44.
39. Ramzy A, Thompson DM, Ward-Hartstonge KA, Ivison S, Cook L, Garcia RV, et al. Implanted pluripotent stem-cell-derived pancreatic endoderm cells secrete glucose-responsive C-peptide in patients with type 1 diabetes. Cell Stem Cell. 2021;28(12):2047–61 e5.
40. Shapiro AMJ, Thompson D, Donner TW, Bellin MD, Hsueh W, Pettus J, et al. Insulin expression and C-peptide in type 1 diabetes subjects implanted with stem cell-derived pancreatic endoderm cells in an encapsulation device. Cell Rep Med. 2021;2(12):100466.
41. Kelly OG, Chan MY, Martinson LA, Kadoya K, Ostertag TM, Ross KG, et al. Cell-surface markers for the isolation of pancreatic cell types derived from human embryonic stem cells. Nat Biotechnol. 2011;29(8):750–6.
42. Rezania A, Bruin JE, Riedel MJ, Mojibian M, Asadi A, Xu J, et al. Maturation of human embryonic stem cell-derived pancreatic progenitors into functional islets capable of treating pre-existing diabetes in mice. Diabetes. 2012;61(8):2016–29.
43. Schulz TC, Young HY, Agulnick AD, Babin MJ, Baetge EE, Bang AG, et al. A scalable system for production of functional pancreatic progenitors from human embryonic stem cells. PLoS One. 2012;7(5):e37004.
44. Kirk K, Hao E, Lahmy R, Itkin-Ansari P. Human embryonic stem cell derived islet progenitors mature inside an encapsulation device without evidence of increased biomass or cell escape. Stem Cell Res. 2014;12(3):807–14.
45. Cuesta-Gomez N, Verhoeff K, Jasra IT, Pawlick R, Dadheech N, Shapiro AMJ. Characterization of stem-cell-derived islets during differentiation and after implantation. Cell Rep. 2022;40(8):111238.
46. Ameri J, Borup R, Prawiro C, Ramond C, Schachter KA, Scharfmann R, et al. Efficient Generation of Glucose-Responsive Beta Cells from Isolated GP2(+) Human Pancreatic Progenitors. Cell Rep. 2017;19(1):36–49.
47. Weng C, Xi J, Li H, Cui J, Gu A, Lai S, et al. Single-cell lineage analysis reveals extensive multimodal transcriptional control during directed beta-cell differentiation. Nat Metab. 2020;2(12):1443–58.
48. Alvarez-Dominguez JR, Donaghey J, Rasouli N, Kenty JHR, Helman A, Charlton J, et al. Circadian Entrainment Triggers Maturation of Human In Vitro Islets. Cell Stem Cell. 2020;26(1):108–22.e10.
49. Mahaddalkar PU, Scheibner K, Pfluger S, Ansarullah, Sterr M, Beckenbauer J, et al. Generation of pancreatic beta cells from CD177(+) anterior definitive endoderm. Nat Biotechnol. 2020;38(9):1061–72.
50. Augsornworawat P, Maxwell KG, Velazco-Cruz L, Millman JR. Single-Cell Transcriptome Profiling Reveals beta Cell Maturation in Stem Cell-Derived Islets after Transplantation. Cell Rep. 2020;32(8):108067.
51. Velazco-Cruz L, Goedegebuure MM, Millman JR. Advances Toward Engineering Functionally Mature Human Pluripotent Stem Cell-Derived beta Cells. Front Bioeng Biotechnol. 2020;8:786.
52. Jiang W, Sui X, Zhang D, Liu M, Ding M, Shi Y, et al. CD24: a novel surface marker for PDX1-positive pancreatic progenitors derived from human embryonic stem cells. Stem Cells. 2011;29(4):609–17.
53. Ramond C, Glaser N, Berthault C, Ameri J, Kirkegaard JS, Hansson M, et al. Reconstructing human pancreatic differentiation by mapping specific cell populations during development. Elife. 2017;6.
54. Cogger KF, Sinha A, Sarangi F, McGaugh EC, Saunders D, Dorrell C, et al. Glycoprotein 2 is a specific cell surface marker of human pancreatic progenitors. Nat Commun. 2017;8(1):331.

55. Aghazadeh Y, Sarangi F, Poon F, Nkennor B, McGaugh EC, Nunes SS, et al. GP2-enriched pancreatic progenitors give rise to functional beta cells in vivo and eliminate the risk of teratoma formation. Stem Cell Reports. 2022;17(4):964–78.
56. Draper JS, Smith K, Gokhale P, Moore HD, Maltby E, Johnson J, et al. Recurrent gain of chromosomes 17q and 12 in cultured human embryonic stem cells. Nat Biotechnol. 2004;22(1):53–4.
57. Inzunza J, Sahlen S, Holmberg K, Stromberg AM, Teerijoki H, Blennow E, et al. Comparative genomic hybridization and karyotyping of human embryonic stem cells reveals the occurrence of an isodicentric X chromosome after long-term cultivation. Mol Hum Reprod. 2004;10(6):461–6.
58. Imreh MP, Gertow K, Cedervall J, Unger C, Holmberg K, Szoke K, et al. In vitro culture conditions favoring selection of chromosomal abnormalities in human ES cells. J Cell Biochem. 2006;99(2):508–16.
59. Baker DE, Harrison NJ, Maltby E, Smith K, Moore HD, Shaw PJ, et al. Adaptation to culture of human embryonic stem cells and oncogenesis in vivo. Nat Biotechnol. 2007;25(2):207–15.
60. Lefort N, Feyeux M, Bas C, Feraud O, Bennaceur-Griscelli A, Tachdjian G, et al. Human embryonic stem cells reveal recurrent genomic instability at 20q11.21. Nat Biotechnol. 2008;26(12):1364–6.
61. Wu H, Kim KJ, Mehta K, Paxia S, Sundstrom A, Anantharaman T, et al. Copy number variant analysis of human embryonic stem cells. Stem Cells. 2008;26(6):1484–9.
62. Yang S, Lin G, Tan YQ, Deng LY, Yuan D, Lu GX. Differences between karyotypically normal and abnormal human embryonic stem cells. Cell Prolif. 2010;43(3):195–206.
63. Mayshar Y, Ben-David U, Lavon N, Biancotti JC, Yakir B, Clark AT, et al. Identification and classification of chromosomal aberrations in human induced pluripotent stem cells. Cell Stem Cell. 2010;7(4):521–31.
64. Gore A, Li Z, Fung HL, Young JE, Agarwal S, Antosiewicz-Bourget J, et al. Somatic coding mutations in human induced pluripotent stem cells. Nature. 2011;471(7336):63–7.
65. Hussein SM, Batada NN, Vuoristo S, Ching RW, Autio R, Narva E, et al. Copy number variation and selection during reprogramming to pluripotency. Nature. 2011;471(7336):58–62.
66. Laurent LC, Ulitsky I, Slavin I, Tran H, Schork A, Morey R, et al. Dynamic changes in the copy number of pluripotency and cell proliferation genes in human ESCs and iPSCs during reprogramming and time in culture. Cell Stem Cell. 2011;8(1):106–18.
67. The International Stem Cell Initiative I, Amps K, Andrews PW, Anyfantis G, Armstrong L, Avery S, et al. Screening ethnically diverse human embryonic stem cells identifies a chromosome 20 minimal amplicon conferring growth advantage. Nat Biotechnol. 2011;29(12):1132–44.
68. Martins-Taylor K, Nisler BS, Taapken SM, Compton T, Crandall L, Montgomery KD, et al. Recurrent copy number variations in human induced pluripotent stem cells. Nat Biotechnol. 2011;29(6):488–91.
69. Taapken SM, Nisler BS, Newton MA, Sampsell-Barron TL, Leonhard KA, McIntire EM, et al. Karotypic abnormalities in human induced pluripotent stem cells and embryonic stem cells. Nat Biotechnol. 2011;29(4):313–4.
70. Avery S, Hirst AJ, Baker D, Lim CY, Alagaratnam S, Skotheim RI, et al. BCL-XL mediates the strong selective advantage of a 20q11.21 amplification commonly found in human embryonic stem cell cultures. Stem Cell Reports. 2013;1(5):379–86.
71. Weissbein U, Ben-David U, Benvenisty N. Virtual karyotyping reveals greater chromosomal stability in neural cells derived by transdifferentiation than those from stem cells. Cell Stem Cell. 2014;15(6):687–91.
72. Narva E, Autio R, Rahkonen N, Kong L, Harrison N, Kitsberg D, et al. High-resolution DNA analysis of human embryonic stem cell lines reveals culture-induced copy number changes and loss of heterozygosity. Nat Biotechnol. 2010;28(4):371–7.

73. Garitaonandia I, Amir H, Boscolo FS, Wambua GK, Schultheisz HL, Sabatini K, et al. Increased risk of genetic and epigenetic instability in human embryonic stem cells associated with specific culture conditions. PLoS One. 2015;10(2):e0118307.
74. Baker D, Hirst AJ, Gokhale PJ, Juarez MA, Williams S, Wheeler M, et al. Detecting Genetic Mosaicism in Cultures of Human Pluripotent Stem Cells. Stem Cell Reports. 2016;7(5):998–1012.
75. Merkle FT, Ghosh S, Kamitaki N, Mitchell J, Avior Y, Mello C, et al. Human pluripotent stem cells recurrently acquire and expand dominant negative P53 mutations. Nature. 2017;545(7653):229–33.
76. Amir H, Touboul T, Sabatini K, Chhabra D, Garitaonandia I, Loring JF, et al. Spontaneous Single-Copy Loss of TP53 in Human Embryonic Stem Cells Markedly Increases Cell Proliferation and Survival. Stem Cells. 2017;35(4):872–85.
77. Assou S, Girault N, Plinet M, Bouckenheimer J, Sansac C, Combe M, et al. Recurrent Genetic Abnormalities in Human Pluripotent Stem Cells: Definition and Routine Detection in Culture Supernatant by Targeted Droplet Digital PCR. Stem Cell Reports. 2020;14(1):1–8.
78. Avior Y, Lezmi E, Eggan K, Benvenisty N. Cancer-Related Mutations Identified in Primed Human Pluripotent Stem Cells. Cell Stem Cell. 2021;28(1):10–1.
79. Halliwell JA, Baker D, Judge K, Quail MA, Oliver K, Betteridge E, et al. Nanopore Sequencing Indicates That Tandem Amplification of Chromosome 20q11.21 in Human Pluripotent Stem Cells Is Driven by Break-Induced Replication. Stem Cells Dev. 2021;30(11):578–86.
80. Lezmi E, Benvenisty N. Identification of cancer-related mutations in human pluripotent stem cells using RNA-seq analysis. Nat Protoc. 2021;16(9):4522–37.
81. Merkle FT, Ghosh S, Genovese G, Handsaker RE, Kashin S, Meyer D, et al. Whole-genome analysis of human embryonic stem cells enables rational line selection based on genetic variation. Cell Stem Cell. 2022;29(3):472–86 e7.
82. Bar S, Benvenisty N. Epigenetic aberrations in human pluripotent stem cells. EMBO J. 2019;38(12).
83. Maitra A, Arking DE, Shivapurkar N, Ikeda M, Stastny V, Kassauei K, et al. Genomic alterations in cultured human embryonic stem cells. Nat Genet. 2005;37(10):1099–103.
84. Mitalipova MM, Rao RR, Hoyer DM, Johnson JA, Meisner LF, Jones KL, et al. Preserving the genetic integrity of human embryonic stem cells. Nat Biotechnol. 2005;23(1):19–20.
85. Garber K. RIKEN suspends first clinical trial involving induced pluripotent stem cells. Nat Biotechnol. 2015;33(9):890–1.
86. Ben-David U, Benvenisty N. The tumorigenicity of human embryonic and induced pluripotent stem cells. Nat Rev Cancer. 2011;11(4):268–77.
87. Steichen C, Hannoun Z, Luce E, Hauet T, Dubart-Kupperschmitt A. Genomic integrity of human induced pluripotent stem cells: Reprogramming, differentiation and applications. World J Stem Cells. 2019;11(10):729–47.
88. Kosanke M, Osetek K, Haase A, Wiehlmann L, Davenport C, Schwarzer A, et al. Reprogramming enriches for somatic cell clones with small-scale mutations in cancer-associated genes. Mol Ther. 2021;29(8):2535–53.
89. Enver T, Soneji S, Joshi C, Brown J, Iborra F, Orntoft T, et al. Cellular differentiation hierarchies in normal and culture-adapted human embryonic stem cells. Hum Mol Genet. 2005;14(21):3129–40.
90. Harrison NJ, Baker D, Andrews PW. Culture adaptation of embryonic stem cells echoes germ cell malignancy. Int J Androl. 2007;30(4):275–81; discussion 81.
91. Werbowetski-Ogilvie TE, Bosse M, Stewart M, Schnerch A, Ramos-Mejia V, Rouleau A, et al. Characterization of human embryonic stem cells with features of neoplastic progression. Nat Biotechnol. 2009;27(1):91–7.
92. Olariu V, Harrison NJ, Coca D, Gokhale PJ, Baker D, Billings S, et al. Modeling the evolution of culture-adapted human embryonic stem cells. Stem Cell Res. 2010;4(1):50–6.
93. Na J, Baker D, Zhang J, Andrews PW, Barbaric I. Aneuploidy in pluripotent stem cells and implications for cancerous transformation. Protein Cell. 2014;5(8):569–79.

94. Barbaric I, Biga V, Gokhale PJ, Jones M, Stavish D, Glen A, et al. Time-lapse analysis of human embryonic stem cells reveals multiple bottlenecks restricting colony formation and their relief upon culture adaptation. Stem Cell Reports. 2014;3(1):142–55.
95. Kim YJ, Go YH, Jeong HC, Kwon EJ, Kim SM, Cheong HS, et al. TPX2 prompts mitotic survival via the induction of BCL2L1 through YAP1 protein stabilization in human embryonic stem cells. Exp Mol Med. 2023;55(1):32–42.
96. Price CJ, Stavish D, Gokhale PJ, Stevenson BA, Sargeant S, Lacey J, et al. Genetically variant human pluripotent stem cells selectively eliminate wild-type counterparts through YAP-mediated cell competition. Dev Cell. 2021;56(17):2455–70 e10.
97. Hindley C, Philpott A. The cell cycle and pluripotency. Biochem J. 2013;451(2):135–43.
98. Liu L, Michowski W, Kolodziejczyk A, Sicinski P. The cell cycle in stem cell proliferation, pluripotency and differentiation. Nat Cell Biol. 2019;21(9):1060–7.
99. Ballabeni A, Park IH, Zhao R, Wang W, Lerou PH, Daley GQ, et al. Cell cycle adaptations of embryonic stem cells. Proc Natl Acad Sci U S A. 2011;108(48):19252–7.
100. Nguyen HT, Geens M, Mertzanidou A, Jacobs K, Heirman C, Breckpot K, et al. Gain of 20q11.21 in human embryonic stem cells improves cell survival by increased expression of Bcl-xL. Mol Hum Reprod. 2014;20(2):168–77.
101. Fazeli A, Liew CG, Matin MM, Elliott S, Jeanmeure LF, Wright PC, et al. Altered patterns of differentiation in karyotypically abnormal human embryonic stem cells. Int J Dev Biol. 2011;55(2):175–80.
102. Abyzov A, Mariani J, Palejev D, Zhang Y, Haney MS, Tomasini L, et al. Somatic copy number mosaicism in human skin revealed by induced pluripotent stem cells. Nature. 2012;492(7429):438–42.
103. Caisander G, Park H, Frej K, Lindqvist J, Bergh C, Lundin K, et al. Chromosomal integrity maintained in five human embryonic stem cell lines after prolonged in vitro culture. Chromosome Res. 2006;14(2):131–7.
104. Catalina P, Montes R, Ligero G, Sanchez L, de la Cueva T, Bueno C, et al. Human ESCs predisposition to karyotypic instability: Is a matter of culture adaptation or differential vulnerability among hESC lines due to inherent properties? Mol Cancer. 2008;7:76.
105. Cowan CA, Klimanskaya I, McMahon J, Atienza J, Witmyer J, Zucker JP, et al. Derivation of embryonic stem-cell lines from human blastocysts. N Engl J Med. 2004;350(13):1353–6.
106. Elliott AM, Elliott KA, Kammesheidt A. High resolution array-CGH characterization of human stem cells using a stem cell focused microarray. Mol Biotechnol. 2010;46(3):234–42.
107. Gertow K, Cedervall J, Unger C, Szoke K, Blennow E, Imreh MP, et al. Trisomy 12 in HESC leads to no selective in vivo growth advantage in teratomas, but induces an increased abundance of renal development. J Cell Biochem. 2007;100(6):1518–25.
108. Guo CW, Kawakatsu M, Idemitsu M, Urata Y, Goto S, Ono Y, et al. Culture under low physiological oxygen conditions improves the stemness and quality of induced pluripotent stem cells. J Cell Physiol. 2013;228(11):2159–66.
109. Hasegawa K, Fujioka T, Nakamura Y, Nakatsuji N, Suemori H. A method for the selection of human embryonic stem cell sublines with high replating efficiency after single-cell dissociation. Stem Cells. 2006;24(12):2649–60.
110. Herszfeld D, Wolvetang E, Langton-Bunker E, Chung TL, Filipczyk AA, Houssami S, et al. CD30 is a survival factor and a biomarker for transformed human pluripotent stem cells. Nat Biotechnol. 2006;24(3):351–7.
111. Kanchan K, Iyer K, Yanek LR, Carcamo-Orive I, Taub MA, Malley C, et al. Genomic integrity of human induced pluripotent stem cells across nine studies in the NHLBI NextGen program. Stem Cell Res. 2020;46:101803.
112. Kilpinen H, Goncalves A, Leha A, Afzal V, Alasoo K, Ashford S, et al. Common genetic variation drives molecular heterogeneity in human iPSCs. Nature. 2017;546(7658):370–5.
113. Lagarkova MA, Volchkov PY, Philonenko ES, Pfannkuche K, Prokhorovich MA, Zabotina T, et al. CD 30 is a marker of undifferentiated human embryonic stem cells rather than a biomarker of transformed hESCs. Cell Cycle. 2008;7(22):3610–2.

114. Ludwig TE, Levenstein ME, Jones JM, Berggren WT, Mitchen ER, Frane JL, et al. Derivation of human embryonic stem cells in defined conditions. Nat Biotechnol. 2006;24(2):185–7.
115. Luo L, Kawakatsu M, Guo CW, Urata Y, Huang WJ, Ali H, et al. Effects of antioxidants on the quality and genomic stability of induced pluripotent stem cells. Sci Rep. 2014;4:3779.
116. Ma H, Morey R, O'Neil RC, He Y, Daughtry B, Schultz MD, et al. Abnormalities in human pluripotent cells due to reprogramming mechanisms. Nature. 2014;511(7508):177–83.
117. Mills JA, Wang K, Paluru P, Ying L, Lu L, Galvao AM, et al. Clonal genetic and hematopoietic heterogeneity among human-induced pluripotent stem cell lines. Blood. 2013;122(12):2047–51.
118. Spits C, Mateizel I, Geens M, Mertzanidou A, Staessen C, Vandeskelde Y, et al. Recurrent chromosomal abnormalities in human embryonic stem cells. Nat Biotechnol. 2008;26(12):1361–3.
119. Ware CB, Nelson AM, Blau CA. A comparison of NIH-approved human ESC lines. Stem Cells. 2006;24(12):2677–84.
120. Russnes HG, Vollan HKM, Lingjaerde OC, Krasnitz A, Lundin P, Naume B, et al. Genomic architecture characterizes tumor progression paths and fate in breast cancer patients. Sci Transl Med. 2010;2(38):38ra47.
121. Privitera AP, Barresi V, Condorelli DF. Aberrations of Chromosomes 1 and 16 in Breast Cancer: A Framework for Cooperation of Transcriptionally Dysregulated Genes. Cancers (Basel). 2021;13(7).
122. Schmidt TM, Fonseca R, Usmani SZ. Chromosome 1q21 abnormalities in multiple myeloma. Blood Cancer J. 2021;11(4):83.
123. Cortes-Gutierrez EI, Davila-Rodriguez MI, Muraira-Rodriguez M, Said-Fernandez S, Cerda-Flores RM. Association between the stages of cervical cancer and chromosome 1 aneusomy. Cancer Genet Cytogenet. 2005;159(1):44–7.
124. Wang L, Guo ZY, Zhang R, Xin B, Chen R, Zhao J, et al. Pseudogene OCT4-pg4 functions as a natural micro RNA sponge to regulate OCT4 expression by competing for miR-145 in hepatocellular carcinoma. Carcinogenesis. 2013;34(8):1773–81.
125. Oosterhuis JW, Looijenga LHJ. Human germ cell tumours from a developmental perspective. Nat Rev Cancer. 2019;19(9):522–37.
126. Abruzzo LV, Herling CD, Calin GA, Oakes C, Barron LL, Banks HE, et al. Trisomy 12 chronic lymphocytic leukemia expresses a unique set of activated and targetable pathways. Haematologica. 2018;103(12):2069–78.
127. Muleris M, Laurent-Puig P, Salmon RJ, Thomas G, Dutrillaux B. Chromosome 12 alterations and c-Ki-ras mutations in colorectal tumors. Cancer Genet Cytogenet. 1993;69(2):161–2.
128. Ben-David U, Arad G, Weissbein U, Mandefro B, Maimon A, Golan-Lev T, et al. Aneuploidy induces profound changes in gene expression, proliferation and tumorigenicity of human pluripotent stem cells. Nat Commun. 2014;5:4825.
129. Zhao S, Xu J, Liu S, Cui K, Li Z, Liu N. Dppa3 in pluripotency maintenance of ES cells and early embryogenesis. J Cell Biochem. 2019;120(4):4794–9.
130. Looijenga LH, Zafarana G, Grygalewicz B, Summersgill B, Debiec-Rychter M, Veltman J, et al. Role of gain of 12p in germ cell tumour development. APMIS. 2003;111(1):161–71; discussion 72–3.
131. Zhang W, Yu Y. The important molecular markers on chromosome 17 and their clinical impact in breast cancer. Int J Mol Sci. 2011;12(9):5672–83.
132. Lee YK, Lee TS, Song IH, Jeong HY, Kang SJ, Kim MW, et al. Inhibition of pulmonary cancer progression by epidermal growth factor receptor-targeted transfection with Bcl-2 and survivin siRNAs. Cancer Gene Ther. 2015;22(7):335–43.
133. Zhang WY, de Almeida PE, Wu JC. Teratoma formation: A tool for monitoring pluripotency in stem cell research. StemBook. Cambridge (MA)2012.
134. Fearon ER, Cho KR, Nigro JM, Kern SE, Simons JW, Ruppert JM, et al. Identification of a chromosome 18q gene that is altered in colorectal cancers. Science. 1990;247(4938):49–56.

135. Tanaka T, Watanabe T, Kazama Y, Tanaka J, Kanazawa T, Kazama S, et al. Chromosome 18q deletion and Smad4 protein inactivation correlate with liver metastasis: A study matched for T- and N- classification. Br J Cancer. 2006;95(11):1562–7.
136. Ando T, Ishiguro H, Kimura M, Mitsui A, Mori Y, Sugito N, et al. Frequent loss of the long arm of chromosome 18 in esophageal squamous cell carcinoma. Oncol Rep. 2007;17(5):1005–11.
137. Kluth M, Graunke M, Moller-Koop C, Hube-Magg C, Minner S, Michl U, et al. Deletion of 18q is a strong and independent prognostic feature in prostate cancer. Oncotarget. 2016;7(52):86339–49.
138. Hodgson JG, Chin K, Collins C, Gray JW. Genome amplification of chromosome 20 in breast cancer. Breast Cancer Res Treat. 2003;78(3):337–45.
139. Scotto L, Narayan G, Nandula SV, Arias-Pulido H, Subramaniyam S, Schneider A, et al. Identification of copy number gain and overexpressed genes on chromosome arm 20q by an integrative genomic approach in cervical cancer: potential role in progression. Genes Chromosomes Cancer. 2008;47(9):755–65.
140. Maffei M, Mongera S, Terpstra L, Donadini A, Voorham QJ, Meijer GA, et al. Chromosome 20 aberrations at the diploid-aneuploid transition in sporadic colorectal cancer. Cytogenet Genome Res. 2014;144(1):9–14.
141. Voutsadakis IA. Chromosome 20q11.21 Amplifications in Colorectal Cancer. Cancer Genomics Proteomics. 2021;18(3 Suppl):487–96.
142. Zhang J, Hirst AJ, Duan F, Qiu H, Huang R, Ji Y, et al. Anti-apoptotic Mutations Desensitize Human Pluripotent Stem Cells to Mitotic Stress and Enable Aneuploid Cell Survival. Stem Cell Reports. 2019;12(3):557–71.
143. Markouli C, Couvreu De Deckersberg E, Regin M, Nguyen HT, Zambelli F, Keller A, et al. Gain of 20q11.21 in Human Pluripotent Stem Cells Impairs TGF-beta-Dependent Neuroectodermal Commitment. Stem Cell Reports. 2019;13(1):163–76.
144. Poh AR, O'Donoghue RJ, Ernst M. Hematopoietic cell kinase (HCK) as a therapeutic target in immune and cancer cells. Oncotarget. 2015;6(18):15752–71.
145. Di Oto E, Monti V, Cucchi MC, Masetti R, Varga Z, Foschini MP. X chromosome gain in male breast cancer. Hum Pathol. 2015;46(12):1908–12.
146. Ropke A, Erbersdobler A, Hammerer P, Palisaar J, John K, Stumm M, et al. Gain of androgen receptor gene copies in primary prostate cancer due to X chromosome polysomy. Prostate. 2004;59(1):59–68.
147. Cerami E, Gao J, Dogrusoz U, Gross BE, Sumer SO, Aksoy BA, et al. The cBio cancer genomics portal: an open platform for exploring multidimensional cancer genomics data. Cancer Discov. 2012;2(5):401–4.
148. Gao J, Aksoy BA, Dogrusoz U, Dresdner G, Gross B, Sumer SO, et al. Integrative analysis of complex cancer genomics and clinical profiles using the cBioPortal. Sci Signal. 2013;6(269):pl1.
149. Friedman NB, Moore RA. Tumors of the testis; a report on 922 cases. Mil Surg. 1946;99(5):573–93.
150. Kleinsmith LJ, Pierce GB, Jr. Multipotentiality of Single Embryonal Carcinoma Cells. Cancer Res. 1964;24:1544–51.
151. Damjanov I, Solter D. Experimental teratoma. Curr Top Pathol. 1974;59:69–130.
152. Andrews PW. From teratocarcinomas to embryonic stem cells. Philos Trans R Soc Lond B Biol Sci. 2002;357(1420):405–17.
153. Andrews PW, Matin MM, Bahrami AR, Damjanov I, Gokhale P, Draper JS. Embryonic stem (ES) cells and embryonal carcinoma (EC) cells: opposite sides of the same coin. Biochem Soc Trans. 2005;33(Pt 6):1526–30.
154. Zwaka TP, Thomson JA. A germ cell origin of embryonic stem cells? Development. 2005;132(2):227–33.
155. Solter D. From teratocarcinomas to embryonic stem cells and beyond: a history of embryonic stem cell research. Nat Rev Genet. 2006;7(4):319–27.

156. Cunningham JJ, Ulbright TM, Pera MF, Looijenga LH. Lessons from human teratomas to guide development of safe stem cell therapies. Nat Biotechnol. 2012;30(9):849–57.
157. Stevens LC, Little CC. Spontaneous Testicular Teratomas in an Inbred Strain of Mice. Proc Natl Acad Sci U S A. 1954;40(11):1080–7.
158. Pierce GB. Teratocarcinoma: model for a developmental concept of cancer. Curr Top Dev Biol. 1967;2:223–46.
159. Stevens LC. The biology of teratomas. Adv Morphog. 1967;6:1–31.
160. Stevens LC. The development of transplantable teratocarcinomas from intratesticular grafts of pre- and postimplantation mouse embryos. Dev Biol. 1970;21(3):364–82.
161. Solter D, Skreb N, Damjanov I. Extrauterine growth of mouse egg-cylinders results in malignant teratoma. Nature. 1970;227(5257):503–4.
162. Brinster RL. Embryo development. J Anim Sci. 1974;38(5):1003–12.
163. Finch BW, Ephrussi B. Retention of multiple developmental potentialities by cells of a mouse terticular teratocarcinoma during prolonged culture *in vitro* and their extinction upon hybridization with cells of permanent lines. Proc Natl Acad Sci U S A. 1967;57(3):615–21.
164. Kahan BW, Ephrussi B. Developmental potentialities of clonal in vitro cultures of mouse testicular teratoma. J Natl Cancer Inst. 1970;44(5):1015–36.
165. Martin GR, Evans MJ. The morphology and growth of a pluripotent teratocarcinoma cell line and its derivatives in tissue culture. Cell. 1974;2(3):163–72.
166. Evans MJ. The isolation and properties of a clonal tissue culture strain of pluripotent mouse teratoma cells. J Embryol Exp Morphol. 1972;28(1):163–76.
167. Damjanov I, Andrews PW. Teratomas produced from human pluripotent stem cells xenografted into immunodeficient mice - a histopathology atlas. Int J Dev Biol. 2016;60(10-11-12):337–419.
168. Damjanov I, Andrews PW. The terminology of teratocarcinomas and teratomas. Nat Biotechnol. 2007;25(11):1212; discussion
169. Atkin NB, Baker MC. Specific chromosome change, i(12p), in testicular tumours? Lancet. 1982;2(8311):1349.
170. Cheng L, Albers P, Berney DM, Feldman DR, Daugaard G, Gilligan T, et al. Testicular cancer. Nat Rev Dis Primers. 2018;4(1):29.
171. Sillars-Hardebol AH, Carvalho B, van Engeland M, Fijneman RJ, Meijer GA. The adenoma hunt in colorectal cancer screening: defining the target. J Pathol. 2012;226(1):1–6.
172. Rhee I, Bachman KE, Park BH, Jair KW, Yen RW, Schuebel KE, et al. DNMT1 and DNMT3b cooperate to silence genes in human cancer cells. Nature. 2002;416(6880):552–6.
173. Hu XM, Lin T, Huang XY, Gan RH, Zhao Y, Feng Y, et al. ID1 contributes to cell growth invasion and migration in salivary adenoid cystic carcinoma. Mol Med Rep. 2017;16(6):8907–15.
174. Hsu CW, Chen YC, Su HH, Huang GJ, Shu CW, Wu TT, et al. Targeting TPX2 Suppresses the Tumorigenesis of Hepatocellular Carcinoma Cells Resulting in Arrested Mitotic Phase Progression and Increased Genomic Instability. J Cancer. 2017;8(8):1378–94.
175. Jeong HC, Go YH, Shin JG, Kim YJ, Cho MG, Gwon D, et al. TPX2 Amplification-Driven Aberrant Mitosis in Culture Adapted Human Embryonic Stem Cells with gain of 20q11.21. Stem Cell Rev Rep. 2023.
176. Zhao L, Liu J, Chen S, Fang C, Zhang X, Luo Z. Prognostic significance of NANOG expression in solid tumors: a meta-analysis. Onco Targets Ther. 2018;11:5515–26.
177. Kubara K, Yamazaki K, Ishihara Y, Naruto T, Lin HT, Nishimura K, et al. Status of KRAS in iPSCs Impacts upon Self-Renewal and Differentiation Propensity. Stem Cell Reports. 2018;11(2):380–94.
178. Keller A, Spits C. The Impact of Acquired Genetic Abnormalities on the Clinical Translation of Human Pluripotent Stem Cells. Cells. 2021;10(11).
179. Bouma MJ, van Iterson M, Janssen B, Mummery CL, Salvatori DCF, Freund C. Differentiation-Defective Human Induced Pluripotent Stem Cells Reveal Strengths and Limitations of the Teratoma Assay and In Vitro Pluripotency Assays. Stem Cell Reports. 2017;8(5):1340–53.

180. Orsetti B, Nugoli M, Cervera N, Lasorsa L, Chuchana P, Ursule L, et al. Genomic and expression profiling of chromosome 17 in breast cancer reveals complex patterns of alterations and novel candidate genes. Cancer Res. 2004;64(18):6453–60.
181. Lane DP. Cancer. p53, guardian of the genome. Nature. 1992;358(6381):15–6.
182. Shaw PH. The role of p53 in cell cycle regulation. Pathol Res Pract. 1996;192(7):669–75.
183. Kandoth C, McLellan MD, Vandin F, Ye K, Niu B, Lu C, et al. Mutational landscape and significance across 12 major cancer types. Nature. 2013;502(7471):333–9.
184. Chen J. The Cell-Cycle Arrest and Apoptotic Functions of p53 in Tumor Initiation and Progression. Cold Spring Harb Perspect Med. 2016;6(3):a026104.
185. Petitjean A, Mathe E, Kato S, Ishioka C, Tavtigian SV, Hainaut P, et al. Impact of mutant p53 functional properties on TP53 mutation patterns and tumor phenotype: lessons from recent developments in the IARC TP53 database. Hum Mutat. 2007;28(6):622–9.
186. Baugh EH, Ke H, Levine AJ, Bonneau RA, Chan CS. Why are there hotspot mutations in the TP53 gene in human cancers? Cell Death Differ. 2018;25(1):154–60.
187. Stein Y, Aloni-Grinstein R, Rotter V. Mutant p53 oncogenicity: dominant-negative or gain-of-function? Carcinogenesis. 2020;41(12):1635–47.
188. Azuhata T, Scott D, Takamizawa S, Wen J, Davidoff A, Fukuzawa M, et al. The inhibitor of apoptosis protein survivin is associated with high-risk behavior of neuroblastoma. J Pediatr Surg. 2001;36(12):1785–91.
189. Blum B, Bar-Nur O, Golan-Lev T, Benvenisty N. The anti-apoptotic gene survivin contributes to teratoma formation by human embryonic stem cells. Nat Biotechnol. 2009;27(3):281–7.
190. Rouhani FJ, Zou X, Danecek P, Badja C, Amarante TD, Koh G, et al. Substantial somatic genomic variation and selection for BCOR mutations in human induced pluripotent stem cells. Nat Genet. 2022;54(9):1406–16.
191. International Society for Stem Cell Research I. Guidelines for Stem Cell Research and Clinical Translation. Internet]. ISSCR; June 05 2023, Available from: https://www.isscr.org/standards-document.
192. Tate JG, Bamford S, Jubb HC, Sondka Z, Beare DM, Bindal N, et al. COSMIC: the Catalogue Of Somatic Mutations In Cancer. Nucleic Acids Res. 2019;47(D1):D941–D7.
193. Astolfi A, Fiore M, Melchionda F, Indio V, Bertuccio SN, Pession A. BCOR involvement in cancer. Epigenomics. 2019;11(7):835–55.
194. Wang Z, Gearhart MD, Lee YW, Kumar I, Ramazanov B, Zhang Y, et al. A Non-canonical BCOR-PRC1.1 Complex Represses Differentiation Programs in Human ESCs. Cell Stem Cell. 2018;22(2):235–51 e9.
195. Nishino K, Toyoda M, Yamazaki-Inoue M, Fukawatase Y, Chikazawa E, Sakaguchi H, et al. DNA methylation dynamics in human induced pluripotent stem cells over time. PLoS Genet. 2011;7(5):e1002085.
196. Lister R, Pelizzola M, Kida YS, Hawkins RD, Nery JR, Hon G, et al. Hotspots of aberrant epigenomic reprogramming in human induced pluripotent stem cells. Nature. 2011;471(7336):68–73.
197. Lund RJ, Narva E, Lahesmaa R. Genetic and epigenetic stability of human pluripotent stem cells. Nat Rev Genet. 2012;13(10):732–44.
198. Weissbein U, Plotnik O, Vershkov D, Benvenisty N. Culture-induced recurrent epigenetic aberrations in human pluripotent stem cells. PLoS Genet. 2017;13(8):e1006979.
199. Poetsch MS, Strano A, Guan K. Human Induced Pluripotent Stem Cells: From Cell Origin, Genomic Stability, and Epigenetic Memory to Translational Medicine. Stem Cells. 2022;40(6):546–55.
200. Tompkins JD, Hall C, Chen VC, Li AX, Wu X, Hsu D, et al. Epigenetic stability, adaptability, and reversibility in human embryonic stem cells. Proc Natl Acad Sci U S A. 2012;109(31):12544–9.
201. Hu Q, Rosenfeld MG. Epigenetic regulation of human embryonic stem cells. Front Genet. 2012;3:238.

202. Kim K, Doi A, Wen B, Ng K, Zhao R, Cahan P, et al. Epigenetic memory in induced pluripotent stem cells. Nature. 2010;467(7313):285–90.
203. Simpson DJ, Olova NN, Chandra T. Cellular reprogramming and epigenetic rejuvenation. Clin Epigenetics. 2021;13(1):170.
204. Bhattacharjee D, Shenoy S, Bairy KL. DNA Methylation and Chromatin Remodeling: The Blueprint of Cancer Epigenetics. Scientifica (Cairo). 2016;2016:6072357.
205. French R, Pauklin S. Epigenetic regulation of cancer stem cell formation and maintenance. Int J Cancer. 2021;148(12):2884–97.
206. Nebbioso A, Tambaro FP, Dell'Aversana C, Altucci L. Cancer epigenetics: Moving forward. PLoS Genet. 2018;14(6):e1007362.
207. Kim I, Park JW. Hypoxia-driven epigenetic regulation in cancer progression: A focus on histone methylation and its modifying enzymes. Cancer Lett. 2020;489:41–9.
208. Nishizawa M, Chonabayashi K, Nomura M, Tanaka A, Nakamura M, Inagaki A, et al. Epigenetic Variation between Human Induced Pluripotent Stem Cell Lines Is an Indicator of Differentiation Capacity. Cell Stem Cell. 2016;19(3):341–54.
209. Keller A, Dziedzicka D, Zambelli F, Markouli C, Sermon K, Spits C, et al. Genetic and epigenetic factors which modulate differentiation propensity in human pluripotent stem cells. Hum Reprod Update. 2018;24(2):162–75.
210. Liu LP, Zheng YW. Predicting differentiation potential of human pluripotent stem cells: Possibilities and challenges. World J Stem Cells. 2019;11(7):375–82.
211. de Boni L, Gasparoni G, Haubenreich C, Tierling S, Schmitt I, Peitz M, et al. DNA methylation alterations in iPSC- and hESC-derived neurons: potential implications for neurological disease modeling. Clin Epigenetics. 2018;10:13.
212. Diaz Perez SV, Kim R, Li Z, Marquez VE, Patel S, Plath K, et al. Derivation of new human embryonic stem cell lines reveals rapid epigenetic progression in vitro that can be prevented by chemical modification of chromatin. Hum Mol Genet. 2012;21(4):751–64.
213. Wainwright EN, Scaffidi P. Epigenetics and Cancer Stem Cells: Unleashing, Hijacking, and Restricting Cellular Plasticity. Trends Cancer. 2017;3(5):372–86.
214. Feinberg AP, Vogelstein B. Hypomethylation distinguishes genes of some human cancers from their normal counterparts. Nature. 1983;301(5895):89–92.
215. Bert SA, Robinson MD, Strbenac D, Statham AL, Song JZ, Hulf T, et al. Regional activation of the cancer genome by long-range epigenetic remodeling. Cancer Cell. 2013;23(1):9–22.
216. Wilson AS, Power BE, Molloy PL. DNA hypomethylation and human diseases. Biochim Biophys Acta. 2007;1775(1):138–62.
217. Ehrlich M, Lacey M. DNA hypomethylation and hemimethylation in cancer. Adv Exp Med Biol. 2013;754:31–56.
218. Hur K, Cejas P, Feliu J, Moreno-Rubio J, Burgos E, Boland CR, et al. Hypomethylation of long interspersed nuclear element-1 (LINE-1) leads to activation of proto-oncogenes in human colorectal cancer metastasis. Gut. 2014;63(4):635–46.
219. Geens M, Chuva De Sousa Lopes SM. X chromosome inactivation in human pluripotent stem cells as a model for human development: back to the drawing board? Hum Reprod Update. 2017;23(5):520–32.
220. Kim JS, Choi HW, Arauzo-Bravo MJ, Scholer HR, Do JT. Reactivation of the inactive X chromosome and post-transcriptional reprogramming of Xist in iPSCs. J Cell Sci. 2015;128(1):81–7.
221. Brenes AJ, Yoshikawa H, Bensaddek D, Mirauta B, Seaton D, Hukelmann JL, et al. Erosion of human X chromosome inactivation causes major remodeling of the iPSC proteome. Cell Rep. 2021;35(4):109032.
222. Bar S, Seaton LR, Weissbein U, Eldar-Geva T, Benvenisty N. Global Characterization of X Chromosome Inactivation in Human Pluripotent Stem Cells. Cell Rep. 2019;27(1):20–9 e3.
223. Looijenga LH, Gillis, A. J., van Gurp, R. J., Verkerk, A. J., Oosterhuis, J. W. X inactivation in human testicular tumors. XIST expression and androgen receptor methylation status. The American journal of pathology. 1997;151(2):581–90.

224. Lobo J, Nunes SP, Gillis AJM, Barros-Silva D, Miranda-Goncalves V, Berg AVD, et al. XIST-Promoter Demethylation as Tissue Biomarker for Testicular Germ Cell Tumors and Spermatogenesis Quality. Cancers (Basel). 2019;11(9).
225. Kawakami T, Okamoto K, Ogawa O, Okada Y. XIST unmethylated DNA fragments in male-derived plasma as a tumour marker for testicular cancer. Lancet. 2004;363(9402):40–2.
226. Cloutier M, Kumar S, Buttigieg E, Keller L, Lee B, Williams A, et al. Preventing erosion of X-chromosome inactivation in human embryonic stem cells. Nat Commun. 2022;13(1):2516.
227. Barakat TS, Ghazvini M, de Hoon B, Li T, Eussen B, Douben H, et al. Stable X chromosome reactivation in female human induced pluripotent stem cells. Stem Cell Reports. 2015;4(2):199–208.
228. Lengner CJ, Gimelbrant AA, Erwin JA, Cheng AW, Guenther MG, Welstead GG, et al. Derivation of pre-X inactivation human embryonic stem cells under physiological oxygen concentrations. Cell. 2010;141(5):872–83.
229. Xie P, Ouyang Q, Leng L, Hu L, Cheng D, Tan Y, et al. The dynamic changes of X chromosome inactivation during early culture of human embryonic stem cells. Stem Cell Res. 2016;17(1):84–92.
230. Park H, Kim D, Kim CH, Mills RE, Chang MY, Iskow RC, et al. Increased genomic integrity of an improved protein-based mouse induced pluripotent stem cell method compared with current viral-induced strategies. Stem Cells Transl Med. 2014;3(5):599–609.
231. Schlaeger TM, Daheron L, Brickler TR, Entwisle S, Chan K, Cianci A, et al. A comparison of non-integrating reprogramming methods. Nat Biotechnol. 2015;33(1):58–63.
232. Kang X, Yu Q, Huang Y, Song B, Chen Y, Gao X, et al. Effects of Integrating and Non-Integrating Reprogramming Methods on Copy Number Variation and Genomic Stability of Human Induced Pluripotent Stem Cells. PLoS One. 2015;10(7):e0131128.
233. Bhutani K, Nazor KL, Williams R, Tran H, Dai H, Dzakula Z, et al. Whole-genome mutational burden analysis of three pluripotency induction methods. Nat Commun. 2016;7:10536.
234. Turinetto V, Orlando L, Giachino C. Induced Pluripotent Stem Cells: Advances in the Quest for Genetic Stability during Reprogramming Process. Int J Mol Sci. 2017;18(9).
235. Forsyth NR, Musio A, Vezzoni P, Simpson AH, Noble BS, McWhir J. Physiologic oxygen enhances human embryonic stem cell clonal recovery and reduces chromosomal abnormalities. Cloning Stem Cells. 2006;8(1):16–23.
236. Lim HJ, Han J, Woo DH, Kim SE, Kim SK, Kang HG, et al. Biochemical and morphological effects of hypoxic environment on human embryonic stem cells in long-term culture and differentiating embryoid bodies. Mol Cells. 2011;31(2):123–32.
237. Mathieu J, Zhou W, Xing Y, Sperber H, Ferreccio A, Agoston Z, et al. Hypoxia-inducible factors have distinct and stage-specific roles during reprogramming of human cells to pluripotency. Cell Stem Cell. 2014;14(5):592–605.
238. Thompson O, von Meyenn F, Hewitt Z, Alexander J, Wood A, Weightman R, et al. Low rates of mutation in clinical grade human pluripotent stem cells under different culture conditions. Nat Commun. 2020;11(1):1528.
239. Nakamura N, Shi X, Darabi R, Li Y. Hypoxia in Cell Reprogramming and the Epigenetic Regulations. Front Cell Dev Biol. 2021;9:609984.
240. Millman JR, Tan JH, Colton CK. Mouse Pluripotent Stem Cell Differentiation Under Physiological Oxygen Reduces Residual Teratomas. Cell Mol Bioeng. 2021;14(6):555–67.
241. Molina-Ruiz FJ, Introna C, Bombau G, Galofre M, Canals JM. Standardization of Cell Culture Conditions and Routine Genomic Screening under a Quality Management System Leads to Reduced Genomic Instability in hPSCs. Cells. 2022;11(13).
242. Hentze H, Soong PL, Wang ST, Phillips BW, Putti TC, Dunn NR. Teratoma formation by human embryonic stem cells: evaluation of essential parameters for future safety studies. Stem Cell Res. 2009;2(3):198–210.
243. International Stem Cell Banking Initiative I. Consensus guidance for banking and supply of human embryonic stem cell lines for research purposes. Stem Cell Rev Rep. 2009;5(4):301–14.

244. The International Stem Cell Initiative I. Assessment of established techniques to determine developmental and malignant potential of human pluripotent stem cells. Nat Commun. 2018;9(1):1925.
245. Wesselschmidt RL. The teratoma assay: an in vivo assessment of pluripotency. Methods Mol Biol. 2011;767:231–41.
246. Bulic-Jakus F, Katusic Bojanac A, Juric-Lekic G, Vlahovic M, Sincic N. Teratoma: from spontaneous tumors to the pluripotency/malignancy assay. Wiley Interdiscip Rev Dev Biol. 2016;5(2):186–209.
247. The International Stem Cell Initiative I, Adewumi O, Aflatoonian B, Ahrlund-Richter L, Amit M, Andrews PW, et al. Characterization of human embryonic stem cell lines by the International Stem Cell Initiative. Nat Biotechnol. 2007;25(7):803–16.
248. Buta C, David R, Dressel R, Emgard M, Fuchs C, Gross U, et al. Reconsidering pluripotency tests: do we still need teratoma assays? Stem Cell Res. 2013;11(1):552–62.
249. Percie du Sert N, Ahluwalia A, Alam S, Avey MT, Baker M, Browne WJ, et al. Reporting animal research: Explanation and elaboration for the ARRIVE guidelines 2.0. PLoS Biol. 2020;18(7):e3000411.
250. Muller FJ, Goldmann J, Loser P, Loring JF. A call to standardize teratoma assays used to define human pluripotent cell lines. Cell Stem Cell. 2010;6(5):412–4.
251. Montilla-Rojo J, Bialecka M, Wever KE, Mummery CL, Looijenga LHJ, Roelen BAJ, et al. Teratoma Assay for Testing Pluripotency and Malignancy of Stem Cells: Insufficient Reporting and Uptake of Animal-Free Methods-A Systematic Review. Int J Mol Sci. 2023;24(4).
252. Gropp M, Shilo V, Vainer G, Gov M, Gil Y, Khaner H, et al. Standardization of the teratoma assay for analysis of pluripotency of human ES cells and biosafety of their differentiated progeny. PLoS One. 2012;7(9):e45532.
253. Plaks V, Kong N, Werb Z. The cancer stem cell niche: how essential is the niche in regulating stemness of tumor cells? Cell Stem Cell. 2015;16(3):225–38.
254. Junttila MR, de Sauvage FJ. Influence of tumour micro-environment heterogeneity on therapeutic response. Nature. 2013;501(7467):346–54.
255. Dressel R. Effects of histocompatibility and host immune responses on the tumorigenicity of pluripotent stem cells. Semin Immunopathol. 2011;33(6):573–91.
256. Zeevaert K, Elsafi Mabrouk MH, Wagner W, Goetzke R. Cell Mechanics in Embryoid Bodies. Cells. 2020;9(10).
257. Pekkanen-Mattila M, Pelto-Huikko M, Kujala V, Suuronen R, Skottman H, Aalto-Setala K, et al. Spatial and temporal expression pattern of germ layer markers during human embryonic stem cell differentiation in embryoid bodies. Histochem Cell Biol. 2010;133(5):595–606.
258. Hidalgo Aguilar A, Smith L, Owens D, Quelch R, Przyborski S. Recreating Tissue Structures Representative of Teratomas In Vitro Using a Combination of 3D Cell Culture Technology and Human Embryonic Stem Cells. Bioengineering (Basel). 2022;9(5).
259. Sheridan SD, Surampudi V, Rao RR. Analysis of embryoid bodies derived from human induced pluripotent stem cells as a means to assess pluripotency. Stem Cells Int. 2012;2012:738910.
260. Du F, Zhao X, Fan D. Soft Agar Colony Formation Assay as a Hallmark of Carcinogenesis. Bio Protoc. 2017;7(12):e2351.
261. Borowicz S, Van Scoyk M, Avasarala S, Karuppusamy Rathinam MK, Tauler J, Bikkavilli RK, et al. The soft agar colony formation assay. J Vis Exp. 2014(92):e51998.
262. Hamburger AW, Salmon SE. Primary bioassay of human tumor stem cells. Science. 1977;197(4302):461–3.
263. Kuroda T, Yasuda S, Kusakawa S, Hirata N, Kanda Y, Suzuki K, et al. Highly sensitive in vitro methods for detection of residual undifferentiated cells in retinal pigment epithelial cells derived from human iPS cells. PLoS One. 2012;7(5):e37342.
264. Hanahan D. Hallmarks of Cancer: New Dimensions. Cancer Discov. 2022;12(1):31–46.

265. Henry MP, Hawkins JR, Boyle J, Bridger JM. The Genomic Health of Human Pluripotent Stem Cells: Genomic Instability and the Consequences on Nuclear Organization. Front Genet. 2018;9:623.
266. McIntire E, Taapken S, Leonhard K, Larson AL. Genomic Stability Testing of Pluripotent Stem Cells. Curr Protoc Stem Cell Biol. 2020;52(1):e107.
267. Tokunaga K, Saitoh N, Goldberg IG, Sakamoto C, Yasuda Y, Yoshida Y, et al. Computational image analysis of colony and nuclear morphology to evaluate human induced pluripotent stem cells. Sci Rep. 2014;4:6996.
268. Nishimura K, Ishiwata H, Sakuragi Y, Hayashi Y, Fukuda A, Hisatake K. Live-cell imaging of subcellular structures for quantitative evaluation of pluripotent stem cells. Sci Rep. 2019;9(1):1777.
269. Issa J, Abou Chaar M, Kempisty B, Gasiorowski L, Olszewski R, Mozdziak P, et al. Artificial-Intelligence-Based Imaging Analysis of Stem Cells: A Systematic Scoping Review. Biology (Basel). 2022;11(10).
270. Coronnello C, Francipane MG. Moving Towards Induced Pluripotent Stem Cell-based Therapies with Artificial Intelligence and Machine Learning. Stem Cell Rev Rep. 2022;18(2):559–69.
271. D'Hulst C, Parvanova I, Tomoiaga D, Sapar ML, Feinstein P. Fast quantitative real-time PCR-based screening for common chromosomal aneuploidies in mouse embryonic stem cells. Stem Cell Reports. 2013;1(4):350–9.
272. Wang R, Carter J, Lench N. Evaluation of real-time quantitative PCR as a standard cytogenetic diagnostic tool for confirmation of microarray (aCGH) results and determination of inheritance. Genet Test Mol Biomarkers. 2013;17(11):821–5.
273. Weissbein U, Schachter M, Egli D, Benvenisty N. Analysis of chromosomal aberrations and recombination by allelic bias in RNA-Seq. Nat Commun. 2016;7:12144.
274. Wang N. Methodologies in cancer cytogenetics and molecular cytogenetics. Am J Med Genet. 2002;115(3):118–24.
275. Dal Cin P. Metaphase harvest and cytogenetic analysis of malignant hematological specimens. Curr Protoc Hum Genet. 2003;Chapter 10:Unit 10 2.
276. Huang H, Chen J. Chromosome Bandings. Methods Mol Biol. 2017;1541:59–66.
277. Hu Q, Maurais EG, Ly P. Cellular and genomic approaches for exploring structural chromosomal rearrangements. Chromosome Res. 2020;28(1):19–30.
278. Imataka G, Arisaka O. Chromosome analysis using spectral karyotyping (SKY). Cell Biochem Biophys. 2012;62(1):13–7.
279. Ben-David U, Mayshar Y, Benvenisty N. Virtual karyotyping of pluripotent stem cells on the basis of their global gene expression profiles. Nat Protoc. 2013;8(5):989–97.
280. Valli R, Marletta C, Pressato B, Montalbano G, Lo Curto F, Pasquali F, et al. Comparative genomic hybridization on microarray (a-CGH) in constitutional and acquired mosaicism may detect as low as 8% abnormal cells. Mol Cytogenet. 2011;4:13.
281. Li D, Liu J, Yang X, Zhou C, Guo J, Wu C, et al. Chromatin Accessibility Dynamics during iPSC Reprogramming. Cell Stem Cell. 2017;21(6):819–33 e6.
282. J Borstlap AK, G Stacey, A Elstner, A Damaschun, B Arán, J C Gerlach, J C Izpisúa, A Veiga. Development of a European human embryonic stem cell registry. Regen Med. 2008;3(6):945–51.
283. Seltmann S, Lekschas F, Muller R, Stachelscheid H, Bittner MS, Zhang W, et al. hPSCreg--the human pluripotent stem cell registry. Nucleic Acids Res. 2016;44(D1):D757–63.
284. Danecek P, McCarthy SA, HipSci C, Durbin R. A Method for Checking Genomic Integrity in Cultured Cell Lines from SNP Genotyping Data. PLoS One. 2016;11(5):e0155014.
285. Muller FJ, Schuldt BM, Williams R, Mason D, Altun G, Papapetrou EP, et al. A bioinformatic assay for pluripotency in human cells. Nat Methods. 2011;8(4):315–7.
286. Bock C, Kiskinis E, Verstappen G, Gu H, Boulting G, Smith ZD, et al. Reference Maps of human ES and iPS cell variation enable high-throughput characterization of pluripotent cell lines. Cell. 2011;144(3):439–52.

287. Tsankov AM, Akopian V, Pop R, Chetty S, Gifford CA, Daheron L, et al. A qPCR ScoreCard quantifies the differentiation potential of human pluripotent stem cells. Nat Biotechnol. 2015;33(11):1182–92.
288. Kirkegaard JS, inventor; Novo Nordisk A/S, assignee. Enrichment of NKX6.1 and C-peptide co-expressing cells derived in vitro from stem cells2020.
289. Cossarizza A, Chang HD, Radbruch A, Abrignani S, Addo R, Akdis M, et al. Guidelines for the use of flow cytometry and cell sorting in immunological studies (third edition). Eur J Immunol. 2021;51(12):2708–3145.
290. Rai N, Singh AK, Singh SK, Gaurishankar B, Kamble SC, Mishra P, et al. Recent technological advancements in stem cell research for targeted therapeutics. Drug Deliv Transl Res. 2020;10(4):1147–69.
291. Ben-David U, Benvenisty N. Chemical ablation of tumor-initiating human pluripotent stem cells. Nat Protoc. 2014;9(3):729–40.
292. Lee MO, Moon SH, Jeong HC, Yi JY, Lee TH, Shim SH, et al. Inhibition of pluripotent stem cell-derived teratoma formation by small molecules. Proc Natl Acad Sci U S A. 2013;110(35):E3281–90.
293. Ben-David U, Gan QF, Golan-Lev T, Arora P, Yanuka O, Oren YS, et al. Selective elimination of human pluripotent stem cells by an oleate synthesis inhibitor discovered in a high-throughput screen. Cell Stem Cell. 2013;12(2):167–79.
294. Halliwell JA, Frith TJR, Laing O, Price CJ, Bower OJ, Stavish D, et al. Nucleosides Rescue Replication-Mediated Genome Instability of Human Pluripotent Stem Cells. Stem Cell Reports. 2020;14(6):1009–17.
295. Zhong C, Liu M, Pan X, Zhu H. Tumorigenicity risk of iPSCs in vivo: nip it in the bud. Precis Clin Med. 2022;5(1):pbac004.
296. Itakura G, Kawabata S, Ando M, Nishiyama Y, Sugai K, Ozaki M, et al. Fail-Safe System against Potential Tumorigenicity after Transplantation of iPSC Derivatives. Stem Cell Reports. 2017;8(3):673–84.
297. Bedel A, Beliveau F, Lamrissi-Garcia I, Rousseau B, Moranvillier I, Rucheton B, et al. Preventing Pluripotent Cell Teratoma in Regenerative Medicine Applied to Hematology Disorders. Stem Cells Transl Med. 2017;6(2):382–93.
298. Liang Q, Monetti C, Shutova MV, Neely EJ, Hacibekiroglu S, Yang H, et al. Linking a cell-division gene and a suicide gene to define and improve cell therapy safety. Nature. 2018;563(7733):701–4.
299. Martin RM, Fowler JL, Cromer MK, Lesch BJ, Ponce E, Uchida N, et al. Improving the safety of human pluripotent stem cell therapies using genome-edited orthogonal safeguards. Nat Commun. 2020;11(1):2713.

Part V
Beta Cell Replacement: Clinical Horizon

An Ethical Perspective on the Social Value of Cell-Based Technologies in Type 1 Diabetes

Dide de Jongh and Eline M. Bunnik

1 Introduction

In recent years, significant progress has been made within the multidisciplinary field of regenerative medicine field in the development of stem cell-based technologies for the treatment of type 1 diabetes. Several cell-based products are currently being developed for transplantation into patients, with the aim of restoring the pancreatic function of insulin production and secretion and regaining glycemic control [1–3]. As these products are composed of biological materials, designed and assembled in laboratories, and harbor the functional capacities of the pancreas, they are referred to as 'bioartificial organs'. Ideally, bioartificial organs should consist of biological materials derived from unlimited, renewable sources, such as stem cells [4]. This would help overcome pressing shortages of islet of Langerhans tissue derived from deceased donors, and make 'transplantation' of bioartificial organs for the treatment of type 1 diabetes available for larger groups of patients. Preferably, these cells should be patient-derived, so that immunosuppression is not required, and patients can be spared the burdens and long-term health complications of taking life-long medication to prevent graft rejection [5]. In laboratories around the world, researchers are currently combining cutting-edge technologies of regenerative medicine, including tissue engineering, induced pluripotent stem cell and organoid

D. de Jongh
Department of Medical Ethics, Philosophy and History of Medicine and the Erasmus MC Transplant Institute, Department of Internal Medicine, University Medical Centre, Rotterdam, The Netherlands
e-mail: d.dejongh.1@erasmusmc.nl

E. M. Bunnik (✉)
Department of Medical Ethics, Philosophy and History of Medicine, Erasmus MC, University Medical Centre, Rotterdam, The Netherlands
e-mail: e.bunnik@erasmusmc.nl

L. Piemonti et al. (eds.), *Pluripotent Stem Cell Therapy for Diabetes*,
https://doi.org/10.1007/978-3-031-41943-0_19

technologies and gene editing, in an attempt to revolutionize the treatment of type 1 diabetes. For the promise of personalized cell-based technologies in the treatment of type 1 diabetes to materialize, however, researchers will have to solve not only biological and technical issues, but also ethical issues.

Various public and private research projects and programs are underway, mostly at the pre-clinical stage of research and development. For example, the European Commission-funded VANGUARD project aims at generating a vascularized, immunoprotected bioartificial pancreas for transplantation, using cells derived from patients, gene-modified cells from donated placental tissue, and islets from deceased human donors or from pigs [2]. Another example is Diabetes Moonshot of Reg Med XB, a Dutch foundation focused on bringing a cure to patients with chronic disease which is developing a beta cell replacement implants containing insulin-producing stem cells [3]. Vertex, a biotechnology company from the United States of America (US), is developing tissue-engineered transplantable insulin-producing cells made from differentiated allogeneic stem cells [1]. At the moment of writing, Vertex is conducting multi-center, open-label phase 1/2 clinical trials of their products VX-880 and VX-264, in patients with type 1 diabetes, impaired hypoglycemic awareness, and severe hypoglycemia [6]. It is currently enrolling patients in the US and in several European countries, including Norway, Switzerland, and the Netherlands. In the VX-880 trial, insulin-producing islet cells are delivered by infusion through the hepatic portal vein. The therapy requires immunosuppressive medication. In the VX-264 trial, islets are encapsulated in pouches that allow for the exchange of oxygen, hormones and proteins, but protect the insulin-producing cells from direct contact with cells of the immune system so as to avoid an immune response. The device is surgically implanted. While this may seem promising for patients who suffer from poor glycemic control despite best available treatment with insulin pumps, glucose sensors, and even hybrid closed-loop systems, it may take time for cell-based therapies to become routinely available, if at all.

In the meantime, during clinical translation of these technologies, research participants enrolling in early-phase clinical trials are exposed to risks. The implantation of a cell-based therapy requires an invasive and non-reversible surgical procedure [7]. Following implantation, the bioartificial organ will integrate with the recipient's tissue, resulting in permanent alterations of the recipient's body. The cell-based therapy will become functional in desired or undesired ways, and may pose expected as well as unforeseen health risks on the short and long term. To justify the enrolment of human research participants in potentially high-risk trials, a reasonable expectation of the value of the research needs to be present [8]. It is widely acknowledged that for clinical research to be ethical, the associated risks and burdens to research participants should be balanced by the social or scientific value of the research [9]. Striking a positive balance of risks and social value is especially challenging for early-phase clinical trials. By definition, in early-phase clinical trials, direct benefits are not expected for individual research participants, and the risks for participants are generally considered high. However, phase 1/2 clinical trials may have *scientific* value, largely by establishing facts and insights that may be useful in other research projects [10], as well as *social* value, which may help offset a negative balance of risks and benefits for individual research participants.

Ultimately, it is the application of scientific knowledge that will generate value for society. As early-phase clinical trials are inevitable first steps towards establishing the safety and efficacy of new treatment options that may improve the health of future generations of patients, they may have social value. The social value of research is defined as its clinical benefit to future patients, *relative* to alternatives that may already be approved for marketing, or that are being developed in parallel, and may become available in the near future [8]. That is, depends not only on the safety and efficacy of the newly developed treatment option, taken in isolation, it also depends on the scientific and societal environment in which the therapy is being developed, notably on the presence or absence of safe, effective, acceptable, and accessible alternative treatment options. If a perfectly satisfactory treatment option exists for patients with type 1 diabetes, the clinical translation of cell-based technologies for the treatment of type 1 diabetes has little social value.

This means that the anticipated social value of stem cell-based technologies for the treatment of type 1 diabetes should be assessed *in comparison to* established dominant treatment options. For patients with type 1 diabetes, two types of treatment modalities are currently available: device-based treatments (i.e. insulin pumps, glucose sensors, and combinations thereof) and beta cell mass replacement-based treatments (i.e. transplantation of either pancreas or islets of Langerhans). Although both treatment modalities are effective, they are associated with various limitations, as well, ranging from medical risks to psychosocial burdens, and from scarcity constraints to justice concerns. To assess the anticipated social value of cell-based treatments for type 1 diabetes, it is important to know the advantages and disadvantages of future cell-based treatments vis-à-vis existing treatment options. An understanding of whether and how cell-based treatments may change the lives of (future) patients is important for researchers and sponsor to help decide whether and how to move forward responsibly in the field of regenerative medicine as applied to type 1 diabetes.

In this chapter, we will conduct a comparative analysis from an ethical point of view of the value of stem cell-based treatments for type 1 diabetes as compared to that of advanced device-based options and beta cell mass transplantation-based options. Moreover, we will present recommendations to help achieve social value with the clinical translation and implementation of stem cell-based treatments.

2 Existing Treatment Modalities

At present, there are two prominent treatment modalities in diabetes care: device-based treatments and islet or pancreas transplantation. In this section, we briefly describe these two treatment modalities, and some of their strengths and limitations (see Tables 1 and 2).

Table 1 Strengths and limitations of device-based treatments for type 1 diabetes

Ethical value	Strengths	Limitations
Beneficence	Improved blood glucose values (compared to insulin injection therapy)	It remains a tool to control the disease, not a cure
	No invasive surgical intervention needed	Risk of infusion site infection and dermatological complications
	No toxic immunosuppressive medication required	Burden of maintenance, frequent alarms and self-management
		Sleep problems and psychological harm (e.g. diabetic distress)
		Patients may develop a sense of dependency on their device
		Delay in sensor blood glucose measuring
		Latency of pharmacological action of insulin
		Technical problems (e.g. calibration, software, insulin leaks, device connectivity, or system defects)
Liberty and autonomy	Fine-grained insight in fluctuating blood glucose values may give patients a sense of control of their disease	High level of self-management required
	Feeling of reassurance and safety during the night	Burden of constantly having to make adjustments for dietary intake or physical activity
		Impact on flexibility (e.g. physical and social activities) in daily life
		Burden of anticipation of potential hypoglycemic or hyperglycemic events (e.g. bringing high-glucose food or extra pumps)
		No possibility to interfere with the system
		Issues of trust and accuracy
Privacy	Real-time data sharing with physicians and family members can improve care and is helpful in notifying when hypoglycemic episodes occur	Concerns about surveillance and loss of control
	Alarms can give a sense of reassurance (when they do not go off)	Data leaks may result in unauthorized third parties accessing personal patient data
		Concerns about commercialization of personal data
		Visibility of the device is associated with stigmatization and social impacts
		Frequent alarms can be burdensome
		Alarm fatigue

(continued)

Table 1 (continued)

Ethical value	Strengths	Limitations
Justice		High level of health literacy and self-management skills required
		Affordability
		Psychosocial burdens, worries about cost and insurance
		Strict reimbursement criteria
		Disparities in utilization along the lines of race/ethnicity, geographical region and socioeconomic status

Table 2 Strengths and limitations of beta cell replacement-based treatments for type 1 diabetes

Ethical value	Strengths	Limitations
Beneficence	Patients may become insulin independent	Taking of toxic immunosuppressive medication is required
	Higher efficacy in reducing severe hypoglycemia	Potentially life-threatening complications (e.g. risks of infections and cancer)
	Reduction of diabetes-related complications	
Autonomy and liberty	More flexibility in daily life (e.g. physical and social activities)	Social impact of having to take immunosuppressive medication and avoid infection
Justice		Shortage of donor organs
		Available only for select patient groups

2.1 Device-Based Treatment

For the past 100 years, the most common treatment option for type 1 diabetes has been administration of insulin. Until recently, insulin treatment has required glucose readings and exogenous insulin (pen) injections multiple times daily. Over the past decades, medical devices for the treatment of type 1 diabetes have rapidly evolved. Today, several continuous glucose monitoring devices and insulin pumps are available to patients, which have improved not only glycemic control (fewer episodes of hypo- and hyperglycemia), and clinical outcomes [11–14], but also health-related quality of life, including daily functioning [15, 16], sleep quality [17, 18], stress, anxiety [19], and work performance [20]. Patients are less likely to miss work or to experience interruptive episodes at work [21]. Furthermore, better glycemic control is associated with lower risks of psychiatric conditions, such as depression and anxiety [22–24]. More advanced technologies have recently entered the market, including so-called artificial pancreases and hybrid closed-loop systems, in which a continuous glucose monitoring device is combined with one or two hormone infusion pumps, allowing for dynamic infusion of insulin and glucagon. When the system detects hypoglycemia, glucagon is administered subcutaneously to help raise blood glucose concentrations and reverse the chance of severe hypoglycemia. Artificical pancreases come with additional catheters and infusion sites for glucagon. Such combined systems are supported by dosing algorithms, installed on the infusion pumps, allowing for continuous, autonomous, and real-time control of glucose levels [25]. Closed-loop systems have proven to be more effective in preventing hypoglycemic and hyperglycemia events than events than conventional therapy [26]. The magnitude of effects of these advancements depends on the algorithm developed for the software and the functionalities the system offers [27, 28]. Even with recent technological advances, however, it is nearly impossible with medical devices to mimic the body's natural regulation of insulin and blood glucose. This means that in some patients who use these devices, severe hypoglycemic events still occur. For those patients whose problematic hypoglycemia persists, islet or pancreas transplant could be considered [29].

2.2 Transplantation

Transplantation is the other dominant treatment option for patients with type 1 diabetes. It is suitable, however, only for small groups of patients. Two subgroups that are eligible for transplantation are, first, patients who need a simultaneous pancreas-kidney transplantation, and second, patients who are suffering of "brittle diabetes", more specifically "problematic hypoglycemia". These patients have frequent, unpredictable, uncontrollable, and sharp changes in blood glucose, without an obvious cause. They also have a greater likelihood of experiencing severe hypoglycemia or ketoacidosis [13, 29]. Through transplantation of either a whole pancreas or isolated islets of Langerhans derived from (one or more) deceased donor(s) [30–32], patients with type 1 diabetes can be cured – albeit, often, for limited periods of time. Pancreas transplantation is an invasive procedure that is associated with significant morbidity and mortality. Islet transplantation, by contrast, offers a less invasive alternative with fewer complications and adequate functional results [33, 34]. Following islet transplantation, most patients with type 1 diabetes achieve insulin independence for at least a year [35, 36]. However, transplantation-based treatments options are faced with two major challenges: first, the enduring shortages of donor organs [32, 37], and second, the need for transplant recipients to undergo lifelong treatment with immunosuppressive medication. These challenges hinder the widespread application of pancreas and islet transplantation as a treatment option for type 1 diabetes. Only for a small number of patients with type 1 diabetes, the benefits of transplantation will outweigh the risks of taking lifelong immunosuppression medication. In the Netherlands, for instance, only 2–25 patients receive an islet transplant each year, whereas approximately 120,000 patients are diagnosed with type 1 diabetes [38]. Thus, there is a global need to develop alternative treatment options for type 1 diabetes patients over and above existing device- and transplantation-based treatment options.

3 Social Value of Stem Cell Technologies

The social value of stem cell technologies in the treatment of diabetic patients – as compared to the currently dominant device-based and transplantation-based treatment modalities – can be evaluated in four ethical dimensions: beneficence, autonomy, privacy, and justice.

3.1 Beneficence

Cell-based treatments hold the promise of improving health outcomes and the quality of life of patients. If they succeed in doing so, their clinical development and implementation is in line with the medical-ethical principle of beneficence [39]. Compared to the two existing device-based and transplantation-based treatment modalities, cell-based approaches may have at least four benefits: (1) the treatment may involve a less invasive procedure as compared to whole-pancreas transplantation; (2) it may come with fewer treatment-related health concerns (e.g. as toxic immunosuppressive medication is not required, as there are no risks of infusion site infection); (3) it may enhance health outcomes by ensuring better glucose regulation as compared advanced medical devices; and (4) it may be a less practically, psychologically, and socially burdensome therapy for patients, as it may not require the level of self-management associated with the technically complex and customizable medical devices.

First, whole-pancreas transplantation is a major surgical procedure with a high risk of complications associated with morbidity and mortality, also as compared to other organ transplantations [31, 33, 40]. Transplantation of islets of Langerhans is surgically less burdensome, but also less effective, mainly due to the loss of beta cell mass during the process of isolation, transplantation, and engraftment [41, 42]. Therefore, islet transplantation for one individual requires pancreatic tissue derived from more than one deceased donors. Furthermore, over time, transplanted islets usually deteriorate and lose their function to produce insulin. Often, patients need a retransplant or must resort again to exogenous insulin after 2–3 years [42, 43]. As said, given the enduring shortage of donor organs, transplantation of islets derived from deceased donors is not a viable treatment approach for the majority of patients with type 1 diabetes.

Second, as with all transplantation, recipients have to take immunosuppressive medication, often for the rest of their lives, to prevent graft rejection. These drugs are toxic and have side effects, which may eventually lead to life-threatening complications. Immunosuppressive regimens required after transplantation make patients more prone to minor problems, such as mouth ulcers, diarrhea, ovarian cysts, and acne, and to major problems in the long term, such as malignancies and serious infections [44]. The requirement of immunosuppression limits the current indication for islet transplantation to type 1 diabetes patients with impaired hypoglycemia awareness who experience severe hypoglycemia events. Less than 10% of the type 1 diabetes patient population meets these criteria [45].

Third, although advanced medical devices have as a major advantage that they do not require chronic immunosuppression, they are far from being able to achieve the same levels of glycemic control as is (recent) islet transplantation. Multiple studies have shown better metabolic outcomes of islet transplantation compared to the administration, multiple times daily, of insulin [46], and to the use of pumps with or without glucose sensors [47]. Further, patients who have had an islet transplantation also had a significantly reduced risk of progression of retinopathy compared to patients who take insulin [48], and better diabetic peripheral sensory neuropathy outcomes [49]. Some of these differences may be explained by delays between detection of changes in glucose levels by medical devices, and their actions. Subcutaneously inserted sensors measure glucose in the interstitial fluid, and there is a time lag in the rising and falling of glucose levels between blood and interstitial fluid, causing delays in detection. Further, there are delays between subcutaneous injection of insulin and its pharmacological action. This latency may vary depending on inter alia skin temperature, physical activity, and location of inserted cannula [50–52]. Newly developed algorithms try to take into account these latency factors; however, despite these improvements, delays cannot be prevented fully, and can still hamper glycemic control [51]. Cell-based treatment modalities, by contrast, will measure and adapt blood glucose levels more immediately inside the body, and might help solve this problem.

In addition, the use of sensors and pumps may lead to dermatological complications, as it involves the insertion of cannulas or filaments under the skin, which results in the maintaining of openings between the body and the outside world. Cannulas or filaments are often secured with adhesives, which may irritate the skin. Continuous monitoring devices are approximately left in place for 7 to 14 days, depending on the device, and insulin pumps are typically exchanged every 2 to 3 days. Skin complications, such as infections, scarring and irritations at infusion sites pose significant concerns for patients and their treating physicians [53–57]. In a study by Berg et al., the prevalence of dermatological complications (i.e. having currently visible dermatological conditions at one site at least) among pediatric and adolescent type 1 diabetic patients using pumps and sensors, was found to be between 46% and 69% [53]. Dermatological complications have been cited as a reason for patients to discontinue the use of advanced devices [58]. In the future, medical devices for the treatment of type 1 diabetes may further advance to 'close the loop' by using faster-acting insulins, increasing accuracy and reducing lag-time of continuous glucose monitoring, and using machine-learning algorithms) [59]. However, they cannot be seen as a cure, and will remain technological tools to help control – not overcome – the disease.

Moreover, clinicians are worried that when patients use medical devices for longer periods of time, their habitual self-management skills in controlling their disease might decline [60]. Patients may develop a sense of dependency on their medical device, when they learn to rely on the system for clinical decision-making, which may become problematic when technical problems arise [61, 62]. These patients could become more vulnerable to malfunctions of their medical devices, and less able to solve problems independently. Unfortunately, technical problems are expected to continue to occur in the future, and device-based treatment modalities will continue to expose patients to health risks associated with these problems.

Fourth, cell-based approaches may reduce some of the psychological and social burdens associated with device-based diabetes management. Hybrid closed-loop systems and artificial pancreases were developed with the intention of simplifying disease management – vis-á-vis finger prick-based blood glucose monitoring and manual glucose injection – and improving care for patients with type 1 diabetes. Still, however, advanced device-based treatments for type 1 diabetes come with limitations. While patients have reported that these new systems make disease management easier [63], it remains challenging for patients to successfully learn how to handle these customizable medical devices. Several participants in a hybrid closed-loop system trial, for instance, reported that they spent more time thinking about their diabetes while using this system than while undergoing standard treatment [64]. In fact, the additional function of glucagon-infusion of the artificial pancreases renders the handling of the device is even more complex for patients, and require additional catheters and infusion sites [25, 50, 65]. In addition, artificial pancreases still require patients to manually adjust the parameters for basal infusions and prandial boluses in accordance with lifestyle factors, such as physical exercise, lifestyle factors, and dietary intake, using exact carbohydrate counting [50, 65]. Thus, patients must be health literate, and skilled, to make effective and safely use of these devices [50, 66–68]. Having to constantly adjust parameters is experienced as a stressful task for some patients with type 1 diabetes [64]. Moreover, devices require maintenance, and, as said, most continuous glucose monitoring and insulin infusion devices need to be replaced every few days to prevent infections. Some devices also have to be calibrated a few times a day with traditional capillary blood glucose values to provide reliable readings, and have a short battery life [17, 69–71]. The burden of maintenance, frequent alarms, and self-managing the treatment of type 1 diabetes may result in sleep problems and psychological harm, such as diabetic distress, especially among parents of children with type 1 diabetes and patients who struggle to attain their target glycemia levels [17, 69]. Some patients and parents are constantly worrying, and in fear of both acute (e.g. severe hypoglycemia) and long-term complications of type 1 diabetes (e.g. retinopathy, neuropathy, nephropathy and cardiovascular disease) [69, 72, 73]. The responsibility of individual patients for managing diabetes care may also have negative impact on patients' social relationships and functioning. Patients with type 1 diabetes often believe that they themselves are fully responsible for the control of their illness. Thus, their inability to attain target glycemia levels may represent for them a form of moral failure and a sign of unworthiness, and may make them feel guilty towards their loved ones, and put even higher pressures on themselves [74]. With future cell-based therapies, patients might ideally stop having to self-manage their disease to achieve acceptable – or optimal – glycemia levels.

3.2 Liberty and Autonomy

The hope is that cell-based therapies can be implanted in patients and function for life, without the need of taking immunosuppressive medication or administering extra insulin, and can thus have a major impact on patients' liberty and autonomy. While recipients of beta cells derived from deceased donors need to take immunosuppressive medication to prevent graft rejection, recipients of personalized, patient-derived cell-based therapies might not. Patients taking immunosuppressive medication are more vulnerable for infection by bacteria causing illness [44]. Patients should, therefore, be more careful in their daily lives to reduce their chances of infection. This affects their liberty and ability to participate fully in society. Ideally, cell-based approaches might restore in patients the liberty to take part in social activities.

Also, in comparison to device-based treatment modalities, cell-based treatments may have positive effects on patients' liberty and autonomy. Device-based treatment modalities have traditionally required high levels of health literacy and disease management skills in patients, as well as time, effort and energy from patients [65]. Advanced medical devices, such as continuous glucose monitoring systems, and especially hybrid closed-loop systems and artificial pancreases, are able to control blood glucose levels relatively autonomously, without the need for constant oversight or intervention on the part of the patients, giving them more time and freedom for activities other than disease management. Also, these systems give patients more fine-grained insights in their fluctuating glucose levels, and may strengthen users' autonomy and sense of control of their disease [60, 64]. Patients using hybrid closed-loop systems reported feeling more self-assured in their ability to manage their disease, as well as experiencing a stronger sense of safety [17]. Yet at the same time, some patients felt that hybrid closed-loop systems were not managing hyperglycemia aggressively enough [64]. A few users mentioned that they would input higher carbohydrate quantities than they had actually consumed, in order to cheat the algorithm and produce a stronger effect (they felt was better) that was closer to what they would have achieved with their previous diabetes management therapy [64]. Systems with non-user-modifiable algorithms, which take full control of blood glucose, can even be experienced by patients as resulting in a loss of autonomy [74]. It can be frustrating for patients when the device overrules their decisions. At the moment, despite the improved user-friendliness of these medical devices aimed at making disease control easier, it seems that these systems are not responsive enough to efficiently and reliably control glycemia without additional self-management by the patient [50, 65]. Thus, their use is still experienced as a burden by patients [60, 64].

If cell-based therapies succeed in providing a cure for type 1 diabetes, they will liberate patients completely from the burdens of disease management. They will have time and head space to participate in society and live healthy lives. They will no longer require a regular and stable lifestyle. They will be free from having to make adjustments for dietary intake or physical activity, and free from the risk of severe hypo- and hyperglycemia events, for example, if they make last-minute decisions to go to restaurants with friends, family or colleagues. They no longer have to anticipate events, for example, by bringing high-glucose food in their bags or extra insulin pumps/pens when they go out or on holidays. They can try out new physical activities without stress and anxiety, and become more flexible and willing to try new experiences.

This does mean that patients need to learn to let go of control. The functioning of an implanted cell-based therapy cannot be adjusted by the patient. This implies a radical turnaround for type 1 diabetes, for they are often habituated to continuously exercising control over their bodies and micromanaging their glucose levels. That this may be difficult for type 1 diabetes patients, is evident from user reports of advanced medical devices. In the beginning, while the algorithm is learning, adjusting to the patient's body, and improving itself, the system may not yet result in optimal glucose values. During that learning phase, patients find it difficult to not interfere with the system when they notice high glucose levels on the screen [64]. In the beginning, patients feel the urge to check and correct the system, and feel hesitant about yielding agency to the device. Ultimately, many patients will become familiar with the device and its capabilities, and will gain confidence in the technology to partially manage their type 1 diabetes. However, technical problems (e.g. problems with equipment, software, insulin leaks, device connectivity, or system defects) may occur. In addition, software updates of medical devices come with a risk of disruption of the workings of the system, which could lead to hypo- or hyperglycemic peaks. Currently, system updates are not always communicated to patients in advance. Therefore, patients always have to be alert for system updates to happen. This reduces their liberty to live without worries, to trust the system fully, and be able to let go of control. Likewise, patients will need to learn to trust future cell-based treatment modalities for a new sense of freedom to arise.

3.3 Privacy

Cell-based treatment modalities can have positive effects on type 1 diabetes patients' privacy in three ways: it does not require the generation, processing, storing or sharing of personal data; it is implanted in the body and, therefore, not visible; it does not require any activities on the part of the patient, and so it does not invade their personal space. Privacy is often defined as the right to be let alone, or the right not to be observed or disturbed by others.

3.3.1 Personal Data

When patients make use of advanced medical devices, they generate data. These data include glucose readings and insulin infusion doses, and are often shared with the manufacturers of the devices and sometimes with treating physicians of family members [75]. Real-time data sharing can be useful in involving physicians in guiding optimal disease management, and helpful in notifying family members or caregivers during hypoglycemic episodes [75, 76]. Especially in pediatric and elderly patients, data sharing can help improve care. Yet sharing of personal data, including medical data, is not without risks to the privacy of patients. Data leaks could result in unauthorized third parties accessing patient data, leading, possibly, to misuse, discrimination and stigmatization. Some patients are afraid of being observed, tracked or spied upon, or of losing control of their medical devices, e.g. through hacks [71, 77]. In addition, sharing patients' clinical data with healthcare professionals or family members could also leave the patient with the impression of being surveilled and observed, especially if this is done without the patient's consent [18, 77]. Also, when data on periods of poor type 1 diabetic management are stored, can be seen, and cannot be deleted from the system, patients may experience a loss of control of their personal data. Finally, preliminary results from an interview study with type 1 diabetes patients showed concerns about the (re)use of personal data by manufacturers, not only to help improve software and technical aspects of the devices, but also to earn money. Because they are biological in nature, cell-based therapies do not require the generation of data. Thus, they are not associated with the privacy implications and risks of surveillance that come with data-driven device-based systems.

3.3.2 Visibility

A major limitation of device-based treatment modalities is that patients have to wear insulin pumps on or outside the body, which makes type 1 diabetes a disease that is readily visible for others. To the dismay of patients, the visibility of the devices often leads strangers to ask question and start conversations [74]. In these moments, patients felt that their identity was being reduced to that of 'the diabetic person'. Patients mentioned that they dislike unsolicited attention, and having to explain their illness to others, and that they would rather be treated as 'normal' [74]. Studies have shown that young type 1 diabetes patients who wore pumps expressed concerns about their visibility [62, 78]. Some patients even refrained from wearing their pumps in public, so as to hide their disease from others, to preserve their self-image and to prevent the stigmatization associated with having a disease [63]. Stigmatization occurs when someone with poor knowledge of the illness falsely attributes negative characteristics to the person with type 1 diabetes, such as blaming them for their illness and labeling them as sick [60]. Further, medical devices located outside the body impose restrictions or require adjustments on clothing and could pose challenges for starting or developing intimate relationships [79]. By contrast, cell-based therapies are not likely to be very large in size, and will be implanted under the skin or inside the body, such that only a surgical scar may be visible. The relative invisibility of cell-based therapies compared to device-based could help increase the feeling of 'normality' in patients with type 1 diabetes and allow them to develop other aspects of their identity.

3.3.3 Alarms

Advanced devices usually have an alarm function that alerts patients and possibly close others (e.g. parents or partners) when blood glucose levels are high, low, or when they fluctuate quickly. On the one hand, these alarms are experienced by patients as useful in providing insight and understanding of fluctuating glucose levels [19], and can give them a sense of reassurance and safety [80]. Knowing that an alarm will wake one up, and/or that a loved one will receive an alarm should a dangerous situation develop, can be a major source of reassurance [81]. On the other hand, alarms are accompanied by a variety of challenges. Users may experience alarms as burdensome [64]. For instance, parents of children with type 1 diabetes report frequent disruption of sleep related to (false) alarms or fear of hypoglycemic events [82, 83]. In addition, alarm fatigue (desensitization to alarms) may occur, leading patients to ignore or fail to act upon alarms. In one study, alarm fatigue is cited as the most common barrier to the use of the device [84]. For example, some children find alarms disruptive during school. Consequently, they may turn off the alarms, leading to worse diabetes management [80]. For some parents, alarms were perceived as a sign of their own failure to achieve optimal glycemic control for their child [80]. By young patients, too, alarms going off and drawing attention at inconvenient time, was perceived as embarrassing [71], and a constant reminder of living with type 1 diabetes [84, 85]. They do not allow patients to escape or take a break from their disease. Alarms may thus disturb and violate the personal space of patients. Cell-based treatment modalities may not have this disadvantage.

3.4 *Justice*

Cell-based therapies might hold the promise of reducing the disease burden associated with type 1 diabetes, but only if they can be accessed and used by the patients who need them.

3.4.1 Affordability

Accessibility requires not only availability, but also affordability. The material costs of the development of cell-based products need not be very high, because cell-based products will likely be composed mainly of patient-derived and donor-derived material. In case of the VANGUARD bioartificial pancreas, for instance, a main component is derived from placentas that are considered medical waste and can be procured at low cost. However, even with low material costs of the source material, cell-based technologies are likely to be expensive. First, they require highly specialized personnel and laboratory facilities for tissue engineering and gene editing, which come at great cost. Second, the costs of clinical development are often very high. The technology, like many other cell-based technologies, is currently being developed in pre-clinical research settings by research groups in academic centers that are subsidized by research grants. Unfortunately, these centers often lack the capacities and the funds to set up and conduct clinical trials of their cell-based innovative products. Major financial investments by large (commercial) manufacturers are often needed to gather clinical evidence, meet regulatory requirements for advanced therapy medicinal products (ATMPs), and bring the product to market [86]. When commercial manufacturers take on the financial risk associated with the clinical development of cell-based products, they need to recover incurred costs and reward their investors within the often limited time frame between marketing authorization and patent expiration, which may drive up the price of the treatment and hamper its accessibility for patients [87]. Given the expected high prices of future cell-based therapies for type 1 diabetes, there are concerns about just distribution of – and equal access to – these technologies. Justice requires that "societies meet healthcare needs fairly under reasonable resource constraints" [88]. It is not yet clear whether it will be possible for societies in resource-constrained settings to offer cell-based therapies to patients with type 1 diabetes.

Concerns about accessibility of best available treatments are not new to the field of type 1 diabetes, nor unique for cell-based treatment modalities. At present, existing device-based treatment modalities, such as continuous glucose monitoring devices, hybrid closed-loop systems and artificial pancreases, are not reimbursed for all eligible patients with type 1 diabetes [67, 89]. In the Netherlands, for instance, many patients cannot make use of advanced device-based treatments, due to strict national reimbursement policies. In fact, continuous glucose monitoring devices are reimbursed only for the following subgroups of patients with type 1 diabetes: children <18 years old, adults with poorly managed disease (in whom despite standard treatment, blood glucose levels remain high with an HbA1c of >8% or >64 mmol/mol), or who suffer from serious episodes of hypoglycemia and/or are unable to detect hypoglycemia (i.e. have hypo-unawareness), and pregnant women and women with a pregnancy wish and a history of gestational diabetes [90]. These eligibility criteria for the provision of continuous glucose monitoring devices are stringently applied by hospitals, and qualification tends to be temporary: as soon as a patient no longer meets the criteria (e.g. for pregnant women, after delivery), the treatment is withdrawn. As payment for medical care by patients themselves is not common – and not deemed morally acceptable – in the Netherlands, this means that patients who do not meet the criteria for reimbursement by the healthcare system, have no way to access the technology. Stringent criteria for reimbursement of advanced device-based treatment modalities apply in other European countries, too [67, 89].

In countries in which (co-)payments by patients themselves are possible, the high costs of the technology often pull up a barrier to access. Many, if not most, patients are simply unable to pay for the treatment. Ideally, continuous glucose monitoring devices should be accessible for all patients with type 1 diabetes, as is recommended by the Endocrine Society [91]. Use of continuous glucose monitoring devices is consistently associated with reduced hypoglycemia [92], and improved glycemic control [93, 94], also for those patients who have already excellent disease control using finger prick-based blood glucose monitoring and insulin injections [95]. At the moment, the main obstacle to the comprehensive implementation of continuous glucose monitoring devices for type 1 diabetes patients appears to be the cost [81]. Yet multiple studies have assessed the cost-effectiveness of continuous glucose monitoring devices for patients with type 1 diabetes, and have found that their use can be cost-effective, also because it reduces the use of consumables such as daily blood test strips, and, most importantly, improves glucose rates, prevents non-severe hypoglycemia events and secondary complications of diabetes, and reduces the indirect cost associated with reduced work productivity and daily hours devoted to diabetes care [96–98]. It can, therefore, be argued that the provision of continuous glucose monitoring devices is not only beneficial for the individual patient, but also for society. For example, Wan et al. found that for adults with type 1 diabetes living in the United States, as compared to self-monitoring of blood glucose with multiple times daily finger pricks, the use of continuous glucose monitoring devices was associated with an incremental cost-effectiveness of approximately $98,108 [€89,140] per quality-adjusted life year (QUALY) for lifetime use of this device [96]. By extending the duration of the use of the sensor from 7 to 10 days, the incremental cost-effectiveness could be further improved to approximately $33,459 [€30,327] per QALY [96]. An additional consideration is that, just like any other technology, continuous glucose monitoring devices will become cheaper after a period of time, when patents expire and monopolies are broken, which will further increase their cost-effectiveness and thus their relative value for society.

Lack of access to best available care, even when it is proven to be cost-effective, is a major concern for many patients in western European countries even [70, 99, 100]. Individuals with type 1 diabetes or parents of children with type 1 diabetes report barriers to the uptake of device-based treatment modalities [71, 84, 89, 101–103]. A study by Rashotte et al. describes parents of pediatric patients and adolescents with type 1 diabetes as "living worried," due to the stress associated with having to shoulder the everyday financial costs of diabetes care [71]. They report medical devices for the treatment of type 1 diabetes being too expensive, largely because of the costs of consumables [71]. Some users of continuous glucose monitoring devices reported using sensors or cannulas longer than was recommended, in order to save money [84].

More serious justice concerns arise in countries in which even basic healthcare needs cannot be met, and in which less costly, basic materials, such as finger-prick blood glucose tests, syringes, and insulin are in short supply. In 2021, the World Health Organization drew attention to limited access to insulin for millions of patients living with diabetes around the globe. Experts fear that future cell-based treatment modalities may not become accessible to patients around the world who may need them most [104]. They are concerned that when research groups have to rely on large commercial manufacturers for the clinical development of these technologies, the resulting therapies may not be affordable and, as a result, may not be used to revolutionize diabetic care for all patients in need, not only in developed countries, but also elsewhere [104].

3.4.2 Health Disparities

The use of medical device-based treatment modalities requires knowledge, skills and adherence. As not all patients have the capacities for adequate disease self-management using advanced device-based treatments, in practice, not all patients may be able to receive this type of care. For example, in most European countries, in order to be eligible for coverage of insulin pumps, patients must undergo various assessments over a period of 3–6 months [74]. This is to ensure patient safety and to monitor treatment effects, to ensure appropriate use of the expensive device as a scarce healthcare resource. Patients must also be educated on the use of the advanced medical device and also, for instance, on data processing and carbohydrate counting, for which they require basic language and numeracy skills. Patients will have to visit the hospital frequently, and not all patients have the time and the resources to do so. Lower literacy or numeracy skills, and the inability to take time off work could serve as barriers for the uptake of advanced device-based technologies [105–107]. This may lead to relatively worse health outcome in patients of lower socio-economic status. The use of stringent eligibility criteria and the requirement of frequent hospital visits may thus increase the health equality gap between patients with health literacy, digital skills, and financial means, and patients without such personal resources.

This health equity gap in type 1 diabetes has been studied widely. Type 1 diabetes patients with more deprived socioeconomic backgrounds have an increased chance to develop diabetic complications earlier than those with less deprived socioeconomic backgrounds [108]. A recent Swedish study demonstrated that a lower socioeconomic status triples the risks of diabetic complications in patients, such as cardiovascular disease and mortality [109]. Social support, flexible work schedules and regularly attending follow-up care appointments are factors that are associated with better health outcomes, but are unfortunately more difficult to achieve for patients with a lower socioeconomic status, which may explain some of the differences in health outcomes [105, 106, 110]. Over the last couple of years, utilization of advanced medical devices has overall been increased for type 1 diabetic patients; however, this increase has been observed preferentially in patients with a higher socioeconomic status [67, 110]. This inequality in uptake has been demonstrated across all age groups in various Western countries, including the United States, Germany and the United Kingdom [105, 110–112]. In addition, studies in the United States have indicated large race/ethnic inequities in the use of diabetes technology, disease-related distress, and disease self-management, regardless of socioeconomic status, in young non-Hispanic black adults [110, 113, 114]. Data consistently shows that non-Hispanic whites and female youths are most likely to use insulin pumps compared with other groups. Other characteristics associated with diabetes technology utilization in children and adolescents, includes having private health insurance, coming from a higher-income household, living in a two-parent household, and having parents with higher education [58, 101, 105, 108, 111,

113, 115, 116]. Disparities in access to technology have been associated with worse health outcomes [105, 108, 109, 115]. In a German study on access to diabetes technology, lower health literacy skills were associated with lower socioeconomic status, resulting in a lower use of diabetes technology in more deprived regions [111]. The necessity to apply for reimbursement by health insurers of device-based therapies and uncertainties surrounding this application process, may have discouraged families in more deprived socioeconomic situations from trying. A persistent association between pump use and area deprivation was found in this study [111]. This access gap in technology utilization for type 1 diabetic patients with ethnic minority or less privileged socioeconomic backgrounds is currently leading to unjust disparities in health outcomes [115].

In theory, cell-based therapies could have major health benefits precisely for patients from ethnic minority groups, populations with lower socioeconomic status, and other patient groups, such as children and patients with cognitive impairment, for whom the level of self-management that is required for current (advanced) device-based treatments is difficult to attain. By taking the – for patients from less privileged backgrounds sometimes insurmountable – burden of self-management away from patients, cell-based therapies could be used to help overcome the current health equity gap in diabetes care. Thus, they can be employed as instruments of change.

4 Concluding Remarks

In conclusion, the potential social value of cell-based therapies for the treatment of type 1 diabetes lies in better health and psychosocial outcomes for patients (beneficence), more freedom and flexibility in patients' daily lives (autonomy), advantages in terms of visibility, surveillance and stigmatization (privacy), and suitability for patients who lack the self-management skills required for device-based treatments (justice). To achieve these various forms of social value, researchers, research ethics review committees, manufacturers, regulatory agencies, clinicians and other stakeholder groups should bear in mind the following points of consideration. Ideally, cell-based therapies should be made from renewable materials, such as (genetically modified) patient-derived stem cells, which come at the additional benefit of personalization and freedom from having to use immunosuppressive medications and their associated toxicity. If this works, the 'transplantation' of insulin-producing organs will no longer be restrained by the enduring shortage of donor organs.

To optimize the social value of cell-based technologies, patient selection will be key. Major benefits of cell-based therapies – if it proves safe and effective – are expected especially for three groups of patients. First, younger patients, who may achieve more health benefit from cell-based therapies if these succeed in preventing secondary complications in the long term, and may benefit the longest from these – potentially expensive – potential future treatment options, thus improving their cost-effectiveness. Also, the disease itself and currently available device-based treatments have especially strong adverse impact on the social functioning and relationships of children and adolescents, and their visibility is experienced as more problematic by younger patients. Second, cell-based therapies might be important for patients who meet the criteria for islet and pancreas transplantation because of impaired hypoglycemia awareness or recurrent episodes of severe hypoglycemia, do not have any effective or suitable treatment options available to them, and may be waiting – in vain – for a transplant. Third, cell-based therapies might be appropriate for patients who have limited knowledge and capacity to use and manage advanced technological devices at the moment. This way, cell-based therapies can be used to help close the current equity gap in health outcomes for type 1 diabetes patients from less privileged socioeconomic backgrounds.

Further, if cell-based treatments prove to be safe and effective, it is important for research groups and manufacturers to make every effort to design and develop these technologies in such a way that they will be affordable and accessible to individual patients and to public healthcare systems [117, 118]. It will be challenging to use cell-based technologies to address global current disparities in access to adequate diabetes care, also because in many countries around the world, there are few, if any, specialized laboratory facilities and clinical care settings available for the production and administration of (personalized) cell-based therapies. It is not yet clear how, if at all, cell-based therapies could benefit those who are most in need, and how focusing our research efforts on cell-based therapies can be made compatible with global justice requiring us to help those who are currently unable to access basic diabetes and general health care (first).

Finally, if cell-based therapies are implemented in clinical care of type 1 diabetes, it will be important for healthcare professionals to ensure that patients understand how to detect problems and when to seek medical assistance in the event of graft failure, and remain knowledgeable of conventional management strategies (e.g. calculating insulin doses according to blood glucose readings) to avoid jeopardizing their health.

It is important for researchers to take into account ethical values associated with the application of regenerative medicine to the field of type 1 diabetes. Cell-based technologies may impact patient populations in various ethical dimensions, including beneficence, liberty, privacy and justice. Researchers should be aware of these potential future impacts during the subsequent phases of clinical development and implementation, and incorporate these in the designs of their trials and products, to help optimize the balance of potential benefits, risks and burdens of cell-based therapies for current and future patients with type 1 diabetes. This will strengthen the social value of cell-based technologies in healthcare systems around the world, and help ensure that these technologies are translated into tools that may bring not only medical benefits, but also psychological, social and existential advantages, to patients with type 1 diabetes.

References

1. Vertex Pharmaceuticals | R&D Pipeline | Type 1 Diabetes. https://www.vrtx.com/our-science/pipeline/type-1-diabetes/. Accessed 12 Feb 2023
2. VANGUARD. https://vanguard-project.eu/. Accessed 12 Feb 2023
3. RegMed XB | Regenerative Medicine Crossing Borders. https://regmedxb.com/. Accessed 12 Feb 2023
4. Khosravi-Maharlooei M, Hajizadeh-Saffar E, Tahamtani Y, et al (2015) THERAPY OF ENDOCRINE DISEASE: Islet transplantation for type 1 diabetes: so close and yet so far away. Eur J Endocrinol 173:R165–83.
5. Edgar L, Pu T, Porter B, et al (2020) Regenerative medicine, organ bioengineering and transplantation. Br J Surg 107:793–800.
6. ClinicalTrials.gov. https://clinicaltrials.gov/. Accessed 3 Feb 2023
7. de Jongh D, Massey EK, Cronin AJ, et al (2022) Early-phase clinical trials of bio-artificial organ technology: a systematic review of ethical issues. Transpl Int. 35:10751.
8. Habets MGJL, van Delden JJM, Bredenoord AL (2017) The unique status of first-in-human studies: strengthening the social value requirement. Drug Discov Today 22:471–475
9. Emanuel EJ, Wendler D, Grady C (2000) What makes clinical research ethical? J Am Med Assoc 283:2701–2711
10. Kimmelman J (2012) Gene transfer and the ethics of first-in-human research: Lost in translation. Gene Transf Ethics First-in-Human Res Lost Transl 1–205
11. Pala L, Dicembrini I, Mannucci E (2019) Continuous subcutaneous insulin infusion vs modern multiple injection regimens in type 1 diabetes: an updated meta-analysis of randomized clinical trials. Acta Diabetol 56:973–980
12. Scott ES, McGrath RT, Januszewski AS, et al (2019) HbA1c variability in adults with type 1 diabetes on continuous subcutaneous insulin infusion (CSII) therapy compared to multiple daily injection (MDI) treatment. BMJ Open. https://doi.org/10.1136/BMJOPEN-2019-033059
13. Pickup JC, Sutton AJ (2008) Severe hypoglycaemia and glycaemic control in Type 1 diabetes: meta-analysis of multiple daily insulin injections compared with continuous subcutaneous insulin infusion. Diabet Med 25:765–774
14. Monami M, Lamanna C, Marchionni N, et al (2010) Continuous subcutaneous insulin infusion versus multiple daily insulin injections in type 1 diabetes: a meta-analysis. Acta Diabetol 1:77-81 https://doi.org/10.1007/S00592-009-0132-5
15. Iturralde E, Tanenbaum ML, Hanes SJ, et al (2017) Expectations and Attitudes of Individuals With Type 1 Diabetes After Using a Hybrid Closed Loop System. Diabetes Educ 43:223–232
16. Lawton J, Blackburn M, Rankin D, et al (2019) Participants' experiences of, and views about, daytime use of a day-and-night hybrid closed-loop system in real life settings: Longitudinal qualitative study. Diabetes Technol Ther 21:119–127
17. Barnard KD, Wysocki T, Allen JM, et al (2014) Closing the loop overnight at home setting: psychosocial impact for adolescents with type 1 diabetes and their parents. BMJ Open Diabetes Res Care 2:e000025
18. Brew-Sam N, Chhabra M, Parkinson A, et al (2021) Experiences of Young People and Their Caregivers of Using Technology to Manage Type 1 Diabetes Mellitus: Systematic Literature Review and Narrative Synthesis. JMIR Diabetes. https://doi.org/10.2196/20973
19. Pickup JC, Holloway MF, Samsi K (2015) Real-time continuous glucose monitoring in type 1 diabetes: A qualitative framework analysis of patient narratives. Diabetes Care 38:544–550
20. Young AJ, Thabit H, Heller SR, et al (2015) Holistic Impact of Closed-Loop Technology on People With Type 1 Diabetes. J Diabetes Sci Technol 9:932–933
21. Jendle J, Buompensiere MI, Holm AL, et al (2021) The Cost-Effectiveness of an Advanced Hybrid Closed-Loop System in People with Type 1 Diabetes: a Health Economic Analysis in Sweden. Diabetes Ther 12:2977–2991

22. Northam EA, Matthews LK, Anderson PJ, et al (2005) Psychiatric morbidity and health outcome in Type 1 diabetes--perspectives from a prospective longitudinal study. Diabet Med 22:152–157
23. Anderson RJ, Grigsby AB, Freedland KE, et al (2002) Anxiety and poor glycemic control: A meta-analytic review of the literature. Int J Psychiatry Med 32:235–247
24. Lustman PJ, Anderson RJ, Freedland KE, et al (2000) Depression and poor glycemic control: a meta-analytic review of the literature. Diabetes Care 23:934–942
25. Russell SJ, El-Khatib FH, Sinha M, et al (2014) Outpatient glycemic control with a bionic pancreas in type 1 diabetes. N Engl J Med 371:313–325
26. McAuley SA, Lee MH, Paldus B, et al (2020) Six Months of Hybrid Closed-Loop Versus Manual Insulin Delivery With Fingerprick Blood Glucose Monitoring in Adults With Type 1 Diabetes: A Randomized, Controlled Trial. Diabetes Care 43:3024–3033
27. Doyle FJ, Huyett LM, Lee JB, et al (2014) Closed-loop artificial pancreas systems: engineering the algorithms. Diabetes Care 37:1191–1197
28. El-Khatib FH, Balliro C, Hillard MA, et al (2017) Home use of a bihormonal bionic pancreas versus insulin pump therapy in adults with type 1 diabetes: a multicentre randomised crossover trial. Lancet (London, England) 389:369–380
29. Choudhary P, Rickels MR, Senior PA, et al (2015) Evidence-Informed Clinical Practice Recommendations for Treatment of Type 1 Diabetes Complicated by Problematic Hypoglycemia. Diabetes Care 38:1016–1029
30. Pathak V, Pathak NM, O'Neill CL, et al (2019) Therapies for Type 1 Diabetes: Current Scenario and Future Perspectives. Clin Med Insights Endocrinol Diabetes. https://doi.org/10.1177/1179551419844521
31. Meirelles Júnior RF, Salvalaggio P, Pacheco-Silva A (2015) Pancreas transplantation: review. Einstein (Sao Paulo) 13:305–309
32. (2016) Global Report on Diabetes. WHO 978:6–86
33. Troppmann C (2010) Complications after Pancreas Transplantation. Curr Opin Organ Transplant. https://doi.org/10.1097/MOT.0b013e3283355349
34. Tamburrini R, Odorico JS (2021) Pancreas transplant versus islet transplant versus insulin pump therapy: in which patients and when? Curr Opin Organ Transplant 26:176–183
35. Vantyghem M-C, Chetboun M, Gmyr V, et al (2019) Ten-Year Outcome of Islet Alone or Islet After Kidney Transplantation in Type 1 Diabetes: A Prospective Parallel-Arm Cohort Study. 2042 Diabetes Care. https://doi.org/10.2337/dc19-0401
36. Hering BJ, Clarke WR, Bridges ND, et al (2016) Phase 3 Trial of Transplantation of Human Islets in Type 1 Diabetes Complicated by Severe Hypoglycemia. Diabetes Care 39:1230–1240
37. Anonymous (2018) Organ Procurement and Transplantation Network (OPTN). National Data Reports. US Dept. of Health and Human Services. 2018:1–62
38. Eurotransplant Eurotransplant – Statistics. In: Eurotransplant Stat. https://statistics.eurotransplant.org/. Accessed 7 Feb 2023
39. Beauchamp TL, Childress JF (2008) Principles of Biomedical Ethics (Principles of Biomedical Ethics. Oxford Univ Press 6:39–64
40. Kopp WH, Verhagen MJJ, Blok JJ, et al (2015) Thirty Years of Pancreas Transplantation at Leiden University Medical Center: Long-term Follow-up in a Large Eurotransplant Center. Transplantation 99:e145–e151
41. Ryan EA, Paty BW, Senior PA, et al (2005) Five-year follow-up after clinical islet transplantation. Diabetes 54:2060–2069
42. Vantyghem MC, de Koning EJP, Pattou F, et al (2019) Advances in β-cell replacement therapy for the treatment of type 1 diabetes. Lancet 394:1274-1285
43. Froud T, Ricordi C, Baidal DA, et al (2005) Islet Transplantation in Type 1 Diabetes Mellitus Using Cultured Islets and Steroid-Free Immunosuppression: Miami Experience. Am J Transplant 5:2037–2046
44. Ryan EE, Paty BW, Senior PA, et al (2004) Risks and side effects of islet transplantation. Curr Diab Rep 4:304–309

45. Geddes J, Schopman JE, Zammitt NN, et al (2008) Prevalence of impaired awareness of hypoglycaemia in adults with Type 1 diabetes. Diabet Med 25:501–504
46. Thompson DM, Meloche M, Ao Z, et al (2011) Reduced progression of diabetic microvascular complications with islet cell transplantation compared with intensive medical therapy. Transplantation 91:373–378
47. Holmes-Walker DJ, Gunton JE, Hawthorne W, et al (2017) Islet Transplantation Provides Superior Glycemic Control With Less Hypoglycemia Compared With Continuous Subcutaneous Insulin Infusion or Multiple Daily Insulin Injections. Transplantation 101:1268–1275
48. Benomar K, Chetboun M, Espiard S, et al (2018) Purity of islet preparations and 5-year metabolic outcome of allogenic islet transplantation. Am J Transplant 18:945–951
49. Fensom B, Harris C, Thompson SE, et al (2016) Islet cell transplantation improves nerve conduction velocity in type 1 diabetes compared with intensive medical therapy over six years. Diabetes Res Clin Pract 122:101–105
50. Gingras V, Taleb N, Roy-Fleming A, et al (2018) The challenges of Achieving Postprandial Glucose Control using Closed-Loop Systems in Patients with Type 1 Diabetes. Diabetes Obes Metab 20:245–256
51. Bekiari E, Kitsios K, Thabit H, et al (2018) Artificial pancreas treatment for outpatients with type 1 diabetes: systematic review and meta-analysis. BMJ 361:1310
52. Zinman B, Ruderman N, Campaigne BN, et al (2003) Physical activity/exercise and diabetes mellitus. Diabetes Care 1:73-77 https://doi.org/10.2337/diacare.26.2007.s73
53. Berg AK, Olsen BS, Thyssen JP, et al (2018) High frequencies of dermatological complications in children using insulin pumps or sensors. Pediatr Diabetes 19:733–740
54. Messer LH, Berget C, Beatson C, et al (2018) Preserving Skin Integrity with Chronic Device Use in Diabetes. Diabetes Technol Ther 20:254–264
55. Forlenza GP, Messer LH, Berget C, et al (2018) Biopsychosocial Factors Associated With Satisfaction and Sustained Use of Artificial Pancreas Technology and Its Components: a Call to the Technology Field. Curr Diab Rep 18:114
56. Hilliard ME, Levy W, Anderson BJ, et al (2019) Benefits and Barriers of Continuous Glucose Monitoring in Young Children with Type 1 Diabetes. Diabetes Technol Ther 21:493
57. Binder E, Lange O, Edlinger M, et al (2015) Frequency of dermatological side effects of continuous subcutaneous insulin infusion in children and adolescents with type 1 diabetes. Exp Clin Endocrinol Diabetes 123:260–264
58. Wong JC, Foster NC, Maahs DM, et al (2014) Real-Time Continuous Glucose Monitoring Among Participants in the T1D Exchange Clinic Registry *A complete list of members of the T1D Ex-change Clinic Network can be found in the Sup-plementary Data. 2702 Diabetes Care. https://doi.org/10.2337/dc14-0303
59. Peyser T, Dassau E, Breton M, et al (2014) The artificial pancreas: current status and future prospects in the management of diabetes. Ann N Y Acad Sci 1311:102–123
60. Farrington C, Murphy HR, Hovorka R (2020) A qualitative study of clinician attitudes towards closed-loop systems in mainstream diabetes care in England. Diabet Med 37:1023
61. Quintal A, Messier V, Rabasa-Lhoret R, et al (2019) A critical review and analysis of ethical issues associated with the artificial pancreas. Diabetes Metab 45:1
62. Saarinen T, Fernström L, Brorsson A-L, et al (2014) Insulin pump therapy is perceived as liberating, but to many it can imply a sense of the diabetes made visible. Eur Diabetes Nurs 11:38–42
63. Barnard KD, Wysocki T, Ully V, et al (2017) Closing the Loop in Adults, Children and Adolescents With Suboptimally Controlled Type 1 Diabetes Under Free Living Conditions: A Psychosocial Substudy. J Diabetes Sci Technol 11:1080–1088
64. Iturralde E, Tanenbaum ML, Hanes SJ, et al (2017) Expectations and Attitudes of Individuals With Type 1 Diabetes After Using a Hybrid Closed Loop System HHS Public Access. Diabetes Educ 43:223–232

65. Boscari F, Avogaro A (2021) Current treatment options and challenges in patients with Type 1 diabetes: Pharmacological, technical advances and future perspectives. Rev Endocr Metab Disord 22:217
66. Association AD (2021) 7. Diabetes Technology: Standards of Medical Care in Diabetes—2021. Diabetes Care 44:S85–S99
67. Janez A, Battelino T, Klupa T, et al (2021) Hybrid Closed-Loop Systems for the Treatment of Type 1 Diabetes: A Collaborative, Expert Group Position Statement for Clinical Use in Central and Eastern Europe. Diabetes Ther 12:3107
68. Reidy C, Bracher M, Foster C, et al (2018) The process of incorporating insulin pumps into the everyday lives of people with Type 1 diabetes: A critical interpretive synthesis. Health Expect 21:714–729
69. Luo X, Pan J, Lu H, et al (2022) Parents' experiences on the combined use of continuous subcutaneous insulin infusion and real-time continuous glucose monitoring to manage Type 1 diabetes in their children: A systematic review and meta-synthesis of qualitative studies. Nurs Open 9:2532–2551
70. Wysocki T, Hirschfeld F, Miller L, et al (2016) Consideration of Insulin Pumps or Continuous Glucose Monitors by Adolescents With Type 1 Diabetes and Their Parents: Stakeholder Engagement in the Design of Web-Based Decision Aids. Diabetes Educ 42:395–407
71. Rashotte J, Tousignant K, Richardson C, et al (2014) Living with sensor-augmented pump therapy in type 1 diabetes: Adolescents' and parents' search for harmony. Can J Diabetes 38:256–262
72. Balfe M, Doyle F, Smith D, et al (2013) What's distressing about having type 1 diabetes? A qualitative study of young adults' perspectives. BMC Endocr Disord 13:1–14
73. Gonder-Frederick L (2014) Lifestyle Modifications in the Management of Type 1 Diabetes: Still Relevant After All These Years? Diabetes Technol Ther 16:695
74. Quintal A, Messier V, Rabasa-Lhoret R, et al (2019) A critical review and analysis of ethical issues associated with the artificial pancreas HHS Public Access. Diabetes Metab 45:1–10
75. Lawton J, Hart RI, Kimbell B, et al (2021) Data Sharing while Using a Closed-Loop System: Qualitative Study of Adolescents' and Parents' Experiences and Views. Diabetes Technol Ther 23:500–507
76. Hart RI, Kimbell B, Rankin D, et al (2022) Parents' experiences of using remote monitoring technology to manage type 1 diabetes in very young children during a clinical trial: Qualitative study. Diabet Med 39:14828
77. Messer LH, Johnson R, Driscoll KA, et al (2018) Best friend or spy: a qualitative meta-synthesis on the impact of continuous glucose monitoring on life with Type 1 diabetes. Diabet Med 35:409–418
78. Payk M, Robinson T, Davis D, et al (2018) An integrative review of the psychosocial facilitators and challenges of continuous subcutaneous insulin infusion therapy in type 1 diabetes. J Adv Nurs 74:528–538
79. Racine E, Quintal A, Sample M (2017) Neuroessentialism in Discussions About the Impact of Closed-Loop Technologies on Agency and Identity. AJOB Neurosci 8:81
80. Lawton J, Blackburn M, Allen J, et al (2018) Patients' and caregivers' experiences of using continuous glucose monitoring to support diabetes self-management: qualitative study. BMC Endocr Disord 18:12 https://doi.org/10.1186/S12902-018-0239-1
81. Kraaijeveld SR (2021) Continuous Glucose Monitoring as a Matter of Justice. Hec Forum 33:345
82. Barnard K, Crabtree V, Adolfsson P, et al (2016) Impact of Type 1 Diabetes Technology on Family Members/Significant Others of People With Diabetes. J Diabetes Sci Technol 10:824–830
83. Oser TK, Oser SM, McGinley EL, et al (2017) A Novel Approach to Identifying Barriers and Facilitators in Raising a Child With Type 1 Diabetes: Qualitative Analysis of Caregiver Blogs. JMIR diabetes 2:e27 https://doi.org/10.2196/DIABETES.8966

84. Mohamed M, Soulodre C, Kaulback K, et al (2018) Continuous Monitoring of Glucose for Type 1 Diabetes: A Health Technology Assessment. Ont Health Technol Assess Ser 18:1
85. McCarthy GM, Rodríguez Ramírez ER, Robinson BJ (2019) Letters to medical devices: a case study on the medical device user requirements of adolescents with type 1 diabetes. Behav and Info Tech 40:39–48 https://doi.org/101080/0144929X20191606939
86. Bock AJ, Johnson D (2018) Regenerative medicine venturing at the university-industry boundary: Implications for institutions, entrepreneurs, and industry. Adv Exp Med Biol 1098:213–236
87. Cossu G, Birchall M, Brown T, et al (2018) Lancet Commission: Stem cells and regenerative medicine. Lancet (London, England) 391:883–910
88. Daniels N (2001) Justice, health, and healthcare. Am J Bioeth 1:2–16
89. Dos Santos TJ, Dave C, Macleish S, et al (2021) Diabetes technologies for children and adolescents with type 1 diabetes are highly dependent on coverage and reimbursement: results from a worldwide survey. BMJ Open Diabetes Res Care 9:2537
90. Formulieren – ZN https://znformulieren.nl/337936417/Formulieren?folderid=338591748&title=Hulpmiddelen. Accessed 12 Feb 2023
91. Peters AL, Ahmann AJ, Battelino T, et al (2016) Diabetes Technology-Continuous Subcutaneous Insulin Infusion Therapy and Continuous Glucose Monitoring in Adults: An Endocrine Society Clinical Practice Guideline. J Clin Endocrinol Metab 101:3922–3937
92. Rodbard D (2017) Continuous Glucose Monitoring: A Review of Recent Studies Demonstrating Improved Glycemic Outcomes. Diabetes Technol Ther 19:S25–S37
93. Foster NC, Miller KM, Tamborlane W V, et al (2016) Continuous Glucose Monitoring in Patients With Type 1 Diabetes Using Insulin Injections. Diabetes Care 39:e81–e82
94. Beck RW, Riddlesworth T, Ruedy K, et al (2017) Effect of Continuous Glucose Monitoring on Glycemic Control in Adults With Type 1 Diabetes Using Insulin Injections: The DIAMOND Randomized Clinical Trial. JAMA 317:371–378
95. Group JDRFCGMS (2009) The Effect of Continuous Glucose Monitoring in Well-Controlled Type 1 Diabetes. Diabetes Care 32:1378
96. Wan W, Skandari MR, Minc A, et al (2018) Cost-effectiveness of Continuous Glucose Monitoring for Adults With Type 1 Diabetes Compared With Self-Monitoring of Blood Glucose: The DIAMOND Randomized Trial. Diabetes Care 41:1227–1234
97. McQueen RB, Ellis SL, Campbell JD, et al (2011) Cost-effectiveness of continuous glucose monitoring and intensive insulin therapy for type 1 diabetes. Cost Eff Resour Alloc 9:13 https://doi.org/10.1186/1478-7547-9-13
98. Huang ES, O'Grady M, Basu A, et al (2010) The cost-effectiveness of continuous glucose monitoring in type 1 diabetes. Diabetes Care 33:1269–1274
99. Jendle J, Pöhlmann J, De Portu S, et al (2019) Cost-Effectiveness Analysis of the MiniMed 670G Hybrid Closed-Loop System Versus Continuous Subcutaneous Insulin Infusion for Treatment of Type 1 Diabetes. https://home.liebertpub.com/dia 21:110–118
100. Roze S, Smith-Palmer J, De Portu S, et al (2019) Cost-effectiveness of sensor-augmented insulin pump therapy vs continuous subcutaneous insulin infusion in patients with type 1 diabetes in The Netherlands. Clinicoecon Outcomes Res 11:73–82
101. Commissariat PV, Boyle CT, Miller KM, et al (2017) Insulin Pump Use in Young Children with Type 1 Diabetes: Sociodemographic Factors and Parent-Reported Barriers. Diabetes Technol Ther 19:363–369
102. Tanenbaum ML, Adams RN, Hanes SJ, et al (2017) Optimal Use of Diabetes Devices: Clinician Perspectives on Barriers and Adherence to Device Use. J Diabetes Sci Technol 11:484–492
103. James S, Perry L, Gallagher R, et al (2016) Diabetes Educators: Perceived Experiences, Supports and Barriers to Use of Common Diabetes-Related Technologies. J Diabetes Sci Technol 10:1115–1121
104. Bunnik EM, Jongh D de, Massey E (2022) Ethics of Early Clinical Trials of Bio-Artificial Organs. Transpl Int 35:10621 https://doi.org/10.3389/TI.2022.10621

105. Fallon C, Jones E, Oliver N, et al (2022) The impact of socio-economic deprivation on access to diabetes technology in adults with type 1 diabetes. Diabet Med 39:14906
106. Harris SM, Shah P, Mulnier H, et al (2017) Factors influencing attendance at structured education for Type 1 diabetes in south London. Diabet Med 34:828–833
107. Lawton J, Kimbell B, Rankin D, et al (2020) Health professionals' views about who would benefit from using a closed-loop system: a qualitative study. Diabet Med 37:1030–1037
108. Low L, Law JP, Hodson J, et al (2015) Impact of socioeconomic deprivation on the development of diabetic retinopathy:A population-based, cross-sectional and longitudinal study over 12 years. BMJ Open 5:007290 https://doi.org/10.1136/BMJOPEN-2014-007290
109. Rawshani A, Svensson AM, Rosengren A, et al (2015) Impact of Socioeconomic Status on Cardiovascular Disease and Mortality in 24,947 Individuals With Type 1 Diabetes. Diabetes Care 38:1518–1527
110. Pauley ME, Berget C, Messer LH, et al (2021) Barriers to Uptake of Insulin Technologies and Novel Solutions. Med Devices (Auckl) 14:339–354
111. Auzanneau M, Rosenbauer J, Maier W, et al (2021) Heterogeneity of Access to Diabetes Technology Depending on AreaDeprivation and Demographics Between 2016 and 2019 in Germany. J Diabetes Sci Technol 15:1059–1068
112. Barnard-Kelly KD, Cherñavvsky D (2020) Social Inequality and Diabetes: A Commentary. Diabetes Ther 11:803–811
113. Lin MH, Connor CG, Ruedy KJ, et al (2013) Race, Socioeconomic Status, and Treatment Center Are Associated with Insulin Pump Therapy in Youth in the First Year Following Diagnosis of Type 1 Diabetes. Diabetes Technol Ther 15:929–934
114. Agarwal S, Kanapka LG, Raymond JK, et al (2020) Racial-Ethnic Inequity in Young Adults With Type 1 Diabetes. J Clin Endocrinol Metab 105:e2960
115. Addala A, Auzanneau M, Miller K, et al (2021) A Decade of Disparities in Diabetes Technology Use and HbA1c in Pediatric Type 1 Diabetes: A Transatlantic Comparison. Diabetes Care 44:133–140
116. Powell PW, Chen R, Kumar A, et al (2013) Sociodemographic effects on biological, disease care, and diabetes knowledge factors in youth with type 1 diabetes. J Child Health Care 17:174–185
117. Wassmer CH, Lebreton F, Bellofatto K, et al (2020) Generation of insulin-secreting organoids: a step toward engineering and transplanting the bioartificial pancreas. Transpl Int 33:1577–1588
118. Iacovacci V, Ricotti L, Menciassi A, et al (2016) The bioartificial pancreas (BAP): Biological, chemical and engineering challenges. Biochem Pharmacol 100:12–27

Beta Cell Replacement Cellular Products: Emerging Regulatory Perspectives and Considerations for Program Development

Bruce S. Schneider

1 Introduction

In planning research programs for the development of cellular products for the replacement of pancreatic beta cells, it is critical at the outset to determine whether the work is intended to produce a marketable product or whether the goal is limited to research. If the former, it is important to consider the entire development pathway prior to initiating studies. But unless the research is not intended to include clinical investigation, the requirements of the US FDA and other regulatory agencies will need to be met at all stages of development, including those for CMC, pharm/tox, and clinical components of the program. This chapter describes some of the pertinent aspects of current FDA regulation of beta cell replacement products, focusing on considerations for clinical trials. Meetings and other interactions with FDA are also discussed. Emphasis on specific topics and inclusion of suggested approaches to planning reflect the author's experience, both within FDA and subsequently as a consultant. This chapter is intended for readers at any level of familiarity with regulatory affairs.

This therapeutic area is advancing rapidly, and one would expect regulatory guidelines to change accordingly, to protect the safety of subjects at all stages of product development and ensure that new cellular products are effective clinically. There is also great interest, among sponsors and investigators, in finding novel approaches to measuring clinically meaningful effectiveness especially during later-phase studies. New technologies, including electronic recording and analysis of data from CGM devices and other sources, hold the promise of providing more comprehensive information than previously afforded by conventional clinical monitoring procedures. FDA and other regulatory agencies will certainly evaluate these approaches as more information becomes available and new techniques are

B. S. Schneider (✉)
B. Schneider BIO Consultancy LLC, New York, NY, USA

L. Piemonti et al. (eds.), *Pluripotent Stem Cell Therapy for Diabetes*,
https://doi.org/10.1007/978-3-031-41943-0_20

proposed by sponsors. It should also be noted that background or standard therapies for the treatment of diabetes, including closed-loop devices, are also evolving. As improved background therapies become part of standard care, or at least additions to treatment options, indications for islet and beta cell replacement may also change, not only in clinical practice but also as part of a regulatory framework.

2 Planning Overall Product Development: The Labeling Concept

The goal of research and development in this field is development of a cellular product that can be used to treat patients suffering from Type 1 diabetes mellitus or perhaps even severely insulinopenic Type 2 diabetes. Due to the nature of cellular and gene therapies, including known and unknown safety considerations associated with the product, the route of administration, and ancillary treatments such as immunosuppression, all stakeholders – investigators, sponsors, regulators, and potential patients – expect that these therapies will have substantial beneficial treatment effects. This issue is discussed in more detail below, but it is worth considering in planning development programs for cellular products.

Possibly the best way to design a comprehensive development program, including planning for clinical, CMC, and pharm/tox studies, is to leap ahead and write the product label first, concentrating on the specific intended indication(s) and claims, as well as other key elements of a product label. Working back from the proposed label, one can help ensure that the preclinical and clinical studies can support claims for the safety and effectiveness of the product, as well as all other required components of a product label. Furthermore, following a focused pathway helps to obviate expenditure of resources on unnecessary studies, improving the efficiency of a development program.

As a valuable aid to this approach, FDA has issued a draft Guidance for writing a Target Product Profile (TPP) [1]. As defined by the FDA Guidance, a TPP is *"a format for a summary of a drug development program described in terms of labeling concepts."* As indicated above, the overall concept is that sponsors *begin* planning of product development with the desired goal in mind – product labeling. The end of the process, which is the labeling for the final approved product, is written first.

The TPP, which is entirely voluntary, may be shared with FDA to enhance the value of important communications with the Agency and make meetings with the review division more productive. Within the TPP template, each specific labeling concept (e.g., indications and usage, dose and administration, warnings and precautions) is accompanied by documentation of the present and planned studies intended to support each concept. The TPP can be modified as development proceeds.

Use of a TPP can help avoid omissions that can lead to delay in product approval. In a hypothetical example, following a successful phase 3 trial that met its primary effectiveness endpoint, a sponsor wishes to add a claim that their stem cell-derived product, administered in a subcutaneous capsule device, contains a given proportion of beta cells along with other islet cell elements, with no evidence of teratoma formation, at periods ranging from 3 to 12 months after administration. In addition, the newly modified label would claim that no clinically important histological abnormalities are found at the administration sites adjacent to the capsules. The supporting data are based on sampling and analytical methods that had not been discussed in advance and that had been performed on a small subset of subjects. At an FDA advisory committee meeting, the panel suggests the need for extensive histological analysis of the capsule contents as well as examination of the tissue adjacent to the capsules – both at 12 months following administration. Methodologies for establishing these claims are discussed in detail. FDA agrees with the panel's recommendation, but the sponsor has not obtained data based on recommended timing and methods. Depending on the degree to which FDA considers these studies important for safety, approval could be delayed, or the product may be approved without the added labeling claims. Using a TPP that stipulates these claims at the outset and discussing the adequacy of studies to support them may have avoided delays in approval.

A TPP template is provided in the above-cited FDA TPP Guidance. The template can be modified to fit specific circumstances and as development proceeds.

3 Which Product?

A detailed discussion of product (CMC) regulation is beyond the scope of this chapter. FDA has not published specific guidance on cellular products for treatment of diabetes aside from the Agency's 2009 Guidance [2], which discusses intact allogeneic islet products. However, a few commonly encountered CMC-related issues in cell and gene therapies are worth considering at the beginning of product development.

Cell and gene-therapy development programs, which often originate in relatively small academic or biotech research laboratories, may initially study a product that will undergo modifications during clinical development. As an example, a gene-therapy program may have used fresh locally prepared transfected autologous CD34+ cells in a phase ½ clinical trial, but as the program expands, the sponsor finds it necessary to ship frozen cells between manufacturing and treatment sites, with the intention of using frozen transfected cells as the marketed product. Occasionally, such changes are made after considerable clinical data have accrued, even as late as phase 3, and FDA may require some form of comparability testing to assess the cogency or even relevance of outcomes of studies using the original product. Decisions regarding the intended marketed cellular product should be made as early as possible, to prevent unnecessary delays in clinical development and approval.

For sponsors who may wish to study more than one version of a cellular or gene-therapy product in a single early-phase trial, FDA has recently issued a new Guidance on study of multiple product versions in an "umbrella" trial [3]. The purpose of such umbrella trials is to enable sponsors to determine which of multiple product versions is best to pursue in subsequent later-phase trials. Studying multiple versions in a single trial should provide information more efficiently than performing several studies sequentially. Ideally, in a single umbrella trial, subjects are randomized to each product version or to a control group. Importantly, use of a single common control group makes more efficient use of a limited number of subjects and may facilitate recruitment and retention of subjects, as fewer would be assigned to a control group, compared to studies performed sequentially.

Several aspects of "umbrella" trials are worth noting: Each version of the product will have its own IND. Unlike "basket trials," in which one product is tested in more than one indication, in umbrella trials, multiple product versions are tested in a single indication. These trials are conducted early in development and are not intended to provide conclusive evidence of effectiveness of a given product version, nor are they likely to be powered to demonstrate statistically significant differences in efficacy (or safety) among product versions.

In this Guidance, FDA provides suggestions for proceeding with umbrella trials, including approaches to organizing the INDs, as well as submitting information, reporting adverse events, and other aspects relevant to conducting these trials.

Release criteria and potency assays: Establishing release criteria and potency assays is essential to development of cellular and gene-therapy products, including stem cell-derived beta cells. FDA requirements and suggestions for the development of release criteria in early-phase trials and potency assays for phase 3 cell and gene-therapy trials are described in a Guidance [4].

4 Clinical Trials: From First-in-Human to Registration

In general, clinical development of novel cell and gene therapies does not strictly follow pathways typical of those for small-molecule drugs (i.e., orderly progression through phase 1, 2, and 3, often with multiple mid- and later-phase trials). Numbers of patients for cell and gene-therapy trials are usually smaller than for trials of small molecules, either because of disease rarity, or inability to manufacture sufficient quantities of experimental product, or owing to the nature of the administration procedure and requirements for extensive monitoring of subjects. Even though Type 1 diabetes is not rare, with prevalence over 1.6 million in the USA (CDC data) [5], numbers of subjects in clinical trials of cells or intact islets have been relatively small, due to the above considerations. However, the general expectation is that a large treatment effect size will permit demonstration of effectiveness even with relatively small numbers of enrolled subjects, particularly when the primary effectiveness outcome is insulin independence, which does not occur in patients with established T1D. Some statistical simulations of effectiveness outcomes for allogeneic islet transplantation have been published [6]. Requirements for size of safety databases for novel products will vary according to emerging and anticipated risks, based on the nature of the product, preclinical pharm/tox data, and accumulated safety data as clinical trials proceed.

Enrollment Population Unlike trials of small-molecule drugs, many of which may enroll normal volunteers in phase 1, cell and gene-therapy trials typically enroll only patients with the target disease. Generally, the population of patients enrolled in trials should match the intended target population at some point in development and in sufficient numbers to support a labeled indication. For example, a trial of a beta cell replacement product may initially enroll very brittle patients who might stand to benefit the most, given the potential risks of the product, the administration procedure, and use of ancillary medications, such as immunosuppressives. This approach has been standard for allogeneic islet transplantation, where the most robust outcome is elimination of severe hypoglycemic events, allowing achievement of acceptable HbA1c levels with reduced or simplified insulin dose regimens. Also, this stringent enrollment criterion facilitates use of a single-arm trial, with its necessary comparison of outcomes to baseline. Alternatively, the first-in-human trial may enroll T1D patients who are more metabolically stable if there are other potential safety concerns about a particular cell product, its method of administration, or requirement for ancillary medications and treatments. As an example, if a stem cell-derived cellular product requires a few months to mature into fully functioning beta cells after in vivo administration, such function will occur well after the subject has left the administration site and may not be able to be monitored adequately. Potential risks of hypoglycemia may be mitigated in metabolically stable T1D patients who have some remaining autonomic function. As development proceeds, the population characteristics might be expanded, based on favorable emerging safety and efficacy outcomes, and ultimately any T1D patient may become a candidate, provided that the benefits justify real and potential risks. The labeled indication will be based on the populations studied in clinical trials that demonstrate the safety and effectiveness of the cellular product.

The other major issue in selecting clinical trial subjects for treatment with a beta cell replacement product is defining a specific benefit that is both clinically meaningful in that patient group and measurable in a clinical trial. This issue has become increasingly important in the past few years as treatment regimens, including newer insulins, use of CGM, and closed-loop systems have evolved. As background therapeutic and monitoring options improve, it is reasonable to assume that FDA will increasingly require justification for use of an experimental agent, in terms of potential benefit over available options. The paradigm for allogeneic islet transplantation via intraportal infusion with immunosuppression – inability to maintain acceptable HbA1c levels without incurring multiple episodes of severe hypoglycemic events despite at least 6 months of intensive insulin therapy by a multidiscipline team – may shift as newer approaches reduce the incidence of hypoglycemia and improve metabolic control. Given the rapid advances in this field, it is not clear which current therapeutic options and approaches FDA will require a subject to have "failed," in order to permit enrollment into a trial of a cellular therapy. On the other hand, recognizing that a large proportion of patients do not achieve acceptable HbA1c levels, with or without hypoglycemia, attainment of insulin independence (or the composite endpoint of HbA1c <6.5% and absence of hypoglycemia with use of some exogenous insulin) could justify potential risks, so the anticipated treatment effect also factors into determination of a patient's suitability to enter a clinical trial. Choice of enrollment criteria requires extensive discussion with the FDA review division, given ongoing evolution of treatment options.

Studies in pediatric patients: Type 1 diabetes can occur at any age including infancy, with peak onset around 13 years. Given the devastating effect this diagnosis exerts on children and their families, often with emotional and behavioral problems limiting ability to achieve good metabolic control, with early onset of microvascular complications, it would seem ideal to be able to treat such young individuals with beta cell replacement therapies. One limiting factor is use of immunosuppressive drugs, with their associated long-term risks. For children in the USA, studies of investigational therapies that pose more than minimal risk must follow 21 CFR Subpart D regulations: *Additional Safeguards for Children in Clinical Investigations*. A key element in these regulations is that there must be the prospect of direct benefit to participants in the trial. Such prospect should ordinarily come from prior studies of the investigational therapy in adults with the same disease. If this is not feasible, either because few affected children reach adulthood (as in some genetic diseases) or the presumably milder form of the disease is substantially different from the childhood form, the prospect of direct benefit may be derived from preclinical studies or other sources. Ordinarily, studies in children begin with adolescents, if they are capable of providing assent as well as informed consent from parents or caregivers. Proposed studies in children should be discussed with the review division at FDA.

Other populations: A small subset of Type 2 diabetics are severely insulinopenic, require insulin to maintain metabolic control, and may experience episodes of hypoglycemia as well. This patient group lacks the autoimmunity present in T1D patients, while retaining allo-immunity. This condition may have some impact on viability of transplanted beta cell products, particularly if they are derived from autologous sources, such as IPS.

4.1 Clinical Trial Design

Considerations and principles underlying clinical trial enrollment criteria are described above. Cell and gene-therapy first-in-human trials often include a dose-escalation evaluation based on safety and, if feasible, some indication of efficacy or at least of pharmacodynamic action. For beta cell replacement therapies, evaluation of the latter should be easily afforded by measurement of C-peptide responses, among other parameters related to glucose metabolic control and required insulin dose. The dose-escalation paradigms for cell and gene therapies differ from those employed in small-molecule phase 1 trials in that, for the former, the starting dose (expressed as cell number or vector genomes/kg) should be predicted to be capable of conveying an acceptable level of clinical benefit to the subject, because the investigational product may remain in the body for a considerable period and given the associated risks as well. Determination of the starting dose can be based on preclinical data (allometric scaling), inferred by comparison to total number of insulin-producing cells in a normal individual, and informed by published results of allogenic islet transplantation (e.g., data published by the Collaborative Islet Transplant Registry in their Annual Report), and clinical studies from other sources. For stem cell-derived cellular products, safety concerns that FDA will likely focus on include tumorigenicity, route of administration, and ancillary medications such as immunosuppressives. Regarding higher doses of cells, hypoglycemia should not be a problem if all the cells respond appropriately to low glucose levels. However, this may need to be determined during the dose-escalation phases, as well as from in vivo preclinical studies. Risks of ectopic tissue and tumor formation may be increased with increasing cell dose.

Under 21 CFR § 314.126 (Adequate and Well-Controlled Studies), approval to market a new drug must be based on data submitted from one or more "adequate and well-controlled" studies. Regulations permit a variety of trial designs that may be "adequate and well controlled," as long as the design permits a valid comparison with the control, in order to provide a quantitative assessment of the effect of the drug.

A randomized control trial (RCT) design is universally recognized as providing the highest standard of evidence of effectiveness and safety as well. Randomization can overcome inequalities in baseline covariates, both known and unrecognized, and reduce or eliminate bias. FDA will almost always recommend this design, especially for phase 3 trials. Due to the need for special, invasive, procedures for administration of a cellular product, as well as use of either transient or permanent immunosuppression, it will usually not be feasible to conduct a placebo-controlled RCT, in which subjects and investigators are blinded to treatment. In beta cell replacement trials, this drawback may be mitigated by using objective outcomes as measures of effectiveness.

In some situations, a RCT is neither feasible nor necessary. FDA has agreed that for allogeneic islet transplantation via intraportal infusion with immunosuppression, a single-arm study may be capable of providing adequate evidence of effectiveness [2]. To make this design acceptable, the treatment effect size must be very large, preferably excluding outcomes that are rarely, if ever, encountered in the course of the disease. A prime example of such an endpoint is proportion of subjects achieving insulin independence following transplantation.

Some options for alternative designs and approaches are listed in the above-cited regulations.

4.2 Endpoints

The array of clinical and metabolic endpoints for Type 1 diabetes trials is large. However, sponsors should recognize that an acceptable endpoint for a phase 3 trial to support product registration must be clinically meaningful – that is, affecting how a patient feels, functions, or survives (or remains out of the hospital) – demonstrated either directly or through an accepted surrogate. The utility of using C-peptide levels as a guide to transplant function is widely recognized, and C-peptide responses to mixed meal stimulation can be helpful at all phases of development. However, in the context of beta cell replacement, FDA has not accepted C-peptide levels as a surrogate for a clinically meaningful outcome. (One reason is that there is not a recognized quantitative relationship between a given change in C-peptide levels and clinical benefit, within the context of islet or beta cell transplantation.)

For allogeneic islet transplantation, FDA has accepted either insulin independence or the composite endpoint of HbA1c $\leq 6.5\%$ (or $\leq 7\%$) and absence of severe hypoglycemic events (as a responder analysis of the proportion of subjects who achieve this composite) [2]. The former endpoint, which does not occur in a patient with established Type 1 diabetes, would be expected a priori to confer a substantial benefit to any patient with this disease, although presumably the magnitude of benefit would vary directly with the degree of baseline metabolic instability. Insulin independence should be specifically defined, in terms of timing and duration of good metabolic control in the absence of exogenous insulin. The latter endpoint, the composite, would represent a clinical benefit in patients who cannot maintain adequate metabolic control without experiencing repeated severe hypoglycemic events (the eligible population for allogeneic islet transplantation), justifying the potential risks of immunosuppression and the intraportal infusion procedure. This endpoint is a composite, not a co-primary – both outcomes must be achieved in the same subject, and the results are analyzed in terms of proportion of responders. If this composite endpoint is to be used in a single-arm clinical trial using the subject's baseline condition as the control, it is essential that: (a) the baseline distribution depicting a hypothetical responder rate, with its mean (point estimate) and 95% confidence intervals, be accepted by FDA, (b) the lower bound of the 95% confidence interval for the point estimate for the proportion of observed responders in the trial lies well above the corresponding upper boundary for the hypothetical controls, and (c) the incidence and severity of baseline hypoglycemia be firmly documented for every subject in the trial. Simulations for power calculations for such single-arm trials have been published (Tiwari et al., *op cit*).

It is reasonable to assume that FDA will have similar policies and considerations for use of these two effectiveness endpoints in trials of stem cell-derived beta cell products, although depending on the trial population and the intended indication, trial designs other than single-arm may be required for the composite endpoint and certainly for other potential endpoints. (Adequate size of the safety population is another matter.) Again, designation of a primary effectiveness endpoint for phase 3 should be discussed with FDA early in product development.

Other primary effectiveness endpoints may be considered, based on the enrolled population, but as the magnitude of the treatment effect becomes smaller, the greater the probability that the outcome may fall within the course of the disease as currently treated, especially as therapeutic modalities advance. For example, achievement of acceptable metabolic control (HbA1c) alone may occur with improved background therapy; and some changes in HbA1c levels may be too small to justify potential risks of stem cell replacement therapy. In addition, for adequate statistical power to detect and separate the effect of the investigational product from that of background therapy, trials with relatively small treatment effects are likely to be large and necessarily employ a RCT design. FDA can be expected to weigh the magnitude of clinical benefit against risks of tumorigenicity, the administration procedure, immunosuppression, and other potential concerns.

Use of time-in-range (TIR), based on output data from CGM devices, can provide near-continuous readouts of glucose levels over time, with an added advantage afforded by remote monitoring, with the potential to reduce the amount of missing data in a clinical trial. An example of this approach applied to a clinical study of a hybrid closed-loop device system has recently been published [7]. It is possible that such data could support endpoints not based on traditional landmark analyses, yielding more clinical and metabolic information over time, compared to that obtained using a traditional approach. For example, analysis of CITR data from allogeneic islet transplantation trials (Tiwari et al. AJT 2012, *op cit*) showed that the driver behind loss of the above-cited composite endpoint at the landmark time point of 1 year was HbA1c, not reappearance of SHE. So even if a subject had maintained the composite endpoint throughout most of the trial period, an elevation of HbA1c > 6.5% at the time of the landmark analysis would categorize that subject as a non-responder, substantially underestimating the treatment effect of the transplant.

At the time of this writing, FDA has not accepted TIR as the primary endpoint for phase 3 trials. One possible reason is inadequate knowledge of a quantitative relationship between any increase in time-in-range and clinical benefit. For example, if, in a RCT, the active-treatment arm demonstrates a mean TIR that is 20% greater than that of controls, it may not be possible to infer a corresponding clinical benefit. Or there may be a clinical benefit but only for the subgroup of subjects who had the widest glucose excursions at baseline. Another issue may relate to concerns about possible errors in CGM readings. This is an active and promising area of development, and FDA policy may change as more information becomes available.

Product Administration and Retrieval: Ancillary Treatments Beta cell product administration poses the greatest current challenge in this field. Several groups have successfully produced stem cell-derived products that are differentiated into functional beta cells, either at the time of implantation or after maturation in vivo. The remaining problem is getting the product into the body in a way that the cells are protected from the body's immune system and the patient is protected from ectopic tissue formation and tumors. To date, the two major approaches to beta cell product administration are encapsulation with implantation of the capsules subcutaneously or into the peritoneal cavity or omentum, or via direct intraportal administration of unencapsulated cells into the liver. Encapsulated cells may be regulated as combination products, but because the prime mechanism of action is via the cells, such products will be regulated primarily by CBER with input from CDRH. Sponsors often express concern that combination-product designation adds complexity to the review and approval process, but this is not the case, and interactions with FDA generally proceed through the primary review division at CBER.

For encapsulated cell products, safety issues include the administration procedure itself, including considerations related to anatomical location, responses of the body to the capsule, and concerns about liberation of undifferentiated cells should the capsule break down. It is anticipated that the capsules will become sufficiently vascularized to permit the aerobic metabolism required to support adequate beta cell viability and function. Concomitant use of immunosuppressive drugs may be required short term or longer, but depending on the structure of the capsule, especially pore size, it may be possible to withdraw these medications. FDA may request information regarding ability to retrieve capsules if this becomes necessary, for example, due to evidence of tumor formation.

There has been substantial research and development aimed at designing capsule material and structure. There is also the need to develop preclinical models, either in vivo or in vitro, that can predict human responses to implanted capsules, to avoid failure in clinical trials. Until capsule design develops to the point of human applicability, unencapsulated beta cells may provide better functionality if they are implanted in a supportive anatomical site. One such site is the liver, with implantation accomplished via intraportal infusion with immunosuppression. The major concern here is tumorigenicity, especially since a teratoma in the liver would present a greater problem than one occurring within a subcutaneous capsule. Accordingly, FDA will likely require plans for monitoring and treating any ectopic tissue or tumor. As immunosuppression will likely be required for unencapsulated products, one potential "kill switch" could be provided by withdrawal of these drugs. Other approaches should be considered as well, as this issue is very likely to come up during review.

5 Expedited Programs: Fast Track, Regenerative Medicine Advanced Therapies (RMAT), Breakthrough, Accelerated Approval, Priority Review

An important part of the current regulatory landscape is CBER's series of expedited programs, details of which are provided in a recent CBER Guidance [8] and in an earlier FDA Guidance [9].

Type 1 diabetes is recognized as a serious condition with unmet medical need, and beta cell replacement products (including genetically modified beta cell products) fall clearly into the category of regenerative medicines; accordingly, developers of any such cellular products should be able to avail themselves of the benefits of these programs, provided the investigational product meets performance criteria required for the specific designation – Fast Track, RMAT, Breakthrough, Accelerated Approval. (Accelerated Approval is based on a biomarker or intermediate endpoint that is "reasonably likely" to predict clinical benefit. Following granting of such approval, sponsors are required to conduct a confirmatory trial, the details of which are worked out with FDA.) The most recent designation, RMAT, is unique to CBER, and it is important to note that, like Breakthrough, granting of RMAT requires cogent clinical data. The CBER *Guidance* [8] provides detailed criteria for qualification for each of these designations.

Probably the most important benefit derived from Fast Track, RMAT, or Breakthrough designation is the opportunity for greater interaction with FDA review staff via increased frequency of meetings.

Regarding potential eligibility for Accelerated Approval, for products granted any of these Expedited Program designations, the standards for granting Accelerated Approval (or, for that matter, traditional approval) are no less stringent than for products granted Accelerated Approval without RMAT designation, except for requirements for confirmatory data after approval. As specified in Section 506(g)(7) of the FD&C Act, for products granted RMAT and Accelerated Approval, post-approval requirements may be fulfilled with clinical data from sources other than traditional clinical trials. These sources may include "real-world" evidence (e.g., electronic medical records), other clinical data sets as agreed upon with FDA, or post-approval monitoring of all patients treated prior to approval. The content, timing, and format of post-approval studies should be agreed upon with FDA. It should also be emphasized that for stem cell-derived cellular products, potential safety issues may require longer trial periods prior to any approval.

These issues should be discussed with FDA during product development. Presumably, sponsors of products granted RMAT will have sufficient opportunity to interact with FDA during this period.

6 Conducting Trials During the COVID Pandemic

One of the few predictable consequences of any pandemic is that, in addition to the morbidity and mortality caused by the infection itself, the normal course of everyone's life will be severely disrupted. The ability to conduct clinical trials is not an exception to this. COVID has affected almost every aspect of clinical trial management, from patient screening and recruitment, to missing clinic visits, and to effects of the virus on patients and investigators during the trial. To address this problem, FDA has issued a Guidance [10] that addresses multiple potential problems associated with conduct of clinical trials during the COVID-19 pandemic. As stated in the Guidance, these problems include "unanticipated quarantines; site closures; travel limitations for subjects, investigators, trial monitors; interruptions in supply of investigational product; and inability of a subject to continue in the trial."

Although the Guidance was issued just prior to widespread availability of COVID vaccines and newer antiviral medications, the pandemic is still with us, with ongoing mutations that have varied effects on transmissibility and clinical severity. Therefore, consideration of principles that may help prevent disruption of a trial or increase in amount of missing data seems prudent. In addition, these principles apply to other widespread contagious diseases and perhaps other unforeseen events that have the potential to disrupt the conduct of clinical investigations, particularly multicenter and multinational trials.

Among the key recommendations outlined in the *Guidance:*

- Sponsors, investigators, and IRBs should establish policies and procedures to protect subjects and manage study conduct during COVID pandemic.
- Sponsors should consider use of electronic data submission/electronic monitoring.
- Protocols should include fewer patient visits.
- "Hybrid trials," should be considered.
- Design trials to minimize occurrence of missing data (also see *The Prevention and Treatment of Missing Data in Clinical Trials* National Research Council. National Academies Press 2010 for more details on addressing the problem of missing data in clinical trials).
- The clinical study report should describe contingency measures employed, including a listing of affected subjects by unique number identifier, nature of study alteration, and analysis of how alternative procedures affected overall safety and effectiveness outcomes.

It is worth noting that some of these recommendations may apply even without a pandemic, especially if patients are unlikely to be concentrated in proximity to treatment sites.

7 Meetings with FDA

Frequent communication with the review division at CBER/FDA can be of great assistance at all stages of program development, from the earliest (INTERACT) meetings through pre-BLA discussions. The format and content of traditional meeting types (A, B, and C) are well known. Very recently, FDA has issued its *SOPP 8101.1: Regulatory Meetings with Sponsors and Applicants for Drugs and Biological Products.* This communication describes types of formal meetings and provides readers with details of responsibilities and procedures *within* FDA. Sponsors should also take note of the new Type D meetings, which focus on one or two topics and do not require input from more than three disciplines or divisions. New PDUFA timelines, which are being phased in, include FDA responses to Type D meeting requests within 14 days and that meetings are to be scheduled within 50 days. Accelerated timelines are being phased in for INTERACT meetings as well. Sponsors are advised to follow developments regarding face-to-face meetings, in terms of evolving schedules and venues.

Finally, as noted in previous sections, meetings can be expedited by obtaining product designation under Expedited Programs. Particularly for stem cell-derived products, focusing on RMAT designation when clinical data become available would be very helpful in this regard. Writing a TPP as early as possible will enhance meeting focus and productivity.

8 Summary

This chapter presents considerations that, in the opinion of the author, will be helpful to the clinical development of stem cell-derived beta cell products. Every aspect of this field, including scientific development, standard therapy, and product regulation, is changing rapidly. Sponsors are encouraged to maintain ongoing communication with the review division at FDA via meetings that occur early in development and as frequently as required and as permitted by FDA policy and resources.

Disclaimer The opinions expressed in this chapter, as well as choice of content, are the author's alone and do not necessarily represent those of any federal agency or other organization.

References

1. FDA *Guidance for Industry and Review Staff: Target Product Profile – a Strategic Development Process Tool.*
2. FDA *Guidance for Industry: Considerations for Allogeneic Pancreatic Islet Cell Products Guidance for Industry.*
3. FDA *Guidance for Industry: Studying Multiple Versions of a Cellular or Gene Therapy Product in an Early-Phase Clinical Trial.*
4. FDA *Guidance for Industry: Potency Tests for Cellular and Gene Therapy Products.*
5. Center for Disease Control and Prevention, *Report on Prevalence of Diabetes in the US*, 2019.
6. Tiwari, J, Schneider, B, and Barton, F, *Islet Cell Transplantation in Type I Diabetes Patients: Analysis of Efficacy Outcomes from a Registry and Considerations for Trial Design.* American Journal of Transplantation (12)2012; 1898–1907.
7. Wadwa, R, et al, *Trial of Hybrid Closed-Loop Control in Young Children with Type 1 Diabetes,* N Engl J Med 2023; 388:991–1001.
8. FDA/CBER Guidance, *Expedited Programs for Regenerative Medicine Therapies for Serious Conditions Guidance for Industry.*
9. FDA Guidance for Industry: Expedited Programs for Serious Conditions – Drugs and Biologics.
10. FDA *Guidance for Industry, Investigators, and Institutional Review Boards: Conduct of Clinical Trials of Medical Products During the COVID-19 Public Health Emergency.*

Lessons Learned from Clinical Trials of Islet Transplantation

Thierry Berney, Lionel Badet, Ekaterine Berishvili, Fanny Buron, Philippe Compagnon, Fadi Haidar, Emmanuel Morelon, Andrea Peloso, and Olivier Thaunat

1 Introduction: The Landmark Impact of the Edmonton Protocol

Thanks to spectacular improvements in functional outcomes, islet transplantation has evolved to become a clinical reality and is part of the standard-of-care for patients with complicated type 1 diabetes in several countries [1]. Advances made

T. Berney (✉)
Division of Transplantation, Department of Surgery, University of Geneva Hospitals, Geneva, Switzerland

Department of Transplantation, Nephrology and Clinical Immunology, Hospices Civils de Lyon, Lyon, France

School of Natural Sciences and Medicine, Ilia State University, Tbilisi, Georgia

Faculty Diabetes Center, University of Geneva, Geneva, Switzerland
e-mail: thierry.berney@unige.ch

L. Badet
Department of Urology and Transplantation Surgery, Hospices Civils de Lyon, Lyon, France

E. Berishvili
Division of Transplantation, Department of Surgery, University of Geneva Hospitals, Geneva, Switzerland

School of Natural Sciences and Medicine, Ilia State University, Tbilisi, Georgia

Faculty Diabetes Center, University of Geneva, Geneva, Switzerland

F. Buron · E. Morelon · O. Thaunat
Department of Transplantation, Nephrology and Clinical Immunology, Hospices Civils de Lyon, Lyon, France

P. Compagnon · F. Haidar · A. Peloso
Division of Transplantation, Department of Surgery, University of Geneva Hospitals, Geneva, Switzerland

L. Piemonti et al. (eds.), *Pluripotent Stem Cell Therapy for Diabetes*,
https://doi.org/10.1007/978-3-031-41943-0_21

by active islet transplantation programs over the last two decades have translated into improved functional outcomes of islet transplantation [2]. Current data show that islet transplantation can deliver excellent functional outcomes with >5 years follow-up. In >1000 islet transplant recipients reported to the Collaborative Islet Transplant Registry (CITR) by 38 centers worldwide, 5-year insulin independence and graft function rates currently range from 20 to 30% and 50–60%, respectively [3]. Importantly, there is a consensus that insulin-independence is not necessarily the sole criterion for defining success [4, 5].

The modern era of islet transplantation has arguably started with the landmark publication in 2000 of what is widely known as the "Edmonton protocol" [6]. Prior to this, the functional outcomes of islet transplantation were rather mediocre, with 1-year insulin independence rates around 10% reported to the International Islet Transplantation Registry [7]. The Edmonton paper reported seven consecutive islet graft recipients in which 100% insulin independence was achieved thanks to a combination of strategies, of which the relative contributions are difficult to assess. The novelties of the protocol thought to have had an impact included (i) an immunosuppressive regimen associating an anti-IL-2R mAb (daclizumab) and a sirolimus-based, "islet-sparing" maintenance, in which steroids were avoided and tacrolimus dosage was minimized and (ii) a strategy of sequential injections of islets, isolated from at least two donors, in order to increase the transplanted endocrine mass to >10,000 islet equivalents (IEQ) per kilogram body weight [6]. A strategy of immediate transplantation of freshly isolated islets, as opposed to maintaining them in culture for 24–48 h was part of the protocol, but was later found to have no real impact on clinical outcomes [8]. The immunosuppressive protocol associated an anti-IL2R mAb (daclizumab) for the induction and a sirolimus–tacrolimus combination for maintenance. Interestingly, the Edmonton protocol specifically targeted patients with problematic hypoglycemia without renal function impairment who were selected for an islet transplant alone (ITA) procedure [6]. Until then, similar to whole pancreas transplantation, the main indication had been simultaneous islet-kidney transplantation (SIK) in patients with type 1 diabetes with end-stage renal failure [9]. Since then, ITA overwhelmingly remains the primary indication for islet transplantation, at least in North America [1, 10].

In this chapter, we will review the designs, results, and conclusions of the clinical trials conducted and reported in the past 2 decades, which have followed the initial publication of the Edmonton series. To better focus on an already significant material, the islet transplantation procedure itself and its associated complications will not be discussed. Neither will the indications for islet transplantation relative to other modalities such as whole pancreas transplantation or technological solutions (artificial insulin delivery systems, closed loops, …) [11], or the difficulties of monitoring the islet graft for timely management of dysfunction [12].

2 Clinical Trials Utilizing the Edmonton Protocol

The perception of the 2000 Edmonton report as a major breakthrough led to the widespread adoption of the Edmonton protocol by existing programs and to the multiplication of new islet transplant centers [1, 3] eager to apply what seemed to be an infallible recipe for a successful islet transplantation program.

An international multicenter trial (NIS01) was almost immediately launched and sponsored by the Immune Tolerance Network (ITN), and co-funded by the National Institutes of Health (NIH) and the Juvenile Diabetes Research Foundation (JDRF), with the aim to explore the feasibility and reproducibility of islet transplantation using the Edmonton protocol [13]. Nine centers participated to this trial (6 in North America and 3 in Europe) and enrolled 36 patients with problematic hypoglycemia for ITA procedures, for a maximum of 3 islet injections. The level of experience in islet transplantation varied from large to naught. A standardized islet isolation protocol was utilized, including common standard operating procedures and identical batches of reagents and digestion enzymes. This was designed as a single-arm phase 1–2 clinical trial with a primary endpoint of "insulin independence with adequate glycemic control". Overall, 58% of study subjects reached insulin independence, but the primary endpoint (at 1 year) was achieved in 44%, which some observers perceived as a relative disappointment. An additional 28% achieved partial graft function, defined by measurable levels of circulating C-peptide, and 70% still had detectable C-peptide at 2 years, indicating at least partial graft function. All patients with full or partial graft function experienced significant improvements in glycemic control, as assessed by HbA1c levels, blood glucose time in range and mean amplitude of glucose excursions [13]. Important lessons were learned from the ITN trial. First, the progressive attrition of islet graft function, reported the previous year in the 5-year update of the Edmonton series [14] was a seemingly inexorable feature of islet transplantation, at least with the initial Edmonton protocol. Second, and importantly, although the Edmonton protocol was indeed reproducible by some of the participating centers, none of the inexperienced centers were able to achieve the primary endpoint in any of their patient, highlighting the fact that "although the dissemination of immunosuppressive protocols is quite easy, transferring the knowledge and expertise required to isolate a large number of quality human islets for transplantation is a far greater challenge" [15]. These were, by and large, also the conclusions of what had been in fact the first clinical trial, which aimed to reproduce the results of the Edmonton protocol. The study conducted at the NIH in 6 patients, achieved 50% insulin independence at 1 year, but serious procedure-related complications (bleeding) were observed in 2/6 subjects, as well as side effects of the immunosuppressive regimen so severe that medication had to be discontinued [16]. In effect, after this initial experience, the NIH clinical center decided to stop their islet transplant program. Factually, the year 2005 marked a peak in the number of islet isolation/transplantation centers in the USA and has seen a steady decline since then [3].

The difficulty of setting up an islet isolation facility was made clear by the ITN and the NIH experience and had indeed been anticipated by several institutions who elected either to join forces with an experienced islet isolation center in a bilateral collaboration, such as the Miami-Houston collaboration [15], or to regroup in islet transplantation networks, pioneered by the Swiss-French GRAGIL consortium [17, 18]. The GRAGIL model was emulated by other national or multinational networks such as the Nordic Network in Scandinavia [19], the UK islet transplant consortium (UKITC) [20] and the Australian Islet Transplant Consortium [21].

A few centers or networks embarked on their own clinical trials, essentially utilizing the Edmonton immunosuppressive regimen and the repeated islet infusion strategy. At the same time, the University of Alberta published a 5-year follow-up of their first 65 patients, of which 47 had completed the islet transplantation protocol. At 5 years, the majority (80%) had detectable C-peptide, but only a minority (approximately 10%) had maintained insulin independence. The median duration of insulin independence was 15 months. The investigators also reported the positive impact of an even partial graft function on HbA1c and control of severe hypoglycemia. They also cautioned that the immunosuppressive regimen was associated with significant toxicity [14].

The GRAGIL network, who had focused its activities on islet-after-kidney (IAK) transplantation, modified its existing immunosuppressive regimen (steroids, anti-IL-2R, ciclosporin, mycophenolate (MMF)) to an "Edmonton-like" regimen (basiliximab–everolimus–ciclosporin instead of daclizumab–sirolimus–tacrolimus), for a new ITA (GRAGIL 2) and a redesigned IAK trial (GRAGIL 1b) [22, 23]. At the same time, the University of Lille launched an ITA and IAK trial of the Edmonton protocol [24]. The first reports were similar in terms of graft function (defined as detectable circulating C-peptide) but differed for the insulin-independence endpoint, which was reached at 1 year in about 30% of patients in the GRAGIL 2 trial and 71% in the Lille trials. There was however a marked difference of practice in the islet infusion strategy: 1–2 infusions in GRAGIL (median 10,700 IEQ/kg) versus 2–3 in Lille (median 12,500 IEQ/kg), which probably contributed to the differences observed [24]. This point was confirmed by the GRAGIL group in a report of 15 IAK recipients, in which those who had received a double infusion achieved insulin independence (67%) more rapidly and lost it less frequently (approximately 50% at 2 years) than recipients of a single-islet transplant [22]. This was further emphasized by the observation by the Lille group that what they called "primary graft function", defined by the β-score [25] at 1 month after last islet infusion, was the main determinant of long-term insulin-independence and graft survival [24]. Another interesting observation reported by the Lille group relates to the level of graft function required to achieve certain objectives. Using continuous glucose monitoring (CGM) data, they could demonstrate that suboptimal function (β-score > 5) was necessary to significantly improve glucose excursions and time in hyperglycemia and optimal function (β-score > 7) was necessary to normalize them. However, partial function (β-score > 3) was sufficient to abrogate hypoglycemia [26].

The ability of a functional islet graft to abolish hypoglycemia, regardless of the achievement of insulin-independence, was also reported in a series of 31 islet transplant recipients. Remarkably, the authors reported that restoration of glycemia awareness persisted even after graft failure, suggesting that patients with autonomic neuropathy could have regained adrenergic symptoms after even a brief period of islet endocrine function [27].

3 The Edmonton Protocol in Islet-After-Kidney and Simultaneous Islet-Kidney Transplantation

As discussed above, the Edmonton protocol was rapidly adapted to the islet-after-kidney (IAK) situation by the GRAGIL network (GRAGIL 1b trial) [22] and the Lille group [24, 28] with results similar to those of ITA. The first report of an IAK transplantation trial utilizing the Edmonton protocol was published by the Geneva group [29]. In this protocol, patients were switched from the current immunosuppressive regimen administered for kidney graft maintenance to the Edmonton regimen. For patients on steroids, they were tapered to 5 mg prednisone daily before proceeding to the first islet injection; this was the only modification to the Edmonton protocol. Seven patients received a median 12,328 IEQ/kg body weight in 2 injections. All came off insulin for at least 3 months, with a 1-year insulin-independence rate of 74% and normalized HbA1c in all patients [29]. One patient went on to lose his kidney graft in a multifactorial context (declining kidney function at the time of inclusion, nephrotoxicity of the sirolimus–tacrolimus combination, sirolimus-induced pulmonitis). The University of Miami, developed a similar trial, with the difference that the immunosuppressive protocol was switched 6 months before the anticipated date of transplant. Seven patients received 14,779 IEQ/kg body weight in 1–3 islet injections. All patients but one came off insulin, with 30% insulin-independence at 1 year, 100% graft function and HbA1c < 6.5% in all but one [30]. These studies confirmed the feasibility and efficacy of IAK transplantation, but gave a word of caution about the need for careful patient selection and a clear definition of objectives [29].

Simultaneous islet-kidney transplantation (SIK) has been performed much less frequently since the turn of the millennium. The University of Zurich pioneered the field of SIK on the Edmonton protocol and reported the 5-year outcome results of 13 SIK patients in comparison to recipients of a simultaneous pancreas-kidney transplant (SPK) or a kidney transplant alone (KTA). The Zurich group applied a different philosophy: lower numbers of islets were transplanted and the focus was placed on metabolic impact rather than insulin independence. In their series, in spite of an insulin independence rate of 9.3% at 5 years, the mean HbA1c was normal and not significantly different from that of SPK, in sharp contrast with outcomes of KTA [31].

4 Adaptations of the Immunosuppressive Edmonton Protocol

The relative disappointment caused by the real-life reports of insulin independence and the seemingly inevitable attrition of islet graft function was met with extensive research and debate on the factors responsible for suboptimal islet engraftment and graft survival. Several factors were identified [32] and tackled whenever possible. Some of these factors were related to the islet isolation procedure (loss of endocrine tissue during the digestion process, loss of extracellular matrix, disruption of the vascular supply, loss of innervation); some to the recipient's adaptive immune response (rejection, recurrence of autoimmunity) which are also occurring after whole pancreas transplantation; some to toxicity of immunosuppression, with the interesting observation that islets implanted in the liver were exposed to elevated portal blood levels of the drugs [33]; some to the glucotoxicity or lipotoxicity sustained within the liver [32]. Perhaps the most significant factor was the exposure to an intense proinflammatory reaction, caused by inflammatory mediators in the graft microenvironment [34], but mostly elicited by contact of the islets with tissue factor expressed by the portal endothelial cells at the time of infusion and termed IBMIR (Instant Blood-Mediated Inflammatory Reaction) [32, 35].

These considerations led to several adaptations of the initial immunosuppressive Edmonton regimen. The rejection issue was addressed by implementing a more powerful induction, replacing anti-IL2R antibodies by T-cell depleting antibodies, for the first islet injection [20, 21, 36–40].

IBMIR and other proinflammatory phenomena were tackled by the inclusion of potent anti-inflammatory drugs, initially anti-TNF-α monoclonal antibodies (infliximab) [8, 40] or TNF-α soluble receptors (etanercept) [37, 41, 42], to which were subsequently added an IL-1 receptor antagonist (anakinra) [40, 43, 44].

The first center to progressively implement these two strategies, with the primary aim to reach success, defined by insulin independence, with single-donor transplantation, was the University of Minneapolis, reporting excellent results in 3 small trials. In the first trial, T-cell depletion was achieved with the experimental humanized monoclonal antibody hOKT3γ1(Ala-Ala), with 4/6 recipients (67%) achieving insulin independence and perfect metabolic control at 1 year [36]. In the second trial, 100% insulin independence was achieved in all subjects and maintained for >1 year in 5/8 ITA recipients [37]. In the third trial, prolonged (3 years) insulin independence in 4 of 6 patients. Five of the 6 study subjects had to receive a second islet injection (median total islet mass: 11, 264 IEQ/kg) because they had not achieved insulin independence after the first transplant [41]. In a further registry analysis, the inclusion of a T-cell-depleting agent in the induction protocol demonstrated its superiority over IL-2R blockade for the achievement of insulin independence and prolonged graft survival [38].

5 Sirolimus in Islet Transplantation: Friend or Foe?

Sirolimus was one of the cornerstones of the original Edmonton regimen, but with the major issue of a toxicity profile characterized by well-known and poorly tolerated side effects (mouth ulcers, diarrhea, nephrotic proteinuria, hypercholesterolemia, etc.). This has led to dose reductions or switches to another agent in a large proportion of patients on the drug [16, 45, 46].

From a more scientific standpoint, there were several arguments in favor or against the use of sirolimus. On the pro side, sirolimus is synergistic with calcineurin inhibitors (CNI) and could have tolerogenic effects; on the contra side, it could impair β-cell proliferation/regeneration, β-cell function thereby causing insulin resistance and contributing to the attrition of graft function [45]. These important limitations led most islet programs to move away from sirolimus or other m-TOR inhibitors for a tacrolimus–MMF combination. Sirolimus-free immunosuppressive regimens represent more than two-thirds of the current practice [3], and the Edmonton group no longer uses the original "Edmonton protocol" immunosuppressive regimen [40].

Relative "newcomers" in the field in Europe (UK, the Netherlands) and Australia adopted a sirolimus-free approach from the start of their programs. ITA or IAK programs were started with an initial tacrlimus–MMF combination. T-cell depleting induction was utilized by all, either with ATG or alemtuzumab, but no TNF-α blockade was administered [20, 21, 39]. Interestingly, the Australian protocol include a switch from MMF to sirolimus at 6 months post-transplant, but, because of toxicity issues, this was only actually done in about half the cohort and maintained in one third [21]. The UK and Australian trials took place in a multicenter national consortium setting. Metabolic goals (graft function, abolishment of severe hypoglycemic events, HbA1c < 7%) were achieved in all three trials in a large majority of study subjects [20, 21, 39].

A somewhat reverse approach, based on the aspiration to eliminate nephrotoxicity, was adopted first by the Brussels group [47] and second, in a bicenter clinical trial performed in Milan and Geneva [48]. The Brussels group performed a non-randomized controlled trial of tacrolimus–sirolimus maintenance versus sirolimus alone, under an ATG induction protocol. Sirolimus target trough levels were similar in both groups (11–15 ng/ml) and tacrolimus trough levels were 3–6 ng/ml in the control group. At 6 months post-transplant, graft function and insulin independence rates were strikingly lower in the sirolimus alone group. Of note, sirolimus had to be discontinued in 80% of patients because of toxicity or side effects [47].

The primary objective of the Milan-Geneva trial was to design an immunosuppressive protocol for islet transplantation that would be compatible with the survival, function, and expansion of subsequently transferred regulatory T-cells (Treg), with the ultimate goal of developing of an adoptive therapy with tolerogenic donor-specific Tregs for patients with type 1 diabetes undergoing islet transplantation [48]. For this reason, the immunosuppression would have to avoid IL-2R blockade [49] and to rely on mTOR inhibition for maintenance, because of its beneficial impact on

Treg biology [45], in contrast to calcineurin inhibitors [50]. Accordingly, 10 patients were enrolled in a phase 1–2 ITA clinical trial and received a median of 12,390 IEQ/kg body weight in 1–3 infusions. Patients were started on sirolimus 17–392 days (median 129 days) before islet transplantation, with target trough levels of 10–12 ng/ml. The immunosuppression protocol included ATG induction for the first islet infusion, and maintenance with sirolimus (target trough levels 12–15 ng/ml) and MMF (2 g/day). The primary endpoint was percentage of patients insulin-free at 3 years post-transplant. Unfortunately, only 1 patient could receive the full ATG dose because of side effects. Also because of side effects, 3 patients had to be switched to CNI (tacrolimus or ciclosporin). At 3 years, the rate of insulin independence was 40%, all other study subjects lost their grafts. Interestingly, 3 of the 4 insulin-free patients were those who had been switched to CNI [48]. Although the 40% insulin independence rate at 4 years was quite good, the rate of complete graft loss was sobering. One conclusion of the study was that the lack of a full induction in most subjects was probably detrimental. Another important message from both trials was the importance of CNI in the immunosuppressive regimen, at least in the peri-transplant period.

6 Immunosuppressive Regimens Including Co-stimulatory Blockade

Because T-cells play a central role in the adaptive immune responses, interfering with their activation appears as an attractive strategy to prolong allograft survival. Activation of a naïve T-cell requires distinct signals. Signal 1 determines the specificity of the immune response and involves the cognate interaction between clonal-specific T-cell receptor (TCR) and MHC-peptide complexes on antigen presenting cells (APC). To trigger the full activation of specific T cells, the first signal need to be complemented with second signals, delivered by multiple surface molecular interactions, known as "costimulation" (Table 1) [51].

While most IS protocols in islet transplantation to date have involved various combinations of "conventional" induction and maintenance agents, the University of California at San Francisco (UCSF) and Emory groups have explored co-stimulatory blockade as the main component of IS regimen. At UCSF, the combination of ATG induction with maintenance with sirolimus and the anti-LFA-1 mAb efalizumab [52], later replaced by CTLA4-Ig (belatacept) when efalizumab was withdrawn from market [53], produced excellent functional results. Efalizumab ($N = 5$ patients) was given subcutaneously at a dose of 1 mg/kg/week for the first 3 months and then tapered to 0.5 mg/kg/week. Belatacept ($N = 5$ patients) was given intravenously at a dose of 10 mg/kg on days 0, 4, 14 and every 2 weeks until day 75, decreased to 5 mg/kg every 4 weeks until 18 months, and then further decreased to 5 mg/kg every 8 weeks [53]. All patients on belatacept, and all but one on efalizumab, achieved insulin-independence after a single transplant. Five patients received a second or third transplant (7 supplementary transplants in total: 4 islets

Table 1 Immunosuppressive regimens in cited clinical trials

	Edmonton (original)	**Edmonton (evolution)**[a]	**Geneva**	**Zurich**	**Lille**	**GRAGIL 1b/2**	**Minneapolis 1**	**Minneapolis 2**	**Minneapolis 3**	**Brussels**
Indication	**ITA**	**ITA/IAK**	**IAK**	**SIK**	**ITA/ IAK**	**IAK**	**ITA**	**ITA**	**ITA**	**ITA**
Induction										
ATG		X						X	X	X
hOKT3γ_1 (Ala-Ala)							X			
Alemtuzumab		X								
Anti-IL2R	X	X	X	X	X	X			X[d]	
Anti-TNF		X							X	
Anti-IL1		X								
Steroids										X
Maintenance										
Sirolimus	X	X	X	X	X		X	X		X[f]
Everolimus						X			X[e]	
Tacrolimus	X	X	X	X	X		X[c]	X		X[f]
Ciclosporin						X			X	
MMF		X					X[c]		X[e]	
Efalizumab										
Belatacept										
Steroids			(X)[b]			(X)[b]				
References	[6, 13]	[40]	[29]	[98, 99]	[24, 26, 28]	[22, 23, 86]	[36]	[37]	[41]	[47]

(continued)

Table 1 (continued)

	UCSF 1	UCSF 2	Emory	Milan/GE	CIT-07	CIT-06	TRIMECO	UK[a]	Australia	Leiden
Indication	**ITA**	**ITA**	**ITA**	**ITA**	**ITA**	**IAK**	**ITA/IAK**	**ITA/IAK**	**ITA**	**IAK**
Induction										
ATG	X	X		X	X	X	X	X	X[l]	
hOKT3γ_1 (Ala-Ala)										
Alemtuzumab								X		X
Anti-IL2R	X[d]	X[d]	X		X[d]	X[d]	X[d]	X	X[m]	X[d]
Anti-TNF					X	X	X			
Anti-IL1				X						
Steroids				X						
Maintenance						X[k]				
Sirolimus	X	X		X[j]	X			X	X[n]	
Everolimus										
Tacrolimus			X		X		X	X	X[n]	X
Ciclosporin								X		
MMF	(X)[g]	(X)[g]	X	X			X	X	X	X
Efalizumab	X		X							
Belatacept/ Abatacept	X[h]	X	X[i]							
Steroids						(X)[b]	(X)[b]			X
References	[5, 52, 53]	[5, 52, 53]	[54]	[48]	[82, 85]	[83, 85]	[84]	[20]	[21]	[39]

[a]Various combinations of immunossuppressants have been used as first-line treatment over time
[b]In IAK recipients, minimal steroid dose was allowed
[c]Tacrolimus progressively replaced by MMF

[d]For subsequent islet injections, ATG is replaced by IL-2R blockade
[e]Everolimus progressively replaced by MMF
[f]Comparison of sirolimus alone vs sirolimus–tacrolimus
[g]Switch to MMF if sirolimus not tolerated (majority of patients)
[h]Switch from efalizumab to belatacept after withdrawal of efalizumab from market
[i]Switch from efalizumab to abatacept after withdrawal of efalizumab from market
[j]Sirolimus started pre-transplant (median 129 days)
[k]CNI-based regimen used for kidney transplant maintained
[l]ATG given for first and second islet infusion
[m]IL-2R blockade given for third islet infusion
[n]Per protocol switch from tacrolimus to sirolimus at 6 months, only done in half of study subjects

and 3 pancreas) because of a return to exogenous insulin. Remarkably, after a 10–14-year follow-up, 4 recipients of a single islet transplant are still off insulin, one of them after complete cessation of all immunosuppression for medication tolerance issues in a remarkable example of operational tolerance [5].

Similar results were reported by the Emory group with efalizumab, combined with IL-2R blockade for induction therapy and tacrolimus/MMF for maintenance [54]. When efalizumab ceased being available, patients were converted to abatacept, another CTLA4-Ig molecule that binds to CD80 and CD86 with lower avidity than belatacept [55]. In a series of 12 patients, 4 subjects on the efalizumab-based regimen were compared to 8 on the Edmonton protocol. All 4 came off insulin after a single-islet transplant and maintained insulin independence at least until the switch to abatacept (follow-up 9–24 months). In contrast, 6 of 8 patients became insulin-independent in the Edmonton protocol group, 3 of which return to exogenous insulin after 9–24 months. Subjects on the efalizumab regimen experienced significantly less side effects, but unfortunately 3 of them came back on exogenous insulin shortly after the switch to abatacept [54].

It is impossible to draw firm conclusions from this initial experience obtained in small cohorts. However, costimulatory blockade is clearly an approach that deserves further investigation in larger controlled trials, notably in view of the prolonged graft success achieved with minimal chronic immunosuppression.

7 Protocols Including Engraftment-Enhancing Agents

Glucagon-like peptide-1 (GLP-1) is an insulinotropic hormone produced in the gut. Exenatide is a synthetic agonist of the GLP-1 receptor able to increase insulin secretion in patients with peripheral insulin resistance and decrease glucagon release. In animal models, exenatide has been shown to directly impact β-cell mass by promoting neogenesis and decreasing apoptosis [56]. Exenatide became available in 2005 and was rapidly included in the peri-transplant protocols of a few centers.

The first center to include exenatide in its protocol was the University of Illinois-Chicago (UIC). After an initial group of 4 patients receiving an ITA on the Edmonton protocol, the investigators added etanercept and exenatide (5 μg subcutaneously twice daily for 1 week) to the following 6 patients [42]. All patients came off insulin, but the exenatide group required 1 islet infusion only, a significantly lower number to 2–3 in their Edmonton protocol group. However, the primary endpoint (insulin independence at 15-month) was achieved in all patients in the Edmonton protocol group and in 4/6 patients in the exenatide group [42]. Three of the 6 exenatide patients required 1 or 2 additional islet injections during the 5-year follow-up, suggesting a transient beneficial impact of exenatide. The overall 5-year insulin independence rate was a remarkable 60% [57].

The Niguarda Hospital in Milan treated 10 patients (ITA, 3 IAK) with peri-transplant exenatide at a dose of 5 μg subcutaneously twice daily for 1 month. In addition, some patients received 2 doses of everolimus for anti-inflammatory purposes, but no TNF-α blockade. Maintenance was with a tacrolimus–MMF combination. Insulin independence was achieved in 80% of patients, and maintained in 50% after a 1 to 5-year follow-up, with excellent metabolic results [58]. The absence of a control group and the heterogeneity of the protocol make it difficult to clearly assign the good results of this small series to exenatide or everolimus.

GLP-1 or dipeptidyl peptidase-4 (PPP4) inhibition were also part of the routine peri-transplant medication in the Leiden or Australian protocols, but a specific impact is impossible to appraise [21, 39].

Peri-transplant exenatide was also used by the Miami group in a series of patients retransplanted for islet graft dysfunction (defined by elevated fasting blood glucose and/or HbA1c). This protocol used a prolonged exenatide treatment strategy, starting subcutaneous exenatide injection when graft dysfunction was detected. Four patients received a supplementary islet transplant after a median of 171 days on exenatide, at the highest tolerated dose (10–25 μg per day). Exenatide was continued indefinitely after the supplementary transplant [59]. These patients were compared to 5 historic controls who had received a supplementary islet transplant for the same reason but without exenatide. All patients came off insulin, but after 18 months, all exenatide-treated patients were still insulin-independent compared to only 1/5 in the control group [59]. These results suggest a beneficial and prolonged effect of exenatide on islet graft function when administered chronically, but the small sample size and the non-randomized nature of the study warrant a word of caution.

The Miami and Vancouver group also reported their experience of treating patients with islet graft dysfunction with exenatide injections in an attempt to restore insulin secretion to higher levels. Both groups reported a positive but transient effect [60, 61].

The San Raffaele group in Milan has been working on reparixin, a pharmacological inhibitor of the CXCR1/2 chemokine receptor, since several years. They demonstrated that its ligand, the proinflammatory chemokine CXCL8, is released in high quantities by islets after isolation and immediately before transplantation. CXCL8 levels were also highly elevated in the circulating blood of islet transplant recipients, independently of the immunosuppressive regimen [62]. These findings led to a phase 2 open-label pilot study in 9 patients receiving a single-donor islet transplant and randomized to receive reparixin or no additional treatment; one patient dropped out of the study. The standard protocol included ATG induction and tacrolimus–MMF maintenance, tacrolimus being switched to sirolimus after 3 months. Reparixin was administered as a continuous intravenous injection (2.772 mg/kg/h) for 7 days [62]. The results showed that all patients in the control group ($N = 3$) lost all graft function after 1 month, in contrast to the treatment group ($N = 5$) in which all patients had persistent graft function with a marked improvement of glycemic control. Three of these patients went on to have a second islet infusion, and 2 of them came off insulin [62].

These encouraging findings led to a multicenter, phase 3, double-blind, randomized, placebo-controlled trial, exploring the efficiency of reparixin, in promoting islet engraftment. Fifty-one patients, recruited by 9 centers in Europe and the USA, and candidates for ITA were randomized 2:1 to receive reparixin or placebo. The complex primary endpoint was the area under the curve [63] for the serum C-peptide level during 2 h of a mixed-meal tolerance test, normalized by the number of IEQ/

kg at day 75 after the first islet infusion and day 365 after the last islet infusion. Several secondary endpoints were also assessed, including the composite endpoint of HbA1c < 7% and absence of SHE. An important issue was the absence of a standardized IS protocol. Each participating center administered their preferred induction and maintenance regimens, which probably resulted in an important center effect in functional results. Of particular relevance, anti-TNF-α agents were not allowed in the study, and 37.5% patients did not receive ATG induction. No difference was observed for the primary endpoint between the study and control groups, with insulin independence rates at 1 year at 32% and 31.3%, respectively. A trend toward higher rates of insulin independence was observed in the subset of patients receiving ATG induction. Overall, >70% subjects were free of SHE and had HbA1c < 7%. Although not reaching statistical significance, reparixin was more effective than placebo in reaching single-donor insulin independence (approximately 20% vs. 6%) [63].

A post-hoc analysis from the University of Chicago analyzed separately the patients enrolled at their center. Twelve recipients were randomized to receive reparixin ($N = 8$) or placebo ($N = 4$). All patients received an ATG induction. Five patients in the reparixin group came off insulin after a single transplant, compared to none of the controls. This analysis suggested that reparixin might have been associated with better results than placebo when a sufficient islet mass is transplanted and when standardized ATG induction is used [64].

Since IBMIR is elicited within the portal vasculature and is triggered by the coagulation cascade [35], anticoagulation strategies by systemic heparinization in the peri-transplant phase were implemented with some success in improving the rate of engraftment [65]. The Nordic Network conducted a multicenter phase 2 randomized controlled trial to examine whether peri-transplant anticoagulation with low molecular weight dextran-sulfate was superior to heparin in controlling IBMIR [19]. Twenty-four candidates for an ITA were enrolled, 10 in the experimental arm and 14 in the control arm. Dextran-sulfate and heparin were both administered to confer full-systemic anticoagulation immediately at the time of islet infusion and for a duration of 5 h. Immunosuppression was based on ATG induction for the first-islet injection, and maintenance on MMF or sirolimus combined with a CNI. The primary endpoint was stimulated C-peptide (mixed meal tolerance test) at 75 days after initial transplant. The study failed to demonstrate an advantage of dextran over heparin, either in the primary endpoint, or in insulin requirements at 75 or 365 days. It is noteworthy that the blood levels of thrombin-antithrombin complexes were identical in both groups in the hours following islet infusion [19].

8 Islet Transplantation in Extra-Hepatic Sites

As previously described, intrahepatic islet transplantation may result in the loss of a large proportion of the infused islets, mainly due to the IBMIR reaction, the exposition to toxic agents coming from the portal blood flow (e.g., immunosuppressive drugs) [33] and the proinflammatory microenvironment of the liver [34, 66]. For these reasons, the search for an extra-hepatic transplantation site has been a significant research goal in the field. An additional feature of a new site should be ease of access for implantation and realization of biopsies.

A large number of potential sites have been explored in animal models such as the subcutaneous space, the muscle, the kidney capsule, the gastric submucosa, the spleen, the pancreas, the epididymal fat pad, the testis, the peritoneum, the omentum, the bone marrow, the anterior chamber of the eye, the cerebral ventricle and the thymus [67], but few have been tested clinically in the modern era.

The striated muscle is an attractive site because it is easily accessible and well vascularized. Promising results were obtained in animals [68, 69] and the muscle has been used in islet autotransplantation with acceptable results [70]. Islet allotransplantation into the skeletal muscle of the forearm has only been performed at the Niguarda hospital in 4 patients in whom intrahepatic transplantation was contraindicated [71]. None of the patients came off insulin and all had a transient elevation of C-peptide, which went back to trace levels between 6 and 12 months after transplantation [71].

Three centers have reported islet transplantation in an omental pouch. It has attracted interest as an extra-hepatic islet implantation site because it displays extensive vascularization and portal venous drainage due to its location in the peritoneal cavity [72]. The first reported case was a type-1 diabetic patient who received an allotransplant with a single-islet infusion (11′300 IEQ/kg) incorporated into a bioartificial thrombin scaffold and embedded into the omentum by a laparoscopic approach [73]. The patient came off insulin after 17 days and remained insulin-independent at 1 year with good metabolic control [73]. This case was reported as part of an ongoing clinical trial, but no further report has been published to date.

The Brussels group has published the results of a pilot study of 7 patients who received an islet graft implanted laparoscopically into the omentum using scaffolds of various compositions. Five different scaffolds were used. None of the omental implants achieved an adequate metabolic effect, with at best low and transient levels of measurable circulating C-peptide [74].

Finally, the Prague group recently reported 2 cases of islet transplantation in an omental pouch. In total, 12,350 and 5,350 IEQ/kg were mixed with autologous plasma, seeded during a laparoscopic procedure onto the omentum, overlaid with human thrombin solution, and fixed by flapping the omentum over. Partial function was reported in both cases, with measurable C-peptide, abolishment of hypoglycemic events and a 50% decrease in insulin requirements, all maintained at least 1 year post-transplant [75].

The gastric submucosa is easily accessible by a minimally invasive endoscopic approach. It is well vascularized with portal venous drainage, and has shown promising results in experimental models [76]. Transplantation into the gastric submucosa has been performed by a Warsaw group for autotransplantation after total pancreatectomy [77], but never in an allotransplantation setting.

The San Raffaele group in Milan has worked on the bone marrow site for several years and demonstrated results superior to intrahepatic transplantation in a murine model [78]. These results made them embark on two randomized clinical trials [79, 80], one of them in the allotransplantation setting. Islets were infused in the iliac crest in patients not eligible for intrahepatic transplantation. In total, 7 patients with type 1 diabetes received an allogeneic islet transplant into the bone marrow.

These studies demonstrated the feasibility and the safety of islet infusion into the bone marrow.

However, despite successful graft implantation, islet function was lost within 6 months after transplantation, partially because of autoimmunity recurrence. Monitoring biopsies were performed by bone marrow aspiration and were able to demonstrate an immune reaction against the islets. Finally, despite having a dense vascular network, the bone marrow has low pO2 and severe hypoxic stress to the islets most probably occurred [81].

Taken together, the results of human studies in search for alternatives to the liver as islet transplantation sites have mostly demonstrated the discrepancy between experimental and clinical research, and the difficulty to translate promising animal studies into the clinic.

9 Phase 3 Prospective Trials

The last years have seen the publication of important multicenter prospective controlled trials that have provided the most robust data in support of islet transplantation.

The CIT-07 trial from the NIH-sponsored Clinical Islet Transplant Consortium (CITC) enrolled 48 patients at 8 centers in the USA and Canada between 2008 and 2011 [82]. It was a phase 3, prospective, open-label, single-arm study, assessing the impact of islet transplantation alone in patients with hypoglycemia unawareness. The immunosuppression protocol included ATG (daclizumab for subsequent injections) and etanercept induction and tacrolimus–sirolimus maintenance. The composite primary endpoint was eradication of severe hypoglycemic events (SHE) and HbA1c < 7% at 1 year after the first islet transplant. Several secondary endpoints were evaluated up to 2 years post-transplant. Remarkably, the primary endpoint was achieved in 87.5% of subjects, and maintained in 71% at 2 years. Of the secondary endpoints, insulin independence was achieved in 52% at 1 year (insufficient data at 2 years), HbA1c ≤ 6.5% was achieved in 79% and 69% at 1 and 2 years, and the estimated probability of SHE-free survival at 2 years was 93% [82]. The results of the CIT-07 trial have provided a compelling rationale for its consideration as standard-of-care in the management of SHE.

The CIT-06 trial was similar in design to CIT-07, but included patients previously transplanted with a kidney for end-stage diabetic nephropathy and with poor metabolic control (IAK). Twenty-four subjects were enrolled at 10 participating centers in North America over the same period as CIT-07 (see Table 9 for IS protocol). The immunosuppression protocol included ATG (daclizumab for subsequent injections) and etanercept induction. The calcineurin-based maintenance immunosuppression regimen used for the renal transplant was continued after islet transplant. Up to 10 mg of prednisone was allowed as part of maintenance immunosuppression. The primary and secondary endpoints were essentially similar to those of CIT-07. The composite primary endpoint (eradication of severe hypoglycemic events and HbA1c < 7% at 1 year post-first islet transplant) was achieved in 62.5% of subjects, and maintained in 58.3% at 2 years and 45.8% at 3 years. Of the secondary endpoints, insulin independence was achieved in 37.5% at 1 year (29.2% at 2 years), HbA1c ≤ 6.5% was achieved in 62.5% and 50% at 1 and 2 years, eradication of SHE was achieved in 79% and 75% at 1 and 2 years [83]. Although the results did not quite match those of CIT-07, they provided strong evidence for the efficacy of islet transplantation in the IAK approach.

The TRIMECO trial was the first randomized trial comparing transplantation to a reference treatment for any type of organ failure. TRIMECO enrolled 50 patients in 15 institutions in France between 2010 and 2013. Study subjects were either patients with hypoglycemia unawareness (ITA) or patients previously transplanted with a kidney for end-stage diabetic nephropathy with poor metabolic control (IAK). The immunosuppression protocol included ATG (basiliximab for subsequent injections) and etanercept induction and tacrolimus–MMF maintenance. For IAK, up to 0.1 mg/kg of prednisone was allowed as part of maintenance immunosuppression. This was a phase 3, open label, two-arm, prospective randomized trial. Subjects were randomized 1:1 to receive immediate islet transplantation or intensive insulin therapy (control arm). Patients randomized to the control arm were eligible for a subsequent islet transplant after the primary endpoint. This trial was conducted with a network approach, in which 3 centralized facilities in France and Switzerland isolated islets for the 15 participating centers. The primary endpoint was assessment of the beta score at 6 months after islet transplantation. Several secondary endpoints were assessed up to 12 months post-transplant. The median beta score increased from 0 to 6 in the transplant group versus 0 to 1.5 in the control group (the beta score was modified to allow values >0 in type 1 diabetic patients with negative stimulated C-peptide to allow a fairer comparison of metabolic outcomes). Regarding secondary endpoints at 6 months, mean HbA1c was 5.6% vs. 8.2%, fasting blood glucose was 5.2 mmol/l vs. 5.7 mmol/l, and 84% had HbA1c < 7% and abolition of SHE vs. 0%. This indicated excellent results of islet transplantation and showed that the effects of intensive insulin therapy were present but unable to match those of islet transplantation [84]. At 1 year, 59% were off insulin, median beta score was 7, median HbA1c was 5.8%, and 70% of patients had a HbA1c < 7% and absence of SHE [84].

10 Long-Term Outcomes

Publication of long-term outcomes (>5 years) is available from 4 different sources: the CIT trials (8 years), the GRAGIL and Lille trials (both 10 years) and the Edmonton series (20 years).

The CIT-08 trial was a follow-up prospective observational study of the CIT-06 and CIT-07 protocols discussed above, to obtain an 8-year follow-up in actuarial terms. Of the original 72 study subjects (48 ITA and 24 IAK), only 42 entered the trial (34 ITA and 8 IAK). Eighteen patients declined participation or had withdrawn from their original trial, 10 had experienced graft failure and 2 had already completed the CIT-6 trial and could not be included. Four additional patients withdrew from the study before completion. Statistical simulation was performed to ensure that the bias due to the censoring of drop-outs did not significantly affect the survival curves. The primary endpoint was graft survival defined as stimulated C-peptide >0.3 ng/ml after a mixed meal tolerance test [85].

Actuarial graft function was 56% for ITA and 49% for IAK, after 8.3 and 7.3-year follow-up, respectively. Insulin independence was achieved in 74% of patients, of which 57% were still off insulin at the end of follow-up, that is, an insulin-independence rate of 42%. Interestingly, most patients with sustained graft function were still insulin-free. HbA1c <7% was achieved in 49% ITA and 35% IAK at approximately 8 years, with no difference between ITA and IAK. Finally, 93% of patients had been free of severe hypoglycemic events [85]. Significant differences were observed in favor of ITA over IAK, but only for graft function and rate of HbA1c <7%.

The Lille group has published the 10-year outcome results of their two-arm prospective trial in 28 patients (14 ITA and 14 IAK). Median follow-up was 11.5 years (8.9–12.9). The stringent primary outcome was insulin-independence with HbA1C <6.5%. All patients achieved insulin-independence and the primary outcome was reached in 39% at 5 years and 28% at 10 years, without significant difference between ITA and IAK. Graft survival, defined by fasting C-peptide level >0.3 ng/ml, was achieved in 82% and 78% at 5 and 10 years, again without differences between ITA and IAK. The Lille concept that primary graft function determines graft survival was confirmed in this study, with a median graft survival of >10 years for optimal graft function (β-score ≥ 7 at 1 month post last islet infusion) vs. 4.5 years for suboptimal graft function. No severe hypoglycemic events were observed throughout follow-up [28].

Unlike the previous studies, the GRAGIL consortium analyzed the 10-year outcome data of their trials retrospectively, but in actual terms, all patients having completed 10 years follow-up (10–17 years). Only patients included in the GRAGIL 1b and GRAGIL 2 trials, that is, IAK and ITA on the Edmonton immunosuppressive regimen, were reported. Graft survival defined by fasting C-peptide level >0.3 ng/ml, was achieved in 79% and 52% at 5 and 10 years, respectively, with no difference between ITA and IAK. Insulin independence, defined by absence of exogenous insulin and Hba1c < 7%, was achieved in 20% and 5% at 5 and 10 years. Median HbA1c was 6.8% and 7.2% at 5 and 10 years. ITA recipients were free of severe hypoglycemic events in 93% and 84% at 5 and 10 years, respectively [86].

The Edmonton group has reported their 20-year experience in a retrospective cohort study of 255 patients with a median follow-up of 7.4 years. Patients received 1–5 islet infusions, with a median islet graft mass of 14,300 IEQ/kg. The immunosuppression regimen evolved from the original Edmonton protocol (IL2-R blockade, sirolimus, and tacrolimus) to include T-cell-depleting induction for the first islet injection, first with ATG, and then with alemtuzumab. Anti-inflammatory strategy was added to include first TNF-α blockade, soon complemented with IL-1 blockade. For maintenance, the sirolimus–tacrolimus association was progressively replaced by tacrolimus–MMF. All patients have received tacrolimus and >90% have received MMF, either from the start or after a switch from sirolimus, only 43% having received tacrolimus. In this series, using the etanercept–anakinra combination for induction had a positive impact on outcome (sustained graft function); MMF was also favorable and sirolimus had a negative impact [40].

Actuarial patient and graft survival (Kaplan-Meier) were 74% and 48% at 20 years, with a median graft survival of 5.9 years. Insulin-independence occurred in 79%, with an actuarial insulin-independence rate of 8% at 20 years. Median duration of insulin-independence was 2.3 years. Excellent metabolic results were reported for those with sustained graft function, with median HbA1c at 6% and median fasting blood glucose at 6 mmol/l at 20 years. Hypoglycemia was near abolished in almost all patients for the whole follow-up, including for those without sustained graft function, up to 8 years. Hypoglycemia unawareness, assessed by the Clarke score, was abolished for the whole follow-up period in patients with sustained graft function, and even up to 5 years in case of graft loss [40].

11 Impact of Islet Transplantation on Kidney Function

A key decision factor for islet transplantation in ITA and IKA candidates is the impact that chronic immunosuppression will have on kidney function, and to balance this against the rate of progression of diabetic nephropathy. Even in candidates for beta-cell replacement with a normal kidney function at the time of transplant, the presence of some degree of glomerulopathy is frequent, if not universal [87]. In a crossover study from Vancouver, in which ITA recipients were compared to ITA candidates on the waiting list, the rate of decrease of the glomerular filtration rate was 1.49 ml/min/year compared to 3.53 ml/min/year, indicating that nephrotoxicity of IS was able to offset diabetic nephropathy [88].

The long-term evolution of renal function has been examined in some of the series mentioned above. In the Lille series, baseline eGFR went from 68 ml/min/1.73 m^2 at baseline to 64 at 5 years and 54 at 10 years. The decrease was 1.1 ml/min/1.73 m^2/year for ITA, and 0.9 for IAK; this decrease did not reach statistical significance and eGFR was in fact stable or even slightly increased in some patients. None of the patients developed end-stage renal failure [28]. In the GRAGIL series, glomerular filtration rates before islet transplantation and after 10 years of follow-up were significantly different, in both IAK and ITA recipients: median eGFR was 76 ml/min/m^2 at 10 years versus 88 at baseline in ITA recipients, and 47.8 ml/min/m^2 at 10 years versus 58.6 at baseline in IAK recipients [86]. Finally, in the 20-year Edmonton report, data are provided in graphs, and the following figures are approximate. In the whole cohort, eGFR decreased from 85 ml/min/1.73 m^2 at baseline to

60 at 10 years and about 58 at 20 years. Decrease of eGFR was more pronounced and a higher incidence of stage 3 chronic kidney disease was observed in patients with sustained islet graft function, probably on account of higher or longer CNI exposure. On the other hand, progression to end-stage renal failure, which occurred in about15–20% over the 20-year follow-up, was delayed by persistent islet graft function [40].

There is obviously no control group in these cohorts and no valid model can provide insights on the fate of kidney function in the absence of a transplant in the study subjects. One should consider these kidney function data in perspective with the long-term risks of developing end-stage renal failure in patients diagnosed with type 1 diabetes [89], and keep in mind that, at the time of first islet transplantation, study subjects already had 33-year, 30-year, and 28-year history of diabetes in the GRAGIL, Edmonton, and Lille series, respectively [28, 40, 90].

12 Risks of HLA Sensitizitation

As observed with other organ transplants, there is a risk of HLA sensitization after islet transplantation. Since islet infusions from multiples donors are usually necessary, the risk of sensitization is theoretically increased. This is a matter of concern, because HLA sensitization may negatively impact the ability of these individuals to undergo further islet, pancreas or, more problematically, kidney transplantation in case of progression of the nephropathy (see paragraph above).

Screening for HLA sensitization was performed in the 98 first islet transplant recipients in the Edmonton series. Twenty-nine patients (31%) developed de novo donor-specific antibodies (DSA) post-transplant. Ten of 14 recipients (71%) who have discontinued immunosuppression are broadly sensitized with panel reactive antibodies (PRA) ≥50%. The majority of these antibodies have persisted over time [91].

Forty-two pancreatic islet graft recipients from the GRAGIL consortium were enrolled in a similar HLA sensitization screening. The prevalence of DSA was 25% at 3 years post-transplant. The risk of sensitization increased steeply after immunosuppressive drug withdrawal. DSA repertoire diversity correlated with the number of HLA and eplet mismatches. DSA titers were significantly lower from those observed in solid organ transplantation [92]. None of the detected DSA bound the complement fraction C3d. However, in contrast with solid organ transplantation, DSA did not seem to negatively affect islet graft survival [93].

These data seem to be in contradiction with the San Raffaele series, in which the 27 of 59 (46%) islet transplant recipients experiencing an increase in DSA titers during follow-up, had a significantly increased risk of graft failure (HR > 5) [94]. However, 29 patients (49%) had pre-formed DSA, which was not case in the GRAGIL series (1 individual with pre-formed DSA; 8%) [93, 94]. Of note, the presence of pre-formed DSA has been associated with reduced islet graft survival in the

Edmonton cohort [95]. In a report of 16 islet transplant recipients from Newcastle, 5 (19%) developed DSA within 4 weeks after transplantation and experienced rapid or delayed graft failure in the following months despite maintained immunosuppression. In contrast, only 1 of 11 patients who did not develop DSA had graft failure during follow-up [96].

However, it should be remembered that the generation of DSA implies the differentiation of B cells into antibody-producing plasma cells, which requires T cells. Hence, DSA appearance could be a "marker but not a maker" of islet rejection [97]. In other words, DSA appearance could reflect an ongoing alloimmune T-cell response involving not only CD4+ T cells (leading to DSA generation) but also CD8+ cytotoxic T cells, which may be responsible for the destruction of the grafted islets [97].

13 What Are the Lessons Learned?

The evidence gathered from clinical trials, summarized and discussed above, has contributed to a better understanding of the phenomena involved in islet engraftment and function. This has fostered progressive changes or fine tuning of the procedures, and continuously improved islet transplantation outcomes [2].

Lesson no. 1: The ability to abolish severe hypoglycemia is the major achievement of islet transplantation. It has been a ubiquitous finding of all islet transplant trials and a prolonged restoration of hypoglycemia awareness has been shown to occur, even after graft failure.

Lesson no. 2: Insulin-independence is not the only criterion for islet transplantation success. This is especially true for patients transplanted for severe hypoglycemia, but also in others in whom blood sugar levels are challenging to control. The control of glycemic variability, HbA1c, and progression of diabetic nephropathy can be obtained even in the absence of insulin-independence. This has been acknowledged by the development of the Igls consensus criteria for the definition of islet and pancreas graft outcomes. These criteria define islet (or pancreas) graft function, using a 4-tier classification of optimal, good, marginal, or failure, based not only on insulin requirements, but also on detectability of C-peptide, HbA1c and control of hyperglycemic events [4]. A reappraisal of the Igls criteria has proposed to integrate CGM data to obtain a refined classification [5].

Lesson no. 3: Attrition of islet graft function is the general rule.

Lesson no. 4: Multiple-donor transplantation is generally required for islet graft success.

Lesson no. 5: Optimal primary graft function must be obtained to ensure prolonged islet graft survival. All long-term clinical trials have seen a seemingly inescapable progressive loss of insulin-independence an islet graft survival. Engraftment of a sufficient functional islet mass is a determinant of prolonged graft function and survival. It is widely thought that a marginal engrafted islet mass will lead to islet cell exhaustion and accelerated graft loss. Optimal primary graft function can be achieved by the rapid transplantation of a sufficient functional islet mass, using up to 3 donors if necessary [24, 28]. This will be of particular interest in the design of regenerative medicine strategies, when quantifying the necessary functional mass of the implant. Optimal primary graft function also requires the appropriate peri-transplant management of acquired and innate immune phenomena, which will be discussed below.

The Edmonton protocol has initially focused only on ITA recipients, and North American centers have adopted this practice, which they have progressively extended to IAK transplantation. Evidence from a large number of centers, mostly in Europe, has demonstrated that the outcomes of islet transplantation were essentially identical after ITA, IAK, and SIK transplantation.

Lesson no. 6: Islet transplantation can be successfully performed in the ITA, IAK, and SIK indications. This allows to offer a form of beta-cell replacement therapy to every patient with complicated type 1 diabetes, notably when pancreas transplantation is contra-indicated.

Some of the lessons relate to the optimal immunosuppressive strategy. The analysis of clinical outcomes has led to the progressive understanding of the mechanisms involved in islet graft damage and loss. Notably, although their sensitivity to immune effectors (antibodies versus T cells) may not be the same, islets are as prone to rejection as vascularized organs and the universally adopted steroid-free regimens call for a strong induction and adapted maintenance. The inflammatory events, notably IBMIR, occurring at the time of islet implantation must be tackled to optimize engraftment. Strategies addressing these issues have markedly improved graft survival but have overall not been able to match the outcomes of pancreas transplantation.

Lesson no. 7: T-cell-depleting induction is required for optimal long-term islet graft outcomes. ATG or alemtuzumab have been utilized with apparently identical results.

Lesson no. 8: Calcineurin inhibitors are required, at least in the initial post-transplant period. CNI are the best class of immunosuppressants to control T-cell responses. Attempts at utilizing CNI-free regimens have failed.

Lesson no. 9: The role of sirolimus (or other mTOR inhibitors) in the maintenance regimen is unclear. A tolerogenic effect has been postulated and some programs have successfully used, but their overall toxicity profile makes them poorly tolerated drugs. A tacrolimus–MMF association is by far the most common maintenance regimen currently in use.

Lesson no. 10: Co-stimulatory blockade, notably belatacept, is a potentially interesting component of the maintenance regimen. Further studies are required to confirm the promising preliminary results. Long-term insulin independence after single-donor transplantation and a case of operational tolerance have been reported.

Lesson no. 11: An anti-inflammatory strategy is required at the time of implantation. The association of TNF-α blockade by etanercept and IL-1 blockade by anakinra has been associated with improved long-term outcomes. Blockade of the CXCL8 chemokine by reparixin may also have a beneficial impact on islet engraftment, but robust evidence is still lacking in clinical trials.

The potentially harmful consequences of islet transplantation for the long-term prognosis of the recipients are essentially related to the impact of chronic immunosuppression on kidney graft function and the occurrence of HLA sensitization, which could prevent or delay access to a subsequent kidney transplant, should end-stage renal failure develop.

Lesson no. 12: Kidney function decreases slowly and slightly after islet transplantation. The decrease seems to be slower in transplanted and immunosuppressed patients than what the natural history of diabetic nephropathy would predict. However, a small proportion of patients will develop end-stage renal failure in the very long term.

Lesson no. 13: HLA sensitization develops in islet transplant recipients essentially after graft failure and discontinuation of immunosuppression. In spite of conflicting evidence, studies in larger cohorts seem to show that de novo DSA can occur when immunosuppression is maintained, but with little impact on islet graft function, if any. Despite a lack of data, a slow weaning of immunosuppression after graft failure can be recommended to avoid or limit the extent of HLA sensitization.

Finally, the search for an islet implantation site better than the liver has been ongoing. Recent clinical trials have explored the gastric submucosa, the skeletal muscle or the bone marrow as potential extra-hepatic sites but have failed to demonstrate superiority to the hepatic site. The omental pouch may be a suitable alternative site but reports of success have been scanty so far and have required incorporation of the liver inside a bioartificial scaffold.

Lesson no. 14: The liver, by intra-portal injection, remains the optimal islet transplantation site.

Lesson no. 15: Successful islet transplantation in extra-hepatic sites will likely require graft inclusion into a bioartificial scaffold. This is of particular relevance for the field of regenerative medicine, where avoiding the inflammatory phenomena that occur in the liver and easy access and retrieval of the graft will be among the priorities. This will call for the design of a bioartificial pancreas.

Islet of Langerhans transplantation has gained a clear place in the therapeutic armamentarium for type 1 diabetes and is now considered standard-of-care in many countries in which it is available [1]. Depending on the indication, patient profile or goals to be achieved, islet transplantation is an excellent first-line approach or a valid alternative to pancreas transplantation. The several lessons that have been learned and that are summarized above may have an impact, beyond the mere field of islet transplantation, on other modalities of beta-cell replacement therapies, notably in regenerative medicine (Fig. 1).

Fig. 1 A graphic summary of 15 lessons learned from clinical trials of islet transplantation

References

1. Berney T, Andres A, Bellin MD, de Koning EJP, Johnson PRV, Kay TWH, et al. A worldwide survey of activities and practices in clinical islet of Langerhans transplantation. Transpl Int. 2022;35:In press.
2. Barton FB, Rickels MR, Alejandro R, Hering BJ, Wease S, Naziruddin B, et al. Improvement in outcomes of clinical islet transplantation: 1999–2010. Diabetes Care. 2012;35(7):1436–45.
3. (CITR) CITR. Eleventh Annual Report 2022. Available from: https://citregistry.org/system/files/11th%20Allograft%20report%20May%2031%202022.pdf.
4. Rickels MR, Stock PG, de Koning EJP, Piemonti L, Pratschke J, Alejandro R, et al. Defining outcomes for beta-cell replacement therapy in the treatment of diabetes: a consensus report on the Igls criteria from the IPITA/EPITA opinion leaders workshop. Transpl Int. 2018;31(4):343–52.
5. Landstra CP, Andres A, Chetboun M, Conte C, Kelly Y, Berney T, et al. Examination of the Igls Criteria for Defining Functional Outcomes of beta-cell Replacement Therapy: IPITA Symposium Report. J Clin Endocrinol Metab. 2021;106(10):3049–59.
6. Shapiro AM, Lakey JR, Ryan EA, Korbutt GS, Toth E, Warnock GL, et al. Islet transplantation in seven patients with type 1 diabetes mellitus using a glucocorticoid-free immunosuppressive regimen. N Engl J Med. 2000;343(4):230–8.
7. International Islet Transplant Registry (ITR). Newsletter #9 2001. Available from: https://www.med.uni-giessen.de/itr/newsletter/no_9/news_9.pdf.
8. Froud T, Ricordi C, Baidal DA, Hafiz MM, Ponte G, Cure P, et al. Islet transplantation in type 1 diabetes mellitus using cultured islets and steroid-free immunosuppression: Miami experience. Am J Transplant. 2005;5(8):2037–46.

9. Bretzel RG, Jahr H, Eckhard M, Martin I, Winter D, Brendel MD. Islet cell transplantation today. Langenbecks Arch Surg. 2007;392(3):239–53.
10. (CITR) CITR. Tenth Annual Report. 2017.
11. Wojtusciszyn A, Branchereau J, Esposito L, Badet L, Buron F, Chetboun M, et al. Indications for islet or pancreatic transplantation: Statement of the TREPID working group on behalf of the Societe francophone du diabete (SFD), Societe francaise d'endocrinologie (SFE), Societe francophone de transplantation (SFT) and Societe francaise de nephrologie - dialyse - transplantation (SFNDT). Diabetes Metab. 2019;45(3):224–37.
12. Ward C, Odorico JS, Rickels MR, Berney T, Burke GW, 3rd, Kay TWH, et al. International Survey of Clinical Monitoring Practices in Pancreas and Islet Transplantation. Transplantation. 2022;106(8):1647–55.
13. Shapiro AM, Ricordi C, Hering BJ, Auchincloss H, Lindblad R, Robertson RP, et al. International trial of the Edmonton protocol for islet transplantation. N Engl J Med. 2006;355(13):1318–30.
14. Ryan EA, Paty BW, Senior PA, Bigam D, Alfadhli E, Kneteman NM, et al. Five-year follow-up after clinical islet transplantation. Diabetes. 2005;54(7):2060–9.
15. Goss JA, Schock AP, Brunicardi FC, Goodpastor SE, Garber AJ, Soltes G, et al. Achievement of insulin independence in three consecutive type-1 diabetic patients via pancreatic islet transplantation using islets isolated at a remote islet isolation center. Transplantation. 2002;74(12):1761–6.
16. Hirshberg B, Rother KI, Digon BJ, 3rd, Lee J, Gaglia JL, Hines K, et al. Benefits and risks of solitary islet transplantation for type 1 diabetes using steroid-sparing immunosuppression: the National Institutes of Health experience. Diabetes Care. 2003;26(12):3288–95.
17. Berney T, Benhamou PY, Kessler L, Morel P. Islet transplantation in multicenter networks – The GRAGIL example. Curr Opin Organ Transplant 2004;9:72–6.
18. Kempf MC, Andres A, Morel P, Benhamou PY, Bayle F, Kessler L, et al. Logistics and transplant coordination activity in the GRAGIL Swiss-French multicenter network of islet transplantation. Transplantation. 2005;79(9):1200–5.
19. von Zur-Muhlen B, Lundgren T, Bayman L, Berne C, Bridges N, Eggerman T, et al. Open Randomized Multicenter Study to Evaluate Safety and Efficacy of Low Molecular Weight Sulfated Dextran in Islet Transplantation. Transplantation. 2019;103(3):630–7.
20. Brooks AM, Walker N, Aldibbiat A, Hughes S, Jones G, de Havilland J, et al. Attainment of metabolic goals in the integrated UK islet transplant program with locally isolated and transported preparations. Am J Transplant. 2013;13(12):3236–43.
21. O'Connell PJ, Holmes-Walker DJ, Goodman D, Hawthorne WJ, Loudovaris T, Gunton JE, et al. Multicenter Australian trial of islet transplantation: improving accessibility and outcomes. Am J Transplant. 2013;13(7):1850–8.
22. Borot S, Niclauss N, Wojtusciszyn A, Brault C, Demuylder-Mischler S, Muller Y, et al. Impact of the number of infusions on 2-year results of islet-after-kidney transplantation in the GRAGIL network. Transplantation. 2011;92(9):1031–8.
23. Badet L, Benhamou PY, Wojtusciszyn A, Baertschiger R, Milliat-Guittard L, Kessler L, et al. Expectations and strategies regarding islet transplantation: metabolic data from the GRAGIL 2 trial. Transplantation. 2007;84(1):89–96.
24. Vantyghem MC, Kerr-Conte J, Arnalsteen L, Sergent G, Defrance F, Gmyr V, et al. Primary graft function, metabolic control, and graft survival after islet transplantation. Diabetes Care. 2009;32(8):1473–8.
25. Ryan EA, Paty BW, Senior PA, Lakey JR, Bigam D, Shapiro AM. Beta-score: an assessment of beta-cell function after islet transplantation. Diabetes Care. 2005;28(2):343–7.
26. Vantyghem MC, Raverdy V, Balavoine AS, Defrance F, Caiazzo R, Arnalsteen L, et al. Continuous glucose monitoring after islet transplantation in type 1 diabetes: an excellent graft function (beta-score greater than 7) Is required to abrogate hyperglycemia, whereas a minimal function is necessary to suppress severe hypoglycemia (beta-score greater than 3). J Clin Endocrinol Metab. 2012;97(11):E2078–83.

27. Leitao CB, Tharavanij T, Cure P, Pileggi A, Baidal DA, Ricordi C, et al. Restoration of hypoglycemia awareness after islet transplantation. Diabetes Care. 2008;31(11):2113–5.
28. Vantyghem MC, Chetboun M, Gmyr V, Jannin A, Espiard S, Le Mapihan K, et al. Ten-Year Outcome of Islet Alone or Islet After Kidney Transplantation in Type 1 Diabetes: A Prospective Parallel-Arm Cohort Study. Diabetes Care. 2019;42(11):2042–9.
29. Toso C, Baertschiger R, Morel P, Bosco D, Armanet M, Wojtusciszyn A, et al. Sequential kidney/islet transplantation: efficacy and safety assessment of a steroid-free immunosuppression protocol. Am J Transplant. 2006;6(5 Pt 1):1049–58.
30. Cure P, Pileggi A, Froud T, Messinger S, Faradji RN, Baidal DA, et al. Improved metabolic control and quality of life in seven patients with type 1 diabetes following islet after kidney transplantation. Transplantation. 2008;85(6):801–12.
31. Gerber PA, Locher R, Zuellig RA, Tschopp O, Ajdler-Schaeffler E, Kron P, et al. Glycemia, Hypoglycemia, and Costs of Simultaneous Islet-Kidney or Islet After Kidney Transplantation Versus Intensive Insulin Therapy and Waiting List for Islet Transplantation. Transplantation. 2015;99(10):2174–80.
32. Harlan DM, Kenyon NS, Korsgren O, Roep BO, Immunology of Diabetes S. Current advances and travails in islet transplantation. Diabetes. 2009;58(10):2175–84.
33. Desai NM, Goss JA, Deng S, Wolf BA, Markmann E, Palanjian M, et al. Elevated portal vein drug levels of sirolimus and tacrolimus in islet transplant recipients: local immunosuppression or islet toxicity? Transplantation. 2003;76(11):1623–5.
34. Delaune V, Berney T, Lacotte S, Toso C. Intraportal islet transplantation: the impact of the liver microenvironment. Transpl Int. 2017;30(3):227–38.
35. Bennet W, Groth CG, Larsson R, Nilsson B, Korsgren O. Isolated human islets trigger an instant blood mediated inflammatory reaction: implications for intraportal islet transplantation as a treatment for patients with type 1 diabetes. Ups J Med Sci. 2000;105(2):125–33.
36. Hering BJ, Kandaswamy R, Harmon JV, Ansite JD, Clemmings SM, Sakai T, et al. Transplantation of cultured islets from two-layer preserved pancreases in type 1 diabetes with anti-CD3 antibody. Am J Transplant. 2004;4(3):390–401.
37. Hering BJ, Kandaswamy R, Ansite JD, Eckman PM, Nakano M, Sawada T, et al. Single-donor, marginal-dose islet transplantation in patients with type 1 diabetes. JAMA. 2005;293(7):830–5.
38. Bellin MD, Barton FB, Heitman A, Harmon JV, Kandaswamy R, Balamurugan AN, et al. Potent induction immunotherapy promotes long-term insulin independence after islet transplantation in type 1 diabetes. Am J Transplant. 2012;12(6):1576–83.
39. Nijhoff MF, Engelse MA, Dubbeld J, Braat AE, Ringers J, Roelen DL, et al. Glycemic Stability Through Islet-After-Kidney Transplantation Using an Alemtuzumab-Based Induction Regimen and Long-Term Triple-Maintenance Immunosuppression. Am J Transplant. 2016;16(1):246–53.
40. Marfil-Garza BA, Imes S, Verhoeff K, Hefler J, Lam A, Dajani K, et al. Pancreatic islet transplantation in type 1 diabetes: 20-year experience from a single-centre cohort in Canada. Lancet Diabetes Endocrinol. 2022;10(7):519–32.
41. Bellin MD, Kandaswamy R, Parkey J, Zhang HJ, Liu B, Ihm SH, et al. Prolonged insulin independence after islet allotransplants in recipients with type 1 diabetes. Am J Transplant. 2008;8(11):2463–70.
42. Gangemi A, Salehi P, Hatipoglu B, Martellotto J, Barbaro B, Kuechle JB, et al. Islet transplantation for brittle type 1 diabetes: the UIC protocol. Am J Transplant. 2008;8(6):1250–61.
43. Onaca N, Takita M, Levy MF, Naziruddin B. Anti-inflammatory Approach With Early Double Cytokine Blockade (IL-1beta and TNF-alpha) Is Safe and Facilitates Engraftment in Islet Allotransplantation. Transplant Direct. 2020;6(3):e530.
44. McCall M, Pawlick R, Kin T, Shapiro AM. Anakinra potentiates the protective effects of etanercept in transplantation of marginal mass human islets in immunodeficient mice. Am J Transplant. 2012;12(2):322–9.
45. Berney T, Secchi A. Rapamycin in islet transplantation: friend or foe? Transpl Int. 2009;22(2):153–61.

46. Berney T, Andres A, Toso C, Majno P, Squifflet JP. mTOR Inhibition and Clinical Transplantation: Pancreas and Islet. Transplantation. 2018;102(2S Suppl 1):S30–S1.
47. Gillard P, Ling Z, Mathieu C, Crenier L, Lannoo M, Maes B, et al. Comparison of sirolimus alone with sirolimus plus tacrolimus in type 1 diabetic recipients of cultured islet cell grafts. Transplantation. 2008;85(2):256–63.
48. Maffi P, Berney T, Nano R, Niclauss N, Bosco D, Melzi R, et al. Calcineurin inhibitor-free immunosuppressive regimen in type 1 diabetes patients receiving islet transplantation: single-group phase 1/2 trial. Transplantation. 2014;98(12):1301–9.
49. Rech AJ, Mick R, Martin S, Recio A, Aqui NA, Powell DJ, Jr., et al. CD25 blockade depletes and selectively reprograms regulatory T cells in concert with immunotherapy in cancer patients. Sci Transl Med. 2012;4(134):134ra62.
50. Miroux C, Morales O, Ghazal K, Othman SB, de Launoit Y, Pancre V, et al. In vitro effects of cyclosporine A and tacrolimus on regulatory T-cell proliferation and function. Transplantation. 2012;94(2):123–31.
51. Muckenhuber M, Wekerle T, Schwarz C. Costimulation blockade and Tregs in solid organ transplantation. Front Immunol. 2022;13:969633.
52. Posselt AM, Bellin MD, Tavakol M, Szot GL, Frassetto LA, Masharani U, et al. Islet transplantation in type 1 diabetics using an immunosuppressive protocol based on the anti-LFA-1 antibody efalizumab. Am J Transplant. 2010;10(8):1870–80.
53. Posselt AM, Szot GL, Frassetto LA, Masharani U, Tavakol M, Amin R, et al. Islet transplantation in type 1 diabetic patients using calcineurin inhibitor-free immunosuppressive protocols based on T-cell adhesion or costimulation blockade. Transplantation. 2010;90(12):1595–601.
54. Turgeon NA, Avila JG, Cano JA, Hutchinson JJ, Badell IR, Page AJ, et al. Experience with a novel efalizumab-based immunosuppressive regimen to facilitate single donor islet cell transplantation. Am J Transplant. 2010;10(9):2082–91.
55. Larsen CP, Pearson TC, Adams AB, Tso P, Shirasugi N, Strobert E, et al. Rational development of LEA29Y (belatacept), a high-affinity variant of CTLA4-Ig with potent immunosuppressive properties. Am J Transplant. 2005;5(3):443–53.
56. Drucker DJ. Glucagon-like peptides: regulators of cell proliferation, differentiation, and apoptosis. Mol Endocrinol. 2003;17(2):161–71.
57. Qi M, Kinzer K, Danielson KK, Martellotto J, Barbaro B, Wang Y, et al. Five-year follow-up of patients with type 1 diabetes transplanted with allogeneic islets: the UIC experience. Acta Diabetol. 2014;51(5):833–43.
58. Bertuzzi F, Marazzi M, De Carlis LG, Rampoldi AG, Bonomo M, Antonioli B, et al. Sustained Islet Allograft Function After Peritransplant Treatment Using Exenatide With and Without Everolimus. Transplantation. 2016;100(11):e117–e8.
59. Faradji RN, Tharavanij T, Messinger S, Froud T, Pileggi A, Monroy K, et al. Long-term insulin independence and improvement in insulin secretion after supplemental islet infusion under exenatide and etanercept. Transplantation. 2008;86(12):1658–65.
60. Froud T, Faradji RN, Pileggi A, Messinger S, Baidal DA, Ponte GM, et al. The use of exenatide in islet transplant recipients with chronic allograft dysfunction: safety, efficacy, and metabolic effects. Transplantation. 2008;86(1):36–45.
61. Ghofaili KA, Fung M, Ao Z, Meloche M, Shapiro RJ, Warnock GL, et al. Effect of exenatide on beta cell function after islet transplantation in type 1 diabetes. Transplantation. 2007;83(1):24–8.
62. Citro A, Cantarelli E, Maffi P, Nano R, Melzi R, Mercalli A, et al. CXCR1/2 inhibition enhances pancreatic islet survival after transplantation. J Clin Invest. 2012;122(10):3647–51.
63. Maffi P, Lundgren T, Tufveson G, Rafael E, Shaw JAM, Liew A, et al. Targeting CXCR1/2 Does Not Improve Insulin Secretion After Pancreatic Islet Transplantation: A Phase 3, Double-Blind, Randomized, Placebo-Controlled Trial in Type 1 Diabetes. Diabetes Care. 2020;43(4):710–8.
64. NHS Blood and Transplant. Organ and Tissue Donation and Transplantation Activity Report [Available from: https://www.odt.nhs.uk/statistics-and-reports/annual-activity-report/

65. Koh A, Senior P, Salam A, Kin T, Imes S, Dinyari P, et al. Insulin-heparin infusions peri-transplant substantially improve single-donor clinical islet transplant success. Transplantation. 2010;89(4):465–71.
66. Wassmer CH, Lebreton F, Bellofatto K, Bosco D, Berney T, Berishvili E. Generation of insulin-secreting organoids: a step toward engineering and transplanting the bioartificial pancreas. Transpl Int. 2020.
67. Merani S, Toso C, Emamaullee J, Shapiro AM. Optimal implantation site for pancreatic islet transplantation. Br J Surg. 2008;95(12):1449–61.
68. Svensson J, Lau J, Sandberg M, Carlsson PO. High vascular density and oxygenation of pancreatic islets transplanted in clusters into striated muscle. Cell Transplant. 2011;20(5):783–8.
69. Christoffersson G, Henriksnas J, Johansson L, Rolny C, Ahlstrom H, Caballero-Corbalan J, et al. Clinical and experimental pancreatic islet transplantation to striated muscle: establishment of a vascular system similar to that in native islets. Diabetes. 2010;59(10):2569–78.
70. Rafael E, Tibell A, Ryden M, Lundgren T, Savendahl L, Borgstrom B, et al. Intramuscular autotransplantation of pancreatic islets in a 7-year-old child: a 2-year follow-up. Am J Transplant. 2008;8(2):458–62.
71. Bertuzzi F, Colussi G, Lauterio A, De Carlis L. Intramuscular islet allotransplantation in type 1 diabetes mellitus. Eur Rev Med Pharmacol Sci. 2018;22(6):1731–6.
72. [Available from: https://www.isletstudy.org/CITDocs/CIT-07%20Protocol%20ver%202.0F%20_11%20Oct%202007_.pdf.
73. Baidal DA, Ricordi C, Berman DM, Alvarez A, Padilla N, Ciancio G, et al. Bioengineering of an Intraabdominal Endocrine Pancreas. N Engl J Med. 2017;376(19):1887–9.
74. Van Hulle F, De Groot K, Hilbrands R, Van de Velde U, Suenens K, Stange G, et al. Function and composition of pancreatic islet cell implants in omentum of type 1 diabetes patients. Am J Transplant. 2022;22(3):927–36.
75. Saudek F, Hladikova Z, Hagerf B, Nemetova L, Girman P, Kriz J, et al. Transplantation of Pancreatic Islets Into the Omentum Using a Biocompatible Plasma-Thrombin Gel: First Experience at the Institute for Clinical and Experimental Medicine in Prague. Transplant Proc. 2022;54(3):806–10.
76. Caiazzo R, Gmyr V, Hubert T, Delalleau N, Lamberts R, Moerman E, et al. Evaluation of alternative sites for islet transplantation in the minipig: interest and limits of the gastric submucosa. Transplant Proc. 2007;39(8):2620–3.
77. Wszola M, Berman A, Gorski L, Ostaszewska A, Serwanska-Swietek M, Krajewska M, et al. Endoscopic Islet Autotransplantation Into Gastric Submucosa-1000-Day Follow-up of Patients. Transplant Proc. 2018;50(7):2119–23.
78. Cantarelli E, Melzi R, Mercalli A, Sordi V, Ferrari G, Lederer CW, et al. Bone marrow as an alternative site for islet transplantation. Blood. 2009;114(20):4566–74.
79. Maffi P, Nano R, Monti P, Melzi R, Sordi V, Mercalli A, et al. Islet Allotransplantation in the Bone Marrow of Patients With Type 1 Diabetes: A Pilot Randomized Trial. Transplantation. 2019;103(4):839–51.
80. Maffi P, Balzano G, Ponzoni M, Nano R, Sordi V, Melzi R, et al. Autologous pancreatic islet transplantation in human bone marrow. Diabetes. 2013;62(10):3523–31.
81. Spencer JA, Ferraro F, Roussakis E, Klein A, Wu J, Runnels JM, et al. Direct measurement of local oxygen concentration in the bone marrow of live animals. Nature. 2014;508(7495):269–73.
82. Hering BJ, Clarke WR, Bridges ND, Eggerman TL, Alejandro R, Bellin MD, et al. Phase 3 Trial of Transplantation of Human Islets in Type 1 Diabetes Complicated by Severe Hypoglycemia. Diabetes Care. 2016;39(7):1230–40.
83. Markmann JF, Rickels MR, Eggerman TL, Bridges ND, Lafontant DE, Qidwai J, et al. Phase 3 trial of human islet-after-kidney transplantation in type 1 diabetes. Am J Transplant. 2021;21(4):1477–92.
84. Lablanche S, Vantyghem MC, Kessler L, Wojtusciszyn A, Borot S, Thivolet C, et al. Islet transplantation versus insulin therapy in patients with type 1 diabetes with severe hypoglycaemia or poorly controlled glycaemia after kidney transplantation (TRIMECO): a multicentre, randomised controlled trial. Lancet Diabetes Endocrinol. 2018;6(7):527–37.

85. Rickels MR, Eggerman TL, Bayman L, Qidwai JC, Alejandro R, Bridges ND, et al. Long-term Outcomes With Islet-Alone and Islet-After-Kidney Transplantation for Type 1 Diabetes in the Clinical Islet Transplantation Consortium: The CIT-08 Study. Diabetes Care. 2022;45(12):2967–75.
86. Lablanche S, Borot S, Wojtusciszyn A, Skaare K, Penfornis A, Malvezzi P, et al. Ten-year outcomes of islet transplantation in patients with type 1 diabetes: Data from the Swiss-French GRAGIL network. Am J Transplant. 2021.
87. Fioretto P, Steffes MW, Sutherland DE, Goetz FC, Mauer M. Reversal of lesions of diabetic nephropathy after pancreas transplantation. N Engl J Med. 1998;339(2):69–75.
88. Thompson DM, Meloche M, Ao Z, Paty B, Keown P, Shapiro RJ, et al. Reduced progression of diabetic microvascular complications with islet cell transplantation compared with intensive medical therapy. Transplantation. 2011;91(3):373–8.
89. Helve J, Sund R, Arffman M, Harjutsalo V, Groop PH, Gronhagen-Riska C, et al. Incidence of End-Stage Renal Disease in Patients With Type 1 Diabetes. Diabetes Care. 2018;41(3):434–9.
90. Lablanche S, Borot S, Wojtusciszyn A, Bayle F, Tetaz R, Badet L, et al. Five-Year Metabolic, Functional, and Safety Results of Patients With Type 1 Diabetes Transplanted With Allogenic Islets Within the Swiss-French GRAGIL Network. Diabetes Care. 2015;38(9):1714–22.
91. Campbell PM, Senior PA, Salam A, Labranche K, Bigam DL, Kneteman NM, et al. High risk of sensitization after failed islet transplantation. Am J Transplant. 2007;7(10):2311–7.
92. Chaigne B, Geneugelijk K, Bedat B, Ahmed MA, Honger G, De Seigneux S, et al. Immunogenicity of Anti-HLA Antibodies in Pancreas and Islet Transplantation. Cell Transplant. 2016;25(11):2041–50.
93. Pouliquen E, Baltzinger P, Lemle A, Chen CC, Parissiadis A, Borot S, et al. Anti-Donor HLA Antibody Response After Pancreatic Islet Grafting: Characteristics, Risk Factors, and Impact on Graft Function. Am J Transplant. 2017;17(2):462–73.
94. Piemonti L, Everly MJ, Maffi P, Scavini M, Poli F, Nano R, et al. Alloantibody and autoantibody monitoring predicts islet transplantation outcome in human type 1 diabetes. Diabetes. 2013;62(5):1656–64.
95. Campbell PM, Salam A, Ryan EA, Senior P, Paty BW, Bigam D, et al. Pretransplant HLA antibodies are associated with reduced graft survival after clinical islet transplantation. Am J Transplant. 2007;7(5):1242–8.
96. Brooks AM, Carter V, Liew A, Marshall H, Aldibbiat A, Sheerin NS, et al. De Novo Donor-Specific HLA Antibodies Are Associated With Rapid Loss of Graft Function Following Islet Transplantation in Type 1 Diabetes. Am J Transplant. 2015;15(12):3239–46.
97. Chen CC, Pouliquen E, Kessler L, Dubois V, Thaunat O. Cum hoc sed non propter hoc. Am J Transplant. 2017;17(7):1960–1.
98. Lehmann R, Weber M, Berthold P, Zullig R, Pfammatter T, Moritz W, et al. Successful simultaneous islet-kidney transplantation using a steroid-free immunosuppression: two-year follow-up. Am J Transplant. 2004;4(7):1117–23.
99. Lehmann R, Graziano J, Brockmann J, Pfammatter T, Kron P, de Rougemont O, et al. Glycemic Control in Simultaneous Islet-Kidney Versus Pancreas-Kidney Transplantation in Type 1 Diabetes: A Prospective 13-Year Follow-up. Diabetes Care. 2015;38(5):752–9.

Minimal SC-β-Cell Properties for Transplantation in Diabetic Patients

Veronica Cochrane, Yini Xiao, Hasna Maachi, and Matthias Hebrok

1 Introduction

The primary function of pancreatic β cells is to properly sense blood glucose levels and secrete insulin accordingly. Insulin secretion from β cells is classified into a triggering (first) phase and an amplifying (second) phase. Upon exposure to high glucose levels, the triggering phase of insulin secretion is identified by a large spike in secreted insulin from a pool of readily releasable insulin vesicles followed by a less pronounced but sustained amplifying phase of insulin secretion from both readily releasable vesicles and a reserve vesicle pool. While both phases of insulin secretion require glucose metabolism, only the triggering-phase mechanism is well defined. During the triggering phase, glucose metabolism causes a rise in the intracellular ATP/ADP ratio that leads to the closure of ATP-sensitive potassium (K_{ATP}) channels, membrane depolarization, activation of voltage-gated calcium channels

Veronica Cochrane, Yini Xiao and Hasna Maachi contributed equally with all other contributors.

V. Cochrane
Diabetes Center, Department of Medicine, University of California, San Francisco, CA, USA

Institute for Diabetes Organoid Technology (IDOT), Helmholtz Center Munich, Helmholtz Diabetes Center (HDC), Neuherberg, Germany

Y. Xiao
Diabetes Center, Department of Medicine, University of California, San Francisco, CA, USA

H. Maachi
Diabetes Center, Department of Medicine, University of California, San Francisco, CA, USA

Center for Organoid Systems (COS), Technical University Munich (TUM), Garching, Germany

Institute for Diabetes Organoid Technology (IDOT), Helmholtz Center Munich, Helmholtz Diabetes Center (HDC), Neuherberg, Germany

L. Piemonti et al. (eds.), *Pluripotent Stem Cell Therapy for Diabetes*,
https://doi.org/10.1007/978-3-031-41943-0_22

(VGCCs) and an ensuing influx of calcium (Ca^{2+}) that stimulates insulin vesicle exocytosis. In contrast, the amplifying phase of insulin secretion occurs even when K_{ATP} channels are inhibited and is thus, K_{ATP} channel-independent. The precise mechanisms underlying the amplifying phase remain an intensive area of research. The following sections will describe the critical aspects of glucose-stimulated insulin secretion (GSIS) and discuss our current knowledge about how this process is replicated in SC-β cells.

2 Insulin Secretion from β Cells

2.1 Glucose Sensing and Metabolism

GSIS from pancreatic β cells is critical for maintaining glucose homeostasis. This requires beta cells to efficiently uptake glucose and couple glucose metabolism to insulin secretion. Glucose enters β cells through passive diffusion facilitated by glucose transporters (GLUT). GLUT1 is prominently expressed in human β cells and its high affinity, low Km (~3 mM), allows for rapid equilibration of extracellular and intracellular glucose levels [1]. Once inside, the cell glucose must be phosphorylated to glucose-6-phosphate (G6P) by hexokinases (HK) to initiate glycolysis. There are a total of four hexokinases (HK1-4), but adult β cells only express HK4, also known as glucokinase (GCK). Unlike other hexokinases, GCK activity is not inhibited by the product G6P, and it has a low affinity for glucose (GCK: Km ~8 mM; HK1–3: Km δ1mM) [2, 3]. These distinct characteristics establish GCK as a high glucose sensor and the rate limiting step governing GSIS (Fig. 1).

Notably, GCK expression corresponds with β-cell functional maturity. Functionally mature adult β cells arise from the postnatal development of cellular processes, especially glucose metabolism and insulin processing. Adult β cells secrete greater amounts of insulin and have a higher glucose threshold for maximal insulin secretion compared to immature β cells [4–8]. The expression of GCK and repression of HK1 is crucial for preventing excess insulin secretion and hypoglycemia [9–11]. Importantly, SC-β cells have been reported to downregulate HK1 and upregulate GCK over the course of their in vitro differentiation [7, 12–18].

M. Hebrok (✉)
Diabetes Center, Department of Medicine, University of California, San Francisco, CA, USA

Center for Organoid Systems (COS), Technical University Munich (TUM), Garching, Germany

Institute for Diabetes Organoid Technology (IDOT), Helmholtz Center Munich, Helmholtz Diabetes Center (HDC), Neuherberg, Germany

Munich Institute of Biomedical Engineering (MIBE), TUM, Garching, Germany

German Center for Diabetes Research, DZD, Neuherberg, Germany
e-mail: matthias.hebrok@tum.de

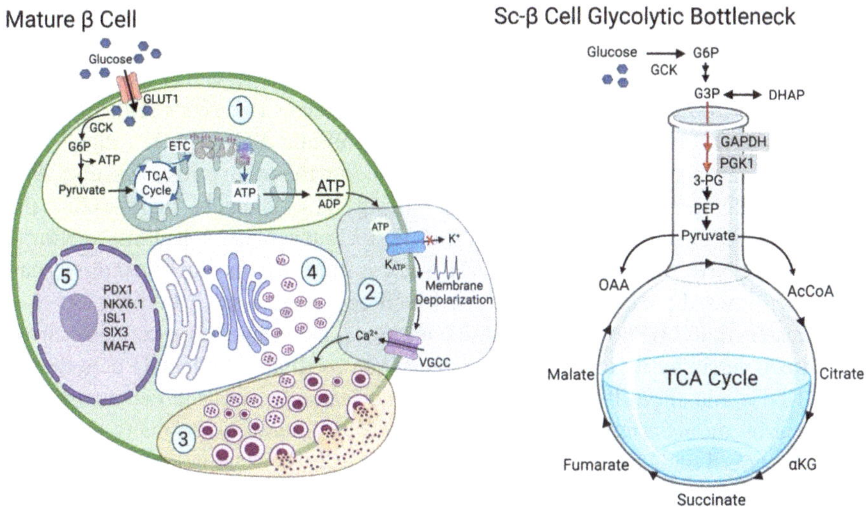

Fig. 1 β-cell schematic. Mature β cell (left). (1) Glucose metabolism in mature β cells starts with glucose uptake via GLUT1 and the phosphorylation of glucose by GCK to generate G6P. Glycolysis proceeds to generate pyruvate which is further metabolized in the mitochondria through the TCA cycle and ETC to produce ATP. This significantly increases the intracellular ATP/ADP ratio. (2) ATP binding to K_{ATP} channels causes the channel closure and inhibition of K^+ efflux. Consequently, the membrane depolarizes and β-cell electrical activity is initiated. This results in increased Ca^{2+} influx through VGCCs. (3) Ca^{2+} oscillations promote the secretion of insulin from readily releasable vesicle primed at the membrane. (4) The intracellular store of insulin is readily replenished by glucose-induced preproinsulin translation which is further processed to mature insulin as it progresses through the ER and Golgi to dense core secretory granules. (5) Transcriptional regulation of β-cell maturation genes maintains β-cell identity and functional maturity. SC-β glycolytic bottleneck (right). The schematic depicts the coupling of glycolysis to mitochondrial metabolism. Red arrows identify GAPDH and PGK1 as the enzymatic steps at which the bottleneck is proposed to occur in SC-β cells

Consistent with this [$U^{13}C6$]-glucose metabolic tracing studies show that as GCK expression increases in SC-β cells, the G6P-to-glucose ratio reduces to levels more similar to those in human islets [15].

From G6P, glycolysis proceeds through a series of enzymatic reactions that generate pyruvate, which is then transported into the mitochondria and further catabolized. Inside the mitochondria matrix, pyruvate is oxidized through the tricarboxylic acid (TCA) cycle to CO_2 [19, 20]. Pyruvate enters the TCA evenly through its conversion to Acetyl-CoA (AcCoA) by pyruvate dehydrogenase (PDH) and oxaloacetate (OAA) by the anaplerotic enzyme pyruvate carboxylase (PC) [20–22]. During pyruvate oxidation, electrons are stored on the electron carriers NADH and $FADH_2$ and then deposited to the electron transport chain (ETC) located in the mitochondrial inner membrane. Electrons are passed down the ETC until they reach the final acceptor oxygen, which then splits to form two molecules of water. The flow of electrons through the ETC pumps protons from the matrix into the intermembrane space creating an electrochemical gradient. Protons fuel the production of ATP as they move down their gradient and pass-through ATP synthase back into the matrix. This process is known as oxidative phosphorylation (OxPhos), and it is extremely efficient at generating ATP. For every molecule of glucose, OxPhos produces a net maximum of 30 ATPs, whereas glycolysis only produces two ATPs [19]. Coupling glycolysis to OxPhos significantly raises the intracellular ATP-to-ADP ratio. ATP is essential for numerous energetic processes, including the production of insulin. ATP fulfills another critical role in GSIS by closing K_{ATP} channels, depolarizing the plasma membrane, and thus permitting Ca^{2+} influx that promotes insulin exocytosis.

Cells that have a high demand for ATP, such as muscle cells during exercise, may be pushed toward anaerobic glycolysis, which occurs at a much faster rate (~100 times faster) than OxPhos. Under these conditions, pyruvate is diverted away from the mitochondria by its interconversion to lactate facilitated by lactate dehydrogenase (LDH) and the diffusion of pyruvate and lactate across the cell membrane mediated by monocarboxylate transporters (MCT). In contrast to many other cell types, mature β cells have very low expression levels of LDH isoform A (LDHA) and MCT1 [23, 24]; thus, pyruvate predominantly enters the mitochondria where it is oxidized. This also serves to prevent insulin secretion in response to exercise-induced circulating pyruvate and lactate. Repression of LDHA and MCT1 and a corresponding decrease of lactate has been reported during the in vitro differentiation of SC-β cells [7, 12–15]. Despite this, there is evidence that SC-β cells exhibit aberrant handling of pyruvate compared to human islets. While lactate levels decline over the course of SC-β-cell differentiation, the lactate-to-pyruvate ratio remains relatively high suggesting that there are unacceptable amounts of LDHA activity [15]. One study also reported that exogenous application of pyruvate led to significant amounts of insulin being secreted from SC-β cells, while human islets had virtually no response [15]. Perhaps one of the more surprising observations from SC-β cells has been that high glucose stimulation induces limited mitochondrial OxPhos, measured by the rate of oxygen consumption, but that exogenous pyruvate markedly escalates mitochondrial respiration [12, 15, 25]. Collectively, these findings indicate that SC-β cells have diminished coupling of glycolysis to oxidative phosphorylation and are metabolically immature.

2.2 Glycolytic Bottleneck in SC-β Cells

SC-β cells are frequently reported to have suboptimal GSIS; for which, perturbed glucose metabolism has been pinpointed as the underlying cause (Fig. 1). Recent studies have utilized [$U^{13}C6$]-glucose metabolic tracing experiments to track the glycolytic fate of glucose in SC-β cells. From this work, it was revealed that SC-β cells have a glycolytic bottleneck [25]. Compared to human islets, SC-β cells have similar levels of the glycolytic metabolite glyceraldehyde-3-phosphate (G3P) and its isomer dihydroxyacetone phosphate (DHAP), but the presence of metabolites downstream of G3P, proceeding from 3-phosphoglycerate (3-PG), phosphoenolpyruvate (PEP), to pyruvate, are significantly reduced [15, 25]. Supplying glycolytic metabolites downstream of G3P, but not upstream of G3P or G3P itself, can greatly enhance insulin secretion [15, 25]. This indicates a disruption at the conversion G3P to 3-PG, which is catalyzed by the enzymes glyceraldehyde-3-phosphate dehydrogenase (GAPDH) and phosphoglycerate kinase 1 (PGK1). Although GAPDH and PGK1 are highly expressed in SC-β cells, their enzymatic activities are lower than that in human islets [25]. Attempts to increase GAPDH and PGK1 activity using small molecule activators have thus far been unsuccessful in overcoming the glycolytic bottleneck [25]. Therefore, it remains unclear whether targeting GAPDH and PGK1 activity will recover glucose metabolism in SC-β cells. The glycolytic bottleneck phenotype is a novel discovery spanning multiple cell lines and differentiation protocols and it presents an outstanding hurdle for generating functionally mature SC-β cells.

The inefficient breakdown of glucose to pyruvate prohibits flux through the TCA cycle and OxPhos. As a result, glucose stimulation is incapable of producing sufficient ATP to shift the intracellular ATP/ADP ratio [15, 25]. The consequence of this is apparent in that nearly all SC-β cells are reported to secrete insulin at a lesser magnitude than human islets [12, 14, 16, 25]. It should be noted that SC-β cells express functional K_{ATP} channels and demonstrate a high capacity for insulin secretion in response to the K_{ATP} channel inhibitor tolbutamide [12, 13, 15, 25, 26]. This indicates that the GSIS phenotype is indeed the result of deficient glucose metabolism and not that of faulty K_{ATP} channels or insulin secretory machinery. In addition to the triggering phase, glucose metabolism produces many glycolytic and mitochondrial metabolites involved in the amplifying phase of insulin secretion. Irregular glucose metabolism likely also explains the lack or reduction of a second phase commonly reported in many SC-β cells [12, 14, 25]. Despite the metabolic immaturity of SC-β cells, there have been some advancement toward generating SC-β cells that are capable of dynamically secreting insulin at levels resembling human islets [13, 15, 26]. The metabolic mechanisms underlying their functionality have not yet been fully characterized, but metabolomics studies suggest that shunting of glucose metabolites to metabolic pathways that run parallel to canonical glycolysis and OxPhos may be key to understanding insulin secretion in these cells [15].

2.3 *Ca^{2+} Influx*

Following the closure of K_{ATP} channels and the consequent inhibition of K^+ efflux, the beta-cell membrane is depolarized. Human beta cells are equipped with three VGCCs consisted of L-type, T-type, and P/Q-type Ca^{2+} channels [27]. After the membrane potential is depolarized to −40 mV, L-type voltage-gated Ca^{2+} channels are activated and in concert with T-type and P/Q-type Ca^{2+} channels to allow an influx of extracellular Ca^{2+} [27]. The increased intracellular Ca^{2+} concentration ($[Ca2+]_i$) triggers insulin granules exocytosis from a pool of readily releasable vesicles located at the plasma membrane in ~7 min [28, 29]. $[Ca^{2+}]_i$ is sensed by multiple Ca^{2+}-binding proteins and tightly controlled by multiple Ca^{2+} exchangers and pumps [30]. The greater Ca^{2+} sensitivity can lead to higher basal insulin secretion that is a marker of immature β cells, thus associating Ca^{2+} signaling with the maturation status of β cells [31]. Glucose-induced Ca^{2+} influx in individual β cells occurs in an oscillatory pattern that reflects β-cell electrical activity. Furthermore, within intact islets, β cells are electrically coupled via gap junctions, which allows for Ca^{2+} influx to be synchronized and for pulsatile secretion of insulin [32].

SC-β cells exhibit a range of calcium responses depending on their differentiated state at the time of the Ca^{2+} assay. Early on in SC-β-cell differentiations, when insulin has started to be expressed but progenitor cells may still be present, high glucose stimulation elicits a slow rise in cytoplasmic Ca^{2+} that does not return to baseline values when the cells are reintroduced to low glucose [12, 18]. SC-β cells at this early stage secrete low basal levels of insulin and are predominantly unresponsive to high glucose concentrations. Conversely, SC-β cells may also transition through an overactive state as they progress along in vitro differentiation [15]. These immature SC-β cells are reminiscent of neonatal islets in that they are unable to suppress insulin secretion at low glucose levels [33]. In addition to high basal insulin secretion, hyperactive SC-β cells show robust Ca^{2+} influx activity [15]. A plausible explanation for this is that HK1 expression in these cells reduces the glucose threshold required to commence glucose metabolism, the subsequent closure of K_{ATP} channels, Ca^{2+} influx and insulin secretion. Evidence for upstream metabolic defects comes from the finding that the K_{ATP} channel opener diazoxide reduces basal insulin secretion from these overactive SC-β cells [15]. At the end stage of several SC-β-cell differentiation protocols, SC-β cells have been shown to respond to glucose stimulation with marked Ca^{2+} flux [12, 15, 25]. However, the changes in SC-β-cell intracellular Ca^{2+} are typically to a lesser extent than that observed in human islets and in some cases, Ca^{2+} flux may only be detectable in subregions of SC-β-cell clusters. The limited degree of Ca^{2+} flux in SC-β cells can likely be attributed to disrupted glucose metabolism being insufficient for inhibiting K_{ATP} channels and initiating β-cell electrical activity [15]. In agreement with this, bypassing the glycolytic bottleneck using the methylated, cell-permeable form of 3-PG (m-3-PG) induces a significant influx of a Ca^{2+} that propagates as a wave across SC-β clusters like that documented in human islets [25]. This finding emphasizes the detrimental consequence of a glycolytic bottleneck for SC-β-cell functional maturity.

2.4 *First-Phase and Second-Phase Insulin Secretion and Their Regulation*

The central parameter for predicting the therapeutic potency of SC-β cell is GSIS [34]. GSIS displays a biphasic pattern that consists of a transient first phase lasting 3–10 min followed by a subsequent sustained second phase lasting 1 h or more until euglycemia is restored. Loss of first-phase secretion and reduced second-phase secretion are characteristics of T2D. As described above, the first phase is regulated by the triggering pathway in which ATP mainly produced via mitochondrial OxPhos induces the closure of the K_{ATP} channels, resulting in the opening of the VGCCs, increase in intracellular Ca^{2+}, and the exocytosis of a "readily releasable" pool of insulin secretory granules located adjacent to the plasma membrane [35]. In addition to glucose, other secretagogues, such as KCl, can also trigger insulin secretion by inducing membrane depolarization that leads to an increase in $[Ca^{2+}]i$. Among the various signals, only "fuel secretagogues" can initiate the second-phase insulin secretion, also named the "metabolic amplifying pathway" [36–38]. This pathway depends on its glucose as a trigger and functions to amplify and sustain insulin secretion. Fuel secretagogues include free fatty acids, amino acids, and glucose metabolites that are calorigenic and whose metabolism yields necessary signals to promote insulin secretion [38]. For instance, glutamate, catalyzed by mitochondrial glutamate dehydrogenase (GDH) to produce α-ketoglutarate (α-KG) and NAD(P)H, can stimulate insulin secretion [39]. In this process, GDH acts as a key enzyme for sensing glutamine and glutamate level in the amplifying pathway. Gain-of-function mutations of GDH in patients lead to "protein-rich meal"-induced hyperinsulinemia [40]. Under glutamine and leucine stimulation, SC-β cells display comparable insulin secretion compared to human islets but show slightly lower mitochondrial respiration rates [15]. In addition, other cytosolic cycling mechanisms, including the Pentose monophosphate shunt, phosphoenolpyruvate cycle, pyruvate–malate cycle, pyruvate–citrate cycle, pyruvate–isocitrate cycle, contribute to the metabolic amplifying pathways [20, 41]. Uncoupling of any of these metabolic cycles will impair insulin secretion [25].

Insulin secretion is further regulated by hormonal factors. Incretins, such as GLP-1 or its analog exendin-4, potentiates insulin secretion by activating cAMP signaling [42]. cAMP is a major second message contributing to hormonal amplification of insulin secretion. The oscillation of cAMP in β cells regulates pulsatile insulin secretion. The action of cAMP is mediated by the cAMP sensor protein kinase A (PKA) and the guanine nucleotide exchange factor Epac, both of which contribute to second-phase secretion [43]. Suppression of glucose-induced cAMP formation impairs both first and second-phase secretion [44, 45]. In contrast, increasing cAMP levels using the compound forskolin activates PKA and enhances amplifying secretion [46]. Of note, the cAMP concentrations measured upon high glucose and exendin-4 stimulation in SC-β cells are similar to those found in adult human islets, indicating that SC-β cells are equipped with appropriate cAMP signaling during insulin secretion [15].

In addition to metabolic regulation, β-cell cytoskeleton is implicated in second-phase insulin secretion. For example, the actin network regulates the exocytosis of insulin granules during the second phase [47]. Glucose stimulation induces transient F-actin remodeling which mobilizes transport of insulin granules to the plasma membrane [47]. The activation and transportation of Cdc42 to secretory granules contributes to this process [48]. Reduction of Cdc42 selectively abolishes second-phase secretion in islets [49]. The CDC42 expression level in SC-β cell is even higher than islet β cells, indicating post-transcriptional regulation of CDC42 might be also important [15].

Overall, the second phase of islet β cells contributes to approximately 75% of total insulin secretion upon glucose stimulation [8]. Modifications of the original differentiation protocols, such as withdrawal of T3 and ALK5 inhibitor in culture media of last stage [26], sorting of mature ENTPD3$^+$ β cells [50], circadian entrainment [51], and targeting cytoskeleton in planer differentiation system [13] have demonstrated improvements in the second-phase secretion of SC-β cells. Nonetheless, there are still considerable functional defects compared to the "gold standard" endogenous islet β cells, most notably the above-mentioned defects in glycolysis caused by the reduced activity of GAPDH [25].

3 Insulin Production and Processing

β cells maintain relatively stable intracellular stores of insulin due to the coordination of insulin secretion and biosynthesis. While both processes are stimulated by glucose, proinsulin biosynthesis has a lower glucose threshold that ensures there is an abundant reservoir of readily available insulin for secretion [52]. Mitochondrial metabolites resulting from glucose metabolism stimulate mRNA translation of preproinsulin comprised of an N-terminal signal peptide and the B- and A-chains joined by a connecting peptide (C-peptide). Upon translocation into the endoplasmic reticulum (ER), the signal peptide is cleaved by a signal peptidase to yield proinsulin, which undergoes folding that is stabilized by three disulfide bonds [53]. The folded proinsulin is then trafficked from the ER, through the Golgi and the trans-Golgi network to immature secretory granules, where it is processed to form insulin and C-peptide. Maturation of proinsulin into insulin and C-peptide occurs in two consecutive steps: (1) C-peptide is cleaved by Ca^{2+}-sensitive protein convertases PC2 and PC1/3 at the B and A chain junction, (2) followed by the cleavage of the dibasic residues R55-R56 at the C-terminal of the B chain by the exopeptidase carboxypeptidase E/H (CPE) [54, 55]. PC1/3 activity is increased by the acidification of the insulin granules through the action of the vesicular ATP-dependent proton pump (V-ATPase) in a glucose-dependent manner [52, 56]. Within insulin secretory granules, mature insulin is stored as zinc insulin hexamers alongside C-peptide, islet amyloid polypeptide (IAPP or amylin), and other less abundant β-cell secretory products [57].

Each secretory granule contains 300,000 or more molecules of insulin, and each β-cell stores 5–10,000 secretory granules [58–60]. Upon acute glucose stimulation, the newly synthesized secretory granules that reside at the plasma membrane are preferentially secreted over the older ones. A prolonged stimulation requires mobilization of the granules from a pool of reserve vesicles. Only a fraction of total insulin granules (~2% per hour) is sufficient to achieve blood glucose levels within the physiological range. The finely tuned replenishment of insulin secretory granules is critical for β cells to secrete sufficient insulin to maintain normoglycemia.

Although insulin processing and granule generation in mature β cells is a well-described cellular process, this remains a largely unexplored area of SC-β-cell biology. The few studies that have examined insulin processing in SC-β cells have done so via transcriptional analysis and electron microscopy (EM) imaging. Much of this work has revealed that over time SC-β cells progressively upregulate insulin processing genes (i.e., PC1/3 and CPE) and acquire dense core insulin granules whose ultrastructure is very similar to that of primary β cells [15, 16, 18, 61]. However, the amount of total insulin content reported in SC-β cells ranges from significantly less than to approximately the same as human islets depending on the protocol and cell line used [12, 13, 15, 25, 26]. Considering the SC-β-cell glycolytic bottleneck, it would be unsurprising for glucose-stimulated insulin translation and processing to also be somewhat disturbed.

4 β-Cell Maturation Genes

Transcription factors are key regulators of gene expression and guarantors of cell identity and maturity. The expression of a specific transcription factor at a specific time can alter or change the fate of a somatic cell, a process called reprogramming. The differentiation of human pluripotent stem cells (hPSCs) into insulin-producing cells offers a great potential for diabetes therapy, hence the importance of identifying the transcription factors that characterize the mature human β cell. Human pancreatic endocrine progenitors are defined by the high expression of NEUROD1, ISL1, NKX2–2, PAX6, PDX1, and NKX6.1 during the period from 8 and 12 weeks (Table 1) [33, 62, 63]. Immature β cells have a high proliferative rate, and a reduced glucose threshold for insulin secretion leading to increased basal insulin secretion and a reduced stimulation index. After birth, β cells undergo a maturation process during which they gain expression of several key regulators of insulin transcription and secretion, including NEUROD1, PAX6, MAFA, INS, SIX2, SIX3, and ESRRG [64–68]. Mature β cells expressing the full complement of these transcription factors are able to respond to high glucose levels with robust insulin release [20, 69]. Transcriptional comparison of SC-β cells to fetal and adult human β cells has revealed that SC-β cells more closely resemble fetal β cells. SC-β cells tend to have reduced expression of genes important for maturation and functionality such as MAFA, UCN3, and SIX3 [7, 15]. Interestingly, RNA-seq analysis of SC-β cells following 6 months of engraftment in mice shows that transplanted cells undergo remarkable transcriptional changes and acquire a gene expression profile that is closer to that of human adult β cells [15, 61].

While the quality of stem cell differentiation is frequently evaluated by the expression of β-cell identity genes (INS, PDX1, NKX6.1, NEUROD1, MAFA, and UCN3), there are also genes that are key for regulating β-cell function. These include genes involved in insulin processing (PCSK1, PCSK2, and CPE), β-cell electrical activity (ABCC8), and glucose transport and metabolism (SLC30A8, GCK, GLUT1, G6PC2, and PFKFB2). Of note, recent studies have described genes that are instrumental for regulating both β-cell identity and β-cell functionality. Among these genes are NEUROD1, MAFA, and ESRRG. NEUROD1 is required

Table 1 Genes involved in β cell maturation, function, and disease. Expression levels in SC- and primary β cells are reported

Gene	Description	β-cell maturation	β-cell function	β-cell defects and/or disease after gene mutation	Expression levels: Sc-β cell	Expression levels: Sc-β-cell post-transplantation	Expression levels: Primary β cells	Ref. (#PMID)
β-cell identity markers								
PDX1	Pancreatic and duodenal homeobox 1	✓	✓	β-cell identity loss, MODY4	++	++	++	24506867, 23863625, 35241836, 32846125, 31068696
PAX6	Paired box 6	✓	✓	Impaired glucose tolerance, early onset T2D	+	++	++	22403172, 11756345, 15842522, 19034419, 31068696
NKX6.1	NK6 homeobox 1	✓	✓	Diabetes, hypoinsulinemia	++	++	++	23863625, 24035389, 32846125, 35241836, 31068696
NKX2.2	NK2 homeobox 2	✓	✓	Bihormonal cells, diabetes	+	++	++	27941248, 32846125
NEUROD1	Neuronal differentiation 1	✓	✓	MODY6	+	++	++	11575290, 20374962, 10545951, 35241836, 32846125
CHGB	Chromogranin B	✓	✓	Impaired secretory granule trafficking, proinsulin processing, and GSIS	+	++	+++	31182646, 31068696, 35241836, 32846125
MAFA	V-maf musculo-aponeurotic fibro-sarcoma oncogene homolog A	✓	✓	T1D, T2D	+	++	++	23863625, 35241836, 2550095, 12368292, 24520122, 32846125
ESRRG	Estrogen-related receptor γ	✓	✓	Impaired OxPhos and GSIS	+	+	+	27076077, 35241836
INS	Insulin	✓	✓	Neonatal diabetes, ER stress, and apoptosis	++	+++	+++	24843467, 30297969, 20938745, 35241836, 32846125, 31068696
ISL1	ISL LIM homeobox 1	✓	✓	Impaired GSIS and glucose tolerance	++	+	+	25028525, 35241836, 31068696
SIX2	SIX homeobox 2	✓	✓	Impaired GSIS	+	++	++	33446570, 32460030, 35241836, 31068696, 32846125

SIX3	SIX homeobox 3	✓	✓	Impaired GSIS	–	–	++	33446570, 32460030, 35241836, 31068696
IAPP	Islet amyloid polypeptide	✓		Increased GSIS, improved glucose tolerance	+	++	++	9753619, 32846125, 35241836, 31068696
UCN3	Urocortin 3	✓		Increased GSIS, glucose tolerance	–	++	++	32369775, 35241836, 25148370, 32846125, 31068696
β-cell functional markers								
SLC2A1	GLUT1		✓	T2D Diabetic nephropathy	++	+	+	22011817, 30297969, 22660720, 32846125, 35164795, 31068696, 35241836
GCK	Glucokinase (low affinity)		✓	MODY2	++	+	+	19790256, 11372010, 9435328, 32846125, 35241836, 31068696
ABCC8	Sulphonylurea receptor 1		✓	MODY12, congenital hyperinsulinism	+	++	++	26246406, 28600547, 31068696
G6PC2	Glucose-6-phosphatase 2		✓	Reduced fasting glycemia T2D	+	++	++	35176280, 32213654, 25625282, 23274894, 35241836, 32846125, 31068696
SLC30A8	Zinc transporter 8		✓	Reduce or gain T2D risk according to mutations	+	++	++	31676859, 24584071, 35842441, 32846125, 35241836, 31068696
β-cell disallowed genes								
HK1	Hexokinase-1 (high affinity)	✓	✓	Congenital hyperinsulinism	+	?	–	23274908, 32846125, 35241836, 31068696
SLC16A1	Monocarboxylate transporter 1	✓	✓	Exercise triggered hyperinsulinism	+	?	–	12502513, 32846125, 35241836, 31068696
LDHA	Lactate dehydrogenase A	✓	✓	Exercise triggered hyperinsulinism	++	+	+	22517647, 33424095, 32846125, 35241836, 31068696

MODY maturity-onset diabetes of the young

for β-cell maturation and maintenance of glucose-responsive capacity [64]. MAFA regulates insulin expression in addition to the expression of genes involved in glucose metabolism (GLUT, GCK), mitochondrial biogenesis (PGC1α), and circadian genes [70]. Estrogen-related receptor gamma (ERRγ) encoded by ESRRG has been identified as a key regulator of the mitochondrial oxidative phosphorylation genes and for MDH1, the gene encoding for the pyruvate–citrate cycle-related enzyme. Overexpression of ERRγ in SC-β cells promotes metabolic maturation by enhancing aerobic oxidative metabolism [68]. Gene expression in addition to functional assays is a valuable tool for assessing the quality and completeness of a SC-β-cell differentiation protocol.

Balboa and colleagues recently described an optimized protocol that allowed for the generation of islets with a transcriptomic signature similar to that of primary human islets. These stem cell-derived islet clusters are characterized by an increase in β-cell maturation markers, an upregulation of glycolysis and TCA cycle-related genes, and a repression of several disallowed genes [15]. They further show a similar architecture to human islets with 40% beta cells, 40–50% alpha cells, and 4% SST cells. Although these clusters show elevated dynamic GSIS with a first-phase and second-phase insulin secretion, the amount of secreted insulin in the second phase is still lower compared to the primary human islets due to the persistent metabolic failure induced by an immature mitochondrial glucose coupling.

5 SC-β Cells in Clinical Trials

Clinical trials using of SC-β cells have recently conducted. The clinical trial (ClinicalTrials Identifier: NCT03163511) performed by Viacyte has used human stem cell-derived pancreatic endoderm cells (PECs) loaded into a non-immunoprotective macrocapsulation device "VC-02" for subcutaneous transplantation into patients with T1D. PECs are precursors that further differentiate into mature β cells within 35–57 weeks following transplantation. Surprisingly, it was found that glucagon-producing α cells prevail as the predominant cell type in these grafts rather than β cells. While the PEC recipients have meal responsive insulin secretion 26 weeks post-transplantation, the circulating C-peptide levels were extremely low and unable to achieve glycemic control (20–40 pM in 5 patients, <10 pM in all other patients) [71–73]. This suggests that the transplantation of a large number of PECs would be required to obtain the necessary dosage of SC-β cells to reverse diabetes. Furthermore, the uncharacteristically high percentage of α cells in grafts compared to primary islets may result in increased circulating glucagon levels affecting the physiological glycemic set point [74]. Another phase 1/2 clinical trial conducted by Vertex Pharmaceuticals (ClinicalTrials Identifier: NCT04786262) recently showed that one T1D patient receiving an allogeneic human pluripotent stem cell-derived islet cell product has achieved insulin independence by day 270. Despite the need for immunosuppression, these findings indicate that SC-β cells can be effective to restore normoglycemia in patients with diabetes.

Additional data and follow up studies will need to be conducted to prove the safety and efficacy of this approach in a larger population.

Summarily, even though SC-β cells exhibit immature features and functionality in vitro, in vivo data shows that insulin secretion from SC-β cells is improved upon transplantation into mice [12, 15, 18]. Furthermore, RNA sequencing studies on transplanted SC-β cells indicate that pronounced transcriptional changes occur. Many of the genes affected are involved in glucose metabolism, the TCA cycle, oxidative phosphorylation, insulin secretion, and insulin processing. The capacity for SC-β cells to further functionally mature is significant and provides promise that refining current differentiation protocols in vitro will achieve metabolically mature SC-β cells that more closely resemble the human islet "gold standard".

Acknowledgements Work in MH's laboratory on stem cells is supported by grants from the NIH (DK129935, DK090570, DK129523, and DK132547) and the JDRF (3-SRA-2022-1216-S-B).

Conflict of Interest MH owns stock/stock options in Encellin, Thymmune, and Minutia. MH is co-founder and SAB member of Minutia.

References

1. De Vos, A. *et al.* Human and rat beta cells differ in glucose transporter but not in glucokinase gene expression. *J Clin Invest* **96**, 2489–2495 (1995). https://doi.org/10.1172/JCI118308
2. Matschinsky, F. M. Regulation of pancreatic beta-cell glucokinase: from basics to therapeutics. *Diabetes* **51 Suppl 3**, S394–404 (2002). https://doi.org/10.2337/diabetes.51.2007.s394
3. Wilson, J. E. Isozymes of mammalian hexokinase: structure, subcellular localization and metabolic function. *J Exp Biol* **206**, 2049–2057 (2003). https://doi.org/10.1242/jeb.00241
4. King, R. A., Smith, R. M. & Dahlenburg, G. W. Long term postnatal development of insulin secretion in early premature infants. *Early Hum Dev* **13**, 285–294 (1986). https://doi.org/10.1016/0378-3782(86)90062-9
5. Arda, H. E. *et al.* Age-Dependent Pancreatic Gene Regulation Reveals Mechanisms Governing Human beta Cell Function. *Cell Metab* **23**, 909–920 (2016). https://doi.org/10.1016/j.cmet.2016.04.002
6. Blum, B. *et al.* Functional beta-cell maturation is marked by an increased glucose threshold and by expression of urocortin 3. *Nat Biotechnol* **30**, 261–264 (2012). https://doi.org/10.1038/nbt.2141
7. Hrvatin, S. *et al.* Differentiated human stem cells resemble fetal, not adult, beta cells. *Proc Natl Acad Sci U S A* **111**, 3038–3043 (2014). https://doi.org/10.1073/pnas.1400709111
8. Henquin, J. C. & Nenquin, M. Dynamics and Regulation of Insulin Secretion in Pancreatic Islets from Normal Young Children. *PLoS One* **11**, e0165961 (2016). https://doi.org/10.1371/journal.pone.0165961
9. Becker, T. C., BeltrandelRio, H., Noel, R. J., Johnson, J. H. & Newgard, C. B. Overexpression of hexokinase I in isolated islets of Langerhans via recombinant adenovirus. Enhancement of glucose metabolism and insulin secretion at basal but not stimulatory glucose levels. *J Biol Chem* **269**, 21234–21238 (1994).
10. Henquin, J. C. *et al.* Congenital hyperinsulinism caused by hexokinase I expression or glucokinase-activating mutation in a subset of beta-cells. *Diabetes* **62**, 1689–1696 (2013). https://doi.org/10.2337/db12-1414
11. Dhawan, S. *et al.* DNA methylation directs functional maturation of pancreatic beta cells. *J Clin Invest* **125**, 2851–2860 (2015). https://doi.org/10.1172/JCI79956

12. Nair, G. G. *et al.* Recapitulating endocrine cell clustering in culture promotes maturation of human stem-cell-derived beta cells. *Nat Cell Biol* **21**, 263–274 (2019). https://doi.org/10.1038/s41556-018-0271-4
13. Hogrebe, N. J., Augsornworawat, P., Maxwell, K. G., Velazco-Cruz, L. & Millman, J. R. Targeting the cytoskeleton to direct pancreatic differentiation of human pluripotent stem cells. *Nat Biotechnol* **38**, 460–470 (2020). https://doi.org/10.1038/s41587-020-0430-6
14. Pagliuca, F. W. *et al.* Generation of functional human pancreatic beta cells in vitro. *Cell* **159**, 428–439 (2014). https://doi.org/10.1016/j.cell.2014.09.040
15. Balboa, D. *et al.* Functional, metabolic and transcriptional maturation of human pancreatic islets derived from stem cells. *Nat Biotechnol* **40**, 1042–1055 (2022). https://doi.org/10.1038/s41587-022-01219-z
16. Russ, H. A. *et al.* Controlled induction of human pancreatic progenitors produces functional beta-like cells in vitro. *EMBO J* **34**, 1759–1772 (2015). https://doi.org/10.15252/embj.201591058
17. Veres, A. *et al.* Charting cellular identity during human in vitro beta-cell differentiation. *Nature* **569**, 368–373 (2019). https://doi.org/10.1038/s41586-019-1168-5
18. Rezania, A. *et al.* Reversal of diabetes with insulin-producing cells derived in vitro from human pluripotent stem cells. *Nat Biotechnol* **32**, 1121–1133 (2014). https://doi.org/10.1038/nbt.3033
19. Nicholls, D. G. The Pancreatic beta-Cell: A Bioenergetic Perspective. *Physiol Rev* **96**, 1385–1447 (2016). https://doi.org/10.1152/physrev.00009.2016
20. Campbell, J. E. & Newgard, C. B. Mechanisms controlling pancreatic islet cell function in insulin secretion. *Nat Rev Mol Cell Biol* **22**, 142–158 (2021). https://doi.org/10.1038/s41580-020-00317-7
21. Alves, T. C. *et al.* Integrated, Step-Wise, Mass-Isotopomeric Flux Analysis of the TCA Cycle. *Cell Metab* **22**, 936–947 (2015). https://doi.org/10.1016/j.cmet.2015.08.021
22. Lu, D. *et al.* 13C NMR isotopomer analysis reveals a connection between pyruvate cycling and glucose-stimulated insulin secretion (GSIS). *Proc Natl Acad Sci U S A* **99**, 2708–2713 (2002). https://doi.org/10.1073/pnas.052005699
23. Schuit, F. *et al.* beta-cell-specific gene repression: a mechanism to protect against inappropriate or maladjusted insulin secretion? *Diabetes* **61**, 969–975 (2012). https://doi.org/10.2337/db11-1564
24. Thorrez, L. *et al.* Tissue-specific disallowance of housekeeping genes: the other face of cell differentiation. *Genome Res* **21**, 95–105 (2011). https://doi.org/10.1101/gr.109173.110
25. Davis, J. C. *et al.* Glucose Response by Stem Cell-Derived beta Cells In Vitro Is Inhibited by a Bottleneck in Glycolysis. *Cell Rep* **31**, 107623 (2020). https://doi.org/10.1016/j.celrep.2020.107623
26. Velazco-Cruz, L. *et al.* Acquisition of Dynamic Function in Human Stem Cell-Derived beta Cells. *Stem Cell Rep* **12**, 351–365 (2019). https://doi.org/10.1016/j.stemcr.2018.12.012
27. Braun, M. *et al.* Voltage-gated ion channels in human pancreatic beta-cells: electrophysiological characterization and role in insulin secretion. *Diabetes* **57**, 1618–1628 (2008). https://doi.org/10.2337/db07-0991
28. Cook, D. L., Ikeuchi, M. & Fujimoto, W. Y. Lowering of pHi inhibits Ca2+-activated K+ channels in pancreatic B-cells. *Nature* **311**, 269–271 (1984). https://doi.org/10.1038/311269a0
29. Satin, L. S. & Cook, D. L. Voltage-gated Ca2+ current in pancreatic B-cells. *Pflugers Arch* **404**, 385–387 (1985). https://doi.org/10.1007/BF00585354
30. Gilon, P., Chae, H. Y., Rutter, G. A. & Ravier, M. A. Calcium signaling in pancreatic beta-cells in health and in Type 2 diabetes. *Cell Calcium* **56**, 340–361 (2014). https://doi.org/10.1016/j.ceca.2014.09.001
31. Huang, C. *et al.* Synaptotagmin 4 Regulates Pancreatic beta Cell Maturation by Modulating the Ca(2+) Sensitivity of Insulin Secretion Vesicles. *Dev Cell* **45**, 347–361 e345 (2018). https://doi.org/10.1016/j.devcel.2018.03.013
32. Idevall-Hagren, O. & Tengholm, A. Metabolic regulation of calcium signaling in beta cells. *Semin Cell Dev Biol* **103**, 20–30 (2020). https://doi.org/10.1016/j.semcdb.2020.01.008

33. Henquin, J. C. & Nenquin, M. Immaturity of insulin secretion by pancreatic islets isolated from one human neonate. *J Diabetes Investig* **9**, 270–273 (2018). https://doi.org/10.1111/jdi.12701
34. Siehler, J., Blochinger, A. K., Meier, M. & Lickert, H. Engineering islets from stem cells for advanced therapies of diabetes. *Nat Rev Drug Discov* **20**, 920–940 (2021). https://doi.org/10.1038/s41573-021-00262-w
35. Gaisano, H. Y. Recent new insights into the role of SNARE and associated proteins in insulin granule exocytosis. *Diabetes Obes Metab* **19 Suppl 1**, 115–123 (2017). https://doi.org/10.1111/dom.13001
36. Henquin, J. C. Triggering and amplifying pathways of regulation of insulin secretion by glucose. *Diabetes* **49**, 1751–1760 (2000). https://doi.org/10.2337/diabetes.49.11.1751
37. Henquin, J. C. Regulation of insulin secretion: a matter of phase control and amplitude modulation. *Diabetologia* **52**, 739–751 (2009). https://doi.org/10.1007/s00125-009-1314-y
38. Prentki, M., Matschinsky, F. M. & Madiraju, S. R. Metabolic signaling in fuel-induced insulin secretion. *Cell Metab* **18**, 162–185 (2013). https://doi.org/10.1016/j.cmet.2013.05.018
39. Stanley, C. A. Regulation of glutamate metabolism and insulin secretion by glutamate dehydrogenase in hypoglycemic children. *Am J Clin Nutr* **90**, 862S–866S (2009). https://doi.org/10.3945/ajcn.2009.27462AA
40. Li, M., Li, C., Allen, A., Stanley, C. A. & Smith, T. J. The structure and allosteric regulation of mammalian glutamate dehydrogenase. *Arch Biochem Biophys* **519**, 69–80 (2012). https://doi.org/10.1016/j.abb.2011.10.015
41. Barsby, T. & Otonkoski, T. Maturation of beta cells: lessons from in vivo and in vitro models. *Diabetologia* **65**, 917–930 (2022). https://doi.org/10.1007/s00125-022-05672-y
42. Fehmann, H. C., Goke, R. & Goke, B. Cell and molecular biology of the incretin hormones glucagon-like peptide-I and glucose-dependent insulin releasing polypeptide. *Endocr Rev* **16**, 390–410 (1995). https://doi.org/10.1210/edrv-16-3-390
43. Henquin, J. C. & Nenquin, M. Activators of PKA and Epac distinctly influence insulin secretion and cytosolic Ca2+ in female mouse islets stimulated by glucose and tolbutamide. *Endocrinology* **155**, 3274–3287 (2014). https://doi.org/10.1210/en.2014-1247
44. Dyachok, O. *et al.* Glucose-induced cyclic AMP oscillations regulate pulsatile insulin secretion. *Cell Metab* **8**, 26–37 (2008). https://doi.org/10.1016/j.cmet.2008.06.003
45. Idevall-Hagren, O., Barg, S., Gylfe, E. & Tengholm, A. cAMP mediators of pulsatile insulin secretion from glucose-stimulated single beta-cells. *J Biol Chem* **285**, 23007–23018 (2010). https://doi.org/10.1074/jbc.M109.095992
46. Tengholm, A. & Gylfe, E. cAMP signalling in insulin and glucagon secretion. *Diabetes Obes Metab* **19 Suppl 1**, 42–53 (2017). https://doi.org/10.1111/dom.12993
47. Seino, S., Shibasaki, T. & Minami, K. Dynamics of insulin secretion and the clinical implications for obesity and diabetes. *J Clin Invest* **121**, 2118–2125 (2011). https://doi.org/10.1172/JCI45680
48. Huang, Q. Y. *et al.* Cdc42: A Novel Regulator of Insulin Secretion and Diabetes-Associated Diseases. *Int J Mol Sci* **20** (2019). https://doi.org/10.3390/ijms20010179
49. Wang, Z., Oh, E. & Thurmond, D. C. Glucose-stimulated Cdc42 signaling is essential for the second phase of insulin secretion. *J Biol Chem* **282**, 9536–9546 (2007). https://doi.org/10.1074/jbc.M610553200
50. Docherty, F. M. *et al.* ENTPD3 Marks Mature Stem Cell-Derived beta-Cells Formed by Self-Aggregation In Vitro. *Diabetes* **70**, 2554–2567 (2021). https://doi.org/10.2337/db20-0873
51. Alvarez-Dominguez, J. R. *et al.* Circadian Entrainment Triggers Maturation of Human In Vitro Islets. *Cell Stem Cell* **26**, 108–122 e110 (2020). https://doi.org/10.1016/j.stem.2019.11.011
52. Omar-Hmeadi, M. & Idevall-Hagren, O. Insulin granule biogenesis and exocytosis. *Cell Mol Life Sci* **78**, 1957–1970 (2021). https://doi.org/10.1007/s00018-020-03688-4
53. Arvan, P. & Halban, P. A. Sorting ourselves out: seeking consensus on trafficking in the beta-cell. *Traffic* **5**, 53–61 (2004). https://doi.org/10.1111/j.1600-0854.2004.00152.x
54. Steiner, D. F., Docherty, K. & Carroll, R. Golgi/granule processing of peptide hormone and neuropeptide precursors: a minireview. *J Cell Biochem* **24**, 121–130 (1984). https://doi.org/10.1002/jcb.240240204

55. Chen, Y. C., Taylor, A. J. & Verchere, C. B. Islet prohormone processing in health and disease. *Diabetes Obes Metab* **20 Suppl 2**, 64–76 (2018). https://doi.org/10.1111/dom.13401
56. Stiernet, P., Guiot, Y., Gilon, P. & Henquin, J. C. Glucose acutely decreases pH of secretory granules in mouse pancreatic islets. Mechanisms and influence on insulin secretion. *J Biol Chem* **281**, 22142–22151 (2006). https://doi.org/10.1074/jbc.M513224200
57. Nishi, M., Sanke, T., Nagamatsu, S., Bell, G. I. & Steiner, D. F. Islet amyloid polypeptide. A new beta cell secretory product related to islet amyloid deposits. *J Biol Chem* **265**, 4173–4176 (1990).
58. Lee, Y. H., Kim, J., Park, K. & Lee, M. S. beta-cell autophagy: Mechanism and role in beta-cell dysfunction. *Mol Metab* **27S**, S92–S103 (2019). https://doi.org/10.1016/j.molmet.2019.06.014
59. Muller, A. *et al.* A Global Approach for Quantitative Super Resolution and Electron Microscopy on Cryo and Epoxy Sections Using Self-labeling Protein Tags. *Sci Rep* **7**, 23 (2017). https://doi.org/10.1038/s41598-017-00033-x
60. Goginashvili, A. *et al.* Insulin granules. Insulin secretory granules control autophagy in pancreatic beta cells. *Science* **347**, 878–882 (2015). https://doi.org/10.1126/science.aaa2628
61. Augsornworawat, P., Maxwell, K. G., Velazco-Cruz, L. & Millman, J. R. Single-Cell Transcriptome Profiling Reveals beta Cell Maturation in Stem Cell-Derived Islets after Transplantation. *Cell Rep* **32**, 108067 (2020). https://doi.org/10.1016/j.celrep.2020.108067
62. Lyttle, B. M. *et al.* Transcription factor expression in the developing human fetal endocrine pancreas. *Diabetologia* **51**, 1169–1180 (2008). https://doi.org/10.1007/s00125-008-1006-z
63. Nair, G. & Hebrok, M. Islet formation in mice and men: lessons for the generation of functional insulin-producing β-cells from human pluripotent stem cells. *Current Opinion in Genetics & Development* **32**, 171–180 (2015). https://doi.org/10.1016/j.gde.2015.03.004
64. Gu, C. *et al.* Pancreatic beta cells require NeuroD to achieve and maintain functional maturity. *Cell Metab* **11**, 298–310 (2010). https://doi.org/10.1016/j.cmet.2010.03.006
65. Gosmain, Y. *et al.* Pax6 is crucial for beta-cell function, insulin biosynthesis, and glucose-induced insulin secretion. *Mol Endocrinol* **26**, 696–709 (2012). https://doi.org/10.1210/me.2011-1256
66. Hang, Y. *et al.* The MafA Transcription Factor Becomes Essential to Islet beta-Cells Soon After Birth. *Diabetes* **63**, 1994–2005 (2014). https://doi.org/10.2337/db13-1001
67. Bevacqua, R. J. *et al.* SIX2 and SIX3 coordinately regulate functional maturity and fate of human pancreatic beta cells. *Genes Dev* **35**, 234–249 (2021). https://doi.org/10.1101/gad.342378.120
68. Yoshihara, E. *et al.* ERRgamma Is Required for the Metabolic Maturation of Therapeutically Functional Glucose-Responsive beta Cells. *Cell Metab* **23**, 622–634 (2016). https://doi.org/10.1016/j.cmet.2016.03.005
69. Lemaire, K., Thorrez, L. & Schuit, F. Disallowed and Allowed Gene Expression: Two Faces of Mature Islet Beta Cells. *Annu Rev Nutr* **36**, 45–71 (2016). https://doi.org/10.1146/annurev-nutr-071715-050808
70. Aguayo-Mazzucato, C. *et al.* Thyroid hormone promotes postnatal rat pancreatic beta-cell development and glucose-responsive insulin secretion through MAFA. *Diabetes* **62**, 1569–1580 (2013). https://doi.org/10.2337/db12-0849
71. Ramzy, A. *et al.* Implanted pluripotent stem-cell-derived pancreatic endoderm cells secrete glucose-responsive C-peptide in patients with type 1 diabetes. *Cell Stem Cell* **28**, 2047–2061 e2045 (2021). https://doi.org/10.1016/j.stem.2021.10.003
72. Shapiro, A. M. J. *et al.* Insulin expression and C-peptide in type 1 diabetes subjects implanted with stem cell-derived pancreatic endoderm cells in an encapsulation device. *Cell Rep Med* **2**, 100466 (2021). https://doi.org/10.1016/j.xcrm.2021.100466
73. de Koning, E. J. P. & Carlotti, F. Stem cell-based islet replacement therapy in diabetes: A road trip that reached the clinic. *Cell Stem Cell* **28**, 2044–2046 (2021). https://doi.org/10.1016/j.stem.2021.11.008

74. Rodriguez-Diaz, R. *et al.* Paracrine Interactions within the Pancreatic Islet Determine the Glycemic Set Point. *Cell Metab* **27**, 549–558 e544 (2018). https://doi.org/10.1016/j.cmet.2018.01.015

Clinical Trials with Stem Cell-Derived Insulin-Producing Cells

Ji Lei and James F. Markmann

1 Introduction

There are two main types of diabetes mellitus (DM): type 1 and type 2. The subset of type 1 diabetes (T1D) represents nearly 10% of this population. T1D is an autoimmune disease in which the insulin-producing β-cells of the pancreas are destroyed, requiring the institution of lifelong insulin treatment. Type 2 diabetes (T2D) is characterized by initial insulin resistance followed by islet dysfunction and islet loss over time, and most of the patients are treated with non-insulin-based medical therapies. The prevalence of DM is accelerating worldwide, with the number of individuals globally affected projected to reach ~700 million from the current ~460 million by 2045 [1]. The global economic burden from this increase is projected to be USD 2.5 trillion annually by 2030 [2, 3]. This chapter focuses on the new developments in the field of T1D, which affects 1.6 million children and adults in the USA alone, with 175 new cases diagnosed per day and a greater than $16 billion annual direct healthcare cost [4, 5].

Insulin was discovered by Frederick Banting and colleagues at the University of Toronto in 1921. It was first used to treat a 14-year-old boy dying from T1D in 1922 [6]. This monumental advance allowed T1D patients to live into adulthood and negated the need for the existing treatments of that time, which primarily relied on ultralow calorie, starvation diets. The discovery of insulin to treat the disease led the way to the development of recombinant human insulin and the subsequent insulin analogs, which drastically refined treatment options and improved patient outcomes

J. Lei (✉) · J. F. Markmann
Division of Transplantation, Department of Surgery, Massachusetts General Hospital, Harvard Medical School, Boston, MA, USA

Center for Transplantation Sciences, Department of Surgery, Massachusetts General Hospital, Harvard Medical School, Boston, MA, USA
e-mail: jlei2@mgh.harvard.edu; JMARKMANN@mgh.harvard.edu

L. Piemonti et al. (eds.), *Pluripotent Stem Cell Therapy for Diabetes*,
https://doi.org/10.1007/978-3-031-41943-0_23

[7, 8]. Due to these developments, T1D has evolved from being a fatal diagnosis into a manageable chronic condition. However, as diabetes therapy continued to rely liberally on the intermittent monitoring of blood glucose and the subcutaneous injection of insulin, it became apparent that exogenous insulin therapy, while averting death from ketoacidosis and dehydration, could not prevent secondary complications such as diabetic nephropathy, retinopathy, and neuropathy. New technologies such as smart wearable continuous glucose monitors and insulin pump delivery devices have improved glycemic control and refined diabetes management. Despite these technological advances, blood monitoring and insulin injections cannot fully replicate the precise physiological glycemic control provided by endogenous pancreatic islet β-cells, and chronic treatment remains burdensome for patients [9, 10]. Even with the use of new continuous glucose sensing devices and insulin pumps that have further advanced the care for T1D patients, it is doubtful that any patient would consider this a cure. Furthermore, standard therapy with exogenous insulin is associated with a significant danger of hypoglycemia and is only partially efficacious in preventing long-term complications. Research has demonstrated the profound health benefits of optimal glycemic control to prevent long-term complications and increase life expectancy [11, 12].

2 Rationale for the Need for Insulin-Producing Cells in Diabetes Therapy

The findings related to diabetes management mentioned above inspired interest to venture into other strategies to gain replacement of physiologic β-cell function. Because of the exquisite glycemic control offered by the replacement of β-cell mass through transplantation, this approach will likely be held as the gold standard for T1D treatment until future therapies such as native β-cell regeneration [13, 14] or use of blastocyst complementation that can grow a patient β-cell in an animal host [15, 16]. In fact, among the earliest proposed treatments by Watson Williams in 1893 was to transplant fragments of the pancreas into a diabetic patient [17, 18] with the goal of achieving an actual cure for the disease. These strategies, during Watson's time, were of course doomed to failure because of having only a rudimentary understanding of the immune system and the lack of β-cell isolation and purification strategies [19]. However, with dramatic progress in solid organ transplantation, developments of protocols in islet isolation [20–22], the successful restoration of euglycemia by islet transplantation in mice by Barker et al. [23], and advances in immunosuppression during the twentieth century, the prospect of whole organ pancreas and islet transplant became a reality.

In 1966, whole pancreas transplant in humans was first achieved clinically at the University of Minnesota [24]. Following that, further innovations in islet isolation and culture led to the first successful clinical islet transplantation in 1990 [25]. The results of these therapies demonstrated meaningful improvements along multiple vectors, including superior glycemic control and reduced glucose variability with better HbA1c levels than that of insulin administration [26]. What is more, islet replacement has the additional benefit of guarding against severe episodes of hypoglycemia, which standard insulin therapy is unable to achieve [27]. Despite the benefits of improved glucose-level regulation and protection from life-threatening episodes of hypoglycemia, pancreas transplantation is primarily performed concurrently or after a renal transplant and reserved only for cases of severe diabetes due to risks of the surgical procedure and complications associated with life-long immunosuppression [28, 29]. These risks are typically thought to outweigh the benefits of β-cell replacement by transplant therapy over the existing solutions of glucose monitoring and insulin administration. While islet transplantation is a substantially safer surgical procedure than whole pancreas transplant, it is hampered by many of the same limitations. Standard immunosuppression treatments were also shown to be toxic to newly transplanted islets, resulting in the only initial, sporadic success of islet transplantation [30]. It was not until the advent of the Edmonton Protocol in the late 1990s, which removed glucocorticoids and decreased calcineurin inhibitors along with multiple doses of islets recovered from different donors, that islet transplantation was able to consistently achieve insulin independence [31, 32]. The modifications also included the use of xeno-protein-free media during islet isolation and the immediate transplantation of the purified islets. A long-term 20-year follow-up study of islet transplantation at the Edmonton single center evaluating 255 patients demonstrated that islet transplantation continued to remain beneficial 20 years later for this subset of patients. In this cohort, 70% of the patients demonstrated positive serum C-peptide, stabilized glycemic control, and corrected HbA1c levels, all denoting persistent islet graft function. Insulin independence was achieved at any time point by 79% of patients, which diminished to 61% by 1 year, 32% by 5 years, 20% by 10 years, and 8% by 20 years after transplantation [33]. While these patients ultimately began to rely on low-dose supplemental insulin, they continued to be protected from severe hypoglycemic episodes [33, 34]. Hering et al. analyzed data from the Collaborative Islet Transplant Registry, which revealed several constructive factors that improve outcomes for the transplantation of primary human islets. The main findings were that transplanting an islet mass constituting at least 325,000 islet equivalents in combination with T-cell depletion and maintaining immunosuppression with rapamycin and calcineurin inhibitors resulted in 73% of patients being free from severe hypoglycemic events and achieving an HbA1c level of <7.0% 5 years after transplantation [35]. These experiences demonstrate that islet transplantation is a safe and effective cellular replacement therapy and could be a routine treatment for severe diabetes mellitus.

3 Rationale for the Need for a Stem Cell-Based Insulin-Producing Cell Therapy in Diabetes

In addition to these developments mentioned, two recent phase 3 studies sponsored by the NIH-Clinical Islet Transplant (CIT) Consortium have concluded that islet transplant is a safe and effective procedure for T1D [36, 37]. These monumental trials laid the groundwork for a Biologic License Application through the FDA to enable islet transplant to be a reimbursable cell-based therapy for patients with severe T1D in the USA. In fact, islet transplantation is already a government-funded, standard-of-care therapy in Europe, Canada, and several countries in Asia [38] for T1D patients who suffer from glycemic lability, hypoglycemia unawareness, and severe hypoglycemic episodes, despite optimal intensive diabetes management. However, the broad application of this therapy is limited by the scarcity of available primary islets from cadaveric donors and the low efficiency of the processing that may require processing four pancreata to achieve two good islet yields to gain robust diabetes reversal in a single patient. Thus, alternative sources of insulin-producing cells, such as porcine islet or stem cell-derived insulin-producing cells, are the next breakthrough therapeutic entities.

During the past two decades, stem cells have become a major focus for regenerative medicine because of the property of self-renewal while maintaining the capacity to differentiate into diverse cell types. Based on the origin and the ability of differentiation, stem cells can be broadly classified into three major categories: somatic stromal stem cells, embryonic pluripotent stem cells, and induced pluripotent stem cells. Somatic stromal stem cells, also termed mesenchymal stem cells (MSCs), are multipotent with bone marrow as the most prevalent source, but MSCs can be derived from almost all organs and tissues [39–41]. In humans, MSCs are referred to as hMSCs, which can reside specifically in vivo niches (endogenous) or be isolated and expanded in vitro (exogenous). Embryonic stem cells (ESCs) are pluripotent stem cells (PSCs) that are derived from embryos at the blastocyst stage. They maintain the genomic stability, self-renewal, and developmental potential to form trophoblast and derivatives of all three embryonic germ layers (endoderm, mesoderm, and ectoderm) [42]. Human blastocyst-derived pluripotent cells are referred to as human embryonic stem cells (hESCs). Induced pluripotent stem cells (iPSCs) are pluripotent stem cells that are generated from reprogrammed somatic cells by ectopic overexpression of specific transcription factors [43] as defined by Shinya Yamanaka, for which a Nobel Prize was awarded in 2012. iPSCs also exhibit the morphology, self-renewal, and differentiation potential as ESCs, while being artificially reproducible. The pluripotent stem cells derived from human somatic cells by reprogramming are referred to as human-induced pluripotent stem cells (hiPSCs). A few studies in both preclinical nonhuman primate models and human clinical trials have demonstrated that both hESCs and hiPSCs can be used to generate insulin-producing, glucose-responsive functional β-cells [44–49]. The schematic stepwise generation of insulin-producing cells from hESCs and hiPSC is shown in Fig. 1 and can occur over seven stages [50].

Because of the properties of self-renewal while maintaining pluripotency after expansion, hESCs and hiPSCs or perhaps hMSCs thus emerged as alternative, unlimited and scalable sources to generate insulin-producing β-cells. The transformative approaches to generate stem cell-derived insulin-producing β-cells (SC-β-cells) or islet-like clusters (SC-Islet) from human stem cells could replace primary islets and enable many more patients to be treated with a more homogeneous product. Stem cell-based approaches have the potential to radically alter the dynamics of diabetes management.

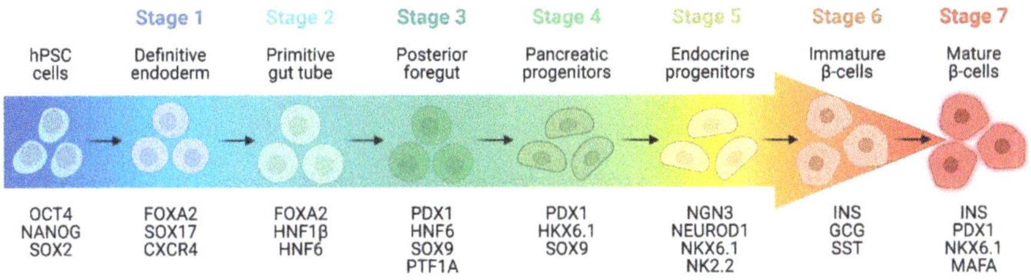

Fig. 1 Pancreatic β-cell differentiation from human pluripotent stem cells (hPSCs) [50]. Schematic illustration of the stepwise differentiation process for generating insulin-producing β-cells from human pluripotent stem cells by recapitulating the ontogeny of the embryonic pancreatic β-cells. The key markers of the distinct developmental stages of pancreatic β-cells are listed

4 Mesenchymal Stem Cell-Based Clinical Trials in Diabetes Therapy

Among the varied types of stem cells used in clinical trials for diabetes or its complications, human mesenchymal stem cells (hMSCs) are the most widely utilized today. A synopsis of selected completed and active mesenchymal stem cells (MSCs) based clinical trials for diabetes from numerous trials registered at www.clinicaltrials.gov since 2010 is listed in Table 1. To date, clinical trials have researched three theories: [1] whether pancreatic progenitors derived from MSCs have the capacity to generate functional β-cells, [2] whether undifferentiated MSCs precursors can transdifferentiate directly in vivo to produce functional insulin-producing cells, and [3] whether MSCs that have not been further differentiated into pancreatic progenitors can provide general health benefits to foster endogenous β-cells survival in vivo. Because of their ease of recovery, many investigators have explored the potential of MSCs as an allogeneic therapeutic therapy (Table 1) [51].

Initial studies suggested MSCs could generate insulin-producing cells [52–55]; however, strong evidence of functional mature β-cells or islet-like organoids, both in vitro and in vivo, has not materialized [56], a finding corroborated by in vivo cell-tracing experiments [57–59]. Despite the lack of evidence, clinical trials testing the hypothesis that MSC-derived pancreatic progenitors generated in vitro can mature into β-cells in vivo continue in T1D, either alone or in combination with bone marrow cell infusion [60, 61]. These trials reported evidence of improved glycemic control with reduced insulin requirements and a fall in A1c, while the evidence that β-cells had been generated from MSCs in vivo still lacks.

More recent trials have diverted from conclusions of in vivo augmentation of the β-cell mass by trans-differentiation and instead favor the hypothesis that administration of MSCs may foster endogenous β-cell health and survival in T1D; however, controlled trials are still needed for confirmation [62, 63]. A meta-analysis performed by Gan et al. in 2018 studied approaches based on hematopoietic stem cells therapy, MSC-based therapy, and co-infusion of hematopoietic and multipotent MSCs. The analysis comprised nine randomized-controlled trials and 14

Table 1 Summary of selected completed and active mesenchymal stem cells (MSCs)-based clinical trials for T1D and T2D (www.clinicaltrials.gov)

Trail ID	Title	Subtype	Interventions
NCT05308836	Evaluate safety of adipose-derived MSC T1D treatment	T1D	Adipose-derived MSCs
NCT05061030	Mesenchymal stromal cells to treat T1D in children and adolescents	T1D	ATMP Protrans
NCT05003908	Safety of cultured allogeneic adult umbilical cord derived MSC intravenous infusion for DM	T1D and T2D	AlloRx
NCT04501341	BM-MNC and UCMSC for T2D patients	T2D	BMMC+UCMSCs
NCT04466007	Safety and efficacy of allogeneic adipose tissue MSCs in DM patients with critical limb ischemia	T1D and T2D	Allogeneic MSCs
NCT04441658	Efficacy and safety of umbilical cord MSC cells transplantation in patients with T2D	T2D	UCMSCs
NCT04078308	MSC cells transplantation in newly diagnosed T1D patients	T1D	Autologous MSCs
NCT04061746	Cellular therapy for T1D using MSCs	T1D	MSCs
NCT04011020	Clinical application of stem cell educator therapy in T1D	T1D	Stem cell educator
NCT03920397	MSCs in patients with T1D	T1D	Adipose MSCs
NCT03912480	Stem cells from human exfoliated teeth in treatment of diabetic patients	T1D	Teeth MSCs
NCT03835312	Sequential transplantation of UCBSCs and islets in children and adolescents with T1D	T1D	UCBSCs
NCT03658655	Stem cells from human exfoliated teeth in treatment of T2D	T2D	Teeth MSCs
NCT03562208	Autologous bone marrow transplant in chronic insulin-dependent diabetic patients	T1D and T2D	BM MSCs
NCT03484741	MSC therapy for T1D patients	T1D	MSCs
NCT03390231	Stem cell educator therapy in diabetes	T1D and T2D	Stem cell educator
NCT03343782	Outcomes of expanded autologous bone marrow-derived MSCs therapy in T2D	T2D	MSCs
NCT02940418	Use of stem cells in T1D	T1D	Adipose MSCs + BMMCs
NCT02893306	MSC administration for the management of T1D	T1D	MSCs
NCT02644759	Transplantation of autologous stem cells for the treatment of T1D	T1D	UCBSCs
NCT02624804	A pilot study of the therapeutic potential of stem cell educator therapy in T1D	T1D	Stem cell educator
NCT02384018	Mesenchymal stem cell and islet co-transplantation	T1D and T2D	Autologous MSCs
NCT02302599	Mesenchymal stem cells to treat type 2 diabetes	T2D	UC-MSCs

(continued)

Table 1 (continued)

Trail ID	Title	Subtype	Interventions
NCT02057211	MSCs to intervene in the development of T1D: a blinded randomized study	T1D	Autologous MSCs
NCT01996228	Reversal of T1D in children by stem cell educator therapy	T1D	Stem cell educator
NCT01954147	Umbilical cord MSCs and liraglutide in diabetes mellitus	T2D	UB-MSCs
NCT01759823	Bone marrow-derived stem cell transplantation in T2D	T2D	MSCs
NCT01576328	Safety study of mesenchymal precursor cells in T2D	T2D	MPCs
NCT01496339	Human menstrual blood-derived MSCs transplantation in treating T1D	T1D	MenMSCs
NCT01453751	Safety and effects of autologous adipose-derived stromal cells delivered in patients with T2D	T2D	AD-SVF
NCT01415726	Stem cell educator therapy in type 2 diabetes	T2D	Stem cell educator
NCT01413035	Safety and efficacy study of umbilical cord/placenta-derived MSCs to treat T2D	T2D	MSCs
NCT01350219	Stem cell educator therapy in type 1 diabetes	T1D	Stem cell educator
NCT01157403	Autologous transplantation of MSCs for treatment of patients with onset of T1D	T1D	MSCs
NCT01143168	Stem cell therapy for type 1 diabetes mellitus	T1D	BMMC-UCMSCs
NCT01142050	Stem cell therapy for type 2 diabetes mellitus	T2D	MSCs
NCT01121029	Hematopoietic stem cell transplantation in T1D	T1D	MSCs
NCT01068951	Treatment of patients with newly onset of T1D with MSCs	T1D	MSCs
NCT00971503	Safety and efficacy of arterial delivery of autologous bone marrow cells in the treatment of T1D	T1D	MSCs

self-controlled trials and concluded that the pooled effect of these therapies resulted in an increased C-peptide level, compared with conventional insulin therapy; whereas, trials based on umbilical cord blood-derived MSCs did not achieve the increase in C-peptide level [64]. In contrast to these findings, a separate meta-analysis conducted by Hwang et al. in 2019, based on six controlled trials for T1D refuted these claims. The study showed no improvement in stimulated and fasting c-peptide levels, indicating a lack of benefits from MSCs therapy for T1D patients [65]. However, more conclusive results can be obtained from ongoing randomized-controlled trials that include larger cohorts of patients and stricter controls, which will provide better clarification regarding the efficacy of MSCs therapy for T1D.

While the benefits of MSCs-based therapies for T1D have remained questionable, MSCs-based clinical trials for T2D have shown consistent and robust efficacy.

T2D MSC-based trials utilize multipotent MSCs derived from the varied sources are listed in Table 1. From 2009 when the first trial by Bhansali et al. [66] was conducted to the most recent trials [67–69], improvements have been observed in HbA1c values, C-peptide levels, and reduction in the required insulin dosage, findings corroborated by a systematic review of ten T2D MSCs-based trials [65]. However, despite the rich in vitro and in vivo preclinical studies to date, the exact mechanism by which MSCs improve outcomes remains to be elucidated. With the prelude of questionable evidence of insulin production by hMSCs, most of this chapter will be devoted to pluripotent stem cell-based insulin-producing clinical trials in T1D therapy.

5 Pluripotent Stem Cell-Based Clinical Trials in Diabetes Therapy

Table 2 provides a summary of ongoing and completed clinical trials testing human pluripotent stem cells (hPSCs) derived β-cells in T1D. Since the first registration of such a trial in 2014, only eight trials have been carried out, sponsored by Vertex Pharmaceuticals Inc., ViaCyte Inc. (acquired by Vertex in July 2022), and CRISPR Therapeutics. For the ease of naming convention, we use the name of the company that sponsored the original trial to describe the findings across time.

5.1 ViaCyte Inc. Trials

In 2014, the first T1D patient was treated with an hESC-based pancreatic progenitor transplant in San Diego. The study was sponsored by ViaCyte Inc. under the supervision of Dr. Robert Henry and his team at UCSD. The purpose of the trial (clinicaltrials.gov: NCT02239354, submitted in 2014) was to investigate the safety, tolerability, and efficacy of ViaCyte product candidate named VC-01™, which is a combination product of hESC-derived pancreatic progenitor cells (named PEC-01) and a macroencapsulation device (Encaptra) for the combination product called PEC-Encap designed for the treatment of T1D. PEC-01 is a stage 4 hESC-derived pancreatic progenitor cell (as shown in Fig. 1) product expected to mature into functional insulin-producing cells upon transplantation in vivo, based on prior mice studies [70, 71]. PEC-Encap is a device with a semipermeable membrane pore that allows small molecule oxygen and nutrient exchange but blocks cell trafficking to protect the pancreatic progenitor cells from harm by the immune system. The latter is especially important for preventing allogeneic and autoimmune rejection. This innovation would eliminate the necessity of immunosuppressive drugs for long-term graft survival. The first transplantation consisted of approximately 40 million pancreatic progenitor cells in each of the larger devices, which were divided into

Table 2 Summary of ongoing and completed clinical trials of human pluripotent stem cells (hPSCs) derived insulin-producing cells in T1D (www.clinicaltrials.gov)

Trail ID	Start date	Completion date	Sponsor collaborators	Trial phase	# of patient	Title	Interventions
NCT02239354	September 2014	Terminated	ViaCyte	Ph.1/2	$N = 19$	A safety, tolerability, and efficacy study of VC-01™ combination product in subjects with T1D	PEC-01 cells loaded into PEC-Encap
NCT04678557	June 2019	November 2021	ViaCyte	Ph.1/2	$N = 31$	A phase 1/2 study to evaluate the safety and efficacy of VC-01™ in subjects with T1D	PEC-01 cells loaded into PEC-Encap
NCT02939118	November 2016	November 2023[a]	ViaCyte	Safety	$N = 51$	A follow-up safety study in subjects previously implanted with VC-01™ combination product	VC-01™ with subsequent explantation of VC-01 units
NCT03162926	July 2017	February 2018	ViaCyte	Ph. 1	$N = 3$	A study evaluating the safety and tolerability of VC-02™ in subjects with T1D	Aka PEC-direct. PEC-01 cells loaded into a delivery device
NCT03163511	July 2017	February 2024[a]	ViaCyte	Ph. 1/2	$N = 49$	A study evaluating the safety and efficacy of VC-02™ combination product in T1D with hypoglycemia unawareness	PEC-01 cells are loaded into PEC-direct
NCT04786262	March 2021	January 2024[a]	Vertex	Ph. 1/2	$N = 17$	A phase 1/2 study to evaluate the safety, tolerability, and efficacy of VX-880 in T1D	VX-880 infused into the hepatic portal vein
NCT05565248	November 2022	August 2025[a]	CRISPR Inc. ViaCyte	Ph. 1/2	$N = 40$	A study evaluating the safety, tolerability, and efficacy of VCTX211in T1D	CRISPR-Cas9 genetically modified PEC211 cells loaded into a perforated delivery device
NCT05210530	January 2022	April 2023	CRISPR Inc. ViaCyte	Ph. 1	$N = 10$	A study evaluating the safety and tolerability of VCTX210A in subjects with T1D	CRISPR-Cas9 genetically modified PEC210A cells loaded into a delivery device

N numbers for ongoing studies.
[a]Estimated dates

two encapsulation devices implanted subcutaneously in the abdomen, along with six smaller encapsulation devices implanted subcutaneously in the arm, serving as sentinels to be removed at different time points to follow cell survival and maturation. This trial evaluated 19 T1D patients over 2 years in an open-label, dose-escalating phase 1/2 study of a planned number of 40 patients but was terminated prematurely due to inconsistencies in cell survival, poor cell engraftment, and the lack of evidence for graft-induced insulin secretion. Additionally, there was evidence indicating a foreign body response (FBR) to the encapsulation devices that likely impeded transit through the semipermeable membrane causing islet death from suffocation. Based on results posted on clinicaltrials.gov, the first 19 patients were reported to tolerate the product well and had few complications related to the product or surgery [72]. This first trial revealed the drawbacks of encapsulation devices in general, namely the foreign body responses and resulting fibrosis which clogs membrane pores preventing vascularization.

In 2017, ViaCyte, in an attempt to mitigate the cell loss due to the device's FBR, an alternative modified encapsulation delivery device named PEC-Direct or the VC-02™ combination product was designed. A phase 1 trial (NCT03162926) aimed to test whether the VC-02™ combination product could be implanted subcutaneously in three participants with T1D and whether it can be maintained safely for up to 4 months. PEC-Direct is a macroencapsulation device with pores designed to enable direct capillary vascular permeation into the device interior space where the pancreatic progenitor cells (PEC-01) are housed. The vascularization property was postulated to promote cell survival and maturation into functional islet-like cells. No modifications were made concerning the hESC-derived pancreatic progenitors used in the NCT02239354 trial. PEC-Direct lacks the ability to keep infiltrating immune cells away from the graft, thus requiring the use of immunosuppressive drugs during this study to protect the graft. Successful outcomes led to the current ongoing phase 1/2 trial (NCT03163511) aimed to study VC-02 safety, tolerability, graft survival, and C-peptide production when implanted subcutaneously in T1D patients with hypoglycemia unawareness for up to 2 years. There are two cohorts in this study design: cohort 1 enrolled up to 15 patients total, and cohort 2 had an additional 47 subjects. Two different types of VC-02 units, a small sentinel unit (VC-02-20) and a larger dose-finding unit (VC-02-300), are implanted in each study subject. The sentinel unit is smaller in size (1.5 cm × 1 cm × 1 mm with 20 μl capacity) and was used to allow serial "sampling" via explant. It is loaded with 6 million PEC-01 cells (a negligible dose) intended not to significantly affect outcomes upon periodic removal but sufficient to reflect the histological characteristics of the larger units. It is explanted and examined periodically to evaluate the status of the dose-finding units (VC-02-300). The larger dose-finding units (VC-02-300, 9 cm × 3 cm × 1 mm with 300 μl capacity) are loaded with approximately 75 million PEC-01 cells and are used to develop the procedures and to assess the safety and efficacy of various cell doses to determine how many cells will eventually be needed for a potential fully therapeutic dose [44]. Up to ten VC-02-20 units (sentinels) were implanted in cohort 1 subjects. Cohort 2 subjects were implanted with up to 12 total units (sentinels). Of the 12 implanted units, no more than ten were VC-02-300 units and the remainder were VC-02-20 units. Each subject in cohort 1 received up to 4 dose-finding units. The first 35 patients implanted in cohort 2 received between 2 and 5 dose-finding units, with all subsequent patients receiving between 8 and 10 dose-finding units based on approval protocol amendments. The dose-finding units are implanted in the subcutaneous tissue of the anterior and lateral abdominal wall.

This trial is still in progress, and preliminary results from 15 cohort 1 patients and 17 cohort 2 patients in their first 12 months of the implant from a single-study center demonstrated that engraftment and insulin expression was observed in 63% of VC-02 units explanted from participates between 3 and 12 months postimplant. In total, 35.3% of patients demonstrated positive C-peptide as early as 6 months postimplant. Subjects who had units with substantial cell engraftment demonstrated increased glucose-responsive C-peptide levels and/or mixed meal-stimulated C-peptide secretion postimplant. Explanted grafts contained a majority of cells presenting an endocrine lineage with cells resembling mature islet cells, which were immunoreactive for insulin, islet amyloid polypeptide, and MAFA. Other ductal and rare exocrine cells were also present in a minority of the graft. Implants were well tolerated, with no teratoma formation or severe graft-related adverse events reported. The most reported adverse events were related to surgical implant or explant procedures (27.9%) or side effects of immunosuppression (33.7%). Protective immunosuppression-related transient increases in circulating regulatory T cells, PD1 high T cells, and IL17A$^+$ CD4$^+$ T cells were detected as well. These initial results suggest that pancreatic progenitor cells derived from pluripotent stem cells can engraft and further differentiate into pancreatic islet-like insulin-producing tissue in patients, achieving meal-regulated insulin secretion transplantation [44, 45]. However, despite the measurable C-peptide observed in 35% of responder subjects, the pilot study has not yet observed a measurable clinical benefit that was statistically meaningful. However, data presented from the ADA oral abstract in 2021, and the 2022 Endocrine Society demonstrated clinical benefit for certain patients. There is notable graft-to-graft variation within and between patients, and every graft explanted also contained substantial regions of non-cellular material close to the pores likely produced by the infiltrating host myofibroblasts, which may impair revascularization and was not observed in the fully encapsulated VC-01 explants [45]. While this trial shows that hESC-derived pancreatic progenitors can engraft into patients, the lack of consistent therapeutic response advocates for the need for further optimization of the device and/or administration method. Perhaps all that is needed is a larger number of hESC-derived pancreatic progenitor cells during implantation, as the inadequate insulin production could simply be due to an insufficient β-cell mass being achieved.

Utilizing immunosuppression drugs to protect grafts is a major barrier impeding the broader application of stem cell-derived β-cell application. To address these challenges, ViaCyte initiated efforts to test two different approaches. First, ViaCyte announced in August 2020 (https://viacyte.com/news-events) that a clinical phase agreement with Gore, a materials science company, aimed for the development of a modified version of the original PEC-Encap, which has the potential to eliminate the need for immunosuppression and address the foreign body response while still allowing vascularization.

5.2 *ViaCyte Inc. and CRISPR Therapeutics Collaboration Trials*

The second approach consists of two clinical trials that use gene editing technology to modify its cells for immune evasiveness and survival. In January 2022, CRISPR Therapeutics collaborating with ViaCyte sponsored a 6-month phase 1 multicenter trial (NCT05210530) aimed to study a combination product named VCTX210A in 10 T1D patients. The VCTX210A combination product (unit) consists of ViaCyte's proprietary allogeneic pancreatic endoderm cells (PEC210A) genetically modified using Cluster Regularly Interspaced Short Palindromic Repeats/CRISPR-associated protein 9 (CRISPR/Cas9) to promote immune evasiveness and survival, together with a durable, removable, and perforated device designed to deliver and retain PEC210A cells. Up to ten sentinel units of the combination product VCTX210A will be implanted in a participant. Its primary outcome is to measure the incidence of adverse events with causality related to VCTX210 A units and/or the surgical procedures required to implant and explant the VCTX210A units from implantation up to 6 months postimplantation. Secondary outcomes include innate and adaptive immune responses to VCTX210A units within the graft, new alloreactive and autorotative antibodies in patient blood, the percentage of viable graft cells per unit, and the percentage of graft cells per unit that have differentiated into endocrine β-cells up to 6 months postimplantation. This study dosed its first patient on February 2, 2022 [73], and is expected to be completed by April 2023.

In November 2022, CRISPR Therapeutics and ViaCyte sponsored an expanded 12-month phase 1/2 multicenter trial (NCT05565248) aimed to study a combination product named VCTX211 in 40 T1D patients. The trial design is the same as (NCT05210530), except for longer study duration, more patient enrollment, use of larger, dose-finding units, and utilizing a different genetically modified pancreatic endoderm cell named PEC211. The studies' primary outcome measurements are the same as VCTX210 but extend the study time and will additionally assess C-peptide increase from the baseline from implantation up to 12 months postimplantation. Secondary outcomes will additionally assess changes in exogenous insulin usage, the number of hypoglycemic events, and changes in hemoglobin A1C levels and evaluate the percentage of time in predefined glycemic ranges as measured by a continuous glucose monitor from baseline. The studies are currently ongoing.

5.3 *Vertex Pharmaceuticals Inc. Trial*

In March 2021, Vertex Pharmaceuticals Inc. initiated a phase 1/2 trial with a planned enrollment of 17 adult participants (NCT04786262), aimed to assess the safety, tolerability, and efficacy of VX-880 up to 5 years in participants with T1D with impaired hypoglycemic awareness and severe hypoglycemic events. VX-880 is an investigational allogeneic human embryonic stem cell-derived, fully differentiated, stage 7 (as shown in Fig. 1) insulin-producing islet-like clusters (SC-islet) manufactured using proprietary technology by Vertex. VX-880 SC-islets are injected into the portal vein, as is the conventional approach for primary islet transplantation done in the clinic with islets recovered from deceased donors. Primary outcomes are [1] to assess adverse events from VX-880 infusion to end of study (up to 5 years); [2] the proportion of participants free of severe hypoglycemic events; [3] either a glycosylated hemoglobin (HbA1c) <7.0% or a ≥1% reduction in HbA1c from baseline 1 year after VX-880 infusion. Secondary outcomes focus on the proportion of participants who are exogenous insulin-independent and improvements in stimulated C-peptide 1 year after VX-880 infusion. The study architecture is composed of three parts: A, B, and C, with 2, 5, and 10 participants each, respectively. The primary endpoint of part A is the safety and tolerability of VX-880, and participants only receive half of the targeted dose of cells as part of a dose escalation schema. In 2021, Dr. J. Markmann and his team at Massachusetts General Hospital supervised the infusion of islet cells to the first patient in the study sponsored by Vertex. Preliminary results of the part A participant demonstrated that VX-880 freed its first participant from the reliance on exogenous insulin. The exciting results were reported at multiple scientific meetings and covered in the *New York Times* on November 27, 2021, with the headline "A cure for type 1 diabetes? For one man, it seems to have worked."

Based on the information presented by Dr. J. Markmann et al. at the 82nd Annual American Diabetes Association Conference in June of 2022, to date, three participants have undergone transplantation, two in part A who have received half the target dose and one participant in part B who received the full target dose of islets (no detailed data as of yet). Participant 1 is a 64-year-old male with a 42-year history of T1D who had undetectable C-peptide at baseline with an A1c of 8.6% and an insulin requirement of 34 units per day. His disease was complicated by impaired awareness of hypoglycemia (IAH) with five severe hypoglycemic events (SHEs) in the year prior to VX-880 infusion. Participant 2 is a 35-year-old female with an over 10-year history of T1D who also had undetectable C-peptide at baseline with A1c of 7.5% and required about 26 units of insulin per day. Her disease was also complicated by IAH and with three SHEs in the year prior to VX-880 infusion. Both patients were infused with VX-880 at half the target dose. Participant 1, at day 90 post-transplant, had regained robust glucose-stimulated C-peptide production, which reached a peak of 560 pmol/L during the mixed meal tolerance test (MMTT) that was associated with markedly improved glucose disposal. At the day 180 visit, the MMTT showed maximum C-peptide levels of 1146 pmol/L with continuing improvements in glycemic control. Graft function improved even further at his most recent visit on day 270. Collectively, these MMTTs reveal remarkable recovery of physiologic regulation of glucose-stimulated insulin release. Participant 1 had a baseline A1c of 8.6% that steadily declined over time to 5.2% at visit day 270, well within the normal range. Patient 1 was receiving 34 units of insulin per day at baseline, which was reduced to approximately 17 units by day 29. Incredibly as of day 210, the patient did not require any exogenous insulin and was deemed by prespecified study criteria as insulin independent at study day 270. Continuous glucose monitoring (CGM) showed marked improvement in dynamic glycemic swing post-transplant. Baseline CGM showed marked swings, both high and low, with only a 40% time in the target glucose range of 70–180 pretransplant. In contrast, during the period between days 121 and 150, time in range improved dramatically to 81.4% of readings in the target range, and during days 241–270, the participant spent a remarkable 99.9% of time in range, with no insulin administered during this period. An October 2023 update from IPITA meeting revealed that this patient remains exogenous insulin-independent even after more than two years post-transplant.

Participant 2 at the day 90 visit MMTT revealed clear production of C-peptide, which reached a peak of 202 pmol/L, with corresponding improvements in glucose disposal, indicating glucose-responsive insulin production and improved glucose regulation as early as 3 months post VX-880 infusion. Patient 2 had a baseline HbA1c of 7.5% that declined to less than 7% by day 29 of the study, further decreasing to 6.4% by day 57. Participant 2 required 25.9 units of insulin at baseline, which dropped to 18.7 units per day by day 60 postinfusion and has remained stable through the day 150 study. Patient 2 also experienced significant improvements comparing the baseline CGM monitoring window with the day 121–150 interval. Time in range improved from a baseline of 35.9% to 51.9% by days 121 through 150 and was associated with a 30% decrease in insulin utilization.

No SHEs in either participant after day 35 post-transplant were reported. Participant 1 had two serious adverse events (SAEs): On day 4, a rash was reported and considered to be grade 1 in severity and related to antithymocyte globulin and enoxaparin. This led to a one-day delay in hospital discharge and was resolved on day 5. On day 186, the participant was hospitalized for dehydration. Both SAEs were resolved. Only one event was considered related or possibly related to VX-880 – participant 1 had a transient increase in liver function testing (LFT) post-VX-880 infusion that was non-serious. This is consistent with what is commonly seen after cadaveric islet infusion [36, 37]. All the adverse events (AEs) for participant 2 were mild or moderate in severity, none of which were related to VX-880. The most common AEs were headaches and hypomagnesemia. Overall, the safety profile of VX-880 was very favorable and generally consistent with the immunosuppressive regimen and the perioperative period common complications.

In summary, VX-880 infusion at half the target dose in these two participants with long-standing T1D, who had no endogenous insulin production, and had a history of IAH and SHEs, resulted in engraftment of the transplant. This led to glucose-regulated increases in both fasting and simulated C-peptide levels, elimination of recurrent severe hypoglycemia, improved glycemic control as demonstrated by improved HbA1c and time in range, and exogenous insulin independence for patient 1 and significant reductions in insulin requirements for patient 2. The safety profile of VX-880 is consistent with the immunosuppression regimen used in the study and the perioperative period. This is the first demonstration of a patient with T1D achieving insulin independence with stem-cell-derived islet cell therapy. Collectively, these data demonstrate the tremendous potential of fully differentiated stem cell-derived islets to benefit T1D patients with poor glycemic control, IAH, and SHEs.

5.4 Other Developmental Trials

In October 2019, Orgenesis Ltd. registered a trial (NCT04038931) to collect human liver biopsy and whole blood samples from total or partial pancreatectomy patients, with the aim to use the sample for potential use as an autologous source for insulin-producing cells in future clinical studies for T1D.

In May 2018, Sernova Corp launched a phase 1/2 trial with 13 patients (NCT03513939) to develop a novel implantable Cell Pouch™ that is transplanted with therapeutic cells such as insulin-producing islets. Upon implantation, the Cell Pouch is designed to form a natural environment rich in microvessels via a scaffold made of non-degradable polymers, formed into small cylindrical chambers which, when implanted against the abdominal muscle, become incorporated with vascularized tissue to the circumference of the device. The Cell Pouch does not protect against immune infiltration and/or rejection.

On March 9, 2023, Vertex announced the FDA's clearance of an IND application for its product, VX-264, which utilizes stem cell-derived, fully differentiated pancreatic islets, the same as its predecessor VX-880. However, VX-264 is equipped with a proprietary immunoprotective device that eliminates the need for immunosuppression after implantation. The clinical trial for VX-264 is a phase 1/2, single-arm, open-label study conducted in multiple sequential parts to assess the safety, tolerability, and efficacy of the treatment. The global clinical trial is expected to enroll approximately 17 patients and be initiated in the first half of 2023.

5.5 Patient Selection-Key Inclusion and Exclusion Criteria

All trials target T1D in adults ages 18–65 years old, but there are variations in patient selection in terms of the severity of the disease. Vertex (NCT04786262) enrolls subjects with a history of T1D with greater than 5 years of duration and at least two episodes of documented severe hypoglycemia in the 12 months prior to enrollment. CRISPR Therapeutics and ViaCyte collaboration trials (NCT05565248, NCT05210530), and ViaCyte trial (NCT04678557) require patients for a minimum of 3 years of history and preclude subjects with the occurrence of two or more severe unexplained hypoglycemic events within 6 months prior to enrollment and advanced complications associated with diabetes. All trials require patients to maintain stable diabetic treatment and the use of continuous glucose monitor (CGM) before screening and for the duration of the study. Other key exclusion criteria include prior islet cell transplant, organ transplant, gene or gene-edited cell therapy, and/or immunosuppressive therapy required for other preexisting diseases.

5.6 Transplant Site

In the first Vertex trial of VX-8880 requiring immunosuppression, hESCs-derived islets are delivered via infusion into the patient's hepatic portal vein, similar to the approach used for primary islets from deceased donors. This differs from ViaCyte, which relies on subcutaneous implantation within its designed chambers. Implantation of hESCs or hiPSCs carries the risk of teratoma or other malignant tumor formation [74–76]. For this reason, implantation of SC-derived islets should be performed so that the transplant is convenient to monitor and easily accessible should its removal be required. Simultaneously, the locale must have a capable vasculature to support graft survival.

The current approaches do not fulfill the desired option. The intraportal infusion approach suffers from an immediate over 50% loss of islet mass post-transplant when deceased donor islets are infused, progressive graft dysfunction over time, and precludes recovery of the graft should there be serious complications. Subcutaneous delivery has high clinical potential in terms of accessibility, convenience, and ease of retrieval, but the limited oxygenation and vascularization can seriously restrict graft survival and function [77]. A few preclinical NHP studies may provide insights to address these issues. Yu et al. reported that in an NHP study, long-term survival of islets is achieved when islets are mixed with a matrix, termed islet viability matrix, consisting of collagen, glutamine, fetal bovine serum, sodium bicarbonate, and medium and transplanted into the subcutaneous space [77]. Our group is currently exploring a strategy with injectable and fully degradable vasculogenic hydrogels to address these limitations by providing a delivering vehicle that generates a stable and functional local vascular bed in the subcutaneous space for transplant engraftment. Other efforts made to identify alternative sites such as the recent reports by Lei et al. of successful islet graft survival and glycemic control by intrapleural space transplant in a diabetic NHP model [78] and by bioengineering the omental site to achieve successful glycemic control in both diabetic NHPs and humans could address the transplant site as well [79–82].

5.7 Immunosuppression Protocols

In theory, hESCs insulin-producing cells, especially the stage 7 mature SC-islet used in the current Vertex clinical trials, might be less immunogenic than deceased donor primary human islets preparation due to the absence of a large exocrine component and the absence of antigen-presenting passenger leukocytes and endothelial cells. However, the cells are still allogenic, and immunosuppressants are needed to protect the implanted cells from allo- and auto-immunity. In Table 2 listing of trials, patients are treated with an immunosuppression regimen that is similar to the one administrated for cadaveric human islet transplants. This includes induction therapy using Thymoglobulin (ATG: up to 6 mg/kg i.v. over 5 days) and tumor necrosis factor-alpha blockade by Etanercept (50 mg i.v. preimplant followed by 25 mg subcutaneous x 3 doses out to 10 days postimplant; maintenance immunosuppression using Tacrolimus (target 10–12 ng/mL for the first 3 months; 7–9 ng/mL thereafter), and mycophenolate mofetil (MMF up to 2 g per day as tolerated) [45], while the Vertex trial used Sirolimus instead of MMF. Anticoagulants such as Enoxaparin or Heparin are administered transiently postcell infusion to prevent portal venous thrombosis. As to primary human islet transplantation, the need for chronic immunosuppression to protect the graft from rejection is a serious barrier to the current stem cell-derived islet trials as well. A few recent innovative tolerance approaches in nonhuman primate islet studies can be translational to SC-islet transplantation. Lei et al. reported that administration of apoptotic donor lymphocytes on the day of transplant combined with a regimen of induction treatment with ATG and anti-interleukin-6 receptor antibody followed by Rapamycin alone for the first 30 days post-transplant achieved prolonged immunosuppression free graft survival in a full-MHC mismatched diabetic NHP islet transplant model [83]. Singh et al. refined this approach by using two doses of apoptotic donor leukocytes infusion under an induction regimen of rapamycin, anti-CD40, and IL-6R antibodies, and soluble TNFR achieved 1-year immunosuppression free-islet survival in MHC II DRB allele-matched diabetic NHPs [84]. In 2022 Lei et al. also reported a local immunomodulation strategy whereby co-transplantation of islets and FasL-presenting biomaterial onto the omentum under transient (1 month) rapamycin monotherapy resulted in immunosuppression-free long-term graft survival in MHC mismatched diabetic NHPs [81]. Other exciting and innovative approaches are emerging in stem-cell islet transplant in murine as well. In 2022, Wang et al. reported that co-transplantation of protective MSCs engineered to express high levels of PDL1 and CTLA4 immunoglobulin fusion proteins together with islets under the kidney capsule resulted in sufficient local immunotolerance to allow immunosuppression-free allogeneic islet survival [85]. In 2023, Gerace et al. showed that targeting HLAs and PD-L1 along with genetically engineering SC-islet to secrete the cytokines IL-10, TGF-β, and IL-2 resulted in a tolerogenic local microenvironment by recruiting Tregs to the SC-islets when they are transplanted under the kidney capsule [86]. These approaches in murine could be translational if confirmed in NHPs and may offer alternative strategies to mitigate early inflammation and immune activation while conferring immuno-evasion and localized immune tolerance.

6 Preclinical Nonhuman Primate Pilot Studies

Most of the human stem cell-derived insulin-producing cells' efficacy and safety studies were carried out in immune-deficient murine models [87]. While the NHP study is thought to be the most robust model for translational preclinical study, it involves xeno immune responses for hPSCs, which are hard to conduct studies and may result in being less informative. As a result, one can make a strong argument to go directly from mice to humans for the hPSCs-β-cells since there is no proper allotransplant preclinical model to allow insight for clinical guidance. As a result, there is a dearth of safety and efficacy studies on human-like, nonhuman primate models before initiating a clinical trial, except for a short-term NHP study done by Vertex. At the 2019 International Society for Stem Cell Research (ISSCR) annual meeting, Vertex reported that embryonic stem cell-derived mature islets generated by its proprietary technology successfully engrafted in the liver and were functional over a period of 6 weeks with a 60% in reduction of insulin usage in a nonhuman primate model. Two recent studies demonstrated that human chemically induced pluripotent stem cell islets (hCiPSC-islets) transplanted either intraportal or beneath the abdominal anterior rectus sheath of diabetic NHPs at a dose of about 39,600 IEQ per kg on average, achieved effective and functional engraftment. The average exogenous insulin requirement was reduced by 49% 15 weeks after transplantation in the liver model and an average 44% reduction of HbA1c from baseline at 3 months after transplantation in the sub-anterior rectus sheath model [46, 48]. No teratoma or other malignancy was observed in all animals, including the three longest 180 days survivors. While it is exciting to show the safety and feasibility of SC-islet for diabetic treatment in human-like settings, insulin independence is not achieved in these studies. In addition, the study durations are too short for meaningful interpretation, and the number of samples is small with most animals terminated by 3 months post-transplant due to either study design, loss of graft function or death due to complications of immunosuppressive therapy to demonstrate study safety.

In our opinion, studies in larger cohorts with adjustment of SC-islet dose and long-term follow-up of transplants are needed to demonstrate hPSCs efficacy and safety in NHP models before SC-islet clinical trials. These stringent preclinical NHP model can also provide a few insights for further investigations on stem cell-derived islet transplants before clinical trials such as: (1) What are the kinetics of diabetes reversal in a human-like setting? (2) What degree of glycemic control can be achieved? (3) What cell dose is needed to achieve the full reversal of diabetes? (4) Is there a normal counter-regulatory hormone response, or is insulin secretion appropriately regulated with fasting? (5) Where is a more hospitable site for SC-islet transplant? (6) Can any biomarkers be used to monitor off-target growth such as teratoma formation or other tumors development? (7) What is the proper immunosuppression regimen, and does immunosuppression withdrawal lead to graft or resulting tumor rejection if it happens? (8) Does hPSC-β-cell islet elicit immunity in vivo? and (9) Does the transcriptional profile of the β-cells change over time post-transplant? In our opinion, not only do stem cell derived islets or β-cells require NHP studies, but also for an any given stem cell-derived product, NHP studies should be conducted prior to the initiation of a clinical trial. In addition, there is a need to develop an equivalent version of nonhuman primate stem cell-derived products for use in allogeneic NHP study settings to guide human stem cell-derived cell clinical trials in general; this would be particularly valuable in designing and testing gene edits that will promote stem cell-derived islets or new cells survival in the absence of immunosuppression.

7 Conclusions and Future Directions

7.1 *The Achievements of Stem Cell-Based Therapies for Diabetes*

Stem cell-based therapies in diabetes continue to evolve rapidly and have made impressive strides over the past two decades. Numerous hMSCs-based clinical trials for both T1D and T2D patients have been conducted, which resulted in inconclusive evidence that hMSCs can differentiate into functional mature β-cells or islet-like organoids to benefit T1D patients. However, in T2D, a systematic review confirmed that a significant increase in the levels of stimulated and fasting C-peptide and improvement of HbA1c values were observed, demonstrating hMSCs' ability to support islet health and survival in T2D patients. In 2014, the first hESCs-derived islet-like organoids were transplanted in T1D patients. This feat was surpassed by the remarkable step of transplanting highly differentiated hESCs into a patient in 2021, initiating Vertex's trial of an allogeneic embryonic stem-cell islet product termed VX-880. Despite receiving only half of the intended target dose for a cure, the transplant freed the first patient from reliance on exogenous insulin after a single infusion of half of the target dose of VX-880. Although a single case, this event marked a pivotal step in the application of stem cell-based insulin-producing cells application in diabetes management, indicating that an unlimited source of cells for β-cell replacement has finally become an actual reality. With these proofs of principle successes, a wide range of stringent human-like nonhuman primate setting tests and clinical trials are now necessary to investigate all aspects of this field to move this therapy forward.

7.2 *A Debate of hESCs Versus hiPSCs*

Before diving into a deeper discussion about future directions, the first question to ask is which type of PSCs will be the likely focus in the near future. There is an ongoing debate over hESCs-derived universal cells reservoir versus hiPSCs-derived autologous personalized cells for their clinical applications. The universal cell approach has advantages such as easier government regulated scale-up production with standard cGMP to yield consistent quality to avoid line and clone variations and an off-the-shelf product to ship across the globe, but it has the major hurdles of allogenic and auto immunities that need to be addressed with approaches such as immuno-isolation, cell genetic engineering, immune suppression and conditioning are needed. The iPSCs-derived patient-specific approach is conceptually more attractive because it is isogenic, having no alloimmune reactivity and fewer ethical concerns. However, recurrent autoimmunity still needs to be addressed and will likely require cell engineering or recipient conditioning. Also, per-patient manufacturing of cells is time-consuming, complex and costly, harder to achieve consistent product quality. Furthermore, a number of other barriers need to be addressed for the patient-specific approach as well, such as (1) their genomic stability of iPSCs is still questionable [88]; (2) risks of oncogenic transformation in residual undifferentiated hiPSCs are safety concerns [89]; (3) their ability to differentiate into mature pancreatic endocrine cells has not yet reached the same quality observed with ESC protocols [90]. In the near foreseeable future, allogenic stem cell-derived β-cells transplant may still be the dominant modality for potential T1D application.

7.3 Immune Protection

Efforts from multiple fronts are ongoing to address the graft immune rejection barrier. Recent efforts have focused on genetic engineering or genome editing approaches that generate insulin-producing cells capable of evading immune rejection [85, 86, 91, 92]. However, these approaches run the risks of generating cells that cannot be detected and or eliminated by the host immune monitoring system should the graft become infected by a virus or should oncogenic transformation occur. On this front, the potential tolerance inductions or drug-eluting scaffolds to modulate the immune reactions may address many alloimmunity hurdles as discussed earlier [81, 83, 84]. Alternatively, immuno-isolation has long been explored as a method to protect the graft from rejection, with significant efforts made to design a variety of types of either microencapsulation or macroencapsulation strategies. While these encapsulation approaches are able to confer a certain degree of immune protection, they are generally transient due to a vigorous foreign body reaction [87, 93–96], including such difficulties as were encountered in ViaCyte trials discussed above.

7.4 Other Considerations

For future trials, not only does the functional property and safety of the insulin-producing cells or organoids need to be extensively studied, but many aspects related to this approach as the followings warrant further intensive investigation (1) What is the optimal immunosuppression or immune isolation or evasion strategies? (2) What is the therapeutic cell dose for a curative effect? (3) Where is the best hospitable transplant site? (4) Efforts to understand the SC-islets' cell composition and study the cell mixtures' quality required to achieve optimal metabolic control are crucial. What is the best composite of endocrine cell types in the cluster? Should it mimic primary islets in terms of percentile of alpha, beta, and delta cells, or enriched to a higher percentile of beta cells or even enhanced efficacy of cells to allow lower minimal efficacious dose? (5) A detailed evaluation of how the implanted endocrine cells evolve in vivo over time would provide insights for further optimization of stem cell-derived β-cell applications. (6) Which stage of cells is the right cells to implant, stage 4 (proliferative) versus stage 7 (mature) as defined in Fig. 1? (7) How to ensure safety, including genomic stability and eliminate harmful off-target growth should be a major effort. To address the risk of teratoma formation due to contaminating undifferentiated stem cells, a strategy that uses CRISPR-Cas9 for in-frame insertion of a suicide gene, iC9, into the endogenous SOX2 locus in ESC can specifically eradicate undifferentiated cells without affecting differentiated cells may be one of the potential solutions to address this concern [97, 98] (8) Are preclinical nonhuman primate investigations needed, including the development of an equivalent version of a nonhuman primate stem cell-derived product for use in allogeneic NHP studies to guide human stem cell-derived cells for clinical application? (9) Can technologies such as robotics, artificial intelligence, and big data analyses expedite the discovery and improve the manufacturing and efficacy of SC-β-cell therapies? The rich experiences gained with primary human islet transplantation are instructive for stem cell-based insulin-producing cell therapy, and many preclinical studies can be translational for its application in T1D as well. The following 5 years of study will likely reveal the answers to many of these questions.Compliance with Ethical Standards

Conflict of Interest Ji Lei declares that he has no conflicts of interest. James Markmann declares that he is on the Scientific Advisory Boards of iTolerance Inc.; eGenesis Inc.; QihanBio Inc.; and a Consultant of Vertex Pharmaceuticals.

Human and Animal Rights and Informed Consent The unpublished data contained in this chapter are from studies with informed consent by Massachusetts General Hospital IRB and IACUC.

References

1. Cho NH, Shaw JE, Karuranga S, et al. IDF Diabetes Atlas: Global estimates of diabetes prevalence for 2017 and projections for 2045. *Diabetes research and clinical practice.* 2018;138:271–281.
2. Bommer C, Sagalova V, Heesemann E, et al. Global Economic Burden of Diabetes in Adults: Projections From 2015 to 2030. *Diabetes care.* 2018;41(5):963–970.
3. Riddle MC, Herman WH. The Cost of Diabetes Care-An Elephant in the Room. *Diabetes care.* 2018;41(5):929–932.
4. National Diabetes Statistics Report, 2020. Estimates of diabetes and its burden in the United States. *https://wwwcdcgov/diabetes/data/statistics/statistics-reporthtml.* 2020.
5. Type 1 Diabetes Facts 2020. JDRF. 2020; https://www.jdrf.org/t1d-resources/about/facts/.
6. White JR, Jr. A Brief History of the Development of Diabetes Medications. *Diabetes Spectr.* 2014;27(2):82–86.
7. Hartman I. Insulin analogs: impact on treatment success, satisfaction, quality of life, and adherence. *Clin Med Res.* 2008;6(2):54–67.
8. Sims EK, Carr ALJ, Oram RA, DiMeglio LA, Evans-Molina C. 100 years of insulin: celebrating the past, present and future of diabetes therapy. *Nat Med.* 2021;27(7):1154–1164.
9. Schuetz C, Markmann JF. Islet cell transplant: Update on current clinical trials. *Curr Transplant Rep.* 2016;3(3):254–263.

10. Latres E, Finan DA, Greenstein JL, Kowalski A, Kieffer TJ. Navigating Two Roads to Glucose Normalization in Diabetes: Automated Insulin Delivery Devices and Cell Therapy. *Cell metabolism.* 2019;29(3):545–563.
11. Schuetz C, Markmann JF. Immunogenicity of beta-cells for autologous transplantation in type 1 diabetes. *Pharmacological research: the official journal of the Italian Pharmacological Society.* 2015;98:60–68.
12. Blonde L. Benefits and risks for intensive glycemic control in patients with diabetes mellitus. *Am J Med Sci.* 2012;343(1):17–20.
13. Bonner-Weir S, Baxter LA, Schuppin GT, Smith FE. A second pathway for regeneration of adult exocrine and endocrine pancreas. A possible recapitulation of embryonic development. *Diabetes.* 1993;42(12):1715–1720.
14. Xu G, Stoffers DA, Habener JF, Bonner-Weir S. Exendin-4 stimulates both beta-cell replication and neogenesis, resulting in increased beta-cell mass and improved glucose tolerance in diabetic rats. *Diabetes.* 1999;48(12):2270–2276.
15. Kobayashi T, Yamaguchi T, Hamanaka S, et al. Generation of rat pancreas in mouse by interspecific blastocyst injection of pluripotent stem cells. *Cell.* 2010;142(5):787–799.
16. Yamaguchi T, Sato H, Kato-Itoh M, et al. Interspecies organogenesis generates autologous functional islets. *Nature.* 2017;542(7640):191–196.
17. Bhatt S, Fung JJ, Lu L, Qian S. Tolerance-inducing strategies in islet transplantation. *Int J Endocrinol.* 2012;2012:396524.
18. Shapiro J. Eighty years after insulin: parallels with modern islet transplantation. *CMAJ.* 2002;167(12):1398–1400.
19. Ricordi C, Strom TB. Clinical islet transplantation: advances and immunological challenges. *Nature reviews Immunology.* 2004;4(4):259–268.
20. Sutherland DE, Matas AJ, Najarian JS. Pancreas and islet transplantation. *World journal of surgery.* 1977;2(1):185–195.
21. Ricordi C, Lacy PE, Scharp DW. Automated islet isolation from human pancreas. *Diabetes.* 1989;38 Suppl 1:140–142.
22. Lacy PE, Kostianovsky M. Method for the isolation of intact islets of Langerhans from the rat pancreas. *Diabetes.* 1967;16(1):35–39.
23. Reckard CR, Ziegler MM, Barker CF. Physiological and immunological consequences of transplanting isolated pancreatic islets. *Surgery.* 1973;74(1):91–99.
24. Kelly WD, Lillehei RC, Merkel FK, Idezuki Y, Goetz FC. Allotransplantation of the pancreas and duodenum along with the kidney in diabetic nephropathy. *Surgery.* 1967;61(6):827–837.
25. Tzakis AG, Ricordi C, Alejandro R, et al. Pancreatic islet transplantation after upper abdominal exenteration and liver replacement. *Lancet.* 1990;336(8712):402–405.
26. Warnock GL, Thompson DM, Meloche RM, et al. A multi-year analysis of islet transplantation compared with intensive medical therapy on progression of complications in type 1 diabetes. *Transplantation.* 2008;86(12):1762–1766.
27. Leitao CB, Tharavanij T, Cure P, et al. Restoration of hypoglycemia awareness after islet transplantation. *Diabetes care.* 2008;31(11):2113–2115.
28. Barton FB, Rickels MR, Alejandro R, et al. Improvement in outcomes of clinical islet transplantation: 1999–2010. *Diabetes care.* 2012;35(7):1436–1445.
29. Ryan EA, Paty BW, Senior PA, Shapiro AM. Risks and side effects of islet transplantation. *Current diabetes reports.* 2004;4(4):304–309.
30. Niclauss N, Bosco D, Morel P, Giovannoni L, Berney T, Parnaud G. Rapamycin impairs proliferation of transplanted islet beta cells. *Transplantation.* 2011;91(7):714–722.
31. Shapiro AM, Lakey JR, Ryan EA, et al. Islet transplantation in seven patients with type 1 diabetes mellitus using a glucocorticoid-free immunosuppressive regimen. *N Engl J Med.* 2000;343(4):230–238.
32. Shapiro AM, Ricordi C, Hering BJ, et al. International trial of the Edmonton protocol for islet transplantation. *N Engl J Med.* 2006;355(13):1318–1330.
33. Marfil-Garza BA, Imes S, Verhoeff K, et al. Pancreatic islet transplantation in type 1 diabetes: 20-year experience from a single-centre cohort in Canada. *Lancet Diabetes Endocrinol.* 2022;10(7):519–532.

34. Brennan DC, Kopetskie HA, Sayre PH, et al. Long-Term Follow-Up of the Edmonton Protocol of Islet Transplantation in the United States. *Am J Transplant.* 2016;16(2):509–517.
35. Hering BJ, Ballou CM, Bellin MD, et al. Factors associated with favourable 5 year outcomes in islet transplant alone recipients with type 1 diabetes complicated by severe hypoglycaemia in the Collaborative Islet Transplant Registry. *Diabetologia.* 2023;66(1):163–173.
36. Hering BJ, Clarke WR, Bridges ND, et al. Phase 3 Trial of Transplantation of Human Islets in Type 1 Diabetes Complicated by Severe Hypoglycemia. *Diabetes care.* 2016;39(7):1230–1240.
37. Markmann JF, Rickels MR, Eggerman TL, et al. Phase 3 trial of human islet-after-kidney transplantation in type 1 diabetes. *Am J Transplant.* 2021;21(4):1477–1492.
38. Ng NH, Tan W, Koh YX, Teo A. Human Islet Isolation and Distribution Efforts for Clinical and Basic Research. *OBM Transplantation.* 2019;3(2):31.
39. Crisan M, Yap S, Casteilla L, et al. A perivascular origin for mesenchymal stem cells in multiple human organs. *Cell Stem Cell.* 2008;3(3):301–313.
40. Jin HJ, Bae YK, Kim M, et al. Comparative analysis of human mesenchymal stem cells from bone marrow, adipose tissue, and umbilical cord blood as sources of cell therapy. *International journal of molecular sciences.* 2013;14(9):17986–18001.
41. Kern S, Eichler H, Stoeve J, Kluter H, Bieback K. Comparative analysis of mesenchymal stem cells from bone marrow, umbilical cord blood, or adipose tissue. *Stem cells.* 2006;24(5):1294–1301.
42. Thomson JA, Itskovitz-Eldor J, Shapiro SS, et al. Embryonic stem cell lines derived from human blastocysts. *Science.* 1998;282(5391):1145–1147.
43. Takahashi K, Yamanaka S. Induction of pluripotent stem cells from mouse embryonic and adult fibroblast cultures by defined factors. *Cell.* 2006;126(4):663–676.
44. Ramzy A, Thompson DM, Ward-Hartstonge KA, et al. Implanted pluripotent stem-cell-derived pancreatic endoderm cells secrete glucose-responsive C-peptide in patients with type 1 diabetes. *Cell Stem Cell.* 2021;28(12):2047–2061 e2045.
45. Shapiro AMJ, Thompson D, Donner TW, et al. Insulin expression and C-peptide in type 1 diabetes subjects implanted with stem cell-derived pancreatic endoderm cells in an encapsulation device. *Cell Rep Med.* 2021;2(12):100466.
46. Du Y, Liang Z, Wang S, et al. Human pluripotent stem-cell-derived islets ameliorate diabetes in non-human primates. *Nat Med.* 2022.
47. Shapiro AMJ, Verhoeff K. A spectacular year for islet and stem cell transplantation. *Nat Rev Endocrinol.* 2022.
48. Liang Z, Sun D, Lu S, et al. Implantation underneath the abdominal anterior rectus sheath enables effective and functional engraftment of stem-cell-derived islets. *Nat Metab.* 2023.
49. Pullen LC. Stem cell-derived islets take a leap toward patients. *Am J Transplant.* 2022;22(3):677–678.
50. Sui L, Leibel RL, Egli D. Pancreatic Beta Cell Differentiation From Human Pluripotent Stem Cells. *Curr Protoc Hum Genet.* 2018;99(1):e68.
51. Song N, Scholtemeijer M, Shah K. Mesenchymal Stem Cell Immunomodulation: Mechanisms and Therapeutic Potential. *Trends in pharmacological sciences.* 2020;41(9):653–664.
52. Guo QS, Zhu MY, Wang L, et al. Combined transfection of the three transcriptional factors, PDX-1, NeuroD1, and MafA, causes differentiation of bone marrow mesenchymal stem cells into insulin-producing cells. *Experimental diabetes research.* 2012;2012:672013.
53. Shivakumar SB, Lee HJ, Son YB, et al. In vitro differentiation of single donor derived human dental mesenchymal stem cells into pancreatic beta cell-like cells. *Biosci Rep.* 2019;39(5).
54. Wu LF, Wang NN, Liu YS, Wei X. Differentiation of Wharton's jelly primitive stromal cells into insulin-producing cells in comparison with bone marrow mesenchymal stem cells. *Tissue engineering Part A.* 2009;15(10):2865–2873.
55. Phadnis SM, Joglekar MV, Dalvi MP, et al. Human bone marrow-derived mesenchymal cells differentiate and mature into endocrine pancreatic lineage in vivo. *Cytotherapy.* 2011;13(3):279–293.
56. Hansson M, Tonning A, Frandsen U, et al. Artifactual insulin release from differentiated embryonic stem cells. *Diabetes.* 2004;53(10):2603–2609.

57. Lechner A, Yang YG, Blacken RA, Wang L, Nolan AL, Habener JF. No evidence for significant transdifferentiation of bone marrow into pancreatic beta-cells in vivo. 2004;53:616–623.
58. Choi JB, Uchino H, Azuma K, et al. Little evidence of transdifferentiation of bone marrow-derived cells into pancreatic beta cells. *Diabetologia.* 2003;46(10):1366–1374.
59. Ezquer F, Ezquer M, Contador D, Ricca M, Simon V, Conget P. The antidiabetic effect of mesenchymal stem cells is unrelated to their transdifferentiation potential but to their capability to restore Th1/Th2 balance and to modify the pancreatic microenvironment. *Stem cells.* 2012;30(8):1664–1674.
60. Dave SD, Vanikar AV, Trivedi HL, Thakkar UG, Gopal SC, Chandra T. Novel therapy for insulin-dependent diabetes mellitus: infusion of in vitrogenerated insulin-secreting cells. *Clin Exp Med.* 2015;15(1):41–45.
61. Thakkar UG, Trivedi HL, Vanikar AV, et al. Insulin-secreting adiposederived mesenchymal stromal cells with bone marrow-derived hematopoietic stem cells from autologous and allogenic sources for type 1 diabetes mellitus. *Cytotherapy.* 2015;17(7):940–947.
62. Mesples A, Majeed N, Zhang Y, Hu X. Early immunotherapy using autologous adult stem cells reversed the effect of anti-pancreatic islets in recently diagnosed type 1 diabetes mellitus: preliminary results. *Medical science monitor: international medical journal of experimental and clinical research.* 2013;19:852–857.
63. Carlsson PO, Schwarcz E, Korsgren O, Le Blanc K. Preserved b-cell function in type 1 diabetes by mesenchymal stromal cells. *Diabetes.* 2015;64(2):587– 592.
64. Gan J, Wang Y, Zhou X. Stem cell transplantation for the treatment of patients with type 1 diabetes mellitus: A meta-analysis. *Experimental and therapeutic medicine.* 2018;16(6):4479–4492.
65. Hwang G, Jeong H, Yang HK, Kim HS, Hong H, Kim NJ. Efficacies of Stem Cell Therapies for Functional Improvement of the b Cell in Patients with Diabetes: A Systematic Review of Controlled Clinical Trials. *Int J Stem Cells.* 2019;12(2):195–205.
66. Bhansali A, Upreti V, Khandelwal N, Marwaha N, Gupta V, Sachdeva N. Efficacy of autologous bone marrow-derived stem cell transplantation in patients with type 2 diabetes mellitus. *Stem cells and development.* 2009;18(10):1407– 1416.
67. Jiang R, Han Z, Zhuo G, Qu X, Li X, Wang X. Transplantation of placenta-derived mesenchymal stem cells in type 2 diabetes: a pilot study. *Front Med.* 2011;5(1):94–100.
68. Bhansali A, Asokumar P, Walia R, et al. Efficacy and safety of autologous bone marrow-derived stem cell transplantation in patients with type 2 diabetes mellitus: a randomized placebo-controlled study. *Cell transplantation.* 2014;23(9):1075–1085.
69. Bhansali S, Dutta P, Kumar V, et al. Efficacy of Autologous Bone Marrow-Derived Mesenchymal Stem Cell and Mononuclear Cell Transplantation in Type 2 Diabetes Mellitus: A Randomized, Placebo-Controlled Comparative Study. *Stem cells and development.* 2017;26(7):471–481.
70. Kroon E, Martinson LA, Kadoya K, et al. Pancreatic endoderm derived from human embryonic stem cells generates glucose-responsive insulin-secreting cells in vivo. *Nature biotechnology.* 2008;26(4):443–452.
71. Schulz TC. Concise Review: Manufacturing of Pancreatic Endoderm Cells for Clinical Trials in Type 1 Diabetes. *Stem cells translational medicine.* 2015;4(8):927–931.
72. Henry RR, Pettus J, Wilensky JON, Shapiro AMJ, Senior PA, Roep B. Initial Clinical Evaluation of VC-01TM Combination Product—A Stem Cell–Derived Islet Replacement for Type 1 Diabetes (T1D. *Diabetes.* 2018;67(Supplement 1):138.
73. CRISPR Therapeutics and ViaCyte, Inc. Announce First Patient Dosed in Phase 1 Clinical Trial of Novel Gene-Edited Cell Replacement Therapy for Treatment of Type 1 Diabetes (T1D) [press release]. GLOBE NEWSWIRE: CRISPR Therapeutics and ViaCyte, Inc., Feb 2 2022.
74. Han L, He H, Yang Y, et al. Distinctive Clinical and Pathologic Features of Immature Teratomas Arising from Induced Pluripotent Stem Cell-Derived Beta Cell Injection in a Diabetes Patient. *Stem cells and development.* 2022;31(5–6):97–101.
75. Fujikawa T, Oh SH, Pi L, Hatch HM, Shupe T, Petersen BE. Teratoma formation leads to failure of treatment for type I diabetes using embryonic stem cell-derived insulin-producing cells. *The American journal of pathology.* 2005;166(6):1781–1791.

76. Knoepfler PS. Deconstructing stem cell tumorigenicity: a roadmap to safe regenerative medicine. *Stem cells.* 2009;27(5):1050–1056.
77. Yu M, Agarwal D, Korutla L, et al. Islet transplantation in the subcutaneous space achieves long-term euglycaemia in preclinical models of type 1 diabetes. *Nat Metab.* 2020;2(10):1013–1020.
78. Lei J, Zhang A, Deng H, et al. Intrapleural transplantation of allogeneic pancreatic islets achieves glycemic control in a diabetic non-human primate. *Am J Transplant.* 2021.
79. Berman DM, O'Neil JJ, Coffey LC, et al. Long-term survival of nonhuman primate islets implanted in an omental pouch on a biodegradable scaffold. *Am J Transplant.* 2009;9(1):91–104.
80. Baidal DA, Ricordi C, Berman DM, et al. Bioengineering of an Intraabdominal Endocrine Pancreas. *N Engl J Med.* 2017;376(19):1887–1889.
81. Lei J, Coronel MM, Yolcu ES, et al. FasL microgels induce immune acceptance of islet allografts in nonhuman primates. *Sci Adv.* 2022;8(19):eabm9881.
82. Deng H, Zhang A, Pang DRR, et al. Bioengineered omental transplant site promotes pancreatic islet allografts survival in non-human primates. *Cell Rep Med.* 2023:100959.
83. Lei J, Kim JI, Shi S, et al. Pilot Study Evaluating Regulatory T Cell-Promoting Immunosuppression and Nonimmunogenic Donor Antigen Delivery in a Nonhuman Primate Islet Allotransplantation Model. *Am J Transplant.* 2015.
84. Singh A, Ramachandran S, Graham ML, et al. Long-term tolerance of islet allografts in nonhuman primates induced by apoptotic donor leukocytes. *Nature communications.* 2019;10(1):3495.
85. Wang X, Wang K, Yu M, et al. Engineered immunomodulatory accessory cells improve experimental allogeneic islet transplantation without immunosuppression. *Sci Adv.* 2022;8(29):eabn0071.
86. Gerace D, Zhou Q, Kenty JH, et al. Engineering human stem cell-derived islets to evade immune rejection and promote localized immune tolerance. *Cell Rep Med.* 2023;4(1):100879.
87. Vegas AJ, Veiseh O, Gurtler M, et al. Long-term glycemic control using polymer-encapsulated human stem cell-derived beta cells in immune-competent mice. *Nat Med.* 2016;22(3):306–311.
88. Gerace D, Martiniello-Wilks R, Habib R, Ren B, Nassif NT, O'Brien BA. Ex Vivo Expansion of Murine MSC Impairs Transcription FactorInduced Differentiation into Pancreatic b-Cells. *Stem Cells Int.* 2019:2019 1395301.
89. Lee AS, Tang C, Rao MS, Weissman IL, Wu JC. Tumorigenicity as a clinical hurdle for pluripotent stem cell therapies. *Nat Med.* 2013;19(8):998–1004.
90. Kondo Y, Toyoda T, Inagaki N, Osafune K. iPSC technology-based regenerative therapy for diabetes. *J Diabetes Investig.* 2018;9(2):234–243.
91. Koga K, Wang B, Kaneko S. Current status and future perspectives of HLA-edited induced pluripotent stem cells. *Inflamm Regen.* 2020;40:23.
92. Lee J, Sheen JH, Lim O, et al. Abrogation of HLA surface expression using CRISPR/Cas9 genome editing: a step toward universal T cell therapy. *Scientific reports.* 2020;10(1):17753.
93. Bochenek MA, Veiseh O, Vegas AJ, et al. Alginate encapsulation as long-term immune protection of allogeneic pancreatic islet cells transplanted into the omental bursa of macaques. *Nat Biomed Eng.* 2018;2(11):810–821.
94. Sremac M, Lei J, Penson MFE, et al. Preliminary Studies of the Impact of CXCL12 on the Foreign Body Reaction to Pancreatic Islets Microencapsulated in Alginate in Nonhuman Primates. *Transplant Direct.* 2019;5(5):e447.
95. Stock AA, Manzoli V, De Toni T, et al. Conformal Coating of Stem Cell-Derived Islets for beta Cell Replacement in Type 1 Diabetes. *Stem Cell Reports.* 2020;14(1):91–104.
96. Wang T. Successful diabetes management without immunosuppressivedrugs in NHP model has been demonstrated. Encapsulation system with taperednanopore conduits achieved normal glycaemia with regulated insulin release. *Artif Cells Nanomed Biotechnol.* 2018;46(sup3):S1162–S1168.
97. Wu Y, Chang T, Long Y, Huang H, Kandeel F, Yee JK. Using Gene Editing to Establish a Safeguard System for Pluripotent Stem-Cell-Based Therapies. *iScience.* 2019;22:409–422.
98. Yagyu S, Hoyos V, Del Bufalo F, Brenner MK. An Inducible Caspase-9 Suicide Gene to Improve the Safety of Therapy Using Human Induced Pluripotent Stem Cells. *Molecular therapy: the journal of the American Society of Gene Therapy.* 2015;23(9):1475–1485.

Modelling of Beta Cell Pathophysiology Using Stem Cell-Derived Islets

Tom Barsby, Hossam Montaser, Väinö Lithovius, Hazem Ibrahim, Eliisa Vähäkangas, Sachin Muralidharan, Vikash Chandra, Jonna Saarimäki-Vire, and Timo Otonkoski

1 Introduction

Insulin is a hormone produced by pancreatic beta cells that regulates blood glucose homeostasis. Beta cells sense circulating glucose levels and constantly adjust their insulin secretion to meet the physiological demand. A family of related metabolic disorders, termed diabetes mellitus (DM), arise from the loss or dysfunction of beta cells, causing an imbalance between insulin requirement and secretion, resulting in chronic hyperglycaemia [95].

DM encompasses a heterogeneous pool of clinical phenotypes that greatly differ in the age of disease onset, the mechanism of beta-cell loss, and the genetic factors that contribute to the disease. The vast majority of DM cases are classified into either type 1 (T1D) or type 2 (T2D), which both share the common feature of loss of pancreatic beta-cell mass (and therefore insulin secretion), but differ in the pathophysiology and timing of disease progression. T1D manifests through autoimmune-mediated destruction of beta cells, with clinical presentation of the disease typically appearing in childhood [5]. Conversely, T2D pathologies manifest later in life, incorporate more heterogeneous phenotypes, and are driven through mechanisms of beta-cell stress and exhaustion related to insulin resistance [16]. In both cases, genetic factors influence disease outcomes.

In rare cases, DM presents within the neonatal period, either as permanent or transient neonatal diabetes (PNDM and TNDM). Neonatal diabetes is invariably caused by mutations within single genes that play a central role in pancreatic development and beta-cell formation [41]. More commonly, monogenic diabetes

T. Barsby · H. Montaser · V. Lithovius · H. Ibrahim · E. Vähäkangas · S. Muralidharan · V. Chandra · J. Saarimäki-Vire · T. Otonkoski (✉)
Stem Cells and Metabolism Research Program, Faculty of Medicine, University of Helsinki, Helsinki, Finland
e-mail: timo.otonkoski@helsinki.fi

L. Piemonti et al. (eds.), *Pluripotent Stem Cell Therapy for Diabetes*,
https://doi.org/10.1007/978-3-031-41943-0_24

presents later, in older children or young adults, as maturity onset diabetes of the young (MODY). MODY is primarily caused by dominant mutations in *HNF1A, HNF1B, HNF4A,* or *GCK* genes [109]. Although monogenic forms of diabetes are rare, their identification has significantly improved our understanding of beta-cell development, physiology, and function. Most importantly, the correct diagnosis of monogenic diabetes is important for the choice of optimal treatment strategies [138].

To gain further insights into the pathogenesis of specific forms of diabetes, pancreatic islets (and beta cells in particular) would need to be directly studied. This is challenging, because human pancreatic tissue (especially from diabetic patients) is not readily available for research. Animal models, particularly mice and rats, have been extensively used in research of T1D/T2D and monogenic forms of diabetes. However, it has become increasingly evident that both the physiology of the endocrine pancreas and the pathogenetic mechanisms leading to diabetes are markedly different between rodents and humans [15, 82]. Alternatively, immortalized human beta-cell lines, such as EndoC-bH1, have been widely used in modelling of beta-cell pathologies [98, 125]. Although they are derived from human beta cells and are able to maintain fairly reasonable glucose stimulated insulin secretion responses in vitro, their low clonal efficiency and aneuploidy nature may limit their use in genetic studies [11].

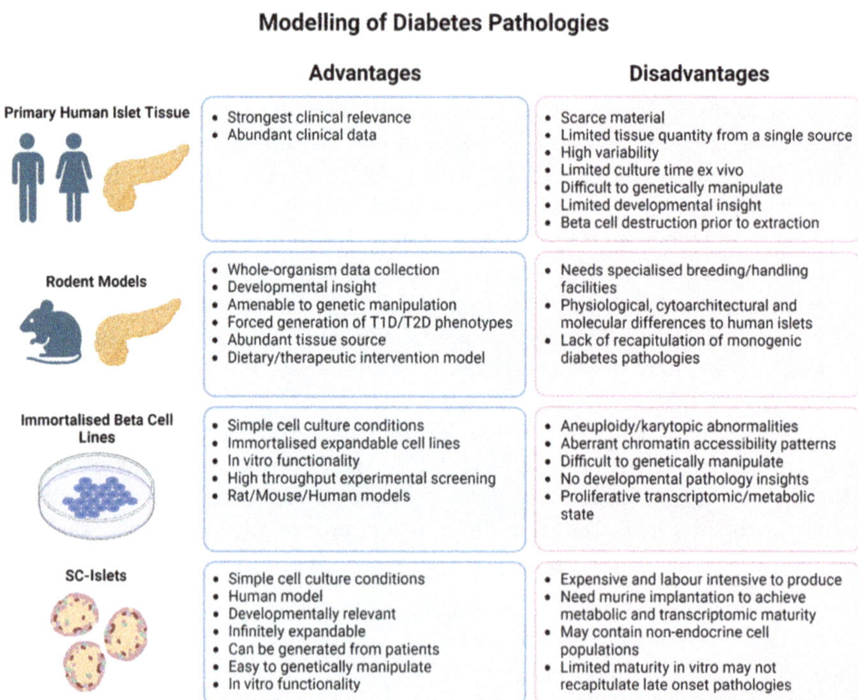

Fig. 1 Models of beta-cell biology and diabetes pathologies. An overview of the related advantages and disadvantages present in current beta-cell modelling systems

Stem cell-derived islets (SC-islets) provide many advantages over these other disease-modelling options (Fig. 1). For example, they can be generated in unlimited numbers from individuals carrying specific disease-associated genotypes and can be efficiently manipulated genetically. Furthermore, SC-islets allow biological processes to be studied throughout development which is not possible with any other human cell model. In this chapter, we review the current possibilities of SC-islet-based modelling of beta-cell pathophysiology and envision how this may impact the development of new therapeutic approaches for rare and common forms of diabetes.

2 Possibilities and Limitations of SC-Islets

Over the last two decades, numerous protocols for generating functional beta cells from pluripotent stem cells (PSCs) have been developed, leveraging the tremendous body of knowledge from developmental biology. Several growth factors and small molecules, temporally activating or inhibiting key pathways, are used to mimic in vivo pancreatic development. The first milestone towards endocrine differentiation in vitro was the development of a robust protocol for the induction of definitive endoderm (DE) [29]. Soon after, several laboratories described protocols for generating pancreatic progenitors and hormone-producing endocrine cells (SC-islets) [91, 99]. However, these cells were mainly polyhormonal and displayed poor glucose-responsive insulin secretion. Recently, more optimized protocols have succeeded in generating monohormonal endocrine cell populations, with glucose-responsiveness similar to primary human islets [8, 126]. Intriguingly, these in vitro SC-derived islets still possess certain metabolic and transcriptomic differences compared to their primary human analogues [8]. Even so, these next generation protocols for the derivation of SC-islets still represent a powerful tool by which to dissect human islet organogenesis and endocrine cell differentiation, as well as specific genetic forms of beta-cell dysfunction that will be discussed in more detail below [12].

One of the major advantages of the SC-islet model is the possibility for effective genome editing of the stem cells (Fig. 1). Advances in genome editing tools have enabled selective targeting of specific genes by which to generate knock-out models, correct/introduce specific mutations, activate/inactivate specific genes, and allow for the production of transgenic reporter lines [9]. Such genetically modified models have been successfully used within SC-islet based modelling of monogenic diabetes and will be expanded on in later sections. Some examples of extremely useful reporter lines used within SC-islet studies include the MEL-INS GFP PSC line, which has been used to purify the beta cells within SC-islet cultures [85], as well as form the basis of a dual reporter system by which to study metabolic enzymes within SC-islet-derived beta cells [47]. This also includes reporter lines that contain fluorescently labelled beta cell-associated transcription factors (such as MNX1), to allow the real-time visualization of developmental gene networks [118]. More functional aspects of beta-cell metabolism have also recently been undertaken using

PSC-derived SC-islet reporter lines that carry fluorescent calcium sensitive probes as a proxy for real-time insulin release [89].

3 Monogenic Beta-Cell Defects

Monogenic beta-cell diseases represent a logical starting point for the application of SC-islet-based diabetes modelling. Due to each pathology being derived from a single-gene mutation allows for both the modelling of aberrant beta-cell development within stem cell-derived endocrine tissue and the opportunity to correct each mutation within the same experimental model. The vast majority of genes responsible for monogenic diabetes have a major role in the islet, and beta cells in particular (Fig. 2). Many of the same genes, the best example being the K-ATP channel genes (*ABCC8, KCNJ11*), can cause either diabetes or congenital hyperinsulinism (CHI), depending on the functional consequence of the mutation (Fig. 2, Tables 1 and 2).

It is also worth noting that the division between clinical PNDM and MODY classifications and the underlying causative mutations are not consistently distinct. Therefore, identical mutations may elicit differences in the age of onset or severity of the disease phenotype. While an argument could be made that the lines between monogenic diabetes classifications based on age of onset are an artefact of clinical presentation, in this chapter, we will retain current nomenclature of PNDM and MODY-related phenotypes as a way of separating and collating research data as it has been classified in the literature. However, as ongoing research utilizes more

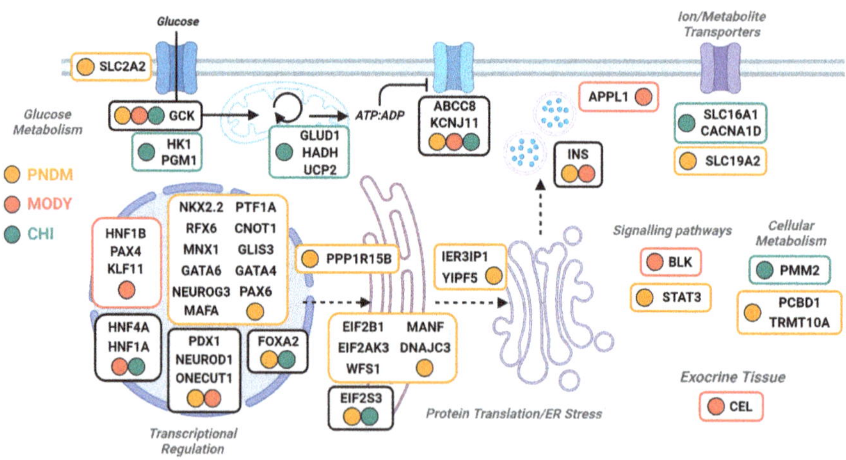

Fig. 2 Genes associated with monogenic forms of diabetes. Schematic of monogenic diabetes causative mutations and localization/function of the expressed proteins. Genes that can result in multiple clinical disease outcomes are grouped accordingly

Table 1 PNDM-related genes and associated SC-Islet studies

PNDM gene	Stem cell study	Related pathologies
ABCC8 (SUR1)	None	MODY, CHI
ARX	Zhu et al. [140]	
CNOT1	None	
DNAJC3	Lytrivi et al. [68]	
EIF2AK3 (PERK)	None	
EIF2B1	None	
EIF2S3	None	CHI
FOXA2	Elsayed et al. [35]	CHI
GATA4	Shi et al. [106]	
GATA6	Shi et al. [106]; Tiyaboonchai et al. [122]	
GCK	Aqel et al. [4]	MODY, CHI
GLIS3	Zhu et al. [140]; Amin et al. [3]	
HES1	Zhu et al. [140]	
KCNJ11	Zeng et al. [137]	MODY, CHI
IER3IP1	None	
INS	Balboa et al. [7]; Ma et al. [70]	MODY
MANF	Montaser et al. [84]	
MNX1	Zhu et al. [140]	
NEUROD1	Romer et al. [100]	MODY
NEUROG3	McGrath et al. [77]; Zhu et al. [140]	
NKX2–2	None	
ONECUT1	Philippi et al. [94]	MODY
PAX6	None	
PCBD1	None	
PDX1	Wang et al. [130–132]; Zhu et al. [140]	MODY
PPP1R15B	None	
PTF1A	Zhu et al. [140]; Nair and Odorico [86]; Miguel-Escalada et al. [78]	
RFX6	Zhu et al. [140]; Trott et al. [124]	
SLC19A2	None	
STAT3	Saarimäki-Vire et al. [101]	
TRMT10A	Cosentino et al. [26]	
WFS1	Maxwell et al. [75]; Shang et al. [105]; Kitamura et al. [59]	
YIPF5	De Franco et al. [40]	

refined genomic association screening techniques, it may in the future be more fruitful to apply the term 'monogenic diabetes' to all early onset causative mutations and resultant pathologies.

Table 2 MODY-related genes and associated SC-Islet studies

Gene (MODY type)	Stem cell study	PSC generation	Further characterization
HNF4A (MODY1)	Carrasco et al. [19]	+	+
	Ghila et al. [43]	+	+
	Ng et al. [88]	+	+
	Braverman-Gross et al. [14]	+	+
	Stoffel and Duncan [115]	+	+
	Teo et al. [119]	+	–
GCK (MODY2)	Teo et al. [119]	+	–
	Aqel et al. [4]	+	–
HNF1A (MODY3)	Cujba et al. [28]	+	+
	Low et al. [66]	+	+
	Cardenas-Diaz et al. [18]	+	+
	Yabe et al. [135]	+	+
	Griscelli et al. [50]	+	–
	González et al. [47]	+	+
	Hermann et al. [55]	+	+
	Teo et al. [119]	+	–
PDX1 (MODY4)	Wang et al. [131]	+	–
	Wang et al. [130]	+	–
	Wang et al. [132]	+	+
HNF1B (MODY5)	El-Khairi et al. [34]	+	+
	Teo et al. [120]	+	+
	Yabe et al. [134]	+	+
	Teo et al. [119]	+	–
NEUROD1 (MODY6)	None	–	–
KLF11 (MODY7)	None	–	–
CEL (MODY8)	Pellegrini et al. [93]	+	+
	Teo et al. [119]	+	–
PAX4 (MODY9)	None	–	–
INS (MODY10)	None	–	–
BLK (MODY11)	None	–	–
ABCC8 (MODY12)	None	–	–
KCNJ11 (MODY13)	Griscelli et al. [49]	+	–
APPL1 (MODY14)	None	–	–

4 Neonatal Diabetes Associated with Developmental Defects

Monogenic mutations that result in insulin deficiency early in the neonatal period, or even in utero, are likely to represent pathologies with the strongest direct genetic effects. These defects can be classified based on the level of pathology, affecting either the development, survival, or secretory function of the beta cells. There is therefore already a large interest in applying SC-islet technology to the modelling of these early onset diabetes pathologies (Table 1) [9, 13, 15]. Pancreatic agenesis or hypoplasia are the most obvious indications of severely arrested pancreatic development. Mutations in at least seven genes encoding transcription factors have been shown to cause pancreatic agenesis in humans (*PDX1, PTF1A, HNF1B, RFX6, GATA6, GATA4,* and *CNOT1*), while at least five transcription factor genes have been linked with the failure of beta-cell development without pancreatic agenesis (*GLIS3, NEUROD1, NEUROG3, NKX2.2,* and *MNX1*) [41].

In a pioneering study, many of these genes were inactivated within a stem cell model of beta-cell development [140], with the expected clear phenotype of arrested islet cell development found in SC-islets carrying mutations within *PDX1* and *RFX6*. These gene knockout models are also in agreement with patient-derived SC-islet models which show that homozygous point mutations within the *PDX1* and *RFX6* genes impede early pancreatic cell differentiation, mirroring key aspects of the clinical presentation of pancreatic agenesis leading to PNDM [124, 132]. SC-islet models of *GATA6* mutations also demonstrate a defect in early pancreatic progenitor formation, and show gene dosage effects and interactions with *GATA4* (another PNDM-associated gene) [106]. This finding highlights that stem cell-based models are a viable method for tracking the interplay of genes and genetic pathways that lead to PNDM pathologies. Early PNDM-related phenotypes were also recorded in SC-islets containing homozygous and heterozygous mutations in *FOXA2* and *ONECUT1* genes, leading to impeded pancreatic and endocrine progenitor formation [35, 94]. Again, reinforcing that the strengths of PNDM-related phenotypes seen in vitro are related to the severity of each mutation as well as aspects of gene dosage present within SC-islet models, as is the case in patients.

Less severe PNDM pathologies that present without pancreatic agenesis have also successfully been recapitulated in current SC-islet models. These include patient-derived SC-islets harbouring homozygous *TRMT10A* mutations which resulted in heightened oxidative stress markers and apoptosis of differentiated beta cells with limited effects on endocrine cell composition [26]. Conversely, the transcription factor *NEUROD1* was shown to be essential for efficient endocrine cell differentiation within an SC-islet model, but with no measurable alteration in proliferation or cell survival [100]. Gene knockout of the key pro-endocrine transcription factor *NEUROG3* in SC-islets completely abolished endocrine progenitor formation, with the authors further showing that only a small percentage of *NEUROG3* transcription was enough to rescue this phenotype [77]. Genes associated with more mature beta-cell function, such as the K_{ATP} channel subunit gene *KCNJ11*, has also been shown to display aberrant beta-cell phenotypes within SC-islet models, although intriguingly the type of mutation used within this study would imply a congenital hyperinsulinism phenotype [137].

However, it should be noted that not all PNDM-related gene mutations display measurable phenotypes within SC-islet models. Homozygous knockout of *PTF1A*, *MNX1,* or *GLIS3* within a beta-cell differentiation model did not uncover any defects in endocrine cell formation [140]. However, studies utilizing overexpression of *PTF1A* within a stem cell-based model did uncover enhanced foregut progenitor differentiation, suggesting a developmentally early effect of *PTF1A* function [86].

A finding echoed in a recent study of *PTF1A*-related enhancer regions in models of pancreatic progenitor patterning [78].

Another possible confounding factor in SC-islet models of PNDM may be the exogenous factors utilized within the in vitro differentiation process acting to obscure or modify the effects of underlying disease-causing mutations. An example of this is seen in SC-islet studies of *GLIS3* knockouts which despite not displaying a measurable phenotype in one study [140], did demonstrate a pancreatic progenitor phenotype when the protocol was modified slightly [3]. This is also true of *GATA6* mutation modelling, which although did show a disease-related phenotype in one study [106], the mutation phenotype could be enhanced through modulation of the beta-cell differentiation conditions [122]. In agreement with this, activating mutations within the *STAT3* gene only displayed an endocrine progenitor phenotype through specific modulation of early pancreatic differentiation protocol steps [101]. Tailoring particular steps of the SC-islet differentiation protocol (both 'early' or 'late' stages) may therefore be necessary to discern particular disease phenotypes as an overly strong differentiation strategy may mask or override subtle effects of developmental gene mutations. Even in the complete absence of *NEUROG3*, insulin positive cells were able to be differentiated in vitro (albeit in low numbers) demonstrating that highly refined differentiation systems can act to overcome even key developmental control points in beta-cell generation [140]. Such modifications must be considered in future modelling applications of SC-islets where developmentally early effects or weaker disease-related phenotypes are expected [42].

5 Neonatal Diabetes Associated with Beta-Cell Stress Mechanisms

The endoplasmic reticulum (ER) is a central organelle responsible for protein folding and quality control [96]. Beta cells possess a highly developed ER in order to cope with the dynamic demand of insulin production in response to elevated glycaemic levels [39]. However, increased stress due to accumulation of misfolded proteins or decreased folding capacity of the ER leads to beta-cell dysfunction and apoptosis [25]. Recent technological advances in SC-islets have enabled the modelling of several PNDM gene mutations that are related to this pathway and have revealed the essential role of ER homeostasis in beta-cell function and survival (Table 1).

Insulin missense mutations are the second most common cause of PNDM [33]. Insulin production comprises a complex series of molecular biosynthetic events that begins with pre-pro-insulin translocation into the ER through its signal peptide sequence [133]. Mutations in this ER signalling peptide (such as L13R) have been reported to cause PNDM [57]. After the cleavage of the signal sequence, proinsulin is folded to its native state by the formation of three intramolecular disulphide bonds which are critical for the subsequent insulin processing steps. Mutations which alter proinsulin folding lead to retention of the misfolded protein in the ER resulting in elevated ER stress. Interestingly, 60% of the pathogenic *INS* mutations are due to the disruption of disulphide bridge formation through the addition of another cysteine to the protein (such as R89C) or abolishing the bond formation by removing one of the native cysteine residues (such as the Akita model C96Y and the Munich model C95S) [33]. SC-islet modelling has shown that *INS*^{C96Y} mutations lead to accumulation of misfolded proinsulin in the ER, resulting in elevated ER stress and reduced beta-cell proliferation [7]. After implantation under the kidney capsule of immunocompromised mice, the INS mutant SC-islet grafts displayed impaired insulin secretion, further elevated ER stress levels, reduced beta-cell size, and altered mTORC signalling and mitochondrial respiration. This indicates that this INS mutation does not only alter beta-cell function, but also affects beta-cell development and mass expansion, at least through impaired mTORC activity. *INS* mutations that abolish insulin formation altogether have also been modelled within SC-islets, although this phenotype does not directly tie to ER functionality [70].

Other common causes of beta-cell dysfunction in PNDM are due to mutations in genes necessary for mediating accurate protein folding and maintenance of the unfolded protein response (UPR) within the ER [25]. One such gene is *WFS1*, which encodes an essential component of the UPR and is a regulator of ER homeostasis [38]. Mutations within this gene lead to Wolfram syndrome and early onset diabetes. SC-islets derived from diabetic patients with *WFS*1 mutations showed elevated ER stress and reduced insulin content [105]. scRNA-seq analyses of stem cell-derived pancreatic progenitors demonstrated that *WFS1* mutations lead to misdirection of the cell fate choice from pancreatic endocrine cells to exocrine or non-pancreatic lineages [75]. Additionally, *WFS1* mutant-beta cells showed decreased insulin content and elevated levels of ER stress markers, mitochondrial stress, and apoptosis [75, 105]. Another ER-resident protein implicated in childhood-onset diabetes is mesencephalic astrocyte-derived neurotrophic factor (MANF), which modulates UPR signalling and ER stress. As seen in the *WFS1* mutation studies, a *MANF* knockout SC-islet model showed that beta cells displayed a trend towards elevated ER stress, and a significant decrease in insulin content. After implantation, MANF-KO SC-islet grafts demonstrated severely impaired functionality and elevated ER stress [84]. The more severe phenotype observed after murine implantation, which exposed the SC-islets to a more metabolically challenging environment, is in agreement with the relatively late onset of diabetes observed in patients. Furthermore, an iPSC model derived from diabetic patient with *DNAJC3* mutations, which is a heat shock protein essential for promoting translational recovery in later phases of the UPR, showed increased beta-cell death by triggering mitochondrial apoptosis via the pro-apoptotic Bcl-2 proteins BIM and PUMA [68]. This highlights the crosstalk between the ER and mitochondria, and the necessity of an intact UPR to maintain the integrity of other organelles.

After proinsulin folding in the ER, it is transported to the Golgi for further processing and packaging into insulin granules. Vesicle trafficking between ER and Golgi is facilitated by an array of proteins such as COPII and COPI which mediate forward and retrograde trafficking, respectively. Disruption of ER-to-Golgi trafficking has been reported to cause at least 10 different neurological disorders [129], and has also been implicated in multiple diabetes pathologies. A recent study reported mutations in *YIPF5* as a new cause of neonatal diabetes, alongside microcephaly and epilepsy, due to disrupted ER-to-Golgi trafficking [40]. SC-islets derived from a patient carrying a YIPF5^{I98S} mutation showed a moderate phenotype of increased apoptosis upon chemical induction of ER stress. However, a stronger phenotype was observed in a *YIPF5* knockout SC-islet model which demonstrated increased proinsulin accumulation in the ER, elevated ER stress markers, and reduced insulin content. Furthermore, murine implanted YIPF5-KO and YIPF5^{I98S} SC-islets displayed reduced C-peptide secretion and beta-cell number, which was more severe in the YIPF5-KO grafts. This cleanly demonstrates the ability of SC-derived diabetes models to recapitulate the phenotypic spectrum of various mutations affecting ER function.

In addition to modelling monogenic forms of ER stress leading to PNDM, SC-islets offer a promising platform for individualized drug screening and discovery. As an example, SC-islets carrying *WFS1* pathogenic variant mutations have been used to test the therapeutic potential of chemical chaperones, such as 4-phenylbutyric acid (4-PBA) and Tauroursodeoxycholic acid (TUDCA). Current studies show that addition of 4-PBA alleviated the detrimental effect induced by *WFS1* mutations in SC-islets by increasing their insulin content and reducing ER stress [59, 105]. Further testing of other promising therapeutic compounds such as recombinant human MANF and oestrogen remain to be explored in the SC-derived islets platform.

In summary, utilizing human stem cell models and the current advances in genome editing shows a great potential in recapitulating PNDM patients' phenotype, and enables us to dissect the mechanistic aspects of stress-induced beta-cell failure. It also opens the venue for targeted drug screening and the discovery of new beta-cell protective agents. Additionally, it will enable the modelling of several other ER stress-related mutations, such as the mutations in the PERK branch genes: *EIF2AK3* [103], *EIF2S3* [110], and *PPP1R15B* [1]; and the ER-to-Golgi trafficking protein IER3IP1 [104], which will deepen our understanding of insulin synthesis and beta-cell homeostasis.

6 SC-Islet Models of Diabetes Presenting After the Neonatal Period (MODY)

Maturity onset diabetes of the young (MODY) is an established term to define monogenic diabetes presenting at young age and resulting from autosomal-dominant mutations that affect pancreatic development or function [109]. It is not possible to categorically discriminate between MODY and neonatal diabetes, since many MODY genes are also causative in neonatal diabetes, such as *GCK*, *PDX1*, *HNF1B*,

NEUROD1, *INS*, *ABCC8*, and *KCNJ11*, depending on the type and severity of the mutation (Fig. 2, Tables 1 and 2).

MODY is thought to account for 1–5% of all diabetes cases worldwide [15] and presently encompasses mutations within 14 genes (Table 2). However, the vast majority of identified cases (approximately 99%) are linked to mutations in just four of these genes: *HNF4A* (MODY1), *GCK* (MODY2), *HNF1A* (MODY3), and *HNF1B* (MODY5) [109]. It is worth noting that the current MODY gene list is not exhaustive and through ongoing genetic screening efforts other forms of MODY are being proposed, as well as more in depth profiling of global patterns and biases of causative mutations [83].

Due to the similar age of onset, many MODY cases are currently misdiagnosed as T1D [138], which restricts the application of more targeted insulin-alternative therapies, as well as limiting the pool of patient material available for phenotypic characterization. Studies in MODY pathologies have also been impeded by the difficulties in recapitulating the diabetic phenotypes in rodent models [10]. SC-islets offer a unique solution to this problem by allowing in depth molecular and functional characterization of MODY mutations in a reproducible human islet model. That being said, the later clinical onset of MODY pathologies is indicative of a less severe phenotype than is seen in the PNDM models discussed above, which may cause challenges for SC-islet modelling. Even so, valuable data on the molecular mechanisms of MODY mutations within pancreatic endocrine development have already been demonstrated in SC-islets (Table 2).

A very early study utilizing an embryonic stem cell (ESC) model of a mutation within the transcription factor *HNF4A* (MODY1) was able to discern a downregulation of glycolysis-related genes at the developmentally early endodermal stage of differentiation [115], findings that were echoed in a more recent induced pluripotent stem cell (iPSC) model that found that 3D encapsulation of MODY1 SC-islets amplified the mutation-causing phenotype of aberrantly expressed glycolytic genes, albeit differing ones to those seen within the endodermal model [19]. Other findings within stem cell-based models have implicated *HNF4A* mutations in the alteration of available HNF4A binding sites, miRNA patterns regulating cell cycle arrest and proliferation, as well as the direct regulation of genes relevant to pancreatic and hepatic differentiation [14, 43, 88]. Stem cell-based models have therefore in principle shown that multiple core aspects of MODY1 pathology can be studied in vitro, and that the phenotype is one that has far reaching consequences for endocrine differentiation, progenitor proliferation, and mature beta-cell function, in agreement with current animal models of HNF4A function [81, 109].

Despite being the second most prevalent form of MODY [109] and having the highest reported number of causal variants in the current literature [97], almost no stem cell-based models of *GCK*-MODY (MODY2) exist. This is perhaps explained by the quite mild clinical presentation of the MODY2 phenotype which generally results in modest hyperglycaemia that does not require pharmacological intervention [138]. The only currently available stem cell-based studies of MODY2 mutations are focused on iPSC generation from patients, with limited or no phenotypic profiling [4, 119]. Hopefully, the advent of improved stem cell-based beta-cell functional modelling will shed further light on the metabolic mechanism of glucose homeostasis set-points within the beta cell and how they relate directly to the function of the GCK enzyme.

Mutations within the *HNF1A* (MODY3) gene constitute the most prevalent form of MODY worldwide [109]. As such, the majority of current stem cell-based models of MODY are focused on *HNF1A* mutations (Table 2). These span from the simple generation of edited or patient-derived PSC lines [50, 114, 119] through to more in-depth pancreatic progenitor and beta-cell profiling [18, 28, 47, 55, 66, 135]. Outcomes from these studies implicate a multifaceted effect of *HNF1A* mutations ranging from the reduced formation of pancreatic progenitors during stem cell differentiation, a loss of alpha cell gene repression, reduced glucose transporter expression/glucose uptake, and even alterations in long non-coding RNA (lncRNA) patterns linked to mitochondrial function. Intriguingly, mutant *HNF1A* also reduced the expression of another MODY gene; *HNF1B* (MODY5), with overexpression of *HNF1B* partially rescuing the MODY3 phenotype [28]. A finding that reinforces potential core overlapping features of many MODY pathologies. Further studies of MODY3 mutations in SC-islets also uncovered defects in glucose-sensitive insulin secretion, insulin granule morphology, and calcium-related exocytotic machinery genes, the knockout of which recapitulated elements of the MODY3 phenotype [47]. Intriguingly, a recent study also demonstrated a hyper-secretory phenotype in mutant *HNF1A* SC-islets, uncovering the possible overlapping mechanisms of this MODY mutation seen within congenital hyperinsulinism (CHI) and pre-diabetic beta-cell states [55]. Stem cell models of *HNF1A* mutations therefore bring into focus elements of dysfunction that occur early on in pancreatic lineage specification as well as functional elements in more mature endocrine cells.

HNF1B (MODY5) mutations represent the last of the ‘big 4’ MODY genes that represent the vast majority of clinical cases. Stem cell-based modelling studies of MODY5 have shown both positive and negative findings. An iPSC model harbouring an R177X mutation did not alter pancreatic progenitor markers compared to healthy controls [134]. Whereas S148L mutant and heterozygous knockout models did report alterations in pancreatic progenitor proliferation and marker gene expression profiles [34, 120]. Intriguingly, it was suggested that these shifts in progenitor gene expression patterns may act to compensate for MODY5-related developmental pancreatic hypoplasia, and explain why *HNF1B* mutations result in primarily a MODY clinical presentation, rather than PNDM [120].

Outside of these core MODY genes, very few stem cell-based models appear in the literature. The exceptions primarily relate to the generation of mutation-harbouring PSC lines, as is the case for *PDX1* (MODY4) [130, 131] and *KCNJ11* gain-of-function (MODY13) [49] mutants. The utilization of the PDX1 point mutations in PSCs and SC-islets revealed a defect in beta-cell differentiation and function, tied to the downregulation of PDX1-driven expression of genes such as *MEG3*, *NNAT*, and *MNX1* (a gene associated with PNDM phenotypes) [132]. A recent study also modelled *CEL* (MODY8) mutations within SC-islets, and found that in the absence of exocrine tissue, no alterations in endocrine cell phenotypes were observed [93].

In summary, the modelling of MODY phenotypes within stem cell-based models shows great promise, but is as of yet still in its infancy. Ongoing and future research utilizing more robust islet differentiation protocols will enable more stable and reproducible cell sources from which to dissect less severe developmental phenotypes, as is seen in many MODY pathologies. These breakthroughs will not only offer a greater understanding in the principles of beta-cell development, maturation, and dysfunction, but also act as models for more targeted detection and therapies for MODY patients.

7 SC-Islet Models of Congenital Hyperinsulinism (CHI)

In addition to modelling defects related to monogenic and common forms of diabetes, SC-islets can also be used to model monogenic congenital hyperinsulinism (CHI). CHI is a genetic disorder of the beta cells, characterized by inappropriately high insulin secretion. Excessive insulin secretion can lead to life-threatening hypoglycaemia and cause damage in the developing brain, especially since hyperinsulinemia inhibits the production of ketone bodies, depriving the central nervous system of both its primary and secondary fuel.

At least 15 genes have been clinically identified as causative for CHI [53], but still around 20–30% of CHI patients remain without a definitive genetic diagnosis [73, 111]. These genes have variable functions, which in turn leads to slightly varied clinical phenotypes. These genes can be broadly grouped into three categories based on their proposed function.

The most common form of CHI (40–60% of all cases) [58, 73, 111], termed K_{ATP}HI, is caused by homozygous loss-of-function [121] or heterozygous-dominant negative mutations [56] in either of the genes encoding the subunits of the K_{ATP}-channel, *ABCC8* and *KCNJ11*. Closing of the K_{ATP}-channel is the critical trigger of insulin secretion that links the increase in ATP/ADP-ratio upon elevated blood glucose to membrane depolarization, calcium influx, and insulin secretion. Loss of K_{ATP}-channel function causes excessive activation of this triggering pathway of insulin secretion, as the K_{ATP}-channel cannot hyperpolarize the beta cells as it would in basal glucose conditions.

Currently, K_{ATP}HI is the only form of CHI that has been modelled using SC-islets [52, 65]. In both studies, homozygous loss-of-function of *ABCC8* was studied in an isogenic setting. The differentiated K_{ATP}HI SC-islets demonstrated two to three-fold higher insulin secretion in low glucose than healthy controls and failed to respond to K_{ATP}-channel acting pharmaceuticals. Following implantation and maturation under the kidney capsule of immunocompromised mice, the *ABCC8*-mutant SC-islet grafts secreted seven-fold higher levels of human C-peptide, and caused a 38% reduction in fasting blood glucose, than control SC-islet grafts [65]. K_{ATP}HI SC-islets were therefore able to replicate the cardinal insulin secretory phenotype of K_{ATP}HI both in vitro and in vivo. Additionally, it was shown that K_{ATP}-channel inactivation biased the development of endocrine cells towards beta-cell lineages (at the expense of alpha cells), possibly due to an increased proliferation rate of *ABCC8*-mutant beta-like cells. Other defects, that conceivably act on a similar mechanism as K_{ATP}HI, that is, by increasing the activity of the triggering pathway, include gain-of-function mutations in *HK1* [128] and *GCK* [45]. Specifically, these two defects cause increased flux of glucose metabolism leading to closure of K_{ATP}-channels in basal glucose concentrations. CHI-causing mutations identified in *CACNA1D* [37], encoding a voltage gated Ca^{2+}-channel in beta cells, would conceivably also act through increasing activity of the triggering pathway.

The second group of genes linked to CHI are ones which encode enzymes and transporters involved in metabolic pathways normally suppressed or inactive in beta cells. Examples of this group include promoter region mutations of *SLC16A1* which lead to the inappropriate expression of the disallowed pyruvate transporter MCT-1. This enables circulating lactate to enter beta cells, which is then metabolized, resulting in HI and hypoglycaemia during strenuous exercise and prolonged fasting [90]. Another example is activating mutations in *GLUD1,* encoding glutamate dehydrogenase (GDH), where glutamate-derived alpha-ketoglutarate production and in turn ATP production are increased, again leading to hyperinsulinism through aberrant activation of the triggering pathway [113]. This is mirrored in loss-of-function mutations in *HADH* result in a loss of interaction between GDH and HADH, increasing the activity of GDH and causing CHI via the same mechanism [24]. Other CHI-genes in this group with predicted function in metabolism include *PGM1* [117] and *UCP2* [46], although the exact mechanism between their mutations and increased insulin secretion remains poorly understood.

The third group of CHI genes are those encoding beta cell-related transcription factors and protein translation machinery. These include HNF4A [36] and *HNF1A* [112], which are also identified as MODY genes (Table 1, Fig. 2), as well as *FOXA2* [44] and *EIF2S3* [48], which can also lead to PNDM pathologies. Mutations in *PMM2*, involved in N-glycosylation, have also been linked to CHI [17]. The exact molecular mechanisms connecting these mutations to insulin hypersecretion remain largely unexplored. However, there is a clear general concept of hyper-excitability of the beta cell as the primary defect in early life, with eventual development of impaired insulin secretion and diabetes later in life.

As outlined above, some clinical and experimental insight exists for the molecular mechanism of the commonest forms of CHI, but much remains to be uncovered. Since scarcity of patient material remains a critical limiting factor in pursuing these mechanistic studies, SC-islets serve as a particularly enticing modelling system for the study of CHI. Studying these mechanisms could reveal precise targets for the development of CHI therapeutics and further model how specific mutations within core beta-cell developmental genes act to generate CHI, PNDM, and MODY pathologies.

8 In Vivo SC-Islet Models

Some of the limitations of SC-islet-based modelling of monogenic pathologies are related to a retained transcriptional and metabolic immaturity of the derived endocrine cells [8]. Mature human beta cells do not only demonstrate elements of beta-cell identity (monohormonal insulin expression, expression of beta-cell typical transcription factors, and expression of insulin release machinery), but also display other maturation-related factors [12]. These include glucose-stimulated insulin secretion (GSIS) at physiologically relevant glucose concentrations, glucose-responsive mitochondrial oxidative metabolism, functional metabolic amplification pathways, insensitivity to pyruvate/lactate-stimulated insulin release, and the absence of disallowed gene expression. Several lines of evidence suggest that the SC-islet cells do indeed mature further after they have been implanted in vivo to a status more closely resembling primary islet tissue [6, 8, 27]. The most common method of SC-islet implantation is into the renal subcapsular space of immunodeficient mice. Implanted human SC-islets, which then mature to levels beyond those obtainable in vitro, allow for the modelling of monogenic diabetes pathologies that may manifest later in development or are contingent on the expression of more mature beta-like phenotypes before the pathology can manifest, some of which have been described above.

For efficient engraftment and maturation, SC-islets require interaction with surrounding cells and tissues such as vasculature, neurons, and immune cells, as is the case with primary islets [127]. In addition, the typical in vivo islet niche is one surrounded by exocrine pancreatic tissue and current research suggests a beneficial effect of interaction between the two tissue types [87]. Recent studies have shown that SC-islet function post-implantation can be greatly augmented through the co-implantation of vascular tissue [2] or embedded in hydrogels containing vascular endothelial cells [89]. The signalling interplay between endocrine tissue and the surrounding niche can work both ways, by which islet-derived signals (such as insulin, VEGFs, and ephrins) can trigger neovascularization [87], whereas signals from the endothelial cells and pericytes (such as BMP4), can have positive effects on SC-islet maturation [102].

Although the subrenal capsule implantation site is most commonly used SC-islet-based humanized mouse model, other implantation sites have been tested. These include subcutaneous, intramuscular, omental, and hepatic implantation sites. Even the ocular anterior chamber has been utilized as an implantation site for intravital imaging [64, 76, 127, 139]. Each site has distinct advantages and disadvantages, and further research will uncover which elements within each implantation site show the greatest benefit for implanted SC-islet maturation.

As has been mentioned above, these ‘humanized’ SC-islet mouse models are especially useful in modelling monogenic pathologies that may not be captured in the developmentally immature SC-islets in vitro. Following implantation, SC-islets are subjected to dynamic shifts in blood glucose levels as well as an array of systemic hormonal and nutritional signalling. Although it is currently unclear that the relative effect of particular elements of the in vivo environment are on SC-islets, it allows for the development of much stronger disease-related phenotypes, as has been seen in studies of monogenic mutations within the insulin gene, *MANF* and *YIPF5* [7, 40, 84].

Models of type 1 diabetes (T1D) can also be generated within mouse models, either through the genetic or chemical ablation of mouse beta cells [30, 63, 107]. Within these models, the implantation of SC-islets would act as a model of T1D-cell therapy and can be used to predict cell dosage as well as possible side effects of the engraftment process. These include studies on how SC-islet grafts may need to be genetically altered to avoid immune rejection in long-term recipients [108, 136].

In summary, the use of SC-islets within murine models is helping to shed light on the processes governing beta-cell maturation, as well as aiding the modelling of later onset diabetes pathologies that are currently impossible in vitro. Furthermore, the use of T1D mouse models can help in the refinement of clinical engraftment processes and the development of novel therapeutics for diabetic patients.

9 Functional Validation of Genetic Variants Associated with Polygenic Diabetes

Type 1 diabetes (T1D) manifests through a specific autoimmune attack on beta cells by cytotoxic $CD8^+$ T cells [5]. Many of the genes associated with T1D have a known function in beta cells as well as T cells, raising the notion that beta cells may play an active role in the disease progression and are not solely a passive factor [51, 80, 116]. So far, there are more than 60 risk loci associated with ≈80% of T1D heritability [60]. These genetic risk factors require sophisticated models to allow the investigation of their impact on the pathophysiology of T1D. Since the seminal studies reporting successful generation of SC-islets from T1D patient material [79], the concept of modelling the more complex polygenic and multi-tissue nature of T1D with SC-islets has become more prevalent [61]. Indeed even alone SC-islets have been used to model the effects of pro-inflammatory cytokines in a model akin to an autoimmune response seen in T1D [31].

Since there is more than one cell type involved in developing T1D, SC-islets provide an outstanding model for studying the crosstalk between the generated islet cells and isolated T-cells derived from the same T1D donor in an autologous setting. Interestingly, T-cells cultured with autologous SC-islets from T1D donors did not show activation nor beta-cell cytotoxicity unless endoplasmic reticulum stress was induced chemically [62]. This indicates the increasing appreciation of the active role of beta cells and their stress response in regulating their autoimmune vulnerability. Genetically engineered SC-islets can be utilized to elucidate the vulnerability or protection potential of the genetic variants in question. Correspondingly, loss or inhibition of the T1D-associated protein TYK2 in SC-islets showed diminished interferon-α response and partially protected beta cells from T-cell-mediated cytotoxicity [22]. SC-islets closely mimic adult human islets in their response to pro-inflammatory cytokines by upregulating the expression of HLA class I molecules and PD-L1 [20, 136], and even present epitopes such as pre-proinsulin peptides to CD8[+] T-cells [123]. Furthermore, since they can be generated from patient-derived pluripotent stem cells (PSCs), SC-islets allow a variety of selection of the different haplotypes from the respective donors. This facilitates the studying of beta-cell/T-cell crosstalk using haplotype-matched T-cell lines. Understanding the immune response against beta cells opens new venues of screening for drugs counteracting T1D progression as well as generating hypoimmunogenic SC-islets that can escape immune recognition in beta-cell replacement therapy [23, 54, 92, 108].

Modern genetic analysis tools have enabled us to gain valuable pathophysiological insights into the mechanisms of complex polygenic diseases such as type 2 diabetes (T2D). The latest meta-analysis of T2D genome-wide association studies (GWAS) have identified 243 genome wide significant loci which comprised 403 distinct association signals implicated in T2D predisposition [71, 72]. Human SC-islet differentiation models along with advanced gene editing tools have already been used to further our understanding of the molecular mechanisms for some of these genetic signals.

An early SC-islet study was carried out by creating isogenic human ESC lines with mutations in GWAS-identified T2D susceptibility genes *CDKAL1*, *KCNJ11*, and *KCNQ1*, leading to impaired glucose secretion in vitro and in vivo [137]. However, such homozygous mutations in *KCNJ11* would in essence also mirror those leading to congenital hyperinsulinism. Another successful example is the modelling of *SLC30A8* (zinc transporter 8) mutations that act to protect against T2D. Several cellular models, including SC-islets, showed that reduced *SLC30A8* expression leads to increased insulin secretion and reduced metabolic stress [32, 69]. More recently, a knockout model of the T2D risk gene *RREB1* was also generated within SC-islets, and demonstrated a regulatory effect on the developmentally relevant *RFX* family of genes [74]. These studies show that despite the later onset forms of diabetes that such mutations may cause, particular genes may also be developmentally relevant with phenotypes that may be studied within the SC-islet model.

The functional study of many T2D genetic associations is challenging as the majority of these loci are located in intronic or intergenic noncoding regions [21]. It is likely that non-coding T2D-associated SNPs exert their effect by regulating the expression of corresponding genes in a cell-type and developmental stage-specific manner. Additionally, T2D is highly heterogeneous and for many genes, the key mechanism may be in insulin action in the target tissue rather than insulin secretion from the beta cell. For example, the most replicated and strongest single genetic risk associated with T2D is the *TCF7L2* SNP rs79031846 located in the non-coding intronic region [67]. Being a transcription factor with a role in the canonical Wnt/beta-catenin signalling pathway, it is likely important for pancreatic and hepatic development. Thus, the effects of this SNP may be based on the regulation of

TCF7L2 expression during early development, which could be studied in PSC-derived endodermal models.

It can be anticipated that PSC-based modelling systems for pancreatic development, as well as insulin-sensitive tissues, will play an increasingly important role in the coming years in order to translate these genetic discoveries into disease mechanisms, with the potential for the development of precision therapies.

10 Conclusion

SC-islets represent a powerful tool by which to not only understand many principles of human islet organogenesis, but also as a controlled and reproducible system to study the mechanisms of beta-cell dysfunction underlying multiple genetic causes of diabetes. SC-islets offer an inexhaustible supply of human islet material with many advantages over immortalized cell lines and rodent models, with the added benefit of the relative ease of genetic manipulation and/or mutation correction. The modern renaissance of improved SC-islet differentiation protocols has encouraged a wave of monogenic diabetes modelling studies, which have uncovered novel mechanistic insights into both early onset and late onset diabetes pathologies. Ongoing research will further refine and develop these models and be applied to more complex polygenic and multi-tissue causes of beta-cell failure, as well candidate T2D susceptibility genes.

Acknowledgements The authors are grateful to Diego Balboa, Ph.D., for enlightening discussions and support. The work in the authors' laboratory was supported by the Academy of Finland, the Novo Nordisk Foundation, the Sigrid Jusélius Foundation, The European Union IMI2 initiative, the Juvenile Diabetes Research Foundation, and the Wellcome Trust.

References

1. Abdulkarim, B. *et al.* (2015) 'A Missense Mutation in PPP1R15B Causes a Syndrome Including Diabetes, Short Stature, and Microcephaly', *Diabetes*, 64(11), pp. 3951–3962. https://doi.org/10.2337/DB15-0477.
2. Aghazadeh, Y. *et al.* (2021) 'Microvessels support engraftment and functionality of human islets and hESC-derived pancreatic progenitors in diabetes models.', *Cell stem cell*, 28(11), pp. 1936–1949.e8. https://doi.org/10.1016/j.stem.2021.08.001.
3. Amin, S. *et al.* (2018) 'Discovery of a drug candidate for GLIS3-associated diabetes', *Nature communications*, 9(1). https://doi.org/10.1038/S41467-018-04918-X.
4. Aqel, Y.W.A. *et al.* (2020) 'Generation of two human iPSC lines from patients with maturity-onset diabetes of the young type 2 (MODY2) and permanent neonatal diabetes due to mutations in the GCK gene', *Stem Cell Research*, 48(May), p. 101991. https://doi.org/10.1016/j.scr.2020.101991.
5. Atkinson, M.A., Eisenbarth, G.S. and Michels, A.W. (2014) 'Type 1 diabetes.', *Lancet (London, England)*, 383(9911), pp. 69–82. https://doi.org/10.1016/S0140-6736(13)60591-7.

6. Augsornworawat, P. *et al.* (2020) 'Single-Cell Transcriptome Profiling Reveals β Cell Maturation in Stem Cell-Derived Islets after Transplantation', *Cell Reports*, 32(8), p. 108067. https://doi.org/10.1016/j.celrep.2020.108067.
7. Balboa, D. *et al.* (2018) 'Insulin mutations impair beta-cell development in a patient-derived iPSC model of neonatal diabetes', *eLife*, 7, pp. 1–35. https://doi.org/10.7554/elife.38519.
8. Balboa, D. *et al.* (2022) 'Functional, metabolic and transcriptional maturation of human pancreatic islets derived from stem cells.', *Nature biotechnology*, 40(7), pp. 1042–1055. https://doi.org/10.1038/s41587-022-01219-z.
9. Balboa, D., Iworima, D.G. and Kieffer, T.J. (2021) 'Human Pluripotent Stem Cells to Model Islet Defects in Diabetes', *Frontiers in Endocrinology*, 12(March), pp. 1–19. https://doi.org/10.3389/fendo.2021.642152.
10. Balboa, D. and Otonkoski, T. (2015) 'Human pluripotent stem cell based islet models for diabetes research.', *Best practice & research. Clinical endocrinology & metabolism*, 29(6), pp. 899–909. https://doi.org/10.1016/j.beem.2015.10.012.
11. Balboa, D., Saarimäki-Vire, J. and Otonkoski, T. (2019) 'Concise Review: Human Pluripotent Stem Cells for the Modeling of Pancreatic β-Cell Pathology.', *Stem cells (Dayton, Ohio)*, 37(1), pp. 33–41. https://doi.org/10.1002/stem.2913.
12. Barsby, T. and Otonkoski, T. (2022) 'Maturation of beta cells: lessons from in vivo and in vitro models', *Diabetologia*, 65(6), pp. 917–930. https://doi.org/10.1007/s00125-022-05672-y.
13. Bartolomé, A. (2022) 'Stem cell-derived β cells: A versatile research platform to interrogate the genetic basis of β cell dysfunction', *International Journal of Molecular Sciences*, 23(1). https://doi.org/10.3390/ijms23010501.
14. Braverman-Gross, C. *et al.* (2018) 'Derivation and molecular characterization of pancreatic differentiated MODY1-iPSCs', *Stem Cell Research*, 31(September 2017), pp. 16–26. https://doi.org/10.1016/j.scr.2018.06.013.
15. Burgos, J.I., Vallier, L. and Rodríguez-Seguí, S.A. (2021) 'Monogenic Diabetes Modeling: In Vitro Pancreatic Differentiation From Human Pluripotent Stem Cells Gains Momentum', *Frontiers in Endocrinology*, 12(July), pp. 1–19. https://doi.org/10.3389/fendo.2021.692596.
16. Butler, A.E. *et al.* (2003) 'Beta-cell deficit and increased beta-cell apoptosis in humans with type 2 diabetes.', *Diabetes*, 52(1), pp. 102–110. https://doi.org/10.2337/diabetes.52.1.102.
17. Cabezas, O.R. *et al.* (2017) 'Polycystic kidney disease with hyperinsulinemic hypoglycemia caused by a promoter mutation in phosphomannomutase 2', *Journal of the American Society of Nephrology*, 28(8), pp. 2529–2539. https://doi.org/10.1681/ASN.2016121312.
18. Cardenas-Diaz, F.L. *et al.* (2019) 'Modeling Monogenic Diabetes using Human ESCs Reveals Developmental and Metabolic Deficiencies Caused by Mutations in HNF1A', *Cell Stem Cell*, 25(2), pp. 273–289.e5. https://doi.org/10.1016/j.stem.2019.07.007.
19. Carrasco, M. *et al.* (2022) 'Spatial Environment Affects HNF4A Mutation-Specific Proteome Signatures and Cellular Morphology in hiPSC-Derived β-Like Cells', *Diabetes*, 71(4), pp. 862–869. https://doi.org/10.2337/db20-1279.
20. Castro-Gutierrez, R. *et al.* (2021) 'Protecting Stem Cell Derived Pancreatic Beta-Like Cells From Diabetogenic T Cell Recognition.', *Frontiers in endocrinology*, 12, p. 707881. https://doi.org/10.3389/fendo.2021.707881.
21. Cebola, I. and Pasquali, L. (2016) 'Non-coding genome functions in diabetes.', *Journal of molecular endocrinology*, 56(1), pp. R1–R20. https://doi.org/10.1530/JME-15-0197.
22. Chandra, V. *et al.* (2022) 'The type 1 diabetes gene TYK2 regulates β-cell development and its responses to interferon-α', *Nature Communications*, 13(1), p. 6363. https://doi.org/10.1038/s41467-022-34069-z.
23. Chimienti, R. *et al.* (2022) 'Engineering of immune checkpoints B7-H3 and CD155 enhances immune compatibility of MHC-I(-/-) iPSCs for β cell replacement.', *Cell reports*, 40(13), p. 111423. https://doi.org/10.1016/j.celrep.2022.111423.
24. Clayton, P.T. *et al.* (2001) 'Hyperinsulinism in short-chain L-3-hydroxyacyl-CoA dehydrogenase deficiency reveals the importance of β-oxidation in insulin secretion', *Journal of Clinical Investigation*, 108(3), pp. 457–465. https://doi.org/10.1172/JCI200111294.

25. Cnop, M. *et al.* (2017) 'Endoplasmic reticulum stress and eIF2α phosphorylation: The Achilles heel of pancreatic β cells', *Molecular Metabolism*. Elsevier GmbH, pp. 1024–1039. https://doi.org/10.1016/j.molmet.2017.06.001.
26. Cosentino, C. *et al.* (2018) 'Pancreatic -cell tRNA hypomethylation and fragmentation link TRMT10A deficiency with diabetes', *Nucleic Acids Research*, 46(19), pp. 10302–10318. https://doi.org/10.1093/nar/gky839.
27. Cuesta-Gomez, N. *et al.* (2022) 'Characterization of stem-cell-derived islets during differentiation and after implantation', *Cell Reports*, 40(8), p. 111238. https://doi.org/10.1016/j.celrep.2022.111238.
28. Cujba, A.M. *et al.* (2022) 'An HNF1α truncation associated with maturity-onset diabetes of the young impairs pancreatic progenitor differentiation by antagonizing HNF1β function', *Cell Reports*, 38(9), p. 110425. https://doi.org/10.1016/j.celrep.2022.110425.
29. D'Amour, K.A. *et al.* (2005) 'Efficient differentiation of human embryonic stem cells to definitive endoderm.', *Nature biotechnology*, 23(12), pp. 1534–1541. https://doi.org/10.1038/nbt1163.
30. Dai, C. *et al.* (2016) 'Stress-impaired transcription factor expression and insulin secretion in transplanted human islets.', *The Journal of clinical investigation*, 126(5), pp. 1857–70. https://doi.org/10.1172/JCI83657.
31. Demine, S. *et al.* (2020) 'Pro-inflammatory cytokines induce cell death, inflammatory responses, and endoplasmic reticulum stress in human iPSC-derived beta cells', *Stem Cell Research and Therapy*, 11(1), pp. 1–15. https://doi.org/10.1186/s13287-019-1523-3.
32. Dwivedi, O.P. *et al.* (2019) 'Loss of ZnT8 function protects against diabetes by enhanced insulin secretion.', *Nature genetics*, 51(11), pp. 1596–1606. https://doi.org/10.1038/s41588-019-0513-9.
33. Edghill, E.L. *et al.* (2008) 'Insulin Mutation Screening in 1,044 Patients With DiabetesMutations in the INS Gene Are a Common Cause of Neonatal Diabetes but a Rare Cause of Diabetes Diagnosed in Childhood or Adulthood', *Diabetes*, 57(4), pp. 1034–1042. https://doi.org/10.2337/DB07-1405.
34. El-Khairi, R. *et al.* (2021) 'Modeling HNF1B-associated monogenic diabetes using human iPSCs reveals an early stage impairment of the pancreatic developmental program', *Stem Cell Reports*, 16(9), pp. 2289–2304. https://doi.org/10.1016/j.stemcr.2021.07.018.
35. Elsayed, A.K. *et al.* (2021) 'Aberrant development of pancreatic beta cells derived from human iPSCs with FOXA2 deficiency', *Cell Death and Disease*, 12(1). https://doi.org/10.1038/s41419-021-03390-8.
36. Flanagan, S.E. *et al.* (2010) 'Diazoxide-responsive hyperinsulinemic hypoglycemia caused by HNF4A gene mutations', *European Journal of Endocrinology*, 162(5), pp. 987–992. https://doi.org/10.1530/EJE-09-0861.
37. Flanagan, S.E. *et al.* (2017) 'A CACNA1D mutation in a patient with persistent hyperinsulinaemic hypoglycaemia, heart defects, and severe hypotonia', *Pediatric Diabetes*, 18(4), pp. 320–323. https://doi.org/10.1111/pedi.12512.
38. Fonseca, S.G. *et al.* (2005) 'WFS1 Is a Novel Component of the Unfolded Protein Response and Maintains Homeostasis of the Endoplasmic Reticulum in Pancreatic β-Cells', *Journal of Biological Chemistry*, 280(47), pp. 39609–39615. https://doi.org/10.1074/JBC.M507426200.
39. Fonseca, S.G., Gromada, J. and Urano, F. (2011) 'Endoplasmic reticulum stress and pancreatic β-cell death', *Trends in Endocrinology and Metabolism*. Elsevier Inc., pp. 266–274. https://doi.org/10.1016/j.tem.2011.02.008.
40. De Franco, E. *et al.* (2020) 'YIPF5 mutations cause neonatal diabetes and microcephaly through endoplasmic reticulum stress.', *The Journal of clinical investigation*, 130(12), pp. 6338–6353. https://doi.org/10.1172/JCI141455.
41. De Franco, E. (2021) 'Neonatal diabetes caused by disrupted pancreatic and β-cell development', *Diabetic Medicine*, 38(12), pp. 1–10. https://doi.org/10.1111/dme.14728.
42. De Franco, E. *et al.* (n.d.) 'Primate-specific ZNF808 is essential for pancreatic development in humans'. https://doi.org/10.1038/s41588-023-01565-x.

43. Ghila, L. *et al.* (2020) 'Bioinformatic Analyses of miRNA – mRNA Signature during hiPSC Differentiation towards Insulin-Producing Cells upon HNF4 α Mutation', pp. 1–20. https://doi.org/10.3390/biomedicines8070179.
44. Giri, D. *et al.* (2017) 'Novel FOXA2 mutation causes Hyperinsulinism, Hypopituitarism with Craniofacial and Endodermderived organ abnormalities', *Human Molecular Genetics*, 26(22), pp. 4315–4326. https://doi.org/10.1093/hmg/ddx318.
45. Glaser, B. *et al.* (1998) 'Familial hyperinsulinism caused by an activating glucokinase mutation', *New England Journal of Medicine*, 338(4), pp. 226–230.
46. González-Barroso, M.M. *et al.* (2008) 'Mutations in UCP2 in congenital hyperinsulinism reveal a role for regulation of insulin secretion', *PLoS ONE*, 3(12). https://doi.org/10.1371/journal.pone.0003850.
47. González, B.J. *et al.* (2022) 'Reduced calcium levels and accumulation of abnormal insulin granules in stem cell models of HNF1A deficiency', *Communications Biology*, 5(1). https://doi.org/10.1038/s42003-022-03696-z.
48. Gregory, L.C. *et al.* (2019) 'Impaired EIF2S3 function associated with a novel phenotype of X-linked hypopituitarism with glucose dysregulation', *EBioMedicine*, 42, pp. 470–480. https://doi.org/10.1016/j.ebiom.2019.03.013.
49. Griscelli, F. *et al.* (2017) 'Generation of an induced pluripotent stem cell (iPSC) line from a patient with maturity-onset diabetes of the young type 13 (MODY13) with a the potassium inwardly-rectifying channel, subfamily J, member 11 (KCNJ11) mutation', *Stem Cell Research*, 23, pp. 178–181. https://doi.org/10.1016/j.scr.2017.07.023.
50. Griscelli, F. *et al.* (2018) 'Generation of an induced pluripotent stem cell (iPSC) line from a patient with maturity-onset diabetes of the young type 3 (MODY3) carrying a hepatocyte nuclear factor 1-alpha (HNF1A) mutation', *Stem Cell Research*, 29, pp. 56–59. https://doi.org/10.1016/j.scr.2018.02.017.
51. Groop, L. and Pociot, F. (2014) 'Genetics of diabetes – Are we missing the genes or the disease?', *Molecular and Cellular Endocrinology*, 382(1), pp. 726–739. https://doi.org/10.1016/j.mce.2013.04.002.
52. Guo, D. *et al.* (2017) 'Modeling Congenital Hyperinsulinism with ABCC8 – Deficient Human Embryonic Stem Cells Generated by CRISPR / Cas9', *Scientific Reports*, 7(3156), pp. 1–8. https://doi.org/10.1038/s41598-017-03349-w.
53. Güemes, M. *et al.* (2020) 'Hyperinsulinemic hypoglycemia in children and adolescents: Recent advances in understanding of pathophysiology and management', *Reviews in Endocrine and Metabolic Disorders*, 21(4), pp. 577–597. https://doi.org/10.1007/s11154-020-09548-7.
54. Han, X. *et al.* (2019) 'Generation of hypoimmunogenic human pluripotent stem cells.', *Proceedings of the National Academy of Sciences of the United States of America*, 116(21), pp. 10441–10446. https://doi.org/10.1073/pnas.1902566116.
55. Hermann, F.M. *et al.* (2023) 'An insulin hypersecretion phenotype precedes pancreatic β cell failure in MODY3 patient-specific cells', *Cell Stem Cell*, 30(1), pp. 38–51.e8. https://doi.org/10.1016/j.stem.2022.12.001.
56. Huopio, H. *et al.* (2000) 'Dominantly inherited hyperinsulinism caused by a mutation in the sulfonylurea receptor type 1', 106(7), pp. 897–906.
57. Hussain, S. *et al.* (2013) 'Permanent neonatal diabetes due to a novel insulin signal peptide mutation', *Pediatric Diabetes*, 14(4), pp. 299–303. https://doi.org/10.1111/PEDI.12011.
58. Kapoor, R.R. *et al.* (2013) 'Clinical and molecular characterisation of 300 patients with congenital hyperinsulinism', *European Journal of Endocrinology*, 168(4), pp. 557–564. https://doi.org/10.1530/EJE-12-0673.
59. Kitamura, R.A. *et al.* (2022) 'Multidimensional analysis and therapeutic development using patient iPSC–derived disease models of Wolfram syndrome', *JCI Insight*, 7(18). https://doi.org/10.1172/JCI.INSIGHT.156549.
60. Klak, M. *et al.* (2020) 'Type 1 diabetes: genes associated with disease development.', *Central-European journal of immunology*, 45(4), pp. 439–453. https://doi.org/10.5114/ceji.2020.103386.

61. Leavens, K.F. *et al.* (2022) 'Stem cell-based multi-tissue platforms to model human autoimmune diabetes', *Molecular Metabolism*, 66(October 2022), p. 101610. https://doi.org/10.1016/j.molmet.2022.101610.
62. Leite, N.C. *et al.* (2020) 'Modeling Type 1 Diabetes In Vitro Using Human Pluripotent Stem Cells', *Cell Reports*, 32(2), p. 107894. https://doi.org/10.1016/j.celrep.2020.107894.
63. Lenzen, S. (2008) 'The mechanisms of alloxan- and streptozotocin-induced diabetes.', *Diabetologia*, 51(2), pp. 216–26. https://doi.org/10.1007/s00125-007-0886-7.
64. Liang, Z. *et al.* (2023) 'Implantation underneath the abdominal anterior rectus sheath enables effective and functional engraftment of stem-cell-derived islets', *Nature Metabolism* [Preprint]. https://doi.org/10.1038/s42255-022-00713-7.
65. Lithovius, V. *et al.* (2021) 'SUR1-mutant iPS cell-derived islets recapitulate the pathophysiology of congenital hyperinsulinism.', *Diabetologia*, 64(3), pp. 630–640. https://doi.org/10.1007/s00125-020-05346-7.
66. Low, B.S.J. *et al.* (2021) 'Decreased GLUT2 and glucose uptake contribute to insulin secretion defects in MODY3/HNF1A hiPSC-derived mutant β cells.', *Nature communications*, 12(1), p. 3133. https://doi.org/10.1038/s41467-021-22843-4.
67. Lyssenko, V. *et al.* (2007) 'Mechanisms by which common variants in the TCF7L2 gene increase risk of type 2 diabetes.', *The Journal of clinical investigation*, 117(8), pp. 2155–2163. https://doi.org/10.1172/JCI30706.
68. Lytrivi, M. *et al.* (2021) 'DNAJC3 deficiency induces β-cell mitochondrial apoptosis and causes syndromic young-onset diabetes', *European Journal of Endocrinology*, 1(aop). https://doi.org/10.1530/eje-20-0636.
69. Ma, Q. *et al.* (2022) 'ZnT8 loss-of-function accelerates functional maturation of hESC-derived β cells and resists metabolic stress in diabetes.', *Nature communications*, 13(1), p. 4142. https://doi.org/10.1038/s41467-022-31829-9.
70. Ma, S. *et al.* (2018) 'β Cell Replacement after Gene Editing of a Neonatal Diabetes-Causing Mutation at the Insulin Locus', *Stem Cell Reports*, 11(6), pp. 1407–1415. https://doi.org/10.1016/j.stemcr.2018.11.006.
71. Mahajan, A. *et al.* (2018) 'Fine-mapping type 2 diabetes loci to single-variant resolution using high-density imputation and islet-specific epigenome maps.', *Nature genetics*, 50(11), pp. 1505–1513. https://doi.org/10.1038/s41588-018-0241-6.
72. Mahajan, A. *et al.* (2022) 'Multi-ancestry genetic study of type 2 diabetes highlights the power of diverse populations for discovery and translation.', *Nature genetics*, 54(5), pp. 560–572. https://doi.org/10.1038/s41588-022-01058-3.
73. Männistö, J.M.E. *et al.* (2020) 'Clinical and genetic characterization of 153 patients with persistent or transient congenital hyperinsulinism', *Journal of Clinical Endocrinology and Metabolism*, 105(4), pp. E1686–E1694. https://doi.org/10.1210/clinem/dgz271.
74. Mattis, K.K. *et al.* (2023) 'Loss of RREB1 in pancreatic beta cells reduces cellular insulin content and affects endocrine cell gene expression.', *Diabetologia* [Preprint]. https://doi.org/10.1007/s00125-022-05856-6.
75. Maxwell, K.G. *et al.* (2020) 'Gene-edited human stem cell – derived β cells from a patient with monogenic diabetes reverse preexisting diabetes in mice', 9106(April).
76. Maxwell, K.G. *et al.* (2022) 'Differential Function and Maturation of Human Stem Cell-Derived Islets After Transplantation', *Stem cells translational medicine*, 11(3). https://doi.org/10.1093/stcltm/szab013.
77. McGrath, P.S. *et al.* (2015) 'The basic helix-loop-helix transcription factor neurog3 is required for development of the human endocrine pancreas', *Diabetes*, 64(7), pp. 2497–2505. https://doi.org/10.2337/db14-1412.
78. Miguel-Escalada, I. *et al.* (2022) 'Pancreas agenesis mutations disrupt a lead enhancer controlling a developmental enhancer cluster', *Developmental Cell*, 57(16), pp. 1922–1936.e9. https://doi.org/10.1016/J.DEVCEL.2022.07.014.
79. Millman, J.R. *et al.* (2016) 'Generation of stem cell-derived β-cells from patients with type 1 diabetes', *Nature Communications*, 7(May). https://doi.org/10.1038/ncomms11463.

80. Mirza, A.H. *et al.* (2014) 'Effects of GWAS-Associated Genetic Variants on lncRNAs within IBD and T1D Candidate Loci', *PLOS ONE*, 9(8), p. e105723.
81. Miura, A. *et al.* (2006) 'Hepatocyte nuclear factor-4α is essential for glucose-stimulated insulin secretion by pancreatic β-cells', *Journal of Biological Chemistry*, 281(8), pp. 5246–5257. https://doi.org/10.1074/jbc.M507496200.
82. Miyachi, Y., Miyazawa, T. and Ogawa, Y. (2022) 'HNF1A Mutations and Beta Cell Dysfunction in Diabetes.', *International journal of molecular sciences*, 23(6). https://doi.org/10.3390/ijms23063222.
83. Mohan, V. *et al.* (2018) 'Comprehensive genomic analysis identifies pathogenic variants in maturity-onset diabetes of the young (MODY) patients in South India', *BMC Medical Genetics*, 19(1), pp. 1–10. https://doi.org/10.1186/s12881-018-0528-6.
84. Montaser, H. *et al.* (2021) 'Loss of MANF Causes Childhood-Onset Syndromic Diabetes Due to Increased Endoplasmic Reticulum Stress', *Diabetes*, 70(4), pp. 1006–1018. https://doi.org/10.2337/db20-1174.
85. Nair, G.G. *et al.* (2019) 'Recapitulating endocrine cell clustering in culture promotes maturation of human stem-cell-derived β cells', *Nature Cell Biology*, 21(2), pp. 263–274. https://doi.org/10.1038/s41556-018-0271-4.
86. Nair, G.G. and Odorico, J.S. (2016) 'PTF1a Activity in Enriched Posterior Foregut Endoderm, but Not Definitive Endoderm, Leads to Enhanced Pancreatic Differentiation in an in Vitro Mouse ESC-Based Model', *Stem Cells International*, 2016. https://doi.org/10.1155/2016/6939438.
87. Narayanan, S. *et al.* (2017) 'Intra-islet endothelial cell and β-cell crosstalk: Implication for islet cell transplantation', *World Journal of Transplantation*, 7(2). https://doi.org/10.5500/wjt.v7.i2.117.
88. Ng, N.H.J. *et al.* (2019) 'HNF4A Haploinsufficiency in MODY1 Abrogates Liver and Pancreas Differentiation from Patient-Derived Induced Pluripotent Stem Cells', *iScience*, 16, pp. 192–205. https://doi.org/10.1016/j.isci.2019.05.032.
89. Nguyen-Ngoc, K.-V. *et al.* (2022) 'Engineered vasculature induces functional maturation of pluripotent stem cell-derived islet organoids', *bioRxiv*, p. 2022.10.28.513298. https://doi.org/10.1101/2022.10.28.513298.
90. Otonkoski, T. *et al.* (2007) 'Physical Exercise–Induced Hypoglycemia Caused by Failed Silencing of Monocarboxylate Transporter 1 in Pancreatic β Cells', *The American Journal of Human Genetics*, 81(3), pp. 467–474. https://doi.org/10.1086/520960.
91. Pagliuca, F.W. *et al.* (2014) 'Generation of functional human pancreatic β cells in vitro', *Cell*, 159(2), pp. 428–439. https://doi.org/10.1016/j.cell.2014.09.040.
92. Parent, A. V *et al.* (2021) 'Selective deletion of human leukocyte antigens protects stem cell-derived islets from immune rejection.', *Cell reports*, 36(7), p. 109538. https://doi.org/10.1016/j.celrep.2021.109538.
93. Pellegrini, S. *et al.* (2021) 'Generation of β Cells from iPSC of a MODY8 Patient with a Novel Mutation in the Carboxyl Ester Lipase (CEL) Gene', *Journal of Clinical Endocrinology and Metabolism*, 106(5), pp. E2322–E2333. https://doi.org/10.1210/clinem/dgaa986.
94. Philippi, A. *et al.* (2021) 'Mutations and variants of ONECUT1 in diabetes', *Nature Medicine*, 27(11), pp. 1928–1940. https://doi.org/10.1038/s41591-021-01502-7.
95. Polonsky, K.S. (2012) 'The Past 200 Years in Diabetes', *New England Journal of Medicine*, 367(14), pp. 1332–1340. https://doi.org/10.1056/nejmra1110560.
96. Preissler, S. and Ron, D. (2019) 'Early events in the endoplasmic reticulum unfolded protein response', *Cold Spring Harbor Perspectives in Biology*, 11(4), p. a033894. https://doi.org/10.1101/cshperspect.a033894.
97. Rafique, I. *et al.* (2021) 'Causal variants in Maturity Onset Diabetes of the Young (MODY) – A systematic review', *BMC Endocrine Disorders*, 21(1), pp. 1–6. https://doi.org/10.1186/s12902-021-00891-7.

98. Ravassard, P. *et al.* (2011) 'A genetically engineered human pancreatic β cell line exhibiting glucose-inducible insulin secretion.', *The Journal of clinical investigation*, 121(9), pp. 3589–3597. https://doi.org/10.1172/JCI58447.
99. Rezania, A. *et al.* (2014) 'Reversal of diabetes with insulin-producing cells derived in vitro from human pluripotent stem cells', *Nature Biotechnology*, 32(11), pp. 1121–1133. https://doi.org/10.1038/nbt.3033.
100. Romer, A.I. *et al.* (2019) 'Murine perinatal β-cell proliferation and the differentiation of human stem cell-derived insulin-expressing cells require NEUROD1', *Diabetes*, 68(12), pp. 2259–2271. https://doi.org/10.2337/db19-0117.
101. Saarimäki-Vire, J. *et al.* (2017) 'An Activating STAT3 Mutation Causes Neonatal Diabetes through Premature Induction of Pancreatic Differentiation', *Cell Reports*, 19(2), pp. 281–294. https://doi.org/10.1016/j.celrep.2017.03.055.
102. Sakhneny, L. *et al.* (2021) 'The postnatal pancreatic microenvironment guides β cell maturation through BMP4 production', *Developmental Cell*, 56(19), pp. 2703–2711.e5. https://doi.org/10.1016/j.devcel.2021.08.014.
103. Senée, V. *et al.* (2004) 'Wolcott-Rallison SyndromeClinical, Genetic, and Functional Study of EIF2AK3 Mutations and Suggestion of Genetic Heterogeneity', *Diabetes*, 53(7), pp. 1876–1883. https://doi.org/10.2337/DIABETES.53.7.1876.
104. Shalev, S.A. *et al.* (2014) 'Microcephaly, Epilepsy and Neonatal Diabetes Due to Compound Heterozygous Mutations in IER3IP1: Insights into the Natural History of a Rare Disorder', *Pediatr Diabetes*, 15(3), pp. 252–256. https://doi.org/10.1111/pedi.12086.
105. Shang, L. *et al.* (2014) 'β-cell dysfunction due to increased ER stress in a stem cell model of wolfram syndrome', *Diabetes*, 63(3), pp. 923–933. https://doi.org/10.2337/db13-0717.
106. Shi, Z.D. *et al.* (2017) 'Genome Editing in hPSCs Reveals GATA6 Haploinsufficiency and a Genetic Interaction with GATA4 in Human Pancreatic Development', *Cell Stem Cell*, 20(5), pp. 675–688.e6. https://doi.org/10.1016/j.stem.2017.01.001.
107. Shirakawa, J. *et al.* (2014) 'Effects of the Antitumor Drug OSI-906, a Dual Inhibitor of IGF-1 Receptor and Insulin Receptor, on the Glycemic Control, β-Cell Functions, and β-Cell Proliferation in Male Mice.', *Endocrinology*, 155(6), pp. 2102–11. https://doi.org/10.1210/en.2013-2032.
108. Sintov, E. *et al.* (2022) 'Whole-genome CRISPR screening identifies genetic manipulations to reduce immune rejection of stem cell-derived islets', *Stem Cell Reports*, 17(9), pp. 1976–1990. https://doi.org/10.1016/j.stemcr.2022.08.002.
109. Skoczek, D., Dulak, J. and Kachamakova-Trojanowska, N. (2021) 'Maturity onset diabetes of the young – new approaches for disease modelling', *International Journal of Molecular Sciences*, 22(14). https://doi.org/10.3390/ijms22147553.
110. Skopkova, M. *et al.* (2017) 'EIF2S3 Mutations Associated with Severe X-Linked Intellectual Disability Syndrome MEHMO', *Human Mutation*, 38(4), pp. 409–425. https://doi.org/10.1002/HUMU.23170.
111. Snider, K.E. *et al.* (2013) 'Genotype and phenotype correlations in 417 children with congenital hyperinsulinism', *Journal of Clinical Endocrinology and Metabolism*, 98(2), pp. 355–363. https://doi.org/10.1210/jc.2012-2169.
112. Stanescu, D.E. *et al.* (2012) 'Novel presentations of congenital hyperinsulinism due to mutations in the MODY genes: HNF1A and HNF4A', *Journal of Clinical Endocrinology and Metabolism*, 97(10), pp. 2026–2030. https://doi.org/10.1210/jc.2012-1356.
113. Stanley, C.A. *et al.* (1998) 'Hyperinsulinism and Hyperammonemia in Infants with Regulatory Mutations of the Glutamate Dehydrogenase Gene', *New England Journal of Medicine*, 338(19), pp. 1352–1357. https://doi.org/10.1056/NEJM199805073381904.
114. Stepniewski, J. *et al.* (2015) 'Induced pluripotent stem cells as a model for diabetes investigation', *Scientific Reports*, 5, pp. 1–14. https://doi.org/10.1038/srep08597.
115. Stoffel, M. and Duncan, S.A. (1997) 'The maturity-onset diabetes of the young (MODY1) transcription factor HNF4α regulates expression of genes required for glucose transport

and metabolism', *Proceedings of the National Academy of Sciences of the United States of America*, 94(24), pp. 13209–13214. https://doi.org/10.1073/pnas.94.24.13209.
116. Størling, J. and Pociot, F. (2017) 'Type 1 Diabetes Candidate Genes Linked to Pancreatic Islet Cell Inflammation and Beta-Cell Apoptosis.', *Genes*, 8(2). https://doi.org/10.3390/genes8020072.
117. Tegtmeyer, L.C. *et al.* (2014) 'Multiple Phenotypes in Phosphoglucomutase 1 Deficiency', *New England Journal of Medicine*, 370(6), pp. 533–542. https://doi.org/10.1056/nejmoa1206605.
118. Temocin, O. *et al.* (2021) 'Generation of an hiPSC-1 knock-in line expressing TY1-tagged MNX1-protein together with mScarlet', *Stem Cell Research*, 56(June), p. 102522. https://doi.org/10.1016/j.scr.2021.102522.
119. Teo, A.K.K. *et al.* (2013) 'Derivation of human induced pluripotent stem cells from patients with maturity onset diabetes of the young', *Journal of Biological Chemistry*, 288(8), pp. 5353–5356. https://doi.org/10.1074/jbc.C112.428979.
120. Teo, A.K.K. *et al.* (2016) 'Early Developmental Perturbations in a Human Stem Cell Model of MODY5/HNF1B Pancreatic Hypoplasia', *Stem Cell Reports*, 6(3), pp. 357–367. https://doi.org/10.1016/j.stemcr.2016.01.007.
121. Thomas, P.M. *et al.* (1995) 'Mutations in the sulfonylurea receptor gene in familial persistent hyperinsulinemic hypoglycemia of infancy', *Science*, 268(April), pp. 426–429.
122. Tiyaboonchai, A. *et al.* (2017) 'GATA6 Plays an Important Role in the Induction of Human Definitive Endoderm, Development of the Pancreas, and Functionality of Pancreatic β Cells.', *Stem cell reports*, 8(3), pp. 589–604. https://doi.org/10.1016/j.stemcr.2016.12.026.
123. van der Torren, C.R. *et al.* (2017) 'Immunogenicity of human embryonic stem cell-derived beta cells.', *Diabetologia*, 60(1), pp. 126–133. https://doi.org/10.1007/s00125-016-4125-y.
124. Trott, J. *et al.* (2020) 'Mitchell-Riley syndrome iPSCs exhibit reduced pancreatic endoderm differentiation due to a mutation in RFX6', *Development (Cambridge)*, 147(21), pp. 1–12. https://doi.org/10.1242/dev.194878.
125. Tsonkova, V.G. *et al.* (2018) 'The EndoC-βH1 cell line is a valid model of human beta cells and applicable for screenings to identify novel drug target candidates', *Molecular Metabolism*, 8(December 2017), pp. 144–157. https://doi.org/10.1016/j.molmet.2017.12.007.
126. Velazco-Cruz, L. *et al.* (2019) 'Acquisition of Dynamic Function in Human Stem Cell-Derived β Cells', *Stem Cell Reports*, 12(2), pp. 351–365. https://doi.org/10.1016/j.stemcr.2018.12.012.
127. Wagner, L.E. *et al.* (2022) 'Mouse models and human islet transplantation sites for intravital imaging', *Frontiers in Endocrinology*, 13(October), pp. 1–10. https://doi.org/10.3389/fendo.2022.992540.
128. Wakeling, M.N. *et al.* (2022) 'Non-coding variants disrupting a tissue-specific regulatory element in HK1 cause congenital hyperinsulinism', *Nature Genetics*, 54(11), pp. 1615–1620. https://doi.org/10.1038/s41588-022-01204-x.
129. Wang, B., Stanford, K.R. and Kundu, M. (2020) 'ER-to-Golgi Trafficking and Its Implication in Neurological Diseases', *Cells*, 9(2). https://doi.org/10.3390/CELLS9020408.
130. Wang, X. *et al.* (2016) 'Generation of a human induced pluripotent stem cell (iPSC) line from a patient carrying a P33T mutation in the PDX1 gene', *Stem Cell Research*, 17(2), pp. 273–276. https://doi.org/10.1016/j.scr.2016.08.004.
131. Wang, X. *et al.* (2016) 'Generation of a human induced pluripotent stem cell (iPSC) line from a patient with family history of diabetes carrying a C18R mutation in the PDX1 gene', *Stem Cell Research*, 17(2), pp. 292–295. https://doi.org/10.1016/j.scr.2016.08.005.
132. Wang, X. *et al.* (2019) 'Point mutations in the PDX1 transactivation domain impair human β-cell development and function', *Molecular Metabolism*, 24(March 2019), pp. 80–97. https://doi.org/10.1016/j.molmet.2019.03.006.
133. Weiss, M., Steiner, D.F. and Philipson, L.H. (2000) *Insulin Biosynthesis, Secretion, Structure, and Structure-Activity Relationships*, *Endotext*. MDText.com, Inc. Available at: http://www.ncbi.nlm.nih.gov/pubmed/25905258.

134. Yabe, S.G. *et al.* (2015) 'Establishment of maturity-onset diabetes of the young-induced pluripotent stem cells from a Japanese patient', *Journal of Diabetes Investigation*, 6(5), pp. 543–547. https://doi.org/10.1111/jdi.12334.
135. Yabe, S.G. *et al.* (2019) 'Expression of mutant mRNA and protein in pancreatic cells derived from MODY3- iPS cells', *PLoS ONE*, 14(5), pp. 1–17. https://doi.org/10.1371/journal.pone.0217110.
136. Yoshihara, E. *et al.* (2020) 'Immune-evasive human islet-like organoids ameliorate diabetes', *Nature*, 586(7830), pp. 606–611. https://doi.org/10.1038/s41586-020-2631-z.
137. Zeng, H. *et al.* (2016) 'An Isogenic Human ESC Platform for Functional Evaluation of Genome-wide-Association-Study-Identified Diabetes Genes and Drug Discovery', *Cell Stem Cell*, 19(3), pp. 326–340. https://doi.org/10.1016/j.stem.2016.07.002.
138. Zhang, H. *et al.* (2021) 'Monogenic diabetes: a gateway to precision medicine in diabetes.', *The Journal of clinical investigation*, 131(3). https://doi.org/10.1172/JCI142244.
139. Zhu, H. *et al.* (2018) 'Selection of Implantation Sites for Transplantation of Encapsulated Pancreatic Islets', *Tissue Engineering – Part B: Reviews*. https://doi.org/10.1089/ten.teb.2017.0311.
140. Zhu, Z. *et al.* (2016) 'Genome Editing of Lineage Determinants in Human Pluripotent Stem Cells Reveals Mechanisms of Pancreatic Development and Diabetes', *Cell Stem Cell*, 18(6), pp. 755–768. https://doi.org/10.1016/j.stem.2016.03.015.

Index

A
Autoimmunity, 5, 97, 169, 284, 286, 291, 292, 309, 319, 320, 322, 323, 333, 343, 364, 365, 490, 504, 514, 565
Autonomy, 464, 465, 467, 470–471, 477
Avascular phase, 152, 156

B
Beta cell, 3, 107, 151, 179, 195, 223, 251, 283, 344, 410, 462, 485, 517, 529, 566, 573
β-cell maturation, 210, 531, 537–540, 588
β-cell replacement, 4–5, 101, 113, 114, 144, 158, 159, 161–163, 166, 195, 240, 251, 252, 254, 258, 283, 285, 286, 291, 292, 311, 364, 382, 462, 465, 485–497, 517, 520, 521, 549, 565, 589
Bioartificial organs, 224, 461, 462
Biodegradation, 109
Bioengineering, 98, 101, 113, 123, 152, 154–155, 157, 161–166, 383, 420, 562
Bioengineering 3-D Organoids of Islet and Mesenchymal Stromal Cells, 186–188
Biofunctionalization, 265
Biomaterials, 97–114, 125, 128, 129, 133–134, 138, 144, 155, 158–160, 162, 206–213, 221–224, 234, 240, 259, 261, 264, 265, 268, 271, 319–343, 379, 380, 563
Biotherapies of Type I Diabetes mellitus, 184

C
Ca^{2+} signaling, 534
Cell fate, 14–16, 26, 30, 32, 34, 49–60, 70, 71, 74, 77, 87, 210, 581
Cell product development, 67, 386, 404, 413–415, 420, 421, 435, 437, 440–442, 489, 490, 492–494, 540
Cell therapy, 16, 26, 57, 86, 91, 185, 292, 293, 303, 309, 404, 549–551, 561, 567, 588
Clinical grade fabrication techniques, 104, 159
Clinical translation, 128, 129, 166, 259, 382, 462, 463
Clinical translation of medical devices, 379
Clinical trial, 5, 6, 17, 50, 57, 68, 73, 77, 100, 112, 125, 128–129, 144, 160, 169, 182–185, 226, 242, 255, 370, 374, 380–384, 419–421, 462, 474, 485, 488–494, 499–522, 540–541, 547–567

D
Dependence on anti-rejection drugs, 185
Development, 4, 26, 49, 67, 86, 100, 138, 154, 185, 197, 223, 252, 302, 319, 364, 404, 419, 461, 485, 519, 530, 547, 573

L. Piemonti et al. (eds.), *Pluripotent Stem Cell Therapy for Diabetes*,
https://doi.org/10.1007/978-3-031-41943-0

Diabetes, 3, 25, 49, 67, 86, 97, 126, 151, 179, 201, 221, 251, 284, 310, 317, 363, 415, 461, 486, 499, 537, 547, 573
Donor Scarcity, 180, 185

E
Edmonton protocol, 4, 97, 334, 499–504, 510, 516, 520, 549
Embryonic stem cell (ESC), 156, 287, 383, 564, 565, 583
Endocrine cell, 5, 6, 9–13, 25, 34, 49, 51, 52, 54, 70–72, 186, 403, 566, 575, 579, 584
Endocrine microenvironment, 153, 157, 269
Engraftment, 76, 108, 134, 151–157, 159, 162–164, 167, 168, 184, 210, 212, 252, 254, 265, 286, 326, 367, 372, 374, 377, 379–381, 383, 384, 504, 510–512, 520, 537, 556, 557, 560, 564, 587
Enterochromaffin cells, 10–12, 16, 37, 54, 420
Epidermal growth factor (EGF), 87–90, 288, 291, 294
Epigenetic aberrations, 433, 442
Ethics, 477
Ex vivo islet engraftment, 151–154
Extracellular matrix (ECM), 89, 90, 98, 102, 110, 151, 153, 195–214, 225, 252, 325, 327, 437, 504
Extrahepatic, 98–100, 123, 126, 129–132, 143, 184–185, 265, 319, 322, 364, 370, 374–383

F
Fate specification, 15, 49–60
FDA regulation, 485
Fibroblast growth factor (FGF), 69, 87, 90, 268, 347
First and second phase insulin secretion, 534–536, 540

G
Gene editing, 183, 303, 305, 306, 308, 310–311, 343–347, 385, 386, 405, 412, 462, 473, 558, 589
Gene regulatory network, 52, 53
Genetic regulatory network, 25–37
Glucose homeostasis, 3, 4, 6, 9, 25, 55, 58, 72, 112, 169, 365, 367, 530, 573, 584
Growth factors, 86, 87, 142, 155, 210, 268, 347, 575

H
HLA sensitizitation, 518–519
hPSC-based therapies, 406, 415, 420, 421, 431, 434–438, 441–443
Human leukocyte antigens (HLA), 166, 283–293, 299–311, 319, 343–345, 385, 518–519, 563, 589
Human pluripotent stem cells (hPSCs), 3–18, 50, 67, 86, 213, 421–434, 537, 551, 554, 555

I
Immune isolation, 155, 161, 183–184, 242, 566
Immune rejection, 4, 168, 169, 227, 252, 300, 302, 303, 310, 346, 566, 588
Immunoengineering, 317–350
Immunomodulation, 252, 327, 334, 563
Immunosuppression, 5, 16, 33–34, 125, 128, 160, 167, 169, 252, 265, 270, 271, 302, 311, 318, 319, 321, 348, 350, 378, 379, 381, 382, 384, 385, 406, 461, 466, 467, 486, 489, 491–494, 504, 510, 512, 514–519, 540, 548, 557, 560–566
Implantation site, 76, 77, 98, 109, 113, 123, 127, 129–132, 142, 184–185, 235, 270, 283, 292, 323, 364, 365, 367, 370, 374–380, 382–383, 521, 587
Induced pluripotent stem cell (iPSC), 26, 74, 100, 101, 142, 166, 271, 319, 346, 419, 461, 550, 583
Insulin independence, 49, 97, 125, 126, 134, 136, 152, 183, 185, 252, 292, 318, 334, 372, 374, 375, 377, 381, 382, 466, 488, 490–492, 500–506, 510–517, 519, 520, 540, 549, 560, 564
Insulin-producing β-cell, 49, 50, 113, 365, 547, 550, 551
Insulin secretion, 3, 35, 68, 98, 132, 154, 196, 229, 257, 309, 366, 510, 529, 557, 573
Intraportal, 112, 169, 185, 266, 283, 364, 365, 367–386, 489, 491–494, 564
Islet, 3, 25, 49, 67, 97, 123, 151, 180, 195, 221, 251, 283, 309, 317, 363, 404, 420, 461, 486, 499, 531, 547, 574

Islet delivery device, 101, 102, 127, 129, 131–133, 137
Islet development, 3–18, 25–37, 50, 154, 201–203, 205
Islet-like cluster, 156, 365, 550, 559
Islet microenvironment, 152, 200, 204, 205, 214, 265, 268–269
Islet transplantation, 4, 49, 67, 75, 97–98, 112, 113, 128, 133–135, 143, 151, 169, 182, 187, 211, 235, 252, 264, 266, 270, 283, 292, 318–323, 333–343, 365, 367–371, 386, 466–468, 489, 491, 499–522, 549, 559, 563, 567
Islets of Langerhans, 49, 51, 221, 222, 238, 463, 466, 467

K
Kidney function, 503, 517–518, 521

L
Lineage reprogramming, 49–60
Local immunosuppression, 252, 266

M
Macroencapsulation, 68, 73, 76–78, 125, 129–132, 136–139, 141–143, 224, 251–271, 384, 554, 556, 566
Malignant potential, 437, 439, 440
Mechanical properties, 102–104, 107–108, 113, 158, 208, 224, 231, 241, 252, 268, 269, 327, 331
Mesenchymal stem cell (MSC), 156, 168, 182, 213, 239, 347, 348, 550–554
Metabolism, 25, 35, 49, 264, 291, 309, 310, 319, 366, 494, 529–537, 540, 575, 586, 587
Microarrangement, 26, 208, 440
Microencapsulation, 99, 125–129, 142, 143, 159, 161, 223, 224, 226, 227, 238, 239, 241, 265, 269, 384, 566
MODY, 574–578, 582–586
Monogenic diabetes, 573–577, 582, 587, 590
Multiomics technology, 27, 28, 33, 36

N
Nano-encapsulation, 99, 155, 162
Natural killer (NK) cells., 224, 289, 300, 304, 325, 345, 364
Nonhuman primate, 550, 563–565, 567
Non-invasive islet monitoring, 113, 114, 164, 210, 378
Notch, 7, 31–33, 87, 90, 91

O
Omental engraftment, 180
Oncogenicity, 17, 70, 71, 184, 566
Oxidative stress, 105, 106, 369, 579
Oxygenation, 137, 156, 159, 162, 185, 251–271, 367, 379, 562

P
Pancreas, 4, 25, 49, 67, 86, 97, 126, 152, 180, 195, 221, 252, 283, 365, 461, 500, 547, 574
Pancreas development, 30, 37, 50–54, 67–70, 73, 86, 90, 91, 201
Pancreatic development, 5, 11, 30–32, 34, 36, 50, 53, 54, 90, 383, 573, 575, 579, 582, 590
Pancreatic progenitors (PP), 5, 29, 50, 51, 67–77, 86, 169, 306, 383, 414, 551, 556, 557, 575, 581, 584
Permanent neonatal diabetes (PNDM), 573, 576, 577, 579–584, 586
Physiological Insulin and other Islet Hormone Delivery, 185
Pluripotent stem cell (PSC), 5, 26, 50, 74, 86, 100, 142, 166, 252, 302, 319, 383, 419, 461, 550, 575
Porosity, 103, 104, 106–107, 113, 329, 331
Proliferation, 31, 32, 36, 57, 69, 73, 85–114, 134, 168, 196, 211, 229, 240, 306, 327, 332, 334, 346, 348, 420, 431, 505, 581, 583, 584, 586

R
Regenerative medicine, 166, 201, 206, 213–214, 318, 364, 404, 421, 434–441, 461, 463, 478, 494–495, 520, 521, 550
Retrievability, 103, 109, 127, 140, 156, 184–185, 187, 379

S
Safety switch, 403–415
Scaffold, 153, 156, 157, 159, 160, 162, 163, 165, 200, 205, 210, 212, 231, 234, 236, 241, 254, 263, 268, 292, 337, 341, 347, 513, 521, 561

SC-beta cells, 8, 9, 12–18, 529–541, 550, 567
scRNA-seq, 12, 14, 16, 27–30, 32, 36, 52, 54, 59, 581
Sex differences, 75
Signaling, 6, 7, 9, 12–15, 17, 26, 32, 36, 50, 52, 54, 69, 87, 89–91, 167, 203, 214, 252, 267, 268, 324, 334, 346, 370, 407, 534, 535
Single cell genomics, 27, 28, 37, 53, 58, 439
Single-cell technology, 28, 29, 52
Small molecules, 14, 89, 224, 404, 407, 410, 412, 413, 441, 442, 488, 575
Social value, 461–478
Stealth beta cells, 285
Stem cell derived islets, 10, 16–18, 196, 202, 203, 212, 214, 266, 267, 318, 320, 325, 343–347, 560, 564, 573–590
Stem cell islet, 232, 299–311, 420, 563–565
Stem cells, 3, 26, 50, 67, 86, 100, 123, 156, 182, 196, 227, 252, 285, 302, 318, 364, 403, 419, 461, 487, 537, 549, 575
Stem cell technology, 169
Suicide gene, 14, 17, 184, 404–416, 442, 567

T

Teratoma, 101, 136, 183, 184, 288, 302, 304–306, 308, 404, 409, 410, 420, 423, 434–438, 441, 442, 487, 494, 557, 562, 564
Topography, 103, 107–108, 113, 327, 332
Transcription factors (TF), 7, 11, 13, 26, 69–72, 166, 537, 550, 575, 579, 586, 587
Transplant site, 4, 183, 210, 318, 319, 324, 325, 335, 349, 365, 562, 566
Transplantation, 4, 25, 49, 67, 86, 97, 123, 151, 182, 196, 221, 252, 283, 300, 317, 363, 403, 420, 461, 489, 499, 540, 548
Treatment of diabetic mice and dogs with 3-D Organoids, 187
Tumorigenicity, 166, 169, 185, 415, 491, 493, 494
Tumour formation, 420, 421, 431, 435, 437, 438, 442
Type 1 diabetes (T1D), 3, 49, 67, 97, 142, 151, 179, 221, 265, 294, 317, 363, 461, 486, 499, 547, 588
Type 2 diabetes (T2D), 49, 74, 75, 201, 420, 486, 534, 547, 552, 553, 565, 573, 589, 590

U

Upscaling of medical devices, 136

V

Vascularization, 97, 100, 107, 133, 134, 153, 155, 157, 163, 168, 183–185, 210, 213, 214, 222, 235, 240, 242, 253–255, 260, 264, 266–270, 288, 293, 366, 374, 376, 382, 384, 513, 556, 557, 562
Vascularized endocrine niche, 151, 153, 154
Vascularized scaffolds, 154

W

WNT, 13, 35, 89–91, 589

Printed in the USA
CPSIA information can be obtained
at www.ICGtesting.com
CBHW052321051124
16972CB00003B/35

9 783031 419423